SCHAUM'S SOLVED PROBLEMS SERIES

800 SOLVED PROBLEMS IN

VECTOR MECHANICS FOR ENGINEERS
Volume I: STATICS

by

Joseph F. Shelley, Ph.D., P.E.
Trenton State College

McGraw-Hill, Inc.
New York   St. Louis   San Francisco   Auckland   Bogotá
Caracas   Lisbon   London   Madrid   Mexico City   Milan
Montreal   New Delhi   San Juan   Singapore
Sydney   Tokyo   Toronto

Joseph F. Shelley, Ph.D., P.E, *Professor of Mechanical Engineering Technology at Trenton State College.*
Dr. Shelley earned the Ph.D. at The Polytechnic University of New York. He has authored a three-volume set of textbooks in Engineering Mechanics, published by McGraw-Hill in 1980.

To
Gabrielle
and
Stefanie, Suzanne,
Matthew, and Meredith

Your sweet love
such wealth brings,
I scorn to change
my state with kings

Project supervision was done by The Total Book.

 This book is printed on recycled paper containing a minimum of 50% total recycled fiber with 10% postconsumer de-inked fiber. Soybean based inks are used on the cover and text.

Library of Congress Cataloging-in-Publication Data

Shelley, Joseph F.
  800 solved problems in vector mechanics for engineers; statics /
Joseph F. Shelley.
  p.   cm. – (Schaum's solved problems series)
ISBN 0-07-056582-1
  1. Mechanics, Applied – Problems, exercises, etc.   2. Vector
analysis – Problems, exercises, etc.   I. Title.   II. Title: Eight
hundred solved problems in vector mechanics for engineers.
III. Series.
TA350.7.S54   1990
620.1'04'076 – dc20
                                                                89-7951
                                                                CIP

1 2 3 4 5 6 7 8 9 BAW BAW 9 4

ISBN 0-07-056835-9 (Formerly published under ISBN 0-07-056582-1.)

CONTENTS

TO THE STUDENT

Engineering mechanics is the study of the effects that forces produce on bodies. It has two major subdivisions: statics, in which the bodies are at rest or are moving with constant velocity; and dynamics, in which the bodies may possess any type of motion. Thus, acceleration is a necessary part of the description of dynamics problems. It is the absence of acceleration effects that distinguishes statics from dynamics.

Statics is one of the beginning courses in the fields of aeronautical, civil, and mechanical engineering and is required of all engineering students. A thorough understanding of its fundamental principles is a prerequisite for further study in dynamics, strength (or mechanics) of materials, structural engineering, stress analysis, and mechanical design and analysis.

This book is the first volume of a two-volume set in statics and dynamics. It is a completely self-contained treatment. All the fundamental definitions, concepts, and problem-solving techniques of static force analysis are introduced through questions. All groups of problems with numerical solutions are preceded by presentation of the particular definition, concept, or technique required for the solution of those problems. The material in each chapter is arranged in grouped sections of topics, and, within each section, the problems are arranged in a generally increasing order of difficulty.

The last question in each chapter is a review of the fundamental definitions, concepts, and techniques introduced and used in that chapter. The final chapter in this book is a self-study review of all the earlier questions on fundamental definitions, concepts, and techniques. This chapter contains 224 review questions, which are referenced by problem number to the original question in order to make it easy for the reader to refer to the answer. These review questions are not counted as part of the 800 Solved Problems.

Many problems are presented as easily recognizable mechanical or structural systems. The intent is to give a physical, real-world flavor to these problems with which the reader can readily identify. Additional commentary on the solutions is frequently provided at the end of problems to clarify or point out a particular characteristic or limitation of the solution. In many problems a comparison is made between the solutions when certain conditions of the original problem are varied. This gives a general engineering design flavor to these problems. The units used in this text are equally divided between U.S. Customary (USCS) units and International System (SI) units.

There is a carefully developed index, by problem number, at the end of the text. All problems involving definitions, concepts, or techniques are cross-referenced by topic. All problems in the text are listed in this index, and those which are more advanced, or have unusually lengthy solutions, are identified. The reader is encouraged to review this index and become familiar with its use, and thus be able to rapidly identify specific problems in any desired area of statics.

This book may be used with any textbook in statics. It may also be used by itself. A cross-reference of this book, by topics, with the three leading textbooks in statics is included in the appendices. These three texts are Beer and Johnston, *Vector Mechanics for Engineers: Statics and Dynamics*, 5th ed.; Hibbeler, *Engineering Mechanics: Statics and Dynamics*, 5th ed.; and Meriam and Kraige, *Engineering Mechanics*, *volume 1*, *Statics*, 2d ed.

Preparation of a work such as this is a very subjective exercise in creativity. It reflects many judgments on the part of the author with respect to organization of material and emphasis of topics. As with any other book, it receives its ultimate review by the readers. The author welcomes comments and suggestions on any matters of content, organization, or emphasis, and such information may be sent to the Schaum Division, McGraw-Hill Publishing Company, 1221 Avenue of the Americas, New York, N.Y. 10020. Every effort will be made to reply to this correspondence.

As a final note, the author wishes to thank Meredith Ann Shelley for her yeoman service in helping to perform the myriad tasks required to bring this work to its final form.

LIST OF SYMBOLS

a	Radius of a circle, scalar multiplier, acceleration, points, lengths
$\mathbf{A}$	General vector quantity
A	Magnitude of $\mathbf{A}$, area, mass element
A_x, A_y, A_z	xyz components of $\mathbf{A}$
A_{ab}	Component of $\mathbf{A}$ along line ab
$A_i, i = 1, 2, \ldots$	Area element
$b, c, d, \ldots$	Points, lengths
$\mathbf{B}, \mathbf{C}, \ldots$	General vector quantities
$B, C, \ldots$	Magnitudes of $\mathbf{B}, \mathbf{C}, \ldots$ mass elements
B_x, B_y, B_z	xyz components of $\mathbf{B}$
d_x, d_y	Separation distances between xy axes and centroidal x_0, y_0 axes
d_x, d_y, d_z	Lengths
$\mathbf{F}, \mathbf{F}_1, \mathbf{F}_2, \ldots$	Forces
$F, F_1, F_2, \ldots$	Magnitudes of $\mathbf{F}, \mathbf{F}_1, \mathbf{F}_2, \ldots$
F_x, F_y, F_z	xyz components of $\mathbf{F}$
$F_{x'}, F_{y'}, \ldots$	Components of $\mathbf{F}$ along x' and y' axes
$F_{1x}, F_{2x}, \ldots$	x components of forces $\mathbf{F}_1, \mathbf{F}_2, \ldots$
$F_{1y}, F_{2y}, \ldots$	y components of forces $\mathbf{F}_1, \mathbf{F}_2, \ldots$
$F_{1z}, F_{2z}, \ldots$	z components of forces $\mathbf{F}_1, \mathbf{F}_2, \ldots$
F_{ab}	Component of $\mathbf{F}$ along line ab
$F_{ij}, i, j = a, b, c, \ldots$	Forces in truss members
$F_{\min}$	Minimum value of $\mathbf{F}$
$F_{\max}$	Maximum value of $\mathbf{F}$
g	Acceleration of gravity
i	$1, 2, 3, \ldots$
$\mathbf{i}, \mathbf{j}, \mathbf{k}$	Unit vectors in x, y, and z directions, respectively
$\mathbf{i}_n$	Unit vector normal to a plane
$\mathbf{i}_{ab}$	Unit vector along line ab
I_x, I_y	Moment of inertia of a plane area or curve about x and y axes, respectively
I_{xy}	Product of inertia of a plane area or curve about x and y axes
I_{0x}, I_{0y}	Centroidal moment of inertia of a plane area or curve about x and y axes, respectively
J	Polar moment of inertia of a plane area or curve
J_0	Centroidal polar moment of inertia of a plane area or curve
k_x, k_y	Radii of gyration
k_p	Polar radius of gyration
l	Length of a straight or curved line
l_i	Length element of a plane curve
m	Mass, number of members in a truss
$\mathbf{M}, \mathbf{M}_1, \mathbf{M}_2, \ldots$	Moments, or couples
M	Magnitude of $\mathbf{M}$
$M_1, M_2, \ldots$	Magnitudes of $\mathbf{M}_1, \mathbf{M}_2$
M_x, M_y, M_z	xyz components of $\mathbf{M}$
$M_a, M_b, \ldots$	Moments about points $a, b, \ldots$
M_0	Applied moment
$\mathbf{M}_R$	Reaction moment
M_R	Magnitude of $\mathbf{M}_R$
M_{ab}	Component of $\mathbf{M}$ along line ab
n	Number of forces in system, number of joints in a truss
$N, N_a, N_b, \ldots$	Compressive normal reaction forces
p	Pressure
p_0	Pressure above a liquid
$\mathbf{P}, \mathbf{P}_1, \mathbf{P}_2, \ldots$	Forces
$P, P_1, P_2, \ldots$	Magnitudes of $\mathbf{P}, \mathbf{P}_1, \mathbf{P}_2, \ldots$
Q_x, Q_y	First moments of a plane area about x and y axes, respectively
r_i	Radial coordinate of a plane area element or a plane curve element
$\mathbf{R}, \mathbf{R}_a, \mathbf{R}_b, \ldots$	Reaction forces

$R, R_a, R_b, \ldots$	Magnitudes of $\mathbf{R}, \mathbf{R}_a, \mathbf{R}_b, \ldots$
R_x, R_y, R_z	xyz components of $\mathbf{R}$
$R_{ax}, R_{bx}, \ldots$	x components of reaction forces $\mathbf{R}_a, \mathbf{R}_b, \ldots$
$R_{ay}, R_{by}, \ldots,$	y components of reaction forces $\mathbf{R}_a, \mathbf{R}_b, \ldots$
$R_{az}, R_{bz}, \ldots$	z components of reaction forces $\mathbf{R}_a, \mathbf{R}_b, \ldots$
SG	Specific gravity
t	Thickness of a plate
T	Torque, or moment
T_1, T_2	Belt tensile forces
w	Force per unit length
W	Weight
W_i	Weight element
x, y, z	Coordinate axes
x', y'	Coordinate axes
x_c, y_c	Coordinates of centroid of a plane area
x_c', y_c'	Coordinates of centroid of a plane curve
x_i, y_i	xy coordinates of a weight, area, or length element

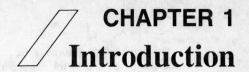

CHAPTER 1
Introduction

1.1 BASIC UNITS OF MECHANICS, FORCE AND MASS, CONVERSION OF UNITS, ANGULAR UNITS

1.1 Define the fundamental quantities *force* and *mass*.

▮ A force is an effect of one body on another body resulting from an attractive effect, such as a gravitational field, or from contact of the bodies. A familiar example of force is the weight associated with a body. The mass of a body, by comparison, is a measure of its resistance to being accelerated. The mass of a body is a constant and unchanging property which is the same at all points in the universe.

Mass and force are related by Newton's second law. This law and its effects are studied in detail in dynamics. For the purposes of the present discussion, the second law will be presented simply as

$$F = ma$$

where m is the mass, F is the force that acts on the mass, and a is the acceleration of the mass. The units of the acceleration term are length divided by time squared.

1.2 Describe the four basic units of mechanics.

▮ The science of mechanics depends on the use of four units:

Length
Time
Force
Mass

Three of these units may be chosen to be *fundamental* units, and the fourth unit must then be expressed in terms of the other three. Units may be treated as algebraic quantities. That is, they may be added, subtracted, multiplied, divided, or raised to a power, in any combination.

1.3 What is the difference between a *gravitational* system of units and an *absolute* system of units?

▮ The four basic units of mechanics are length, time, force, and mass. If length, time, and force are chosen to be fundamental units, then the derived units of mass may be found from Newton's second law. Such a system is called a *gravitational system of units*. If, on the other hand, length, time, and mass are chosen to be fundamental units, then the units of a force may be derived from Newton's second law. This type of system of units is called an *absolute system of units*.

1.4 Describe the *U.S. Customary* units.

▮ The U.S. Customary System (USCS) is a gravitational system, with the units length, time, and force chosen to be fundamental units. The length unit is the foot, the force unit is the pound, and the time unit is the second. The symbols used to represent these quantities are

Foot, ft
Pound, lb
Second, s

The value of the gravitational acceleration g in USCS units is $32.17 \, \text{ft/s}^2 = 386.0 \, \text{in/s}^2 \approx 32.2 \, \text{ft/s}^2$.
The mass unit is defined to be the *slug*. It is that mass which, when acted on by a force of one pound, will experience an acceleration of one foot per second squared.

1.5 Describe the *International System* units.

▮ SI (International System) units are an absolute system of units with the three fundamental units chosen to be length, time, and mass. The unit of length is the meter, the unit of time is the second, and the unit of mass is the kilogram. The symbols used to represent these quantities are

Meter, m
Second, s
Kilogram, kg

The value of the gravitational acceleration g in SI units is $9.807 \, \text{m/s}^2 \approx 9.81 \, \text{m/s}^2$.

The derived unit of force is the newton, with the symbol N. A force of one newton gives a mass of one kg an acceleration of one meter per second squared.

1.6 Give the conversion factors for length, force, and mass between the USCS and SI units.

❚ The factors for conversion between the USCS units and SI units are

$$1 \text{ ft} = 0.3048 \text{ m} \approx 0.305 \text{ m} \qquad 1 \text{ m} = 3.281 \text{ ft} \approx 3.28 \text{ ft}$$
$$1 \text{ lb} = 4.448 \text{ N} \approx 4.45 \text{ N} \qquad 1 \text{ N} = 0.2248 \text{ lb} \approx 0.225 \text{ lb}$$
$$1 \text{ slug} = 14.59 \text{ kg} \approx 14.6 \text{ kg} \qquad 1 \text{ kg} = 0.06854 \text{ slug} \approx 0.0685 \text{ slug}$$

As an estimate of the relative sizes of the above length and force units, it may be noted that 1 m is approximately 3 ft, while 1 N is approximately 0.25 lb.

1.7 Define the basic units of angular measurement.

❚ A radian, *rad*, is the angle subtended by an arc of length equal to the radius of a circle, as shown in Fig. 1.7. The unit of angular measurement is the degree, with the symbol °, defined by

$$1 \text{ revolution} = 1 \text{ r} = 360° = 2\pi \text{ radians} = 2\pi \text{ rad}$$
$$1 \text{ rad} = 57.30°$$

Angular measurement in SI units is the same as in the U.S. Customary System.

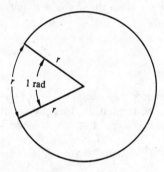

Fig. 1.7

1.8 Express the fundamental relationship between the weight W and the mass m of the body shown in Fig. 1.8.

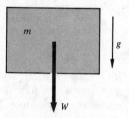

Fig. 1.8

❚ Newton's second law is

$$F = ma$$

The weight force W of the body in a gravitational field with acceleration g is then

$$W = mg \qquad \text{or} \qquad m = \frac{W}{g}$$

1.9 Derive the units of a slug of mass.

❚ Using Newton's second law, we have

$$F = ma$$

$$1 \text{ lb} = (1 \text{ slug})\left(1 \frac{\text{ft}}{\text{s}^2}\right)$$

$$1 \text{ slug} = 1 \frac{\text{lb} \cdot \text{s}^2}{\text{ft}}$$

From the above equations, the derived units of mass in the USCS are pound-seconds squared per foot. It may be seen that this mass unit is expressed in terms of the three fundamental units, force, length, and time.

1.10 Derive the units of a newton of force.

❚ Using Newton's second law, we get

$$F = ma$$

$$1\,\text{N} = 1\,\text{kg}\left(1\,\frac{\text{m}}{\text{s}^2}\right) = 1\,\frac{\text{kg}\cdot\text{m}}{\text{s}^2}$$

From the above equations, the derived units of the force are seen to be kilogram-meters per second squared. This result is in terms of the three fundamental units, length, time, and mass.

1.11 Find the weight of a mass of 1 slug in USCS and SI units.

❚

$$W = mg$$

$$= (1\,\text{slug})\left(32.17\,\frac{\text{ft}}{\text{s}^2}\right)$$

$$= \left(1\,\frac{\text{lb}\cdot\text{s}^2}{\text{ft}}\right)32.17\,\frac{\text{ft}}{\text{s}^2} = 32.17\,\text{lb}$$

$$W = 32.17\,\text{lb}\left(\frac{4.448\,\text{N}}{1\,\text{lb}}\right) = 143.1\,\text{N}$$

1.12 Find the weight of a mass of 1 kilogram in USCS and SI units.

❚ From Prob. 1.5, $g = 9.807\,\text{m/s}^2$.

$$W = mg$$

$$W = (1\,\text{kg})9.807\,\frac{\text{m}}{\text{s}^2} = 9.807\,\text{N}$$

$$W = 9.087\,\text{N}\left(\frac{1\,\text{lb}}{4.448\,\text{N}}\right) = 2.205\,\text{lb}$$

1.13 Define the terms *pressure* and *stress*.

❚ Pressure is a force-induced quantity. The units of pressure are force per unit area. Typical USCS units of pressure are pounds per square inch and pounds per square foot, written as psi and psf.

The fundamental SI pressure unit is the pascal, abbreviated Pa, with units newtons per meter squared and written as

$$1\,\text{Pa} = 1\,\frac{\text{N}}{\text{m}^2}$$

A term with the same units as pressure is stress. The term pressure is generally used to describe the force effects caused by a liquid or a gas. The term stress is used to describe the intensity of loading in the material in a body, as a result of the forces acting on the body. Stress is a basic term used to characterize problems in strength, or mechanics, of materials.

1.14 Define the material property *specific weight*, and explain how it relates to the weight of a body.

❚ The specific weight, with the symbol γ, is defined to be the weight per unit volume of the material. Typical units of specific weight are pounds per cubic foot or newtons per cubic meter. Some average values of specific weights of common materials are $62.4\,\text{lb/ft}^3$ for water at 60°F, $489\,\text{lb/ft}^3$ for steel, and $173\,\text{lb/ft}^3$ for aluminum.

The weight of a homogeneous body is the product of the volume of the body and the specific weight of the material of the body.

1.15 Define the material property *density*, and explain how it relates to the mass of a body.

❚ The density ρ is defined to be the mass per unit volume of the material. The units of density are slugs per cubic foot or kilograms per cubic meter. Some average values of densities of common materials are $1,000\,\text{kg/m}^3$ for water at 60°F, $7,830\,\text{kg/m}^3$ for steel, and $2,770\,\text{kg/m}^3$ for aluminum.

The mass of a homogeneous body is the product of the volume of the body and the density of the material of the body.

1.16 What is the relationship between specific weight and density?

▌ The relationship between specific weight and density, from Newton's second law, is

$$\gamma = \rho g \qquad \rho = \frac{\gamma}{g}$$

It should be observed that the specific weight in USCS units is the *fundamental* quantity, because of its expression in terms of force units. The density, by comparison, is a *derived* quantity, found by using Newton's second law. In SI units the density is the fundamental quantity and the specific weight is the derived quantity.

1.17 What is the *specific gravity* of a material?

▌ The specific gravity (SG) of a material is defined to be the ratio of the density of the material to that of water at 60° F. From this definition, it follows that specific gravity is a dimensionless quantity. Three typical values are shown below:

$$
\begin{array}{lll}
\text{Water at } 60°\text{F:} & \text{SG} = 1 \\
\text{Steel:} & \text{SG} = 7,830/1,000 = 7.830 \\
\text{Aluminum:} & \text{SG} = 2,770/1,000 = 2.770
\end{array}
$$

The magnitude of the specific gravity tells how much "heavier" the material is than water. Since the specific gravity is a dimensionless quantity, it has the same value in USCS and SI units.

1.18 Find the density of steel in USCS units.

▌
$$\rho = \frac{\gamma}{g}$$

$$= \frac{489 \text{ lb/ft}^3}{32.17 \text{ ft/s}^2} = 15.20 \frac{\text{lb} \cdot \text{s}^2}{\text{ft}^4} = 15.20 \frac{\text{slugs}}{\text{ft}^3}$$

1.19 Assume that the density of steel is $7,830 \text{ kg/m}^3$. Find the specific weight of steel in SI units.

▌
$$\gamma = \rho g$$

$$= \left(7,830 \frac{\text{kg}}{\text{m}^3}\right)\left(9.807 \frac{\text{m}}{\text{s}^2}\right) = 76,790 \frac{\text{kg} \cdot \text{m}}{\text{s}^2} \Big/ \text{m}^3$$

$$\gamma = 76,790 \frac{\text{N}}{\text{m}^3} = 76.79 \frac{\text{kN}}{\text{m}^3}$$

1.20 A heavy steel roller is 2.5 ft in diameter and 8 ft long. Given that the specific weight of steel is 489 lb/ft^3, find the weight of this roller.

▌ The volume of the roller is $V = \pi \dfrac{d^2}{4} h$, where d is the diameter and h is the length.

$$V = \frac{\pi(2.5)^2}{4} \text{ ft}^2(8 \text{ ft}) = 39.27 \text{ ft}^3$$

The weight of the roller is then

$$W = \gamma V$$

$$= \left(489 \frac{\text{lb}}{\text{ft}^3}\right)(39.27 \text{ ft}^3) = 19.200 \text{ lb} = 19.2 \text{ kip}$$

1.21 The bar shown in Fig. 1.21 is made of aluminum, which has a specific weight of 173 lb/ft^3. Find the (a) weight and (b) mass of the bar in USCS and SI units.

▌ (a) $\quad W = \gamma V$

$$= [2(4)30 \text{ in}^3]\left(\frac{1 \text{ ft}}{12 \text{ in}}\right)^3 173 \frac{\text{lb}}{\text{ft}^3} = 24.03 \text{ lb}$$

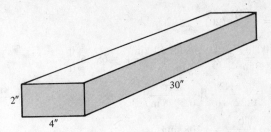

Fig. 1.21

$$W = 24.03 \text{ lb}\left(\frac{4.448 \text{ N}}{1 \text{ lb}}\right) = 106.9 \text{ N}$$

(b) $\quad m = \dfrac{W}{g} = \dfrac{24.03 \text{ lb}}{32.17 \text{ ft/s}^2} = 0.7470 \text{ slug}$

$$m = 0.7470 \text{ slug}\left(\frac{14.59 \text{ kg}}{1 \text{ slug}}\right) = 10.90 \text{ kg}$$

1.22 The disk shown in Fig. 1.22 is made of a material with SG = 9.2. Find the mass and weight of the disk in SI and USCS units.

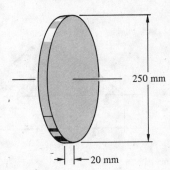

250 mm

20 mm

Fig. 1.22

❚ $\quad m = \text{SG} \cdot \rho_{\text{water}} \cdot V \quad$ and $\quad V = \pi \dfrac{d^2}{4} h$

$$m = 9.2\left(1,000 \frac{\text{kg}}{\text{m}^3}\right)\left[\frac{\pi (250)^2}{4}(20) \text{ mm}^3\right]\left(\frac{1 \text{ m}}{1,000 \text{ mm}}\right)^3 = 9.032 \text{ kg}$$

$$m = 9.032 \text{ kg}\left(\frac{0.06854 \text{ slug}}{1 \text{ kg}}\right) = 0.6191 \text{ slug}$$

$$W = mg = 9.032 \text{ kg}\left(9.807 \frac{\text{m}}{\text{s}^2}\right) = 88.58 \text{ N}$$

$$W = mg = 0.6191 \text{ slug}\left(32.17 \frac{\text{ft}}{\text{s}^2}\right) = 19.92 \text{ lb}$$

1.23 The bar in Fig. 1.23 weighs 100 lb. Find the (a) specific weight, (b) mass, (c) density, and (d) specific gravity of the bar in USCS and SI units.

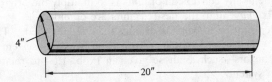

4″

20″

Fig. 1.23

▌ (a) $V = \pi \dfrac{d^2}{4} h = \dfrac{\pi(4)^2}{4} (20) = 251.3 \text{ in}^3$

$\gamma = \dfrac{W}{V} = \dfrac{100 \text{ lb}}{251.3 \text{ in}^3} = 0.3979 \dfrac{\text{lb}}{\text{in}^3} \left(\dfrac{12 \text{ in}}{1 \text{ ft}} \right)^3 = 687.6 \dfrac{\text{lb}}{\text{ft}^3}$

$\gamma = 687.6 \dfrac{\text{lb}}{\text{ft}^3} \left(\dfrac{1 \text{ N}}{0.2248 \text{ lb}} \right) \left(\dfrac{1 \text{ ft}}{0.3048 \text{ m}} \right)^3 = 1.080 \times 10^5 \dfrac{\text{N}}{\text{m}^3} = 108 \dfrac{\text{kN}}{\text{m}^3}$

(b) $m = \dfrac{W}{g} = \dfrac{100 \text{ lb}}{32.17 \text{ ft/s}^2} = 3.108 \text{ slug}$

$m = 3.108 \text{ slug} \left(\dfrac{1 \text{ kg}}{0.06854 \text{ slug}} \right) = 45.35 \text{ kg}$

(c) $\rho = \dfrac{\gamma}{g} = \dfrac{687.6}{32.17} = 21.37 \dfrac{\text{slug}}{\text{ft}^3}$

$\rho = 21.37 \dfrac{\text{slug}}{\text{ft}^3} \left(\dfrac{1 \text{ kg}}{0.06854 \text{ slug}} \right) \left(\dfrac{1 \text{ ft}}{0.3048 \text{ m}} \right)^3 = 11,010 \dfrac{\text{kg}}{\text{m}^3}$

(d) $SG = \dfrac{687.6}{62.4} = \dfrac{11,010}{1,000 \text{ kg/m}^3} = 11.01$

1.24   The pile of earth shown in Fig. 1.24 forms a cone-shaped mound.   An average value of 95 lb/ft³ for earth is assumed.   Find the (a) weight and (b) mass, in USCS and SI units, of the mound of earth.   [The volume of a cone is $\frac{1}{3}$(base area)(height).]

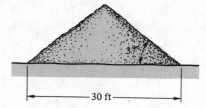

—— 30 ft ——   **Fig. 1.24**

▌ (a)   The height h of the cone is

$h = 15 \tan 37° = 11.30 \text{ ft}$

$V = \dfrac{1}{3} \pi r^2 h$

$\quad = \dfrac{1}{3} \left[\dfrac{\pi(30)^2}{4} \right] 11.30 = 2,662 \text{ ft}^3$

$W = \gamma V = 95 \dfrac{\text{lb}}{\text{ft}^3} (2,662 \text{ ft}^3) = 2.529 \times 10^5 \text{ lb} = 126.5 \text{ ton}$

$W = 2.529 \times 10^5 \text{ lb} \left(\dfrac{4.448 \text{ N}}{1 \text{ lb}} \right) = 1.125 \times 10^6 \text{ N} = 1,125 \text{ kN} = 1.125 \text{ MN}$

(b) $m = \dfrac{W}{g} = \dfrac{2.529 \times 10^5 \text{ lb}}{32.17 \text{ ft/s}^2} = 7,861 \text{ slug}$

$m = 7,861 \text{ slug} \left(\dfrac{14.59 \text{ kg}}{1 \text{ slug}} \right) = 1.147 \times 10^5 \text{ kg}$

1.25   State the basic technique used in the conversion of units and show how to convert a speed of 60 mi/h to inches per second.

▌ The basic technique in any conversion of units is to multiply the original quantity by a succession of factors, *each of whose value is unity*, until the desired units are obtained.

   Two types of problems in the conversion of units may be identified.   The first type is a conversion within the same system of units.   The second type is a conversion from one system of units to another.   This latter type requires equivalence factors—such as those given in Prob. 1.6—to relate the fundamental length, force, and mass units of the two systems.

As basic definitions,

$$5{,}280 \text{ ft} = 1 \text{ mi} \qquad \frac{5{,}280 \text{ ft}}{1 \text{ mi}} = 1$$

$$12 \text{ in} = 1 \text{ ft} \qquad \frac{12 \text{ in}}{1 \text{ ft}} = 1$$

$$60 \text{ min} = 1 \text{ h} \qquad \frac{60 \text{ min}}{1 \text{ h}} = 1$$

$$60 \text{ s} = 1 \text{ min} \qquad \frac{60 \text{ s}}{1 \text{ min}} = 1$$

The original term 60 mi/h is now multiplied by the appropriate factors, and the units are canceled after each successive multiplication. The first step in this operation is

$$60 \frac{\text{mi}}{\text{h}} \left(\frac{5{,}280 \text{ ft}}{1 \text{ mi}} \right)$$

The factor in parentheses has a value of 1, and the units of the product are feet per hour. The next step in the conversion appears as

$$60 \frac{\text{mi}}{\text{h}} \left(\frac{5{,}280 \text{ ft}}{1 \text{ mi}} \right) \left(\frac{12 \text{ in}}{1 \text{ ft}} \right)$$

After this step, the units of the product are inches per hour. This procedure is continued, and the final appearance of the conversion equation is

$$60 \frac{\text{mi}}{\text{h}} \left(\frac{5{,}280 \text{ ft}}{1 \text{ mi}} \right) \left(\frac{12 \text{ in}}{1 \text{ ft}} \right) \left(\frac{1 \text{ h}}{60 \text{ min}} \right) \left(\frac{1 \text{ min}}{60 \text{ s}} \right) = 1{,}056 \frac{\text{in}}{\text{s}}$$

The single factor to convert mi/h to in/s can be found from the above equation as

$$\left(\frac{5{,}280 \text{ ft}}{1 \text{ mi}} \right) \left(\frac{12 \text{ in}}{1 \text{ ft}} \right) \left(\frac{1 \text{ h}}{60 \text{ min}} \right) \left(\frac{1 \text{ min}}{60 \text{ s}} \right) = 17.60 \frac{\text{in/s}}{\text{mi/h}}$$

As a check on the original calculation

$$60 \frac{\text{mi}}{\text{n}} \left(17.60 \frac{\text{in/s}}{\text{mi/h}} \right) \overset{?}{=} 1{,}056 \frac{\text{in}}{\text{s}} \qquad 1{,}056 \frac{\text{in}}{\text{s}} \equiv 1{,}056 \frac{\text{in}}{\text{s}}$$

1.26 Express 1.72 yd^2 in square inches.

❚ The conversion equation has the form

$$1.72 \text{ yd}^2 \left(\frac{3 \text{ ft}}{1 \text{ yd}} \right)^2 \left(\frac{12 \text{ in}}{1 \text{ ft}} \right)^2 = 2{,}229 \text{ in}^2$$

1.27 Find the equivalent in cubic millimeters of 0.110 in^3.

❚ This is an example of a conversion between two different systems of units, and the conversion equation appears as

$$0.110 \text{ in}^3 \left(\frac{1 \text{ ft}}{12 \text{ in}} \right)^3 \left(\frac{0.3048 \text{ m}}{1 \text{ ft}} \right)^3 \left(\frac{1{,}000 \text{ mm}}{1 \text{ m}} \right)^3 = 1{,}803 \text{ mm}^3$$

In this problem, the conversion was effected by multiplying by a succession of primary conversion factors. It is possible, however, to construct a single factor to convert cubic inches directly to cubic millimeters. From the above equation,

$$\left(\frac{1 \text{ ft}}{12 \text{ in}} \right)^3 \left(\frac{0.3048 \text{ m}}{1 \text{ ft}} \right)^3 \left(\frac{1{,}000 \text{ mm}}{1 \text{ m}} \right)^3 = 1.639 \times 10^4 \frac{\text{mm}^3}{\text{in}^3}$$

By using this factor, the original 0.110 in^3 could have been converted directly:

$$0.110 \text{ in}^3 \left(1.639 \times 10^4 \frac{\text{mm}^3}{\text{in}^3} \right) = 1{,}803 \text{ mm}^3$$

1.28 The density of a certain metal in USCS units is 19.2 slugs/ft^3. Express this density in kilograms per cubic meter.

▌ The conversion equation is

$$19.2 \frac{\text{slugs}}{\text{ft}^3} \left(\frac{14.59 \text{ kg}}{1 \text{ slug}}\right)\left(\frac{1 \text{ ft}}{0.3048 \text{ m}}\right)^3 = 9,893 \frac{\text{kg}}{\text{m}^3}$$

1.29 Find the result of multiplying the given quantities in Table 1.1 by a succession of primary conversion factors. In each case, compute the single conversion factor that will convert the given quantity directly to the units of the desired result.

TABLE 1.1

	given	units of desired result		given	units of desired result
(a)	13 mi	yds	(k)	1100 ton/ft^2	psi
(b)	155 yds	in	(l)	30,000 psi	psf
(c)	2800 in^2	yd^2	(m)	735°	rad
(d)	0.25 mi^2	yd^2	(n)	38 km	mm
(e)	3 yd^3	ft^3	(o)	113 mm^2	m^2
(f)	520 in^3	ft^3	(p)	4 km^2	m^2
(g)	0.489 lb/in^3	lb/ft^3	(q)	1250 mm^3	m^3
(h)	62.4 lb/ft^3	lb/yd^3	(r)	131 kN/m^3	N/mm^3
(i)	22 slug/ft^3	lb·s^2/in^4	(s)	1.2×10^4 kg/m^3	kg/mm^3
(j)	1940 in/s	mi/h	(t)	875 kPa	N/mm^2

▌ (a) $13 \text{ mi}\left(\frac{5,280 \text{ ft}}{1 \text{ mi}}\right)\left(\frac{1 \text{ yd}}{3 \text{ ft}}\right) = 22,880 \text{ yd}$

$1,760 \text{ yd/mi}$

(e) $3 \text{ yd}^3\left(\frac{3 \text{ ft}}{1 \text{ yd}}\right)^3 = 81 \text{ ft}^3$

$27 \text{ ft}^3/\text{yd}^3$

(b) $155 \text{ yd}\left(\frac{3 \text{ ft}}{1 \text{ yd}}\right)\left(\frac{12 \text{ in}}{1 \text{ ft}}\right) = 5,580 \text{ in}$

36 in/yd

(f) $520 \text{ in}^3\left(\frac{1 \text{ ft}}{12 \text{ in}}\right)^3 = 0.3009 \text{ ft}^3$

$5.787 \times 10^{-4} \text{ ft}^3/\text{in}^3$

(c) $2,800 \text{ in}^2\left(\frac{1 \text{ ft}}{12 \text{ in}}\right)^2\left(\frac{1 \text{ yd}}{3 \text{ ft}}\right)^2 = 2.160 \text{ yd}^2$

$7.716 \times 10^{-4} \text{ yd}^2/\text{in}^2$

(g) $0.489 \frac{\text{lb}}{\text{in}^3}\left(\frac{12 \text{ in}}{1 \text{ ft}}\right)^3 = 845.0 \frac{\text{lb}}{\text{ft}^3}$

$1,728 \frac{\text{lb/ft}^3}{\text{lb/in}^3}$

(d) $0.25 \text{ mi}^2\left(\frac{5,280 \text{ ft}}{1 \text{ mi}}\right)^2\left(\frac{1 \text{ yd}}{3 \text{ ft}}\right)^2 = 7.745 \times 10^5 \text{ yd}^2$

$3.098 \times 10^6 \text{ yd}^2/\text{mi}^2$

(h) $62.4 \frac{\text{lb}}{\text{ft}^3}\left(\frac{3 \text{ ft}}{1 \text{ yd}}\right)^3 = 1,685 \frac{\text{lb}}{\text{yd}^3}$

$27 \frac{\text{lb/yd}^3}{\text{lb/ft}^3}$

(i) $22 \frac{\text{slug}}{\text{ft}^3}\left(\frac{1 \text{ lb·s}^2/\text{ft}}{1 \text{ slug}}\right)\left(\frac{1 \text{ ft}}{12 \text{ in}}\right)^4 = 0.001061 \text{ lb·s}^2/\text{in}^4$

$4.823 \times 10^{-5} \frac{\text{lb·s}^2/\text{in}^4}{\text{slug/ft}^3}$

(*j*) $1,940 \dfrac{\text{in}}{\text{s}} \left(\dfrac{1\text{ ft}}{12\text{ in}} \right) \left(\dfrac{1\text{ mi}}{5,280\text{ ft}} \right) \left(\dfrac{60\text{ s}}{1\text{ min}} \right) \left(\dfrac{60\text{ min}}{1\text{ h}} \right) = 110.2\text{ mi/h}$

$$0.05682\,\dfrac{\text{mi/h}}{\text{in/s}}$$

(*k*) $1,100 \dfrac{\text{ton}}{\text{ft}^2} \left(\dfrac{2,000\text{ lb}}{\text{ton}} \right) \left(\dfrac{1\text{ ft}}{12\text{ in}} \right)^2 = 15,280\text{ lb/in}^2$

$$13.89\,\dfrac{\text{lb/in}^2}{\text{ton/ft}^2}$$

(*n*) $38\text{ km} \left(\dfrac{1,000\text{ m}}{1\text{ km}} \right) \left(\dfrac{1,000\text{ mm}}{1\text{ m}} \right) = 3.8 \times 10^7\text{ mm}$

$$1 \times 10^6\text{ mm/km}$$

(*l*) $30,000\text{ lb/in}^2 \left(\dfrac{12\text{ in}}{1\text{ ft}} \right)^2 = 4.320 \times 10^6\text{ lb/ft}^2$

$$144\,\dfrac{\text{lb/ft}^2}{\text{lb/in}^2}$$

(*o*) $113\text{ mm}^2 \left(\dfrac{1\text{ m}}{1,000\text{ mm}} \right)^2 = 1.13 \times 10^{-4}\text{ m}^2$

$$1 \times 10^{-6}\text{ m}^2/\text{mm}^2$$

(*m*) $735° \left(\dfrac{2\pi\text{ rad}}{360°} \right) = 12.83\text{ rad}$

$$0.01745\text{ rad}/°$$

(*p*) $4\text{ km}^2 \left(\dfrac{1,000\text{ m}}{1\text{ km}} \right)^2 = 4 \times 10^6\text{ m}^2$

$$1 \times 10^6\text{ m}^2/\text{km}^2$$

(*q*) $1,250\text{ mm}^3 \left(\dfrac{1\text{ m}}{1,000\text{ mm}} \right)^3 = 1.250 \times 10^{-6}\text{ m}^3$

$$1 \times 10^{-9}\text{ m}^3/\text{mm}^3$$

(*r*) $131 \dfrac{\text{kN}}{\text{m}^3} \left(\dfrac{1,000\text{ N}}{1\text{ kN}} \right) \left(\dfrac{1\text{ m}}{1,000\text{ mm}} \right)^3 = 1.31 \times 10^{-4}\text{ N/mm}^3$

$$1 \times 10^{-6}\,\dfrac{\text{N/mm}^3}{\text{kN/m}^3}$$

(*s*) $1.2 \times 10^4 \dfrac{\text{kg}}{\text{m}^3} \left(\dfrac{1\text{ m}}{1,000\text{ mm}} \right)^3 = 1.2 \times 10^{-5}\text{ kg/mm}^3$

$$1 \times 10^{-9}\,\dfrac{\text{kg/mm}^3}{\text{kg/m}^3}$$

(*t*) $875\text{ kPa} \left(\dfrac{1,000\text{ Pa}}{1\text{ kPa}} \right) \left(\dfrac{1\text{ N/m}^2}{1\text{ Pa}} \right) \left(\dfrac{1\text{ m}}{1000\text{ mm}} \right)^2 = 0.875\text{ N/mm}^2$

$$0.001\,\dfrac{\text{N/mm}^2}{\text{kPa}}$$

1.30 Convert all the results obtained in USCS units in Prob. 1.29 to answers in terms of meters, newtons, kilograms, and seconds.

▌ (a) $22{,}880 \text{ yd}\left(\dfrac{3 \text{ ft}}{1 \text{ yd}}\right)\left(\dfrac{0.3048 \text{ m}}{1 \text{ ft}}\right) = 2.092 \times 10^4 \text{ m}$

(b) $5{,}580 \text{ in}\left(\dfrac{1 \text{ ft}}{12 \text{ in}}\right)\left(\dfrac{0.3048 \text{ m}}{1 \text{ ft}}\right) = 141.7 \text{ m}$

(c) $2.16 \text{ yd}^2\left(\dfrac{3 \text{ ft}}{1 \text{ yd}}\right)^2\left(\dfrac{0.3048 \text{ m}}{1 \text{ ft}}\right)^2 = 1.806 \text{ m}^2$

(d) $7.745 \times 10^5 \text{ yd}^2\left(\dfrac{3 \text{ ft}}{1 \text{ yd}}\right)^2\left(\dfrac{0.3048 \text{ m}}{1 \text{ ft}}\right)^2 = 6.476 \times 10^5 \text{ m}^2$

(e) $81 \text{ ft}^3\left(\dfrac{0.3048 \text{ m}}{1 \text{ ft}}\right)^3 = 2.294 \text{ m}^3$

(f) $0.3009 \text{ ft}^3\left(\dfrac{0.3048 \text{ m}}{1 \text{ ft}}\right)^3 = 0.008521 \text{ m}^3$

(g) $845.0 \dfrac{\text{lb}}{\text{ft}^3}\left(\dfrac{1 \text{ ft}}{0.3048 \text{ m}}\right)^3\left(\dfrac{4.448 \text{ N}}{1 \text{ lb}}\right) = 1.327 \times 10^5 \text{ N/m}^3$

(h) $1{,}685 \dfrac{\text{lb}}{\text{yd}^3}\left(\dfrac{1 \text{ yd}}{3 \text{ ft}}\right)^3\left(\dfrac{1 \text{ ft}}{0.3048 \text{ m}}\right)^3\left(\dfrac{4.448 \text{ N}}{1 \text{ lb}}\right) = 9{,}803 \text{ N/m}^3$

(i) $0.001061 \dfrac{\text{lb}\cdot\text{s}^2}{\text{in}^4}\left(\dfrac{12 \text{ in}}{1 \text{ ft}}\right)\left(\dfrac{1 \text{ slug}}{1 \text{ lb}\cdot\text{s}^2/\text{ft}}\right)\left(\dfrac{12 \text{ in}}{1 \text{ ft}}\right)^3\left(\dfrac{1 \text{ ft}}{0.3048 \text{ m}}\right)^3\left(\dfrac{14.59 \text{ kg}}{1 \text{ slug}}\right) = 11{,}340 \text{ kg/m}^3$

(j) $110.2 \dfrac{\text{mi}}{\text{n}}\left(\dfrac{5{,}280 \text{ ft}}{1 \text{ mi}}\right)\left(\dfrac{0.3048 \text{ m}}{1 \text{ ft}}\right)\left(\dfrac{1 \text{ h}}{3{,}600 \text{ s}}\right) = 49.26 \text{ m/s}$

(k) $15{,}280 \dfrac{\text{lb}}{\text{in}^2}\left(\dfrac{12 \text{ in}}{1 \text{ ft}}\right)^2\left(\dfrac{1 \text{ ft}}{0.3048 \text{ m}}\right)^2\left(\dfrac{4.448 \text{ N}}{1 \text{ lb}}\right) = 1.053 \times 10^8 \text{ N/m}^2 = 1.053 \times 10^8 \text{ Pa} = 105.3 \text{ MPa}$

(l) $4.320 \times 10^6 \dfrac{\text{lb}}{\text{ft}^2}\left(\dfrac{1 \text{ ft}}{0.3048 \text{ m}}\right)^2\left(\dfrac{4.448 \text{ N}}{1 \text{ lb}}\right) = 2.068 \times 10^8 \text{ N/m}^2 = 2.068 \times 10^8 \text{ Pa} = 206.8 \text{ MPa}$

1.31 Convert all results obtained in SI units in Prob. 1.29 to answers in terms of feet, pounds, slugs, and seconds.

▌ (n) $3.8 \times 10^7 \text{ mm}\left(\dfrac{1 \text{ m}}{1{,}000 \text{ mm}}\right)\left(\dfrac{3.281 \text{ ft}}{1 \text{ m}}\right) = 1.247 \times 10^5 \text{ ft}$

(o) $1.13 \times 10^{-4} \text{ m}^2\left(\dfrac{3.281 \text{ ft}}{1 \text{ m}}\right)^2 = 0.001216 \text{ ft}^2$

(p) $4 \times 10^6 \text{ m}^2\left(\dfrac{3.281 \text{ ft}}{1 \text{ m}}\right)^2 = 4.306 \times 10^7 \text{ ft}^2$

(q) $1.250 \times 10^{-6} \text{ m}^3\left(\dfrac{3.281 \text{ ft}}{1 \text{ m}}\right)^3 = 4.415 \times 10^{-5} \text{ ft}^3$

(r) $1.31 \times 10^{-4} \dfrac{\text{N}}{\text{mm}^3}\left(\dfrac{1{,}000 \text{ mm}}{1 \text{ m}}\right)^3\left(\dfrac{1 \text{ m}}{3.281 \text{ ft}}\right)^3\left(\dfrac{0.2248 \text{ lb}}{1 \text{ N}}\right) = 833.8 \text{ lb/ft}^3$

(s) $1.2 \times 10^{-5} \dfrac{\text{kg}}{\text{mm}^3}\left(\dfrac{1 \text{ kg}}{1{,}000 \text{ g}}\right)\left(\dfrac{0.06854 \text{ slug}}{1 \text{ kg}}\right)\left(\dfrac{1{,}000 \text{ mm}}{1 \text{ m}}\right)^3\left(\dfrac{1 \text{ m}}{3.281 \text{ ft}}\right)^3 = 23.29 \text{ slug/ft}^3$

(t) $0.875 \dfrac{\text{N}}{\text{mm}^2}\left(\dfrac{1{,}000 \text{ mm}}{1 \text{ m}}\right)^2\left(\dfrac{1 \text{ m}}{3.281 \text{ ft}}\right)^2\left(\dfrac{0.2248 \text{ lb}}{1 \text{ N}}\right) = 1.827 \times 10^4 \text{ lb/ft}^2$

1.2 FUNDAMENTAL CONCEPTS AND DEFINITIONS

1.32 Define the term *magnitude*.

▌ The magnitude of a quantity tells "how much" there is. A magnitude may be expressed only by a number, and there is no other way to describe this term. Typical examples of magnitude are \$100, 33°, $1\frac{1}{2}$ lb, 3 kg, and

10 mi.   In the above examples, the type, or kind, of quantity is indicated by the descriptive information which accompanies the term, while the magnitude is expressed solely by the number.

1.33   Define the term *direction*.

▌ In Fig. 1.33*a* straight line *aa* and axis *bb* lie in the plane of the paper.

This line possesses an orientation with respect to the horizontal *bb* axis.   This characteristic of orientation is referred to as direction.   The term direction should *not* be thought of in terms of movement along line *aa*, but rather as describing the *position* of line *aa* with respect to the axis *bb*.

The orientation of the line *aa* may be specified in one of two ways.   The first way is shown in Fig. 1.33*b*.   Angle θ measures the orientation of line *aa* with respect to axis *bb*, and this *angle* is defined to be the direction of line *aa*.

The second method of specifying direction is shown in Fig. 1.33*c*.   Here, the small right triangle has legs of known length *c* and *d*, and *d* is parallel to the *bb* axis.   The ratio *c/d* is defined to be the *slope* of line *aa*.   Thus, the direction of a line may be defined by specifying its slope.

It may be seen that the above two methods of specifying the direction of a line are not independent.   From Fig. 1.33*c* it may be concluded that

$$\tan \theta = \frac{c}{d}$$

Thus, the slope of a line is the tangent of the angle that defines the direction of the line.

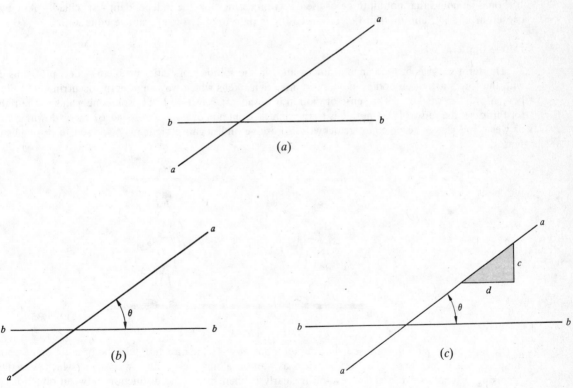

Fig. 1.33

1.34   How is the direction of a line defined in three-dimensional space?

▌ The definition of a slope, or of an included angle, given in Prob. 1.33 is sufficient to determine the direction of a line that lies in a plane.   For the more general case, where the straight line must be referenced to a three-dimensional space, the definition of direction becomes somewhat more elaborate.   The line *aa*, now shown in Fig. 1.34*a* as occupying some position in space, is to be referenced to the *xyz* coordinate axes.   This may be accomplished by specifying the three angles between line *aa* and each of the three coordinate axes, as is shown in Fig. 1.34*b*.   Each of the angles is measured in the plane that is defined by line *aa* and the

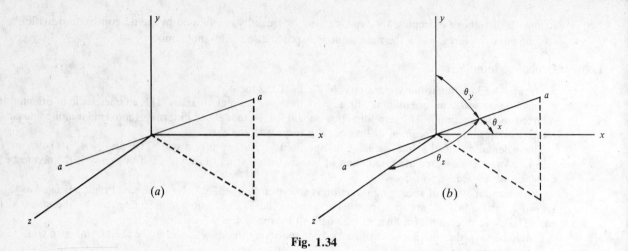

Fig. 1.34

corresponding coordinate axis. The three angles θ_x, θ_y, and θ_z are called *direction angles*. It can be shown that all possible values of a direction angle lie between 0 and 180°. The cosines of the direction angles are called *direction cosines*. These direction cosines must satisfy the identity

$$\cos^2 \theta_x + \cos^2 \theta_y + \cos^2 \theta_z = 1$$

It should be noted that not all three of the direction cosines may be independently specified. Any two may be chosen arbitrarily, and the third direction cosine is then found from the above equation.

1.35 Define the term *sense*.

❚ The term sense is of fundamental importance in the solution of statics problems—i.e., problems concerned with the forces acting on bodies in equilibrium. Sense tells which way an effect is occurring. In Fig. 1.35a, a person walks to the left. The sense of motion is to the left. In Fig. 1.35b, someone walks to the right so that, for this case, the sense of the motion is to the right. Common examples of sense are up or down, in or out, east or west, and clockwise or counterclockwise. Positive and negative signs may be used to distinguish opposite senses.

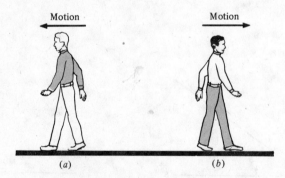

Fig. 1.35

The term *sense* is often confused with the term *direction*. The direction tells the orientation of a straight line in space, while the sense of an effect, such as motion along such a line, would tell which way the effect is occurring. Readers are urged to establish clearly in their minds the distinction between direction and sense.

1.36 Find the direction θ of the line *aa* shown in Fig. 1.36.

❚
$$\tan \theta = \frac{2}{5} \qquad \theta = 21.80°$$

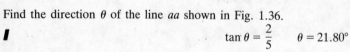

Fig. 1.36

1.37   Figure 1.37 shows that the line *aa* with direction θ lies in the *xy* plane.   Find the three direction cosines.

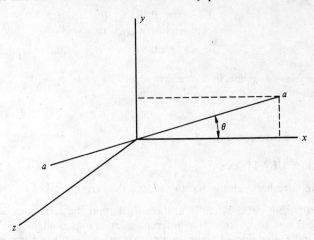

Fig. 1.37

❚ From the basic definition of a direction cosine,

$$\theta_x = \theta \qquad \cos \theta_x = \cos \theta$$

The angle between line *aa* and the *y* axis is $90° - \theta$.   Thus

$$\theta_y = 90° - \theta \qquad \text{and} \qquad \cos \theta_y = \sin \theta$$

Line *aa* is perpendicular to the *z* axis, so that

$$\theta_z = 90° \qquad \cos \theta_z = 0$$

The formal statement of the three direction cosines that define the position of line *aa* is then

$$\cos \theta_x = \cos \theta \qquad \cos \theta_y = \sin \theta \qquad \cos \theta_z = 0$$

1.38   Two direction angles of a line are $\theta_x = 65°$ and $\theta_y = 110°$.   Find θ_z and show the orientation of the line with respect to the *xyz* axes.

❚
$$\cos^2 \theta_x + \cos^2 \theta_y + \cos^2 \theta_z = 1$$
$$\cos^2 65° + \cos^2 110° + \cos^2 \theta_z = 1$$
$$\cos^2 \theta_z = 0.7044 \qquad \cos \theta_z = \pm 0.8393$$
$$\theta_z = 32.94°, \ 147.1°$$

Both of the above solutions for θ_z are correct. $\theta_z = 32.94° \approx 33°$ measures the acute angle between line *aa* and the *z* axis, while $\theta_z = 147.1° \approx 147°$ measures the obtuse angle between these two lines.   The orientation of line *aa* is shown in Fig. 1.38.

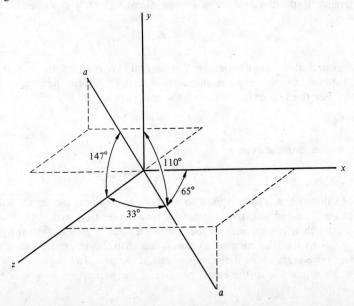

Fig. 1.38

1.39 Two direction angles of a line in space are $\theta_x = 40°$ and $\theta_y = 70°$. Find the third direction angle.

$$\cos^2 \theta_x + \cos^2 \theta_y + \cos^2 \theta_z = 1$$
$$\cos^2 40° + \cos^2 70° + \cos^2 \theta_z = 1$$
$$\cos \theta_z = \pm 0.5442 \qquad \theta_z = 57.03°, \ 123.0°$$

1.40 Two direction angles of a line in space are $\theta_x = 20°$ and $\theta_z = 70°$. Find the third direction angle, and discuss the solution.

$$\theta_y = 90°$$

and the line lies in the xz plane.

1.3 NUMERICAL CALCULATIONS

1.41 What is the difference between the term *error* and the term *mistake*?

▮ The term error should be carefully distinguished from the term mistake. An *error* is defined to be the difference between two numbers. This difference, for example, might be between a value which is measured experimentally and the computed magnitude of that value. An error may also occur in an answer because of the rounding off of numbers during the calculations. A *mistake*, by comparison, results from an incorrect step in a solution or an incorrect computation. A mistake results in a wrong answer. Mistakes can be avoided only by using care in performing the calculations. Errors, by comparison, are inherent in the system of numbers which are manipulated to obtain a final answer.

1.42 Show two methods of comparing the magnitudes of two different numbers, and give numerical examples.

▮ The first method is to state a *percent difference*, abbreviated %D. If there are two numbers A and B and it is desired to compare number A with number B, the percent difference between these two numbers is defined by

$$\%D = \frac{A - B}{B} \ 100$$

In this equation, B is the reference value with which A is to be compared.

Suppose, for example, that a calculated value of a force is 175 N while the experimentally measured value of this quantity is 184 N. For this case, the calculated value would be used as the reference, and the error would be expressed as

$$\%D = \frac{184 - 175}{175} \ 100 = 5.14\%$$

The interpretation of this result is that the *difference* between the two values is 5.14 percent of the reference value. Using the above value for the percent difference, we can write this difference as

$$0.0514 \ (175) = 184 - 175 = 9 \ \text{N}$$

In the above example, the compared value of 184 is greater than the reference value of 175. Thus, the percent difference is positive. If, in a second measurement, the value 172 N were recorded, the percent difference would be

$$\%D = \frac{172 - 175}{175} \ 100 = -1.71\%$$

The minus sign indicates that the compared value of 172 N is less than the reference value of 175 N.

A second method of comparing two numerical values is to form their ratio, with the reference value in the denominator. For the first example above, the ratio would be

$$\frac{184}{175} = 1.05$$

while in the second case the ratio is

$$\frac{172}{175} = 0.983$$

It may be seen that, when a ratio is used to compare two values, the results will never be negative, since it is assumed that positive and negative numbers are not compared with one another. If the ratio of the two numbers is greater than 1, it means that the compared value is greater than the reference value. If this ratio is less than 1, it means that the compared value is less than the reference value.

Both of the above methods of describing error are useful. The use of the %D generally results in a better "feeling" for the magnitude of the error, because of the percentage nature of this term.

1.43 Describe the technique of rounding off a number to three significant figures.

I The first step is to make a partition between the third and fourth significant figures:

$$673|42 \qquad 0.0164|679$$

A decimal point is imagined to exist between the partition line and the fourth significant figure. If the fraction to the right of the partition line is less than 0.5, the third digit is retained with its original value and all digits to the right of the third digit are discarded. If the fraction to the right of the partition line is greater than 0.5, the third digit is increased by 1 and the digits to the right of the third digit are discarded. For the two numbers above, the application of the above rules results in

$$67,342 \approx 67,300$$
$$0.0164679 \approx 0.0165$$

If the fraction to the right of the partition is exactly 0.5, one of two possible roundoff techniques may be used. In the first, the digit in the third place will be increased by 1 if the result of this operation is to make the digit in the third place an even number. The rationale for this particular roundoff technique is that the rounded number may be divided by 2 to obtain a result with no more significant figures than the rounded number has. In the second technique, the digit in the third place is increased by 1 regardless of whether this operation makes the third place digit odd or even. For the number

$$0.978|500$$

the first technique rounds off this value to 0.978, while the use of the second technique yields 0.979.

The second technique will be used consistently throughout this book. That is, if the imagined fraction to the right of the partition line is 0.5 or greater, the digit in the third place will be increased by 1. It should be observed that the technique described above may be used to round off numbers to any number of significant figures.

It is generally sufficient to state an answer to a problem in engineering mechanics in terms of *three* significant figures. Since an electronic pocket calculator will almost invariably be used for the calculations, the rounding off is usually done when the final answer is obtained.

All of the answers in the following problems, unless otherwise indicated, will be presented in terms of three significant figures.

1.44 Find the maximum error in a number that was rounded off to 237, and the maximum possible error when rounding a number to three significant figures.

I For the number 237, the digit in the third place will be assumed to be uncertain, but is known to be between 6 and 8. That is, the error in the digit in the third place is assumed to be no greater than 1. Thus the *maximum* percent difference is

$$\%D = \frac{1}{237} 100 = 0.42\%$$

Note that the same conclusion would be reached if the original number were 0.000237, or 237×10^6. The error inherent in a number is a function of *only* the number of significant figures.

The range of magnitudes of three-significant-figure numbers is 100 to 999. The corresponding range of percent difference is

$$\frac{1}{100} 100 = 1\% \qquad \text{to} \qquad \frac{1}{999} 100 = 0.1\%$$

Thus, when three-significant-figure numbers are used, the maximum error to be expected is less than or equal to 1 percent. This accuracy is usually more than sufficient for engineering calculations.

1.45 Shown are the numerical values recorded in 8 runs of an experiment in which the computed value is 150. (*a*) Find the percent difference and the comparison ratio for each run. (*b*) Which piece of experimental data appears to be unrepresentative of the experiment, and what criterion is used in making this determination?

155	151
157	150
143	130
144	156

▎(a) The definitions of the percent difference and comparison ratio given in Prob. 1.42 are used, with the results shown.

value	percent difference	comparison ratio
155	$\dfrac{155 - 150}{150}(100) = 3.3\%$	$\dfrac{155}{150} = 1.033$
157	$\dfrac{157 - 150}{150}(100) = 4.7\%$	$\dfrac{157}{150} = 1.047$
143	$\dfrac{143 - 150}{150}(100) = -4.7\%$	$\dfrac{143}{150} = 0.953$
144	$\dfrac{144 - 150}{150}(100) = -4\%$	$\dfrac{144}{150} = 0.960$
151	$\dfrac{151 - 150}{150}(100) = 0.7\%$	$\dfrac{151}{150} = 1.007$
150	$\dfrac{150 - 150}{150}(100) = 0\%$	$\dfrac{150}{150} = 1.000$
130	$\dfrac{130 - 150}{150}(100) = -13.3\%$	$\dfrac{130}{150} = 0.867$
156	$\dfrac{156 - 150}{150}(100) = 4\%$	$\dfrac{156}{150} = 1.040$

(b) The value of 130 seems to be unrepresentative of the experiment because the percent difference between this value and the computed value is much larger than any of the other percent differences.

1.46 Round off the following numbers to three, two, and one significant figures and find the percent difference between each rounded-off number and the original number, using the original number as the reference.

(a)	643.02	(f)	0.049984
(b)	13,270	(g)	2.0565×10^{-3}
(c)	1.9432×10^{6}	(h)	0.0079996
(d)	0.26985	(i)	0.44505
(e)	4.445×10^{4}	(j)	0.30005

▎Use the percent difference definition

$$\%D = \frac{A - B}{B} \, 100$$

where A is the rounded value and B is the original value.

(a) For the first value of 643.02

$$643.02 \approx 643 \qquad (3 \text{ significant figures})$$

$$\%D = \frac{643 - 643.02}{643.02} \, 100 = -0.003\%$$

$$643.02 \approx 640 \quad \text{(2 significant figures)}$$

$$\%D = \frac{640 - 643.02}{643.02}\,100 = -0.5\%$$

$$643.02 \approx 600 \quad \text{(1 significant figure)}$$

$$\%D = \frac{600 - 643.02}{643.02}\,100 = -6.7\%$$

(b) For $13{,}270 \approx 13{,}300$, $\%D = 0.2\%$; $13{,}270 \approx 13{,}000$, $\%D = -2\%$; $13{,}270 \approx 10{,}000$, $\%D = -24.6\%$.

(c) For $1.9432 \times 10^6 \approx 1.94 \times 10^6$, $\%D = -0.2\%$; $1.9432 \times 10^6 \approx 1.9 \times 10^6$, $\%D = -2.2\%$; $1.9432 \times 10^6 \approx 2 \times 10^6$, $\%D = 2.9\%$.

(d) For $0.26985 \approx 0.270$, $\%D = 0.06\%$; $0.26985 \approx 0.27$, $\%D = 0.06\%$; $0.26985 \approx 0.3$, $\%D = 11.2\%$.

(e) For $4.445 \times 10^4 \approx 4.45 \times 10^4$, $\%D = 0.1\%$; $4.445 \times 10^4 \approx 4.5 \times 10^4$, $\%D = 1.2\%$; $4.445 \approx 5 \times 10^4$, $\%D = 12.5\%$.

(f) For $0.049984 \approx 0.0500$, $\%D = 0.03\%$; $0.049984 \approx 0.050$, $\%D = 0.03\%$; $0.049984 \approx 0.05$, $\%D = 0.03\%$.

(g) For $2.0565 \times 10^{-3} \approx 2.06 \times 10^{-3}$, $\%D = 0.2\%$; $2.0565 \times 10^{-3} \approx 2.1 \times 10^{-3}$, $\%D = 2.1\%$; $2.0565 \times 10^{-3} \approx 2 \times 10^{-3}$, $\%D = -2.7\%$.

(h) For $0.0079996 \approx 0.00800$, $\%D = 0.005\%$; $0.0079996 \approx 0.0080$, $\%D = 0.005\%$; $0.0079996 \approx 0.008$, $\%D = 0.005\%$.

(i) For $0.44505 \approx 0.445$, $\%D = -0.01\%$; $0.44505 \approx 0.45$, $\%D = 1.1\%$; $0.44505 \approx 0.5$, $\%D = 12.3\%$.

(j) For $0.30005 \approx 0.300$, $\%D = -0.02\%$; $0.30005 \approx 0.30$, $\%D = -0.02\%$; $0.30005 \approx 0.3$, $\%D = -0.02\%$.

1.47 A pair of simultaneous algebraic equations is given as

$$x \cos 36° - 0.35y = 0$$
$$x \sin 36° - 2{,}940 + y = 0$$

Solve for x and y and verify that these two values are the correct solutions to the original set of equations.

▮ When an algebraic equation or a set of algebraic equations is being solved, the answers can be checked by substituting these solutions back into the original equations, to see whether these results satisfy the equations. The results of this substitution, however, must be interpreted with care. The following example shows one type of problem which may arise.

The first equation is solved for y, with the result

$$y = 2.31x$$

This value of y is then substituted in the second equation to get

$$x \sin 36° - 2{,}940 + 2.31x = 0$$
$$x = 1{,}010$$

Then y is found as

$$y = 2.31\,(1{,}010) = 2{,}330$$

As a check on the calculations, the values of x and y are substituted into the original equations. The results are

$$1{,}010 \cos 36° - 0.35\,(2{,}330) \overset{?}{=} 0 \qquad 1.61 \overset{?}{=} 0$$
$$1{,}010 \sin 36° - 2{,}940 + 2{,}330 \overset{?}{=} 0 \qquad -16.3 \overset{?}{=} 0$$

Inspection of the above two equations reveals that testing of the equality signs is inconclusive, since zero has no magnitude. A closer inspection reveals that both equations express the *difference* of terms as equal to zero. If the sum of positive terms is retained on the left side of the equals sign and the sum of the negative terms is moved to the right side of the equals sign, then the results are

$$1{,}010 \cos 36° \overset{?}{=} 0.35(2{,}330) \qquad 817 \overset{?}{=} 816$$
$$1{,}010 \sin 36° + 2{,}330 \overset{?}{=} 2{,}940 \qquad 2{,}920 \overset{?}{=} 2{,}940$$

The percent difference between the two terms in the first equation is 0.1 percent, while for the second equation this value is 0.7 percent. These errors are within the range of accuracy of three significant figures, and thus the computed values of x and y may be accepted as being correct.

1.4 REVIEW OF PLANE TRIGONOMETRY

1.48 A right triangle is shown in Fig. 1.48. Discuss the relationships among the sides A, B, and C and the angle θ.

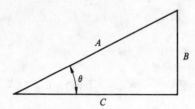

Fig. 1.48

▌ A, B, and C are the lengths of the three sides, and θ is the interior angle opposite leg B. The lengths of the three sides must satisfy the Pythagorean theorem, given by

$$A^2 = B^2 + C^2$$
$$A = +\sqrt{B^2 + C^2}$$

The basic trigonometric relationships among the three sides and angle θ are

$$\sin \theta = \frac{B}{A} \qquad \cos \theta = \frac{C}{A} \qquad \tan \theta = \frac{B}{C}$$

If A is eliminated between the first two equations, and the result is used with the third equation, it can be shown that

$$\tan \theta = \frac{\sin \theta}{\cos \theta}$$

Angle θ may be found from any of the above equations in the forms

$$\theta = \arcsin \frac{B}{A} = \sin^{-1} \frac{B}{A}$$

$$\theta = \arccos \frac{C}{A} = \cos^{-1} \frac{C}{A}$$

$$\theta = \arctan \frac{B}{C} = \tan^{-1} \frac{B}{C}$$

Throughout this book a situation will occur repeatedly in which side A and angle θ of a right triangle are known and it is desired to find the lengths of sides B and C. These two lengths may be expressed as

$$B = A \sin \theta \qquad C = A \cos \theta$$

The interpretation of the above two equations is that legs B and C are projections of length A on the vertical and horizontal axes. These equations may be used, together with $A^2 = B^2 + C^2$, to derive the trigonometric identity

$$\sin^2 \theta + \cos^2 \theta = 1$$

1.49 Express the *law of cosines* for the general triangle shown in Fig. 1.49.

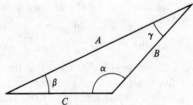

Fig. 1.49

▌ The law of cosines is written as

$$A^2 = B^2 + C^2 - 2BC \cos \alpha$$

This equation may be used to find side A when sides B and C and angle α are known, or to find angle α when sides A, B, and C are known. The above equation includes angle α, the angle opposite side A. It is left as an exercise for the reader to write the forms of the law of cosines in terms of angle β and angle γ.

1.50 Express the *law of sines* for the triangle in Prob. 1.49.

❚ The second basic equation that may be used with the general triangle is the law of sines. This equation has the form

$$\frac{A}{\sin \alpha} = \frac{B}{\sin \beta} = \frac{C}{\sin \gamma}$$

It may be observed that when $\alpha = 90°$, the law of cosines and the law of sines reduce to the relationships for the right triangle of Prob. 1.48.

1.51 For the right triangle in Fig. 1.51, find the length of side A, the sine, cosine, and tangent of angles θ_1 and θ_2, and the values of θ_1 and θ_2.

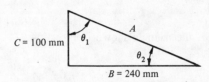

Fig. 1.51

❚
$$A = \sqrt{B^2 + C^2} = \sqrt{240^2 + 100^2} = 260 \text{ mm}$$

$$\sin \theta_1 = \frac{B}{A} = \cos \theta_2 = \frac{240}{260} = 0.923$$

$$\theta_1 = 67.4° \qquad \theta_2 = 22.6°$$

$$\cos \theta_1 = \frac{C}{A} = \sin \theta_2 = \frac{100}{260} = 0.385$$

$$\tan \theta_1 = \frac{B}{C} = \frac{240}{100} = 2.4$$

$$\tan \theta_2 = \frac{C}{B} = \frac{100}{240} = 0.417$$

1.52 The direction of line *aa* in Fig. 1.52 is given as $122°$. Find the length h.

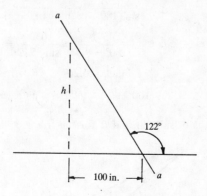

Fig. 1.52

❚
$$\tan(180° - 122°) = \frac{h}{100} \qquad h = 160 \text{ in}$$

1.53 Find the length of side A and the angles β and γ in the triangle shown in Fig. 1.53.

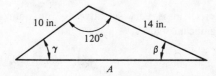

Fig. 1.53

❚ Using the law of cosines, we get

$$A^2 = B^2 + C^2 - 2BC \cos \alpha$$
$$= 10^2 + 14^2 - 2(10)14 \cos 120°$$
$$A = 20.9 \text{ in}$$

The law of sines (see Prob. 1.50) now appears as

$$\frac{20.9}{\sin 120°} = \frac{10}{\sin \beta} = \frac{14}{\sin \gamma}$$

$$\sin \beta = 0.414 \qquad \beta = 24.5°$$
$$\sin \gamma = 0.580 \qquad \gamma = 35.5°$$

As a check on the above calculations, the sum of the three interior angles is

$$120° + 24.5° + 35.5° = 180°$$

which is the expected result.

1.54 Find angles θ_1, θ_2, and θ_3 for the triangle shown in Fig. 1.54.

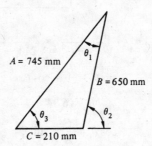

A = 745 mm   B = 650 mm   C = 210 mm   **Fig. 1.54**

❚ Using the law of cosines,

$$B^2 = A^2 + C^2 - 2AC \cos \theta_3 \qquad 650^2 = 745^2 + 210^2 - 2(745)210 \cos \theta_3$$
$$\cos \theta_3 = 0.564 \qquad \theta_3 = 55.7°$$

The law of sines has the form

$$\frac{A}{\sin(180° - \theta_2)} = \frac{B}{\sin \theta_3} = \frac{C}{\sin \theta_1}$$

$$\frac{745}{\sin(180° - \theta_2)} = \frac{650}{\sin 55.7°} = \frac{210}{\sin \theta_1}$$

$$\sin \theta_1 = 0.267 \qquad \theta_1 = 15.5°$$

$$\sin(180° - \theta_2) = 0.947 = \sin \theta_2 \qquad \theta_2 = 71.2°$$

1.55 Find angle θ_1 and the lengths of sides A and C for the triangle shown in Fig. 1.55.

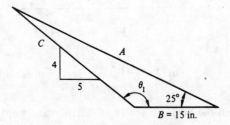

C   A   4   5 θ_1 25°   B = 15 in.   **Fig. 1.55**

❚ The solution to this problem is very sensitive to the number of significant figures used in the computations. Using *three* significant figures,

$$\theta_1 + \tan^{-1}\left(\frac{4}{5}\right) = 180° \qquad \theta_1 = 141°$$

The angle opposite side B is designated α, and

$$\alpha + 25° + \theta_1 = 180° = \alpha + 25° + 141° \qquad \alpha = 14°$$

The law of sines has the form

$$\frac{A}{\sin \theta_1} = \frac{B}{\sin \alpha} = \frac{C}{\sin 25°}$$

$$\frac{A}{\sin 141°} = \frac{15}{\sin 14°} = \frac{C}{\sin 25°}$$

$$A = 39.0 \text{ in} \qquad C = 26.2 \text{ in}$$

When using *four* significant figures,

$$\theta_1 + \tan^{-1}\left(\frac{4}{5}\right) = 180° \qquad \theta_1 = 141.3°$$

$$\alpha + 25° + 141.3° = 180° \qquad \alpha = 13.7°$$

$$\frac{A}{\sin 141.3°} = \frac{15}{\sin 13.7°} = \frac{C}{\sin 25°}$$

$$A = 39.6 \text{ in} \qquad C = 26.8 \text{ in}$$

The %D is now found, using the four-significant-figure values as the reference.

$$\text{Length } A: \qquad \%D = \frac{39.6 - 39.0}{39.6}(100) = 1.5\%$$

$$\text{Length } C: \qquad \%D = \frac{26.8 - 26.2}{26.8}(100) = 2.2\%$$

1.56   Find angles θ_1 and θ_2, and the length of side A, for the triangle shown in Fig. 1.56.

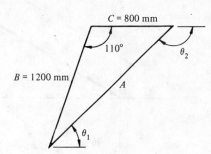

Fig. 1.56

▮ Using the law of cosines,

$$A^2 = B^2 + C^2 - 2BC \cos \alpha$$

$$= 1{,}200^2 + 800^2 - 2(1200)800 \cos 110°$$

$$A = 1{,}650 \text{ mm}$$

The angle between legs A and B is designated α, and the angle between legs A and C is β.   The law of sines has the form

$$\frac{A}{\sin 110°} = \frac{1{,}650}{\sin 110°} = \frac{1{,}200}{\sin \beta} = \frac{800}{\sin \alpha}$$

$$\alpha = 27.1° \qquad \beta = 43.1°$$

$$\theta_1 = 70° - \alpha = 42.9°$$

$$\theta_2 = 180° - \beta = 137°$$

1.5   SCALAR AND VECTOR QUANTITIES

1.57   What is a *scalar* quantity?

▮ A scalar quantity requires for its complete description only an algebraic statement of magnitude.   Typical examples of scalar quantities are mass, the radius of a circle, area, volume, energy, power, and temperature.   Certain scalar quantities, such as mass, radius, or volume, may have only positive values, while other scalar quantities may have both positive and negative values, as in the case of temperature.   A very important set of scalar quantities is the system of real numbers, since these values form the basis of all arithmetic operations.

1.58   What is a *vector* quantity?

❚   A vector is a quantity that requires, for its complete definition, a statement of three descriptive characteristics:

    Magnitude
    Direction
    Sense

A common example of a vector quantity is force   Other vector quantities that appear in engineering mechanics include moment, displacement, velocity, and acceleration.

Vectors may be added and subtracted, and they may be multiplied and divided by scalar quantities.   One vector *may not* be divided by another vector, since this operation is undefined.

A vector quantity may be represented graphically by an arrow, and this arrow may be either straight or curved.   The use of a curved arrow to represent a vector will be shown in Chap. 3 for the graphical representation of a term called moment.   In either case, the end of the arrow on which the arrowhead is drawn is referred to as the *tip*, while the other end is called the *tail*.

The *magnitude* of a vector describes the amount of the quantity.   The magnitude may also be interpreted as the *length* of an arrow, with reference to some arbitrarily chosen scale, which is selected to represent the vector.   As an example, suppose that the magnitude of a vector is given as 100 N.   Using a graphical scale of $1\,\text{N} = 1\,\text{mm}$, the length of the arrow that represents the vectors will be 100 mm.   This effect is shown in Fig. 1.58.

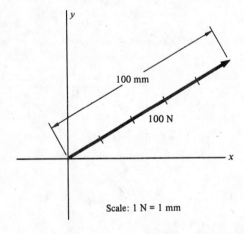

Scale: $1\,\text{N} = 1\,\text{mm}$

Fig. 1.58

The *direction* of a vector is the position, with respect to a fixed reference line, of the line along which the arrow representing the vector is drawn.   The position line along which a force acts is also referred to as the *line of action* of the force.   The direction may be described by the angle between a fixed reference line and the line along which the force acts.   The alternative method of describing the direction is to specify the slope of the line of action of the force.

Finally, the *sense* of a vector indicates which way the effect is acting.   If an arrow is used to graphically represent a vector, then the sense of the vector indicates the end of the arrow on which the tip is to be placed.   From this definition, the sense indicates which way the arrow is pointing.

Some authors use the term *direction* to describe both the direction, as defined above, and the sense.   With this type of description, a vector would be fully defined by stating its magnitude and direction.

Throughout this book all vector quantities will be represented by boldface type, such as the vector **A**.   The magnitude of the vector will be designated by an uppercase italic letter in regular type.   A typical statement, then, would be that A is the magnitude of the vector **A**.   The usual technique in writing or typing a symbol to represent a vector is to place a bar above or beneath the symbol, to represent the vector.   The vector **A** would then appear as $\bar{A}$ *or* $\underline{A}$.

1.59   Show the graphical construction of a vector.

❚   In the usual graphical construction of a vector, the direction is first established, as shown in Fig. 1.59*a*.   Using a convenient scale, the line segment that represents the magnitude is drawn next, as shown in

Fig. 1.59b. The third (and final) step is to place the tip on the end of the line segment in such a way as to properly represent the sense. It may be observed that there are two, and only two, possible choices for the sense of a vector. These two locations are the ends of the line segment that represents the vector, and this effect is shown in Fig. 1.59c.

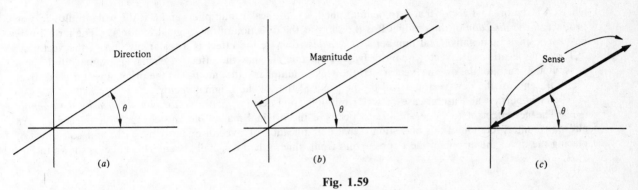

Fig. 1.59

1.60 The force shown in Fig. 1.60 is described as 100 N, acting through the origin of the coordinate system and up and to the right at an angle of 35° with the x axis. Give formal statements of the (a) magnitude, (b) direction, and (c) sense of the force.

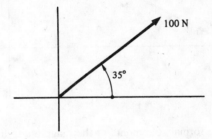

Fig. 1.60

▌ (a) The magnitude of the force is 100 N.

(b) The direction is 35°, measured in a counterclockwise sense from the positive x axis.

(c) The sense is up and to the right.

1.61 A force of 500 N acts in the xy plane, down and to the right at an angle of 50° with the vertical y axis. Sketch the force.

▌ In Fig. 1.61a both lines ab and cd have directions of 50° with the vertical. However, since the sense of the force is described as "down and to the right," this force must act along line cd. In addition, the statement of sense indicates that the tip of the arrow must be placed at the lower end of the line segment that represents the force. If a scale of 10 N = 1 mm is chosen, the final result is as shown in Fig. 1.61b.

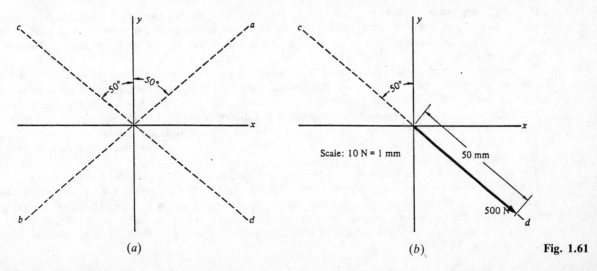

(a)　　　　　　　　　　　　　　　　　(b)　　　　　　　Fig. 1.61

1.62 Define the scalar multiplication operation.

❚ The scalar multiplication operation is defined by

$$\mathbf{B} = a\mathbf{A}$$

where **A** is the original vector that is to be multiplied, a is the scalar multiplier, and **B** is the scalar multiplication product. The *direction* of the product **B** is the same as the direction of the original vector **A**. The term a may be any positive or negative real number. If a is less than 1, the effect is a scalar division of the vector **A**.

The multiplication of a vector quantity by a scalar quantity has the effect of changing the magnitude, and possibly the sense, of the original vector. If the scalar multiplier a that multiplies the original vector is positive, then the sense of the new vector **B** will be the same as that of the original vector.

A very important and useful case of scalar multiplication occurs when *a vector is multiplied by the quantity* -1. The net effect of this operation is to preserve the original magnitude and change *only the sense*. An alternative interpretation of this operation is that multiplication of a vector quantity by -1 has the sole effect of placing the tip of the arrow on the opposite end from which it had originally been. This effect is shown in Fig. 1.62.

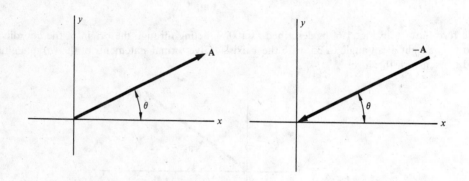

Fig. 1.62

1.63 Refer to Prob. 1.60 and give formal statements of the magnitude, direction, and sense of the original force of 100 N if this force is multiplied by (*a*) a factor of 8.8 and (*b*) a factor of -3.2.

❚ (*a*) The magnitude of the force is 880 N, and the direction and sense are the same as in Prob. 1.60.

(*b*) The magnitude of the force is 320 N. The direction of the force is the same as in Prob. 1.60, and the sense is down and to the left.

1.64 Give a summary of the basic concepts of mechanics presented in this chapter.

❚ The basic units of mechanics are length, time, force, and mass. If length, time, and force are chosen to be fundamental units, the resulting system is defined to be a gravitational system of units. The U.S. Customary System is such a system of units. The units of mass in this system may be derived from Newton's second law. The fundamental units in this system are the foot, second, and pound, and the derived mass unit is the slug.

If length, time, and mass are chosen to be fundamental units, the resulting system is defined to be an absolute system. The SI units form such a system. The units of force in this system may be derived from Newton's second law. The fundamental units in this system are the meter, second, and kilogram, and the derived force unit is the newton.

The specific weight of a material is defined to be the weight per unit volume of the substance. The density is defined to be the mass per unit volume. The specific gravity is the ratio of the density of the material to the density of water at 60°F. Pressure, or stress, is a force-induced quantity with the units force per unit area.

The units of a particular term may be converted to different units by multiplying the original term by a succession of factors, each of whose magnitude is 1, until the desired units are obtained.

The magnitude of a quantity tells how much of the quantity there is. A magnitude may be expressed by a number only.

The direction of a straight line defines its position. If the line lies in a plane, the direction may be defined either by the included angle between the line and a reference axis or by the slope of the line with respect to the reference axis. If the line has a general orientation in space, the direction may be defined by specifying the three angles—called direction angles—between the line and a set of mutually perpendicular coordinate axes. The cosines of these angles are called direction cosines. The sum of the squares of the three direction cosines is always 1.

The sense of an effect tells which way the effect is occurring.   Positive and negative signs may be used to distinguish opposite senses.

An error is the difference between two numbers.   The percent difference (%D) is a measure of the magnitude of the error.   The %D between numbers A and B, with respect to B, is given by

$$\%D = \frac{A - B}{B}\,100$$

If this term is positive, the compared term A is greater than the reference value B, while a negative value for %D indicates that A is less than B.   The magnitude of an error may also be described by forming the ratio of the compared value to the reference value.

It is generally adequate to express the results of an engineering computation to three significant figures.   The maximum error to be expected in using this number of significant figures is 1 percent.

A complete description of a scalar quantity requires a statement of its magnitude only.   A vector quantity, by comparison, requires for its complete description a statement of magnitude, direction, and sense.   Force is a vector quantity.   The magnitude of a force defines the amount of the quantity.   The direction describes the position, with respect to some known reference position, of the line of action of the force.   The sense indicates on which end, of the arrow which represents the force, the tip is to be placed.   Thus, the sense indicates which way the force is acting.

A vector may be multiplied by a scalar quantity.   The effect of this operation is to change either the magnitude or the sense, or both, of the original vector while leaving the direction unchanged.   Scalar multiplication of a vector by -1 has the single effect of reversing the sense of the original vector.

CHAPTER 2
Operations with Forces

2.1 ADDITION OF FORCES TO FIND RESULTANT FORCE, VECTOR ADDITION AND SUBTRACTION

2.1 Give a basic definition of the term *force*

❚ Force is defined as the effect of one physical body on another physical body. This force effect between two bodies has the interpretation of a "push" or a "pull" of one of these bodies on the other body. Force is the basic effect in *all* problems in statics, and the importance of a thorough understanding of the operations with forces cannot be overemphasized. In Chap. 3 the operations with a force-induced effect, called *moment*, will be studied in detail. One or both of the effects, designated by the terms *force* or *moment*, enter into the analysis of *all* problems in statics.

A force is a vector quantity which requires, for its complete description, a statement of its magnitude, direction, and sense.

2.2 Figure 2.2a shows two forces $\mathbf{F}_1$ and $\mathbf{F}_2$ that lie in the xy plane. Find the single force $\mathbf{F}$ that produces the same force effect as that produced by the original two forces.

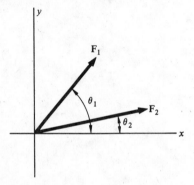

Fig. 2.2a

❚ Two or more forces may be combined, or added, to obtain a single force which produces the same effect as the original system of forces. This single vector is called the *sum*, or the *resultant force*, of the original system of forces.

The forces are redrawn in Fig. 2.2b, and lines parallel to the original two forces are drawn through the tip end of each force. This effect is shown by the dashed lines; this construction produces a parallelogram. A line is now drawn to join points a and b, and this line is the diagonal of the parallelogram. The tip of the arrow is placed at point b, the intersection of the two construction lines. Line segment ab, in Fig. 2.2c, then represents $\mathbf{F}$, the *vector sum*, or *resultant*, of forces $\mathbf{F}_1$ and $\mathbf{F}_2$. The magnitude of this force may be found by comparison of length ab with the scale used to establish the lengths of the original forces $\mathbf{F}_1$ and $\mathbf{F}_2$. The direction of this resultant force is given by angle θ. This single resultant force $\mathbf{F}$ will now produce the same force effect as the original pair of forces $\mathbf{F}_1$ and $\mathbf{F}_2$. The above method of adding two vectors is referred to as the *parallelogram law of force addition*.

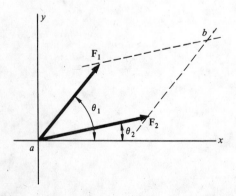

Fig. 2.2b

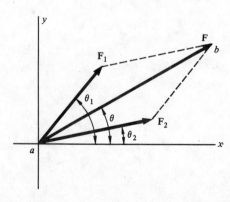

Fig. 2.2c

2.3   Show *the triangle law of force addition* method of summing two forces, and compare this law with the parallelogram law of force addition.

▌ An alternative method of summing two forces is called the triangle law of force addition. This technique is shown in Fig. 2.3a for the case of the two forces in Prob. 2.2. In using this method, the original two forces are redrawn so that the tail end of $\mathbf{F}_1$ is placed at the tip end of $\mathbf{F}_2$. The arrow drawn from the tail end of $\mathbf{F}_2$ to the tip end of $\mathbf{F}_1$ represents the resultant force $\mathbf{F}$. An alternative form of this construction is shown in Fig. 2.3b. Here, the tail end of $\mathbf{F}_2$ is placed at the tip end of $\mathbf{F}_1$. The arrow drawn from the tail end of $\mathbf{F}_1$ to the tip end of $\mathbf{F}_2$ is, then, the representation of the resultant force $\mathbf{F}$. In both cases, the direction of the resultant is given by angle θ. Figure 2.2c is redrawn in Fig. 2.3c, and the two regions A and B are identified. It may be seen that Fig. 2.3a corresponds to region B, and Fig. 2.3b corresponds to region A, of the original parallelogram construction. Thus it may be concluded that the parallelogram and triangle laws are equivalent methods of force addition.

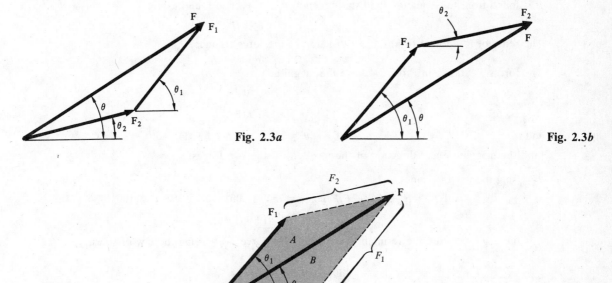

Fig. 2.3a

Fig. 2.3b

Fig. 2.3c

The parallelogram and triangle laws of force addition may be written in equation form as

$$\mathbf{F} = \mathbf{F}_1 \mathbin{+\!\!\!\!\to} \mathbf{F}_2$$

In this equation, $\mathbf{F}$ is the vector sum, or resultant force, of the two forces $\mathbf{F}_1$ and $\mathbf{F}_2$. The sign $+\!\!\!\!\to$ signifies a *vector addition*, in which the directions of the forces must be considered. If the vector nature of the addition defined by the above equation is clearly understood, then this equation may be written with the usual addition sign in the form

$$\mathbf{F} = \mathbf{F}_1 + \mathbf{F}_2$$

The parallelogram and triangle laws of force addition are the fundamental definitions of the operation of vector, or force, addition.

2.4   If vectors $\mathbf{F}_1$ and $\mathbf{F}_2$ in the general case in Prob. 2.2 are known, show how the magnitude, direction, and sense of the resultant force $\mathbf{F}$ may be found.

▌ The system of forces shown in Fig. 2.3a can be represented by the triangle shown in Fig. 2.4.
    The law of cosines may be written as

$$F^2 = F_1^2 + F_2^2 - 2F_1F_2 \cos \alpha$$

where

$$\alpha = 180° - \theta_{12}$$

$$\theta_{12} = \theta_1 - \theta_2$$

In a given problem, the terms in the right-hand side of the first equation are known. This equation may then be used to solve for F, the magnitude of the resultant force.

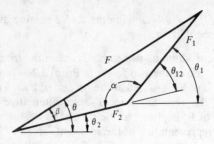

Fig. 2.4

The law of sines has the form

$$\frac{F_1}{\sin \beta} = \frac{F}{\sin \alpha}$$

β is found from this equation, and the direction θ of **F** is then found as

$$\theta = \theta_2 + \beta$$

The sense of **F** is found from the construction of the force triangle.

2.5 The forces F_1 and F_2 in Prob. 2.4 have the magnitudes

$$F_1 = 750 \text{ lb}$$

$$F_2 = 500 \text{ lb}$$

and $\qquad\qquad \theta_1 = 60° \qquad \theta_2 = 24° \qquad \theta_{12} = 60° - 24° = 36°$

Find the magnitude and direction of the resultant of the two forces.

❚ Using the law of cosines,

$$F^2 = F_1^2 + F_2^2 - 2F_1 F_2 \cos \alpha = 750^2 + 500^2 - 2(750)500 \cos (180° - 36°)$$

$$F = 1{,}190 \text{ lb}$$

β is the angle between the resultant force F and the force F_2. Using the law of sines,

$$\frac{F_1}{\sin \beta} = \frac{F}{\sin \alpha}$$

$$\frac{750}{\sin \beta} = \frac{1{,}190}{\sin (180° - 36°)} \qquad \beta = 21.7°$$

$$\theta = \theta_2 + \beta = 24° + 21.7° = 45.7°$$

The direction of the resultant force is 45.7° from the horizontal axis, acting up and to the right.

2.6 Show how the parallelogram law of force addition method may be extended to cases where more than two forces are to be summed to find a resultant force.

❚ The practical use of this technique is limited to force systems in which all the forces lie in the same plane. In using this method to add more than two forces, a pair of forces is first combined to obtain the resultant of these two forces. This resultant is then combined with one of the forces in the original system to obtain a new resultant. This process may be continued until the original system of forces is reduced to a single resultant force. This technique is illustrated in the following example for three forces that lie in the xy plane, as shown in Fig. 2.6a.

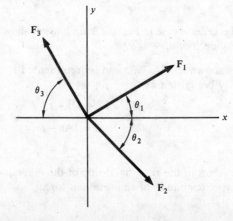

Fig. 2.6a

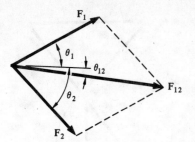

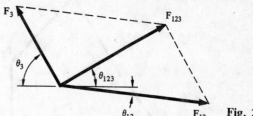

Fig. 2.6b

Fig. 2.6c

The forces $\mathbf{F}_1$ and $\mathbf{F}_2$ are first combined, by using the parallelogram law of force addition, as shown in Fig. 2.6b. The double subscript 12 on the force $\mathbf{F}_{12}$ indicates that it is the resultant of forces $\mathbf{F}_1$ and $\mathbf{F}_2$. The force $\mathbf{F}_{12}$ is next added to force $\mathbf{F}_3$, using the parallelogram law construction, as shown in Fig. 2.6c. The force $\mathbf{F}_{123}$ is the final value of the resultant of the three forces shown in Fig. 2.6a.

2.7   Explain how the triangle law of force addition may also be used to add more than two forces in the same plane.

▌ When this technique is used, the arrows representing the forces are drawn in a tip-to-tail fashion. The order in which these arrows are drawn is a matter of convenience. The arrow drawn from the tail end of the first arrow to the tip end of the last arrow is the resultant force of the system of forces. When more than two forces are added, the triangle law of force addition is referred to as the *polygon law of force addition*.

In practice, it is simpler to use the polygon law to find the sum of the forces than to use the parallelogram law, since the polygon law requires the drawing of fewer construction lines.

The equation form of the vector addition of several forces $\mathbf{F}_1$, $\mathbf{F}_2$, $\mathbf{F}_3$, . . . is

$$\mathbf{F} = \mathbf{F}_1 + \mathbf{F}_2 + \mathbf{F}_3 + \cdots$$

where $\mathbf{F}$ is the vector sum, or resultant force, of the system of forces. It should be remembered that the above equation represents a vector addition, and thus the directions of the forces must be taken into consideration.

2.8   Sketch the resultant force of the system of five forces shown in Fig. 2.8a by using the polygon law of force addition. All five forces lie in the xy plane.

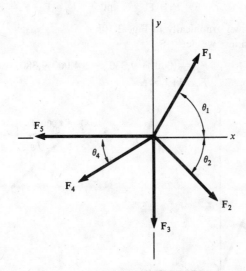

Fig. 2.8a

▌ Two possible graphical solutions are shown in Fig. 2.8b, and $\mathbf{F}$ is the resultant of the five forces. The magnitude F and direction θ of this resultant may be found from the figure if the forces are drawn to scale.

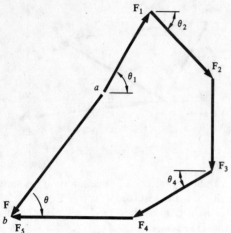

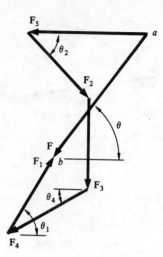

Fig. 2.8b

2.9 Describe the operation of *vector subtraction*.

❚ If **A** and **B** are two vectors, the difference of these two vectors is also a vector. If this vector difference is designated **C**, then

$$C = A - B$$

This equation may be written in the form

$$C = A + (-B)$$

Scalar multiplication of a vector (in this case, the vector **B**) by −1 has the net effect of changing the sense of the vector, as shown in Prob. 1.62. To subtract a vector, then, one simply changes the *sense* of this vector and proceeds as in addition. The operation of vector subtraction is not used frequently in static analysis. However, one application of this technique is shown in the following problem.

2.10 The resultant of two forces is known to be 9,380 N, acting up and to the left at 65° with the x axis. One of the two forces has a known magnitude of 4,000 N, and it acts to the right along the horizontal x axis. Find the magnitude, direction, and sense of the second force.

❚ The resultant force and the 4,000-N force are designated by **F** and F_1, respectively, and the system of these two forces is shown in Fig. 2.10a.
The unknown force is called F_2. From the problem statement,

$$F = F_1 + F_2 \quad \text{and} \quad F_2 = F - F_1 = F + (-F_1)$$

The above result is shown graphically in Fig. 2.10b.
Using the law of cosines, we get

$$F_2^2 = 4,000^2 + 9,380^2 - 2(4,000)9,380 \cos(90° + 25°)$$
$$F_2 = 11,600 \text{ N}$$

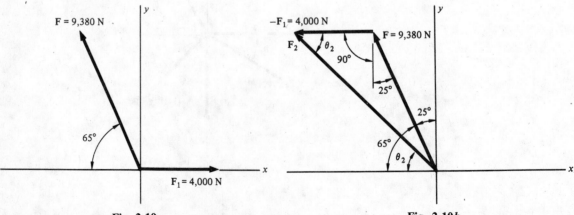

Fig. 2.10a   Fig. 2.10b

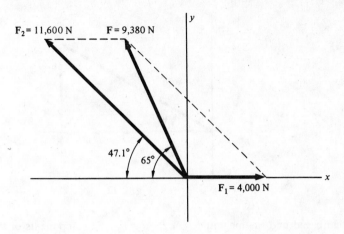

Fig. 2.10c

With the law of sines,

$$\frac{9,380}{\sin \theta_2} = \frac{11,600}{\sin (90° + 25°)} \qquad \theta_2 = 47.1°$$

The formal statement of the solution is that force F_2 has a magnitude of 11,600 N and acts up and to the left at an angle of 47.1° with the negative x axis. The complete force system is shown in Fig. 2.10c.

2.2 COMPONENTS OF A FORCE

2.11 What is meant by the *components* of a force?

❚ The preceding problems showed the techniques for combining two or more forces to obtain a single equivalent force, called the resultant force. The opposite situation occurs when a single force is given and we want to know what other system of forces would have the same effect as this original single force. The forces in the new system are called the *components* of the original force. *The components of a force, by definition, produce the same effect as the original force.* Thus, the vector sum of the components of a force must be equal to this force.

A force may be represented by any number of components along arbitrary directions. Figure 2.11 shows four possible sets of components F_1, F_2, F_3, F_4 of the force F. In the typical problem, the directions along which the components are desired are known. Thus, only the magnitudes and senses of the components are required to fully define these quantities.

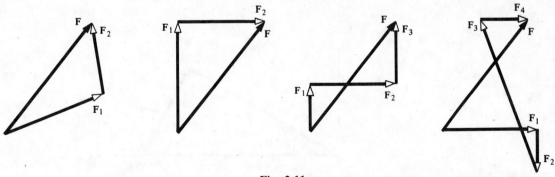

Fig. 2.11

2.12 Find the magnitudes of the two components F_1 and F_2 of the 150-kip force (1 kip = 1000 lb) shown in Fig. 2.12.

❚ The angle opposite side F_2 is 60°. Using the law of sines,

$$\frac{150}{\sin 90°} = \frac{F_1}{\sin 30°} = \frac{F_2}{\sin 60°}$$

$$F_1 = 75 \text{ kip} \qquad F_2 = 130 \text{ kip}$$

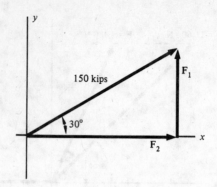

Fig. 2.12

2.13 Refer to the figure and find the magnitudes of F_1 and F_2, the components of the 150-kip force.

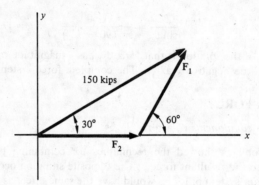

Fig. 2.13

▮ The angle opposite side F_2 is 30°, and the angle opposite the 150-kip force is 120°. Using the law of sines,

$$\frac{150}{\sin 120°} = \frac{F_1}{\sin 30°} = \frac{F_2}{\sin 30°}$$

$$F_1 = F_2 = 86.6 \text{ kip}$$

2.14 Refer to Fig. 2.14. Find the magnitudes of F_1 and F_2.

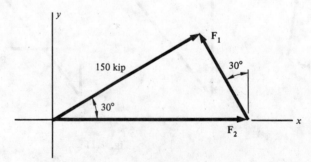

Fig. 2.14

▮ The angle opposite side F_2 is 90°, and the angle opposite the 150-kip force is 60°.

$$\frac{150}{\sin 60°} = \frac{F_1}{\sin 30°} = \frac{F_2}{\sin 90°}$$

$$F_1 = 86.6 \text{ kip} \qquad F_2 = 173 \text{ kip}$$

2.15 Find the magnitudes of components F_1 and F_2 for the force system shown in Fig. 2-15.

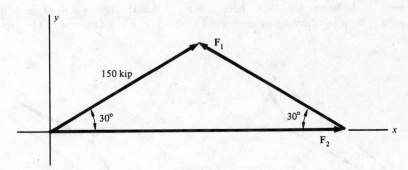

Fig. 2.15

❚ The angle opposite side F_2 is 120°.

$$\frac{150}{\sin 30°} = \frac{F_1}{\sin 30°} = \frac{F_2}{\sin 120°}$$

$$F_1 = 150 \text{ kip} \qquad F_2 = 260 \text{ kip}$$

2.16 Compare the solutions to Probs. 2.12 through 2.15 and discuss the correspondence between a force and its two components.

❚ The given force of 150 kip is the same in all four problems. The components F_1 and F_2 for all cases are listed in the table. It may be seen that there is a large range of values for both F_1 and F_2. There is *no* unique correspondence between a force and its two components. The magnitudes of these components are a function *only* of the specific directions of the components.

problem	F_1, kip	F_2, kip
2.12	75	130
2.13	86.6	86.6
2.14	86.6	173
2.15	150	260

2.17 Refer to Fig. 2.17a and find the components of the force **F** along the directions *aa* and *bb*.

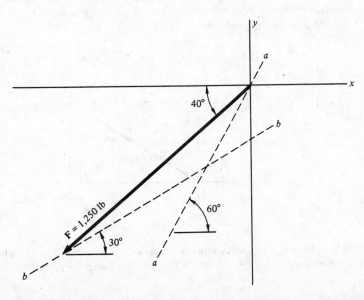

Fig. 2.17a

❚ By using the triangle law of force addition, the relationship between the given force and its components must be as shown in Figs. 2.17b or c.

The senses of the components have been chosen so that the vector sum of these two forces is equal to the original force.   By using the diagram shown in Fig. 2.17b, the force triangle can be represented as shown in Fig. 2.17d.

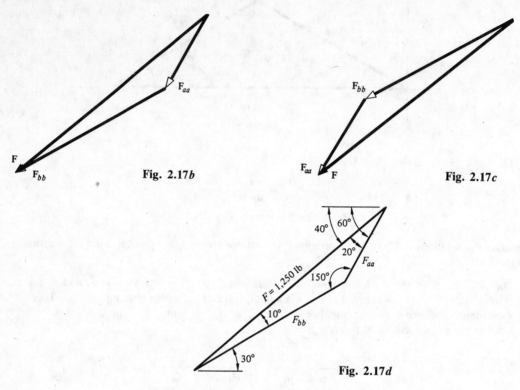

Fig. 2.17b   Fig. 2.17c

Fig. 2.17d

The law of sines is written as

$$\frac{F_{aa}}{\sin 10°} = \frac{F_{bb}}{\sin 20°} = \frac{1,250}{\sin 150°}$$

$$F_{aa} = 434 \text{ lb} \qquad F_{bb} = 855 \text{ lb}$$

2.3 RECTANGULAR COMPONENTS OF A FORCE

2.18   Figure 2.18a shows a force **F** in the xy plane.   The direction of the force is given by angle θ, measured from the x axis.   Find the two components of this force, along the x and y axes, in terms of the known magnitude F and the known direction θ of the original force **F**.

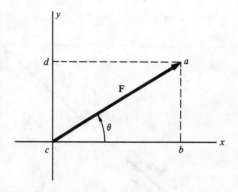

Fig. 2.18a

❚ In practice usually only two components of a force, which are perpendicular to each other, are required.   These two perpendicular components are called *rectangular components*, since they form the sides of a rectangle.

Since the directions of the two components are known to be along the x and y axes, we need to find only the magnitudes and senses of these component forces. Line segment ac is projected onto the x and y axes by dropping perpendiculars from point a to these axes. This construction, shown by the dashed lines in Fig. 2.18a, forms a rectangle. A rectangle is a special form of parallelogram, and the length of line segment ac is proportional to the magnitude of the force $\mathbf{F}$. From the parallelogram law of force addition of two vectors, it then follows that the lengths of the two sides of the rectangle must be proportional to the magnitudes of the two components of the force along the x and y axes.

The x component, represented by line segment cb, is designated F_x, and the y component, line segment cd, is called F_y. The force $\mathbf{F}$ and its two rectangular components are shown in Fig. 2.18b.

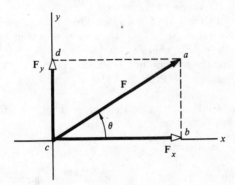

Fig. 2.18b

The given force F has known magnitude and direction, and it is assumed to be an inherently positive term. The components F_x and F_y of this force have known directions along the x and y axes, respectively.

By using the right-triangle trigonometric relationships, the magnitudes of the two rectangular components of the given force F are

$$F_x = F \cos \theta$$
$$F_y = F \sin \theta$$

The above equations express the components F_x and F_y in terms of the magnitude F and the direction θ of the given force. Two additional relationships will now be obtained which express F and θ in terms of the two components F_x and F_y. The two equations are squared, with the result

$$F_x^2 = F^2 \cos^2 \theta$$
$$F_y^2 = F^2 \sin^2 \theta$$

These equations are added to get

$$F_x^2 + F_y^2 = F^2 \cos^2 \theta + F^2 \sin^2 \theta = F^2(\sin^2 \theta + \cos^2 \theta)$$

By using the identity $\sin^2 \theta + \cos^2 \theta = 1$

$$F^2 = F_x^2 + F_y^2$$

the square root is taken of both sides of the equation, and the final result is

$$F = \sqrt{F_x^2 + F_y^2}$$

This equation expresses the magnitude of the given force F in terms of its two components along the x and y axes. In taking the square root, only the positive root is retained, since F was earlier assumed to be positive. The two components are divided, with the result

$$\frac{F_y}{F_x} = \frac{\cancel{F} \sin \theta}{\cancel{F} \cos \theta}$$

Recalling that

$$\tan \theta = \frac{\sin \theta}{\cos \theta}$$

we get for the final result

$$\tan \theta = \frac{F_y}{F_x}$$

This equation gives the direction θ of the original force F, in terms of its components along the x and y axes.

The above equations are very important fundamental results that express the relationships among a given force and its components along two perpendicular axes. Repeated below, these equations are perfectly general results that are true for all values of θ between 0 and 360°, and the term F in these equations is always positive.

$$F_x = F \cos \theta$$
$$F_y = F \sin \theta$$
$$F = \sqrt{F_x^2 + F_y^2}$$
$$\tan \theta = \frac{F_y}{F_x}$$

Finally, a component of a given force is considered to be positive or negative depending on whether it acts in the positive or negative coordinate sense. This effect is shown in Prob. 2.19.

It is always desirable to be able to check an engineering calculation. In using the above equations, only two of them will be required for the solution of a particular problem. The remaining two equations may then be used to check the computations (see Prob. 2.20).

2.19 A force F is given as 100 N. Find the x and y components of this force for (*a*) $\theta = 40°$, (*b*) $\theta = 140°$, (*c*) $\theta = 220°$, and (*d*) $\theta = 320°$.

▌ The configurations of the original system, and of the components, for the above four cases, are shown in Figs. 2.19*a* through *d*.

It may be observed from the results above that if the equations in Prob. 2.18 are used literally, with the direction angle θ always measured counterclockwise from the positive x axis, then the sense of the components *automatically* comes out of the associated sine or cosine function. These senses are positive in the positive coordinate senses. Finally, when a force is sketched, it does not need a minus sign affixed to it. The reason is that the minus sign has already been taken into consideration in determining the sense in which to draw the force.

In the practical solution of problems, it is usually more convenient to *restrict the angle* θ *to values between 0 and 90°* and to always measure this angle from either the positive or negative side of the horizontal x axis. For

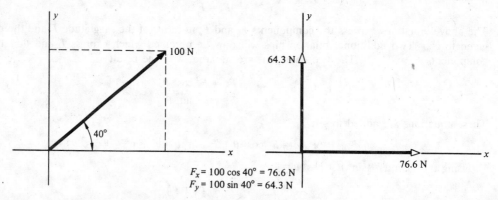

$F_x = 100 \cos 40° = 76.6$ N
$F_y = 100 \sin 40° = 64.3$ N

Fig. 2.19a

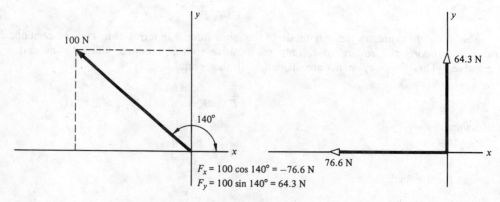

$F_x = 100 \cos 140° = -76.6$ N
$F_y = 100 \sin 140° = 64.3$ N

Fig. 2.19b

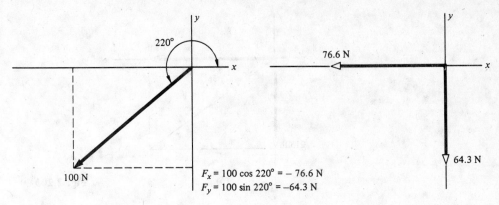

$F_x = 100 \cos 220° = -76.6$ N
$F_y = 100 \sin 220° = -64.3$ N

Fig. 2.19c

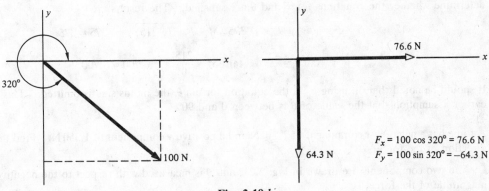

$F_x = 100 \cos 320° = 76.6$ N
$F_y = 100 \sin 320° = -64.3$ N

Fig. 2.19d

this range of values of θ, $\sin \theta$ and $\cos \theta$ are positive; thus the terms F_x and F_y will always be computed as positive quantities and would be used as such in all calculations. It is only in the final statement of the answer for a component that a minus sign would be used to indicate that the force component has a sense opposite that of the positive coordinate axis. If this interpretation were used in the present problem, the angle θ in all four cases would be 40°.

2.20 Find the x and y components of the force shown in Fig. 2.20a.

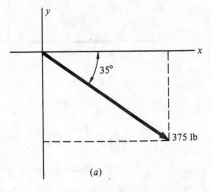

(a)

Fig. 2.20a

❚ The force is projected onto the xy axes, as indicated by the dashed lines in the figure.

$$F_x = 375 \cos 35° = 307 \text{ lb}$$
$$F_y = -375 \sin 35° = -215 \text{ lb}$$

When the equation for F_y is written, the minus sign is included to indicate that the y component of the force acts in the negative y-coordinate sense. The two components of the force are shown in Fig. 2.20b.

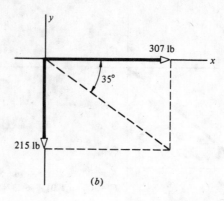

(b)   Fig. 2.20b

As a check on the above calculations, the numerical values obtained for the force components will be used to determine whether the equations for F and θ are satisfied.   The results are

$$F = \sqrt{F_x^2 + F_y^2} \qquad 375 \stackrel{?}{=} \sqrt{307^2 + (-215)^2} \qquad 375 \equiv 375$$

$$\tan\theta = \frac{F_y}{F_x} \qquad \tan 35° = \frac{215}{307} \qquad 0.700 \equiv 0.700$$

It should be noted that, in using F_y in the equation for $\tan\theta$, the minus sign is omitted.   This is because of the earlier assumption that the value of θ is between 0 and 90°.

2.21   A force has a negative x component of 2,740 N and a positive y component of 1,350 N.   Find the magnitude and direction of the force.

▮   The two components are drawn in Fig. 2.21, and θ is measured with respect to the negative x axis. F, the magnitude of the force, is

$$F = \sqrt{F_x^2 + F_y^2} = \sqrt{2,740^2 + 1,350^2} = 3,050 \text{ N}$$

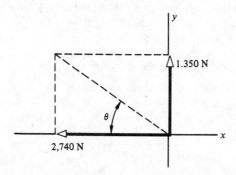

Fig. 2.21

The direction θ is found from

$$\tan\theta = \frac{F_y}{F_x} = \frac{1,350}{2,740} = 0.493$$

$$\theta = 26.2°$$

It should be noted that F_x is entered into the above equation as a positive number, even though it is a negative component in terms of the xy coordinate system.

2.22   The y component of a force is -80 lb.   The direction of the force is 55°, measured counterclockwise from the negative x axis.   Find the magnitude of the force and its x component.

▮   The given data are sketched in Fig. 2.22.

$$\tan\theta = \frac{F_y}{F_x} \qquad \tan 55° = \frac{80}{F_x} \qquad F_x = 56 \text{ lb}$$

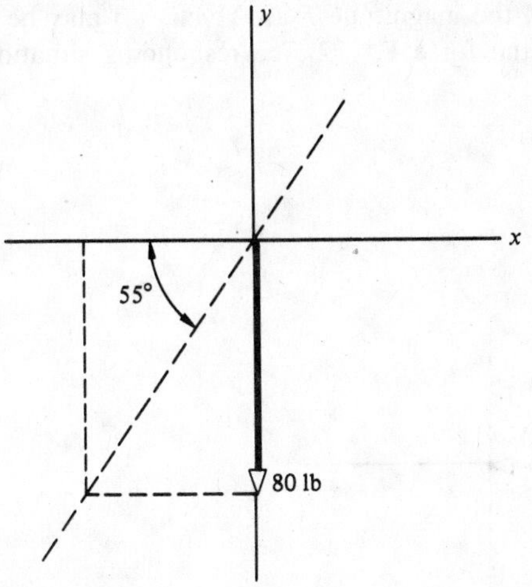

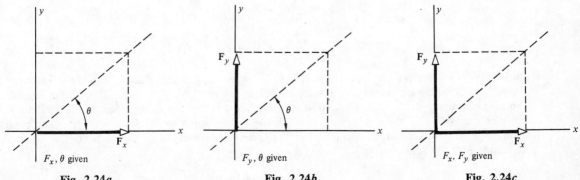

Fig. 2.22

The magnitude of the force is found from

$$F = \sqrt{F_x^2 + F_y^2} = \sqrt{56^2 + (-80)^2} = 97.7 \text{ lb}$$

As a check, we get

$$F_x = F \cos\theta \qquad 56 \stackrel{?}{=} 97.7 \cos 55° \qquad 56 \equiv 56$$
$$F_y = F \sin\theta \qquad 80 \stackrel{?}{=} 97.7 \sin 55° \qquad 80 \equiv 80$$

The formal statement of the solution is

$$F_x = -56 \text{ lb} \qquad F = 97.7 \text{ lb}$$

The first value for F_x is the result of a computation. With the restriction that θ be between 0 and 90°, this term will always be positive. The second value for F_x has the minus sign affixed to it to indicate that this force component actually acts to the left, or in the negative x-coordinate sense, in the figure. Readers should convince themselves that the two apparently contradictory statements for F_x are consistent results.

2.23 The x component of a force is $-2,600$ lb and the direction of the force is 127°, measured in a counterclockwise sense from the positive x axis. Find the y component, and the magnitude, of the resultant force.

❚ The angle between the force and the negative x axis is 53°.

$$F_y = F_x \tan\theta = 2,600 \tan 53° = 3,450 \text{ lb}$$

F_y acts in the positive y sense.

$$F = \frac{F_x}{\cos\theta} = \frac{2,600}{\cos 53°} = 4,320 \text{ lb}$$

2.24 The description of the components of a force along two perpendicular directions involves the four terms F, θ, F_x, and F_y. If any of the pairs F_x, θ; F_y, θ; F_x, F_y are given, the force F is uniquely defined, and these configurations are shown in 2.24a through c. Show that if the magnitude of the force and one component are given, the force is *not* uniquely defined.

Fig. 2.24a Fig. 2.24b Fig. 2.24c

▐ In Fig. 2.24d, F_x and the magnitude F are given. It may be seen that both of the directions 1 and 2 are possible orientations of the force $\mathbf{F}$. The corresponding situation for F_y, and the magnitude of F, given is shown in Fig. 2.24e.

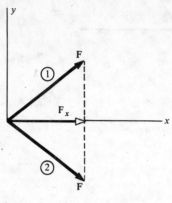

Fig. 2.24d   **Fig. 2.24e**

2.25   The y component of a force of 115 lb is 60 lb. What are the possible values of the x component, and the direction, of the force? The force may act up and to the left, or up and to the right, at an angle θ with the x axis.

$$\sin\theta = \frac{F_y}{F} = \frac{60}{115} \qquad \theta = 31.4°$$

▐ $$F_x = +F\cos\theta = \pm115\cos31.4° = \pm98.2\ \text{lb}$$

2.26   A force of 370 kN has an x component of -260 kN. Discuss the y component, and the direction, of the force. The force may act up and to the left, or down and to the left. The direction of the force is θ, measured either clockwise or counterclockwise from the negative x axis.

$$\cos\theta = \frac{F_x}{F} = \frac{260}{370} \qquad \theta = 45.4°$$

▐ $$F_y = \pm F\sin\theta = \pm370\sin45.4° = \pm263\ \text{kN}$$

2.27   The parallelogram, triangle, and polygon laws of force addition were shown to be equivalent graphical methods for finding the resultant of two or more forces lying in the same plane. Describe the method of summing the components to find the resultant of a system of forces.

▐ The first step in using this method is to find the x and y components of each force. A minus sign is affixed to any component that acts in a negative coordinate sense. The magnitude of the resultant force of the system of forces will be designated by the symbol F, and this force will have the components F_x and F_y. The relationships among the components of the individual forces in the system and the components of the resultant are defined by

$$F_x = \sum_{i=1}^{i=n} F_{ix} = F_{1x} + F_{2x} + \cdots + F_{nx}$$

$$F_y = \sum_{i=1}^{i=n} F_{iy} = F_{1y} + F_{2y} + \cdots + F_{ny}$$

The forces in the system are designated $1, 2, 3, \ldots, n$, so that n represents the number of forces that are summed. In the above equations, typical terms such as F_{1x} and F_{2x} represent the x and y components of forces 1 and 2, respectively. The interpretation of these equations is that the x component of the resultant force is the algebraic summation of the x components of all the forces acting in the system. The y component of the resultant force is the algebraic summation of the y components of all the forces acting in the system. The magnitude F of the resultant force is found from

$$F = \sqrt{F_x^2 + F_y^2}$$

The direction θ of the resultant force is given by

$$\tan \theta = \frac{F_y}{F_x}$$

Again, in using the above equation, F_x and F_y will be entered as positive quantities so that θ will be an angle, with respect to the horizontal x axis, that lies between 0 and 90°.

2.28 Show how the results in Prob. 2.27 can be arranged in a tabular form.

❚ If there are several forces in the system, the terms required are most conveniently handled in a tabular form. Table 2.1 shows a recommended format.

TABLE 2.1				
(1) Force	(2) x Component	(3) Value	(4) y Component	(5) Value
1	785 cos 52°	483 N	−785 sin 52°	−619 N
Sum	F_x		F_y	

Each force will require a single row in the table, and the number of each force is entered in column 1. In columns 2 and 4, the functional forms of the expressions for the x and y components of each force are entered. A minus sign is written in front of a term if the force component acts in the negative coordinate sense. For the sample expressions shown in the table, it would be concluded that force F_1, of magnitude 785 N, acts down and to the right at an angle of 52° with the positive x axis. This effect is shown in Fig. 2.28.

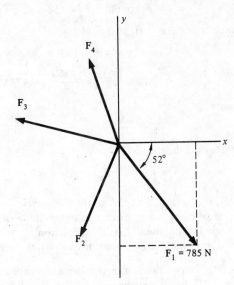

Fig. 2.28

The numerical values of the expressions in columns 2 and 4 are computed and written in columns 3 and 5. The last row of the table is reserved for the components of the resultant force. The numerical values in columns 3 and 5 are summed algebraically, and these quantities are entered into this last row. These two sums are now the x and y components of the resultant force $\mathbf{F}$.

The final steps in the solution are to find the magnitude and direction of this resultant force. Note that the table is a display of all of the terms in the two equations for F_x and F_y.

2.29 The resultant force in Prob. 2.5 was found by using the law of cosines. Check the value of this resultant force by using the method of summation of components.

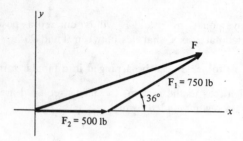

Fig. 2.29

▮ The original system, shown in Prob. 2.4, is redrawn in Fig. 2.29, with force F_2 along the x axis. The components of the resultant force F are then

$$F_x = 500 + 750 \cos 36° = 1,110 \text{ lb}$$
$$F_y = 750 \sin 36° = 441 \text{ lb}$$
$$F = \sqrt{F_x^2 + F_y^2} = \sqrt{1,110^2 + 441^2} = 1,190 \text{ lb}$$

This result agrees with the solution obtained in Prob. 2.5.

2.30 Show how the use of rectangular components of a force would have been a more efficient method of solving Probs. 2.12 through 2.15 than the use of the law of sines.

▮ Prob. 2.12
$$F_1 = 150 \sin 30° = 75 \text{ kip}$$
$$F_2 = 150 \cos 30° = 130 \text{ kip}$$

Prob. 2.13   Projecting forces on the y axis,

$$F_1 \sin 60° = 150 \sin 30° \qquad F_1 = 86.6 \text{ kip}$$

For projection of the forces on the x axis,

$$F_1 \cos 60° + F_2 = 150 \cos 30° \qquad F_2 = 86.6 \text{ kip}$$

Prob. 2.14   Since the force triangle is a right triangle,

$$F_1 = 150 \tan 30° = 86.6 \text{ kip}$$

$$F_2 = \frac{150}{\cos 30°} = 173 \text{ kip}$$

Prob. 2.15   From symmetry,

$$F_1 = 150 \text{ kip}$$

Projecting forces on the x axis,

$$F_2 = 2(150 \cos 30°) = 260 \text{ kip}$$

2.31 A force is given as 3,470 N, with a direction 219° counterclockwise from the positive x axis. It is to be replaced by three components. One of these components has a magnitude of 5,000 N and the direction and sense of the negative y axis. The second component has the direction of the x axis, and the direction of the third component is 40° clockwise from the x axis. Find the magnitudes and senses of the two unknown components.

▮ The given force has a direction of 39° counterclockwise from the negative x axis. The three components are designated F_1, F_2, and F_3. F_1 has a magnitude of 5,000 N, with the direction and sense of the negative y axis. F_2 has the direction of the x axis and is assumed to act in the negative x sense. F_3 is assumed to be acting up and to the left.

Projecting the forces on the y axis gives

$$3,470 \sin 39° + F_3 \sin 40° = 5,000 \qquad F_3 = 4,380 \text{ N}$$

The projection on the x axis yields

$$3,470 \cos 39° = F_3 \cos 40° + F_2 \qquad F_2 = -659 \text{ N}$$

Since F_2 is negative, this force must have a sense opposite to that which was originally assumed. Thus, component F_2 acts in the positive x sense. It is left as an exercise for the reader to make a sketch of the system of forces and confirm the above results.

2.32 Using the method of summation of components, find the magnitude and direction of the resultant force of the system of forces shown in Fig. 2.32a. All the forces lie in the xy plane.

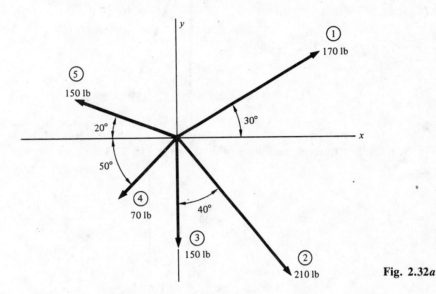

Fig. 2.32a

▌ The x and y components of each of the five forces are listed in Table 2.2. A minus sign is affixed to the x components of F_4 and F_5 and to the y components of F_2, F_3, and F_4, since these components act in the negative coordinate senses.

	TABLE 2.2			
Force	x Component	Value	y Component	Value
1	$170 \cos 30°$	147	$170 \sin 30°$	85
2	$210 \sin 40°$	135	$-210 \cos 40°$	-161
3	—	0	—	-150
4	$-70 \cos 50°$	-45	$-70 \sin 50°$	-54
5	$-150 \cos 20°$	-141	$150 \sin 20°$	51
Sum	F_x	96	F_y	-229

The magnitude and direction of the resultant force are then found from

$$F = \sqrt{F_x^2 + F_y^2} = \sqrt{96^2 + (-229)^2} = 248 \text{ lb}$$

$$\tan \theta = \frac{F_y}{F_x} = \frac{229}{96} = 2.39$$

$$\theta = 67.3°$$

As before, in solving for the direction θ, all minus signs are omitted from F_x or F_y. The resultant force and its components are shown in Fig. 2.32b.

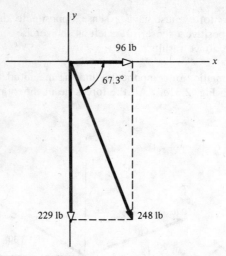

Fig. 2.32*b*

2.33   Find the resultant force of the force system shown in Fig. 2.33.

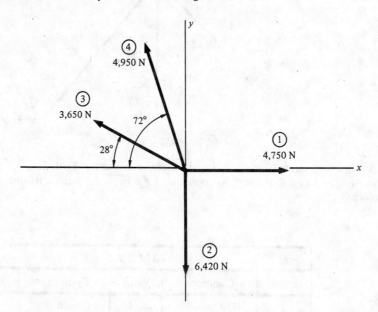

Fig. 2.33

❚ The force components are entered and summed in Table 2.3.   It may be seen that

$$F_x = F_y = 0$$

Since both components of the resultant force are zero, it follows that the resultant force must also be zero.   A force system whose resultant is zero has a very special significance.   Under these conditions, the force system is described as being in *equilibrium*.   This effect will be studied in detail in subsequent chapters of this book.

TABLE 2.3				
Force	x Component	Value	y Component	Value
1	—	4,750	—	0
2	—	0	—	−6,420
3	−3,650 cos 28°	−3,220	3,650 sin 28°	1,710
4	−4,950 cos 72°	−1,530	4,950 sin 72°	4,710
Sum	F_x	0	F_y	0

2.34 The resultant force is to be 160 N at 60° with the negative x axis, as shown in Fig. 2.34.   Find the values of F_4 and θ that will produce this resultant force.

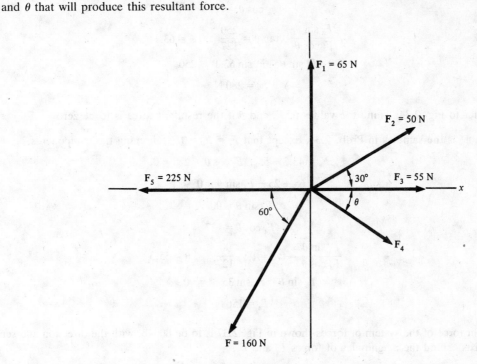

Fig. 2.34

▮ The components of the forces are shown in Table 2.4.   The components are summed to obtain

$$F_x = 43.3 + 55 + F_4 \cos \theta - 225 = -80$$
$$F_y = 65 + 25 - F_4 \sin \theta = -139$$
$$F_4 \sin \theta = 229$$
$$F_4 \cos \theta = 46.7$$

The above two equations are divided, with the result

$$\frac{F_4 \sin \theta}{F_4 \cos \theta} = \tan \theta = \frac{229}{46.7} \quad \theta = 78.4°$$

$$F_4 \sin \theta = F_4 \sin 78.4° = 229 \qquad F_4 = 234\ \text{N}$$

TABLE 2.4				
Force	x Component	Value	y Component	Value
1	—	0	—	65
2	$50 \cos 30°$	43.3	$50 \sin 30°$	25
3	—	55	—	0
4	$F_4 \cos \theta$	$F_4 \cos \theta$	$-F_4 \sin \theta$	$-F_4 \sin \theta$
5	—	−225	—	0
F	$-160 \cos 60°$	−80	$-160 \sin 60°$	−139

2.35 Refer to Fig. 2.34.   If the resultant force of 160 N is to have the direction and sense of the negative y axis, determine the values of F_4 and θ that will produce this force.

▮ The table is the same as in Prob. 2.34, except that $F_x = 0$ and $F_y = -160$ N.   Summing the components,

$$F_x = 43.3 + 55 + F_4 \cos \theta - 225 = 0$$
$$F_y = 65 + 25 - F_4 \sin \theta = -160$$

$$F_4 \sin \theta = 250$$
$$F_4 \cos \theta = 127$$
$$\frac{F_4 \sin \theta}{F_4 \cos \theta} = \tan \theta = \frac{250}{127} \qquad \theta = 63.1°$$
$$F_4 \sin \theta = F_4 \sin 63.1° = 250$$
$$F_4 = 280 \text{ N}$$

2.36 Again, refer to Fig. 2.34. Find the values of F_4 and θ if the resultant force is to be zero.

┃ The table is the same as in Prob. 2.34, except that $F_x = F_y = 0$. Summing the components,

$$F_x = 43.3 + 55 + F_4 \cos \theta - 225 = 0$$
$$F_y = 65 + 25 - F_4 \sin \theta = 0$$
$$F_4 \sin \theta = 90$$
$$F_4 \cos \theta = 127$$
$$\frac{F_4 \sin \theta}{F_4 \cos \theta} = \tan = \theta = \frac{90}{127} \qquad \theta = 35.3$$
$$F_4 \sin \theta = F_4 \sin 35.3° = 90$$
$$F_4 = 156 \text{ N}$$

2.37 The resultant force of the system of forces shown in Fig. 2.37 is to be 800 lb, with the direction and sense of the positive x axis. Find the magnitudes of forces F_1 and F_2.

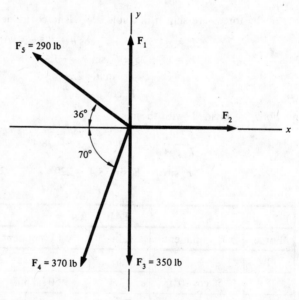

Fig. 2.37

┃ Summing the components,

$$F_x = F_2 - 290 \cos 36° - 370 \cos 70° = 800 \text{ lb}$$
$$F_2 = 1,160 \text{ lb}$$
$$F_y = F_1 + 290 \sin 36° - 370 \sin 70° - 350 = 0$$
$$F_1 = 527 \text{ lb}$$

2.38 Assuming that the value of the resultant force is to be zero, recalculate your results for Prob. 2.37.

┃ Since $F_y = 0$, the value of F_1 is the same as in Prob. 2.37, given as

$$F_1 = 527 \text{ lb}$$
$$F_x = F_2 - 290 \cos 36° - 370 \cos 70° = 0$$
$$F_2 = 361 \text{ lb}$$

2.39   Show how the analytic methods developed for finding the resultant of a system of forces in a single plane may be extended to the case where the forces have arbitrary directions in space.

❚   This arrangement of forces is called a *three-dimensional* force system.   Each of the forces in such a system is referenced to the same *xyz* coordinate system.   The orientation of the lines of action of the forces, with respect to the coordinate axes, may be specified by direction angles.   A direction angle is the angle between the force and a positive coordinate axis, measured in the plane containing the force and this axis.   There will be three such angles for a force that is referenced to a three-dimensional coordinate system.   These direction angles are designated θ_x, θ_y, and θ_z, and they are defined as

$$\cos \theta_x = \frac{F_x}{F} \qquad \cos \theta_y = \frac{F_y}{F} \qquad \cos \theta_z = \frac{F_z}{F}$$

The three direction angles are shown in Fig. 2.39.

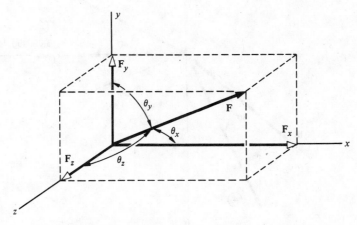

Fig. 2.39

In using the above equations, the three components F_x, F_y, and F_z must be entered with their appropriate algebraic sign.   All possible values of these three direction angles lie between 0 and 180°.   These three angles must also satisfy the relationship

$$\cos^2 \theta_x + \cos^2 \theta_y + \cos^2 \theta_z = 1$$

If any two direction angles are known, the above equation may be used to find the value of the third direction angle.   If the direction of a force is given by three direction angles, then the three rectangular components of this force are

$$F_x = F \cos \theta_x \qquad F_y = F \cos \theta_y \qquad F_z = F \cos \theta_z$$

When these three equations are used, the sense will automatically come from the cosine function, since $\cos \theta$ is negative for $90° < \theta < 180°$.   Thus, no decision must be made as to whether a minus sign should be affixed to a particular component.   This effect of the minus sign coming from the solution is comparable to the effect seen in Prob. 2.19.

    The direction of a force may also be described *in terms of angles with respect to fixed reference planes*.   There is no general solution for these angles, and they must be determined in each case.

    The initial step in finding the resultant force of a three-dimensional force system is to find the *x*, *y*, and *z* components of each force in the system.   These components are obtained by projecting the force onto the coordinate axes.   The resultant force is designated **F**, with the components F_x, F_y, and F_z.   The magnitudes of the components of the resultant force are then found from

$$F_x = \sum_{i=1}^{i=n} F_{ix} = F_{1x} + F_{2x} + \cdots + F_{nx}$$

$$F_y = \sum_{i=1}^{i=n} F_{iy} = F_{1y} + F_{2y} + \cdots + F_{ny}$$

$$F_z = \sum_{i=1}^{i=n} F_{iz} = F_{1z} + F_{2z} + \cdots + F_{nz}$$

As before, *n* represents the total number of forces in the system.   The information required may be recorded in a table, such as the one in Prob. 2.28.

Finally, the relationship between the magnitudes of the resultant force and its components is given by

$$F = \sqrt{F_x^2 + F_y^2 + F_z^2}$$

2.40 Find the magnitude and direction of the resultant force of the three-dimensional force system shown in Fig. 2.40*a*.

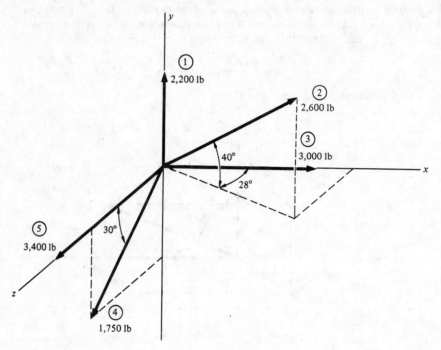

Fig. 2.40*a*

❚ Force F_1 lies along the *y* axis and is completely defined by

$$F_1 = F_{1y} = 2,200 \text{ lb}$$

Force F_2 is redrawn in 2.40*b*. This force lies in a plane that (1) contains the *y* axis and (2) is referenced to the *xy* plane by the 28° angle. Force F_2, in this plane, is referenced to the *zx* plane by the 40° angle. It should be noted that *neither* of these angles is a direction angle, since neither angle is measured directly from the line of action of the force to a coordinate axis. This force is first resolved into two rectangular components along the *y* axis and in the *zx* plane. The component in the *zx* plane is then resolved into rectangular components along the *z* and *x* axes, as shown in the figure. This completes the resolution of force F_2 into components along the *x*, *y*, and *z* axes.

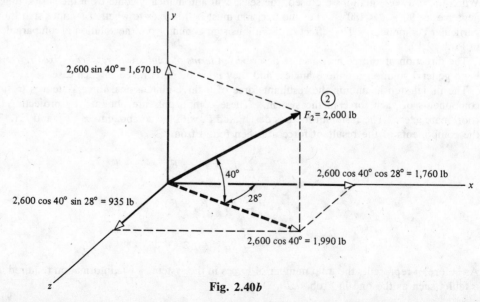

Fig. 2.40*b*

Force F_3 lies along the x axis and has the single component

$$F_3 = F_{3x} = 3,000\,\text{lb}$$

Force F_4 lies in the yz plane, and its x component is zero. This force is shown in true view in Fig. 2.40c. The force is resolved into rectangular components along the y and z axes, as shown in the figure.

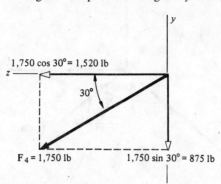

Fig. 2.40c

Force F_5 lies along the z axis and is expressed by

$$F_5 = F_{5z} = 3,400\,\text{lb}$$

Table 2.5 summarizes the above information. Since direction cosines were *not* used to find the force components, minus signs must be affixed to any component that acts in the negative coordinate sense.

TABLE 2.5					
Force	x Component	Value	y Component	Value	z Component	Value
1	—	0	—	2,200	—	0
2	2,600 cos 40° cos 28°	1,760	2,600 sin 40°	1,670	2,600 cos 40° sin 28°	935
3	—	3,000	—	0	—	0
4	—	0	−1,750 sin 30°	−875	1,750 cos 30°	1,520
5	—	0	—	0	—	3,400
Sum	F_x	4,760	F_y	3,000	F_z	5,860

From the table, the three components of the resultant force are

$$F_x = 4,760\,\text{lb}$$
$$F_y = 3,000\,\text{lb}$$
$$F_z = 5,860\,\text{lb}$$

The magnitude of F is found from

$$F = \sqrt{F_x^2 + F_y^2 + F_z^2} = \sqrt{4,760^2 + 3,000^2 + 5,860^2}$$
$$= 8,120\,\text{lb}$$

The resultant force and its components are shown in Fig. 2.40d.

The direction angles θ_x, θ_y, and θ_z are found from

$$\cos\theta_x = \frac{4,760}{8,120} \qquad \theta_x = 54.1°$$

$$\cos\theta_y = \frac{3,000}{8,120} \qquad \theta_y = 68.3°$$

$$\cos\theta_z = \frac{5,860}{8,120} \qquad \theta_z = 43.8°$$

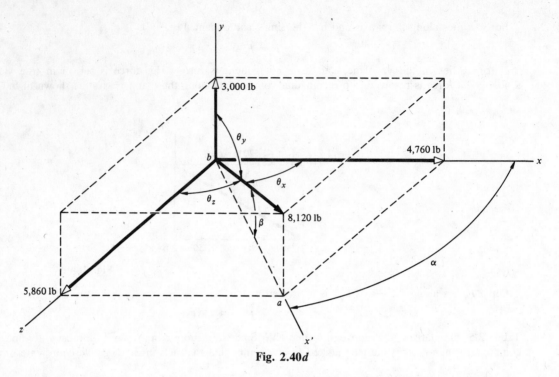

Fig. 2.40d

2.41 Show how the direction of the resultant force in Problem 2.40 may be expressed in terms of the angles α and β shown in Fig. 2.40d.

▮ The angle α measures the location of the plane containing the resultant and the y axis. β measures the location of the resultant in this plane, with respect to the zx plane. These two angles will now be computed. Figure 2.41a shows the zx plane in true view. Line ab is the projection of the resultant force onto this plane, along the x' axis. From triangle abc,

$$\tan \alpha = \frac{5,860}{4,760} = 1.23 \qquad \alpha = 50.9°$$

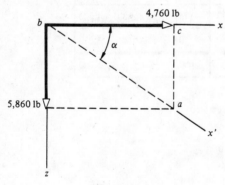

Fig. 2.41a

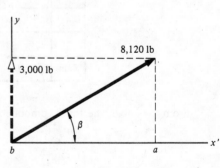

Fig. 2.41b

Figure 2.41b shows the yx' plane in true view. Angle β is found from

$$\sin \beta = \frac{3,000}{8,120} = 0.369 \qquad \beta = 21.7°$$

It may be noted from Fig. 2.40d that a geometric requirement is

$$\theta_y + \beta = 90°$$

Using the value above for β and the value of θ_y from Prob. 2.40, we get

$$68.3° + 21.7° \stackrel{?}{=} 90° \qquad 90° \equiv 90°$$

In general, the use of two angles such as α and β is a more convenient way of expressing the direction of the resultant than the use of the direction angles.

2.42 The direction of a force **F** is given by the two angles α and β shown in Fig. 2.42. Find the general forms for the x, y, and z components of the force, in terms of **F**, α, and β.

▮ The projection of the force on the xz plane is

$$F_{xz} = F \cos \beta$$

The force is now projected onto the x and z axes to obtain the x and z components, with the forms

$$F_x = F_{xz} \cos \alpha = F \cos \beta \cos \alpha$$
$$F_z = F_{xz} \sin \alpha = F \cos \beta \sin \alpha$$

In order to find F_y, the force is projected onto the y axis, with the result

$$F_y = F \sin \beta$$

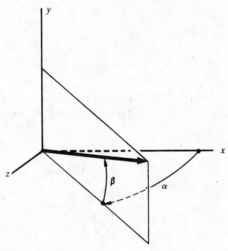

Fig. 2.42

2.43 Check the results found in Prob. 2.42

$$F = \sqrt{F_x^2 + F_y^2 + F_z^2}$$
$$\overset{?}{=} \sqrt{(F \cos \beta \cos \alpha)^2 + (F \sin \beta)^2 + (F \cos \beta \sin \alpha)^2}$$
$$\overset{?}{=} \sqrt{F^2(\cos^2 \beta \cos^2 \alpha + \sin^2 \beta + \cos^2 \beta \sin^2 \alpha)}$$
$$\overset{?}{=} F\sqrt{\cos^2 \beta \underbrace{(\sin^2 \alpha + \cos^2 \alpha)}_{1} + \sin^2 \beta}$$
$$\overset{?}{=} F\sqrt{\underbrace{\sin^2 \beta + \cos^2 \beta}_{1}} \qquad F \equiv F$$

2.44 Refer to Fig. 2.44. Find the magnitude of the resultant force of the system of forces, the direction of this resultant force in terms of the three direction angles and the angles α and β.

$$F = \sqrt{F_x^2 + F_y^2 + F_z^2}$$
$$= \sqrt{3{,}400^2 + 1{,}800^2 + 2{,}600^2}$$
$$= 4{,}640 \text{ lb}$$

$$\cos \theta_x = \frac{F_x}{F} = \frac{3{,}400}{4{,}640} \qquad \theta_x = 42.9°$$

$$\cos \theta_y = \frac{F_y}{F} = \frac{1{,}800}{4{,}640} \qquad \theta_y = 67.2°$$

$$\cos \theta_z = \frac{F_z}{F} = \frac{2{,}600}{4{,}640} \qquad \theta_z = 55.9°$$

$$\tan \alpha = \frac{F_z}{F_x} = \frac{2{,}600}{3{,}400} \qquad \alpha = 37.4°$$

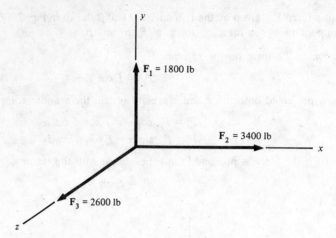

Fig. 2.44

$$\sin \beta = \frac{F_y}{F} = \frac{1,800}{4,640} = \beta = 22.8°$$

2.45 For the system of forces shown in Fig. 2.45, find the magnitude of the resultant force and the direction of the resultant force in terms of the three direction angles and the angles α and β.

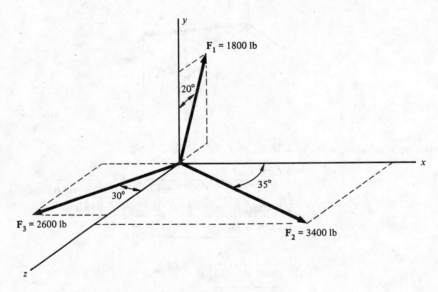

Fig. 2.45

❙ The three forces are projected onto the x, y, and z axes, with the results

$$F_x = 3,400 \cos 35° - 2,600 \sin 30°$$
$$= 1,490 \text{ lb}$$
$$F_y = 1,800 \cos 20° = 1,690 \text{ lb}$$
$$F_z = 2,600 \cos 30° + 3,400 \sin 35° - 1,800 \sin 20° = 3,590 \text{ lb}$$
$$F = \sqrt{F_x^2 + F_y^2 + F_z^2}$$
$$= \sqrt{1,490^2 + 1,690^2 + 3,590^2}$$
$$= 4,240 \text{ lb}$$

$$\cos \theta_x = \frac{F_x}{F} = \frac{1,490}{4,240} \qquad \theta_x = 69.4°$$

$$\cos \theta_y = \frac{F_y}{F} = \frac{1,690}{4,240} \qquad \theta_y = 66.5°$$

$$\cos \theta_z = \frac{F_z}{F} = \frac{3,590}{4,240} \qquad \theta_z = 32.1°$$

$$\tan \alpha = \frac{F_z}{F_x} = \frac{3,590}{1,490} \qquad \alpha = 67.5°$$

$$\sin \beta = \frac{F_y}{F} = \frac{1,690}{4,240} \qquad \beta = 23.5°$$

2.46 Three forces act through the origin of an xyz coordinate system. The information given about the forces is

$$F_1 = 875 \text{ lb} \qquad \theta_{x1} = 60° \qquad \theta_{y1} = 50° \qquad \theta_{z1} = 54.5°$$
$$F_2 = 1,140 \text{ lb} \qquad \theta_{x2} = 90° \qquad \theta_{y2} = 70° \qquad \theta_{z2} = 20°$$
$$F_3 = 720 \text{ lb} \qquad \theta_{x3} = 110° \qquad \theta_{y3} = 120° \qquad \theta_{z3} = 143°$$

Find the components, and the magnitude, of the resultant force and the direction of this resultant force in terms of the three direction angles and the angles α and β.

$$F_x = F_1 \cos \theta_{x1} + F_2 \cos \theta_{x2} + F_3 \cos \theta_{x3}$$
$$= 875 \cos 60° + 1,140 \cos 90° + 720 \cos 110° = 191 \text{ lb}$$
$$F_y = F_1 \cos \theta_{y1} + F_2 \cos \theta_{y2} + F_3 \cos \theta_{y3}$$
$$= 875 \cos 50° + 1,140 \cos 70° + 720 \cos 120° = 592 \text{ lb}$$
$$F_z = F_1 \cos \theta_{z1} + F_2 \cos \theta_{z2} + F_3 \cos \theta_{z3}$$
$$= 875 \cos 54.5° + 1,140 \cos 20° + 720 \cos 143° = 1,000 \text{ lb}$$
$$F = \sqrt{F_x^2 + F_y^2 + F_z^2}$$
$$= \sqrt{191^2 + 592^2 + 1,000^2}$$
$$= 1,180 \text{ lb}$$

$$\cos \theta_x = \frac{F_x}{F} = \frac{191}{1,180} \qquad \theta_x = 80.7°$$

$$\cos \theta_y = \frac{F_y}{F} = \frac{592}{1,180} \qquad \theta_y = 59.9°$$

$$\cos \theta_z = \frac{F_z}{F} = \frac{1,000}{1,180} \qquad \theta_z = 32.1°$$

$$\tan \alpha = \frac{F_z}{F_x} = \frac{1,000}{191} \qquad \alpha = 79.2°$$

$$\sin \beta = \frac{F_y}{F} = \frac{592}{1,180} \qquad \beta = 30.1°$$

2.47 Find the components of the force which, when added to the force system of Prob. 2.46, will make the resultant force equal to zero.

❚ This force has the same magnitude, and opposite sense, of the force in Prob. 2.46, with the components

$$F_x = -191 \text{ lb}$$
$$F_y = -592 \text{ lb}$$
$$F_z = -1,000 \text{ lb}$$

2.48 Two forces act through the origin of an xyz coordinate system. The first force is described as $F_1 = 24 \text{ lb}$, with $\theta_x = 60°$, $\theta_y = 32°$, and $\theta_z = 80°$. The second force is described as $F_2 = 39 \text{ lb}$, with $\alpha = 40°$ and $\beta = 50°$. Find the magnitude of the resultant force and the direction of this resultant force in terms of the three direction angles and the angles α and β.

❚ The forms for F_x, F_y, and F_z, in terms of α and β, obtained in Prob. 2.42, are used to find the components of the second force.

$$F_x = F_1 \cos \theta_x + F_2 \cos \beta \cos \alpha$$
$$= 24 \cos 60° + 39 \cos 50° \cos 40° = 31.2 \text{ lb}$$
$$F_y = F_1 \cos \theta_y + F_2 \sin \beta$$
$$= 24 \cos 32° + 39 \sin 50° = 50.2 \text{ lb}$$
$$F_z = F_1 \cos \theta_z + F_2 \cos \beta \sin \alpha$$
$$= 24 \cos 80° + 39 \cos 50° \sin 40° = 20.3 \text{ lb}$$

$$F = \sqrt{F_x^2 + F_y^2 + F_z^2}$$
$$= \sqrt{31.2^2 + 50.2^2 + 20.3^2} = 62.5 \text{ lb}$$

$$\cos \theta_x = \frac{F_x}{F} = \frac{31.2}{62.5} \qquad \theta_x = 60.0°$$

$$\cos \theta_y = \frac{F_y}{F} = \frac{50.2}{62.5} \qquad \theta_y = 36.6°$$

$$\cos \theta_z = \frac{F_z}{F} = \frac{20.3}{62.5} \qquad \theta_z = 71.0°$$

$$\tan \alpha = \frac{F_z}{F_x} = \frac{20.3}{31.2} \qquad \alpha = 33.0°$$

$$\sin \beta = \frac{F_y}{F} = \frac{50.2}{62.5} \qquad \beta = 53.4°$$

2.49 Do the same as in Prob. 2.48 if the second force is described by $F_2 = 39$ lb, with $\alpha = -40°$ and $\beta = -50°$.

$$F_x = F_1 \cos \theta_x + F_2 \cos \beta \cos \alpha$$
$$= 24 \cos 60° + 39 \cos (-50°) \cos (-40°)$$

The sine function is odd, with

$$\sin (-\theta) = -\sin \theta$$

and the cosine function is even, so that

$$\cos (-\theta) = \cos \theta$$
$$F_x = 24 \cos 60° + 39 \cos 50° \cos 40° = 31.2 \text{ lb}$$
$$F_y = F_1 \cos \theta_y + F_2 \sin \beta$$
$$= 24 \cos 32° + 39 \sin (-50°)$$
$$= 24 \cos 32° - 39 \sin 50° = -9.5 \text{ lb}$$
$$F_z = F_1 \cos \theta_z + F_2 \cos \beta \sin \alpha$$
$$= 24 \cos 80° + 39 \cos (-50°) \sin (-40°)$$
$$= 24 \cos 80° - 39 \cos 50° \sin 40° = -11.9 \text{ lb}$$
$$F = \sqrt{F_x^2 + F_y^2 + F_z^2}$$
$$= \sqrt{31.2^2 + (-9.5)^2 + (-11.9)^2} = 34.7 \text{ lb}$$

$$\cos \theta_x = \frac{F_x}{F} = \frac{31.2}{34.7} \qquad \theta_x = 26.0°$$

$$\cos \theta_y = \frac{F_y}{F} = \frac{-9.5}{34.7} \qquad \theta_y = 106°$$

$$\cos \theta_z = \frac{F_z}{F} = \frac{-11.9}{34.7} \qquad \theta_z = 110°$$

$$\tan \alpha = \frac{F_z}{F_x} = \frac{-11.9}{31.2}$$

$$-\tan \alpha = \frac{11.9}{31.2}$$

The tangent function is odd, with

$$\tan (-\theta) = -\tan \theta$$

Thus,

$$\tan (-\alpha) = \frac{11.9}{31.2} \qquad -\alpha = 20.9° \qquad \alpha = -20.9°$$

$$\sin \beta = \frac{F_y}{F} = \frac{-9.5}{34.7} \qquad -\sin \beta = \frac{9.5}{34.7}$$

Since the sine function is odd,

$$\sin (-\beta) = \frac{9.5}{34.7} \qquad -\beta = 15.9° \qquad \beta = -15.9°$$

2.50 Refer to Fig. 2.50. The force F_2 is perpendicular to the plane containing the forces F_1 and F_3. Find the magnitude of the resultant of the system of forces and the direction of this resultant force in terms of the three direction angles and the angles α and β.

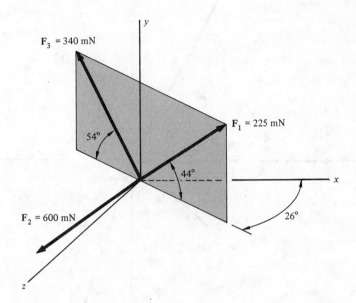

Fig. 2.50

▐
$$F_x = 225 \cos 44° \cos 26° - 340 \cos 54° \cos 26° - 600 \sin 26° = -297 \text{ mN}$$
$$F_y = 225 \sin 44° + 340 \sin 54° = 431 \text{ mN}$$
$$F_z = 225 \cos 44° \sin 26° - 340 \cos 54° \sin 26° + 600 \cos 26° = 523 \text{ mN}$$
$$F = \sqrt{F_x^2 + F_y^2 + F_z^2}$$
$$= \sqrt{(-297)^2 + 431^2 + 523^2} = 740 \text{ mN}$$

$$\cos \theta_x = \frac{F_x}{F} = \frac{-297}{740} \qquad \theta_x = 114°$$

$$\cos \theta_y = \frac{F_y}{F} = \frac{431}{740} \qquad \theta_y = 54.4°$$

$$\cos \theta_z = \frac{F_z}{F} = \frac{523}{740} \qquad \theta_z = 45.0°$$

α' is defined to be the angle between the projection of the resultant force on the xz plane and the negative x axis.

$$\tan \alpha' = \left| \frac{F_z}{F_x} \right| = \left| \frac{523}{-297} \right| \qquad \alpha' = 60.4° \qquad \alpha = 180 - \alpha' = 120°$$

$$\sin \beta = \frac{F_y}{F} = \frac{431}{740} \qquad \beta = 35.6°$$

2.51 A single force is to be added to the system of forces in Prob. 2.50 to make the resultant force equal to zero. Find the x, y, and z components of the force which is to be added.

▐ The force must have the same magnitude, and opposite sense, as the force in Prob. 2.50, so that

$$F_x = 297 \text{ mN}$$
$$F_y = -431 \text{ mN}$$
$$F_z = -523 \text{ mN}$$

2.52 Shown in Fig. 2.52a is a three-dimensional force system. Find the magnitude and direction of the resultant force.

▐ It may be seen that the line of action of force F_1 is defined in terms of the two direction angles $\theta_x = 70°$ and $\theta_z = 140°$. This information by itself is not sufficient for a direct solution of the y component of force F_1. The y component may be found by either of two methods, as follows:

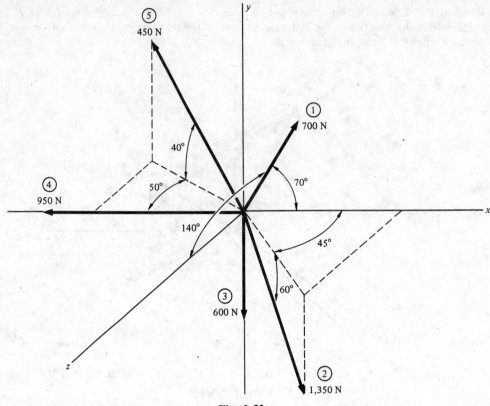

Fig. 2.52a

1. The direction angles must satisfy the relationship

$$\cos^2 \theta_x + \cos^2 \theta_y + \cos^2 \theta_z = 1$$

From the figure,

$$\theta_x = 70° \qquad \theta_z = 140°$$

so that

$$\cos^2 70° + \cos^2 \theta_y + \cos^2 140° = 1$$
$$\cos \theta_y = \pm\sqrt{1 - \cos^2 70° - \cos^2 140°} = \pm 0.544$$
$$\theta_y = 57°, \ 123°$$

From the figure, for the given magnitudes of θ_x and θ_z, we may conclude that

$$\theta_y < 90°$$

Therefore,

$$\theta_y = 57°$$

The y component of force F_1 is

$$F_{1y} = F_1 \cos \theta_y = 700 \cos 57° = 381 \text{ N}$$

2. In this method of finding the y component of force F_1, the x and z components of this force are found first. These quantities are

$$F_{1x} = F_1 \cos \theta_x = 700 \cos 70° = 239 \text{ N}$$
$$F_{1z} = F_1 \cos \theta_z = 700 \cos 140° = -536 \text{ N}$$

The relationship between force F_1 and its components is

$$F_1 = \sqrt{F_{1x}^2 + F_{1y}^2 + F_{1z}^2}$$

This equation may be solved for F_{1y}, with the result

$$F_{1y} = \pm\sqrt{F_1^2 - F_{1x}^2 - F_{1z}^2} = \pm\sqrt{700^2 - 239^2 - (-536)^2} = \pm 381 \text{ N}$$

From consideration of the position of force F_1 in the figure, the positive sign is used. Thus,

$$F_{1y} = 381 \text{ N}$$

The values of the components of the remaining forces are shown in Table 2.6.   The magnitude of the resultant force F is

\multicolumn{7}{c}{TABLE 2.6}
Force	x Component	Value	y Component	Value	z Component	Value
1	700 cos 70°	239	700 cos 57°	381	700 cos 140°	−536
2	1,350 cos 60° cos 45°	477	−1,350 sin 60°	−1,170	1,350 cos 60° sin 45°	477
3	—	0	—	−600	—	0
4	—	−950	—	0	—	0
5	−450 cos 40° cos 50°	−222	450 sin 40°	289	−450 cos 40° sin 50°	−264
Sum	F_x	−456	F_y	−1,100	F_z	−323

$$F = \sqrt{F_x^2 + F_y^2 + F_z^2} = \sqrt{(-456)^2 + (-1,100)^2 + (-323)^2} = 1,230 \text{ N}$$

The direction angles of the resultant force are found to be

$$\cos \theta_x = \frac{F_x}{F} = \frac{-456}{1,230} \qquad \theta_x = 112°$$

$$\cos \theta_y = \frac{F_y}{F} = \frac{-1,100}{1,230} \qquad \theta_y = 153°$$

$$\cos \theta_z = \frac{F_z}{F} = \frac{-323}{1,230} \qquad \theta_z = 105°$$

Figure 2.52b shows the resultant force together with its components.   The definition of the direction of the resultant force in terms of the angles α and β defined in Fig. 2.52b is

$$\tan \alpha = \frac{323}{456} \qquad \alpha = 35°$$

$$\sin \beta = \frac{1,100}{1,230} \qquad \beta = 63°$$

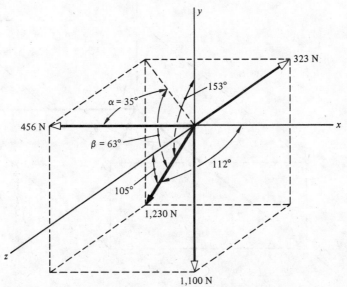

Fig. 2.52b

Again, it should be noted that when the components of a force are drawn, they do not have any minus signs affixed to them. The reason is that these indications of the senses of the components have already been used to establish the orientation in sketching the forces in the figure.

2.4 VECTOR ADDITION WITH UNIT VECTORS, VECTOR DOT PRODUCT

2.53 Define the *unit vectors* of an *xyz* coordinate system.

▌ A system of three vectors, called the unit vectors, is shown in Fig. 2.53.

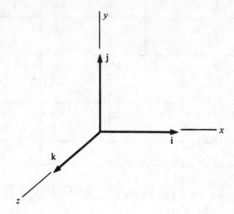

Fig. 2.53

These three vectors are designated **i**, **j**, and **k**, and their *directions* are along the *x*, *y*, and *z* axes, respectively. Each has a dimensionless *magnitude* of 1, and each is positive in the *positive sense* of the three coordinate axes. Any vector may be expressed in terms of these three unit vectors.

2.54 Show how a vector **A** may be expressed in terms of unit vectors.

▌ A general vector **A** is shown in Fig. 2.54. In terms of components along the *x*, *y*, and *z* axes, this vector could be written as

$$\mathbf{A} = \mathbf{A}_x + \mathbf{A}_y + \mathbf{A}_z$$

where the three terms on the right-hand side are the *x*, *y*, and *z* components of **A**.

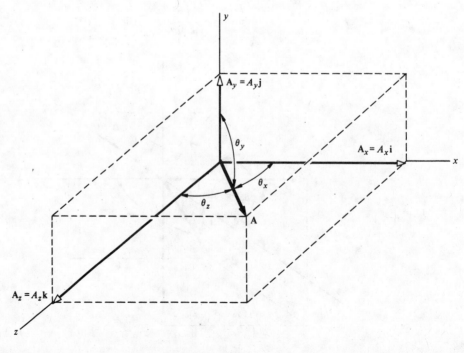

Fig. 2.54

A_x, A_y, and A_z will be chosen to define the magnitude and, depending on whether they are positive or negative, the senses of the three components of **A**. Based on the definition of scalar multiplication of a vector quantity, the vector **A** may then be expressed as

$$\mathbf{A} = A_x\mathbf{i} + A_y\mathbf{j} + A_z\mathbf{k}$$

The three unit vectors describe the *directions* of the components of **A** and thus establish the vector nature of this term. The scalar coefficients A_x, A_y, and A_z define the *magnitudes* of these components, and the *senses* are determined by whether these coefficients are positive or negative. The above equation is represented graphically in the figure. The direction of **A** is given by the three direction angles θ_x, θ_y, and θ_z. If A is the positive magnitude of vector **A**, then this term is related to the magnitudes of the components by

$$A_x = A \cos\theta_x$$
$$A_y = A \cos\theta_y$$
$$A_z = A \cos\theta_z$$

The direction angles are always measured from the positive coordinate axes, and these angles have values between 0 and 180°. In the use of the above three equations the proper sign, which indicates the sense of the component, will come out of the cosine function.

If both sides of the above three equations are squared and the results summed, it can be shown that

$$A = +\sqrt{A_x^2 + A_y^2 + A_z^2}$$

The plus sign in front of the radical indicates that the magnitude A of the vector **A** is an inherently positive term. This equation is the basic relationship between the magnitudes of a vector and its components.

2.55   Express the force shown in Fig. 2.55 in terms of the unit vectors.

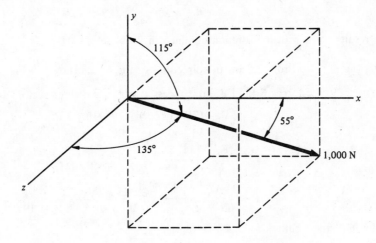

Fig. 2.55

$$A_x = A \cos\theta_x = 1{,}000 \cos 55° = 574\text{ N}$$
$$A_y = A \cos\theta_y = 1{,}000 \cos 115° = -423\text{ N}$$
$$A_z = A \cos\theta_z = 1{,}000 \cos 135° = -707\text{ N}$$

It may be seen that the y and z components have negative senses, the result of using the cosine of an angle between 90° and 180°. The formal statement describing vector **A** is then

$$\mathbf{A} = 574\mathbf{i} - 423\mathbf{j} - 707\mathbf{k} \quad \text{N}$$

2.56   A vector **B** is defined by

$$\mathbf{B} = B_x\mathbf{i} + B_y\mathbf{j} + B_z\mathbf{k}$$

Show the operation of vector addition, where **C** is the sum of **A** and **B**, using the unit vectors.

$$\mathbf{C} = \mathbf{A} + \mathbf{B}$$
$$= A_x\mathbf{i} + A_y\mathbf{j} + A_z\mathbf{k} + B_x\mathbf{i} + B_y\mathbf{j} + B_z\mathbf{k}$$
$$= (A_x + B_x)\mathbf{i} + (A_y + B_y)\mathbf{j} + (A_z + B_z)\mathbf{k}$$

The above equation has a vector on each side of the equals sign. In order for two vectors to be equal, the x, y, and z components of each vector must be *separately* equal to each other. It follows from this equation that the necessary condition of equality is that

$$C_x = A_x + B_x \qquad C_y = A_y + B_y \qquad C_z = A_z + B_z$$

It may be seen that the x component of the sum is the sum of the x components of the vectors that are added, with a similar interpretation for the y and z components of the sum. The result shown above may be generalized to the case where more than two vectors are to be added.

2.57 Solve Prob. 2.5 using unit vectors.

▮ The forces $\mathbf{F}_1$ and $\mathbf{F}_2$, in the figure in Prob. 2.4, are expressed in terms of unit vectors as

$$\mathbf{F}_1 = (F_1 \cos \theta_1)\mathbf{i} + (F_1 \sin \theta_1)\mathbf{j}$$
$$= (750 \cos 60°)\mathbf{i} + (750 \sin 60°)\mathbf{j}$$
$$= 375\mathbf{i} + 650\mathbf{j} \qquad \text{lb}$$
$$\mathbf{F}_2 = (F_2 \cos \theta_2)\mathbf{i} + (F_2 \sin \theta_2)\mathbf{j}$$
$$= (500 \cos 24°)\mathbf{i} + (500 \sin 24°)\mathbf{j}$$
$$= 457\mathbf{i} + 203\mathbf{j} \qquad \text{lb}$$

The resultant force $\mathbf{F}$ is given by

$$\mathbf{F} = \mathbf{F}_1 + \mathbf{F}_2$$
$$= (375 + 457)\mathbf{i} + (650 + 203)\mathbf{j}$$
$$= 832\mathbf{i} + 853\mathbf{j} \qquad \text{lb}$$
$$F = \sqrt{F_x^2 + F_y^2} = \sqrt{832^2 + 853^2} = 1{,}190 \text{ lb}$$

$$\tan \theta = \frac{F_y}{F_x} = \frac{853}{832} \qquad \theta = 45.7°$$

The above results agree with the solution obtained in Prob. 2.5.

2.58 Using unit vectors, solve Prob. 2.5 for the directions of the forces shown in Fig. 2.29.

▮ Using Fig. 2.29, the forces $\mathbf{F}_1$ and $\mathbf{F}_2$ are expressed in terms of unit vectors as

$$\mathbf{F}_1 = (750 \cos 36°)\mathbf{i} + (750 \sin 36°)\mathbf{j}$$
$$= 607\mathbf{i} + 441\mathbf{j} \qquad \text{lb}$$
$$\mathbf{F}_2 = 500\mathbf{i} \qquad \text{lb}$$
$$\mathbf{F} = \mathbf{F}_1 + \mathbf{F}_2 = (607 + 500)\mathbf{i} + 441\mathbf{j}$$
$$= 1{,}110\mathbf{i} + 441\mathbf{j} \qquad \text{lb}$$
$$F = \sqrt{F_z^2 + F_y^2} = \sqrt{1{,}110^2 + 441^2} = 1{,}190 \text{ lb}$$

The direction θ of the resultant force $\mathbf{F}$ is given by

$$\tan \theta = \frac{F_y}{F_x} = \frac{441}{1{,}110} \qquad \theta = 21.7°$$

It should be observed that the above value of θ is the same as the value of β in Prob. 2.5.

2.59 Solve Prob. 2.10 using unit vectors.

▮ The original equation has the form

$$\mathbf{F} = \mathbf{F}_1 + \mathbf{F}_2$$

Forces $\mathbf{F}$ and $\mathbf{F}_1$ are expressed in terms of unit vectors as

$$\mathbf{F} = -(9{,}380 \cos 65°)\mathbf{i} + (9{,}380 \sin 65°)\mathbf{j}$$
$$\mathbf{F}_1 = 4{,}000\mathbf{i} \qquad \text{N}$$

The solution for $\mathbf{F}_2$ is

$$\mathbf{F}_2 = \mathbf{F} - \mathbf{F}_1 = \mathbf{F} + (-\mathbf{F}_1)$$
$$= (-9{,}380 \cos 65° - 4{,}000)\mathbf{i} + (9{,}380 \sin 65°)\mathbf{j}$$
$$= (-7{,}960\mathbf{i} + 8{,}500\mathbf{j}) = F_x\mathbf{i} + F_y\mathbf{j} \qquad \text{N}$$
$$F_2 = \sqrt{F_x^2 + F_y^2} = \sqrt{(-7{,}960)^2 + 8{,}500^2} = 11{,}600 \text{ N}$$

The direction and sense of $\mathbf{F}_2$ are found the same way as in Prob. 2.10.

2.60 Solve Prob. 2.37 using unit vectors.

▮ The forces have the forms

$$\mathbf{F} = 800\mathbf{i} \quad \text{lb}$$
$$\mathbf{F}_1 = F_1\mathbf{j}$$
$$\mathbf{F}_2 = F_2\mathbf{i}$$
$$\mathbf{F}_3 = -350\mathbf{j}$$
$$\mathbf{F}_4 = -(370\cos 70°)\mathbf{i} - (370\sin 70°)\mathbf{j} \quad \text{lb}$$
$$\mathbf{F}_5 = -(290\cos 36°)\mathbf{i} + (290\sin 36°)\mathbf{j} \quad \text{lb}$$

The resultant force $\mathbf{F}$ is given by

$$\mathbf{F} = \mathbf{F}_1 + \mathbf{F}_2 + \mathbf{F}_3 + \mathbf{F}_4 + \mathbf{F}_5$$
$$800\mathbf{i} = (F_2 - 370\cos 70° - 290\cos 36°)\mathbf{i} + (F_1 - 350 - 370\sin 70° + 290\sin 36°)\mathbf{j}$$

The coefficients of the unit vectors are equated, to obtain

$$F_2 - 370\cos 70° - 290\cos 36° = 800 \qquad F_2 = 1{,}160\text{ lb}$$
$$F_1 - 350 - 370\sin 70° + 290\sin 36° = 0 \qquad F_1 = 527\text{ lb}$$

2.61 Use the method of vector addition with unit vectors to solve Prob. 2.40.

▮ From Table 2.5 (Prob. 2.40), the five forces in terms of the unit vectors are

$$\mathbf{F}_1 = 2{,}200\mathbf{j} \quad \text{lb}$$
$$\mathbf{F}_2 = 1{,}760\mathbf{i} + 1{,}670\mathbf{j} + 935\mathbf{k} \quad \text{lb}$$
$$\mathbf{F}_3 = 3{,}000\mathbf{i} \quad \text{lb}$$
$$\mathbf{F}_4 = -875\mathbf{j} + 1{,}520\mathbf{k} \quad \text{lb}$$
$$\mathbf{F}_5 = 3{,}400\mathbf{k} \quad \text{lb}$$

The resultant $\mathbf{F}$ of this force system is then

$$\mathbf{F} = F_x\mathbf{i} + F_y\mathbf{j} + F_z\mathbf{k}$$
$$= (1{,}760 + 3{,}000)\mathbf{i} + (2{,}200 + 1{,}670 - 875)\mathbf{j} + (935 + 1{,}520 + 3400)\mathbf{k}$$
$$F_x\mathbf{i} + F_y\mathbf{j} + F_z\mathbf{k} = 4{,}760\mathbf{i} + 3{,}000\mathbf{j} + 5{,}860\mathbf{k}$$

The coefficients of the unit vectors are equated, with the final result

$$F_x = 4{,}760\text{ lb}$$
$$F_y = 3{,}000\text{ lb}$$
$$F_z = 5{,}860\text{ lb}$$

The above values are those in the last row of Table 2.5.

2.62 Solve Prob. 2.50 using unit vectors.

▮ The forces have the forms

$$\mathbf{F}_1 = (225\cos 44°\cos 26°)\mathbf{i} + (225\sin 44°)\mathbf{j} + (225\cos 44°\sin 26°)\mathbf{k}$$
$$= 145\mathbf{i} + 156\mathbf{j} + 71.0\mathbf{k} \quad \text{mN}$$
$$\mathbf{F}_2 = -(600\sin 26)\mathbf{i} + (600\cos 26°)\mathbf{k}$$
$$= -263\mathbf{i} + 539\mathbf{k} \quad \text{mN}$$
$$\mathbf{F}_3 = -(340\cos 54°\cos 26°)\mathbf{i} + (340\sin 54°)\mathbf{j} - (340\cos 54°\sin 26°)\mathbf{k}$$
$$= -180\mathbf{i} + 275\mathbf{j} - 87.6\mathbf{k} \quad \text{mN}$$

The resultant force $\mathbf{F}$ is then

$$\mathbf{F} = \mathbf{F}_1 + \mathbf{F}_2 + \mathbf{F}_3$$
$$= (145 - 263 - 180)\mathbf{i} + (156 + 275)\mathbf{j} + (71.0 + 539 - 87.6)\mathbf{k}$$
$$= -298\mathbf{i} + 431\mathbf{j} + 522\mathbf{k} \quad \text{mN}$$
$$F = \sqrt{F_x^2 + F_y^2 + F_z^2} = \sqrt{(-298)^2 + 431^2 + 522^2} = 740\text{ mN}$$

The direction of $\mathbf{F}$ is found the same way as in Prob. 2.50.

2.63   Compare the method of finding a resultant force by summing x, y, and z components with the method using formal vector addition with unit vectors.

❚ It may be seen that the techniques of vector addition used to solve Probs. 2.57 through 2.62 are exactly the same as the methods originally used to solve these problems.   Thus, in performing vector addition, there is no economy of effort in stating the problem in terms of unit vectors.   The advantage of using the methods of vector analysis will be seen in the following problems in this chapter, which use the vector dot product multiplication operation, and in Chapter 3, where the vector cross product multiplication operation is used.

Vector analysis is a systematic way of handling operations with vector quantities.   Its use does not constitute a new theory.   Rather, these techniques follow a limited number of general operational rules.   These methods of vector analysis greatly reduce the need for detailed conceptualization of the spatial relationships in statics problems, particularly in the case of three-dimensional force or moment systems.

2.64   Define the *vector dot product* operation.

❚ If **A** and **B** are two vectors, then the dot product C is defined by

$$C = \mathbf{A} \cdot \mathbf{B} = \mathbf{B} \cdot \mathbf{A} = AB \cos \theta$$

In this equation, θ is the angle between the vectors **A** and **B**. A and B are the scalar magnitudes of these two vectors, and the dot product is a *scalar* quantity.   It may also be seen that the dot product operation is commutative.   The above equation may be written as

$$C = (A \cos \theta)B = (B \cos \theta)A$$

This effect is shown in Fig. 2.64*a*.

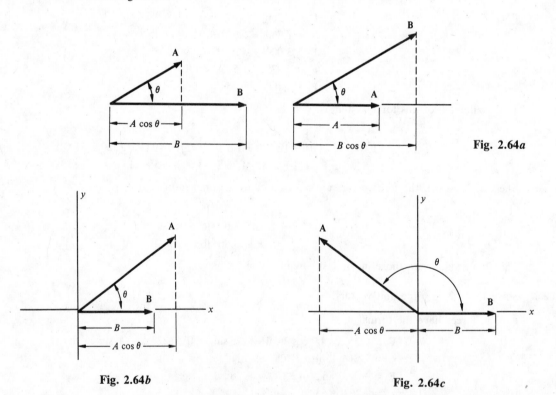

Fig. 2.64*a*

Fig. 2.64*b*

Fig. 2.64*c*

This interpretation of the dot product of two vectors is that one of these vectors is projected onto the direction of the other.   The product of this projected length and the magnitude of the second vector is, then, the dot product.   The dot product has an implicit sign convention, as shown in Figs. 2.64*b* and *c*.

Vector **B** is collinear with the x axis and positive in the positive sense of this axis.   If vector **A** lies in the positive x region, with $-90° < \theta < +90°$, as shown in Fig. 2.64*b*, then $\cos \theta$ is positive and the dot product $\mathbf{A} \cdot \mathbf{B}$ will also be positive.   If vector **A** lies in the negative x region, with $90° < \theta < 270°$, as seen in Fig. 2.64*c*, then $\cos \theta$, and the dot product $\mathbf{A} \cdot \mathbf{B}$, will be negative.

2.65   Show the general form of the solution for the vector dot product.

❚ By using the general forms for **A** and **B**, the dot product may be written as

$$C = \mathbf{A} \cdot \mathbf{B}$$
$$= (A_x\mathbf{i} + A_y\mathbf{j} + A_z\mathbf{k}) \cdot (B_x\mathbf{i} + B_y\mathbf{j} + B_z\mathbf{k})$$
$$C = A_xB_x\mathbf{i}\cdot\mathbf{i} + A_xB_y\mathbf{i}\cdot\mathbf{j} + A_xB_z\mathbf{i}\cdot\mathbf{k} + A_yB_x\mathbf{j}\cdot\mathbf{i} + A_yB_y\mathbf{j}\cdot\mathbf{j} + A_yB_z\mathbf{j}\cdot\mathbf{k} + A_zB_x\mathbf{k}\cdot\mathbf{i} + A_zB_y\mathbf{k}\cdot\mathbf{j} + A_zB_z\mathbf{k}\cdot\mathbf{k}$$

The right-hand side of the above equation contains nine terms involving dot products of the unit vectors. It will now be shown that six of these nine terms are identically zero. Two typical dot products of the unit vectors are shown below.

$$\mathbf{i}\cdot\mathbf{i} = (1)(1)\cos 0° = 1$$
$$\mathbf{i}\cdot\mathbf{j} = (1)(1)\cos 90° = 0$$

The nine possible dot combinations of the unit vectors are then

$$\mathbf{i}\cdot\mathbf{i} = \mathbf{j}\cdot\mathbf{j} = \mathbf{k}\cdot\mathbf{k} = 1$$
$$\mathbf{i}\cdot\mathbf{j} = \mathbf{i}\cdot\mathbf{k} = \mathbf{j}\cdot\mathbf{i} = \mathbf{j}\cdot\mathbf{k} = \mathbf{k}\cdot\mathbf{i} = \mathbf{k}\cdot\mathbf{j} = 0$$

With these results, the general solution appears as

$$\mathbf{A}\cdot\mathbf{B} = A_xB_x + A_yB_y + A_zB_z$$

In this form, the scalar nature of the dot product can be seen clearly.

The dot product has two major applications in the solution of problems in mechanics. The first of these is to define the term *work*, and this is treated in dynamics. The second use of the dot product is to find the component of a vector along an arbitrary direction, and this effect is shown in the following problem.

2.66   Figure 2.66a shows vector **A** and the direction of line *ab*. Show how the dot product operation may be used to find the component of **A** along line *ab*.

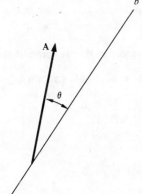

Fig. 2.66a

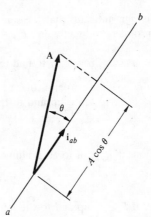

Fig. 2.66b

❚ To find this component, a unit vector $\mathbf{i}_{ab}$ is directed along line *ab*, as shown in Fig. 2.66b. This vector has a magnitude of 1 and an arbitrarily chosen positive sense. The component A_{ab} of **A** along line *ab* is then

$$A_{ab} = \mathbf{A}\cdot\mathbf{i}_{ab} = A(1)\cos\theta = A\cos\theta$$

Since this result is positive, the component of **A** along line *ab* has the same sense as $\mathbf{i}_{ab}$. It is left as an exercise for the reader to show that if $\mathbf{i}_{ab}$ were chosen as positive when acting down and to the left, with the included angle $180° - \theta$, the same result would be obtained.

This second application of the dot product is very useful in the solution of problems in statics where the component of a force or moment along an arbitrary direction is required. This effect is shown in the following problems.

2.67   Refer to Fig. 2.67. Use the definition of the dot product to find the magnitude and sense of the component of the force along line *ab*. The force and the line both lie in the *xy* plane.

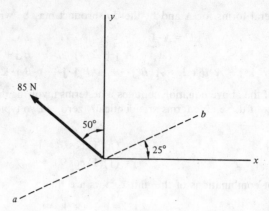

Fig. 2.67

▮ A unit vector $\mathbf{i}_{ab}$ is assumed to act along line ab, with positive sense from a to b. In terms of the unit vectors of the xy coordinates,

$$\mathbf{i}_{ab} = (\cos 25°)\mathbf{i} + (\sin 25°)\mathbf{j}$$
$$= 0.906\mathbf{i} + 0.423\mathbf{j}$$

The force is expressed in vector notation as

$$\mathbf{F} = -(85 \sin 50°)\mathbf{i} + (85 \cos 50°)\mathbf{j}$$
$$= -65.1\mathbf{i} + 54.6\mathbf{j} \quad \text{N}$$

The component F_{ab} of force $\mathbf{F}$ along line ab is then

$$F_{ab} = \mathbf{F} \cdot \mathbf{i}_{ab} = -65.1(0.906) + 54.6(0.423) = -35.9 \text{ N}$$

The minus sign indicates that the sense of F_{ab} is from b to a. For this elementary problem the answer could have been obtained directly as $F_{ab} = -85 \sin 25° = -35.9$ N.

2.68 Do the same as in Prob. 2.67 to find the magnitude and sense of the component that is perpendicular to line ab.

▮ Line $a'b'$ is a line perpendicular to line ab, at an angle of 25° with the y axis. Unit vector $\mathbf{i}_{a'b'}$ lies along this line with a positive sense up and to the left.

$$\mathbf{i}_{a'b'} = -(\sin 25°)\mathbf{i} + (\cos 25°)\mathbf{j} = -0.423\mathbf{i} + 0.906\mathbf{j}$$

The component $F_{a'b'}$ of force $\mathbf{F}$ along line $a'b'$ is

$$F_{a'b'} = \mathbf{F} \cdot \mathbf{i}_{a'b'} = -65.1(-0.423) + 54.6(0.906) = 77.0 \text{ N}$$

The sense of $F_{a'b'}$ is up and to the left. For this elementary case the answer could again have been obtained directly from considering the figure. The result is $F_{a'b'} = 85 \cos 25° = 77.0$ N.

2.69 A force and its components are shown in Fig. 2.69a. The line ab lies in the yz plane, with the direction angle $\theta_y = 40°$. Find the component of the force along line ab.

▮ The yz plane is shown in true view in Fig. 2.69b, and $\mathbf{i}_{ab}$ has the positive sense shown. In terms of the unit vectors,

$$\mathbf{i}_{ab} = (0)\mathbf{i} + 1(\cos 40°)\mathbf{j} + 1(\sin 40°)\mathbf{k}$$

The force $\mathbf{F}$ is written as

$$\mathbf{F} = 450\mathbf{i} + 500\mathbf{j} + 650\mathbf{k}$$

The component F_{ab} of this force along the line ab is

$$F_{ab} = \mathbf{F} \cdot \mathbf{i}_{ab} = 450(0) + 500 \cos 40° + 650 \sin 40° = 801 \text{ N}$$

Since F_{ab} is a positive quantity, this component of the force acts in the positive sense of $\mathbf{i}_{ab}$.

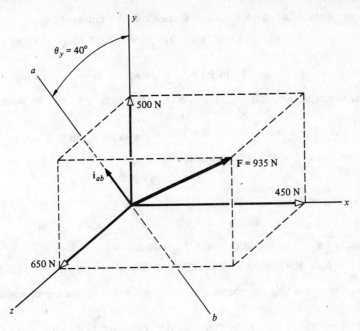

Fig. 2.69*a*

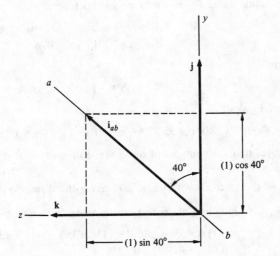

Fig. 2.69*b*

2.70 A line passing through the origin of an *xyz* coordinate system is defined by the direction angles $\theta_x = 75°$ and $\theta_y = 65°$. A force acting through the origin has components of 870, 565, and 610 N in the *x*, *y*, and *z* directions, respectively. Find the magnitude of the component of the force along the line.

❙ The third direction angle of the line is found first.

$$\cos^2 \theta_x + \cos^2 \theta_y + \cos^2 \theta_z = 1$$
$$\cos^2 75° + \cos^2 65° + \cos^2 \theta_z = 1$$
$$\cos \theta_z = \pm 0.869 \qquad \theta_z = 29.7°, 150°$$

Designate $\mathbf{i}_0$ to be a unit vector that lies along the line. The positive sense of this unit vector is chosen to be outward from the origin, so that $\theta_z = 29.7°$ is chosen.

$$\mathbf{i}_0 = [(1) \cos 75°]\mathbf{i} + [(1) \cos 65°]\mathbf{j} + [(1) \cos 29.7°]\mathbf{k}$$
$$= 0.259\mathbf{i} + 0.423\mathbf{j} + 0.869\mathbf{k}$$

The force is expressed in vector notation as

$$\mathbf{F} = 870\mathbf{i} + 565\mathbf{j} + 610\mathbf{k} \qquad \text{N}$$

The magnitude of the component F_0 of this force along the line is then

$$F_0 = \mathbf{F} \cdot \mathbf{i}_0 = 870(0.259) + 565(0.423) + 610(0.869) = 994 \text{ N}$$

2.71 Find the component of the force $\mathbf{F}_1$ in Prob. 2.45 along the direction of force $\mathbf{F}_2$.

❚ $\mathbf{i}_2$ is a unit vector acting along $\mathbf{F}_2$, in the sense of this force. $\mathbf{F}_2$ may be written in terms of the unit vectors as

$$\mathbf{F}_2 = F_2 \mathbf{i}_2 = F_2 \cos 35°\mathbf{i} + F_2 \sin 35°\mathbf{k}$$

The term F_2 cancels out of the above equation, so that

$$\mathbf{i}_2 = (\cos 35°)\mathbf{i} + (\sin 35°)\mathbf{k}$$

Force $\mathbf{F}_1$ is written as

$$\mathbf{F}_1 = (1{,}800 \cos 20°)\mathbf{j} - (1{,}800 \sin 20°)\mathbf{k}$$

The component of $\mathbf{F}_1$ along the direction of $\mathbf{F}_2$ is designated F_{12}, and

$$F_{12} = \mathbf{F}_1 \cdot \mathbf{i}_2 = 0(\cos 35°) + 1{,}800 \cos 20°(0) + (-1{,}800 \sin 20°)\sin 35° = -353 \text{ lb}$$

The minus sign indicates that the component of $\mathbf{F}_1$ along $\mathbf{F}_2$ has a sense opposite that of $\mathbf{F}_2$.

2.72 Do the same as in Prob. 2.71 for the component of the force $\mathbf{F}_2$ in Prob. 2.45 along the direction of $\mathbf{F}_1$.

❚ $\mathbf{i}_1$ is a unit vector acting along $\mathbf{F}_1$, in the sense of this force.

$$\mathbf{F}_1 = F_1 \mathbf{i}_1 = (F_1 \cos 20°)\mathbf{j} - (F_1 \sin 20°)\mathbf{k}$$
$$\mathbf{i}_1 = (\cos 20°)\mathbf{j} - (\sin 20°)\mathbf{k}$$

Force $\mathbf{F}_2$ has the form

$$\mathbf{F}_2 = (3{,}400 \cos 35°)\mathbf{i} + (3{,}400 \sin 35°)\mathbf{k} \qquad \text{lb}$$

The component F_{21} of $\mathbf{F}_2$ along $\mathbf{F}_1$ is then

$$F_{21} = \mathbf{F}_2 \cdot \mathbf{i}_1 = (3{,}400 \cos 35°)0 + 0(\cos 20°) + (3{,}400 \sin 35°)(-\sin 20°) = -667 \text{ lb}$$

The minus sign indicates that the component has a sense opposite that of force $\mathbf{F}_1$.

2.73 Find the component of the force shown in Fig. 2.73 along the direction ab.

❚ The unit vector $\mathbf{i}_{ab}$ lies along line ab, with positive sense from a to b.

$$\mathbf{i}_{ab} = [(1) \cos 35° \cos 40°]\mathbf{i} + (1) \sin 35°\mathbf{j} + [(1) \cos 35° \sin 40°]\mathbf{k}$$
$$= 0.628\mathbf{i} + 0.574\mathbf{j} + 0.527\mathbf{k}$$

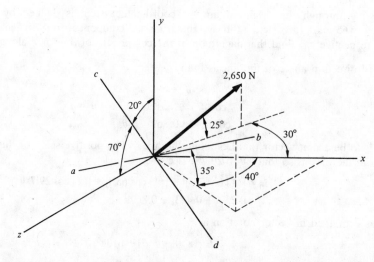

Fig. 2.73

The force is written in vector notation as

$$\mathbf{F} = (2{,}650 \cos 25° \cos 30°)\mathbf{i} + (2{,}650 \sin 25°)\mathbf{j} - (2{,}650 \cos 25° \sin 30°)\mathbf{k}$$
$$= 2{,}080\mathbf{i} + 1{,}120\mathbf{j} - 1{,}200\mathbf{k} \quad \text{N}$$

The component F_{ab} of $\mathbf{F}$ along line ab is

$$F_{ab} = \mathbf{F} \cdot \mathbf{i}_{ab} = 2{,}080(0.628) + 1{,}120(0.574) - 1{,}200(0.527) = 1{,}320 \text{ N}$$

Since F_{ab} is positive, this force component has the same sense as $\mathbf{i}_{ab}$.

2.74 Find the component of the force shown in Fig. 2.73 along the direction cd.

▎ The unit vector $\mathbf{i}_{cd}$ lies along line cd, with sense from d to c.

$$\mathbf{i}_{cd} = [(1) \cos 20°]\mathbf{j} + [(1) \cos 70°]\mathbf{k}$$
$$= 0.940\mathbf{j} + 0.342\mathbf{k}$$

The force is given in Prob. 2.73. The component F_{cd} of $\mathbf{F}$ along line cd is

$$F_{cd} = \mathbf{F} \cdot \mathbf{i}_{cd} = 2{,}080(0) + 1{,}120(0.940) - 1{,}200(0.342) = 642 \text{ N}$$

The force component acts in the positive sense of $\mathbf{i}_{cd}$.

2.75 What is the significance of a dot product that is equal to zero?

▎ If $\mathbf{A} \cdot \mathbf{B} = 0$, then either $\mathbf{A}$ or $\mathbf{B}$, or both, are zero, or the two vectors are perpendicular to each other. This latter situation is the more usual when a dot product is zero. The graphical interpretation is that neither vector has a component along the direction of the other vector.

2.76 Show that, for the system of forces in Prob. 2.50,

$$\mathbf{F}_2 \cdot \mathbf{F}_1 = 0 \qquad \text{and} \qquad \mathbf{F}_2 \cdot \mathbf{F}_3 = 0$$

▎ The forces $\mathbf{F}_1$, $\mathbf{F}_2$, and $\mathbf{F}_3$ were expressed in vector notation in Prob. 2.62.

$$\mathbf{F}_2 \cdot \mathbf{F}_1 \overset{?}{=} 0$$
$$-263(145) + 0(156) + 539(71.0) \overset{?}{=} 0$$
$$263(145) \overset{?}{=} 539(71.0)$$
$$38{,}100 \approx 38{,}300 \qquad (\%D = 0.5\%)$$
$$\mathbf{F}_2 \cdot \mathbf{F}_3 \overset{?}{=} 0$$
$$-263(-180) + 0(275) + 539(-87.6) \overset{?}{=} 0$$
$$263(180) \overset{?}{=} 539(87.6)$$
$$47{,}300 \approx 47{,}200 \qquad (\%D = 0.2\%)$$

The physical interpretation of the above results is that $\mathbf{F}_2$ is perpendicular to $\mathbf{F}_1$ and to $\mathbf{F}_3$.

2.77 Show how the vector dot product may be used to find the angle between two intersecting vectors.

▎ Let the two vectors be $\mathbf{A}$ and $\mathbf{B}$. The vector dot product is defined as

$$\mathbf{A} \cdot \mathbf{B} = AB \cos \theta = A_x B_x + A_y B_y + A_z B_z$$

where θ is the angle between the two vectors. θ is then found from

$$\cos \theta = \frac{A_x B_x + A_y B_y + A_z B_z}{AB}$$

2.78 Use the vector dot product to find the angle between forces $\mathbf{F}_1$ and $\mathbf{F}_3$ in Prob. 2.50.

▎ The forces were expressed in vector notation in Prob. 2.62, and the magnitudes of F_1 and F_3 are shown in Prob. 2.50. θ is the angle between $\mathbf{F}_1$ and $\mathbf{F}_3$, and

$$\mathbf{F}_1 \cdot \mathbf{F}_3 = F_1 F_3 \cos \theta = F_{1x} F_{3x} + F_{1y} F_{3y} + F_{1z} F_{3z}$$

$$\cos \theta = \frac{145(-180) + 156(275) + 71.0(-87.6)}{225(340)} = 0.138 \qquad \theta = 82.1°$$

From inspection of Fig. 2.50,

$$\theta = 180° - 54° - 44° = 82°\cdot \quad \text{and} \quad 82.1° \approx 82°$$

2.79 Find the angle between the line and the force in Prob. 2.70.

❚ θ is the angle between the line and the force. From the definition of the vector dot product,

$$F_0 = \mathbf{F} \cdot \mathbf{i}_0 = F(1) \cos \theta$$

$$\cos \theta = \frac{F_0}{F}$$

Using values for the components and F_0 found in Prob. 2.70,

$$F = \sqrt{F_x^2 + F_y^2 + F_z^2} = \sqrt{870^2 + 565^2 + 610^2} = 1,200 \text{ N}$$

$$F_0 = 994 \text{ N}$$

$$\cos \theta = \frac{994}{1,200} \qquad \theta = 34.1°$$

2.80 Using the vector dot product, find the angle between forces $\mathbf{F}_2$ and $\mathbf{F}_4$ in Prob. 2.40.

❚ The description of the forces in terms of unit vectors, obtained in Prob. 2.61, is

$$\mathbf{F}_2 = 1,760\mathbf{i} + 1,670\mathbf{j} + 935\mathbf{k} \qquad \text{lb}$$
$$F_2 = 2,600 \text{ lb}$$
$$\mathbf{F}_4 = -875\mathbf{j} + 1,520\mathbf{k} \qquad \text{lb}$$
$$F_4 = 1,750 \text{ lb}$$

The angle between $\mathbf{F}_2$ and $\mathbf{F}_4$ is designated θ, and

$$\mathbf{F}_2 \cdot \mathbf{F}_4 = F_2 F_4 \cos \theta = F_{2x}F_{4x} + F_{2y}F_{4y} + F_{2z}F_{4z}$$

$$\cos \theta = \frac{1,760(0) + 1,670(-875) + 935(1,520)}{2,600(1,750)} = -0.00880 \qquad \theta = 90.5°$$

2.81 Find the direction angles of force $\mathbf{F}_2$ in Prob. 2.40.

❚ From Prob. 2.61,

$$\mathbf{F}_2 = 1,760\mathbf{i} + 1,670\mathbf{j} + 935\mathbf{k} \qquad \text{lb}$$
$$F_2 = 2,600 \text{ lb}$$

θ_x is measured in the plane that contains $\mathbf{F}_2$ and the x axis.

$$\mathbf{F}_2 \cdot \mathbf{i} = F_2(1) \cos \theta_x = F_{2x}(1)$$

$$\cos \theta_x = \frac{F_{2x}}{F_2} = \frac{1,760}{2,600} \qquad \theta_x = 47.4°$$

In a similar fashion,

$$\mathbf{F}_2 \cdot \mathbf{j} = F_2(1) \cos \theta_y = F_{2y}(1)$$

$$\cos \theta_y = \frac{F_{2y}}{F_2} = \frac{1,670}{2,600} \qquad \theta_y = 50.0°$$

$$\mathbf{F}_2 \cdot \mathbf{k} = F_2(1) \cos \theta_z = F_{2z}(1)$$

$$\cos \theta_z = \frac{F_{2z}}{F_2} = \frac{935}{2,600} \qquad \theta_z = 68.9°$$

2.82 Refer to Fig. 2.82 and find the angle between the force and line aa.

❚ $\mathbf{i}_a$ is a unit vector acting along line aa, in a sense which is downward in the figure.

$$\mathbf{i}_a = [(1) \cos 22° \cos 35°]\mathbf{i} - [(1) \sin 22°]\mathbf{j} + [(1) \cos 22° \sin 35°]\mathbf{k}$$
$$= 0.760\mathbf{i} - 0.375\mathbf{j} + 0.532\mathbf{k}$$

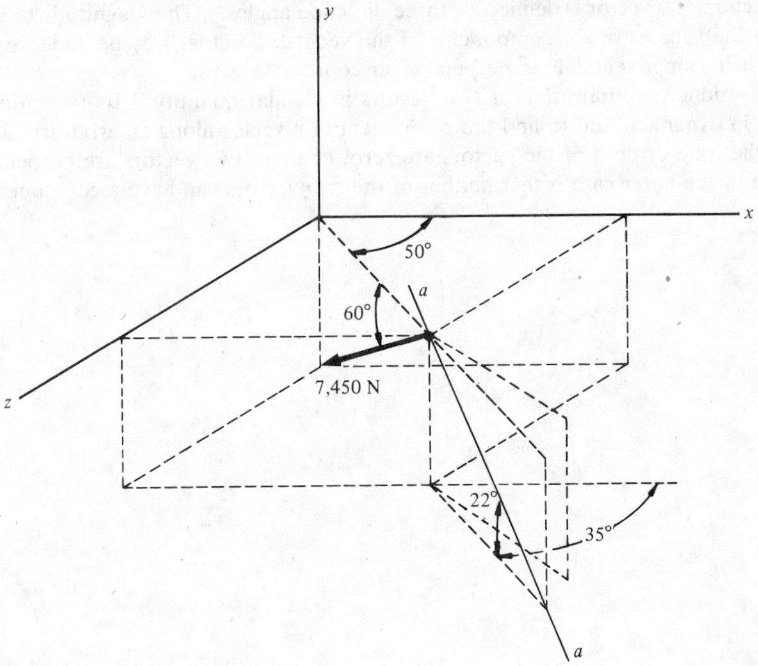

Fig. 2.82

Force **F** has the form

$$\mathbf{F} = -(7{,}450 \cos 60° \cos 50°)\mathbf{i} - (7{,}450 \sin 60°)\mathbf{j} - (7{,}450 \cos 60° \sin 50°)\mathbf{k}$$
$$= -2{,}390\mathbf{i} - 6{,}450\mathbf{j} - 2{,}850\mathbf{k} \quad \text{N}$$

From the figure,

$$F = 7{,}450 \text{ N}$$

The component F_a of **F** along line aa is

$$F_a = \mathbf{F} \cdot \mathbf{i}_a = -2{,}390(0.760) - 6{,}450(-0.375) - 2{,}850(0.532)$$
$$= -914 = F(1) \cos \theta = 7{,}450 \cos \theta$$

where θ is the angle between the force and the line.

$$\cos \theta = \frac{-914}{7{,}450} \qquad \theta = 97.0°$$

2.83 Give a summary of the basic concepts of operations with forces.

❚ Forces may be added graphically by using the parallelogram, triangle, or polygon laws of force addition. The vector sum of a system of forces is called the resultant force. This single force will produce the same effect as that produced by the original set of forces.

A force may be resolved into any number of components along arbitrary directions. The vector sum of these force components is equal to the original force. The most common situation in static analysis is when a force is expressed in terms of components along perpendicular directions. The components of a force along perpendicular directions are called rectangular components. They may be found by projecting the original force onto the known perpendicular directions.

A component of a force is considered to be negative if it acts in a negative coordinate sense. This sense of the component is determined by inspection of the orientation of the force with respect to the coordinate axes. If direction angles are used to specify the direction of a force, the proper sense of the components will be automatically found by using the associated direction cosines.

The resultant of a system of forces may be found analytically by summing the components of the force along perpendicular directions. This is the usual technique used in static analysis. The three unit vectors **i**, **j**, and **k** have the directions of the x, y, and z axes, respectively. They each have a magnitude of 1, and they are positive in the respective positive coordinate sense of the axes. Any vector may be expressed in terms of the unit vectors. The unit vectors establish the directions of the components of the vector. The scalar quantities which multiply the unit vectors define the magnitudes and senses of the components.

The direction of a vector is defined by three direction angles. The magnitude of a vector is the square root of the sum of the squares of the components of the vector. Vectors may be added or subtracted by algebraically summing their components along perpendicular coordinate axes.

The dot product multiplication of two vectors is a scalar quantity. Its two principal uses are to define the term work in dynamics, and to find the component of a vector along an arbitrary direction. If the dot product is zero, either one or both of the vectors are zero, or these two vectors are perpendicular to each other. The implication of the latter case is that neither of the two vectors can have a component along the direction of the other.

CHAPTER 3
Operations with Moments and Couples

3.1 MOMENT OF A FORCE, MOMENT AS THE SUM OF MOMENTS OF COMPONENTS OF FORCE, RESULTANT MOMENT IN A TWO-DIMENSIONAL FORCE SYSTEM

3.1 Define the term *moment of a force*.

▌ A force F lies in the xz plane. The perpendicular distance from the line of action of the force to the y axis is d. From physical considerations, it may be observed that the force F tends to create a turning, or rotation, effect about the y axis. This tendency of the force to rotate about the y axis is referred to as a *moment effect*. The force F in Fig. 3.1 is described as having a moment about, or with respect to, the y axis. Throughout this book the symbol M will be used to designate the magnitude of a moment, and a subscript may be placed on M to indicate the axis about which the moment acts. The y axis intersects the xz plane at point 0, the origin

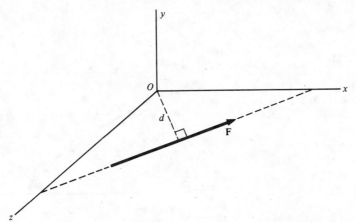

Fig. 3.1

of the coordinate system. In terms of a description of the moment caused by the force F, two interpretations may be made. In the first, the moment is thought of as being about the y axis. *This is the basic description of the effect*, and the more general statement is that the axis about which the moment acts is perpendicular to the plane containing the force. In the second interpretation, the moment is described as being about point 0 in the plane of the force. The *magnitude* of the moment of a force F, with respect to a reference *point* in the plane which contains the force, is defined to be the product of the magnitude of the force and the perpendicular distance d between the line of action of this force and this reference point. The magnitude M of the moment is given by

$$M = Fd$$

Two alternative, equivalent statements for the moment effect shown in the figure are

$$M_y = Fd \qquad M_0 = Fd$$

Here, the subscript y indicates that the moment is about the y axis, while the subscript 0 refers to the origin of the coordinates.

3.2 For what condition will the moment of a force be zero?

▌ From consideration of the equation $M = Fd$, the moment will be zero if, and only if, F or $d = 0$. If $d = 0$, the line of action of the force must intersect the reference axis or point. The very important conclusion may now be drawn that *a force produces zero moment about any axis or reference point which intersects the line of action of the force*. This result will be used subsequently to simplify the computation of moments.

3.3 What are the fundamental units of a moment?

▌ Since a moment is defined by $M = Fd$, it may be seen that the fundamental units of moment are the product of force and length. Typical moment units are then foot-pounds or pound-feet in U.S. Customary System units and newton-meters or meter-newtons in SI units. There is no basic difference in the order of writing these units. Throughout this book, however, the preferred usages will be foot-pounds and newton-meters, abbreviated as ft·lb and N·m.

3.4   Show that a moment is a vector quantity.

▮ The definition of three quantities is required to describe the moment caused by the force in Fig. 3.1. The first is the magnitude of the moment. This quantity is expressed by the equation $M = Fd$ in terms of the magnitudes of the force F which causes the moment and the distance d of this force from the reference axis. The second descriptive quantity required is the orientation of the moment effect. Plane A in Fig. 3.4 contains a force F. It is evident that if this plane is rotated to the position B, then the force F will create a different moment about the y axis than it did in the original position. If plane A were rotated through 90° from its first position, for example, the moment of the force about the y axis would be zero. This characteristic of orientation is called the *direction* of the moment. In defining this direction, one of two points of view may be taken. In the first, the orientation is given of the plane which contains the force. With this definition, the moment of the force F in plane A in Fig. 3.4 would be described as acting in the xz plane. The second method is to specify the direction of an axis which is *perpendicular* to the plane that contains the force. In using this technique, the moment due to the force F in plane A would be described as having the direction of the y

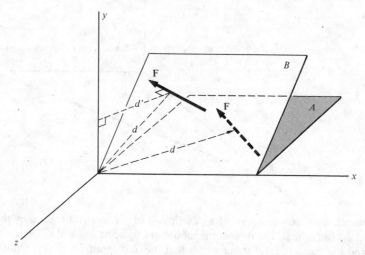

Fig. 3.4

axis. It may be observed, with this latter method of description, that the direction of the moment is the axis about which the moment effect actually takes place. The third element in the description of a moment is a statement of the *sense* of the moment, and this quantity tells whether the rotation effect is clockwise or counterclockwise with respect to some reference. As a consistent usage in this book, the moment caused by any force which lies in the plane of an illustration will be considered *positive* if its sense is counterclockwise.

From the above three requirements to define a moment, it follows that *a moment is a vector quantity*. Thus, all the vector operations which were shown earlier for forces are also applicable to moments. The earlier convention of boldface type to represent a vector will again be used. Thus, **M** is the vector which represents the moment of magnitude M.

3.5   Show the graphical representation of a moment using a curved arrow.

▮ In the first convention, the moment is represented by a curved arrow which encircles the axis about which the moment acts. This effect is shown in Fig. 3.5. The curved arrow emphasizes the fact that the moment is a turning, or rotation, effect. The tip is placed in such a way as to show the sense of this turning effect, and an arbitrary sense of rotation is chosen to represent positive moment. In using this method to show the action of a moment, the length of the curved arrow is *never* drawn to a scale which would represent the magnitude of the moment. The curved arrow may be drawn in the plane of the force or in a plane parallel to the plane of the force.

3.6   Show the graphical representation of a moment using a straight arrow.

▮ The second way of graphically depicting a moment is shown in Fig. 3.6a. In this case, the moment is represented by a straight arrow with a double arrowhead at the tip end. The direction of the double-headed

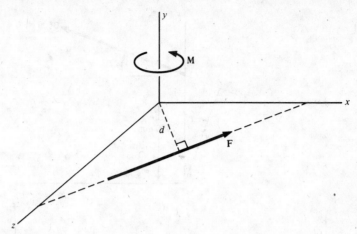

Fig. 3.5

arrow is along the axis about which the moment acts, and this direction is perpendicular to the plane containing the force.

In order to establish the sense of the moment, the thumb of the right hand is imagined to be pointing in the positive sense of the double-headed arrow. The relaxed, curved fingers of the right hand will then be acting in the turning sense of the moment. This effect is shown in Fig. 3.6b. With the use of the double-arrowhead notation, the length of the arrow may be used to represent the magnitude of the moment.

Both conventions for the graphical representation of moments are useful. For simple systems, the curved-arrow technique is probably the more convenient method to use. The double-arrowhead representation of moments finds its greatest utility in complex systems where moments about axes with different directions must be combined. Both conventions will be used in the following problems.

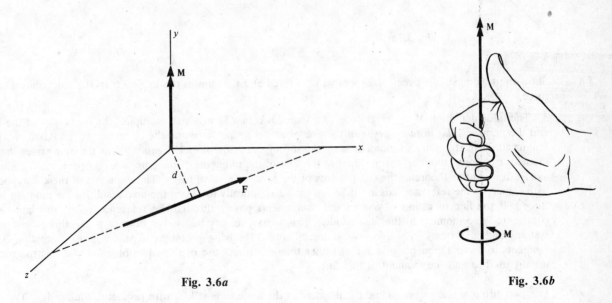

Fig. 3.6a   **Fig. 3.6b**

3.7 The force in Fig. 3.7a lies in the xy plane. Find the moment of this force with respect to the origin 0 of the coordinate system.

❚ The dotted construction lines are drawn in the figure, and

$$d = 1,200 \cos 27° = 1,070 \text{ mm}$$

The magnitude M_0 of the moment of the 3,500-N force about the origin is

$$M_0 = Fd = -3,500 \text{ N} (1,070 \text{ mm})\left(\frac{1 \text{ m}}{1,000 \text{ mm}}\right) = -3,750 \text{ N} \cdot \text{m}$$

This result is negative, since the moment, as viewed in the figure, is clockwise. The above result is shown graphically in Fig. 3.7b. Figure 3.7c shows the same moment in double-arrowhead notation.

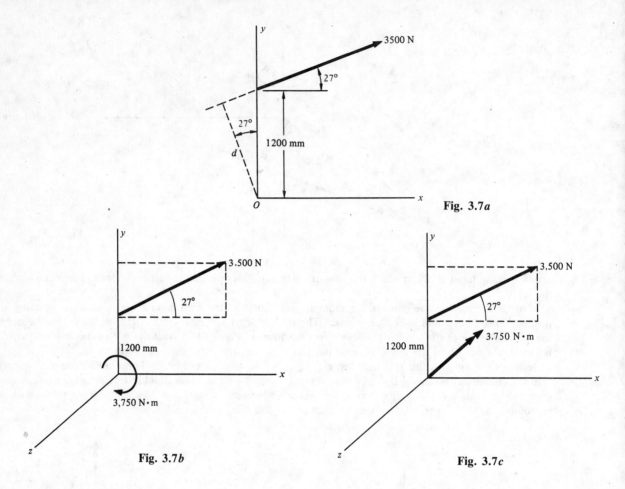

Fig. 3.7a

Fig. 3.7b

Fig. 3.7c

3.8 Show the relationship between the moment of a force and the sum of the moments of the components of the force.

❚ The basic definition of the magnitude of a moment is in a form which multiplies the *magnitude* of the force and the *perpendicular distance* between the line of action of the force and the reference axis. Another way of finding the magnitude of this moment is by the use of the *theorem of Varignon*. This theorem states that the moment of a force about an axis is equal to the sum of the moments of the components of this force about the axis. In using this theorem, the components of the force are found first. The moments of these components with respect to the reference axis are then summed algebraically to obtain the final result for the moment of the force. If the line of action of one of the components passes through the reference point, this component contributes zero moment to the final result. This technique may be used in both two- and three-dimensional systems and in systems in which several forces act. In solving problems in statics, the summation of the moments of the components of a force, rather than the direct use of the equation $M = Fd$, is the method usually used to find the moment of this force.

3.9 Use two different techniques to find the moment of the force in Fig. 3.9a with respect to the z axis. The force lies in the *xy* plane.

❚ From the graphical construction in Fig. 3.9b,

$$d = x_1 \sin \theta$$

The moment of the force about the z axis is designated M_z, and

$$M_z = Fd = F(x_1 \sin \theta) = Fx_1 \sin \theta$$

The actual sense of the moment is counterclockwise. The components of the force are shown in Fig. 3.9c. It may be seen that the line of action of F_x intersects the reference axis, and thus this component produces no moment. From the figure,

$$M_z = (F \sin \theta)x_1 = Fx_1 \sin \theta$$

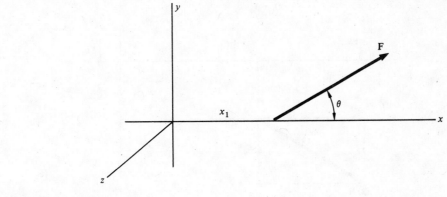

Fig. 3.9a

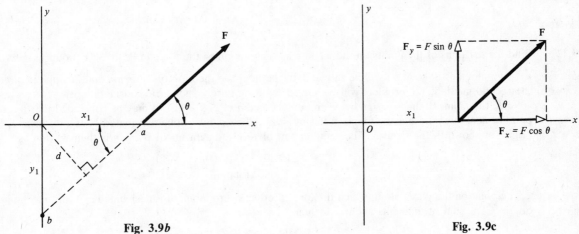

Fig. 3.9b

Fig. 3.9c

3.10 About what points on the x and y axes does the force in Prob. 3.9 produce zero moment?

▌ The moment of the force F will be zero if either of the two reference points a and b is chosen, since the line of action of the force intersects both of these points. The result for the value y_1 which defines the location of point b is found from direct consideration of triangle $0ab$ in Fig. 3.9b. The formal statement of the locations on the x and y axes about which the moment due to the force F is zero is, then,

$$x = x_1 \qquad y = -y_1 = -x_1 \tan \theta$$

3.11 The 300-N force in Fig. 3.11a lies in the xy plane. Find the moment of this force about the origin 0 and about point a.

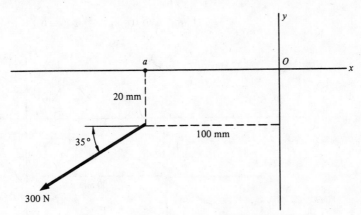

Fig. 3.11a

▌ The components of the force are shown in Fig. 3.11b. The solution is

$$M_0 = -246(20) + 172(100) = 12,300 \text{ N} \cdot \text{mm} = 12.3 \text{ N} \cdot \text{m}$$
$$M_a = -246(20) = -4,920 \text{ N} \cdot \text{mm} = -4.92 \text{ N} \cdot \text{m}$$

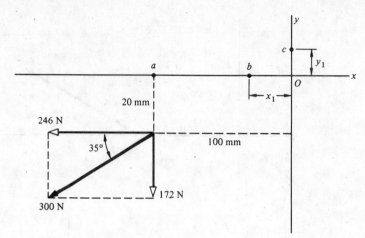

Fig. 3.11b

3.12 Find the locations of the points on the x and y axes about which the force in Prob. 3.11 produces zero moment.

▮ Two methods of solution will be shown. In the first, the theorem of Varignon will be applied directly. In the second method, the problem will be solved by using purely geometric considerations.

Since the moment changes sign between points 0 and a in Fig. 3.11b, it may be concluded that the reference point of zero moment must be found in this interval. This point will be designated b, at a distance x_1 from the origin, as shown in the figure. The requirement of zero moment about point b is then written as

$$M_b = -246(20) + 172(100 - x_1) = 0$$
$$x_1 = 71.4 \text{ mm}$$

The point on the y axis about which the force produces zero moment is designated c, at distance y_1 above the origin. The sum of moments about c is then

$$M_c = -246(y_1 + 20) + 172(100) = 0$$
$$y_1 = 49.9 \text{ mm}$$

In the solution below, the values of x_1 and y_1 are found from purely geometric considerations. If the moment of a force about a point is zero, then the line of action of the force must intersect the point. The system is redrawn in Fig. 3.12 and the line of action of the 300-N force is extended to include points b and c. x_1 and y_1 are then found from

$$\tan 35° = \frac{20}{100 - x_1}$$
$$x_1 = 71.4 \text{ mm}$$
$$\tan 35° = \frac{y_1}{x_1} = \frac{y_1}{71.4}$$
$$y_1 = 50.0 \text{ mm} \approx 49.9 \text{ mm} \qquad (\%D = 0.2\%)$$

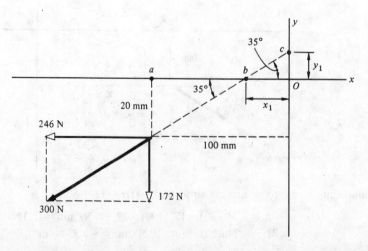

Fig. 3.12

3.13 Refer to Fig. 3.13. Find θ if M_a is to be equal to zero for any value of the force **F**.

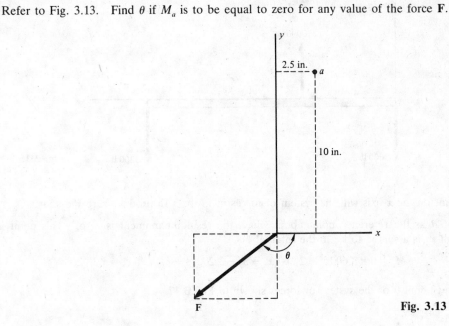

Fig. 3.13

❚ Summing moments about a gives

$$M_a = -[F \cos (180° - \theta)]10 + [F \sin(180° - \theta)]2.5$$

Using

$$\cos (180° - \theta) = -\cos \theta \qquad \sin (180° - \theta) = \sin \theta$$

M_a has the form

$$M_a = (F \cos \theta)10 + (F \sin \theta)2.5 = 0$$

$$10 \cos \theta + 2.5 \sin \theta = 0$$

$$\frac{\sin \theta}{\cos \theta} = \tan \theta = -\frac{10}{2.5} = -4 \qquad \theta = 104°$$

An alternative method of solution is shown below. Let α be the angle between the force and the y axis. Then

$$\tan \alpha = \frac{2.5}{10} \qquad \alpha = 14.0° \qquad \theta = 90° + \alpha = 104°$$

3.14 Describe the general technique for finding the *resultant moment* of a system of several forces.

❚ If there are several forces acting in a system, it is possible to find the combined moment effect of all these forces about a given reference axis. Since a moment caused by a force is a vector quantity, the moment about a reference axis is the vector sum of the moments of each of the forces about this axis. This vector sum of these individual moments is defined to be the resultant moment of the system of forces.

If all the forces in the system lie in the same plane, then the directions of the moments caused by these forces are all the same, and these directions are perpendicular to the plane of the forces. For this case, the resultant moment is found by direct algebraic summation of the moments due to the individual forces. The force system described above is called a two-dimensional force system. The general characteristics of such a system are studied in detail in subsequent chapters of this book.

If all the forces in the system *do not* lie in a common plane, then the resultant moment may be found by using the theorem of Varignon together with the methods of vector addition presented in Chap. 2. For this case, an additional computation is required to determine the *direction* of the resultant moment. These effects will be shown in the following problems.

3.15 Find the moments of the system of forces shown in Fig. 3.15 about points a, b, and c

❚
$$M_a = 2,800(3) - 3,200(8) = -17,200 \text{ ft} \cdot \text{lb}$$
$$M_b = 3,600(3) - 3,200(5) = -5,200 \text{ ft} \cdot \text{lb}$$
$$M_c = 3,600(8) - 2,800(5) = 14,800 \text{ ft} \cdot \text{lb}$$

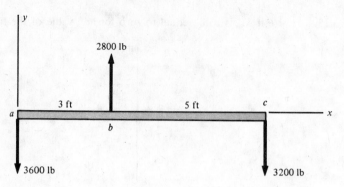

2800 lb

3 ft   5 ft   c

a
b   x

3600 lb   3200 lb   **Fig. 3.15**

3.16 About what point on the x axis will the system of forces in Prob. 3.15 produce zero moment?

▮ Define point d as the reference point about which the resultant moment is zero. This point is located at $x = x_1$, and x_1 is assumed to be in the interval bc.

$$M_d = 0 = 3{,}600x_1 - 2{,}800(x_1 - 3) - 3{,}200(8 - x_1) \qquad x_1 = 4.30 \text{ ft}$$

3.17 Find the moment about 0 of the system of forces shown in Fig. 3.17.

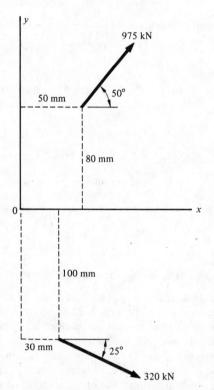

975 kN

50 mm   50°

80 mm

0   x

100 mm

30 mm   25°

320 kN   **Fig. 3.17**

▮ $$M_0 = -(975\cos 50°)\frac{80}{1{,}000} + (975\sin 50°)\frac{50}{1{,}000} + (320\cos 25°)\frac{100}{1{,}000} - (320\sin 25°)\frac{30}{1{,}000}$$

$$= 12.2 \text{ kN} \cdot \text{m}$$

3.18 The force system of Prob. 3.17 is now modified as shown in Fig. 3.18. For what value of F will the resultant moment M_0 be equal to zero?

▮ $$M_0 = -(975\cos 50°)\frac{80}{1{,}000} + (975\sin 50°)\frac{50}{1{,}000} + (F\cos 35°)\frac{100}{1{,}000} - (F\sin 35°)\frac{30}{1{,}000} = 0$$

$$M_0 = 0.0647F - 12.8 = 0 \qquad F = 198 \text{ kN}$$

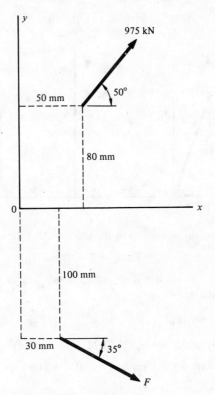

Fig. 3.18

3.19 Find F if the moment about 0 of the system of forces in Prob. 3.18 is equal to $30 \, \text{kN} \cdot \text{m}$.

▮ From Prob. 3.18,

$$M_0 = 0.0647F - 12.8 = 30 \qquad F = 662 \, \text{kN} \cdot \text{m}$$

3.20 Find F if the moment about 0 of the system of forces in Prob. 3.18 is $-30 \, \text{kN} \cdot \text{m}$.

▮ From Prob. 3.18,

$$M_0 = 0.0647F - 12.8 = -30 \qquad F = -266 \, \text{kN}$$

The minus sign indicates that the force F has a sense opposite that shown in Fig. 3.18.

3.21 The forces shown in Fig. 3.21a all lie in the same plane. Using the theorem of Varignon, find the moment of the system of forces about point 0.

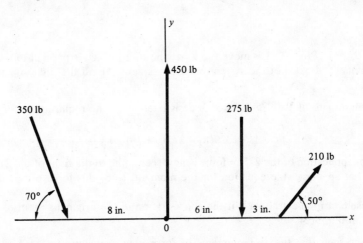

Fig. 3.21a

▮ All the forces are expressed in terms of their x and y components, as shown in Fig. 3.21b. It may be seen that none of the x components of the forces produces a moment, since these forces all pass through point 0. By

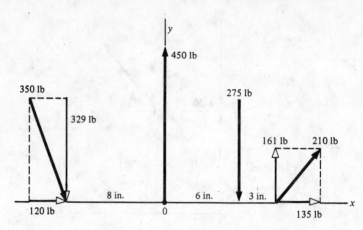

Fig. 3.21*b*

using the theorem of Varignon, the solution for the resultant moment of the system of forces about point 0 is

$$M_0 = 329(8) - 275(6) + 161(9) = 2,430 \text{ in} \cdot \text{lb}$$

It should be noted that the 450-lb force does not enter into any of the above calculations, since this force passes through the reference point 0.

3.22   Check the result of Prob. 3.21 by using the fundamental definition of moment given by $M = Fd$.

▌ Figure 3.21*a* is redrawn as shown in Fig. 3.22.   With direct application of the definition $M = Fd$,

$$M_0 = 350(8 \sin 70°) - 275(6) + 210(9 \sin 50°) = 2,430 \text{ in} \cdot \text{lb}$$

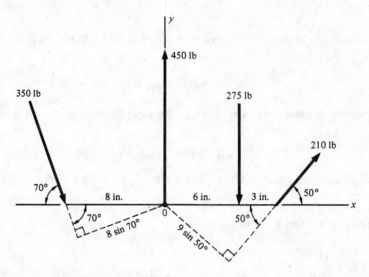

Fig. 3.22

3.23   The 275-lb force in Prob. 3.21 is moved to a new position, which is parallel to the original position, so that the resultant moment about 0 of the system of forces is zero.   Find the x coordinate of the new position of this force.

▌ The x coordinate of the 275-lb force is designated x_1.   The requirement for zero resultant moment about point 0 is

$$M_0 = 329(8) - 275x_1 + 161(9) = 0 \qquad x_1 = 14.8 \text{ in}$$

The point of application of the 275-lb force is now seen to lie to the right of the 210-lb force.   Again, the 450-lb force does not appear in the equation for the moment, since this force passes through the reference point 0.

3.24   Find the resultant moment about 0 if the configuration of the original system of Prob. 3.21 is now changed to that shown in Fig. 3.24.

▌ It may be seen that the x components of force now contribute to the resultant moment about 0.   This moment is

$$M_0 = 329(8) + 120(2.5) - 275(6) + 161(9) - 135(2) = 2,460 \text{ in} \cdot \text{lb}$$

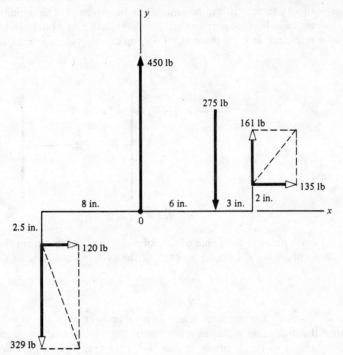

Fig. 3.24

3.2 THE COUPLE, TORQUE, REPLACEMENT OF A FORCE BY A FORCE AND A COUPLE, OR BY A FORCE AND TWO COUPLES

3.25 Define the term *couple*.

▮ A very special situation exists when the configuration of two forces is as shown in Fig. 3.25. The description of this system is that two forces

(*a*) Are parallel to each other and thus lie in the same plane

(*b*) Are of equal magnitude

(*c*) Have opposite senses

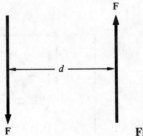

Fig. 3.25

The force system described above is called a couple, and this system gives rise to a *pure turning effect*. The magnitude *M* of this couple is referred to as the *moment of the couple*. It is the product of the force and the perpendicular separation distance, or

$$M = Fd$$

It may be seen that this result has the same form as the equation which expresses the moment of a force about a known reference axis.

3.26 What is the fundamental difference between a *moment* and a *couple*?

▮ There is a significant difference between a moment and a couple. The couple is a pure *turning* effect which may be moved anywhere *in its own plane*, or into a *parallel plane*, without change of its effect on the body. Because of these characteristics, the couple is a *free vector*. The definition of a moment due to a force, by comparison, *must* include a description of the reference axis about which the moment is taken. This independence of a reference axis, in the case of a couple, may be seen from consideration of Fig. 3.26.

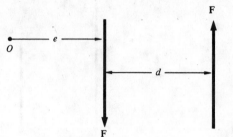

Fig. 3.26

Point 0 is an arbitrary point in the plane of the forces, at a distance e from the line of action of the left-hand force. The moment M_0, with respect to point 0, of the system consisting of the two forces F is, then

$$M_0 = -Fe + F(e + d) = -Fe + Fe + Fd$$
$$= Fd$$

The sense of the couple is counterclockwise. It may be seen that the result for M_0 is *independent* of the distance e. Since the dimension e locates point 0, this point may be at any location in the plane of the pair of forces. By an extension of this analysis, it can be shown that the couple *may be moved to a parallel plane without changing its effect on the system.*

The resultant force of the pair of forces in a couple is always zero, since the forces are equal and opposite. The magnitude of the couple is given by $M = Fd$, and the sense is arbitrarily chosen. Throughout this book, the counterclockwise sense of a couple which acts in the plane of an illustration will be considered positive. If more than one couple in a single plane, or in parallel planes, acts in a system, the resultant couple may be found by direct *algebraic* summation of these couples. The reason is that all these couples have a *common* direction.

3.27   Give examples of a couple.

❙ Figure 3.27 shows two physical examples of a couple. In the first example, the forces **F** represent the forces of a person's hands on a T-shaped handle. In the second example, the forces of a person's fingers on the fluted surface of a bolt rotate this element.

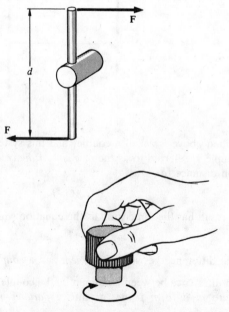

Fig. 3.27

3.28 Find the magnitude and sense of M_B if the resultant couple which acts on the system of pulleys shown in Fig. 3.28 is zero.

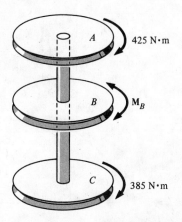

Fig. 3.28

▎ When the system of pulleys in Fig. 3.28 is viewed from the top, the positive sense is assumed to be counterclockwise. M_B is assumed to be positive.

$$M = -425 + M_B - 385 = 0 \qquad M_B = 810 \, \text{N} \cdot \text{m}$$

Since the result for M_B is positive, the actual sense of this couple is counterclockwise.

3.29 Find the resultant of the system of couples shown in Fig. 3.29

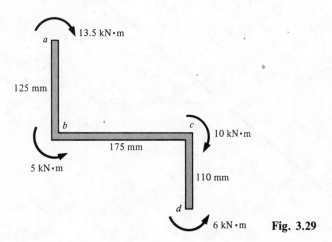

Fig. 3.29

▎
$$M = -13.5 + 5 - 10 + 6 = -12.5 \, \text{kN} \cdot \text{m}$$

3.30 How would the answer in Prob. 3.29 change if the 6-kN · m couple acted at point b instead of at point d?

▎ The result found in Prob. 3.29 would not change, since a couple may be moved anywhere in its own plane without change in its effect on a body.

3.31 The three couples shown in Fig. 3.31a act on a body. The planes of action of the couples are parallel. Find the resultant couple.

▎ The resultant couple is designated M_R, and

$$M_R = -72 - 55 + 150 = 23 \, \text{ft} \cdot \text{lb}$$

This result is seen to have a counterclockwise sense. A graphical representation of the original system, using the double-arrowhead notation, is shown in Fig. 3.31b and Fig. 3.31c shows the resultant couple. It may be seen that the separation distance of 3.35 in shown in Fig. 3.31a does not enter into the problem.

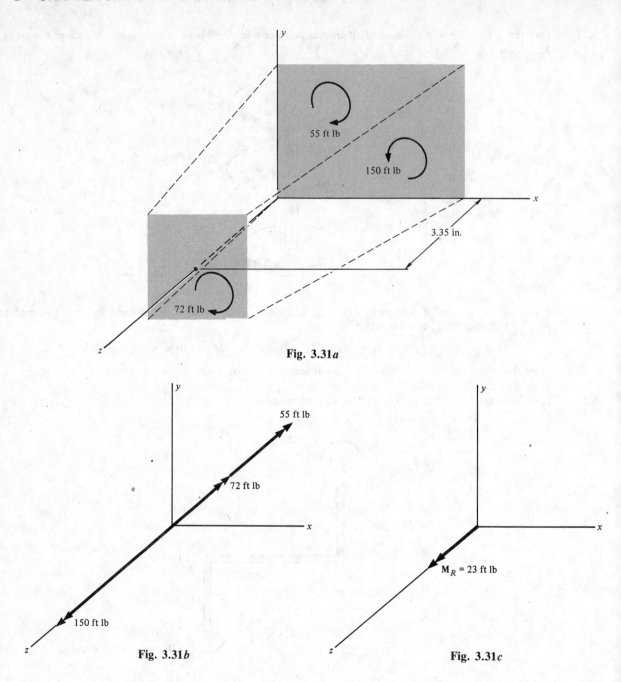

Fig. 3.31a

Fig. 3.31b

Fig. 3.31c

3.32 Define the term *torque*.

❚ Figure 3.32 illustrates a problem which is encountered frequently in machine design. The shaft shown has the gears or pulleys A, B, C, and D mounted on it. Power is transmitted to the shaft through one or more of these gears or pulleys, and this power is taken out through the remaining gears or pulleys. The model of this system is that the gear teeth, or pulley belts, cause the couples M_A, M_B, M_C, and M_D to act on the disks, which represent the gears or pulleys. The effect of each couple is transmitted through the shaft between adjacent disks. Such a transmitted moment effect is called torque. A torque has the identical characteristics of the couple defined in Prob. 3.25. However, this terminology to describe a couple is usually reserved for use in the subjects strength, or mechanics, of materials or machine design, for the case of a moment effect being transmitted through a shaft.

The system shown in the figure is a set of couples which act in parallel planes. Thus, the resultant is found by direct algebraic summation of the individual couples. An arbitrary sense is chosen to represent positive couple. A typical definition of sense for the system shown would be that, when viewed from the left side of the system, counterclockwise couples are considered to be positive.

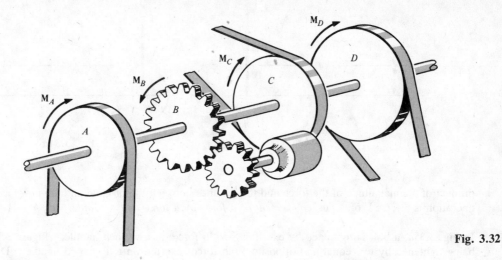

Fig. 3.32

3.33 For the shaft-pulley system of Prob. 3.32, $M_A = 6,700$ N·m, $M_B = -13,200$ N·m, $M_C = 4,500$ N·m, and $M_D = 1,000$ N·m. Find the resultant couple that acts on the system. Assume the definition of sense given in Prob. 3.32.

❚ The resultant couple is designated M_R, and

$$M_R = 6,700 - 13,200 + 4,500 + 1,000 = -1,000 \text{ N·m}$$

The magnitude of the resultant couple is $-1,000$ N·m, and the sense is clockwise when viewed from the left end of the system.

3.34 Show how to replace a force by a force and a couple.

❚ A problem which occurs frequently in the force analysis of machine and structural members is shown in Fig. 3.34a. A force F is applied at a known location a, and it is desired to find the effect of this force at location b, which is at a known distance d from point a.

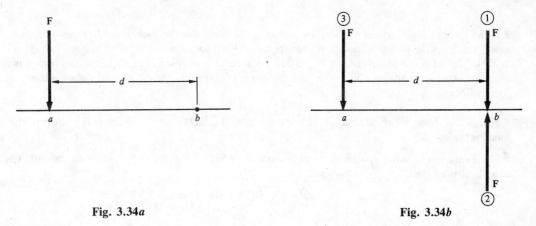

Fig. 3.34a   **Fig. 3.34b**

The system of Fig. 3.34a is redrawn in Fig. 3.34b. Two equal and opposite forces F are placed at point b, with their directions parallel to that of the original force. The forces are numbered 1 through 3. The resultant force of the system of forces 1 and 2 is zero. Thus, the introduction of these two forces has no effect on the original problem. The pair of forces 2 and 3 results in a counterclockwise couple of magnitude M, defined by

$$M = Fd$$

The original system consisted of force F at point a, at a distance d from point b. This original force may now be replaced *by an equal and parallel force F and a counterclockwise couple of magnitude Fd*, both acting at point b. This effect is shown in Fig. 3.34c. Had the original force been moved left a distance d to point c, the equivalent system would appear as shown in Fig. 3.34d.

It may be concluded from the above that the effect of a force at a certain location may be replaced by a parallel force of equal magnitude and sense at a new location, and a couple whose magnitude is equal to the

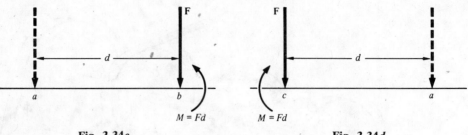

Fig. 3.34c   Fig. 3.34d

product of the magnitude of the force and the distance between the original and the new locations. The above operation is referred to as *the replacement of a force by a force and a couple*.

3.35 In Fig. 3.35a, a bolt is tightened by exerting a 50-lb force on a wrench handle. Figure 3.35b shows the bolt being tightened by two equal and opposite 50-lb forces exerted on a T-shaped handle. Discuss the resulting force effects on the wrench socket for these two cases.

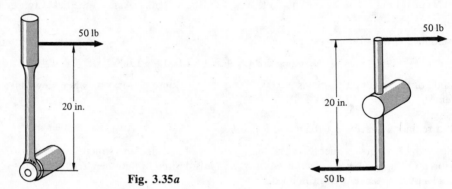

Fig. 3.35a   Fig. 3.35b

▮ The force in Fig. 3.35a is replaced by a force and a couple at the socket. After this operation, the socket is acted on by a 50-lb force acting to the right and a clockwise couple of magnitude $M = Fd = 50(20) = 1,000$ in · lb. The pair of forces in Fig. 3.35b constitutes a couple. For this case the socket is acted on by a clockwise couple of magnitude $M = Fd = 50(20) = 1,000$ in · lb, but the socket *does not* experience a direct force effect.

3.36 Figure 3.36a shows a curved bracket with a vertical strip attached to it. What is the effect of the 25-N force at locations a and b?

▮ The equivalent force system at point a, shown in Fig. 3.36b, is 25 N, acting to the left, and a clockwise couple of magnitude

$$M = Fd = 25(140) = 3,500 \text{ N} \cdot \text{mm} = 3.5 \text{ N} \cdot \text{m}$$

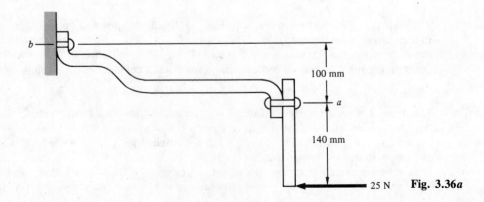

25 N   Fig. 3.36a

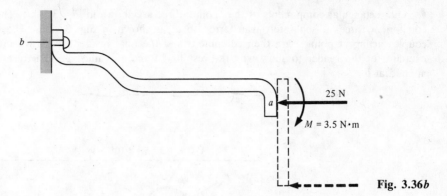

Fig. 3.36b

The equivalent force at point b is again 25 N, acting to the left, and the clockwise couple now has the magnitude

$$M = Fd = 25(140 + 100) = 6{,}000 \text{ N} \cdot \text{mm} = 6 \text{ N} \cdot \text{m}$$

This effect is seen in Fig. 3.36c.

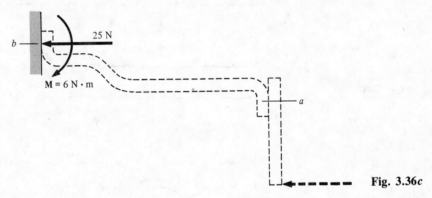

Fig. 3.36c

3.37   Figure 3.37a shows a structural configuration which is referred to as an eccentrically loaded bolted connection. Point a is called the *centroid* of the bolt group. The methods for determining the location of a centroid are presented in Chap. 9. In order to determine the strength of this type of joint, the effect of the applied force at point a must be found first. What force and moment acting at a will produce the same effect at this point as the original force acting at point b does?

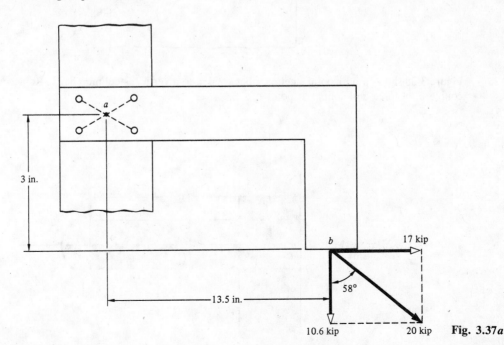

Fig. 3.37a

▌ The rectangular components of the applied force are shown in Fig. 3.37a. The effect of each of these two component forces is separately transferred to *a*, as shown in Fig. 3.37b. The resultants of the forces and couples acting at point *a* are then computed, and the final results are shown in Fig. 3.37c. It is left as an exercise for the reader to show why the resultant force at *a* must be exactly the same as the original force applied at *b*.

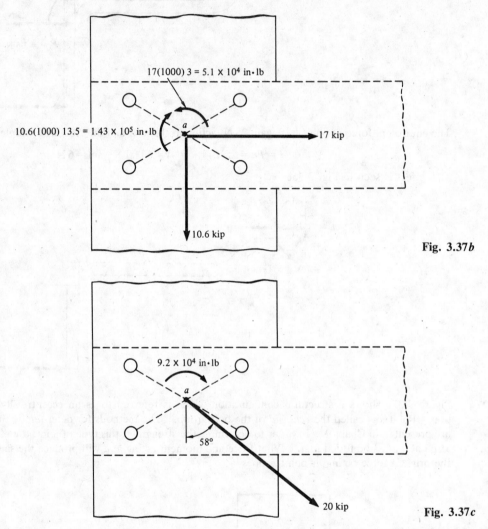

17(1000) 3 = 5.1 × 10⁴ in·lb

10.6(1000) 13.5 = 1.43 × 10⁵ in·lb

17 kip

10.6 kip

Fig. 3.37*b*

9.2 × 10⁴ in·lb

58°

20 kip

Fig. 3.37*c*

3.38 Figure 3.38 shows a horizontal steel member welded to a vertical steel member. Replace the applied force and couple by a force and couple at point *a*.

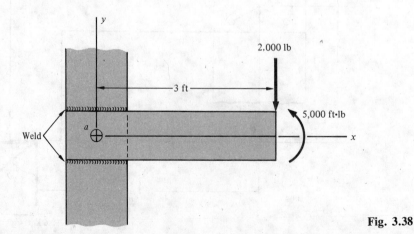

y

2,000 lb

3 ft

5,000 ft·lb

x

a

Weld

Fig. 3.38

▍ The force at point a is 2,000 lb, acting vertically downward. The resultant couple at point a is

$$M_a = -2,000(3) + 5,000 = -1,000 \, \text{ft} \cdot \text{lb}$$

The sense of the resultant couple is clockwise.

3.39 In Prob. 3.38, find the x coordinate of the location where the vertical 2,000-lb force could be placed to give the same effect as the original applied force and couple.

▍ The new location of the 2,000-lb force is given by $x = x_1$. For $M_a = -1,000 \, \text{ft} \cdot \text{lb}$, from Prob. 3.38,

$$M_a = -1,000 = -2,000x_1 \qquad x_1 = 0.5 \, \text{ft}$$

The required location is 0.5 ft to the right of point a.

3.40 Replace the system of forces and couples shown in Fig. 3.40 by a single force and couple at point a.

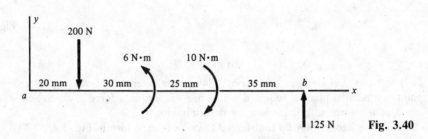

Fig. 3.40

▍ The resultant force at point a is

$$F_a = -200 + 125 = -75 \, \text{N}$$

The minus sign indicates that the sense of the force is downward in the figure. The resultant couple at point a is

$$M_a = -200\left(\frac{20}{1000}\right) + 6 - 10 + 125\left(\frac{20 + 30 + 25 + 35}{1000}\right) = 5.75 \, \text{N} \cdot \text{m}$$

3.41 Do the same as in Prob. 3.40, but for point b.

▍ The resultant force at point b is

$$F_b = -200 + 125 = -75 \, \text{N}$$

and this is the same result as that obtained in Prob. 3.40. The resultant couple at point b is

$$M_b = 200\left(\frac{30 + 25 + 35}{1000}\right) + 6 - 10 = 14 \, \text{N} \cdot \text{m}$$

3.42 Find the magnitude, direction, sense, and location of the single force which would replace the system of forces and couples in Prob. 3.40.

▍ The single force which would replace the system of forces and couples in Prob. 3.40 is designated F. From the results of the last two problems, the magnitude of this force must be 75 N. The force acts in the y direction, and its sense is downward in the figure.

From Prob. 3.40, $M_a = 5.75 \, \text{N} \cdot \text{m}$. In order to replace this 5.75-N·m couple, the force must be moved *leftward* from point a through a distance x_1, so that

$$75x_1 = 5.75 \qquad x_1 = 0.0767 \, \text{m} = 76.7 \, \text{mm}$$

The formal answer to the problem is that $F = 75 \, \text{N}$, acting vertically downward, at the location $x = -76.7 \, \text{mm}$. It is left as an exercise for the reader to show that the same result could have been obtained by starting the problem with the value of M_b found in Prob. 3.41.

3.43 Show how to replace a force by a force and two couples.

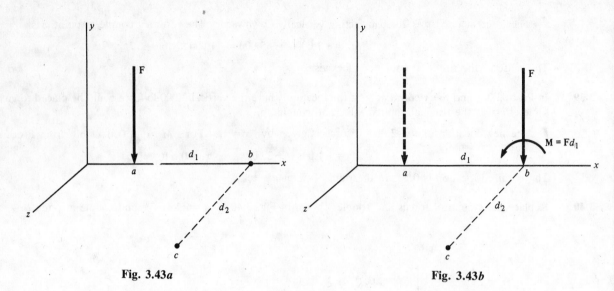

Fig. 3.43a   Fig. 3.43b

❚ The techniques developed in Prob. 3.34 may be extended readily to the case where the effect of a force at a point in a parallel plane is required. Figure 3.43a shows a force F which acts in the xy plane at point a. The effect of this force at point c is to be determined.

The original force may be transferred to point b, as shown in Fig. 3.43b. The couple $M = Fd_1$ may next be transferred to a parallel plane without change in its effect. This plane is chosen to be parallel to the xy plane and contain the point c. The force F is now transferred from point b to point c, and the final result is shown in Fig. 3.43c, in terms of the curved-arrow moment notation. Figure 3.43d shows the same result in terms of the double-arrowhead notation.

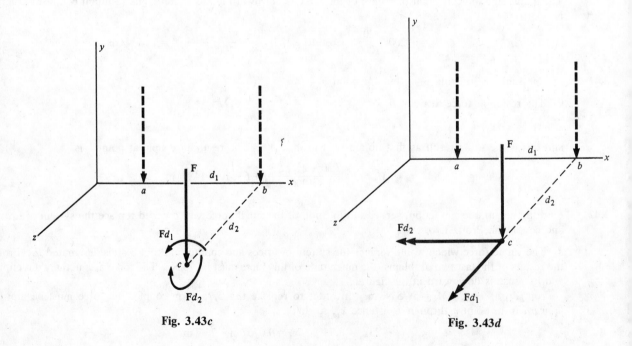

Fig. 3.43c   Fig. 3.43d

3.44   Figure 3.44a shows a structural element which is called an eccentrically loaded column. Find the effect of the 10-kip force at 0, the origin of the coordinate system.

❚ The force is transferred along the x and z directions, and the final result is shown in Fig. 3.44b.

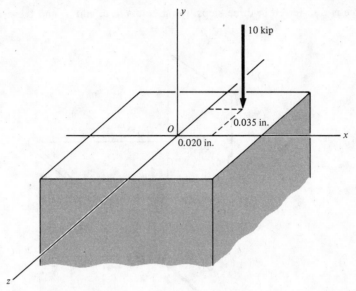

10 kip

0.035 in.

O

0.020 in.

x

y

z

Fig. 3.44a

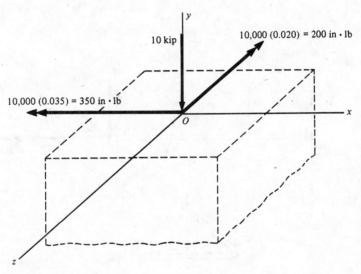

y

10 kip

10,000 (0.020) = 200 in · lb

10,000 (0.035) = 350 in · lb

O

x

z

Fig. 3.44b

3.45 Figure 3.45a shows a crank arm arrangement which lies in a horizontal plane and is attached to a vertical wall at point d. Someone pushes vertically downward, with a force of 165 N, at point a. What force and moment effects are produced at point d?

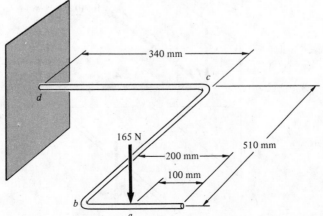

340 mm

c

d

165 N

200 mm

510 mm

100 mm

b

a

Fig. 3.45a

▮ The force is transferred in three steps, using points b, c, and d, and these results are shown in Figs. 3.45b through d.

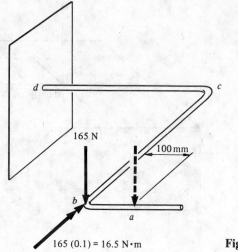

Fig. 3.45b

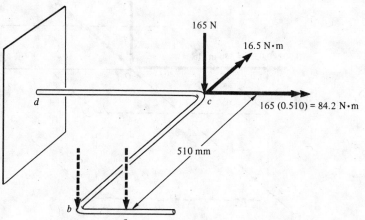

Fig. 3.45c

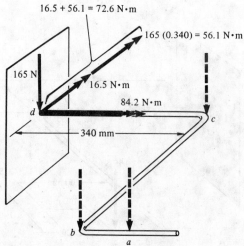

Fig. 3.45d

3.46   Show another form of solution of Prob. 3.45.

    ❚ An alternative construction, using two steps, is shown in Fig. 3.46.

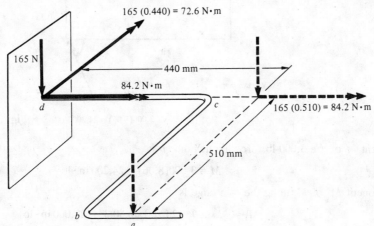

Fig. 3.46

3.47   Refer to Fig. 3.47a. Replace the force by a force and two couples at point b. Sketch the results, using the double-arrowhead representation of couples.

    ❚ The force is first transferred to point a. Then this force and couple is transferred to point b. The replacement of the original force by a force and two couples at point b is shown in Fig. 3.47b.

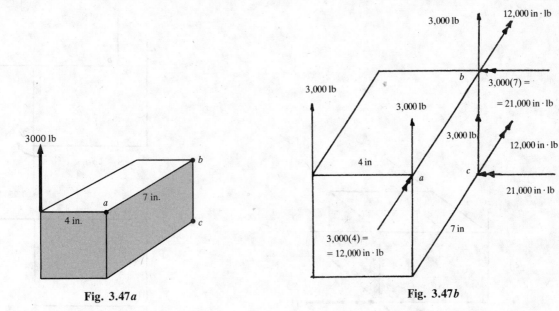

Fig. 3.47a   Fig. 3.47b

3.48   Refer to Fig. 3.47a. Replace the force by a force and two couples at point c. Sketch the results, using the double-arrowhead representation of couples.

    ❚ The force and two couples transfer from point b to point c without change. This effect is shown in Fig. 3.47b.

3.49   Replace the force in Prob. 3.47 by a force and a *single* couple at point b. Using the fact that a couple is a vector quantity, show that the result is equivalent to the result obtained in that problem.

    ❚ A top view of the body of Prob. 3.47 is shown in Fig. 3.49.

$$\tan \theta = \frac{7}{4} \qquad \theta = 60.3°$$

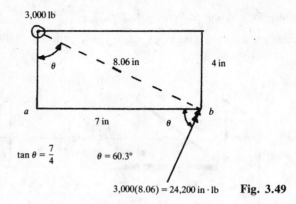

$$\tan \theta = \frac{7}{4} \qquad \theta = 60.3°$$

$$3,000(8.06) = 24,200 \text{ in} \cdot \text{lb} \qquad \textbf{Fig. 3.49}$$

The moment M of the 3,000-lb force about point b is

$$M = 3,000(8.06) = 24,200 \text{ in} \cdot \text{lb}$$

The component M_1 of M along the 4-in edge is

$$M_1 = M \sin \theta = 24,200 \sin 60.3° = 21,000 \text{ in} \cdot \text{lb}$$

The component M_2 of M along the 7-in edge is

$$M_2 = M \cos \theta = 24,000 \cos 60.3° = 12,000 \text{ in} \cdot \text{lb}$$

The values of M_1 and M_2 are the same as the values of the couples in Prob. 3.47.

3.50 In Fig. 3.50a the system of forces and couples at point a is to be replaced by a single force at point b. Find the magnitude, direction, and sense of this force, as well as the coordinates x_1 and y_1 of point b.

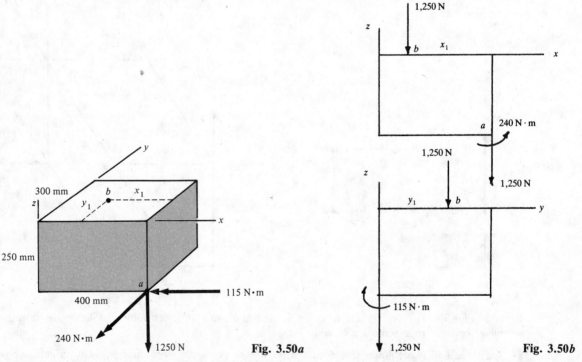

Fig. 3.50a **Fig. 3.50b**

▮ Two side views of the body are shown in Fig. 3.50b. The magnitude of the force at point b is 1,250 N, acting in the negative z sense.

For replacement of the couple and force, in a plane parallel to the xz plane, by a single force,

$$1,250x_1 = 240 \qquad x_1 = 0.192 \text{ m} = 192 \text{ mm}$$

For the plane parallel to the yz plane,

$$1,250y_1 = 115 \qquad y_1 = 0.0920 \text{ m} = 92.0 \text{ mm}$$

3.3 MOMENT IN A THREE-DIMENSIONAL FORCE SYSTEM, RESULTANT MOMENT, AND COMPONENTS OF MOMENT

3.51   Show the technique for representing a moment in a three-dimensional coordinate system.

▮ In the preceding problems, all the forces were in the same plane and this plane was perpendicular to the reference axes about which moments were taken. In the most general case, the force may have an arbitrary direction, and it is referenced to an *xyz* coordinate system. For this case, it is possible for the force to produce components of moment about any of, or all, the three coordinate axes. In describing the sense of moments about the three coordinate axes, the terms *clockwise* and *counterclockwise* are by themselves ambiguous. To avoid this situation, a new sign convention for moment will be introduced. This convention is shown in Fig. 3.51, where the moment **M**, with components $\mathbf{M}_x, \mathbf{M}_y$, and $\mathbf{M}_z$, acts about the origin.

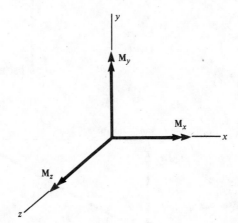

Fig. 3.51

A component of moment about a coordinate axis is considered to be positive if it has the sense shown in the figure. The interpretation of these positive senses of moment is that an observer, positioned at the origin and looking out in the three positive coordinate senses, would define positive moments as having clockwise senses. In solving for the components of moments due to a force which is referenced to a three-dimensional coordinate system, certain components of moment may be determined, by inspection, to be zero. This situation occurs when the line of action of the component of force and the reference axis about which moments are to be taken lie in the same plane. Under these conditions *the line of action of the force either intersects the reference axis or is parallel to it*, and thus no moment is produced by this component of force.

The formal vector description of the moment in the figure is

$$\mathbf{M} = \mathbf{M}_x + \mathbf{M}_y + \mathbf{M}_z$$
$$= M_x \mathbf{i} + M_y \mathbf{j} + M_z \mathbf{k}$$

The following problems illustrate the techniques of solution for the magnitude and direction of the resultant moment of a force in a three-dimensional coordinate system, or of the resultant couple of a system of couples with different directions. A major difference between these two types of problems should be noted. In the case of the resultant moment of a force, *a statement must be included of the reference axis about which the resultant moment acts.* For the case of the resultant couple of a system of couples, no such definition of a reference axis is required. All couples are pure turning effects and thus are *free vectors*. They have no single reference axis about which the moment effect occurs.

3.52   Figure 3.52*a* shows a force referenced to an *xyz* coordinate system. Find the components of moment about the *x*, *y*, and *z* axes and the resultant moment.

▮ The *y* component of moment, M_y, is zero by inspection, since the force and the *y* axis lie in the same plane. Using the sign convention of Prob. 3.51, we have

$$M_x = 750(3) = 2,250 \text{ in} \cdot \text{lb}$$
$$M_z = -750(4.5) = -3,380 \text{ in} \cdot \text{lb}$$
$$\mathbf{M} = 2,250\mathbf{i} - 3,380\mathbf{k} \quad \text{in} \cdot \text{lb}$$

The above equations fully define the components of moment about the three coordinate axes. When the resultant of these three components is found, an additional statement must be made to define the reference

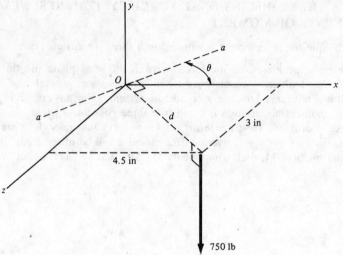

Fig. 3.52a

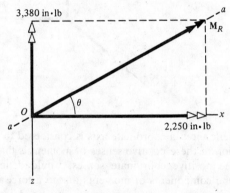

Fig. 3.52b

point about which this resultant moment acts. For the present problem, this point is the origin of the coordinate system, since this is the common point of intersection of the components of moment. The components of moment are shown in true view in Fig. 3.52b.

The axis about which the resultant moment acts is line aa. The magnitude M_R of this moment about line aa is

$$M_R = \sqrt{M_x^2 + M_z^2} = \sqrt{2,250^2 + 3,380^2} = 4,060 \text{ in} \cdot \text{lb}$$

The direction θ is found from

$$\tan \theta = \frac{3,380}{2,250} = 1.50$$

$$\theta = 56.3°$$

The direction cosines of the resultant moment are

$$\theta_x = \theta = 56.3°$$
$$\theta_y = 90°$$

$$\cos \theta_z = \frac{-3,380}{4,060} \qquad \theta_z = 146°$$

For this elementary case, the moment due to the 750-lb force could have been found directly from Fig. 3.52a. The moment arm d is given by

$$d = \sqrt{4.5^2 + 3^2} = 5.41 \text{ in}$$

From the fundamental definition of moment,

$$M_R = Fd = 750(5.41) = 4,060 \text{ in} \cdot \text{lb}$$

3.53 The force in Fig. 3.53a has the three components shown. Find the components of moment of this force about the x, y, and z axes. Specify the magnitude and direction of the resultant moment about the origin of the coordinate system.

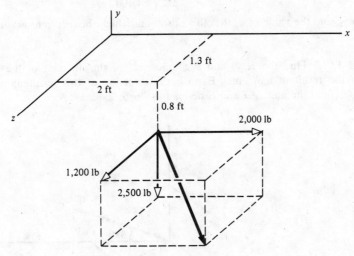

Fig. 3.53a

❙ With the sign convention of Prob. 3.51, these three components of moment are

$$M_x = 2,500(1.3) - 1,200(0.8) = 2,290 \text{ ft} \cdot \text{lb}$$
$$M_y = 2,000(1.3) - 1,200(2) = 200 \text{ ft} \cdot \text{lb}$$
$$M_z = 2,000(0.8) - 2,500(2) = -3,400 \text{ ft} \cdot \text{lb}$$

The formal statement describing the resultant moment is

$$\mathbf{M}_R = 2,290\mathbf{i} + 200\mathbf{j} - 3,400\mathbf{k} \quad \text{ft} \cdot \text{lb}$$

These components are sketched in Fig. 3.53b in their actual senses. The magnitude of the resultant moment is

$$M_R = \sqrt{M_x^2 + M_y^2 + M_z^2} = \sqrt{2,290^2 + 200^2 + (-3,400)^2}$$
$$= 4,100 \text{ ft} \cdot \text{lb}$$

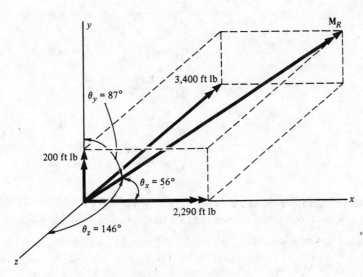

Fig. 3.53b

The direction of this resultant moment is defined by

$$\cos \theta_x = \frac{M_x}{M} = \frac{2,290}{4,100} \qquad \theta_x = 56°$$

$$\cos \theta_y = \frac{M_y}{M} = \frac{200}{4,100} \qquad \theta_y = 87°$$

$$\cos \theta_z = \frac{M_z}{M} = \frac{-3,400}{4,100} \qquad \theta_z = 146°$$

It may be seen from the value of θ_y that the axis about which the resultant acts is very nearly perpendicular to the y axis.

3.54 Refer to Fig. 3.54. Find the x, y, and z components of the moment of the force about point a, and the magnitude of the resultant moment. Express the direction of this resultant moment in terms of the three direction angles, and the angles α and β defined in Prob. 2.42.

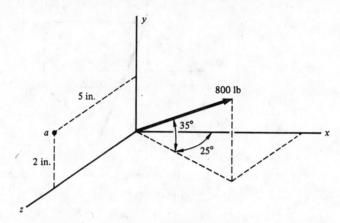

Fig. 3.54

$M_x = (800 \sin 35°)5 - (800 \cos 35°) \sin 25°(2) = 1,740 \text{ in} \cdot \text{lb}$

$M_y = -(800 \cos 35°) \cos 25°(5) = -2,970 \text{ in} \cdot \text{lb}$

$M_z = (800 \cos 35°) \cos 25°(2) = 1,190 \text{ in} \cdot \text{lb}$

$$\mathbf{M} = 1,740\mathbf{i} - 2,970\mathbf{j} + 1,190\mathbf{k} \qquad \text{in} \cdot \text{lb}$$

$$M = \sqrt{M_x^2 + M_y^2 + M_z^2}$$

$$= \sqrt{1,740^2 + (-2,970)^2 + 1,190^2}$$

$$M = 3,640 \text{ in} \cdot \text{lb}$$

$$\cos \theta_x = \frac{M_x}{M} = \frac{1,740}{3,640} \qquad \theta_x = 61.4°$$

$$\cos \theta_y = \frac{M_y}{M} = \frac{-2,970}{3,640} \qquad \theta_y = 145°$$

$$\cos \theta_z = \frac{M_z}{M} = \frac{1,190}{3,640} \qquad \theta_z = 70.9°$$

$$\tan \alpha = \frac{M_z}{M_x} = \frac{1,190}{1,740} \qquad \alpha = 34.4°$$

$$\sin \beta = \frac{M_y}{M} = \frac{-2,970}{3,640} \qquad \beta = -54.7°$$

3.55 Three couples act on the cube-shaped body shown in Fig. 3.55a.
Show the graphical construction required to find the resultant of the three couples.

❚ The original system is shown in Fig. 3.55b. $\mathbf{M}_1$ and $\mathbf{M}_3$ are added vectorially to obtain their resultant $\mathbf{M}_{13}$ along the x' axis. The yx' plane is shown in true view in Fig. 3.55c. $\mathbf{M}_{13}$ and $\mathbf{M}_2$ are added vectorially to obtain the resultant $\mathbf{M}_{123}$ of the original system. From Fig. 3.55b,

$$M_{13} = \sqrt{M_1^2 + M_3^2}$$

and, from Fig. 3.55c,

$$M_{123} = \sqrt{M_{13}^2 + M_2^2} = \sqrt{M_1^2 + M_2^2 + M_3^2}$$

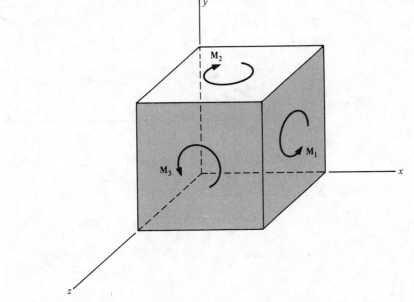

Fig. 3.55a

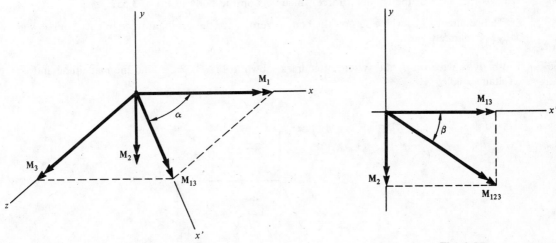

Fig. 3.55b

Fig. 3.55c

3.56 If the magnitudes of M_1, M_2, and M_3 in Prob. 3.55 are 150, 210, and 330 ft · lb, respectively, find the magnitude and direction of the resultant couple which acts on the body.

▌ For the numerical values of this problem,

$$M_{13} = \sqrt{150^2 + 330^2} = 362 \text{ ft} \cdot \text{lb}$$
$$M_{123} = \sqrt{362^2 + 210^2} = 419 \text{ ft} \cdot \text{lb}$$

The direction of the resultant couple is defined by the angles α and β shown in Figs. 3.55b and c. The values of these angles are found from

$$\tan \alpha = \frac{M_3}{M_1} = \frac{330}{150} = 2.20 \qquad \alpha = 65.6°$$

$$\tan \beta = \frac{M_2}{M_{13}} = \frac{210}{362} = 0.580 \qquad \beta = 30.1°$$

The resultant couple of the original system of couples causes a twisting effect in the plane perpendicular to the direction of the vector M_{123} in Fig. 3.55c. This plane of action is shown in Fig. 3.56 as plane *abcd*. Although the resultant couple M_{123} is shown as acting through the origin, it should be emphasized that *this vector may have any position perpendicular to plane abcd*. The reason is the earlier observation that a couple is a free vector which may be moved to any other position parallel to its original direction, without change in its effect on the body on which it acts.

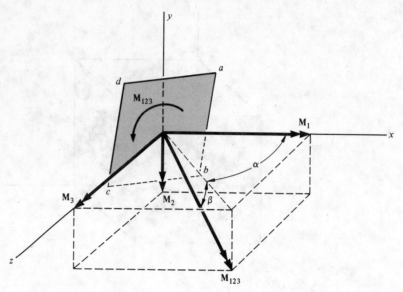

Fig. 3.56

3.57   What is the value of the resultant force which acts on the body in Prob. 3.55?

▮ The resultant force which acts on the body is zero, since the resultant force of each couple which acts on the body is separately zero.

3.58   The body is acted on by the system of couples shown in Fig. 3.58a. Find the magnitude and direction of the resultant couple

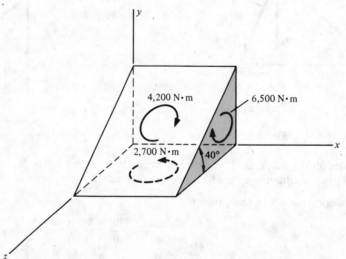

Fig. 3.58a

▮ The original system is shown in Fig. 3.58b. The 4,200-N · m couple is projected onto the y and z axes, as shown in Fig. 3.58b, and the components of the resultant couple are shown in Figs. 3.58c and d. The magnitude M_R of this resultant couple is

$$M_R = \sqrt{(-6,500)^2 + (-520)^2 + (-2,700)^2} = 7,060 \text{ N} \cdot \text{m}$$

The direction of the line of action of the resultant couple will be given in terms of the direction angles. In solving for these angles, it may be noted that all three components shown in Fig. 3.58d are negative. These angles are found as

$$\cos \theta_x = \frac{M_x}{M_R} = \frac{-6,500}{7,060} \qquad \theta_x = 157°$$

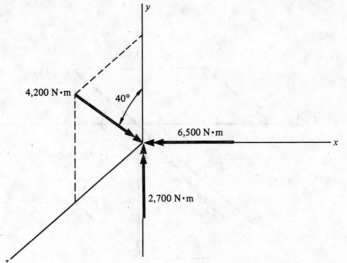

Fig. 3.58b

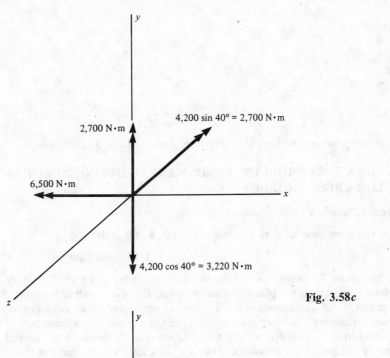

Fig. 3.58c

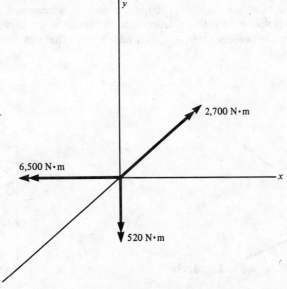

Fig. 3.58d

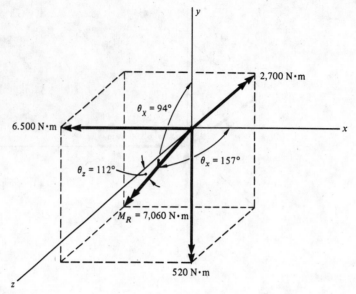

Fig. 3.58e

$$\cos \theta_y = \frac{M_y}{M_R} = \frac{-520}{7,060} \qquad \theta_y = 94°$$

$$\cos \theta_z = \frac{M_z}{M_R} = \frac{-2,700}{7,060} \qquad \theta_z = 112°$$

The components, and the three direction angles, of the resultant couples are shown in Fig. 3.58e.

3.4 VECTOR CROSS PRODUCT, FUNDAMENTAL DEFINITION OF MOMENT AS A VECTOR CROSS PRODUCT

3.59 Define the *vector cross product* operation.

▎ The cross product **C** of the two vectors **A** and **B** is defined by

$$\mathbf{C} = \mathbf{A} \times \mathbf{B} = (AB \sin \theta)\mathbf{i}_n$$

A and B are the magnitudes of the two vectors, and θ is the angle between their directions; $\mathbf{i}_n$ is a unit vector which is normal to the plane formed by **A** and **B**. *The result of the cross product multiplication is a vector*, in contrast to the scalar result in the dot product operation. The cross product is shown graphically in Fig. 3.59a.

In performing the cross product operation, the vector **A** is imagined to be rotated toward **B** in the plane formed by these two vectors. If this rotation is in the sense of the curved fingers of the right hand, then the thumb will extend in the positive sense of $\mathbf{i}_n$. It may be seen that this is exactly the same convention which was

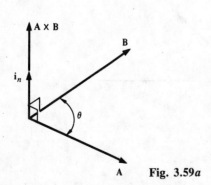

Fig. 3.59a

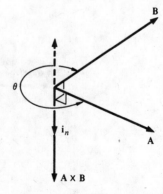

Fig. 3.59b

presented in Prob. 3.6 for the case of the representation of a moment by a double-headed arrow. The angle θ usually is chosen to be the smaller of the two angles between the directions of **A** and **B**, as shown in Fig. 3.59a. In this view, $\theta < 180°$ and the sine function is positive. The actual sense of the cross product **A** × **B**, then, would be upward, as shown in Fig. 3.59a.

In Fig. 3.59b, by using the right-hand rule, the unit vector $\mathbf{i}_n$ is positive when pointing downward. Since $180° < \theta < 360°$, the sine function is negative. The net effect is that the final result for **A** × **B** acts upward, in the sense of the dashed arrow shown in the figure. Thus, the two constructions in Figs. 3.59a and b give identical results.

The vector cross product is not commutative. That is,

$$\mathbf{A} \times \mathbf{B} \neq \mathbf{B} \times \mathbf{A}$$

3.60 Show the general form of the solution for the vector cross product.

▌ The cross product of the two vectors **A** and **B** can be written as

$$\mathbf{A} \times \mathbf{B} = (A_x\mathbf{i} + A_y\mathbf{j} + A_z\mathbf{k}) \times (B_x\mathbf{i} + B_y\mathbf{j} + B_z\mathbf{k})$$

The right-hand side will contain nine terms, and each of these terms will include cross products of pairs of unit vectors. Typical cross products of these unit vectors are

$$\mathbf{i} \times \mathbf{i} = (1)(1) \sin 0° \, \mathbf{k} = 0$$
$$\mathbf{i} \times \mathbf{j} = (1)(1) \sin 90° \, \mathbf{k} = \mathbf{k}$$
$$\mathbf{i} \times \mathbf{k} = (1)(1) \sin 90° \, (-\mathbf{j}) = -\mathbf{j}$$

The remaining six cross products of the unit vectors are

$$\mathbf{j} \times \mathbf{i} = -\mathbf{k} \qquad \mathbf{k} \times \mathbf{i} = \mathbf{j}$$
$$\mathbf{j} \times \mathbf{j} = 0 \qquad \mathbf{k} \times \mathbf{j} = -\mathbf{i}$$
$$\mathbf{j} \times \mathbf{k} = \mathbf{i} \qquad \mathbf{k} \times \mathbf{k} = 0$$

The above equations are combined, with the final result

$$\mathbf{A} \times \mathbf{B} = (A_y B_z - A_z B_y)\mathbf{i} + (A_z B_x - A_x B_z)\mathbf{j} + (A_x B_y - A_y B_x)\mathbf{k}$$

In performing the cross product operation, it is essential that a right-hand xyz coordinate system be used. The reason is that the above equations were obtained on the basis of such a coordinate system. A right-hand coordinate system is shown in Fig. 3.60a. An example of a coordinate system which is *not* right-hand is shown in Fig. 3.60b.

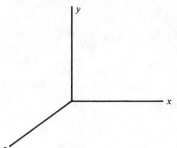

Fig. 3.60a

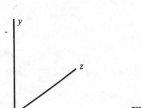

Fig. 3.60b

3.61 Show how the vector cross product **A** × **B** can be compactly represented by a 3 × 3 determinant.

▌ Refer to Fig. 3.61. The first row contains the unit vectors **i**, **j**, and **k**. The second row has the three components A_x, A_y, and A_z of vector **A**. In the third row are the components B_x, B_y, and B_z of vector **B**. All of the above terms must be in the order shown. The order, and associated sign, of multiplication of the elements is indicated by the arrows. In many problems, one or more of the components A_x, A_y, A_z, B_x, B_y, B_z may be zero.

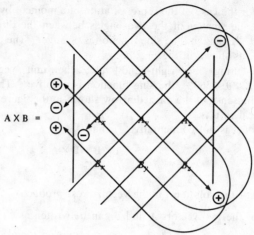

$$A \times B =$$

Fig. 3.61

3.62 Show the *vector cross product definition of the moment of a force.*

▌ The principal use of the cross product in statics is to find the moment due to a force. Figure 3.62 shows a force **F** in space. 0′ is an arbitrary point, and it is desired to find the moment, together with its components, of the force **F** with respect to the point 0′. This moment **M** is defined as

$$\mathbf{M} = \mathbf{r} \times \mathbf{F}$$

In the above equation, **r** is a *position vector* from 0′ to *any* point on the line of action of **F**. This equation is a particularly useful tool for solving for the moment in a three-dimensional force system, since the use of this technique *does not* require a visualization of the moment arms of the forces or of the senses of the components of moment.

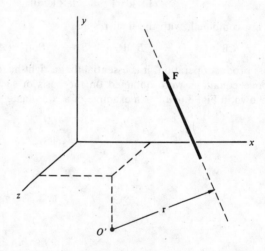

Fig. 3.62

3.63 Solve Prob. 3.52 by using the vector cross product definition of moment.

▌ The force is written as

$$\mathbf{F} = -750\mathbf{j} \qquad \text{lb}$$

and the position vector **r** from point 0 to the line of action of the force is

$$\mathbf{r} = 4.5\mathbf{i} + 3\mathbf{k} \qquad \text{in}$$

The moment $\mathbf{M}_0$ of the force with respect to the origin is then

$$\mathbf{M}_0 = \mathbf{r} \times \mathbf{F} = \begin{vmatrix} \mathbf{i} & \mathbf{j} & \mathbf{k} \\ 4.5 & 0 & 3 \\ 0 & -750 & 0 \end{vmatrix} = [0 - (-750)3]\mathbf{i} + (0 - 0)\mathbf{j} + [-750(4.5) - 0]\mathbf{k}$$

$$= 2{,}250\mathbf{i} - 3{,}380\mathbf{k} \qquad \text{in} \cdot \text{lb}$$

These results were obtained earlier in Prob. 3.52. The direction of $\mathbf{M}_0$ is found as before.

3.64   Solve Prob. 3.53 by using the vector cross product definition of moment.

▐   The position vector **r** and the force **F** are written as

$$\mathbf{r} = 2\mathbf{i} - 0.8\mathbf{j} + 1.3\mathbf{k} \quad \text{ft}$$
$$\mathbf{F} = 2{,}000\mathbf{i} - 2{,}500\mathbf{j} + 1{,}200\mathbf{k} \quad \text{lb}$$

The moment of the force about the origin is then

$$\mathbf{M}_0 = \mathbf{r} \times \mathbf{F} = \begin{vmatrix} \mathbf{i} & \mathbf{j} & \mathbf{k} \\ 2 & -0.8 & 1.3 \\ 2{,}000 & -2{,}500 & 1{,}200 \end{vmatrix}$$

$$= [-0.8(1{,}200) - (-2{,}500)(1.3)]\mathbf{i}$$
$$+ [1.3(2{,}000) - (1{,}200)2]\mathbf{j} + [-2{,}500(2) - 2{,}000(-0.8)]\mathbf{k}$$
$$= 2{,}290\mathbf{i} + 200\mathbf{j} - 3{,}400\mathbf{k} \quad \text{ft} \cdot \text{lb}$$

The components of this moment about the three coordinate axes are

$$M_x = 2{,}290 \text{ ft} \cdot \text{lb}$$
$$M_y = 200 \text{ ft} \cdot \text{lb}$$
$$M_z = -3{,}400 \text{ ft} \cdot \text{lb}$$

These are the results found in Prob. 3.53, and the direction of the moment is found as before.

3.65   Solve Prob. 3.54 by using the vector cross product definition of moment.

▐   The position vector **r** from point *a* to the tail end of the force is

$$\mathbf{r} = -2\mathbf{j} - 5\mathbf{k} \quad \text{in}$$

The force **F** is expressed in terms of the unit vectors as

$$\mathbf{F} = (800 \cos 35° \cos 25°)\,\mathbf{i} + (800 \sin 35°)\,\mathbf{j} + (800 \cos 35° \sin 25°)\,\mathbf{k}$$
$$= (594\mathbf{i} + 459)\mathbf{j} + 277\mathbf{k} \quad \text{lb}$$

The moment **M** of the force about point *a* is then

$$\mathbf{M} = \mathbf{r} \times \mathbf{F} = \begin{vmatrix} \mathbf{i} & \mathbf{j} & \mathbf{k} \\ 0 & -2 & -5 \\ 594 & 459 & 277 \end{vmatrix}$$

$$= [-2(277) - 459(-5)]\mathbf{i} + [-5(594) - 270(0)]\mathbf{j} + [459(0) - 594(-2)]\mathbf{k}$$
$$= 1{,}740\mathbf{i} - 2{,}970\mathbf{j} + 1{,}190\mathbf{k} \quad \text{in} \cdot \text{lb}$$

The magnitude and direction of **M** are found as in Prob. 3.54.

3.66   Compare the method of solving for the moment in a three-dimensional force system by direct application of the fundamental definition $M = Fd$ with the method using the vector cross product definition of moment.

▐   Problems 3.63 through 3.65 show the use of the vector cross product definition of moment to solve problems which had previously been solved by direct use of the definition $M = Fd$ of moment.   In the direct solutions, the magnitudes and senses of the moment components must carefully be determined from inspection of the sketch of the problem.   In addition, a conscious effort must be made to ensure that no moment component has been inadvertently omitted.

When using the vector cross product definition of moment, only two major steps are required.   In the first step, the forces are expressed in terms of unit vectors.   In the second, a position vector *from* the reference point *to any known point* on the line of action of the force is defined.   Using $\mathbf{M} = \mathbf{r} \times \mathbf{F}$,   the solution proceeds in a completely automatic fashion to obtain the final results for the moment components.   These final results express the correct magnitudes and senses, and no separate determination need be made of these quantities.

When solving moment problems with two-dimensional force systems, the usual method is to use Varignon's theorem, with the definition $M = Fd$,   to directly obtain the scalar values of the components of moment.   When solving problems with three-dimensional force systems, the vector cross product method is almost always used.   The magnitudes and senses of the three components found by using the vector cross product method may be confirmed by inspection of the sketch of the problem.

3.67   The force shown in Fig. 3.67 has a magnitude of 1,350 lb.   Find the x, y, and z components of the moment of this force about point a.   Find the magnitude and the direction angles of this moment.

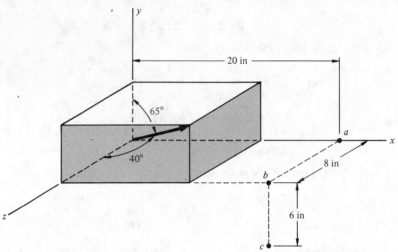

Fig. 3.67

❚ θ_x is found first from

$$\cos^2 \theta_x + \cos^2 65° + \cos^2 40° = 1 \qquad \theta_x = 61.0°, \ 119°$$

From consideration of the figure,

$$\theta_x = 61.0°$$

The position vector $\mathbf{r}$ from point a to the tail end of the force is

$$\mathbf{r} = -20\mathbf{i} \qquad \text{in}$$

The force $\mathbf{F}$ is expressed as

$$\mathbf{F} = (1,350 \cos 61.0°)\mathbf{i} + (1,350 \cos 65°)\mathbf{j} + (1,350 \cos 40°)\mathbf{k}$$
$$= 654\mathbf{i} + 571\mathbf{j} + 1,030\mathbf{k} \qquad \text{lb}$$

$\mathbf{M}_a$ is the moment of the force about point a, given by

$$\mathbf{M}_a = \mathbf{r} \times \mathbf{F} = \begin{vmatrix} \mathbf{i} & \mathbf{j} & \mathbf{k} \\ -20 & 0 & 0 \\ 654 & 571 & 1,030 \end{vmatrix}$$

$$= (0)\mathbf{i} + [0 - 1,030(-20)]\mathbf{j} + [571(-20) - 0]\mathbf{k}$$
$$= 20,600\mathbf{j} - 11,400\mathbf{k} \qquad \text{in} \cdot \text{lb}$$
$$M_a = \sqrt{M_y^2 + M_z^2} = \sqrt{20,600^2 + (-11,400)^2} = 23,500 \text{ in} \cdot \text{lb}$$

The direction of $\mathbf{M}_a$ is given by

$$\cos \theta_x = \frac{M_x}{M_a} = 0 \qquad\qquad \theta_x = 90°$$

$$\cos \theta_y = \frac{M_y}{M_a} = \frac{20,600}{23,500} \qquad \theta_y = 28.8°$$

$$\cos \theta_z = \frac{M_z}{M_a} = \frac{-11,400}{23,500} \qquad \theta_z = 119°$$

3.68   Refer to Fig. 3.67 and again assume a force with a magnitude of 1350 lb.   Find the x, y, and z components of the moment of this force about point a.

❚ The position vector $\mathbf{r}$ from point b to the tail end of the force shown in Prob. 3.67 is

$$\mathbf{r} = -20\mathbf{i} - 8\mathbf{k} \qquad \text{in}$$

The moment $\mathbf{M}_b$ of the force about point b is

$$\mathbf{M}_b = \mathbf{r} \times \mathbf{F} = \begin{vmatrix} \mathbf{i} & \mathbf{j} & \mathbf{k} \\ -20 & 0 & -8 \\ 654 & 571 & 1,030 \end{vmatrix}$$

$$= [0(1,030) - 571(-8)]\mathbf{i} + [-8(654) - 1,030(-20)]\mathbf{j} + [571(-20) - 654(0)]\mathbf{k}$$
$$= 4,570\mathbf{i} + 15,400\mathbf{j} - 11,400\mathbf{k} \quad \text{in} \cdot \text{lb}$$
$$M_b = \sqrt{M_x^2 + M_y^2 + M_z^2} = \sqrt{4,570^2 + 15,400^2 + (-11,400)^2} = 19,700 \text{ in} \cdot \text{lb}$$

The direction of $\mathbf{M}_b$ is defined by

$$\cos \theta_x = \frac{M_x}{M_b} = \frac{4,570}{19,700} \qquad \theta_x = 76.6°$$

$$\cos \theta_y = \frac{M_y}{M_b} = \frac{15,400}{19,700} \qquad \theta_y = 38.6°$$

$$\cos \theta_z = \frac{M_z}{M_b} = \frac{-11,400}{19,700} \qquad \theta_z = 125°$$

3.69 Refer to Fig. 3.67 and find the x, y, and z components of the moment of 1,350-lb force about point c.

❙ The position vector from point c to the tail end of the force is

$$\mathbf{r} = -20\mathbf{i} + 6\mathbf{j} - 8\mathbf{k} \qquad \text{in}$$

The moment $\mathbf{M}_c$ of the force about point c is

$$\mathbf{M}_c = \mathbf{r} \times \mathbf{F} = \begin{vmatrix} \mathbf{i} & \mathbf{j} & \mathbf{k} \\ -20 & 6 & -8 \\ 654 & 571 & 1,030 \end{vmatrix}$$

$$= [6(1,030) - 571(-8)]\mathbf{i} + [-8(654) - 1,030(-20)]\mathbf{j} + [571(-20) - 654(6)]\mathbf{k}$$
$$= 10,700\mathbf{i} + 15,400\mathbf{j} - 15,300\mathbf{k} \quad \text{in} \cdot \text{lb}$$
$$M_c = \sqrt{M_x^2 + M_y^2 + M_z^2} = \sqrt{10,700^2 + 15,400^2 + (-15,300)^2} = 24,200 \text{ in} \cdot \text{lb}$$

The direction angles of $\mathbf{M}_c$ are found from

$$\cos \theta_x = \frac{M_x}{M_c} = \frac{10,700}{24,200} \qquad \theta_x = 63.8°$$

$$\cos \theta_y = \frac{M_y}{M_c} = \frac{15,400}{24,200} \qquad \theta_y = 50.5°$$

$$\cos \theta_z = \frac{M_z}{M_c} = \frac{-15,300}{24,200} \qquad \theta_z = 129°$$

3.70 Refer to Fig. 3.70. Find the x, y, and z components of the moment of the force about point a. Find the magnitude and direction angles of this moment.

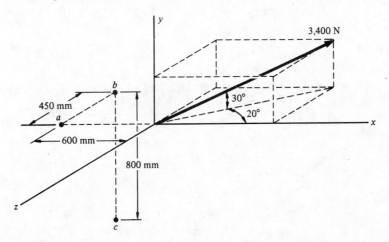

Fig. 3.70

$$\mathbf{F} = (3{,}400 \cos 30° \cos 20°)\mathbf{i} + (3{,}400 \sin 30°)\mathbf{j} - (3{,}400 \cos 30° \sin 20°)\mathbf{k}$$
$$= 2{,}770\mathbf{i} + 1{,}700\mathbf{j} - 1{,}010\mathbf{k} \qquad \text{N}$$

The position vector from point a to the tail end of the force is

$$\mathbf{r} = \frac{600}{1000}\,\mathbf{i} = 0.6\mathbf{i} \qquad \text{m}$$

$$\mathbf{M}_a = \mathbf{r} \times \mathbf{F} = \begin{vmatrix} \mathbf{i} & \mathbf{j} & \mathbf{k} \\ 0.6 & 0 & 0 \\ 2{,}770 & 1{,}700 & -1{,}010 \end{vmatrix}$$

$$= [0(-1{,}010) - 1{,}700(0)]\mathbf{i} + [0(2{,}770) - (-1{,}010)0.6]\mathbf{j} + [1{,}700(0.6) - 2{,}770(0)]\mathbf{k}$$
$$= 606\mathbf{j} + 1{,}020\mathbf{k} \qquad \text{N} \cdot \text{m}$$
$$M_a = \sqrt{M_y^2 + M_z^2} = \sqrt{606^2 + 1{,}020^2} = 1{,}190 \ \text{N} \cdot \text{m}$$

$$\cos \theta_x = \frac{M_x}{M_a} = 0 \qquad\qquad \theta_x = 90°$$

$$\cos \theta_y = \frac{M_y}{M_a} = \frac{606}{1{,}190} \qquad \theta_y = 59.4°$$

$$\cos \theta_z = \frac{M_z}{M_a} = \frac{1{,}020}{1{,}190} \qquad \theta_z = 31.0°$$

3.71 Do the same as in Prob. 3.70, but for point b.

❙ The position vector $\mathbf{r}$ has the form

$$\mathbf{r} = \frac{1}{1{,}000}\,(600\mathbf{i} + 450\mathbf{k}) = 0.6\mathbf{i} + 0.45\mathbf{k} \qquad \text{m}$$

$$\mathbf{M}_b = \begin{vmatrix} \mathbf{i} & \mathbf{j} & \mathbf{k} \\ 0.6 & 0 & 0.45 \\ 2{,}770 & 1{,}700 & -1{,}010 \end{vmatrix}$$

$$= [0(-1{,}010) - 1{,}700(0.45)]\mathbf{i} + [0.45(2{,}770) - (-1{,}010)0.6]\mathbf{j} + [1{,}700(0.6) - 2{,}770(0)]\mathbf{k}$$
$$= -765\mathbf{i} + 1{,}850\mathbf{j} + 1{,}020\mathbf{k} \qquad \text{N} \cdot \text{m}$$
$$M_b = \sqrt{M_x^2 + M_y^2 + M_z^2} = \sqrt{(-765)^2 + 1{,}850^2 + 1{,}020^2} = 2{,}250 \ \text{N} \cdot \text{m}$$

$$\cos \theta_x = \frac{M_x}{M_b} = \frac{-765}{2{,}250} \qquad \theta_x = 110°$$

$$\cos \theta_y = \frac{M_y}{M_b} = \frac{1{,}850}{2{,}250} \qquad \theta_y = 34.7°$$

$$\cos \theta_z = \frac{M_z}{M_b} = \frac{1{,}020}{2{,}250} \qquad \theta_z = 63.0°$$

3.72 Do the same as in Prob. 3.70, but for point c.

❙
$$\mathbf{r} = \frac{1}{1{,}000}\,(600\mathbf{i} + 800\mathbf{j} + 450\mathbf{k}) = 0.6\mathbf{i} + 0.8\mathbf{j} + 0.45\mathbf{k} \qquad \text{m}$$

$$\mathbf{M}_c = \begin{vmatrix} \mathbf{i} & \mathbf{j} & \mathbf{k} \\ 0.6 & 0.8 & 0.45 \\ 2{,}770 & 1{,}700 & -1{,}010 \end{vmatrix}$$

$$= [0.8(-1{,}010) - 1{,}700(0.45)]\mathbf{i} + [0.45(2{,}770) - (-1{,}010)0.6]\mathbf{j} + [1{,}700(0.6) - 2{,}770(0.8)]\mathbf{k}$$
$$= -1{,}570\mathbf{i} + 1{,}850\mathbf{j} - 1{,}200\mathbf{k} \qquad \text{N} \cdot \text{m}$$
$$M_c = \sqrt{M_x^2 + M_y^2 + M_z^2} = \sqrt{(-1{,}570)^2 + 1{,}850^2 + (-1{,}200)^2} = 2{,}710 \ \text{N} \cdot \text{m}$$

$$\cos \theta_x = \frac{M_x}{M_c} = \frac{-1{,}570}{2{,}710} \qquad \theta_x = 125°$$

$$\cos \theta_y = \frac{M_y}{M_c} = \frac{1{,}850}{2{,}710} \qquad \theta_y = 46.9°$$

$$\cos \theta_z = \frac{M_z}{M_c} = \frac{-1{,}200}{2{,}710} \qquad \theta_z = 116°$$

3.73   Find the moment of the force of Prob. 3.53 about the axis *ab* shown in Fig. 3.73*a*.

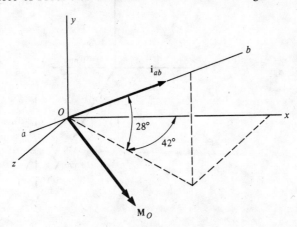

Fig. 3.73*a*

❚ The component of moment about axis *ab* will be found by projecting $\mathbf{M}_R$ onto this axis. The unit vector $\mathbf{i}_{ab}$ is directed along axis *ab*. This vector may be expressed in terms of the unit vectors of the *xyz* coordinate system by

$$\mathbf{i}_{ab} = (1)(\cos 28° \cos 42°)\mathbf{i} + (1)(\sin 28°)\mathbf{j} + (1)(\cos 28° \sin 42°)\mathbf{k}$$
$$= 0.656\mathbf{i} + 0.469\mathbf{j} + 0.591\mathbf{k}$$

From Prob. 3.53,

$$\mathbf{M}_R = 2,290\mathbf{i} + 200\mathbf{j} - 3,400\mathbf{k} \qquad \text{ft} \cdot \text{lb}$$

The component M_{ab} of $\mathbf{M}_R$ along axis *ab* is then found from the dot product

$$M_{ab} = \mathbf{M}_R \cdot \mathbf{i}_{ab}$$
$$= 2,290(0.656) + 200(0.469) - 3,400(0.591)$$
$$= -413 \text{ ft} \cdot \text{lb}$$

The actual sense of this result is shown in Fig. 3.73*b*. It may be seen that the actual sense of a result is determined by comparison with the sense which was arbitrarily chosen for the unit vector $\mathbf{i}_{ab}$.

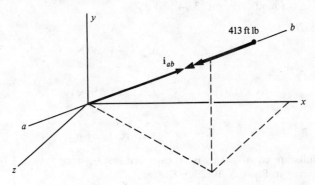

Fig. 3.73*b*

3.74   Find the component of the moment in Prob. 3.70 along a line joining points *a* and *c* shown in the figure in that problem.

❚ The moment of the force about *a* was found in Prob. 3.70 as

$$\mathbf{M}_a = 606\mathbf{j} + 1,020\mathbf{k} \qquad \text{N} \cdot \text{m}$$

A unit vector $\mathbf{i}_{ac}$ is drawn along line *ac*, as shown in Fig. 3.74, and

$$\mathbf{i}_{ac} = -\frac{800}{918}\mathbf{j} - \frac{450}{918}\mathbf{k}$$

The component M_{ac} of $\mathbf{M}_a$ along line *ac* is then found as

$$M_{ac} = \mathbf{M}_a \cdot \mathbf{i}_{ac} = 606\left(-\frac{800}{918}\right) + 1,020\left(-\frac{450}{918}\right) = -1,030 \text{ N} \cdot \text{m}$$

M_{ac} acts from *c* to *a*, with an actual sense which is clockwise when viewed from end *c*.

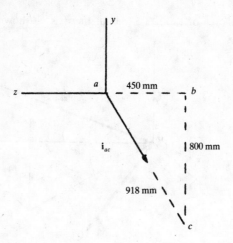

Fig. 3.74

3.75   Do the same as in Prob. 3.74 for the moment found in Prob. 3.72.

▌ From Prob. 3.72,

$$\mathbf{M}_c = -1{,}570\mathbf{i} + 1{,}850\mathbf{j} - 1{,}200\mathbf{k} \qquad \text{N} \cdot \text{m}$$

The component of $\mathbf{M}_c$ along line ac is

$$M_{ac} = \mathbf{M}_c \cdot \mathbf{i}_{ac}$$

where $\mathbf{i}_{ac}$ is the unit vector, obtained in Prob. 3.74, along line ac.

$$M_{ac} = 1{,}850\left(-\frac{800}{918}\right) - 1{,}200\left(-\frac{450}{918}\right) = -1{,}020 \,\text{N} \cdot \text{m}$$

3.76   Compare the results found in Probs. 3.74 and 3.75.

▌ From Prob. 3.74,

$$M_{ac} = -1{,}030 \,\text{N} \cdot \text{m}$$

and, from Prob. 3.75,

$$M_{ac} = -1{,}020 \,\text{N} \cdot \text{m}$$

It may be seen that

$$-1{,}030 \approx -1{,}020$$

and the two results are equal.   This is an expected result, since Probs. 3.74 and 3.75 are *both* solving for the moment of the force in Prob. 3.70 *about line ac*.

3.77   Refer to Fig. 3.77.   Find the moment of the force $\mathbf{F}$ about point a, the intersection of line ab with the x axis.

▌ The position vector $\mathbf{r}$ from a to the tail end of the force is given by

$$\mathbf{r} = (-2\mathbf{i} + 1.5\mathbf{j} + 6.3\mathbf{k}) \,\text{m} \qquad \text{and} \qquad \mathbf{F} = (-340\mathbf{i} + 650\mathbf{j} - 460\mathbf{k}) \,\text{N}$$

The moment $\mathbf{M}_a$ of $\mathbf{F}$ about point a is

$$\mathbf{M}_a = \mathbf{r} \times \mathbf{F} = \begin{vmatrix} \mathbf{i} & \mathbf{j} & \mathbf{k} \\ -2 & 1.5 & 6.3 \\ -340 & 650 & -460 \end{vmatrix}$$

$$= [1.5(-460) - 650(6.3)]\mathbf{i} + [6.3(-340) - (-460)(-2)]\mathbf{j} + [650(-2) - (-340)1.5]\mathbf{k}$$

$$= -4{,}790\mathbf{i} - 3{,}060\mathbf{j} - 790\mathbf{k} \qquad \text{N} \cdot \text{m}$$

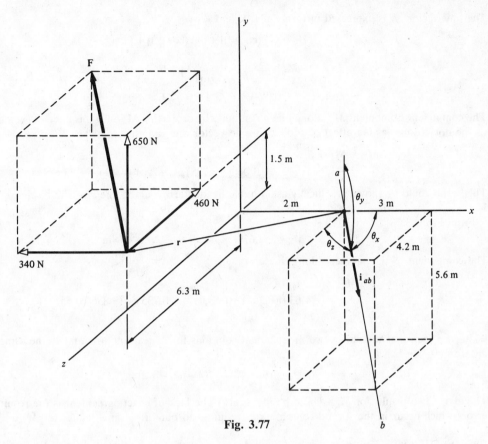

Fig. 3.77

3.78 Do the same as in Prob. 3.77 for the moment of the force about point b, the intersection of line ab with the lower corner of the dashed parallelepiped.

▎ The position vector from point b to the tail end of the force in Fig. 3.77 has the form

$$\mathbf{r} = -(3+2)\mathbf{i} + (5.6+1.5)\mathbf{j} + (-4.2+6.3)\mathbf{k}$$
$$= -5\mathbf{i} + 7.1\mathbf{j} + 2.1\mathbf{k} \quad \text{m}$$

From Prob. 3.77,

$$\mathbf{F} = -340\mathbf{i} + 650\mathbf{j} - 460\mathbf{k} \quad \text{N}$$

The moment $\mathbf{M}_b$ of $\mathbf{F}$ about point b is

$$\mathbf{M}_b = \mathbf{r} \times \mathbf{F} = \begin{vmatrix} \mathbf{i} & \mathbf{j} & \mathbf{k} \\ -5 & 7.1 & 2.1 \\ -340 & 650 & -460 \end{vmatrix}$$

$$= [7.1(-460) - 650(2.1)]\mathbf{i} + [2.1(-340) - (-460)(-5)]\mathbf{j} + [650(-5) - (-340)(7.1)]\mathbf{k}$$
$$= -4{,}630\mathbf{i} - 3{,}010\mathbf{j} - 836\mathbf{k} \quad \text{N} \cdot \text{m}$$

3.79 Show how the results in either Prob. 3.77 or Prob. 3.78 may be used to find the moment of the force in Prob. 3.77 about the line ab.

▎ The unit vector $\mathbf{i}_{ab}$ is directed along line ab, as shown in Fig. 3.77. The direction angles between the direction of line ab and the positive coordinate axes are θ_x, θ_y, and θ_z. The length ab of line ab is found as

$$ab = \sqrt{3^2 + 5.6^2 + 4.2^2} = 7.62 \text{ m}$$

The three direction cosines are then

$$\cos \theta_x = \frac{3}{7.62} = 0.394 \qquad \theta_x = 67°$$

$$\cos \theta_y = \frac{-5.6}{7.62} = -0.735 \qquad \theta_y = 137°$$

$$\cos \theta_z = \frac{4.2}{7.62} = 0.551 \qquad \theta_z = 57°$$

The unit vector $\mathbf{i}_{ab}$ is expressed in terms of $\mathbf{i}$, $\mathbf{j}$, and $\mathbf{k}$ by

$$\mathbf{i}_{ab} = (1)(\cos \theta_x)\mathbf{i} + (1)(\cos \theta_y)\mathbf{j} + (1)(\cos \theta_z)\mathbf{k}$$

$$= \left(\frac{3}{7.62}\right)\mathbf{i} - \left(\frac{5.6}{7.62}\right)\mathbf{j} + \left(\frac{4.2}{7.62}\right)\mathbf{k}$$

$$= 0.394\mathbf{i} - 0.735\mathbf{j} + 0.551\mathbf{k}$$

The component of moment M_{ab} along line ab is found by projecting $\mathbf{M}_a$ onto this line. Using the interpretation of the dot product as the effect of projecting one vector onto another,

$$M_{ab} = \mathbf{M}_a \cdot \mathbf{i}_{ab}$$

$$= -4,790(0.394) - 3,060(-0.735) + (-790)(0.551) = -73.5 \text{ N} \cdot \text{m}$$

This result could be shown graphically as a double-headed arrow along line ab, with a sense opposite to that of $\mathbf{i}_{ab}$.

From Prob. 3.78,

$$\mathbf{M}_b = -4,630\mathbf{i} - 3,010\mathbf{j} - 836\mathbf{k} \qquad \text{N} \cdot \text{m}$$

The component of this moment along line ab is

$$M_{ab} = \mathbf{M}_b \cdot \mathbf{i}_{ab}$$

$$= (-4,630)(0.394) + (-3,010)(-0.735) + (-836)(0.551)$$

$$= -72.5 \qquad \text{N} \cdot \text{m}$$

It may be seen that the above two independent solutions for M_{ab} are approximately the same, with a percent difference given by

$$\%\text{D} = \frac{73.5 - 72.5}{72.5}\,100 = 1.4\%$$

The above two results for M_{ab} *should* be identical. The lack of exact agreement is the result of the roundoff errors which occur in the several computations required to obtain the final values for M_{ab}.

3.80 Refer to Fig. 3.80. Find the magnitude and direction of the moment of the force about point a,

$$\mathbf{F} = -(280 \sin 36° \cos 38°)\mathbf{i} - (280 \cos 36°)\mathbf{j} + (280 \sin 36° \sin 38°)\mathbf{k}$$

$$= -130\mathbf{i} - 227\mathbf{j} + 101\mathbf{k} \qquad \text{lb}$$

The position vector $\mathbf{r}$ from point a to the force is

$$\mathbf{r} = -22\mathbf{i} + 24\mathbf{k} \qquad \text{in}$$

The moment $\mathbf{M}_a$ of the force about point a is then

$$\mathbf{M}_a = \begin{vmatrix} \mathbf{i} & \mathbf{j} & \mathbf{k} \\ -22 & 0 & 24 \\ -130 & -227 & 101 \end{vmatrix}$$

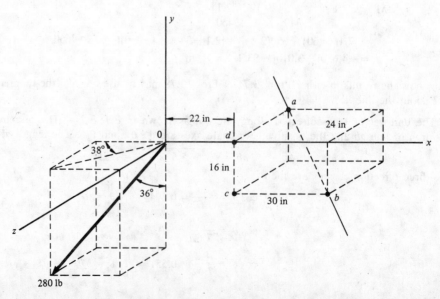

280 lb

Fig. 3.80

$$= [0(101) - (-227)24]\mathbf{i} + [24(-130) - 101(-22)]\mathbf{j} + [-227(-22) - (-130)0]\mathbf{k}$$
$$= 5{,}450\mathbf{i} - 898\mathbf{j} + 4{,}990\mathbf{k}$$
$$= \sqrt{5{,}450^2 + (-898)^2 + 4{,}990^2} = 7{,}440 \text{ in} \cdot \text{lb}$$

$$\cos\theta_x = \frac{M_x}{M_a} = \frac{5{,}450}{7{,}440} \qquad \theta_x = 42.9°$$

$$\cos\theta_y = \frac{M_y}{M_a} = \frac{-898}{7{,}440} \qquad \theta_y = 96.9°$$

$$\cos\theta_z = \frac{M_z}{M_a} = \frac{4{,}990}{7{,}440} \qquad \theta_z = 47.9°$$

3.81　Find the component of the moment in Prob. 3.80 along the line *ab* shown in Fig. 3.80.

❙　A unit vector $\mathbf{i}_{ab}$ acts along line *ab*, with positive sense from *a* to *b*. The distance *d* between points *a* and *b* is

$$d = \sqrt{30^2 + 16^2 + 24^2} = 41.6 \text{ in}$$
$$\mathbf{i}_{ab} = (\cos\theta_x)\mathbf{i} + (\cos\theta_y)\mathbf{j} + (\cos\theta_z)\mathbf{k}$$
$$= \left(\frac{30}{41.6}\right)\mathbf{i} + \left(\frac{-16}{41.6}\right)\mathbf{j} + \left(\frac{24}{41.6}\right)\mathbf{k}$$

The component M_{ab} of the moment along line *ab* is

$$M_{ab} = \mathbf{M}_a \cdot \mathbf{i}_{ab}$$
$$= 5{,}450\left(\frac{30}{41.6}\right) - 898\left(\frac{-16}{41.6}\right) + 4{,}990\left(\frac{24}{41.6}\right)$$
$$= 7{,}150 \text{ in} \cdot \text{lb}$$

The sense of this moment is clockwise when viewed from end *a* of line *ab*.

3.82　Give a summary of the basic concepts of operations with moments and couples.

❙　A moment is a force-induced effect which causes a turning, twisting, or rotation effect about a reference axis. A moment is a vector quantity. Therefore, for its complete definition, it requires a statement of its magnitude, direction, and sense. The magnitude of a moment is defined to be the product of the magnitude of the force and the perpendicular separation distance between the reference axis, about which the moment is taken, and the line of action of the force. The units of a moment are force times length. The moment acts in a plane which is normal to the reference axis. This plane defines the direction of the vector which represents the moment. An alternative definition of the direction of a moment is the direction of the axis about which the moment acts. An arbitrary sense of rotation is chosen to represent positive moment.

　　The theorem of Varignon states that the moment of a force about a reference axis is equal to the sum of the moments of the components of this force about the axis.

　　A couple occurs when two forces are equal in magnitude, opposite in sense, and parallel. The couple is a vector quantity. The magnitude of the couple is equal to the product of the magnitude of one of the forces and the perpendicular separation distance between the two forces. The direction of the couple may be taken to be either the plane in which the two forces lie or the direction of the axis which is perpendicular to this plane. An arbitrary sense of rotation is chosen to represent a positive couple. The couple is a pure turning effect. It may be moved anywhere in its own plane or into a parallel plane without change in its effect on the body on which it acts. The couple is thus a free vector. The resultant force of any couple is zero.

　　Moments and couples both give turning, twisting, or rotation effects. There is a significant difference between these two quantities. The definition of a moment requires a statement of the reference axis about which this effect occurs. A couple, by contrast, is a pure turning effect and thus does not require a reference axis for its definition.

　　Any force may be replaced by a parallel force of equal magnitude and sense, and a couple at a location which is different from that of the original force. Conversely, any system consisting of a force and a couple may be replaced by a single parallel force, of equal magnitude and sense, at a location which is different from that of the original force.

　　The cross product multiplication of two vectors is a vector which is perpendicular to the plane of the original two vectors. The major application of the cross product in statics is to find the moment of a force about an arbitrary point. If the cross product is zero, either one or both of the vectors are zero, or the two vectors are collinear or parallel.

CHAPTER 4
Fundamentals of Force Analysis

4.1 PHYSICAL INTERPRETATION OF FORCE, BODY AND
SURFACE FORCES, TENSILE AND COMPRESSIVE FORCES, TYPES OF FORCE SYSTEMS

4.1   What is the physical interpretation of the term *force*?

I   Force is formally defined as the effect of one physical body on another physical body.   In order to examine this concept, it is necessary to first define the term physical body.   A *physical body* is a body which is composed of matter and which has a volume that is not zero.   The body may be classified as being either rigid or deformable.   A *rigid* body is a body within which two arbitrary points always have the same separation distance, no matter what type of force is applied to the body.   A necessary consequence of this definition is that the exterior shape of a rigid body is always the same.   A *deformable* body, by comparison, may have its shape changed because of the effect of another body acting on it.   Common examples of deformable bodies are a droplet of liquid, a balloon, and a rubber band.

No physical body is totally rigid.   If there are two bodies present, the second body exerts an influence, or force, on the first body.   This force will cause some deformation of the first body, no matter how slight.   The *concept* of a rigid body, however, is a very useful idealization for the solution of problems in engineering mechanics.   From the definition of a force, it may be concluded that *a system must contain at least two bodies if a force is to be produced.*

The basic assumption used in static analysis is that *bodies on which the forces or moments act are perfectly rigid*.   With this assumption, there is no possibility of deformation of the body resulting from the application of forces.   Thus, there can be no change in the relative positions of the forces and moments which act on the body, either with respect to one another or to any arbitrary reference point in the body.

4.2   What is the difference between *body force* and *surface force*?

I   Forces may be divided into two broad classes: body forces and surface forces.   Body forces are the effect of the second body on the *mass particles* of the first body.   Typical examples of body forces are weight forces caused by gravitational attraction, magnetic forces caused by the magnetic field of a second body, and centrifugal forces resulting from the angular velocity of a body.   Two examples of body forces are shown in Fig. 4.2a.

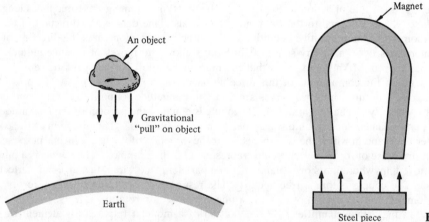

Fig. 4.2a

In the case of surface forces, the force caused by the second body on the first body is a *contact effect* that is transmitted through the surfaces of the bodies that touch one another.   Surface forces are independent of the mass of the body.   It is apparent that for a surface force to exist, the bodies must be in physical contact with one another.   Figure 4.2b shows two examples of surface forces.   Since the moment and couple are basically force-induced effects, a similar definition of surface and body moments could be made.   In practice this is not done, since it is more convenient to identify a particular body or surface force and then determine the moment which this force produces.

A person pulls on handle of lever

A person pushes a door open

Fig. 4.2*b*

In static analysis, no distinction is made between body and surface forces. They are both treated simply as forces. It is only in dynamics, in considering an effect described as the inertia force, that the distinction between body and surface forces becomes significant. It should again be noted that body forces are related to the mass of the body while surface forces are independent of mass.

4.3   What is the difference between *tensile force* and *compressive force*?

❚ The effect of a force may be to either pull or push on a physical body, and this pull or push is the effect of the second body on the first body. Figure 4.3*a* shows a fixed eyebolt, to which a cable is attached. As shown in the figure, a person pulls on the cable, and the cable is seen to "pull" on the eyebolt.

In this case, the first two physical bodies involved in the production of the force are the person and the cable. In a second configuration, the two bodies involved are the cable and the eyebolt. For this particular problem, the cable serves as an intermediate element to transfer the force from the person to the eyebolt. It may be observed that the cable could be eliminated from the problem completely if the person hooked a finger through the eyebolt and exerted a force on this element directly.

The force in this situation is referred to as a tensile force, and the force in the cable is called the cable tensile force. If a force tends to *pull* on the body under consideration, the force is referred to as a tensile force.

In Fig. 4.3*b* a person pushes on the eyebolt. The eyebolt and the person's hand are the two physical bodies which allow the force to be produced. The effect of the person's hand pushing on the eyebolt is referred to as a compressive force. If a force tends to *push* on the body under consideration, the force is referred to as a compressive force.

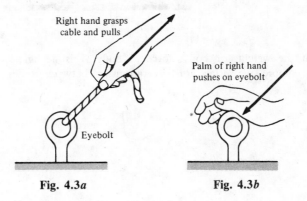

Right hand grasps cable and pulls

Palm of right hand pushes on eyebolt

Eyebolt

Fig. 4.3*a*   **Fig. 4.3*b***

4.4   Describe *weight force*.

❚ A very special force is the force which defines the weight of a body. In this case, the second body that causes the force is the earth, which exerts a gravitational attraction on the mass particles of the body. This weight force acts through a point in the body called the *center of mass, or gravity*. This is a point which has a fixed location within the body, where all the weight may be assumed to be "concentrated." The direction of this weight force is along a line joining the center of mass of the body and the center of the earth. For engineering purposes, the direction of the weight force may be considered to be normal to the surface of the earth.

4.5   What is the basic physical effect in engineering mechanics?

❚ The basic effect in engineering mechanics is force. The moment and couple are force-induced quantities. The magnitude M_0 of the moment $\mathbf{M}_0$ of the force $\mathbf{F}$ shown in Fig. 4.5a, with respect to point 0, is expressed by

$$M_0 = Fd$$

This equation is the product of a length and the magnitude of a force. The length is an inherently arbitrary quantity, since it depends on the choice of the reference point 0. Thus it may be concluded that the basic quantity which causes a moment is the force F.
    The magnitude of the moment of the couple shown in Fig. 4.5b is given by

$$M = Fd$$

Again, the conclusion may be drawn that the basic effect in this expression is the force F.

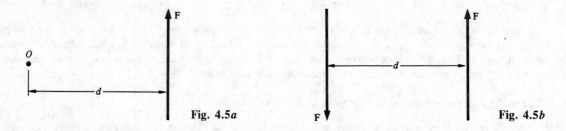

Fig. 4.5a   Fig. 4.5b

In Probs. 4.6 through 4.15, each figure shows a physical situation in which there is a force effect between bodies A and B. This force may be transmitted either directly between the bodies or through intermediate connecting bodies. For each figure, identify the bodies involved in the force transmission between bodies A and B and describe all force transmission effects. In addition, state whether each force is a body or surface force and whether it is a tensile or compressive force.

4.6

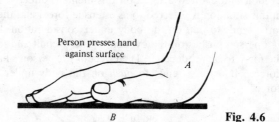

Person presses hand against surface

Fig. 4.6

❚ Assume that the weight, or body, force of body A is negligible compared with the force which is transmitted. A compressive surface force is transmitted between bodies A and B.

4.7

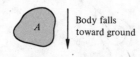

Body falls toward ground

Fig. 4.7

❚ A tensile weight, or body, force is transmitted between bodies A and B.

4.8

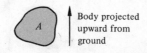

Body projected
upward from
ground

B   **Fig. 4.8**

❚ As in Prob. 4.7, a tensile weight, or body, force is transmitted between bodies *A* and *B*.

4.9

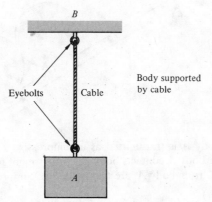

Body supported
by cable

Fig. 4.9

❚ Neglect the weight forces of the eyebolts and cable. The weight, or body, force of body *A* is transmitted as a tensile surface force from body *A* to the lower eyebolt. A compressive surface force is transmitted from the eyebolt to the cable. A tensile surface force is transmitted through the cable, and a compressive surface force is transmitted from the cable to the upper eyebolt. Finally, a tensile surface force is transmitted from the upper eyebolt to body *B*.

4.10

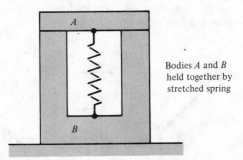

Bodies *A* and *B*
held together by
stretched spring

Fig. 4.10

❚ The weight, or body, force of body *A* is transmitted as a compressive surface force at the interface between bodies *A* and *B*. The spring force is transmitted as a tensile surface force between bodies *A* and *B* at the points where the spring is attached to these bodies.

4.11

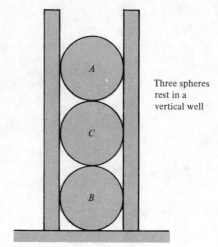

Three spheres
rest in a
vertical well

Fig. 4.11

▮ The weight, or body, force of body A is transmitted as a compressive surface force to body C. This force and the weight force of body C are transmitted as a combined compressive surface force to body B.

4.12

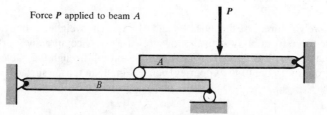

Two cylinders rest
on a hinged arm

Fig. 4.12

▮ The weight force of body A is transmitted as a compressive surface force at the locations where body A contacts body B and where body A contacts body C. The weight force of body C, and the compressive surface force transmitted to body C from body A, are transmitted as compressive surface forces at locations where body C contacts body B.

4.13

Force P applied to beam A

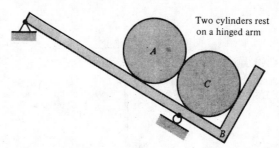

Fig. 4.13

▮ P is a surface force applied to body A. Assume that the weight forces of body A and of the ball support are negligible compared to force P. A compressive surface force is transmitted between the left end of body A and the ball support and between the ball support and body B.

4.14

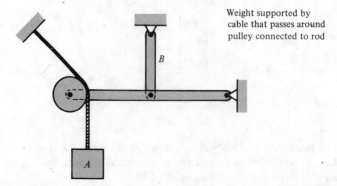

Weight supported by
cable that passes around
pulley connected to rod

Fig. 4.14

❚ Assume that the weight forces of the pulley and members are negligible compared with the weight force of body A. The weight of body A is transmitted as a tensile surface force to the cable. The force in the cable is transmitted as a tensile surface force to the foundation and as a compressive surface force to the pulley. This pulley surface force is transmitted as a compressive surface force from the pulley to the pulley pin and from the pulley pin to the end of the horizontal member. The force is next transmitted, as a compressive surface force, through the horizontal member to the pin at the lower end of body B and from this pin to body B, also as a compressive surface force.

4.15

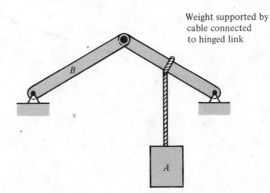

Weight supported by cable connected to hinged link

Fig. 4.15

❚ Assume that the weight forces of the members are negligible compared with the weight force of body A. The weight force of body A is transmitted as a tensile surface force to the cable. The force in the cable is transmitted as a tensile surface force. The force is transmitted from the cable to the right-side member as a compressive surface force. The force is transmitted from the right-side member, through the connecting pin, to body B as a compressive surface force.

4.16 What is the difference between a *fixed* vector and a *free* vector?

❚ Vector quantities, and in particular forces, may be described as either fixed or free. A force which is applied at a particular location on a body is a fixed vector. It is fixed since, if the force were applied at any other point on the body not on the line of action of the force, it would change the problem. A moment is also a fixed vector, because its definition depends on the reference point about which the moment acts. Fixed vectors are also known as bound vectors. A couple, by comparison, is a free vector, since it can be moved anywhere in its own plane, or in a parallel plane, without change in its effect on the body.

4.17 Describe the *principle of transmissibility*?

❚ The left view in Fig. 4.17 shows a force F which is applied to a body at point a. The force can be imagined to be slid along its line of action and be applied at point b in the right view. In terms of its external static effect on the body, there is no difference between the application of the force at points a and b. This concept of sliding a force along its line of action is referred to as the principle of transmissibility of a force.

In terms of the *internal effect* on the body, there would be a difference between application of the force at points a and b. At point a, the force causes a tensile effect on the central portion of the body, while at point b the same force causes a compressive effect in this region. All bodies in statics are assumed to be rigid, however. Thus the observations made above do not alter the earlier conclusion that the force may be moved along its line of action without change in its effect on the system.

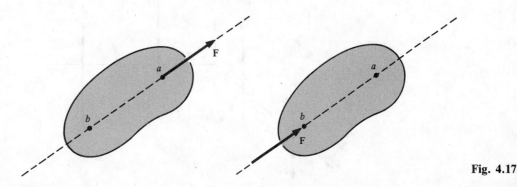

Fig. 4.17

4.18   Show a collinear force system.

▮ The collinear force system is the most elementary type of force system.   It is characterized by all the forces having a common line of action, as shown in Fig. 4.18.

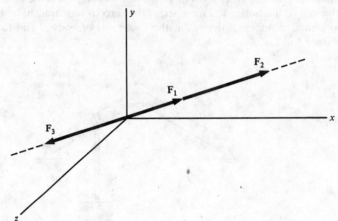

Fig. 4.18

4.19   Show a concurrent force system.

▮ In a concurrent force system the lines of action of all the forces pass through a common point.   This effect is seen in Fig. 4.19.

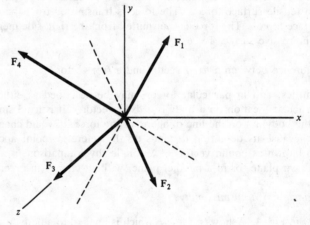

Fig. 4.19

4.20   Show a parallel force system.

▮ In the parallel force system the lines of action of all the forces are parallel, as shown in Fig. 4.20.

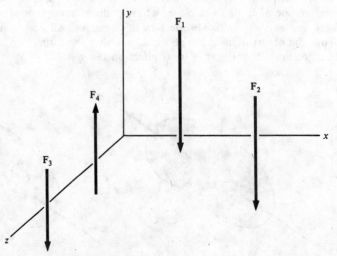

Fig. 4.20

4.21   Show a general force system.

❚ If a force system is not one of the three types shown in Probs. 4.18 through 4.20, it may be called a general force system. A typical example of a general force system is shown in Fig. 4.21. It should be noted that collinear, concurrent, and parallel force systems are all special cases of the general force system.

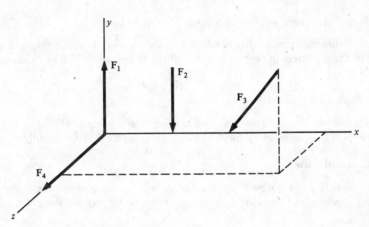

Fig. 4.21

4.22   What is the difference between *coplanar* and *noncoplanar* force systems?

❚ In addition to the classifications in Probs. 4.18 through 4.21, a further distinction may be made. If all the forces in the system lie in the same plane, the force system is referred to as coplanar, or two-dimensional. If the forces have different directions, which are not all in a single plane, the system is referred to as noncoplanar, or three-dimensional. In this book the usage "two- or three-dimensional" will be used in preference to "coplanar or noncoplanar."

4.23   List the types of force systems.

❚ From consideration of Probs. 4.18 through 4.22, seven distinct types of force systems may be identified:

(*a*)   Collinear, one-dimensional

(*b*)   Concurrent, two-dimensional

(*c*)   Parallel, two-dimensional

(*d*)   General, two-dimensional

(*e*)   Concurrent, three-dimensional

(*f*)   Parallel, three-dimensional

(*g*)   General, three-dimensional

4.2   FREE-BODY DIAGRAM, NEWTON'S LAWS OF MOTION, EQUILIBRIUM, FREE-BODY-DIAGRAM SUPPORT CONDITIONS

4.24   What is a *free-body*, or *equilibrium*, *diagram*?

❚ A concept which is fundamental to the solution of all problems in statics is the free-body diagram. A free-body diagram is a view of a body or any portion of that body which is imagined to be removed from its immediate supporting foundation. This immediate supporting foundation may be either the ground supporting the entire body or an adjoining part of the body. On the free-body diagram of the portion of the body under consideration are drawn all the forces and moments acting on this element.

4.25   Explain the difference between the *applied forces and moments* and the *reaction forces and moments* which are shown in a free-body diagram.

❚ The forces and moments which act on the free-body diagram may be one of two general types. The first kind is the *known* forces and moments which act on the body. They are referred to as applied loads, and these quantities are known at the outset of the problem. The second type of forces and moments which are shown in the free-body diagram are the effects *on the body by the material which was imagined to be removed*. These latter effects may be forces, moments, or both. They are referred to as the reaction forces and moments, and they are *unknown* quantities. The complete free-body diagram, then, consists of a view of the portion of interest of a body together with the applied loads and the reaction forces and moments which act on this body.

In terms of the physical effect on the body shown in the free-body diagram, there is no difference between the applied forces and moments and the reaction forces and moments. It is convenient to distinguish between them, however, since the applied loads are *known* and the reaction forces and moments *are to be found.*

4.26 Give the formal statement of Newton's three laws of motion.

❚ All engineering mechanics is based on three principles which were presented by Sir Isaac Newton during the latter part of the seventeenth century. These three principles, which are known as Newton's laws, are written here.

Newton's First Law

A body at rest (or moving with straight-line motion at constant velocity) will remain at rest (or moving with straight-line motion at constant velocity) unless acted on by a resultant force.

Newton's Second Law

A body which is acted on by a resultant force will experience an acceleration *directly proportional* to this force. This acceleration, which is a vector quantity, will have the same direction and sense as the resultant force.

Newton's Third Law

For every force action of a first body on a second body, there is an *equal and opposite* force action of the second body on the first body.

4.27 Interpret Newton's laws of motion.

❚ The resultant force in Newton's first and second laws is the resultant force of the system of forces which acts on the body. The techniques for finding a resultant force were presented in Chap. 2. Newton's first and third laws form the basis of the subject known as *statics.* In the following problems in this text, Newton's first law will be interpreted in terms of a concept called *equilibrium.* Newton's third law may be interpreted as follows. In Fig. 4.27a body B pushes on body A with force F. By using the third law, it would be concluded that body A pushes back on body B with a force *of the same magnitude and direction,* but *opposite sense,* as shown in Fig. 4.27b. An alternative statement of Newton's third law is that *for every force action there is an equal and opposite force reaction.*

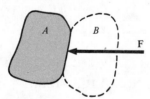

Fig. 4.27a

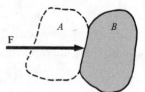

Fig. 4.27b

Newton's second law is the basis of the subject of dynamics. If the acceleration of a body is zero, then the second law is equivalent to the first law.

4.28 Define the term *static equilibrium.*

❚ Newton's first law states that a body at rest (or moving in straight-line motion with constant velocity) will remain so unless it is acted on by a resultant force. A preliminary step in using the first law is to draw a free-body diagram of the body of interest. The resultant force is the resultant of the applied and reaction forces which act on the body. Static equilibrium occurs when Newton's first law is satisfied. Thus, a *necessary* condition for the static equilibrium of a body is that the resultant force which acts on the body be *zero.* In certain types of force systems, an additional requirement is that the resultant moment which acts on the body also be zero.

Statics is concerned primarily with the force effects on bodies which remain at rest. For these cases the resultant force, of *all* the forces which act on the body, must be identically zero. For bodies which move in straight-line motion with constant velocity, the acceleration is zero. From Newton's second law, the requirement of zero acceleration implies that the resultant force which acts on the body must also be zero.   Thus the cases of a body at rest and a body moving in straight-line motion with constant velocity are statically equivalent.

4.29   State three uses for the resultant force of a system of forces.

▌ In Chap. 2 we considered the problem of finding the resultant force of a system of forces.   The resultant force of any system of forces is that *single* force which will produce the same effect as the original system of forces.   A description of this resultant force, as a vector quantity, requires a statement of its magnitude, direction, and sense.   In certain types of force systems an additional statement, one that describes the location of the resultant force within the system of forces, is required.

The concept of the resultant force has three major useful applications.

(*a*)   The resultant force of a system of forces which acts on a body may be used to visualize the *net* effect on the body of this system of forces.   In Fig. 4.29, for example, are shown two cable tensile forces F_1 and F_2 acting on a post.   By finding F, the resultant of the two cable forces, one may visualize the net effect of these two forces on the post.

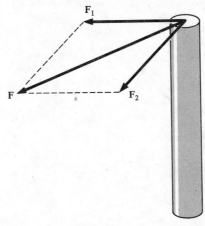

Fig. 4.29

(*b*)   Newton's second law, which is the fundamental law of dynamics, states that the acceleration of a body is directly proportional to the resultant force acting on the body.   This acceleration has the same direction and sense as the resultant force.   The first step in using Newton's second law is to draw a free-body diagram of the body of interest and find the resultant force which acts on this body.   This resultant force will include the effects of both the applied forces which act on the body and the reaction forces which are shown in the free-body diagram.   From the above, it may be concluded that a preliminary, *necessary* condition for finding the acceleration of a body is to first obtain the resultant force which acts on the body.

(*c*)   The resultant force of a system of forces which acts on a body is used in the definition of static equilibrium.   *The concept of static equilibrium is the basis of all problems in static force analysis.*

4.30   Show the free-body diagram condition for support of a body by a *cable*.

▌ Figure 4.30*a* shows a cable which is used to support the end of a body.   A cable is a flexible element used to transmit *tensile force only*, and the direction of the force is collinear with the axis of the cable.

Figure 4.30*b* shows the free-body diagram representation for this case.   The tensile force R is the force *of* the cable *on* the body.   The direction of R is the *known* direction θ of the cable.   The sense of this force is known as a tensile force on the body.   The only unknown in Fig. 4.30*b* is the *magnitude* of the force.

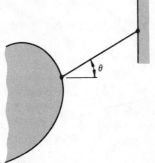

Fig. 4.30*a*

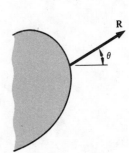

Fig. 4.30*b*

4.31 Show the free-body diagram condition for support of a body by a *smooth plane surface*.

▌ A part of a structure is supported by, or rests against, a smooth plane surface, as shown in Fig. 4.31*a*. The smooth surface in this figure is the *xz* plane. The part of the body which contacts this surface is called a *member*. A smooth surface is one on which the effects of friction are negligible. Typical examples are a sheet of ice and a lubricated polished metal plate. The assumption is made that the tip of the member in Fig. 4.31*a* remains in contact with the smooth surface at all times. The consequence of this assumption is that the force between the member and the surface must be *compressive*, and thus *a tensile force cannot be transmitted between a member and a smooth surface*.

For the case of a body in contact with a smooth plane surface, the line of action of the force transmitted between these two elements is normal to the contacting surface. The immediate consequence is that, for this type of support, the *direction* of the force transmitted between the two surfaces is known and this direction is *normal* to the smooth plane surface. The force exerted *by* the smooth plane surface *on* the lower end of the member is shown in Fig. 4.31*b*. In this figure, the force *R* is the unknown reaction force exerted on the member by the smooth surface, and the direction of this force is that of the normal to the smooth plane surface. The only unknown in this problem is the magnitude of this compressive force.

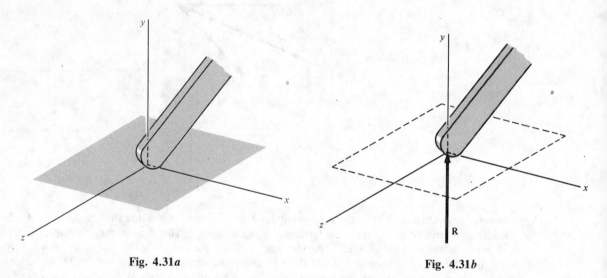

Fig. 4.31*a*   Fig. 4.31*b*

4.32 Show the free-body-diagram condition for the portion of the smooth plane surface that supports the body in Fig. 4.31.

▌ Figure 4.32 shows the free-body diagram of a portion of the smooth plane surface near the lower end of the member. The force *R* in this figure is the force exerted *on* the smooth plane surface *by* the member. The forces *R* in Figs. 4.31*b* and 4.32, which have the same magnitude and direction but *opposite* senses, are a pair of Newton's third law action-reaction forces. The only unknown quantity in this problem is the magnitude of the reaction force.

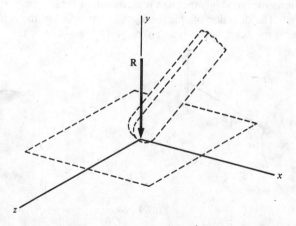

Fig. 4.32

4.33   Show the free-body-diagram condition for support of a body by a *smooth curved surface*.

❚ Figures 4.31 and 4.32 show the structural support for the case of the member contacting a smooth *plane* surface. The same conclusions may be arrived at if the smooth surface is curved. Figures 4.33*a* and *b* show two such configurations. At the point of contact between the member and the smooth curved surface there is one, and only one, tangent plane to the smooth surface. The normal to this plane is the *known* direction of the *unknown* reaction force between the surface and the member. The member is assumed to remain in contact with the smooth curved surface. Thus, the sense of the unknown reaction force is known, and this sense is a compressive force effect between the two bodies. The only unknown in this problem is the magnitude of the reaction force.

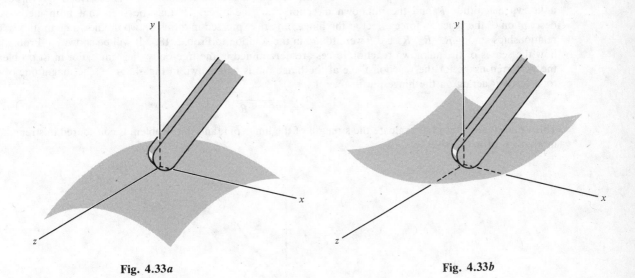

Fig. 4.33*a*   Fig. 4.33*b*

4.34   Show the free-body diagram condition for a body which is supported by a *ball* or *roller*.

❚ A situation which occurs frequently in problems in static analysis is shown in Fig. 4.34*a*. This figure shows a member which is supported by a ball or roller. Both of these elements present smooth, curved contacting surfaces. Thus, the free-body diagram of the end of the member would be as shown in Fig. 4.34*b* where R_1 is the reaction force of the ball or roller on the member. The free-body diagram of the roller is shown in Fig. 4.34*c*, and Fig. 4.34*d* shows the free-body diagram of the ground in the vicinity of the ball or roller. The forces R_2 and R_3 are the forces exerted on the ball or roller by the member and by the ground, respectively. The force R_4 is the force exerted on the ground by the ball or roller. Subsequently it will be seen that if the ball or roller is weightless, the magnitudes of the forces R_1, R_2, R_3, and R_4 are the same.

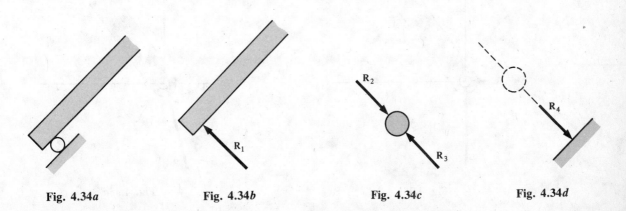

Fig. 4.34*a*   Fig. 4.34*b*   Fig. 4.34*c*   Fig. 4.34*d*

4.35 Show the free-body diagram condition for support of a body by a *hinge pin*.

▮ A common method of physically connecting one body to another is with a hinge pin, as shown in Fig. 4.35*a*. The element which is attached to the ground is called a hinge, and the pin which connects the hinge to the member is called a hinge pin. From physical considerations, it may be seen that the member is constrained to move in the *xy* plane and that the only possible motion of this member is a rotation about the hinge pin.

In a connection such as this, the hinge pin is usually assumed to be frictionless. Thus there is no frictional moment of the hinge pin on the member. The reaction force *of* the ground *on* the member, transmitted through the hinge pin, is a single force of unknown magnitude and direction. There are two ways of showing this reaction force in a free-body diagram. In Fig. 4.35*b*, the reaction force is described by the unknown rectangular force components R_x and R_y. These two force components have the *known directions* shown and unknown magnitudes. The senses are implicitly known, since the reaction force components shown in the figure have been chosen acting up and to the right. In Fig. 4.35*c* this reaction force is described in terms of the unknown magnitude R and the unknown direction θ of the force. It may be seen that both methods of · description of the reaction force effect at the hinge pin are expressed in terms of *two* unknown quantities. The relationships among R_x, R_y, R, and θ were found in the solution to Prob. 2.18. It will be shown, in Prob. 4.40, how the senses of the unknown reaction forces are determined in the free-body-diagram. For most problems, the preferred usage for the reaction force at the hinge pin is that shown in Fig. 4.35*b*. The magnitude of the total force R acting on the hinge pin is given by

$$R = \sqrt{R_x^2 + R_y^2}$$

This value could be used to evaluate the strength of the hinge pin, and this problem is considered in strength, or mechanics, of materials.

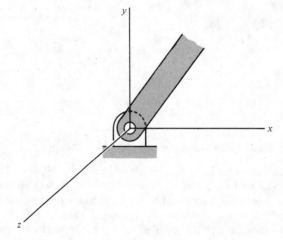

Fig. 4.35*a*

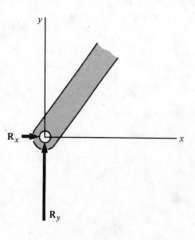

Fig. 4.35*b*

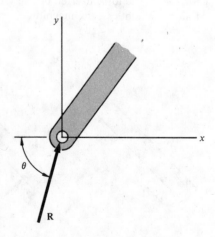

Fig. 4.35*c*

4.36 Show the free-body-diagram condition for support of a body by a *ball joint*.

▮ The three-dimensional equivalent of the hinge shown in Fig. 4.35a is referred to as ball joint. This connection is shown in Fig. 4.36a. The member may have any angular position in space, but the lower end of this element may not move in any direction. The ball joint is assumed to be frictionless. The free-body diagram of the lower end of the member is shown in Fig. 4.36b, and it may be seen that there are three unknown components of the reaction force *of* the ground *on* the member. This type of support occurs in more advanced problems, and it will subsequently be considered in Chap. 12 on three-dimensional force systems.

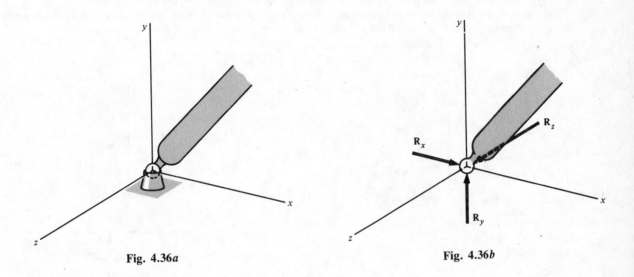

Fig. 4.36a   Fig. 4.36b

4.37 Show the free-body-diagram condition for support of a body by a *rough surface*.

▮ A case which is physically different from, but statically similar to, the preceding two problems is shown in Fig. 4.37. Here, the member rests against a rough surface. A rough surface is a surface on which the friction forces are sufficiently large to prevent motion of the end of the member in the plane of the surface. The major difference between this case and that of the hinge or ball joint support is that *a tensile force cannot be transmitted between the foundation and the member*. If the member were constrained to move in the *xy* plane, then the description of the free-body-diagram reactions would be the same as that shown in Fig. 4.35b or c, with the restriction that the forces R_y and R must be *compressive* force effects between the rough surface and the member. The same conclusions may be arrived at if the supporting rough surface is curved instead of plane. For general, three-dimensional motion of the member, the reaction forces would be as shown in Fig. 4.36b. For this case, the reaction force component R_y is a compressive surface force between the rough surface and the member.

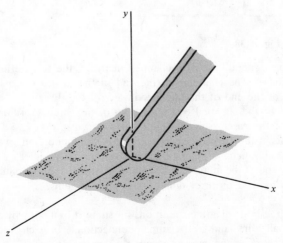

Fig. 4.37

4.38 Show the free-body condition for support of a body by a *clamped connection*.

❚ The final type of connection to be considered is shown in Fig. 4.38a. Here, the end of the member forms an integral part of the foundation structure. This type of connection of the end of the member to the ground is referred to as "clamped," "built in," or "cantilevered."

The case will first be considered where the force effects acting on the member tend to make it move in the xy plane only. The lower end of the member is prevented from moving in this plane, and there can be no rotation of this end about the z axis. The reaction effects *of* the foundation *on* the member are then as shown in Fig. 4.38b.

For this case there are three unknown reaction effects. These are the two components R_x and R_y of the reaction force and the unknown reaction moment M_R. The force effects *of* the member *on* the foundation, using Newton's third law, are shown in Fig. 4.38c.

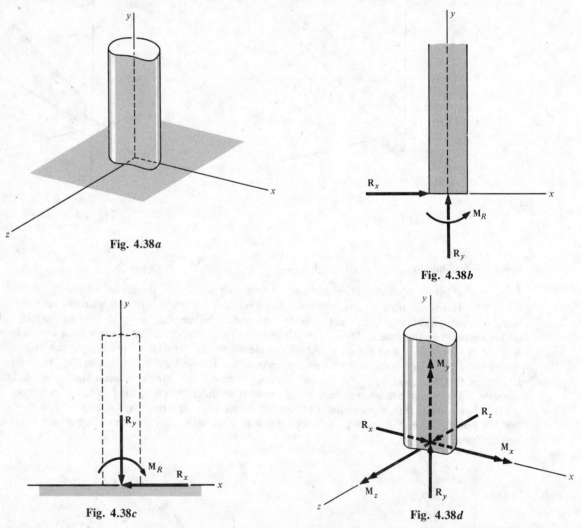

Fig. 4.38a

Fig. 4.38b

Fig. 4.38c

Fig. 4.38d

For the most general case of a clamped member, the force effects acting at the clamped end may have arbitrary directions. The clamping of the end of the member provides two distinct effects. The first is to prevent motion of the end of the member in any of the three coordinate directions. The second effect is to prevent rotation of the clamped end about any of the three coordinate axes. The free-body diagram of the clamped end, for the case of general loading, is shown in Fig. 4.38d. In this figure, R_x, R_y, and R_z are the three components of the unknown reaction force *of* the foundation *on* the member, and M_x, M_y, and M_z are the three unknown components of the moment exerted *on* the member *by* the foundation. This most general case, where the upper end of the member may have any type of applied loading, is an advanced problem which will be investigated in Chap. 12 on three-dimensional force systems.

4.39 Summarize the free-body-diagram conditions for support of a body by a cable, by a smooth or rough surface, by a hinge pin or ball joint, and by a clamped connection. In each case indicate the number of unknown force and/or moment terms at the support location.

I Seven cases are shown in Table 4.1. In the most elementary cases, those of support by a cable or by a smooth surface, there is only one unknown reaction force. For the most complex case, that of clamped support, there are three unknown force components and three unknown moment components.

4.40 How are the senses of the unknown reaction forces and moments in a free-body diagram determined?

TABLE 4.1

Type of support	Physical appearance	Free-body diagram representation	Unknown terms, and restrictions	Number of unknown terms
Cable			Tensile force **R**	1
Plane or curved smooth surface			Compressive force **R**	1
Hinge			R_x, R_y or **R**, Θ	2
Ball joint			R_x, R_y, R_z	3
Plane or curved rough surface			R_x, R_y, R_z R_y is a compressive force	3
Clamped; all force effects act in xy plane			R_x, R_y, M_R	3
Clamped; general force effects			R_x, R_y, R_z, M_x, M_y, M_z	6

❚ In the process of drawing the free-body diagram, the actual senses of the unknown reaction forces and moments may or may not be known. For the case of a member resting against a smooth surface, as shown in Fig. 4.31a, the unknown reaction force must have the sense shown in Fig. 4.31b. The reason is that a smooth contacting surface cannot transmit a tensile force. In the case of support by a cable, it was shown that a cable can transmit only a tensile load. For both of the above cases, the actual senses of the reaction forces are known. In other cases, however, such as with the hinged connection, the senses of the components of the reaction forces are unknown. For those cases where the *actual* sense of the reaction force is unknown, the following technique will be used.

In drawing the free-body diagram of the element of interest, the components of the reaction forces and the reaction moments will be drawn in *assumed* actual senses. If, after the problem is solved, a reaction force or moment has a negative value, it simply means that the sense of this term is opposite to that which was assumed in the original construction of the free-body diagram.

4.41   Devise a notation for reaction forces and moments in a free-body diagram.

❚ Throughout this book, the symbol R will be reserved exclusively for reaction forces which are shown in the free-body diagrams. This symbol may be used by itself or with single or double subscripts.

If there is a single reaction force with a known direction, this force may be designated R. For the case of a single reaction force of unknown direction, as with a hinge pin, the components of this reaction force may be designated R_x and R_y. If there are several support points of the portion of the structure for which the free-body diagram is drawn, these points will be designated $a, b, c, \ldots$. If the directions of these reaction forces are known, as in the case of roller support, these forces may be designated $R_a, R_b, \ldots$. For the case of unknown directions of these reaction forces, typical designations are R_{ax} and R_{ay}, indicating the x and y components of the reaction force at point a. If, in addition to the reaction forces, there are also reaction moments, these may be designated $M_a, M_b, \ldots$. The use of these notations will be seen in the following problems.

4.3   CONSTRUCTION OF FREE-BODY DIAGRAMS

4.42   Figure 4.42a shows a weightless ring which is pulled on by cable cd. The tensile force in this cable is F, and the ring is connected to the ground by cable ab. Draw the free-body diagram of the ring.

❚ Cable ab is imagined to be cut, as indicated by the dashed line in Fig. 4.42a. In terms of the free-body diagram of the ring, cable ab is the foundation which is imagined to be removed. The force F is the applied force. It has the direction of line segment cd, and this force is known. The unknown reaction force is the force of cable ab on the ring. This force is designated by R, and it has the direction of cable ab. The final result for the free-body diagram of the ring is shown in Fig. 4.42b.

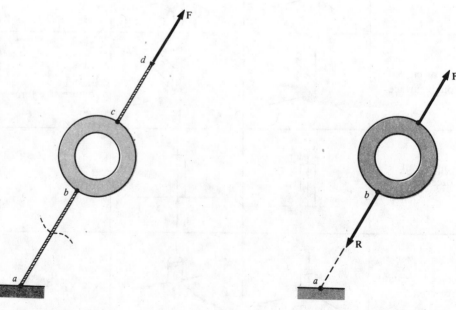

Fig. 4.42a   **Fig. 4.42b**

4.43 A block of weight W rests on a smooth surface, as shown in Fig. 4.43a. Draw the free-body diagram of the block.

▮ The ground is imagined to be removed, and the force of the ground on the block is designated R. Since the contacting surfaces are smooth, this reaction force must be normal to these surfaces. W is the weight force of the block, and the final result for the free-body diagram is shown in Fig. 4.43b. It may be observed that the sense chosen for R shows a *compressive* force acting *on* the block.

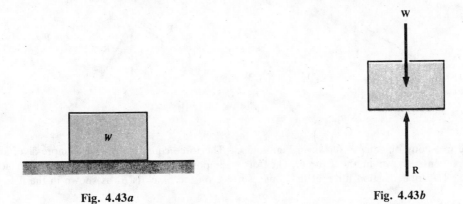

Fig. 4.43a Fig. 4.43b

4.44 The block of the preceding problem now rests on a rough surface and is acted on by a tensile force F, as shown in Fig. 4.44a. The rough surface prevents sliding of the block. Draw the free-body diagram of the block.

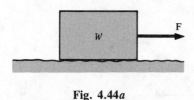

Fig. 4.44a

▮ Since the mating surface between the block and the ground is rough, the conditions of Prob. 4.37 apply. The ground is imagined to be removed, and the free-body diagram is shown in Fig. 4.44b. The reaction force component R_x is arbitrarily assumed to act in the sense shown, and R_x and R_y are the components of the total reaction force R, shown in Fig. 4.44c, of the ground on the block. R_y must be a compressive force, with the sense shown in Fig. 4.44b or Fig. 4.44c.

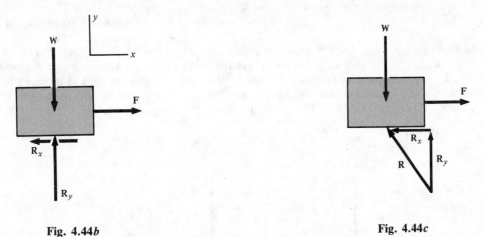

Fig. 4.44b Fig. 4.44c

4.45 The block of Prob. 4.44 now rests on the inclined rough surface shown in Fig. 4.45a. The rough surface prevents sliding of the block. Draw the free-body diagram of the block.

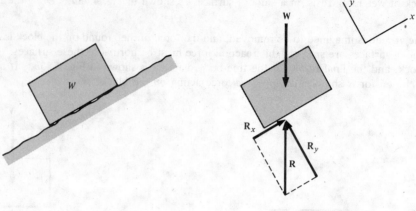

Fig. 4.45a   Fig. 4.45b

❚ The free-body diagram is shown in Fig. 4.45b. The force component R_y is compressive, and thus it must have the sense shown in the figure. The force component R_x is arbitrarily assumed to act in the sense shown. The total reaction force of the rough surface on the block is R, as shown in the figure.

4.46   The ladder of weight W shown in Fig. 4.46a rests against a smooth wall. Draw the free-body diagram of the ladder.

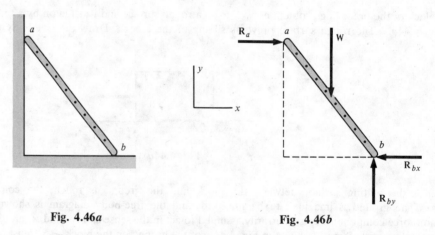

Fig. 4.46a   Fig. 4.46b

❚ The free-body diagram is shown in Fig. 4.46b. Since the wall is smooth, the direction of the reaction force of the wall on the ladder is normal to the wall. The force components R_a and R_{by} must be compressive, and they have the actual senses shown. The force component R_{bx} is assumed to act to the left in the figure.

4.47   The beam shown in Fig. 4.47a is connected to the ground by a roller at a and a hinge at b. A *beam* is a slender structural element that is usually, but not necessarily, loaded by applied forces whose directions are perpendicular to its axis. The characteristics of beams as structural elements are studied in strength, or mechanics, of materials. The known applied loads and moments are F_1, F_2, F_3, and M_1, and these force effects all act in the xy plane. Draw the free-body diagram of the beam.

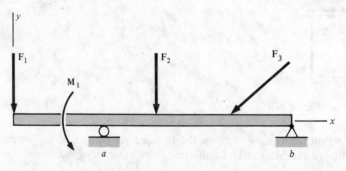

Fig. 4.47a

 The beam is supported by a roller at a, and this is a case of smooth contacting surfaces. The unknown reaction force of the ground on the beam at a must then be perpendicular to the longitudinal axis of the beam. The direction of the reaction force at b of the hinge on the beam is unknown, and this force is represented by the two components R_{bx} and R_{by}. The senses of these two components are arbitrarily chosen. The final form of the free-body diagram is shown in Fig. 4.47b. It should be noted that the unknown reaction forces are subscripted in accordance with the conventions introduced in Prob. 4.41.

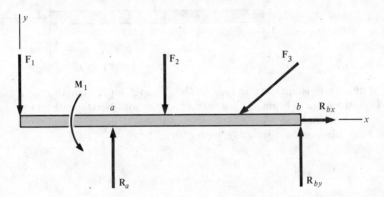

Fig. 4.47b

4.48 The beam shown in Fig. 4.48a is clamped to the ground at its right end. This type of beam is referred to as a *cantilever beam*, and the loads acting on the beam are assumed to be acting in the xy plane. Draw the free-body diagram of the beam.

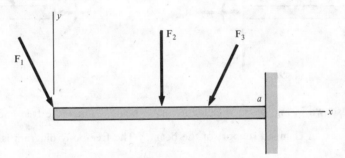

Fig. 4.48a

 The beam is clamped at the right end, and this is the type of support shown in Fig. 4.38b. The foundation is now imagined to be removed, and the free-body diagram is shown in Fig. 4.48b.

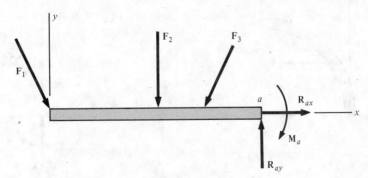

Fig. 4.48b

4.49 A cutting plane is now imagined to be passed through the beam of Prob. 4.48, as shown in Fig. 4.49a. Draw the free-body diagrams of lengths ab and bc of the beam.

 Length bc is imagined to be removed from the beam, and the immediate supporting foundation is the remaining length ab of the beam. This remaining length ab exerts a clamping-type support on length bc. The final appearance of the free-body diagram of length bc is shown in Fig. 4.49b. The components R_{bx} and R_{by} are drawn in assumed actual senses. In drawing the free-body diagram of length ab, the appearance of the

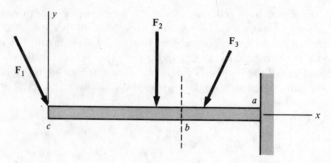

Fig. 4.49a

right end of this length is as shown in Fig. 4.48b. The left end of length *ab* is a mating portion of the right end of length *bc* in the original beam. Thus, in accordance with Newton's third law, any force effects which act on end *b* of length *bc* must have *opposite* senses when shown as acting on end *b* of length *ab*. This effect is shown in the free-body diagram, in Fig. 4.49c, of length *ab*.

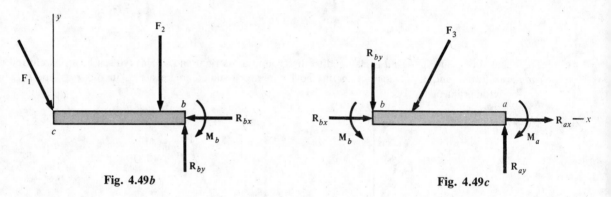

Fig. 4.49b

Fig. 4.49c

4.50 The element shown in Fig. 4.50a is called a curved beam. Draw the free-body diagram of this element.

❚ This is a case of clamped support of the beam. The free-body diagram is shown in Fig. 4.50b.

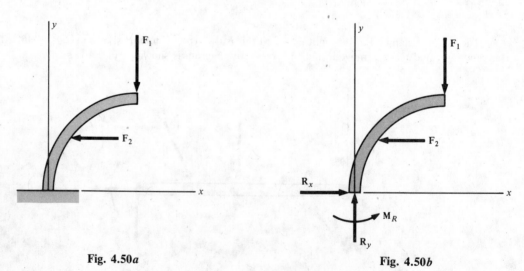

Fig. 4.50a

Fig. 4.50b

4.51 The structure shown in Fig. 4.51a is called a plane framework. It is referred to as "plane," since both the applied forces and the physical structure lie in the plane of the figure. Draw the free-body diagrams of the entire framework and of each member of the framework.

❚ Supports *a* and *c* are hinge pins, and there will be two unknown components of reaction force at each of these locations. The supporting ground is imagined to be removed, and the free-body diagram of the entire framework is shown in Fig. 4.51b, where actual senses have been assumed for the reaction force components.

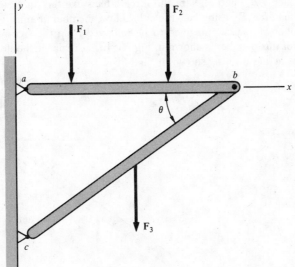

Fig. 4.51a

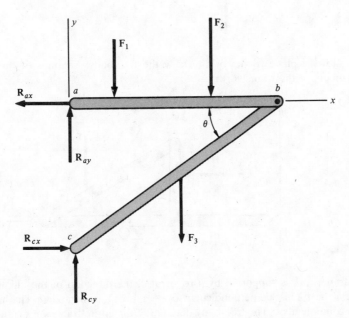

Fig. 4.51b

Member *ab* is now imagined to be removed from the free-body diagram in Fig. 4.51b. The effect of member *bc* on member *ab*, at location *b*, must now be shown. Since the connection at *b* is a hinge pin, it may transmit two unknown components of reaction force. These will be designated R_{bx} and R_{by}, and the final appearance of the free-body diagram of member *ab* is shown in Fig. 4.51c. It should be noted that end *a* of member *ab* has the same forces acting on it as shown in Fig. 4.51b.

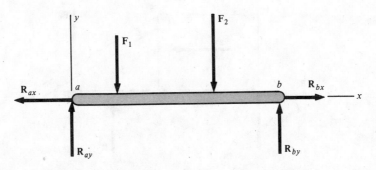

Fig. 4.51c

The force components R_{bx} and R_{by} are the force effects exerted on member *ab* by member *bc*, through the hinge pin connection at *b*. By using Newton's third law, member *ab* must exert an equal and opposite effect on member *bc*. Member *bc* is now imagined to be removed from the free-body diagram of Fig. 4.51*b*, and the free-body diagram of this member is shown in Fig. 4.51*d*. Again, it should be noted that end *c* of this member has the same forces acting on it as shown in Fig. 4.51*b*.

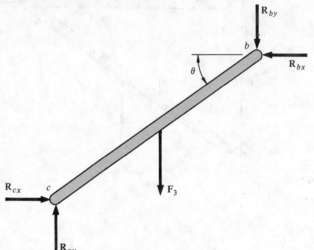

Fig. 4.51*d*

4.52 Figure 4.52*a* shows a plane framework. Draw the free-body diagrams of the entire framework and of the individual members of the framework.

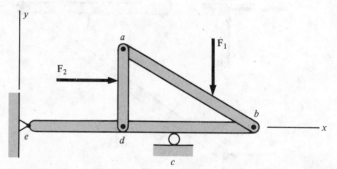

Fig. 4.52*a*

▋ Since the framework is supported by a roller at *c*, the direction of the unknown reaction force at *c* of the ground on member *eb* must be perpendicular to the axis of this member. The hinge pin at *e* has two unknown reaction force components. The free-body diagram of the entire framework is then as shown in Fig. 4.52*b*. In this figure the reaction force components R_{ex} and R_{ey} are drawn in the assumed actual senses shown.

In drawing the free-body diagrams of the members, Newton's third law will be used at each of the hinged joints. This technique was illustrated for the preceding problem. Member *ad*, for example, exerts a certain

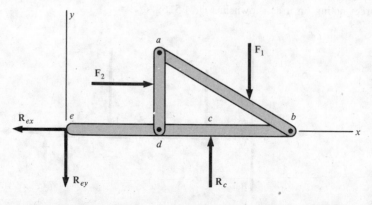

Fig. 4.52*b*

force on member ab, through the physical connection of the hinge pin at a. Member ab, however, must exert an *equal and opposite* force on member ad, because of Newton's third law. The free-body diagrams of the three members of the framework of Fig. 4.52a are shown in Figs. 4.52c through 4.52e. It should again be emphasized that actual senses are *assumed* for unknown force components.

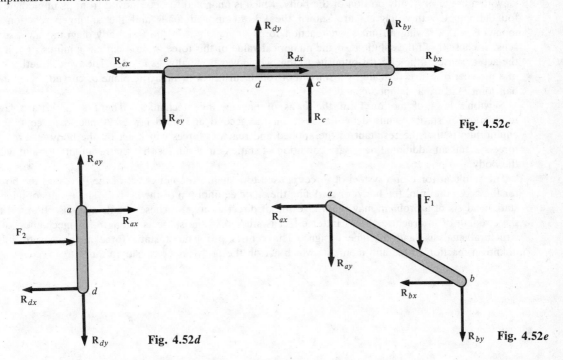

Fig. 4.52c

Fig. 4.52d

Fig. 4.52e

4.53   State two general uses for the solutions to problems in static force analysis.

I   The basic problem of static force analysis is to solve for the unknown reaction forces and moments which act on physical bodies. When the values of these unknown quantities have been obtained, they may be used in two different ways.

In the first way, the answer may be used directly. A typical example would be the solution for a cable force which acts on a body. Another example would be to find the moment which is produced on a bolt by the application of a known force on a wrench of known length. An alternative statement of this direct use of the result is that the forces or moments which are solved for are of interest in themselves.

In the second use, the forces or moments are utilized to compute the intensity of loading in the members comprising the structure. This intensity of loading is called a *stress*, and this quantity has the units of force divided by area. These stresses can then be compared with the material strength of the members to determine whether these elements are strong enough to support the loads. This second use of the results from static analysis is treated in subjects which are referred to as strength, or mechanics, of materials; structural analysis; and machine design. It is probably safe to say that the major application of static force analysis involves this second type of problem.

4.54   What is the *central* problem in static force analysis?

I   *The central problem of static force analysis is to solve for the unknown reaction forces and moments which act on the body under consideration.*

4.55   Give a summary of the basic concepts of force analysis.

I   A force is the effect of one physical body on another. In order for a force to be produced, two physical bodies must be present in the system. A physical body is composed of matter, and it has a nonzero volume. In statics, all bodies are assumed to be rigid. A rigid body is one whose exterior shape does not change under the application of a force or moment. Body forces are due to the effect of a one body on the mass particles of a second body. Surface forces result from the contact of one body with a second body. If a force pulls on a body, the effect is referred to as a tensile force. A compressive force occurs when a force pushes on a body. The weight force of a body is caused by the gravitational attraction between the body and the earth. This weight force is assumed to have a direction which is perpendicular to the surface of the earth.

Force is the basic effect in statics, and the moment and couple are force-induced quantities. Force systems may be identified as being one of seven types: collinear, concurrent, parallel, and general, with the additional description of whether the force system is one-, two-, or three-dimensional.

A fundamental concept in statics and dynamics is that of the free-body diagram. A free-body diagram is a view of a body, or of any portion of the body, which is imagined to be removed from its immediate supporting foundation. On this diagram are shown the known applied loads and the unknown reaction forces and moments. In drawing an unknown reaction force or moment on the free-body diagram, an assumed actual sense is chosen. If the solution for the numerical value of this force or moment has a minus sign, it means that the actual sense of the effect is opposite to that which was originally assumed. The typical methods of support of a member of a structure are by a smooth or rough surface which may be plane or curved, by a hinge pin or a ball joint, and by a clamped connection.

Newton's laws of motion form the basis of engineering mechanics. The first and third laws are the foundation of statics, while dynamics rests on the second and third laws. A necessary condition of static equilibrium is that the resultant of the applied and reaction forces which act on the body is zero. In certain force systems an additional necessary condition of static equilibrium is that the resultant moment which acts on the body also be zero.

The resultant force of a system of forces is used to visualize the net effect of the forces on the body, to solve for the acceleration of the body, or to define the static equilibrium of the body. The results obtained from the static analysis of a problem may themselves be of direct interest. These results may also be used to evaluate the strength of a structure. This latter effect is studied in the subjects strength, or mechanics, of materials, structural analysis, and machine design. The central problem of static force analysis is to solve for the unknown reaction forces and moments which act on the body under consideration.

CHAPTER 5

Analysis of Two-Dimensional Force Systems*

5.1 THE COLLINEAR FORCE SYSTEM—RESULTANT FORCE AND EQUILIBRIUM REQUIREMENT

5.1 Describe the collinear force system and give the general form for the resultant force.

▮ The collinear force system is the most elementary type of force assembly. It is characterized by all the forces having a common line of action.

A common application of the collinear force system is in the analysis of plane truss members, and this subject is treated in Chap. 6.

For the collinear force system, the magnitude F of the resultant force is the *direct algebraic* summation of the forces, or

$$F = \sum_{i=1}^{i=n} F_i = F_1 + F_2 + F_3 + \cdots + F_n$$

where n is the number of forces acting in the system. The direction of this resultant force is along the line of action of the original forces. For brevity, the above equation is written as $F = \Sigma F_i$.

Figure 5.1 shows the appearance of a typical collinear force system in which the forces F_1, F_2, and F_3 act on a stationary bar which is attached to the ground.

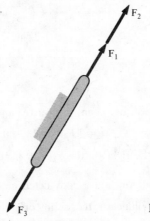

Fig. 5.1

5.2 Find the resultant force which acts on the bar in Fig. 5.1.

▮ A preliminary step is to choose a positive sense for the forces. If the forces F_1 and F_2 in Fig. 5.1 are assumed to be positive, then the resultant force for this case is

$$F = \Sigma F_i \qquad F = F_1 + F_2 - F_3$$

and this result is shown in Fig. 5.2a. Had the force F_3 been chosen to establish the positive sense, then

$$F = \Sigma F_i \qquad F = F_3 - F_2 - F_1$$

and the situation would be as shown in Fig. 5.2b. In both of the above equations, the term F is inherently positive. It should be recalled, from Prob. 1.62, that the net effect of multiplying a vector by -1 is to place the tip of the arrow which represents the vector on the opposite end from where it had originally been. The reader should carefully study Figs. 5.2a and b to be convinced that these figures show identical results for the resultant force F acting on the link.

*NOTE: Before beginning this chapter the reader is urged to thoroughly review Problems 4.26 through 4.28, which define and describe Newton's first and third laws and the requirements for equilibrium of a force system. *The requirements for the equilibrium of a force system are fundamental concepts which undergird all force analysis problems in engineering mechanics.*

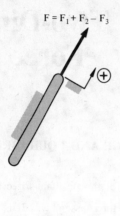

$$F = F_1 + F_2 - F_3$$

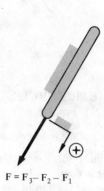

$$F = F_3 - F_2 - F_1$$

Fig. 5.2a   Fig. 5.2b

5.3   Two collinear forces act on the eyebolt shown in Fig. 5.3.   Find the resultant force.

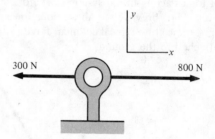

Fig. 5.3

❙ Forces acting to the right will be taken as positive, and

$$F = \Sigma F_i \qquad F = 800 - 300 = 500 \text{ N}$$

The resultant force acting to the right on the eyebolt is 500 N.

5.4   State the requirement for equilibrium of a body acted on by a collinear force system.

❙ In order for the body with collinear forces acting on it to be in static equilibrium, *the resultant force acting on the body must be identically zero.*   For the collinear force system this condition, following the equation in Prob. 5.1, is

$$F = \sum_{i=1}^{i=n} F_i = F_1 + F_2 + F_3 + \cdots + F_n = 0$$

The above relationship will include, in addition to the applied forces in the problem, the reaction forces which result from drawing the free-body diagram of the system.   The above equation is a formal statement of the static equilibrium requirement for the collinear force system.   For brevity, it is written as $\Sigma F = 0$.

5.5   The link shown in Prob. 5.1 is now removed from its support, and it is assumed to be suspended in space under the influence of the forces F_1, F_2, and F_3.   This link is further assumed to be weightless.   Find the relationship among the forces F_1, F_2, and F_3 if the bar is in equilibrium.

❙ The free-body diagram of this element is shown in Fig. 5.5.   For static equilibrium, the three forces which act on the link must satisfy

$$\Sigma F = 0 \qquad F_1 + F_2 - F_3 = 0$$

where the positive sense of force is chosen to be that of force F_1 or F_2.   This result follows from either direct application of the equation in Prob. 5.4, or from setting $F = 0$ in the equation in Prob. 5.1.

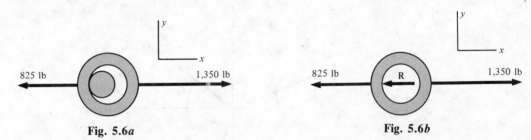

Fig. 5.5

5.6   The ring in Fig. 5.6a fits over a stationary post and is acted on by two collinear forces.   Find the force exerted on the post by the ring.

Fig. 5.6a   **Fig. 5.6b**

I   The free-body diagram of the ring is shown in Fig. 5.6b and R is the unknown reaction force *of* the post *on* the ring.
    Forces acting to the right are considered positive, and the equilibrium requirement for the ring is

$$\Sigma F = 0 \qquad -825 - R + 1{,}350 = 0 \qquad R = 525\ \text{lb}$$

By Newton's third law, a force equal and opposite to R must act on the post.   The final answer for the force exerted *by* the ring *on* the post is 525 lb, acting to the right.

5.7   Six people engage in a tug of war, as shown in Fig. 5.7a.   The compressive contact force between the ring and the fixed post is 15 lb.   With what force is the person on the left end pulling on the rope?

Fig. 5.7a

I   Two methods of solution are shown.

(a)   From consideration of the problem statement, and of the figure, the resultant force on the ring is 15 lb, acting to the left.   Using the definition of the resultant force, with forces acting to the left considered positive,

$$F = \Sigma F_i \qquad 15 = F_1 + 60 + 83 - 75 - 68 - 63 \qquad F_1 = 78\ \text{lb}$$

(**b**) Figure 5.7*b* shows the free-body diagram of the ring. Using the sense definition in part (*a*), the equilibrium requirement is

$$\Sigma F = 0 \qquad F_1 + 60 + 83 - 15 - 75 - 68 - 63 = 0 \qquad F_1 = 78 \text{ lb}$$

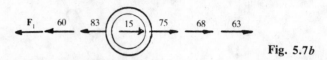

Fig. 5.7*b*

5.2 THE CONCURRENT FORCE SYSTEM—RESULTANT FORCE AND EQUILIBRIUM REQUIREMENTS, ROTATION OF COORDINATE AXES

5.8   Describe the concurrent force system and give the general forms for the resultant force.

▌ In a concurrent force system, the lines of action of all the forces intersect. The resultant force of the concurrent force system also acts through this point of intersection. A complete description of the resultant force of the concurrent force system requires a statement of the magnitude, direction, and sense of this force. An example of a concurrent force system is shown in Fig. 5.8. The angles θ_1, θ_2, and θ_3 give the directions of the forces with respect to the x axis.

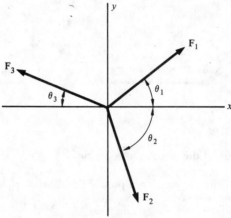

Fig. 5.8

Two methods of solution for the resultant of a concurrent two-dimensional force system were presented in Chap. 2. The first is a graphical method that uses the parallelogram, triangle, or polygon laws of force addition.

In the second method, the components of the resultant are obtained analytically. The forces are referenced to a set of xy coordinates, as seen in Fig. 5.8, to establish positive senses of force components. A component of any force is considered to be positive if it acts in the positive coordinate sense. The components F_x and F_y of the resultant force of the concurrent force system are then

$$F_x = \sum_{i=1}^{i=n} F_{ix} = F_{1x} + F_{2x} + F_{3x} + \cdots + F_{nx}$$

$$F_y = \sum_{i=1}^{i=n} F_{iy} = F_{1y} + F_{2y} + F_{3y} + \cdots + F_{ny}$$

In the above equations, terms such as F_{1x} and F_{1y} are the x and y components, respectively, of the force F_1. The summations are algebraic, and, as before, n is the number of forces acting in the system.

After the components F_x and F_y have been obtained, the value F of the resultant force may be found from

$$F = \sqrt{F_x^2 + F_y^2}$$

The direction θ of the line of action of the resultant force with respect to the x axis is given by

$$\tan \theta = \frac{F_y}{F_x}$$

When using the above equation, the terms F_x and F_y will be treated as positive quantities. Thus, θ will be an acute angle with the x axis.

In formal vector notation, the resultant force has the form

$$\mathbf{F} = F_x \mathbf{i} + F_y \mathbf{j}$$

where F_x and F_y are given by the equations above.

5.9 Find the resultant force of the concurrent force system shown in Fig. 5.9a.

▮ Using the equations in Prob. 5.8,

$$F_x = \Sigma F_{ix} \qquad F_x = 75 \cos 45° + 50 - 100 \cos 60° = 53.0 \text{ N}$$
$$F_y = \Sigma F_{iy} \qquad F_y = 75 \sin 45° - 100 \sin 60° = -33.6 \text{ N}$$

$$F = \sqrt{Fx^2 + Fy^2} = \sqrt{53.0^2 + (-33.6)^2} = 62.8 \text{ N} \qquad \tan \theta = \frac{33.6}{53.0} = 0.634 \qquad \theta = 32.4°$$

These results are shown in Fig. 5.9b.

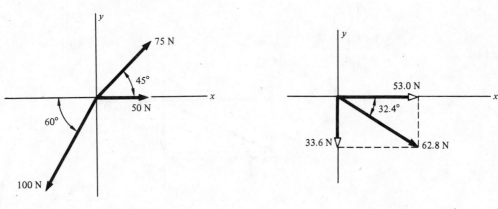

Fig. 5.9a **Fig. 5.9b**

5.10 State the requirement for equilibrium of a body acted on by a concurrent force system.

▮ For static equilibrium of the concurrent force system, the resultant force acting on the system must be *identically zero*. This requirement, following the equations in Prob. 5.8, appears as

$$F_x = \sum_{i=1}^{i=n} F_{ix} = F_{1x} + F_{2x} + F_{3x} + \cdots + F_{nx} = 0$$

$$F_y = \sum_{i=1}^{i=n} F_{iy} = F_{1y} + F_{2y} + F_{3y} + \cdots + F_{ny} = 0$$

where n is the number of forces acting in the system. It should be noted that the above two equations will contain *unknown reaction forces which are introduced in the process of drawing the free-body diagram*.

The above two equations are a formal statement of the static equilibrium requirements for the concurrent force system. For brevity, these equations will be written as $\Sigma F_x = 0$ and $\Sigma F_y = 0$. In formal vector notation, the requirement for equilibrium of the concurrent force system is

$$\mathbf{F} = F_x \mathbf{i} + F_y \mathbf{j} = 0$$

5.11 The ring shown in Fig. 5.11a is mounted on a fixed post. The three forces acting on the ring are cable tensile forces. Find the components of the reaction force of the post on the ring and the magnitude and direction of this reaction force.

▮ The free-body diagram of the ring is shown in Fig. 5.11b, and the x and y axes in the figure establish the positive senses of force. Since the senses of the components R_x and R_y of the reaction force of the post on the ring are unknown, these forces are shown in the figure as acting in the positive coordinate senses. A negative value for either of these forces would merely indicate that the force acts in the opposite sense from that in which it is drawn. The requirements for static equilibrium of the ring are

$$\sum F_x = 0 \qquad 800 \cos 35° - 400 \sin 20° + R_x = 0$$

$$\sum F_y = 0 \qquad R_y + 400 \cos 20° + 800 \sin 35° - 600 = 0$$

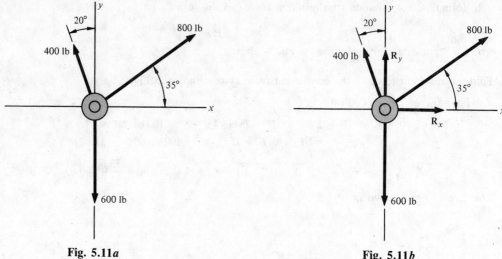

Fig. 5.11a   Fig. 5.11b

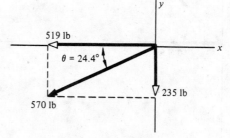

Fig. 5.11c

$$R_x = -519 \text{ lb} \qquad R_y = -235 \text{ lb}$$
$$\mathbf{R} = -519\mathbf{i} - 235\mathbf{j} \qquad \text{lb}$$

These forces are drawn in their actual senses in Fig. 5.11c. The magnitude R of the reaction force of the post on the ring is

$$R = \sqrt{R_x^2 + R_y^2} = \sqrt{(-519)^2 + (-235)^2} = 570 \text{ lb}$$

The angle θ which defines the direction of the resultant force is found from

$$\tan \theta = \frac{235}{519} = 0.453 \qquad \theta = 24.4°$$

5.12   (a) Find the angle θ which defines the contact point between the ring and the fixed post shown in Fig. 5.12a.
       (b) Find the magnitude, direction, and sense of the force exerted by the ring on the post.
       (c) Compare the force found in part (b) with the resultant of the applied forces acting on the ring.

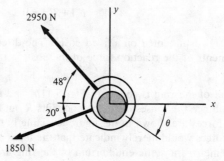

Fig. 5.12a

▌ (a) Figure 5.12b shows the free-body diagram of the ring. The force exerted on the ring by the post is R. For equilibrium,

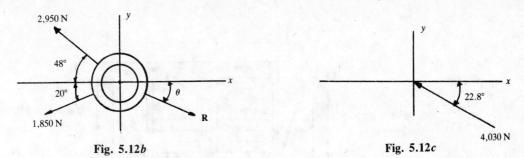

Fig. 5.12b   **Fig. 5.12c**

$$\Sigma F_x = 0 \qquad -2{,}950 \cos 48° - 1{,}850 \cos 20° + R \cos \theta = 0$$
$$\Sigma F_y = 0 \qquad 2{,}950 \sin 48° - 1{,}850 \sin 20° - R \sin \theta = 0$$
$$R \sin \theta = 1{,}560 \qquad R \cos \theta = 3{,}710$$

The first equation is divided by the second equation to obtain

$$\tan \theta = 0.420 \qquad \theta = 22.8°$$

(**b**) From part (*a*),

$$R \sin \theta = 1{,}560 \qquad R \sin 22.8° = 1{,}560 \qquad R = 4{,}030 \text{ N}$$

The direction of force R is given by angle θ in part (*a*).   The sense of this force is down and to the right.   The force exerted on the post by the ring has the magnitude and direction of the force shown in Fig. 5.12*b* and, using Newton's third law, the opposite sense.   This result is shown in Fig. 5.12*c*.

(**c**) The force exerted by the ring on the post is equal to the resultant of the two applied forces that act on the ring.

5.13 Find the value of W which is required to maintain the equilibrium configuration shown in Fig. 5.13*a*.

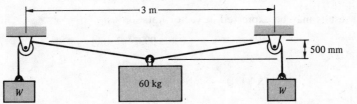

Fig. 5.13a

❚ Each weight W is acted on by the cable tensile force F and the weight force W.   This pair of forces is a collinear force system.   The system of forces acting on the 60-kg mass is a concurrent force system, with the free-body diagram shown in Fig. 5.13*b*.   For equilibrium,

$$\Sigma F_y = 0 \qquad 2\left(\frac{500}{1{,}580}\right)W - 589 = 0 \qquad W = 931 \text{ N}$$

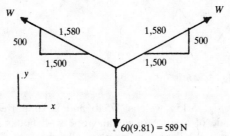

Fig. 5.13b

5.14 A cable of length 6 m is attached to the two posts shown in Fig. 5.14*a*.   The crate has a mass of 500 kg.   Find the x and y components of the forces exerted by the cable on the eyebolts, and find the cable tensile force.

❚ The crate is acted on by the weight force W and the two cable tensile forces F.   These forces form a concurrent force system, since all lines of action intersect at point a in the free-body diagram shown in Fig. 5.14*b*.

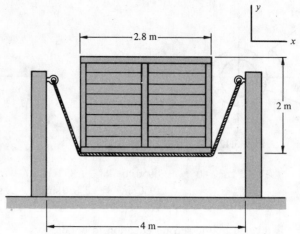

Fig. 5.14a

The length l of cable between the posts and the crate is

$$l = \tfrac{1}{2}(6 - 2.8) = 1.6 \text{ m}$$

and

$$\cos\theta = \frac{0.6}{1.6} \qquad \theta = 68°$$

For equilibrium of the crate,

$$\Sigma F_y = 0 \qquad -4{,}910 + 2F\sin 68° = 0 \qquad F = 2{,}650 \text{ N}$$

The forces acting on the eyebolts are shown in Fig. 5.14c. The values of the force components acting on the eyebolts are

$$F_x = \pm 2{,}650\cos 68° = \pm 993 \text{ N}$$
$$F_y = -2{,}650\sin 68° = -2{,}460 \text{ N}$$

The above results may be expressed in vector notation as

$$\mathbf{F} = \pm 993\mathbf{i} - 2{,}460\mathbf{j} \qquad \text{N}$$

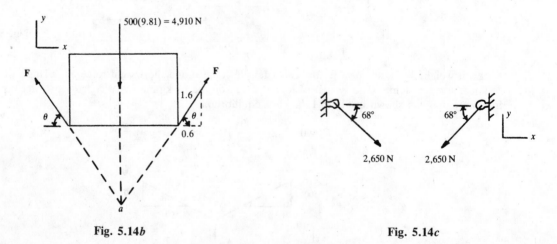

Fig. 5.14b   Fig. 5.14c

5.15 The drum of liquid in Fig. 5.15a weighs 475 lb and is supported by a sling hoist, shown as *abc* in the figure. This hoist is made of a piece of manila rope with hooks on each end. The rope has a rated load capacity of 750 lb. What is the minimum length *abc* of the rope if the load capacity of the rope is not to be exceeded?

∎ The drum is acted on by the two cable tensile forces F and the weight force. These forces form a concurrent force system since all lines of action intersect at point a shown in the free-body diagram in Fig. 5.15b. Using the maximum load capacity of the rope,

$$F = 750 \text{ lb}$$

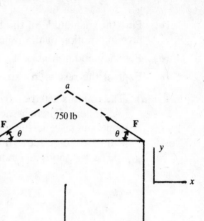

Fig. 5.15a   Fig. 5.15b

For equilibrium,

$$\Sigma F_y = 0 \qquad 2(750 \sin \theta) - 475 = 0 \qquad \theta = 18.5°$$

The configuration of the rope length is shown in Fig. 5.15c. The minimum length l of the rope is found from

$$\cos 18.5° = \frac{2}{l/2} \qquad l = 4.22 \text{ ft}$$

For any length greater than 4.22 ft, angle θ will increase and force F will decrease. Thus, $l = 4.22$ ft is the *minimum* length of rope that can be used if the force is not to exceed 750 lb.

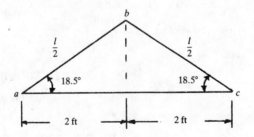

Fig. 5.15c

5.16 The 1,500-kg roller in Fig. 5.16a has a rough surface.

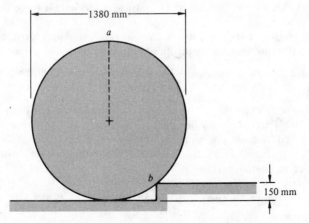

Fig. 5.16a

(a) Find the magnitude of the horizontal force applied at the center of the roller which will cause impending pivoting motion of the roller about the edge of the step.

(b) For the condition at part (a), find the magnitude, direction, and sense of the reaction force at b.

(c) Express the results in parts (a) and (b) in formal vector notation.

❚ (a) The roller is acted on by its weight force, the reaction force at b, and the applied force at the center of the roller. There is no reaction force acting on the bottom of the roller since pivoting motion is impending. The three forces acting on the roller form a concurrent force system since their lines of action all intersect at the center of the roller.

The free-body diagram of the roller is shown in Fig. 5.16b. The dimension a is found as

$$a = \sqrt{690^2 - 540^2} = 430 \text{ mm}$$

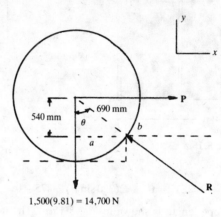

1,500(9.81) = 14,700 N

Fig. 5.16b

Angle θ is given by

$$\cos \theta = \frac{540}{690} \qquad \theta = 38.5°$$

For equilibrium,

$$\Sigma F_x = 0 \qquad P - R \sin \theta = 0 \qquad R = \frac{P}{\sin \theta} = \frac{P}{\sin 38.5°} = 1.61P$$

$$\Sigma F_y = 0 \qquad -14,700 + R \cos \theta = -14,700 + (1.61P) \cos 38.5° = 0 \qquad P = 11,700 \text{ N}$$

(b) Using $R = 1.61P$,

$$R = 1.61P = 1.61(11,700) = 18,800 \text{ N}$$

The direction of force R is 38.5° from the y axis, acting up and to the left.

(c) In formal vector notation the results for P and R are

$$\mathbf{P} = P\mathbf{i} = 11,700\mathbf{i} \qquad \text{N}$$

$$\mathbf{R} = (-R \sin \theta)\mathbf{i} + (R \cos \theta)\mathbf{j} = (-18,800 \sin 38.5°)\mathbf{i} + (18,800 \cos 38.5°)\mathbf{j} = -11,700\mathbf{i} + 14,700\mathbf{j} \qquad \text{N}$$

It may be seen that the x component in the above equation is equal and opposite to the force P and that the y component is equal and opposite to the weight force of the roller.

5.17 Find the magnitude, direction, and sense of the *minimum* required force, applied at the center of the roller in Prob. 5.16, that will cause impending pivoting motion of the roller about the edge of the step.

❚ Two methods of solution are shown. The first uses a graphical technique, and the second uses the definition of the minimum value of a function of one variable.

(a) Figure 5.17a shows the triangle law of force addition (defined in Prob. 2.3) for this problem. The weight force of 14,700 N acts vertically downward. The direction of the reaction force R is along line ab. The force P required to cause impending pivoting motion of the roller may have any of the typical positions shown by P_1, P_2, and P_3 in the figure. Since P must have a minimum value, this force must be perpendicular to line ab. Thus, $P = P_2$ and

$$\sin 38.5° = \frac{P_2}{14,700} \qquad P_2 = 9,150 \text{ N}$$

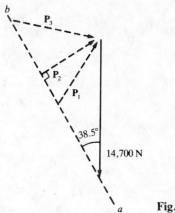

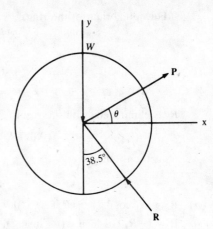

Fig. 5.17a

Fig. 5.17b

(b) Figure 5.17b shows the free-body diagram of the roller, and force P has the arbitrary direction θ. For equilibrium,

$$\Sigma F_x = 0 \qquad P\cos\theta - R\sin 38.5° = 0$$
$$\Sigma F_y = 0 \qquad -W + P\sin\theta + R\cos 38.5° = 0$$

From the first equation,

$$R = \frac{P\cos\theta}{\sin 38.5°}$$

Using this result in the second equation,

$$-W + P\sin\theta + \frac{P\cos\theta}{\sin 38.5°}\cos 38.5° = 0 \qquad P = \frac{W}{\sin\theta + \cos\theta/\tan 38.5°}$$

The necessary condition for P to have an extreme value (for this case, a minimum value) is

$$\frac{dP}{d\theta} = 0 \qquad \cos\theta - \cot 38.5°\sin\theta = 0$$

$$\tan\theta = \tan 38.5° \qquad \theta = 38.5°$$

It may be concluded that force P is perpendicular to the direction of force R. The remainder of the solution is the same as in part (a).

5.18 The thin ring of mass 5 kg shown in Fig. 5.18a is supported in a vertical plane by two rollers. The mean diameter of the ring is 300 mm. (a) Find the forces exerted by the ground on the ring. (b) Express the results in part (a) in formal vector notation.

▌ (a) The free-body diagram of the ring is shown in Fig. 5.18b. The forces which act on the ring form a concurrent force system since all of the lines of action intersect at the center of the ring.

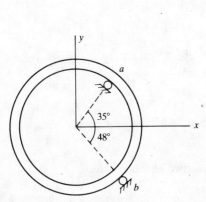

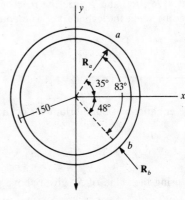

Fig. 5.18a

$W = 5(9.81) = 49.1$ N

Fig. 5.18b

For equilibrium of the ring,

$$\Sigma F_x = 0 \qquad R_a \cos 35° - R_b \cos 48° = 0$$

$$R_a = \frac{\cos 48°}{\cos 35°} R_b = 0.817 R_b$$

$$\Sigma F_y = 0 \qquad R_a \sin 35° + R_b \sin 48° - 49.1 = 0$$

R_a is eliminated, with the result

$$(0.817 R_b) \sin 35° + R_b \sin 48° = 49.1 \qquad R_b = 40.5 \text{ N}$$

R_a is then found as

$$R_a = 0.817 R_b = 0.817(40.5) = 33.1 \text{ N}$$

(b) $\mathbf{R}_a = (R_a \cos 35°)\mathbf{i} + (R_a \sin 35°)\mathbf{j} = (33.1 \cos 35°)\mathbf{i} + (33.1 \sin 35°)\mathbf{j} = 27.1\mathbf{i} + 19.0\mathbf{j}$ N

$\mathbf{R}_b = (-R_b \cos 48°)\mathbf{i} + (R_b \sin 48°)\mathbf{j} = (-40.5 \cos 48°)\mathbf{i} + (40.5 \sin 48°)\mathbf{j} = -27.1\mathbf{i} + 30.1\mathbf{j}$ N

It may be observed that the x components in the equations for $\mathbf{R}_a$ and $\mathbf{R}_b$ are equal and opposite and that the sum of the two y components is equal to the weight of the ring.

5.19   The two smooth cylinders in Fig. 5.19a rest in a well with vertical walls.   All contacting surfaces are assumed to be smooth.   Find the reaction forces of the cylindrical surfaces on each other, and on the walls and floor of the well.   Cylinder A has a mass of 15.3 kg, and cylinder B has a mass of 30.6 kg.

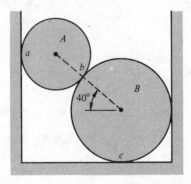

Fig. 5.19a

❚   Since the reaction forces for smooth contacting surfaces have directions which are normal to the surface, it follows that the sets of forces on each of the cylinders in this problem are separate concurrent force systems.   The weights of the cylinders are

$$W_A = 15.3 \text{ kg}\left(9.81 \frac{\text{m}}{\text{s}^2}\right) = 150 \frac{\text{kg} \cdot \text{m}}{\text{s}^2} = 150 \text{ N}$$

$$W_B = 30.6 \text{ kg}\left(9.81 \frac{\text{m}}{\text{s}^2}\right) = 300 \frac{\text{kg} \cdot \text{m}}{\text{s}^2} = 300 \text{ N}$$

The free-body diagrams of the two cylinders are shown in Fig. 5.19b, together with a set of coordinate axes. R_a, R_c, and R_d are the reaction forces of the walls and floor on the cylinders, while R_b is the reaction force between the cylinders.   The equilibrium requirements for cylinder A are

$$\sum F_x = 0 \qquad R_a - R_b \cos 40° = 0$$

$$\sum F_y = 0 \qquad -150 + R_b \sin 40° = 0$$

$$R_b = 233 \text{ N} \qquad R_a = 178 \text{ N}$$

The equations of equilibrium for cylinder B are

$$\sum F_x = 0 \qquad R_b \cos 40° - R_d = 0$$

$$\sum F_y = 0 \qquad -R_b \sin 40° - 300 + R_c = 0$$

By using the value of R_b given above, the solutions to the above two equations are

$$R_d = 178 \text{ N} \qquad R_c = 450 \text{ N}$$

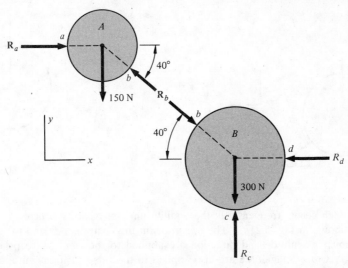

Fig. 5.19b

It may be observed that $R_a = R_d$, which is a necessary condition for horizontal equilibrium of the two cylinders *drawn as a single free-body diagram*. Also, the reaction force R_c of the floor on cylinder B is equal to the combined weight of the two cylinders. The final answer for the reaction forces of the cylinders on the walls and floor of the well, then, would be the *negative* of the values of R_a, R_c, and R_d computed above.

5.20 Find the tensile force in the cable in Fig. 5.20a and the contact forces at points a to d. Each cylinder weighs 30 lb and has a diameter of 8 in.

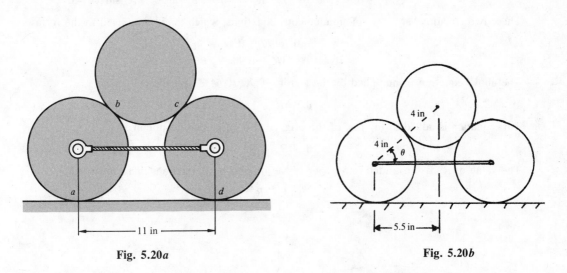

Fig. 5.20a   Fig. 5.20b

❚ From the configuration in Fig. 5.20b,

$$\cos \theta = \frac{5.5}{8} \qquad \theta = 46.6°$$

The free-body diagram of the top cylinder is shown in Fig. 5.20c. For equilibrium,

$$\Sigma F_x = 0 \qquad R_b \cos 46.6° - R_c \cos 46.6° = 0 \qquad R_b = R_c$$
$$\Sigma F_y = 0 \qquad -30 + 2R_b \sin 46.6° = 0 \qquad R_b = R_c = 20.6 \text{ lb}$$

The free-body diagram of the left bottom cylinder is shown in Fig. 5.20d. For equilibrium,

$$\Sigma F_y = 0 \qquad R_a - 20.6 \sin 46.6° - 30 = 0 \qquad R_a = 45.0 \text{ lb}$$
$$\Sigma F_x = 0 \qquad -20.6 \cos 46.6° + T = 0 \qquad T = 14.2 \text{ lb}$$

From the symmetry of the problem,

$$R_d = R_a = 45.0 \text{ lb}$$

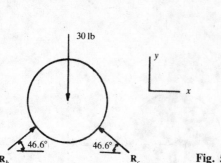

30 lb

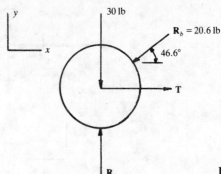

30 lb

$R_b = 20.6$ lb

46.6°

T

R_a

$46.6°$ $46.6°$

R_b R_c

Fig. 5.20c

Fig. 5.20d

5.21 A situation which occurs frequently in the equilibrium analysis of a concurrent force system is typified by the arrangement shown in Fig. 5.21a. The apparatus in this figure is referred to as a force table. Two forces F_1 and F_2 of known magnitude and direction are applied to the ring. Two other forces R_a and R_b of known direction but *unknown* magnitude are also applied to the ring. The magnitudes of these latter two forces are then adjusted so that the ring is in a concentric position about the fixed center post of the force table. Under these conditions the ring is in equilibrium because of the concurrent force system, F_1, F_2, R_a, and R_b. If the free-body diagram of a typical ring is as shown in Fig. 5.21b, find forces R_a and R_b.

▌ The two known forces are 700 N and 850 N, and R_a and R_b may be considered to be unknown reaction forces which are to be found. The equilibrium equations appear as

$$\sum F_x = 0 \qquad -700 \sin 20° - R_a \cos 40° + 850 \cos 30° + R_b \cos 65° = 0$$

$$\sum F_y = 0 \qquad 700 \cos 20° - R_a \sin 40° + 850 \sin 30° - R_b \sin 65° = 0$$

These two equations are a set of simultaneous equations, which may be reduced to the forms

$$-0.766R_a + 0.423R_b + 497 = 0$$
$$-0.643R_a - 0.906R_b + 1,080 = 0$$

The above equation is multiplied by the factor $-0.766/0.643$, with the result

$$0.766R_a + 1.08R_b - 1,290 = 0$$

This equation is added to the first of the original two equations, to obtain

$$1.50R_b = 793 \qquad R_b = 528 \text{ N}$$

The value of R_b is substituted into either of the original two equations, with the result

$$R_a = 940 \text{ N}$$

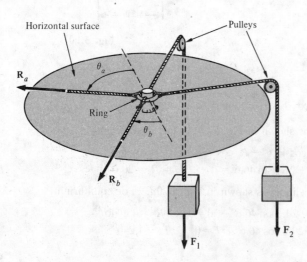

Horizontal surface

Pulleys

R_a

θ_a

Ring

θ_b

R_b

F_1

F_2

Fig. 5.21a

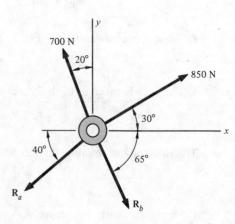

y

700 N

20°

850 N

30°

x

40°

65°

R_a

R_b

Fig. 5.21b

5.22 The computation in Prob. 5.21 required several steps. Show a method of solution which permits R_a or R_b to be found directly in a single step.

❚ The x and y axes in Fig. 5.21b are now rotated to the new positions x' and y' shown in Fig. 5.22. This new location of the coordinate axes is chosen so that one of these axes is collinear with one of the unknown forces. A force summation along the second axis will then involve *only one* unknown force, which may be solved for directly. For the coordinates in Fig. 5.22,

$$\sum F_{y'} = 0 \qquad 700 \sin 70° - R_b \sin 75° - 850 \sin 10° = 0 \qquad R_b = 528 \text{ N}$$

$$\sum F_{x'} = 0 \qquad -R_a - R_b \cos 75° + 700 \cos 70° + 850 \cos 10° = 0$$

$$-R_a - 528 \cos 75° + 700 \cos 70° + 850 \cos 10° = 0$$

$$R_a = 940 \text{ N}$$

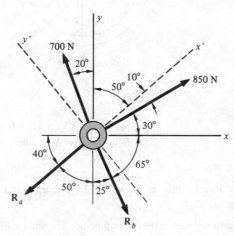

Fig. 5.22

The above method of solution is referred to as *solution by rotation of coordinate axes*. The reader should compare the two methods of solution shown above and be convinced that the computational effort is reduced substantially when the method of rotation of coordinate axes is used.

5.23 The concurrent force system in Fig. 5.23a is in equilibrium. Find the unknown reaction forces, using the method of rotation of coordinate axes.

❚ The x and y axes are rotated to the positions x' and y' shown in Fig. 5.23b. The equilibrium requirements are

$$\sum F_{y'} = 0 \qquad 110 \sin 50° + 85 \cos 50° - R_a \cos 15° = 0 \qquad R_a = 144 \text{ lb}$$

$$\sum F_{x'} = 0 \qquad 110 \cos 50° - 85 \sin 50° - R_a \sin 15° + R_b = 0 \qquad R_b = 31.7 \text{ lb}$$

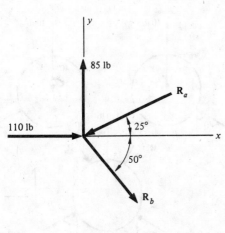

Fig. 5.23a

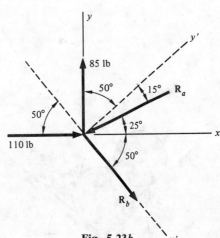

Fig. 5.23b

5.24 Solve Prob. 5.18 by using the method of rotation of coordinate axes.

❚ The free-body diagram of the ring is shown in Fig. 5.24. The x' and y' axes are placed on the ring as shown in the figure.

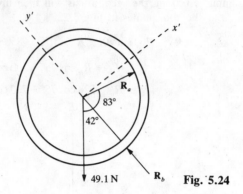

Fig. 5.24

For equilibrium,

$$\Sigma F_{x'} = 0 \qquad R_a \sin 83° - 49.1 \sin 42° = 0 \qquad R_a = 33.1 \text{ N}$$
$$\Sigma F_{y'} = 0 \qquad -49.1 \cos 42° + R_b - R_a \cos 83° = -49.1 \cos 42° + R_b - 33.1 \cos 83° = 0$$
$$R_b = 40.5 \text{ N}$$

5.25 Three 10-lb cylinders rest in a trough, as shown in Fig. 5.25a. Find the magnitudes of the compressive contact forces at points *a* through *f*.

❚ The free-body diagrams of the three cylinders are shown in Fig. 5.25b. The forces which act on each cylinder are systems of concurrent forces. For equilibrium of the left cylinder in the direction of R_a,

$$\Sigma F_{R_a} = 0 \qquad R_a - 10 \cos 40° = 0 \qquad R_a = 7.66 \text{ lb}$$

For equilibrium in the direction of R_b,

$$\Sigma F_{R_b} = 0 \qquad R_b - 10 \sin 40° = 0 \qquad R_b = 6.43 \text{ lb}$$

From symmetry,

$$R_e = R_b = 6.43 \text{ lb} \qquad R_f = R_a = 7.66 \text{ lb}$$

For the lower cylinder, from symmetry, $R_d = R_c$. For equilibrium of this element,

$$\Sigma F_y = 0 \qquad -2R_b \sin 40° - 10 + 2R_c \cos 40° = 0$$
$$R_c = 11.9 \text{ lb} \qquad R_d = R_c = 11.9 \text{ lb}$$

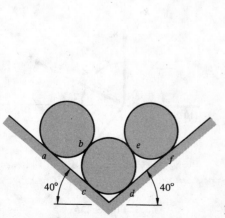

Fig. 5.25a

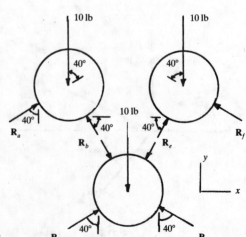

Fig. 5.25b

5.3 THE PARALLEL FORCE SYSTEM—RESULTANT FORCE AND EQUILIBRIUM REQUIREMENTS

5.26 Describe the parallel force system and give the general form for the resultant force.

❚ In a parallel force system, all the forces which act in the system are parallel to one another. Figure 5.26 shows one such system. A common physical application of this type of force system is the case of lateral loading on beams. The magnitude of the resultant force **F** is found from a *direct algebraic* summation of the forces, or

$$F = \sum_{i=1}^{i=n} F_i = F_1 + F_2 + F_3 + \cdots + F_n$$

where n is the number of forces acting in the system. The resultant force F which is defined by the above equation has a direction which is parallel to that of the original system of forces. It may be observed that this equation is identical to the equation, in Prob. 5.1, which defined the resultant force for the collinear force system. In both cases, the force summation is *purely algebraic*. This is so because, in each of these two types of systems, all of the forces involved have the *same* direction. For brevity, the above equation may be written as $F = \Sigma F_i$.

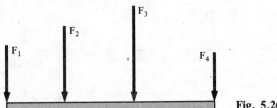

Fig. 5.26

Unlike the problems involving the collinear force and concurrent force systems, no general statement may be made about the location of the resultant force. This location must be computed for each particular problem. The resultant force of any system of forces must provide the same static effect as the original system of forces. One consequence of this definition is that the resultant force must produce, in addition to the same equivalent force as the original system, *an equivalent moment*. This moment may be with respect to any convenient point in the system, and this effect is shown in the following problem.

5.27 Figure 5.27 shows a parallel force system. Find the magnitude and location of the resultant force F.

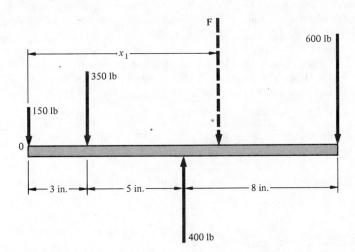

Fig. 5.27

❚ The resultant force F is shown in the figure, and x_1 is the location to be determined. Forces acting downward will be considered positive. The magnitude of the resultant force is

$$F = \Sigma F_i \qquad F = 150 + 350 - 400 + 600 = 700 \, \text{lb}$$

The location 0 of the 150-lb force will be taken as the reference for moments, and counterclockwise moments, as before, are considered positive. The moment M_0 of the original system of forces, with resect to the reference, is

$$M_0 = -350(3) + 400(8) - 600(16) = -7,450 \, \text{in} \cdot \text{lb}$$

The moment M of the resultant force F with respect to the same reference is

$$M = -Fx_1 = -700x_1$$

For equivalence of these two moments,

$$M = M_0 \qquad -700x_1 = -7,450 \text{ in} \cdot \text{lb} \qquad x_1 = 10.6 \text{ in}$$

5.28 The system of Prob. 5.27 is redrawn as shown in Fig. 5.28. In addition to the applied forces which acted in the original system, an applied counterclockwise couple of magnitude 1,200 in · lb is also applied, as shown. Find the magnitude and the location of the resultant force of the new system. Compare these values with the results obtained in Prob. 5.27.

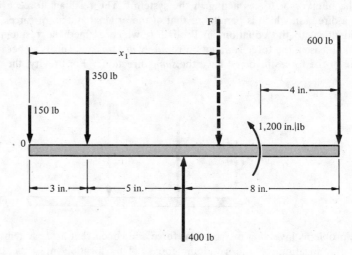

Fig. 5.28

❚ Because of the applied couple, the system is no longer referred to as a parallel force system. This is merely a matter of definition, since the forces in both systems are identical. The magnitude F of the resultant force is still 700 lb, since the resultant force of the couple is identically zero. The moment M_0 of the system of forces and couples shown in Fig. 5.28, with respect to point 0, is

$$M_0 = -350(3) + 400(8) + 1,200 - 600(16) = -6,250 \text{ in} \cdot \text{lb}$$

It may be observed that the value 1,200 in · lb in the above equation, which represents the effect of the couple, has no associated length dimension. This is an expected result, since a couple may be moved to any location in its own plane without change in its effect on the body.

The resultant force F must produce the same moment with respect to point 0 as that produced by the original system of forces, together with the applied couple, with respect to this point. The equation $M = -Fx_i$ is still valid, where x_1 describes the location of the resultant force in this new problem. This equation is now combined with the above value of M_0, with the result

$$-700x_1 = -6,250 \qquad x_1 = 8.93 \text{ in}$$

Comparing this result with the value found in Prob. 5.27, it may be seen that the effect of the applied couple is to move the location of the resultant force closer to the left end of the system.

5.29 Find the magnitude, direction, sense, and location of the resultant of the system of applied forces which act on the beam in Fig. 5.29.

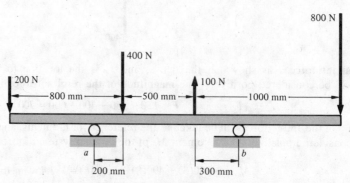

Fig. 5.29

❚ The resultant force F acts vertically downward, at x_1 mm to the right of point a. The magnitude of F is found from

$$F = \Sigma F_i = 4{,}800 + 6{,}500 + 3{,}400 = 14{,}700 \text{ N}$$

For equivalence of the moments about point a,

$$\Sigma M_a = -Fx_1 \qquad -4{,}800(100) - 6{,}500(250) - 3{,}400(850) = -14{,}700x_1 \qquad x_1 = 340 \text{ mm}$$

5.30 State the requirement for equilibrium of a body acted on by a parallel force system.

❚ For static equilibrium of the parallel force system, the resultant force F of the system must be identically zero. From Prob. 5.26, this requirement is

$$F = \sum_{i=1}^{i=n} F_i = F_1 + F_2 + F_3 + \cdots + F_n = 0$$

It should be noted that the above equation will contain *unknown reaction forces which are introduced in the process of drawing the free-body diagram of the system.*

The moment M of the resultant force is of the form

$$M = Fx_1$$

where x_1 is the distance between the resultant force and the reference point for taking moments. If the resultant force F is zero, it follows that

$$M = 0$$

for *any value* of the dimension x_1. This equation now leads to the very important conclusion that, for equilibrium of the parallel force system, *the algebraic sum of moments about any arbitrary reference point is zero.* In practice, this reference point is chosen so as to minimize the computational effort. The formal requirements for static equilibrium of the parallel force system are then

1. The algebraic summation of the forces is zero.
2. The algebraic summation of the moments about *any arbitrary reference point* is zero.

These two requirements may be written in the compact forms $\Sigma F = 0$ and $\Sigma M = 0$. In formal vector notation, the above equilibrium requirements are

$$\mathbf{F} = 0 \qquad \mathbf{M} = 0$$

5.31 A boy and a girl sit on the seesaw in Fig. 5.31a. The girl weighs 105 lb, and the boy weighs 140 lb. The seesaw plank weighs 85 lb.

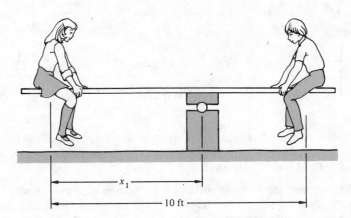

Fig. 5.31a

(*a*) What should the distance x_1 be if the seesaw is to be in equilibrium in the position shown?

(*b*) Find the force exerted by the roller on the seesaw.

❚ (*a*) The free-body diagram of the seesaw is shown in Fig. 5.31b. The set of forces acting on the seesaw is a parallel force system. For moment equilibrium about point 0,

$$\Sigma M_0 = 0 \qquad 105x_1 + 85(x_1 - 5) - 140(10 - x_1) = 0 \qquad x_1 = 5.53 \text{ ft}$$

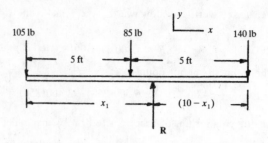

Fig. 5.31*b*

(*b*) For force equilibrium,

$$\Sigma F_y = 0 \qquad -105 - 85 - 140 + R = 0 \qquad R = 330 \text{ lb}$$

It may be seen that R is equal to the combined weight of the boy, the girl, and the plank.

5.32 A strip is welded to a plate, as shown in Fig. 5.32*a*. Find the forces exerted by welds *a* and *b* on the strip.

❚ The free-body diagram of the strip is shown in Fig. 5.32*b*, and F_a and F_b are the forces exerted by the welds on the strip. The set of forces acting on the strip is a parallel force system. For equilibrium,

$$\Sigma M_0 = 0 \qquad -12.5(1.25) + 6F_b = 0 \qquad F_b = 2.60 \text{ kip}$$
$$\Sigma F_y = 0 \qquad F_a - 12.5 + F_b = 0 \qquad F_a = 9.9 \text{ kip}$$

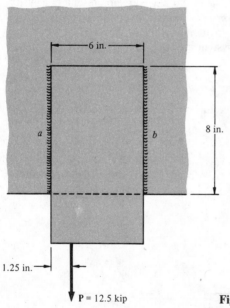

Fig. 5.32*a*

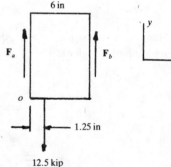

Fig. 5.32*b*

5.33 Figure 5.33*a* shows a laterally loaded beam which is supported by two rollers and is in equilibrium. Find the values of the reaction forces R_a and R_b of the rollers on the beam and the reaction forces exerted by the ground on the rollers. The beam is weightless.

❚ The rollers are assumed to be smooth, so that the normal contact forces between these elements and the beam are perpendicular to the axis of the beam. The system of forces that acts on the beam is a parallel force system. The free-body diagrams are shown in Fig. 5.33*b*, and the weights of the rollers are neglected. Note that the set of forces on each of the two rollers constitutes a collinear force system.

The reference point for taking moments is chosen to be the line of action of R_a, and counterclockwise moments are considered positive. The reason for choosing this point is that force R_a passes through the point and thus contributes zero moment. The statement of moment equilibrium is then

$$\sum M_a = 0 \qquad 200(600) - 400(200) + 100(700) + R_b(1,000) - 800(1,700) = 0$$
$$R_b = 1,250 \text{ N}$$

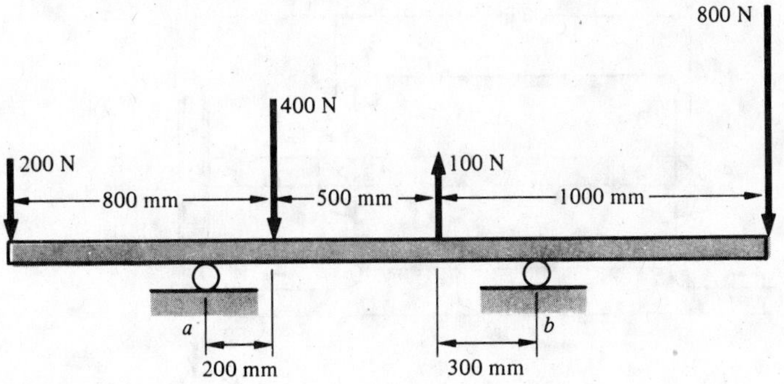

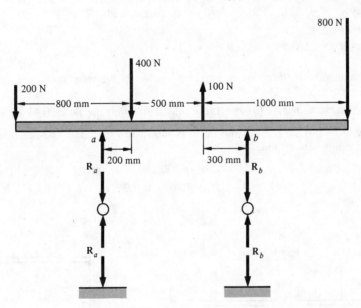

Fig. 5.33a

By taking upward forces as positive, the statement of force equilibrium is

$$\sum F = 0 \qquad -200 + R_a - 400 + 100 + R_b - 800 = 0$$
$$-200 + R_a - 400 + 100 + 1{,}250 - 800 = 0$$
$$R_a = 50 \text{ N}$$

These results may be checked by summing moments about a reference point different from that used in the first moment summation. If point b is chosen as a reference, the moment summation appears as

$$\sum M_b = 0 \qquad 200(1{,}600) - 50(1{,}000) + 400(800) - 100(300) - 800(700) \overset{?}{=} 0$$
$$200(1{,}600) + 400(800) \overset{?}{=} 50(1{,}000) + 100(300) + 800(700)$$
$$6.4 \times 10^5 \equiv 6.40 \times 10^5$$

Fig. 5.33b

5.34 A motor drives a single-vee belt pulley. The belt tensile forces are as shown in Fig. 5.34a.

(a) Draw the free-body diagram of the pulley and find the force exerted on the pulley by the shaft.

(b) Draw the free-body diagram of the shaft and find the forces exerted by the shaft on bearings a and b.

(c) Express the results in parts (a) and (b) in formal vector notation.

▌ (a) The free-body diagram of the pulley is shown in Fig. 5.34b. For equilibrium,

$$\sum F_y = 0 \qquad -40 + R - 15 = 0 \qquad R = 55 \text{ lb}$$

(b) The free-body diagram of the shaft is shown in Fig. 5.34c, and R_a and R_b are the forces exerted by the bearings on the shaft. As an action-reaction force effect, the force R has opposite senses when drawn in Figs. 5.34b and 5.34c. For equilibrium of the shaft,

$$\sum M_a = 0 \qquad -55(14) + R_b(10) = 0 \qquad R_b = 77 \text{ lb}$$
$$\sum F_y = 0 \qquad -R_a + R_b - 55 = 0 \qquad R_a = 22 \text{ lb}$$

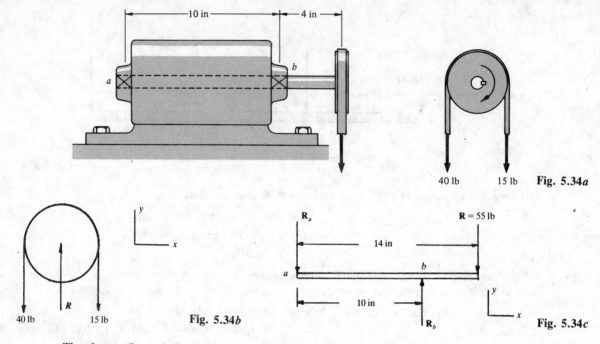

Fig. 5.34a

Fig. 5.34b

Fig. 5.34c

The forces R_a and R_b exerted by the shaft on the bearings have senses opposite those given above. $R_a = 22$ lb acts upward on bearing a, and $R_b = 77$ lb acts downward on bearing b.

(c) $\mathbf{R} = R\mathbf{j} = 55\mathbf{j}$ lb

$\mathbf{R}_a = R_a\mathbf{j} = 22\mathbf{j}$ lb $\qquad \mathbf{R}_b = -R_b\mathbf{j} = -77\mathbf{j}$ lb

5.35 (a) Find the magnitude, direction, sense, and location of the resultant of the reaction forces which act on the beam in Prob. 5.29.

(b) Using the results in part (a) and Prob. 5.29, show that the beam is in equilibrium.

▌ (a) The free-body diagram of the beam is shown in Fig. 5.35a. The reaction forces R_a and R_b are found first. For equilibrium,

$$\Sigma M_a = 0 \qquad -4,800(100) - 6,500(250) + R_b(400) - 3,400(850) = 0 \qquad R_b = 12,500 \text{ N}$$
$$\Sigma F_y = 0 \qquad R_a - 4,800 - 6,500 + R_b - 3,400 = 0 \qquad R_a = 2,200 \text{ N}$$

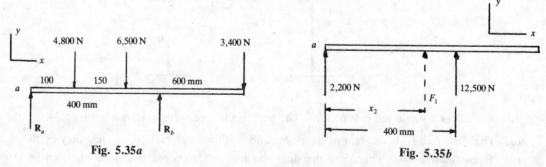

Fig. 5.35a

Fig. 5.35b

Figure 5.35b shows the beam acted on by the two reaction forces R_a and R_b, and F_1 is the resultant of these two forces. The magnitude and location of F_1 are found as

$$F_1 = \Sigma F_i \qquad F_1 = 2,200 + 12,500 = 14,700 \text{ N}$$
$$\Sigma M_a = F_1 x_2 \qquad 12,500(400) = F_1 x_2 = 14,700 x_2 \qquad x_2 = 340 \text{ mm}$$

(b) Figure 5.35c shows the beam acted on by two resultant forces. The downward force of 14,700 N is the resultant of the applied forces which act on the beam. The upward force of 14,700 N is the resultant of the reaction forces which act on the beam. Since these two forces are equal, opposite in sense, and collinear, the resultant force on the beam is zero. This is the necessary condition for equilibrium of the beam.

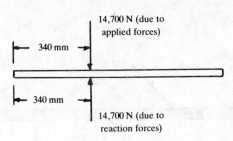

14,700 N (due to applied forces)

340 mm

340 mm

14,700 N (due to reaction forces)

Fig. 5.35c

5.4 THE GENERAL TWO-DIMENSIONAL FORCE SYSTEM—RESULTANT FORCE AND EQUILIBRIUM REQUIREMENTS

5.36 Describe the general two-dimensional force system and give the general form for the resultant force.

❚ As the name implies, this is the most unrestricted form of force system in which all the forces lie in the same plane. Characteristically, the lines of action of the forces in such a system have arbitrary directions and hence do not necessarily intersect at a single point. Figure 5.36 shows a typical view of a general two-dimensional force system.

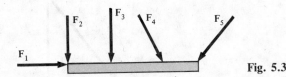

Fig. 5.36

The resultant force in the general two-dimensional force system is of the same form as the resultant force in the concurrent two-dimensional force system. If this resultant force is designated **F**, with the components F_x and F_y, then

$$F_x = \sum_{i=1}^{i=n} F_{ix} = F_{1x} + F_{2x} + F_{3x} + \cdots + F_{nx}$$

$$F_y = \sum_{i=1}^{i=n} F_{iy} = F_{1y} + F_{2y} + F_{3y} + \cdots F_{ny}$$

where n is the number of forces which act in the system. For brevity, the above equations may be written as $\Sigma F_x = 0$ and $\Sigma F_y = 0$.

As in the case of the parallel force system, a complete definition of the resultant force requires a description of its *location*, as well as its magnitude, direction, and sense. This location is found by applying the principle that the moment of the resultant force, with respect to some arbitrary reference point, must be the same as the moment of the original system of forces with respect to the same point. In formal vector notation, the resultant force has the form

$$\mathbf{F} = F_x \mathbf{i} + F_y \mathbf{j}$$

where F_x and F_y are given by the equations above.

5.37 The beam shown in Fig. 5.37a is supported by a hinge and a roller. Find the resultant force of the *applied* loads on the beam.

❚ The x and y axes in the figure establish the positive senses of force. The components of the resultant force are then

$$F_x = \Sigma F_{ix} \qquad F_x = -750 \cos 45° + 1{,}500 \cos 60° - 2{,}000 = -1{,}780 \, \text{lb}$$
$$F_y = \Sigma F_{iy} \qquad F_y = -800 - 750 \sin 45° + 500 - 1{,}500 \sin 60° = -2{,}130 \, \text{lb}$$

The magnitude of the resultant force is

$$F = \sqrt{F_x^2 + F_y^2} = \sqrt{(-1{,}780)^2 + (-2{,}130)^2} = 2{,}780 \, \text{lb}$$

The direction θ is found from

$$\tan \theta = \frac{2{,}130}{1{,}780} = 1.20 \qquad \theta = 50.2°$$

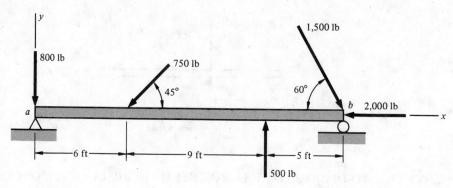

The orientation of the resultant force of the applied loads on the beam is shown in Fig. 5.37b. The resultant force may be expressed in vector notation as

$$\mathbf{F} = F_x\mathbf{i} + F_y\mathbf{j} = -1{,}780\mathbf{i} - 2{,}130\mathbf{j} \quad \text{lb}$$

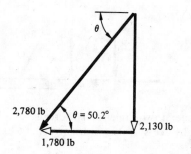

Fig. 5.37b

The system is redrawn in Fig. 5.37c, and x_1 defines the location of the resultant force. The left end of the beam is chosen as the reference point for moments, and counterclockwise moments are considered positive. The moment M_a due to the original system of applied forces is

$$M_a = -(750 \sin 45°)6 + 500(15) - (1{,}500 \sin 60°)20 = -21{,}700 \text{ ft} \cdot \text{lb}$$

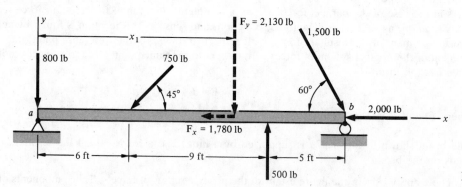

Fig. 5.37c

It may be observed that several forces in Fig. 5.37c have directions which pass through the reference point, and thus these forces produce no moment. The moment M of the resultant force about a is

$$M = -2{,}130x_1$$

The location x_1 is then found from

$$M = M_a \qquad -2{,}130x_1 = -21{,}700 \qquad x_1 = 10.2 \text{ ft}$$

It should be emphasized that a free-body diagram is not required in this problem, since the problem statement calls for only the resultant force of the applied loads that act on the beam.

5.38   State the requirements for equilibrium of a body acted on by a general two-dimensional force system.

❚   For equilibrium of the general two-dimensional force system, the resultant force of all the forces acting in the system must be identically zero.   As before, a free-body diagram of the system is first drawn, which shows both the applied forces in the problem and the unknown reaction forces.   The formal requirements for static equilibrium of the general two-dimensional force system are:

   1.   The algebraic summation of the forces in *two perpendicular directions* is zero.
   2.   The algebraic summation of the moments *about any reference* point is zero.

The above three requirements may be written in equation form as

$$\sum F_x = 0 \qquad \sum F_y = 0 \qquad \sum M = 0$$

In formal vector notation, the above requirements for equilibrium may be written as

$$\mathbf{F} = F_x \mathbf{i} + F_y \mathbf{j} = 0 \qquad \mathbf{M} = 0$$

5.39   Draw the free-body diagram for the beam shown in Fig. 5.37a and find the reaction forces of the foundation on the beam.   The beam is weightless.   Express all results in both scalar and vector forms.

❚   The free-body diagram is shown in Fig. 5.39.   The two unknown force components R_{ax} and R_{ay} at the hinge pin are assumed to act in the positive coordinate senses.   Since the roller can exert only a compressive force, the reaction force R_b is shown in its actual sense.   The hinge pin is taken as the reference for moments, since R_{ax} and R_{ay} act through this point and contribute zero moment.   Counterclockwise moments are considered positive.   For moment equilibrium,

$$\sum M_a = 0 \qquad -(750 \sin 45°)(6) + 500(15) - (1,500 \sin 60°)(20) + R_b(20) = 0 \qquad R_b = 1,080 \text{ lb}$$

$$\mathbf{R}_b = R_b \mathbf{j} = 1,080 \mathbf{j} \qquad \text{lb}$$

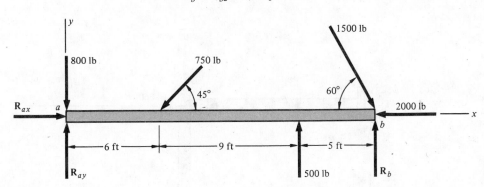

Fig. 5.39

For force equilibrium in the y direction,

$$\sum F_y = 0 \qquad -800 + R_{ay} - 750 \sin 45° + 500 - 1,500 \sin 60° + R_b = 0$$
$$-800 + R_{ay} - 750 \sin 45° + 500 - 1,500 \sin 60° + 1,080 = 0 \qquad R_{ay} = 1,050 \text{ lb}$$

For force equilibrium in the x direction,
$$\sum F_x = 0 \qquad R_{ax} - 750 \cos 45° + 1,500 \cos 60° - 2,000 = 0 \qquad R_{ax} = 1,780 \text{ lb}$$
$$\mathbf{R}_a = R_{ax} \mathbf{i} + R_{ay} \mathbf{j} = 1,780 \mathbf{i} + 1,050 \mathbf{j} \qquad \text{lb}$$

The resultant force R_a at hinge a is
$$R_a = \sqrt{R_{ax}^2 + R_{ay}^2} = \sqrt{1,780^2 + 1,050^2} = 2,070 \text{ lb}$$

5.40   Show that the resultant force of the system of reaction forces R_{ax}, R_{ay}, and R_b in Prob. 5.39 is equal and opposite to the resultant force of the applied loads acting on the beam of Prob. 5.37.

❚   The vector sum of forces R_{ax}, R_{ay}, and R_b is shown in Fig. 5.40.   The x component of the resultant of these three forces is

$$F_x = R_{ax} = 1,780 \text{ lb}$$

The y component is

$$F_y = R_{ay} + R_b = 1,050 + 1,080 = 2,130 \text{ lb}$$

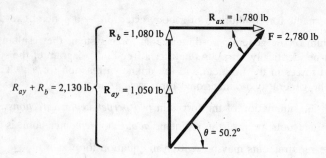

Fig. 5.40

The resultant F of the forces R_{ax}, R_{ay}, and R_b is then

$$F = \sqrt{F_x^2 + F_y^2} = \sqrt{1,780^2 + 2,130^2} = 2,780 \text{ lb}$$

The direction θ of this force is

$$\tan \theta = \frac{2,130}{1,780} = 1.20 \qquad \theta = 50.2°$$

The above results are now compared with Fig. 5.37b, and it may be seen that the resultant force of the applied loads is equal and opposite to the resultant force of the reaction loads which act on the beam. The *net* resultant force acting on the beam is thus zero, which is the expected condition of static equilibrium.

5.41 Find the magnitude, direction, and sense of the reaction forces at a and b for the machine part shown in Fig. 5.41a. Express all results in both scalar and vector forms.

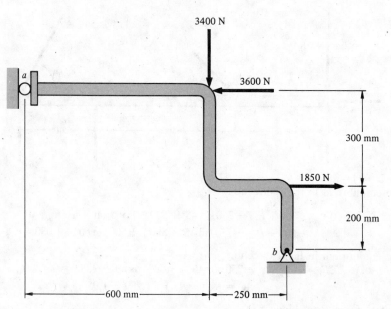

Fig. 5.41a

❙ The free-body diagram of the part is shown in Fig. 5.41b. For equilibrium,

$$\Sigma M_b = 0 \qquad -R_a(500) + 3,400(250) + 3,600(500) - 1,850(200) = 0$$
$$R_a = 4,560 \text{ N} \qquad \mathbf{R}_a = R_a \mathbf{i} = 4,560\mathbf{i} \qquad \text{N}$$
$$\Sigma F_x = 0 \qquad R_a - 3,600 + 1,850 \qquad -R_{bx} = 0 \qquad R_{bx} = 2,810 \text{ N}$$
$$\Sigma F_y = 0 \qquad -3,400 + R_{by} = 0 \qquad R_{by} = 3,400 \text{ N}$$
$$\mathbf{R}_b = -R_{bx}\mathbf{i} + R_{by}\mathbf{j} = -2,810\mathbf{i} + 3,400\mathbf{j} \qquad \text{N}$$
$$R_b = \sqrt{R_{bx}^2 + R_{by}^2} = \sqrt{2,810^2 + 3,400^2} = 4,410 \text{ N}$$

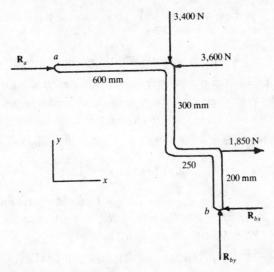

Fig. 5.41b

5.42 (a) Draw the free-body diagram of the curved bracket arrangement shown in Fig. 5.42a, and find the force effects exerted by the wall on the bracket. Express the results in both scalar and vector forms.

(b) Compare these results with the solution to Prob. 3.36.

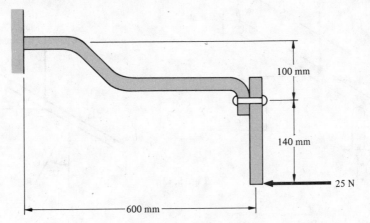

Fig. 5.42a

▌ (a) The free-body diagram is shown in Fig. 5.42b. For equilibrium,

$$\Sigma M_a = 0 \qquad -25\left(\frac{240}{1,000}\right) + M_a = 0$$

$$M_a = 6 \text{ N} \cdot \text{m} \qquad \mathbf{M} = M_a \mathbf{k} = 6\mathbf{k} \qquad \text{N} \cdot \text{m}$$

$$\Sigma F_x = 0 \qquad R_{ax} - 25 = 0 \qquad R_{ax} = 25 \text{ N}$$

$$\Sigma F_y = 0 \qquad R_{ay} = 0$$

$$\mathbf{R}_a = R_{ax}\mathbf{i} + R_{ay}\mathbf{j} = R_{ax}\mathbf{i} = 25\mathbf{i} \qquad \text{N}$$

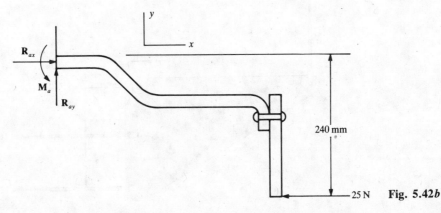

Fig. 5.42b

(*b*) The results in part (*a*) are equal and opposite to the force effects exerted *on* the wall *by* the bracket, found as the solution to Prob. 3.36.

5.43 The smooth rod in Fig. 5.43*a* has a weight $W = 18\,\text{lb}$, and $\theta = 32°$.

(*a*) Find the magnitude of the compressive contact force exerted on the rod by the fixed post.

(*b*) Find the *x* and *y* components and the magnitude of the hinge pin force at *b*.

(*c*) Express the results in parts (*a*) and (*b*) in formal vector notation.

❚ (*a*) The free-body diagram is shown in Fig. 5.43*b*. For equilibrium,

$$\Sigma M_b = 0 \qquad 18(12.5\cos 32°) - R_a\left[2(12.5) - \frac{7}{\cos 32°}\right] = 0 \qquad R_a = 11.4\,\text{lb}$$

(*b*) $\Sigma F_x = 0 \qquad -R_a \sin 32° + R_{bx} = 0 \qquad R_{bx} = R_a \sin 32° = 11.4 \sin 32° = 6.04\,\text{lb}$

$\Sigma F_y = 0 \qquad R_a \cos 32° - 18 + R_{by} = 0 \qquad R_{by} = 18 - 11.4 \cos 32° = 8.33\,\text{lb}$

$R_b = \sqrt{R_{bx}^2 + R_{by}^2} = \sqrt{6.04^2 + 8.33^2} = 10.3\,\text{lb}$

(*c*) $\mathbf{R}_a = (-R_a \sin 32°)\mathbf{i} + (R_a \cos 32°)\mathbf{j} = (-11.4 \sin 32°)\mathbf{i} + (11.4 \cos 32°)\mathbf{j} = -6.04\mathbf{i} + 9.67\mathbf{j} \qquad \text{lb}$

$\mathbf{R}_b = R_{bx}\mathbf{i} + R_{by}\mathbf{j} = 6.04\mathbf{i} + 8.33\mathbf{j} \qquad \text{lb}$

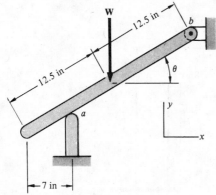

Fig. 5.43*a*

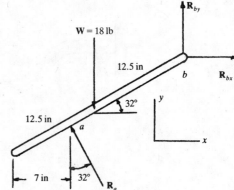

Fig. 5.43*b*

5.44 (*a*) Find the force *P* which must be applied at point *a* to keep the hinged plate in Fig. 5.44*a* in equilibrium in the position shown.

(*b*) Find the magnitude, direction, and sense of the hinge pin force at *b*.

❚ (*a*) The free-body diagram of the plate is shown in Fig. 5.44*b*. For equilibrium of the plate,

$$\Sigma M_b = 0 \qquad -P(12) + 16(3) + 12(6.25) = 0 \qquad P = 10.3\,\text{kip}$$

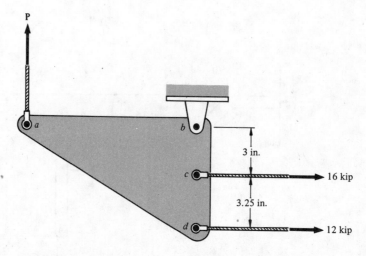

Fig. 5.44*a*

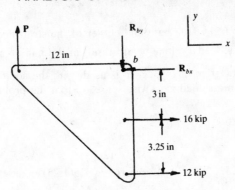

Fig. 5.44b

(b) The x and y components of the pin force at b are found from

$$\Sigma F_x = 0 \qquad -R_{bx} + 16 + 12 = 0 \qquad R_{bx} = 28 \text{ kip}$$
$$\Sigma F_y = 0 \qquad P - R_{by} = 0 \qquad R_{by} = P = 10.3 \text{ kip}$$

The magnitude of R_b is given by

$$R_b = \sqrt{R_{bx}^2 + R_{by}^2} = \sqrt{28^2 + 10.3^2} = 29.8 \text{ kip}$$

The direction θ, with respect to the horizontal axis, of R_b is found as

$$\tan \theta = \frac{R_{by}}{R_{bx}} = \frac{10.3}{28} \qquad \theta = 20.2°$$

The sense of R_b is down and to the left.

5.45 The hinged plate in Prob. 5.44 is to be kept in equilibrium by the application of a couple acting in the plane of the element. For this case, no cable force is exerted at location a.

(a) Find the magnitude and sense of this applied couple, and the magnitude of the hinge pin force at b.

(b) Compare the hinge pin force with the hinge pin force found in Prob. 5.44.

I (a) The free-body diagram of the plate is shown in Fig. 5.45. For equilibrium of the plate,

$$\Sigma M_b = 0 \qquad -M + 16(3) + 12(6.25) = 0 \qquad M = 123 \text{ kip} \cdot \text{in}$$
$$\Sigma F_x = 0 \qquad -R_{bx} + 16 + 12 = 0 \qquad R_{bx} = 28 \text{ kip}$$
$$\Sigma F_y = 0 \qquad R_{by} = 0 \qquad R_b = R_{bx} = 28 \text{ kip}$$

It may be observed that R_{bx} has the same value as R_{bx} in Prob. 5.44.

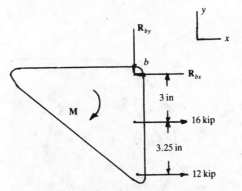

Fig. 5.45

(b) The hinge pin force of 28 kip, when the plate is acted on by the couple, is smaller than the pin force of 29.8 kip, when the plate is acted on by the applied force at point a.

5.46 The loading on the welded strip in Prob. 5.32 is modified, as indicated below. In each case, find the required solution.

(a) Find the couple, acting in the plane of the figure, which would have to be applied to the strip to make the two weld forces equal in magnitude.

(b) If the load P is applied at the left-hand corner of the strip, find the two weld forces.

(c) What position of the load P alone would result in the minimum values of the weld forces?

▌ (a) The free-body diagram of Fig. 5.32b is used, with the addition of a couple of magnitude M acting on the strip with an assumed clockwise sense. From the problem statement,

$$F_a = F_b$$

For equilibrium,

$$\Sigma F_y = 0 \qquad F_a - 12.5 + F_b = 0 \qquad 2F_b = 12.5 \qquad F_b = 6.25 \text{ kip}$$
$$\Sigma M_0 = 0 \qquad -M + F_b(6) - 12.5(1.25) \qquad -M + 6.25(6) - 12.5(1.25) = 0$$
$$M = 21.9 \text{ kip} \cdot \text{in}$$

Since M is positive, the assumed sense is the actual sense.

(b) The free-body diagram of the strip is shown in Fig. 5.46. For equilibrium,

$$\Sigma M_0 = 0 \qquad F_b(6) = 0 \qquad F_b = 0$$
$$\Sigma F_y = 0 \qquad F_a - 12.5 = 0 \qquad F_a = 12.5 \text{ kip}$$

(c) The minimum values of the weld forces will occur when the load $P = 12.5$ kip acts at the midpoint of the width of the strip. For this case,

$$F_a = F_b = \tfrac{1}{2}P = \tfrac{1}{2}(12.5) = 6.25 \text{ kip}$$

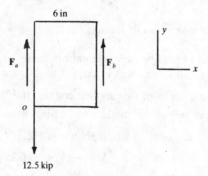

Fig. 5.46

5.47 In Fig. 5.47a belt aa drives a pulley which has an angular velocity of 1,250 r/min. Power is taken out through belts bb and cc.

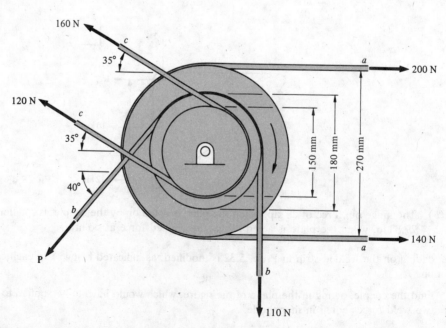

Fig. 5.47a

(a) Find the required belt tensile force P for static equilibrium of the pulley.

(b) Find the magnitude, direction, and sense of the force exerted by the pulley on the hinge pin.

(c) Express the results in parts (a) and (b) in formal vector notation.

▎ (a) The free-body diagram of the pulley is shown in Fig. 5.47b. For moment equilibrium of the pulley,

$$\Sigma M = 0 \qquad -200(135) + 140(135) - 110(90) + P(90) + 160(75) - 120(75) = 0 \qquad P = 167 \text{ N}$$

(b) For force equilibrium of the pulley,

$$\Sigma F_x = 0 \qquad R_x + 200 + 140 - 160 \cos 35° - 120 \cos 35° - 167 \cos 40° = 0 \qquad R_x = 17.3 \text{ N}$$
$$\Sigma F_y = 0 \qquad R_y - 110 + 160 \sin 35° + 120 \sin 35° - 167 \sin 40° = 0 \qquad R_y = 56.7 \text{ N}$$
$$R = \sqrt{R_x^2 + R_y^2} = \sqrt{17.3^2 + 56.7^2} = 59.2 \text{ N}$$

It may be seen that the speed of the pulley does not enter into the problem. The force R is the force exerted *on* the pulley *by* the hinge pin. From Newton's third law, the force exerted *on* the hinge pin *by* the pulley has a sense opposite to that shown in Fig. 5.47b. The force exerted on the hinge pin is shown in Fig. 5.47c. The direction of this force is found from

$$\tan \theta = \frac{56.7}{17.3} \qquad \theta = 73.0°$$

(c) $\mathbf{P} = (-P \cos 40°)\mathbf{i} - (P \sin 40°)\mathbf{j} = (-167 \cos 40°)\mathbf{i} - (167 \sin 40°)\mathbf{j} = -128\mathbf{i} - 107\mathbf{j} \qquad$ N

From Fig. 5.47c,

$$\mathbf{R} = -17.3\mathbf{i} - 56.7\mathbf{j} \qquad \text{N}$$

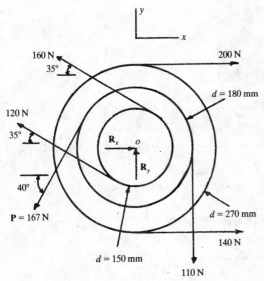

Fig. 5.47b

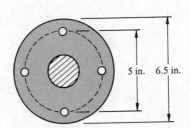

Fig. 5.47c

5.48 Figure 5.48a shows a flanged bolt coupling which transmits a torque of 800 ft · lb. Assuming that the bolts share the load equally, find the magnitude of the force acting on each bolt.

▎ The directions of the four bolt forces are assumed to be perpendicular to the radius of the coupling. The free-body diagram is shown in Fig. 5.48b. For equilibrium,

$$\Sigma M_0 = 0 \qquad 4(P)2.5 - 800(12) \text{ in} \cdot \text{lb} = 0 \qquad P = 960 \text{ lb}$$

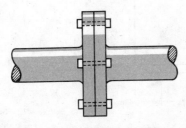

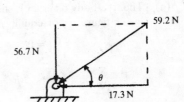

Fig. 5.48a

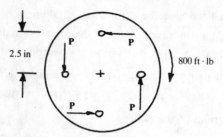

Fig. 5.48b

5.49 (*a*) The bracket shown in Fig. 5.49*a* is contained in a loose-fitting smooth guide. Find the reaction forces which act on the bracket at *a* and *b*.

(*b*) Express the results in part (*a*) in vector forms.

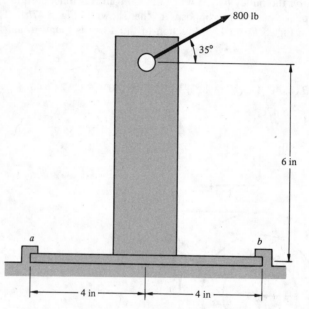

Fig. 5.49a

▌ (*a*) The free-body diagram is shown in Fig. 5.49*b*. Since the bracket is loose-fitting, $R_{ax} = 0$, as shown in the figure. For equilibrium,

$$\Sigma M_b = 0 \qquad -(800 \cos 35°)6 - (800 \sin 35°)4 + R_{ay}(8) = 0 \qquad R_{ay} = 721 \text{ lb}$$

$$\Sigma F_x = 0 \qquad 800 \cos 35° - R_{bx} = 0 \qquad R_{bx} = 655 \text{ lb}$$

$$\Sigma F_y = 0 \qquad -R_{ay} + 800 \sin 35° + R_{by} = 0 \qquad R_{by} = 262 \text{ lb}$$

$$R_b = \sqrt{R_{bx}^2 + R_{by}^2} = \sqrt{655^2 + 262^2} = 705 \text{ lb}$$

(*b*) $\mathbf{R}_a = -R_{ay}\mathbf{j} = -721\mathbf{j}$ lb

$\mathbf{R}_b = -R_{bx}\mathbf{i} + R_{by}\mathbf{j} = -655\mathbf{i} + 262\mathbf{j}$ lb

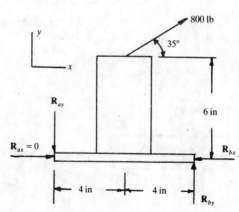

Fig. 5.49b

5.50 Do the same as in Prob. 5.49 if, in addition to the 800-lb force, a 3,000 in · lb clockwise couple, in the plane of the illustration, acts on the bracket.

▌ (a) The free-body diagram is the same as that shown in Fig. 5.49b, with the addition of a 3,000 in · lb clockwise couple. For equilibrium,

$$\Sigma M_b = 0 \qquad -(800\cos 35°)6 - (800\sin 35°)4 - 3,000 + R_{ay}(8) = 0 \qquad R_{ay} = 1,100 \text{ lb}$$
$$\Sigma F_x = 0 \qquad 800\cos 35° - R_{bx} = 0 \qquad R_{bx} = 655 \text{ lb}$$
$$\Sigma F_y = 0 \qquad -R_{ay} + 800\sin 35° + R_{by} = 0 \qquad R_{by} = 641 \text{ lb}$$
$$R_b = \sqrt{R_{bx}^2 + R_{by}^2} = \sqrt{655^2 + 641^2} = 916 \text{ lb}$$

(b) $\mathbf{R}_a = -R_{ay}\mathbf{j} = -1,100\mathbf{j} \qquad$ lb

$\mathbf{R}_b = -R_{bx}\mathbf{i} + R_{by}\mathbf{j} = -655\mathbf{i} + 641\mathbf{j} \qquad$ lb

5.51 Find the magnitude, direction, and sense of the couple which would have to be applied to the bracket in Prob. 5.49 if the reaction force at a is to be zero.

▌ (a) The free-body diagram is the same as that shown in Fig. 5.49b, with the addition of an assumed clockwise couple of magnitude M_0. For equilibrium, with $R_a = 0$,

$$\Sigma M_b = 0 \qquad -(800\cos 35°)6 - (800\sin 35°)4 - M_0 = 0$$
$$M_0 = -5,770 \text{ in · lb}$$

The minus sign indicates that the actual sense of M_0 is counterclockwise.

$$\Sigma F_x = 0 \qquad 800\cos 35° - R_{bx} = 0 \qquad R_{bx} = 655 \text{ lb}$$
$$\Sigma F_y = 0 \qquad 800\sin 35° + R_{by} = 0 \qquad R_{by} = -459 \text{ lb}$$
$$R_b = \sqrt{655^2 + (-459)^2} = 800 \text{ lb}$$

The negative result for R_{by} indicates that the actual sense of this force is downward in Fig. 5.49b. The direction of R_b is given by

$$\tan\theta = \frac{459}{655} \qquad \theta = 35.0°$$

θ is the angle with respect to the x axis, and the sense of R_b is down and to the left. It may be seen that the reaction force R_b and the applied force of 800 lb are equal, opposite, and parallel. These two forces form a couple which is equal and opposite to the applied couple.

(b) The above results may be written in vector notation as

$$\mathbf{R}_a = 0 \qquad \mathbf{R}_b = -R_{bx}\mathbf{i} + R_{by}\mathbf{j} = -655\mathbf{i} - 459\mathbf{j} \qquad \text{lb}$$
$$\mathbf{M} = 5,770\mathbf{k} \qquad \text{in · lb}$$

5.52 Solve Prob. 5.18 by using $\Sigma M_b = 0$ to find R_a directly.

▌ Figure 5.18b is used, and the force R_a is moved along its line of action and is assumed to act at the center of the ring. For equilibrium,

$$\Sigma M_b = 0 \qquad 49.1(150\cos 48°) - (R_a\sin 35°)(150\cos 48°) - (R_a\cos 35°)(150\sin 48°) = 0$$
$$49.1(150)\cos 48° - R_a(150)(\sin 35°\cos 48° + \cos 35°\sin 48°) = 0$$

$$R_a(150)\sin(35° + 48°) = 49.1(150)\cos 48°$$
$$(R_a\sin 83°)\cancel{150} = 49.1(\cancel{150})\cos 48°$$
$$R_a = 33.1 \text{ N}$$

It may be observed that the term $(R_a\sin 83°)$ is the component of force R_a which is perpendicular to force R_b. If the components of R_a along and perpendicular to the line of action of force R_b had been formed, the equation of moment equilibrium about point b could have been written directly as

$$-(R_a\sin 83°)150 + 49.1(150\cos 48°) = 0 \qquad R_a = 33.1 \text{ N}$$

R_b may then be found by using $\Sigma F_x = 0$ or $\Sigma F_y = 0$.

5.53 Solve Prob. 5.16 by using $\Sigma M_b = 0$ to directly find force P.

▌ Using the free-body diagram in Fig. 5.16b,

$$\Sigma M_b = 0 \qquad 14,700(690\sin 38.5°) - P(540) = 0 \qquad P = 11,700 \text{ N}$$

The force R may then be found by using either $\Sigma F_x = 0$ or $\Sigma F_y = 0$.

5.54 Do the same as in Prob. 5.16 if the horizontal force is applied at point *a* on the top of the roller.

▌ (*a*) The free-body diagram of the roller is shown in Fig. 5.54. For moment equilibrium of the roller,

$$\Sigma M_b = 0 \qquad 14{,}700(430) - P(690 + 540) = 0 \qquad P = 5{,}140 \text{ N}$$

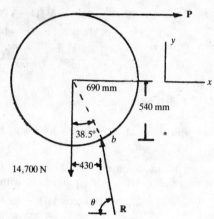

690 mm

540 mm

38.5°

14,700 N ⊢430⊣

θ

R

Fig. 5.54

(*b*) For force equilibrium of the roller,

$$\Sigma F_x = 0 \qquad P - R \cos \theta = 0 \qquad 5{,}140 - R \cos \theta = 0$$
$$\Sigma F_y = 0 \qquad -14{,}700 + R \sin \theta = 0$$

Dividing the second equation by the first equation yields

$$\frac{\sin \theta}{\cos \theta} = \tan \theta = \frac{14{,}700}{5{,}140} = 2.86 \qquad \theta = 70.7°$$

The magnitude of R is then found from

$$-14{,}700 + R \sin 70.7° = 0 \qquad R = 15{,}600 \text{ N}$$

(*c*) In formal vector notation, the results for P and R are

$$\mathbf{P} = P\mathbf{i} = 5{,}140\mathbf{i} \quad \text{N}$$
$$\mathbf{R} = (-R \cos \theta)\mathbf{i} + (R \sin \theta)\mathbf{j} = (-15{,}600 \cos 70.7°)\mathbf{i} + (15{,}600 \sin 70.7°)\mathbf{j} = -5{,}160\mathbf{i} + 14{,}700\mathbf{j} \quad \text{N}$$

When comparing the results for $\mathbf{P}$ and $\mathbf{R}$, with $5{,}140 \approx 5{,}160$, it may be seen that the x component of $\mathbf{R}$ is equal and opposite to $\mathbf{P}$. The y component of $\mathbf{R}$ is equal to the weight force of the roller.

5.55 The rod of weight W in Fig. 5.55a is smooth.

(*a*) Find the reaction forces exerted on the rod at *a* and *b*, in terms of W and θ.
(*b*) Find the angle θ, in terms of x_1 and l, which defines the equilibrium position.
(*c*) Find the numerical values for parts *a* and *b*, if $x_1 = 100$ mm, $l = 350$ mm, and $m = W/g = 1.8$ kg.

▌ (*a*) The free-body diagram of the rod is shown in Fig. 5.55b. For force equilibrium of the rod,

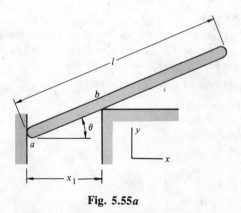

l

b

θ

a

x_1

Fig. 5.55a

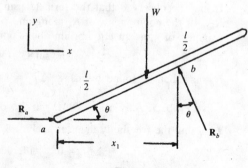

W

$\frac{l}{2}$

b

$\frac{l}{2}$

$\mathbf{R}_a$

θ

a

θ

$\mathbf{R}_b$

x_1

Fig. 5.55b

$$\Sigma F_y = 0 \quad -W + R_b \cos\theta = 0 \quad R_b = \frac{W}{\cos\theta}$$

$$\Sigma F_x = 0 \quad R_a - R_b \sin\theta = 0$$

$$R_a = R_b \sin\theta = \frac{W}{\cos\theta}\sin\theta = W\tan\theta$$

(*b*) For moment equilibrium of the rod,

$$\Sigma M_a = 0 \quad -W\left(\frac{l}{2}\cos\theta\right) + R_b\frac{x_1}{\cos\theta} = 0 \quad \frac{Wl}{2}\cos\theta = R_b\frac{x_1}{\cos\theta} \quad R_b = \frac{Wl}{2x_1}\cos^2\theta$$

Using the result for R_b found in part (*a*),

$$R_b = \frac{W}{\cos\theta} = \frac{Wl}{2x_1}\cos^2\theta \quad \cos^3\theta = \frac{2x_1}{l} \quad \cos\theta = \left(\frac{2x_1}{l}\right)^{1/3}$$

(*c*) Using $x_1 = 100$ mm and $l = 350$ mm,

$$\cos\theta = \left[\frac{2(100)}{350}\right]^{1/3} \quad \theta = 33.9°$$

$$R_a = W\tan\theta = 1.8(9.81)\tan 33.9° = 11.9\text{ N} \quad R_b = \frac{W}{\cos\theta} = \frac{1.8(9.81)}{\cos 33.9°} = 21.3\text{ N}$$

5.56 The beam shown in Fig. 5.56*a* is acted on by only applied couples. Find the reaction forces of the hinge pin and roller on the beam. The beam is assumed to be weightless.

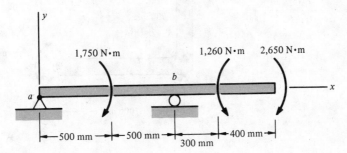

Fig. 5.56*a*

I The free-body diagram of the beam is shown in Fig. 5.56*b*. The reaction components R_{ax} and R_{ay} are assumed to act in the positive coordinate senses. For equilibrium,

$$\Sigma M_a = 0 \quad -1,750 + R_b(1,000\text{ mm})\left(\frac{1\text{ m}}{1,000\text{ mm}}\right) + 1,260 - 2,650 = 0$$

$$R_b = 3,140\text{ N}$$

$$\Sigma F_y = 0 \quad R_{ay} + R_b = R_{ay} + 3,140 = 0 \quad R_{ay} = -3,140\text{ N}$$

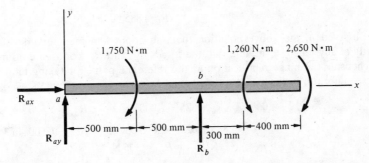

Fig. 5.56*b*

The minus sign in the above result indicates that the actual sense of R_{ay} is opposite to that assumed in Fig. 5.56*b*.

For equilibrium in the *x* direction,

$$\Sigma F_x = 0 \quad R_{ax} = 0$$

It may be observed that R_{ay} and R_b are equal, opposite, and parallel forces. They thus form a *reaction couple* M_R, with magnitude

$$M_R = 3{,}140(1{,}000 \text{ mm})\left(\frac{1 \text{ m}}{1{,}000 \text{ mm}}\right) = 3{,}140 \text{ N} \cdot \text{m}$$

It can be shown that a *system acted on by only applied couples will have a reaction effect which is expressible as a couple.*

5.57 The rigid crank arm in Fig. 5.57a is supported by a hinge and a smooth roller. In addition to the applied forces shown, there is an applied moment of 80 N · m acting on the right end. Draw the free-body diagram and find the reaction forces of the foundation on the crank.

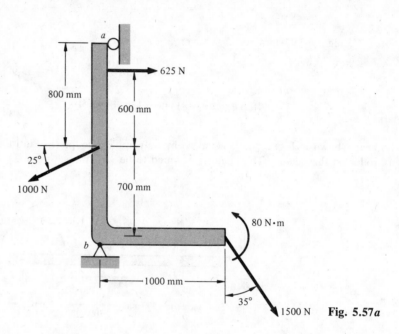

Fig. 5.57a

The free-body diagram is shown in Fig. 5.57b. Since the smooth roller cannot support a tensile force, the reaction force R_a is shown in its actual sense. The unknown components R_{bx} and R_{by} of the reaction force of the foundation on the crank at the hinge pin are assumed to act in the positive coordinate senses.
    For moment equilibrium about the hinge pin

$$\sum M_b = 0$$

$$R_a(1{,}500) - 625(1{,}300) + (1{,}000 \cos 25°)(700) - (1{,}500 \cos 35°)(1{,}000) + 80 \text{ N} \cdot \text{m}\left(\frac{1{,}000 \text{ mm}}{1 \text{ m}}\right) = 0$$

$$R_a = 885 \text{ N}$$

It may be observed in the above equation that the applied moment enters the moment equation directly, independently of any length term, and that the forces R_{bx} and R_{by} contribute no moment effect. This was the reason for choosing the hinge pin location as the reference point for taking moments.
    For force equilibrium in the x and y directions,

$$\sum F_x = 0 \qquad -R_a + 625 - 1{,}000 \cos 25° + R_{bx} + 1{,}500 \sin 35° = 0$$

$$-885 + 625 - 1{,}000 \cos 25° + R_{bx} + 1{,}500 \sin 35° = 0$$

$$R_{bx} = 306 \text{ N}$$

$$\sum F_y = 0 \qquad -1{,}000 \sin 25° + R_{by} - 1{,}500 \cos 35° = 0 \qquad R_{by} = 1{,}650 \text{ N}$$

It should be noted that the applied moment of 80 N · m does not appear in the above two force equations. Finally, the resultant force R_b which acts at the hinge pin is

$$R_b = \sqrt{R_{bx}^2 + R_{by}^2} = 1{,}680 \text{ N}$$

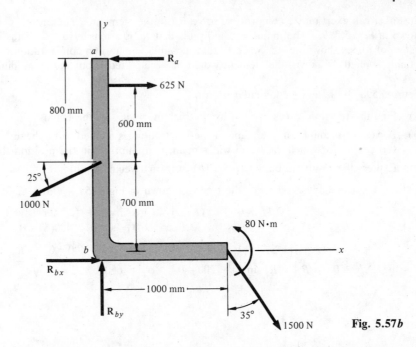

Fig. 5.57b

5.58 A hinged structural element is shown in Fig. 5.58a. It is claimed that the element is in equilibrium with the set of forces shown. How would one verify that the system shown is a possible equilibrium configuration?

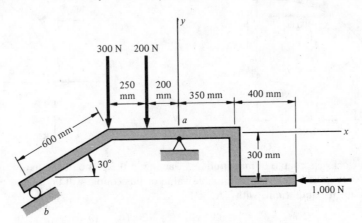

Fig. 5.58a

I The free-body diagram of the element is shown in Fig. 5.58b. The moments are summed about the hinge pin, with the result

$$\sum M_a = 0 \qquad -R_b(600 + 390) + 300(450) + 200(200) - 1,000(300) = 0 \qquad R_b = -126\,\text{N}$$

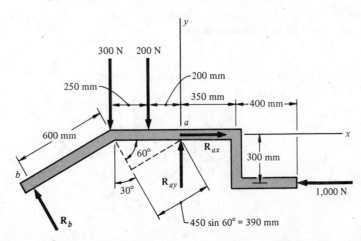

Fig. 5.58b

The roller can exert only a compressive force R_b on the element. Therefore, R_b must be positive in the sense shown in Fig. 5.58b. The minus sign indicates that R_b is a tensile force. Thus it may be concluded that the system of forces shown in Fig. 5.58a is *not* a possible equilibrium configuration. If this system of forces were actually applied as shown, the element would be accelerated about the hinge pin in a clockwise sense.

5.59   Figure 5.59a shows a hinged circular beam.

(a)   Find the reaction forces exerted by the ball and hinge on the beam.

(b)   A clockwise couple, in the plane of the illustration, is applied to the beam. Find the maximum value of this couple for which the beam will remain in equilibrium in the position shown in the figure.

(c)   Express the results in parts (a) and (b) in formal vector notation.

▌ (a)   The free-body diagram of the beam is shown in Fig. 5.59b. For equilibrium of the beam,

$$\Sigma M_a = 0 \qquad -(R_b \cos 40°)(6 \sin 40°) - (R_b \sin 40°)6(1 - \cos 40°)$$
$$+ 300(6 \cos 30°) - 200(6) + 400(6) = 0 \qquad R_b = 715 \text{ lb}$$

$$\Sigma F_x = 0 \qquad R_{ax} + R_b \cos 40° - 300 - 400 = 0 \qquad R_{ax} = 152 \text{ lb}$$

$$\Sigma F_y = 0 \qquad R_{ay} - R_b \sin 40° - 200 = 0 \qquad R_{ay} = 660 \text{ lb} \qquad R_a = \sqrt{R_{ax}^2 + R_{ay}^2} = \sqrt{152^2 + 660^2} = 677 \text{ lb}$$

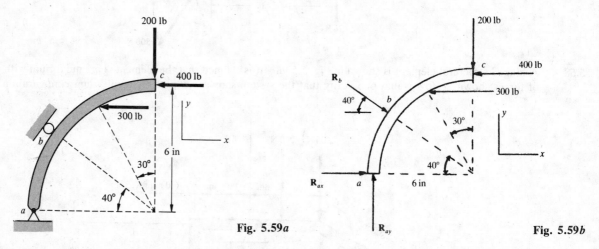

Fig. 5.59a   **Fig. 5.59b**

(b)   A clockwise couple of magnitude M_0, acting in the plane of the figure, is imagined to be applied to the beam. The maximum permissible value of this couple will be limited by the condition $R_b = 0$. For moment equilibrium, with $R_b = 0$,

$$\Sigma M_a = 0 \qquad + 300(6 \cos 30°) - 200(6) + 400(6) - M_0 = 0 \qquad M_0 = 2{,}760 \text{ in} \cdot \text{lb}$$

(c) $\mathbf{R}_a = R_{ax}\mathbf{i} + R_{ay}\mathbf{j} = 152\mathbf{i} + 660\mathbf{j} \qquad \text{lb}$

$\mathbf{R}_b = (R_b \cos 40°)\mathbf{i} - (R_b \sin 40°)\mathbf{j} = (715 \cos 40°)\mathbf{i} - (715 \sin 40°)\mathbf{j} = 548\mathbf{i} - 460\mathbf{j} \qquad \text{lb}$

$\mathbf{M}_0 = -2{,}760\mathbf{k} \qquad \text{in} \cdot \text{lb}$

5.60   Check the results of Prob. 5.25 by drawing the free-body diagram of a system consisting of the three cylinders and showing that this system is in equilibrium.

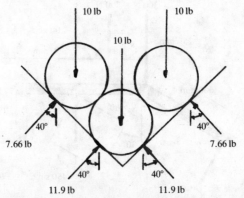

Fig. 5.60

❘ The free-body diagram of the system of three cylinders is shown in Fig. 5.60. By inspection, from symmetry, $\Sigma F_x = 0$ is satisfied. For force equilibrium in the y direction,

$$\Sigma F_y = 0 \qquad 2(7.66 \cos 40°) + 2(11.9 \cos 40°) - 3(10) \overset{?}{=} 0$$
$$2(7.66 \cos 40°) + 2(11.9 \cos 40°) \overset{?}{=} 30$$
$$30 \equiv 30$$

5.61 A cylinder of mass 30 kg rests in the right-angle trough in Fig. 5.61a.

(a) Find the forces exerted by the walls of the trough on the cylinder.

(b) Draw the free-body diagram of the trough. Find the x and y components of the hinge pin force, and the reaction force at the ball support.

❘ (a) The free-body diagram of the cylinder is shown in Fig. 5.61b. The directions of the x and y axes are chosen to permit direct solutions for R_a and R_b. For equilibrium of the cylinder,

$$\Sigma F_x = 0 \qquad R_a - 294 \sin 30° = 0 \qquad R_a = 147 \text{ N}$$
$$\Sigma F_y = 0 \qquad R_c - 294 \cos 30° = 0 \qquad R_c = 255 \text{ N}$$

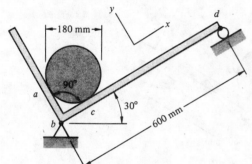

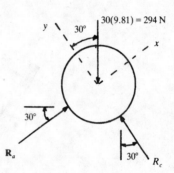

Fig. 5.61a   **Fig. 5.61b**

(b) The free-body diagram of the trough is shown in Fig. 5.61c. From Newton's third law, the reaction forces R_a and R_c are shown with opposite senses in Figs. 5.61b and c. For equilibrium of the trough,

$$\Sigma M_b = 0 \qquad 147(90) - 225(90) + 600R_d = 0 \qquad R_d = 16.2 \text{ N}$$
$$\Sigma F_x = 0 \qquad R_{bx} - 147 = 0 \qquad R_{bx} = 147 \text{ N}$$
$$\Sigma F_y = 0 \qquad R_{by} - 255 + R_d = 0 \qquad R_{by} = 239 \text{ N}$$
$$R_b = \sqrt{R_{bx}^2 + R_{by}^2} = \sqrt{147^2 + 239^2} = 281 \text{ N}$$

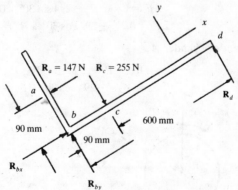

Fig. 5.61c

5.62 (a) Show that the same results as in part (b) of Prob. 5.61 are obtained by drawing the free-body diagram of the combined cylinder-trough assembly and solving the equations of static equilibrium.

(b) Which of the two methods of solution yields more information about the system?

❘ (a) The free-body diagram of the cylinder-trough assembly is shown in Fig. 5.62. For equilibrium,

$$\Sigma M_b = 0 \qquad -(294 \cos 30°)90 + (294 \sin 30°)90 + R_d(600) = 0 \qquad R_d = 16.1 \text{ N}$$
$$\Sigma F_x = 0 \qquad R_{bx} - 294 \sin 30° = 0 \qquad R_{bx} = 147 \text{ N}$$
$$\Sigma F_y = 0 \qquad R_{by} - 294 \cos 30° + R_d = 0 \qquad R_{by} = 239 \text{ N}$$

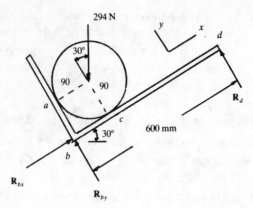

Fig. 5.62

(b) The method used in Prob. 5.61 yields the normal reaction forces between the cylinder and the trough. The method in part (a) above does not produce this information.

5.63 A second cylinder is added to the system of Prob. 5.61, as shown in Fig. 5.63a.

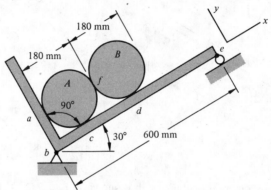

Fig. 5.63a

(a) Find the forces exerted by the walls of the trough on the cylinders, and the force exerted by cylinder A on cylinder B.

(b) Draw the free-body diagram of the trough. Find the x and y components of the hinge pin force and the reaction force at the ball support.

┃ (a) The free-body diagrams of the two cylinders are shown in Fig. 5.63b. For equilibrium of cylinder B,

$$\Sigma F_x = 0 \qquad R_a - 294 \sin 30° = 0 \qquad R_a = 147 \text{ N}$$
$$\Sigma F_y = 0 \qquad -294 \cos 30° + R_d = 0 \qquad R_d = 255 \text{ N}$$

For equilibrium of cylinder A,

$$\Sigma F_x = 0 \qquad R_a - 294 \sin 30° - R_f = 0 \qquad R_a = 294 \text{ N}$$
$$\Sigma F_y = 0 \qquad -294 \cos 30° + R_c = 0 \qquad R_c = 255 \text{ N}$$

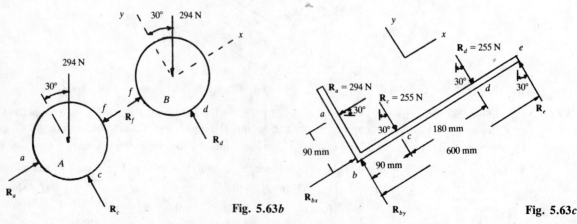

Fig. 5.63b

Fig. 5.63c

(b) The free-body diagram of the trough is shown in Fig. 5.63c. For equilibrium,

$$\Sigma M_b = 0 \qquad 294(90) - 255(90) - 255(270) + R_e(600) = 0 \qquad R_e = 109 \text{ N}$$
$$\Sigma F_x = 0 \qquad R_{bx} - 294 = 0 \qquad R_{bx} = 294 \text{ N}$$
$$\Sigma F_y = 0 \qquad R_{by} - 255 - 255 + 109 = 0 \qquad R_{by} = 401 \text{ N}$$
$$R_b = \sqrt{R_{bx}^2 + R_{by}^2} = \sqrt{294^2 + 401^2} = 497 \text{ N}$$

5.64 (a) Show that the same results as in part (b) of Prob. 5.63 are obtained by drawing the free-body diagram of the combined cylinder-trough assembly and solving the equations of static equilibrium.

(b) Which method of solution yields more information about the system?

▋ (a) The free-body diagram of the combined cylinder-trough assembly is shown in Fig. 5.64. For equilibrium,

$$\Sigma M_b = 0 \qquad -(294 \cos 30°)90 + (294 \sin 30°)90 - (294 \cos 30°)270 + (294 \sin 30°)90 + R_e(600) = 0$$
$$R_e = 109 \text{ N}$$
$$\Sigma F_x = 0 \qquad R_{bx} - 294 \sin 30° - 294 \sin 30° = 0 \qquad R_{bx} = 294 \text{ N}$$
$$\Sigma F_y = 0 \qquad R_{by} - 294 \cos 30° - 294 \cos 30° + R_e = 0 \qquad R_{by} = 400 \text{ N}$$

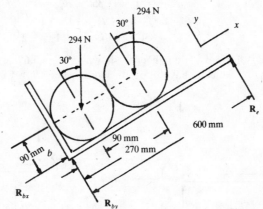

Fig. 5.64

(b) As before, more information is obtained if separate free-body diagrams are drawn for each element in the system.

5.65 (a) Draw the free-body diagram of the beam shown in Fig. 5.65a and find the force effects exerted by the wall on the beam.

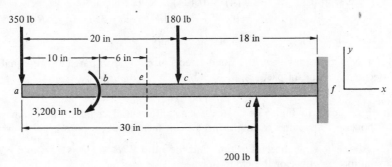

Fig. 5.65a

(b) Draw the free-body diagram of length *ae*, and solve for the unknown reaction force effects.

(c) Draw the free-body diagram of length *ef*, and solve for the unknown reaction force effects.

(d) Discuss the force effects at *e*, found as the solutions to parts *b* and *c*.

Express all results in both scalar and vector forms.

▋ (a) The free-body diagram of the beam is shown in Fig. 5.65b. For equilibrium of this element,

$$\Sigma M_f = 0 \qquad +350(38) - 3,200 + 180(18) - 200(8) - M_f = 0 \qquad M_f = 11,700 \text{ in} \cdot \text{lb}$$

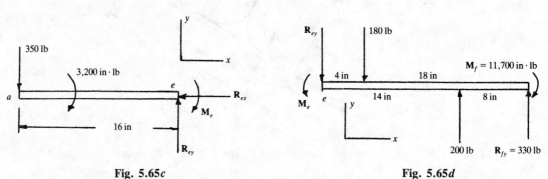

Fig. 5.65b

$$\Sigma F_y = 0 \qquad -350 - 180 + 200 + R_{fy} = 0 \qquad R_{fy} = 330 \text{ lb}$$
$$\Sigma F_x = 0 \qquad R_{fx} = 0 \qquad R_f = R_{fy} = 330 \text{ lb}$$
$$\mathbf{R}_f = 330\mathbf{j} \quad \text{lb} \qquad \mathbf{M}_f = -11{,}700\mathbf{k} \quad \text{in} \cdot \text{lb}$$

(*b*) The free-body diagram of length *ae* is shown in Fig. 5.65*c*. For equilibrium,

$$\Sigma M_e = 0 \qquad 350(16) - 3{,}200 - M_e = 0 \qquad M_e = 2{,}400 \text{ in} \cdot \text{lb}$$
$$\Sigma F_y = 0 \qquad -350 + R_{ey} = 0 \qquad R_{ey} = 350 \text{ lb}$$
$$\Sigma F_x = 0 \qquad R_{ex} = 0 \qquad R_e = R_{ey} = 350 \text{ lb}$$
$$\mathbf{R}_e = 350\mathbf{j} \quad \text{lb} \qquad \mathbf{M}_e = -2{,}400\mathbf{k} \quad \text{in} \cdot \text{lb}$$

(*c*) The free-body diagram of length *ef* is shown in Fig. 5.65*d*. R_{ey} and M_e, as action-reaction force effects, have opposite senses when shown in Figs. 5.65*c* and *d*.

Fig. 5.65*c*   Fig. 5.65*d*

For equilibrium of this segment of the beam,

$$\Sigma M_e = 0 \qquad M_e - 180(4) + 200(14) + 330(22) - 11{,}700 = 0 \qquad M_e = 2{,}360 \approx 2{,}400 \text{ in} \cdot \text{lb}$$
$$\Sigma F_y = 0 \qquad -R_{ey} - 180 + 200 + 330 = 0 \qquad R_{ey} = 350 \text{ lb}$$

(*d*) $\mathbf{R}_e = -350\mathbf{j} \quad \text{lb} \qquad \mathbf{M}_e = 2{,}400\mathbf{k} \quad \text{in} \cdot \text{lb}$

From Newton's third law, the force effects at *e* found in part (*b*) are equal and opposite to the force effects at *e* found in part (*c*).

5.66 The mechanism shown in Fig. 5.66*a* is called a "differential chain hoist." An endless chain traverses the pulleys, as shown. A person exerting a force *P* on the chain can raise a heavy weight *W*.

(*a*) If the system shown is in equilibrium, find the ratio *W*/*P* in terms of r_1 and r_2. (This ratio indicates the force-multiplying effect of the chain hoist.)

(*b*) If $r_1 = 4.5$ in, $r_2 = 5$ in, and the person can exert a force of 100 lb on the chain, what is the heaviest weight that the person can raise with the chain hoist?

(*c*) For the conditions of part (*b*), find the forces which act on the two pulley pins.

▮ (*a*) Chain force *P* is assumed to act in the vertical direction. The free-body diagram of the upper pulley is shown in Fig. 5.66*b*. For moment equilibrium about point *a*,

$$\Sigma M_a = 0 \qquad +\frac{W}{2}r_2 - \frac{W}{2}r_1 - Pr_2 = 0 \qquad \frac{W}{P} = \frac{2r_2}{r_2 - r_1}$$

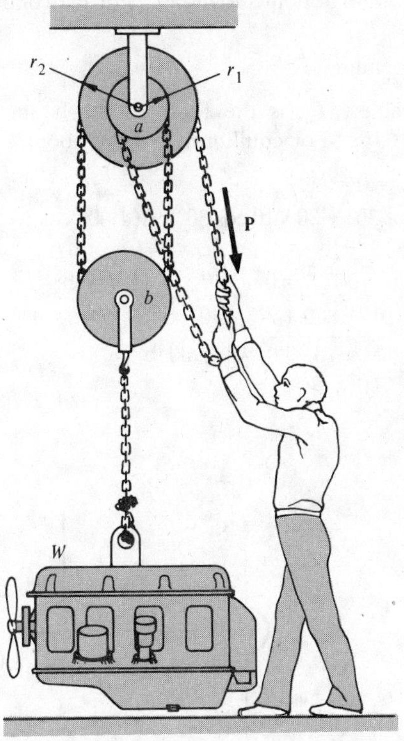

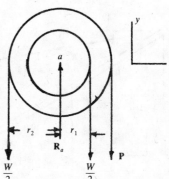

Fig. 5.66a

Fig. 5.66b

(*b*) Using $r_1 = 4.5$ in, $r_2 = 5$ in, and $P = 100$ lb,

$$\frac{W}{100} = \frac{2(5)}{5 - 4.5} \qquad W = 2{,}000 \text{ lb} = 1 \text{ ton}$$

(*c*) The force on pulley pin *b* is

$$R_b = W = 2{,}000 \text{ lb}$$

For pulley pin *a*, from Fig. 5.66*b*,

$$\Sigma F_y = 0 \qquad -\frac{W}{2} + R_a - \frac{W}{2} - P = 0 \qquad R_a = W + P = 2{,}000 + 100 = 2{,}100 \text{ lb}$$

5.67 The crane shown in Fig. 5.67*a* has a total weight of 8,500 lb, acting through point *c*. The boom accounts for 800 lb of this weight. The upper cable in the figure is retracted to hoist the load. The lower cable is used to raise and lower the boom.

(*a*) If $\theta_1 = 30°$, $\theta_2 = 10°$, and the load weighs 4,000 lb, find the two cable forces and the hinge pin forces at *a* and *b*.

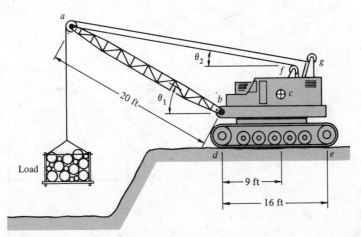

Fig. 5.67a

(*b*) The ground is assumed to exert reaction forces on the crane at points *d* and *e*. For the conditions of part (*a*), find the values of these forces.

(*c*) For what load will tipping motion of the crane be impending?

▌ (*a*) $F_{ag} = 4,000$ lb is the tensile force in the load cable. F_{af} is the force in the boom cable. The free-body diagram of the boom is shown in Fig. 5.67*b*. For equilibrium of the boom,

$$\Sigma M_b = 0$$
$$(4,000 + 4,000 \sin 10° + F_{af} \sin 10°)20 \cos 30° + 800(10 \cos 30°) - (4,000 \cos 10°$$
$$+ F_{af} \cos 10°)20 \sin 30° = 0 \qquad F_{af} = 7,140 \text{ lb}$$
$$\Sigma F_x = 0 \qquad (4,000 + F_{af}) \cos 10° \qquad - R_{bx} = 0 \qquad R_{bx} = 11,000 \text{ lb}$$
$$\Sigma F_y = 0 \qquad -4,000 - (4,000 + F_{af}) \sin 10° - 800 + R_{by} = 0 \qquad R_{by} = 6,730 \text{ lb}$$
$$R_b = \sqrt{R_{bx}^2 + R_{by}^2} = \sqrt{11,000^2 + 6,730^2} = 12,900 \text{ lb}$$

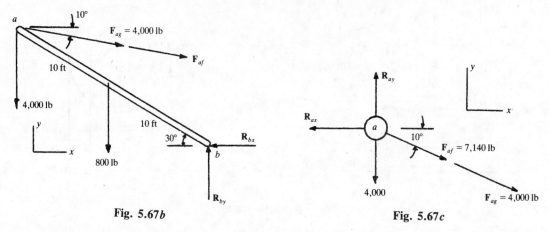

Fig. 5.67*b*   Fig. 5.67*c*

The free-body diagram of pin *a* is shown in Fig. 5.67*c*. R_{ax} and R_{ay} are the components of the force exerted on pin *a* by the boom. For equilibrium of pin *a*,

$$\Sigma F_x = 0 \qquad -R_{ax} + (7,140 + 4,000) \cos 10° = 0 \qquad R_{ax} = 11,000 \text{ lb}$$
$$\Sigma F_y = 0 \qquad R_{ay} - 4,000 - (7,140 + 4,000) \sin 10° = 0 \qquad R_{ay} = 5,930 \text{ lb}$$
$$R_a = \sqrt{R_{ax}^2 + R_{ay}^2} = \sqrt{11,000^2 + 5,930^2} = 12,500 \text{ lb}$$

(*b*) A symbolic free-body diagram of the crane is shown in Fig. 5.67*d*, and R_d and R_e are the ground reaction forces acting on the crane. The set of forces shown in this figure is a parallel force system. For equilibrium of the crane,

$$\Sigma M_d = 0 \qquad 4,000(20 \cos 30°) - 8,500(9) + R_e(16) = 0 \qquad R_e = 451 \text{ lb}$$
$$\Sigma F_y = 0 \qquad -4,000 + R_d - 8,500 + R_e = 0 \qquad R_d = 4,000 + 8,500 - 451 = 12,000 \text{ lb}$$

(*c*) When $R_e = 0$, tipping motion is impending. Using Fig. 5.67*d*, with load *W*,

$$\Sigma M_d = 0 \qquad W(20 \cos 30°) - 8,500(9) = 0 \qquad W = 4,420 \text{ lb}$$

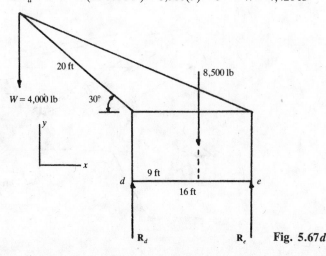

Fig. 5.67*d*

5.68 The mechanical system shown in Fig. 5.68a is referred to as a "cam and follower." A counterclockwise couple of 4,850 in · lb acts on the cam, and this element has a smooth surface.

(a) Find the compressive force R_c transmitted between the cam and the follower, and the magnitude of the couple M_a which must be exerted on the follower to maintain the equilibrium position shown.

(b) Find the hinge pin forces at a and b.

(c) Express the results in parts (a) and (b) in formal vector notation.

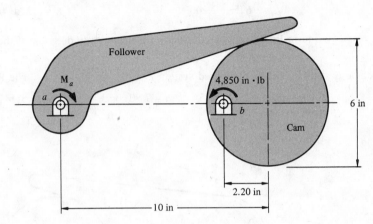

Fig. 5.68a

(a) The free-body diagrams of the cam and follower are shown in Fig. 5.68b and c. From Fig. 5.68b, x_1 and θ are found as

$$\frac{x_1}{9.54} = \frac{2.2}{10} \qquad x_1 = 2.10 \text{ in} \qquad \cos\theta = \frac{3}{10} \qquad \theta = 72.5°$$

The equilibrium requirements are as follows.

Cam: $\qquad \Sigma M_0 = 0 \qquad 4,850 - R_c(2.10) = 0 \qquad R_c = 2,310 \text{ lb}$

Follower: $\qquad \Sigma M_a = 0 \qquad -M_a + R_c(9.54) = 0 \qquad M_a = 22,000 \text{ in} \cdot \text{lb}$

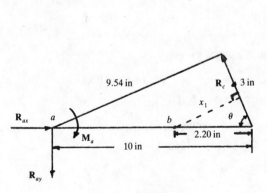

Fig. 5.68b

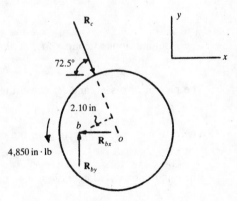

Fig. 5.68c

(b) The pin forces are as follows.

Follower:

$$\Sigma F_x = 0 \qquad R_{ax} - R_c \cos 72.5° = 0 \qquad R_{ax} = 695 \text{ lb}$$
$$\Sigma F_y = 0 \qquad -R_{ay} + R_c \sin 72.5° = 0 \qquad R_{ay} = 2,200 \text{ lb}$$

$$R_a = \sqrt{R_{ax}^2 + R_{ay}^2} = \sqrt{695^2 + 2,200^2} = 2,310 \text{ lb} \qquad \tan\theta = \frac{2,200}{695} \qquad \theta = 72.5°$$

Cam:

$$\Sigma F_x = 0 \qquad R_c \cos 72.5° - R_{bx} = 0 \qquad R_{bx} = 695 \text{ lb}$$
$$\Sigma F_y = 0 \qquad -R_c \sin 72.5° + R_{by} = 0 \qquad R_{by} = 2,200 \text{ lb}$$

$$R_b = \sqrt{R_{bx}^2 + R_{by}^2} = \sqrt{695^2 + 2,200^2} = 2,310 \text{ lb} \qquad \tan\theta = \frac{2,200}{695} \qquad \theta = 72.5°$$

(c) $\mathbf{R}_a = R_{ax}\mathbf{i} - R_{ay}\mathbf{j} = 695\mathbf{i} - 2{,}200\mathbf{j}$ lb
$\mathbf{R}_b = -R_{bx}\mathbf{i} + R_{by}\mathbf{j} = -695\mathbf{i} + 2{,}200\mathbf{j}$ lb
For force $\mathbf{R}_c$ acting on the cam,

$$\mathbf{R}_c = (R_c \cos 72.5°)\mathbf{i} - (R_c \sin 72.5°)\mathbf{j} = (2{,}310 \cos 72.5°)\mathbf{i} - (2{,}310 \sin 72.5°)\mathbf{j}$$
$$= 695\mathbf{i} - 2{,}200\mathbf{j}   \text{lb}$$

For force $\mathbf{R}_c$ acting on the follower,

$$\mathbf{R}_c = -695\mathbf{i} + 2{,}200\mathbf{j}   \text{lb}   \mathbf{M}_a = -M_a\mathbf{k} = -22{,}000\mathbf{k}   \text{in} \cdot \text{lb}$$

5.69   Check the results found in Prob. 5.68 by considering the free-body diagram of the combined cam and follower system.

❚   The free-body diagram of the combined system is shown in Fig. 5.69. The angle α and length d are
$$\alpha = 90° - 72.5 = 17.5°   d = 7.80 \cos \alpha = 7.80 \cos 17.5° = 7.44 \text{ in}$$

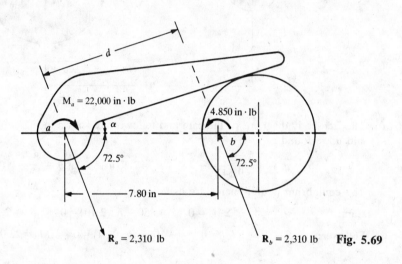

$M_a = 22{,}000 \text{ in} \cdot \text{lb}$ 4.850 in · lb   72.5°   72.5°   7.80 in $R_a = 2{,}310$ lb $R_b = 2{,}310$ lb   **Fig. 5.69**

The resultant applied couple M_1, acting in a clockwise sense, is
$$M_1 = 22{,}000 - 4{,}850 = 17{,}200 \text{ in} \cdot \text{lb}$$
The pair of forces R_a and R_b constitute a counterclockwise couple M_2 of magnitude
$$M_2 = 2{,}310(7.44) = 17{,}200 \text{ in} \cdot \text{lb}$$
The resultant couple M acting on the combined system is then
$$M = M_1 - M_2 = 17{,}200 - 17{,}200 = 0$$
which is the expected condition of static equilibrium.

5.70   The cam and follower arrangement in Fig. 5.70a is in equilibrium.

(a)   Find the forces, at c and d, exerted by the wall on the smooth follower rod, and the contact force between the cam and the follower at point b.

(b)   Find the required value M_0 of the applied couple which acts on the cam.

(c)   Find the hinge pin force at a.

(d)   Express the results in parts (a) through (c) in formal vector notation.

❚   (a)   The free-body diagrams of the cam and follower are shown in Figs. 5.70b and c.
From Fig. 5.70b,

$$\sin \theta = \frac{2.5}{3.5}   \theta = 45.6°$$
$$d = 2.4 \sin \theta = 2.4 \sin 45.6° = 1.71 \text{ in}$$

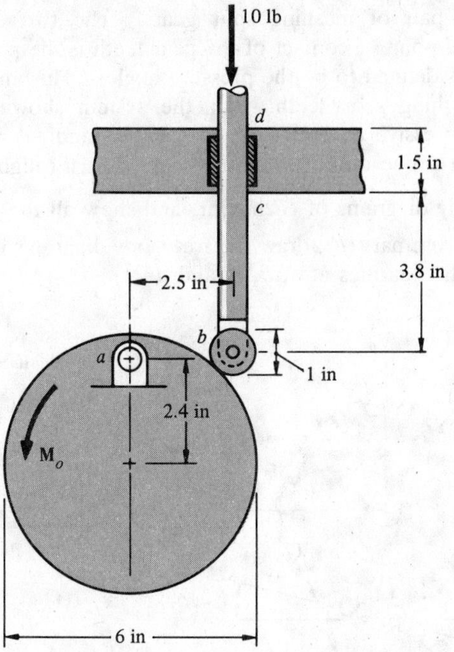

Fig. 5.70a

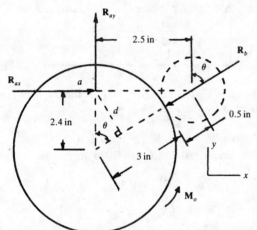

Fig. 5.70b

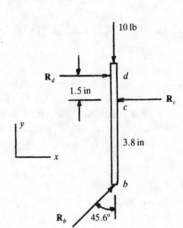

Fig. 5.70c

For equilibrium of the follower, from Fig. 5.70c,

$$\Sigma F_y = 0 \qquad -10 + R_b \cos 45.6° = 0 \qquad R_b = 14.3 \text{ lb}$$
$$\Sigma M_d = 0 \qquad -R_c(1.5) + (R_b \sin 45.6°)(5.3) = 0 \qquad R_c = 36.1 \text{ lb}$$
$$\Sigma F_x = 0 \qquad R_d - R_c + R_b \sin 45.6° = 0 \qquad R_d = 25.9 \text{ lb}$$

(b) For equilibrium of the cam, from Fig. 5.70b,

$$\Sigma M_a = 0 \qquad M_0 - R_b(d) = 0 \qquad M_0 = R_b(d) = 14.3(1.71) = 24.5 \text{ in} \cdot \text{lb}$$

(c) $$\Sigma F_x = 0 \qquad R_{ax} - R_b \sin 45.6° = 0 \qquad R_{ax} = 10.2 \text{ lb}$$
$$\Sigma F_y = 0 \qquad R_{ay} - R_b \cos 45.6° = 0 \qquad R_{ay} = 10.0 \text{ lb}$$
$$R_a = \sqrt{R_{ax}^2 + R_{ay}^2} = \sqrt{10.2^2 + 10^2} = 14.3 \text{ lb}$$

(d) $\mathbf{R}_a = R_{ax}\mathbf{i} + R_{ay}\mathbf{j} = 10.2\mathbf{i} + 10.0\mathbf{j} \qquad \text{lb}$
For force R_b acting on the cam,

$$\mathbf{R}_b = (-R_b \sin \theta)\mathbf{i} - (R_b \cos \theta)\mathbf{j} = (-14.3 \sin 45.6°)\mathbf{i} - (14.3 \cos 45.6°)\mathbf{j} = -10.2\mathbf{i} - 10.0\mathbf{j} \qquad \text{lb}$$

For force $\mathbf{R}_b$ acting on the follower,

$$\mathbf{R}_b = 10.2\mathbf{i} + 10.0\mathbf{j} \qquad \text{lb} \qquad \mathbf{R}_c = -R_c\mathbf{i} = -36.1\mathbf{i} \qquad \mathbf{R}_d = R_d\mathbf{i} = 25.9\mathbf{i}$$

5.71 Figure 5.71*a* shows a pair of meshing spur gears. The two dotted circles in the figure are the pitch circles. The theoretical point of contact of the gear teeth is the point at which the pitch circles are tangent to each other. Angle ϕ is defined to be the pressure angle. This angle is the direction of the total compressive force between the meshing gear teeth. For the system shown, gear A drives gear B with a constant speed. The total compressive contact force is 1,200 lb, and $\phi = 20°$. The torque input M_a to shaft cd is through end c, and the torque output M_B from shaft ab is through end b.

(*a*) Draw the free-body diagrams of each gear, and show all the force effects which act on these gears.

(*b*) Using the results from part (*a*), draw the free-body diagrams of the two shafts and find the forces exerted by the shafts on the bearings at a, b, c, and d.

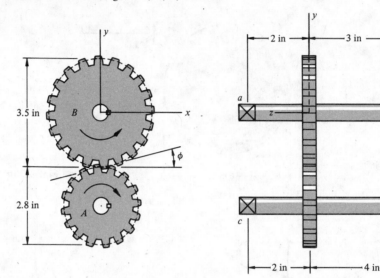

Fig. 5.71*a*

▌ (*a*) The free-body diagrams of the two gears are shown in Fig. 5.71*b* and *c*. M_A is the moment exerted by shaft cd on gear A, and M_B is the moment exerted by shaft ab on gear B.

For force equilibrium of gear A,

$$\Sigma F_x = 0 \qquad R_{Ax} - 1,200 \cos 20° = 0 \qquad R_{Ax} = 1,130 \text{ lb}$$
$$\Sigma F_y = 0 \qquad R_{Ay} - 1,200 \sin 20° = 0 \qquad R_{Ay} = 410 \text{ lb}$$

$$R_A = \sqrt{1,130^2 + 410^2} = 1,200 \text{ lb} \qquad \tan \theta_1 = \frac{410}{1,130} = 0.363 \qquad \theta_1 = 20°$$

It follows that

$$\theta_1 = \phi = 20°$$

For force equilibrium of gear B,

$$\Sigma F_x = 0 \qquad 1,200 \cos 20° - R_{Bx} = 0 \qquad R_{Bx} = 1,130 \text{ lb}$$
$$\Sigma F_y = 0 \qquad 1,200 \sin 20° - R_{By} = 0 \qquad R_{By} = 410 \text{ lb}$$

$$R_B = \sqrt{R_{bx}^2 + R_{by}^2} = \sqrt{1,130^2 + 410^2} = 1,200 \text{ lb} \qquad \tan \theta_2 = \frac{410}{1,130} = 0.363 \qquad \theta_2 = 20°$$

and

$$\theta_2 = \phi = 20°$$

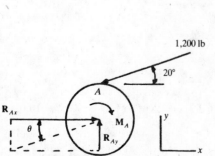

Fig. 5.71*b*

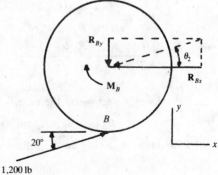

Fig. 5.71*c*

For moment equilibrium of gear A,

$$\Sigma M = 0 \qquad -M_A + (1,200 \cos 20°)\frac{2.8}{2} \qquad M_A = 1,580 \text{ in} \cdot \text{lb}$$

For moment equilibrium of gear B,

$$\Sigma M = 0 \qquad -M_B + (1,200 \cos 20°)\frac{3.5}{2} \qquad M_B = 1,970 \text{ in} \cdot \text{lb}$$

(b)   The free-body diagrams of the two shafts are shown in Figs. 5.71d and e.
For equilibrium of shaft ab,

$$\Sigma M_a = 0 \qquad -1,200(2) + R_b(5) = 0 \qquad R_b = 480 \text{ lb}$$
$$\Sigma F_y = 0 \qquad R_a - 1,200 + R_b = 0 \qquad R_a = 720 \text{ lb}$$

For equilibrium of shaft cd,

$$\Sigma M_c = 0 \qquad -1,200(2) + R_d(6) = 0 \qquad R_d = 400 \text{ lb}$$
$$\Sigma F_y = 0 \qquad R_c - 1,200 + R_d = 0 \qquad R_c = 800 \text{ lb}$$

The forces R_a, R_b, R_c, and R_d exerted *by* the shafts *on* the bearings will have senses opposite to those shown in Figs. 5.71d and e.

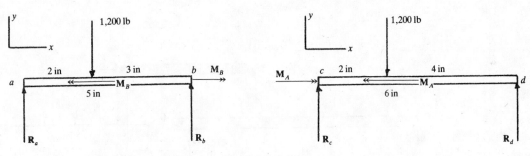

Fig. 5.71d Fig. 5.71e

5.72   The device shown in Fig. 5.72a is referred to as a "trammel plate." The hydraulic cylinder exerts a sufficient force to maintain the equilibrium configuration shown. Find the force exerted by pin c on the trammel plate.

I   The free-body diagram of the trammel plate is shown in Fig. 5.72b. For equilibrium of this element,

$$\Sigma M_c = 0 \qquad -4(6) - (5.5 \sin 40°)6 - (8 \cos 35°)6 + R_d(8) = 0 \qquad R_d = 10.6 \text{ kip}$$
$$\Sigma F_x = 0 \qquad 4 - 5.5 \cos 40° - 8 \cos 35° + R_{cx} = 0 \qquad R_{cx} = 6.77 \text{ kip}$$

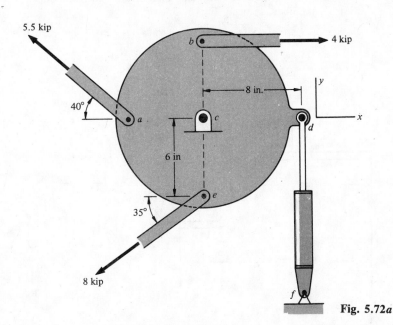

Fig. 5.72a

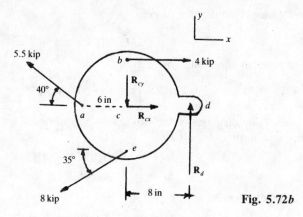

Fig. 5.72b

$$\Sigma F_y = 0 \qquad -R_{cy} + 5.5\sin 40° - 8\sin 35° + R_d = 0 \qquad R_{cy} = 9.55 \text{ kip}$$
$$R_c = \sqrt{R_{cx}^2 + R_{cy}^2} = \sqrt{6.77^2 + 9.55^2} = 11.7 \text{ kip}$$

5.73 Find the forces acting on the pins in the structure shown in Fig. 5.73a.

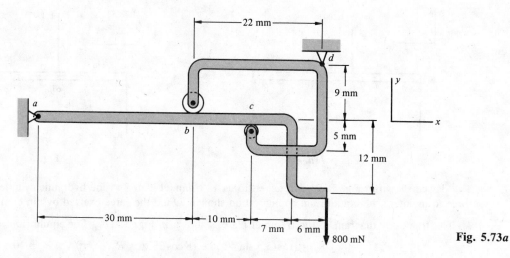

Fig. 5.73a

❚ Figures 5.73b and c show the free-body diagrams of the two members.
For equilibrium of member ac, using Fig. 5.73b,

$$\Sigma M_a = 0 \qquad -R_b(30) + R_c(40) - 800(53) = 0 \qquad (1)$$
$$\Sigma F_x = 0 \qquad R_{ax} = 0$$
$$\Sigma F_y = 0 \qquad -R_{ay} - R_b + R_c - 800 = 0 \qquad (2)$$

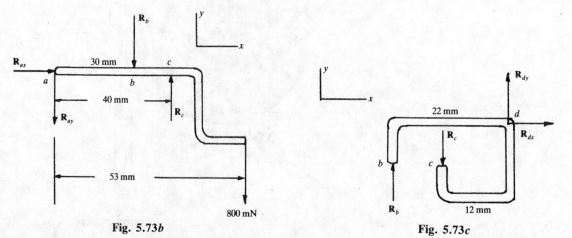

Fig. 5.73b

Fig. 5.73c

For moment equilibrium of member bd, using Fig. 5.73c,

$$\Sigma M_d = 0 \qquad -R_b(22) + R_c(12) = 0 \qquad R_b = 0.545R_c \qquad (3)$$

Equation (3) is substituted into Eq. (1), with the result

$$-30(0.545R_c) + 40R_c - 800(53) = 0 \qquad R_c = 1{,}790 \text{ mN} \qquad (4)$$

Using Eq. (4) in Eq. (3),

$$R_b = 0.545R_c = 0.545(1{,}790) = 976 \text{ mN} \qquad (5)$$

Using Eqs. (5) and (4) in Eq. (2),

$$-R_{ay} - 976 + 1{,}790 - 800 = 0 \qquad R_{ay} = 14 \text{ mN}$$

Using Fig. 5.73c,

$$\Sigma F_x = 0 \qquad R_{dx} = 0$$
$$\Sigma F_y = 0 \qquad R_b - R_c + R_{dy} = 0 \qquad R_{dy} = 814 \text{ mN}$$

A summary of the forces acting on the pins is shown below.

$$R_a = R_{ay} = 14 \text{ mN} \qquad R_b = 976 \text{ mN}$$
$$R_c = 1{,}790 \text{ mN} \qquad R_d = R_{dy} = 814 \text{ mN}$$

5.74 The device shown in Fig. 5.74a is called a Scotch yoke. If the disk rotates with constant angular velocity, the rod will move with simple harmonic motion (i.e., the plot of the position of the rod vs. time will be either a sine or a cosine curve). The system is in equilibrium with the 8-lb yoke rod force and the couple M_0 applied to the disk.

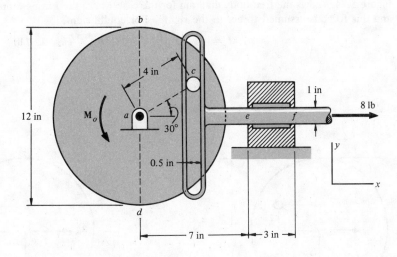

Fig. 5.74a

(a) Find the forces acting on the yoke assembly at c, e, and f; the pin force at a; and the magnitude of M_0.

(b) Do the same as in part (a), if the couple is replaced by a horizontal force acting at point b on the disk, and find the magnitude and sense of this force.

(c) Do the same as in part (b), if the horizontal force acts at point d on the disk.

(d) For which of the above methods of loading is the pin force at point a in Fig. 5.74a maximum?

▮ (a) All surfaces are assumed to be smooth, and the free-body diagram of the yoke assembly is shown in Fig. 5.74b. For equilibrium,

$$\Sigma F_x = 0 \qquad -R_c + 8 = 0 \qquad R_c = 8 \text{ lb}$$
$$\Sigma M_f = 0 \qquad -R_e(3) + R_c(2) = 0 \qquad R_e = 5.33 \text{ lb}$$
$$\Sigma F_y = 0 \qquad R_e - R_f = 0 \qquad R_f = 5.33 \text{ lb}$$

Figure 5.74c shows the free-body diagram of the disk. For equilibrium of this element,

$$\Sigma M_a = 0 \qquad M_0 - R_c(2) = 0 \qquad M_0 = 2R_c = 2(8) = 16 \text{ in} \cdot \text{lb}$$
$$\Sigma F_y = 0 \qquad R_{ay} = 0$$

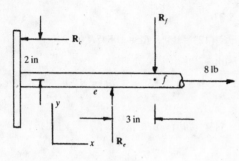

Fig. 5.74b

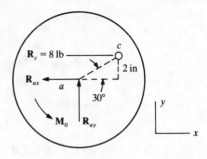

Fig. 5.74c

$$\Sigma F_x = 0 \qquad -R_{ax} + 8 = 0 \qquad R_{ax} = 8\,\text{lb}$$
$$R_a = R_{ax} = 8\,\text{lb}$$

(*b*) R_c is the same as in part (*a*) since this value is an equilibrium requirement of the yoke assembly. Forces R_e and R_f are also unchanged.

Figure 5.74*d* shows the free-body diagram of the disk. The applied force F_b is assumed to act to the left. For equilibrium,

$$\Sigma M_a = 0 \qquad F_b(6) - 8(2) = 0 \qquad F_b = 2.67\,\text{lb}$$
$$\Sigma F_y = 0 \qquad R_{ay} = 0$$
$$\Sigma F_x = 0 \qquad -F_b - R_{ax} + 8 = 0 \qquad R_{ax} = 5.33\,\text{lb}$$
$$R_a = R_{ax} = 5.33\,\text{lb}$$

(*c*) Figure 5.74*e* shows the free-body diagram for the case where the force is applied to the disk at point *d*, and this force is assumed to act to the right. For equilibrium,

$$\Sigma M_a = 0 \qquad F_d(6) - 8(2) = 0 \qquad F_d = 2.67\,\text{lb}$$
$$\Sigma F_y = 0 \qquad R_{ay} = 0$$
$$\Sigma F_x = 0 \qquad F_d - R_{ax} + 8 = 0 \qquad R_{ax} = 10.7\,\text{lb}$$
$$R_a = R_{ax} = 10.7\,\text{lb}$$

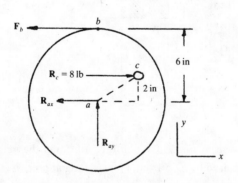

Fig. 5.74d

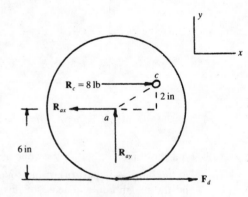

Fig. 5.74e

(*d*) The pin force at *a* has a maximum value of 10.7 lb for the case where the horizontal force acts at point *d*. As a comparison, the minimum value of this pin force of 5.33 lb occurs when the force is applied to the disk at point *b*.

5.75 Give a summary of the basic concepts of the resultant force, and equilibrium, for a two-dimensional force system.

❚ Two concepts used widely in this chapter are the resultant force and equilibrium. It should be emphasized that these are related, but distinct, effects. The resultant force of any force system is that *single* force which will produce the same effect as the original system of forces. A formal description of the resultant force requires a statement of the magnitude, direction, and sense of this force as well as its location within the system.

In establishing the static equilibrium of a body, the first step is to draw a free-body diagram. On this diagram are shown all the forces and moments which act on the body. Two types of such forces may be identified. The first group includes the known forces which are applied to the body. The second type of forces comprise the unknown reaction forces exerted on the body by the supporting structure which was removed. *There is no physical difference between the effects of either of these two types of forces on the body.* It is convenient, however, to distinguish between the two types of forces, since *the basic problem of static equilibrium is to solve for the unknown reaction forces*.

After the free-body diagram is drawn, the requirement of static equilibrium is that the resultant of *all* the forces which act on the free-body diagram must be identically zero. In practice, this requirement is expressed as a set of *equilibrium equations* for the particular system under consideration.

The requirements for the static equilibrium of the four types of systems considered in this chapter are summarized below.

1. Collinear force system
   The algebraic summation of forces along one known direction is zero.

$$\sum F = 0$$

   In vector notation,

$$\mathbf{F} = 0$$

2. Concurrent force system
   The algebraic summation of forces along two perpendicular directions is zero.

$$\sum F_x = 0 \quad , \quad \sum F_y = 0$$

   In vector notation,

$$\mathbf{F} = F_x \mathbf{i} + F_y \mathbf{j} = 0$$

3. Parallel force system
   (a) The algebraic summation of forces along one known direction is zero.

   (b) The algebraic summation of moments about any reference point in the system is zero.

$$\sum F = 0 \qquad \sum M = 0$$

   In vector notation,

$$\mathbf{F} = 0 \qquad \mathbf{M} = 0$$

4. General two-dimensional force system
   (a) The algebraic summation of forces along two perpendicular directions is zero.

   (b) The algebraic summation of moments about any reference point in the system is zero.

$$\sum F_x = 0 \qquad \sum F_y = 0 \qquad \sum M = 0$$

   In vector notation,

$$\mathbf{F} = F_x \mathbf{i} + F_y \mathbf{j} = 0 \qquad \mathbf{M} = 0$$

CHAPTER 6
Force Analysis of Plane Trusses

6.1 TERMINOLOGY, FORCES IN TRUSS MEMBERS, STABILITY FORCE TRANSMISSION THROUGH A JOINT, METHOD OF SUPPORT OF TRUSSES

6.1 (a) Give an example of a *plane truss*.

  (b) What is the difference between the two general types of trusses, plane trusses and *space trusses*?

  (c) What is the general use of trusses?

▎ (a) Figure 6.1*a* shows a typical application of a plane truss. The bridge supports a roadway over which vehicles will pass. The applied forces shown in the figure represent the loads that are transmitted from the roadway structure to the bridge. The structural configuration shown as profile *abcdef* in the figure is called a *plane trusswork*. Throughout this chapter such a configuration, for brevity, will be referred to simply as a plane truss. The *span* of the truss is given by the length *l* in the figure.

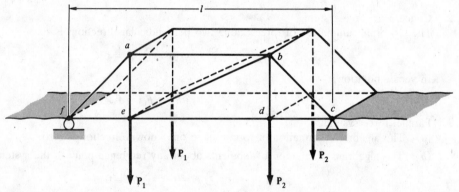

Fig. 6.1*a*

  (b) In plane trusses, both the truss structure and the applied loads lie in the same plane. This situation is typified by the structure shown in Fig. 6.1*a*. In space trusses, either the structure, or the applied loads, or both lie in different planes. Figure 6.1*b* shows a typical space truss. The analysis of space trusses is an advanced problem which is covered in the chapter on three-dimensional force systems. This chapter is devoted to the force analysis of plane trusses.

  (c) The general use of trusses is to support loads over a span length.

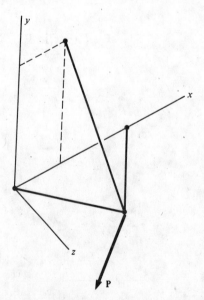

Fig. 6.1*b*

192

6.2 (*a*) Describe the details of truss construction.

(*b*) Describe the method of application of the loads to the truss.

▌ (*a*) A truss is made of structural elements which are referred to as *truss members*. These members are straight and, in the usual application, long and slender. The truss members are connected to one another only at their ends. These junctions, where the ends of two or more truss members are connected, are called *joints*. The joints of a truss will be identified by the lowercase letters $a, b, c, \ldots$.

(*b*) The loads are applied to the truss *only* at the joints. In cases where the weight of the members is not assumed to be negligible, the assumption is made that half of the weight of each member acts as an applied force at the joints at each end of the member.

If, in an actual structure, an applied load acted on a member at a location other than the joint, the member would be classified as a *framework member*. The analysis of this type of structural element is considered in the problems in Chap. 7. Figure 6.2*a* shows typical examples of truss configurations.

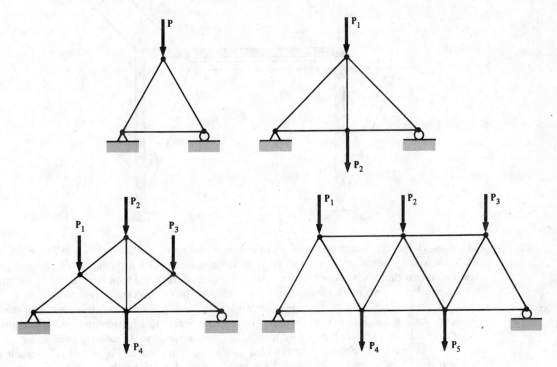

Fig. 6.2*a*

Figures 6.2*b* and *c* show two methods of physically connecting the ends of the members. In Fig. 6.2*b*, a plate is riveted or bolted to the ends of the members which meet at the joint. In Fig. 6.2*c*, the ends of the members have holes in them. A pin is passed through these holes to join the ends of the members. This latter type of connection is referred to as pinned, or hinged. The joint construction shown in Fig. 6.2*b* is assumed to act as a pinned connection.

Fig. 6.2*b*   **Fig. 6.2*c***

6.3 (*a*) Why are plane trusses also called *static*, or *rigid*, trusses?

(*b*) What are the requirements for a stable truss in terms of the number of joints and the number of members?

▮ (*a*) In order to have a plane truss, there must be no possibility of relative motion of the truss members with respect to one another. Such a configuration is called a stable, or rigid, truss.

Figure 6.3*a* shows a configuration of three members which are pin-connected to one another and to the ground. It is apparent that if a force *F* were applied to joint *b*, the shape of the structure would undergo a drastic change, as indicated by the typical dashed outline *da'b'c*. The structure shown in Fig. 6.3*a* is referred to as a *plane linkage*. The characteristic of this assembly is that the links which comprise the structure may have relative rotation with respect to one another. The motion of these links is studied in the subject kinematics. The structure shown in this figure is evidently not suitable for use as a plane truss.

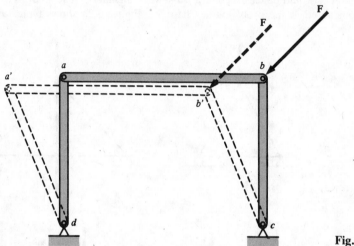

Fig. 6.3*a*

(*b*) The most elementary construction of a plane truss is the triangular form shown in Fig. 6.3*b*. By using the basic assumption of static analysis (that all the elements which are considered are rigid), it may be concluded that there is no possibility of relative motion of the three members in this figure with respect to one another. Two more members may be added to the original truss construction, as shown by the dashed lines in Fig. 6.3*c*. It may again be concluded, from physical considerations, that the new truss configuration *adbc* in Fig. 6.3*c* is stable. The construction in this figure employed the addition of two members and one joint, in a basically triangular configuration, to an existing truss structure. In Fig. 6.3*d* the two members *de* and *be* are added to the configuration of Fig. 6.3*c*. The result is again a stable truss. This technique may now be generalized, and it can be shown that a *necessary* condition for a stable, or rigid, truss is

$$m = 2n - 3$$

where *m* is the number of members and *n* is the number of joints. This equation may be used to determine whether a given truss configuration is stable.

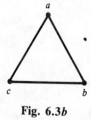

Fig. 6.3*b*

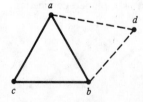

Fig. 6.3*c*

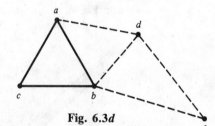

Fig. 6.3*d*

6.4 What are the basic assumptions used in the *force analysis of trusses*?

▮ Two basic assumptions are made in the force analysis of trusses. The first is that the members are pin-connected at their ends, as shown in Fig. 6.2*c*, and that these pins are frictionless. The second assumption is that the loads are applied to the truss *only* at the joints.

The idealized plane truss, then, consists of straight members which are joined to one another, with pin connections, only at their ends. This connection of the ends of two or more truss members is called a joint, and the applied loads on the truss act only at the joints. As a result of this construction, the forces in the members of the truss possess certain special characteristics. These will be described in Probs. 6.5 and 6.6.

6.5 Using the description given in Prob. 6.4, demonstrate that all truss members are *two-force members*.

❚ The members of a truss have loads applied at their ends only, with no other points of load application on the member. A structural element such as this is referred to as a *two-force member*. The special characteristics of such a member will now be established.

Figure 6.5a shows a member with the two forces F_1 and F_2 applied at the ends a and b. The magnitude, direction, and sense of these two forces, as forces on a two-force member, will now be obtained.

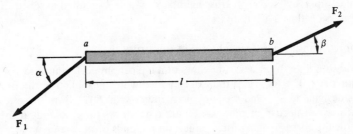

Fig. 6.5a

At the outset of the problem, the forces F_1 and F_2 will have arbitrary directions which are given by the angles α and β. It will be assumed that these two forces lie in the same plane. This system may now be identified as a general two-dimensional force system. The equilibrium requirements for such a system are that the summation of forces in two perpendicular directions be zero, and that the summation of moments about any arbitrary point be zero. For moment equilibrium about point a,

$$\sum M_a = 0 \qquad (F_2 \sin \beta)l = 0 \qquad F_2 l \sin \beta = 0$$

l is the length of the member, and this term may not be zero. If F_2 is zero, it would mean that no force is applied to end b of the member. It then follows that

$$\sin \beta = 0 \qquad \text{or} \qquad \beta = 0°, 180°$$

From the above equation it follows that the direction of force F_2 is collinear with the axis of the member. An additional interpretation which follows from the above equation is that $\beta = 0$ corresponds to F_2 as a tensile force, while $\beta = 180°$ represents the case of F_2 being a compressive force.

The member is redrawn in Fig. 6.5b, using $\beta = 0$. For equilibrium in the direction normal to the axis of the member,

$$F_1 \sin \alpha = 0$$

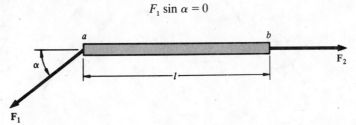

Fig. 6.5b

F_1 may not be zero. If it were, this would indicate that no force is applied to end a of the member. Thus,

$$\sin \alpha = 0 \qquad \text{or} \qquad \alpha = 0°, 180°$$

The above result leads to the conclusion that the direction of the force F_1 acting on the member at end a is collinear with the axis of the member. Again, $\alpha = 0$ represents the case of F_1 being a tensile force, while $\alpha = 180°$ represents the case of F_1 being a compressive force. The free-body diagram of the member, using $\alpha = 0$, now appears as shown in Fig. 6.5c.

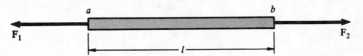

Fig. 6.5c

As a final step, the summation of forces along the axis of the member results in

$$F_1 = F_2$$

It may now be concluded that, *for the case of a member which has forces applied only at its ends, these forces must be equal and opposite to one another, with directions which are collinear with the axis of the member.*

From the discussion of truss construction in Prob. 6.4, it may be concluded that *every truss member is a two-force member.* The forces which act at each end of such members have known directions, and these directions are those of the axes of the respective members. These two forces at the ends of a member are also equal. Then it follows that the force which acts on one end of a truss member is transmitted, without change, to the other end of this member. The two-force truss member is sometimes referred to as a *truss link*.

The forces in a truss member may also be characterized as being either *tensile* or *compressive*. The member forces in a truss are sometimes referred to as the internal forces.

Tensile and compressive forces in truss members may be examined in terms of a concept which is referred to as force transmission through a joint.

6.6   Discuss the force transmission through a joint in a truss.

▌ Figure 6.6a shows the front and side views of a typical physical connection of the ends of two truss members to a connecting pin. The force F is the external load which is applied to the joint through pin a. The force in member ab is designated F_{ab}, and the force in member ac is F_{ac}. Both these forces have the known directions of the axes of their respective members. Figure 6.6b shows the free-body diagram of the hinge pin, and Figs. 6.6c and d show the corresponding force effects on the ends of the members which physically attach to the hinge pin. From Newton's third law, there is a force action-reaction effect between the pin and the mating holes in the members. The pin in Fig. 6.6a thus serves the single function of distributing the applied load F to the ends of the members. If there is no applied force at a particular joint, then the pin at this joint has the function of transmitting the member forces across the joint.

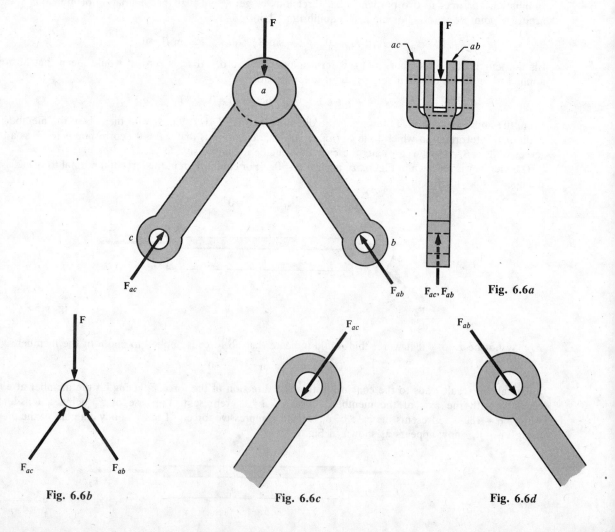

Fig. 6.6a

Fig. 6.6b

Fig. 6.6c

Fig. 6.6d

6.7   State the two basic methods for solving for the forces in truss members.

❚   There are two basic techniques for solving for the forces in truss members. The first method is used when the forces in all the members are desired. This method is referred to as the *method of joints*. In using this technique, the solutions may be found by using the equilibrium requirements for a concurrent force system. The second technique for solving for the forces in truss members is referred to as the *method of sections*. This method is usually used when the values of the forces in only certain members are desired; it is illustrated by the problems in Sec. 6.4.

6.8   (*a*)   State the preliminary step in solving for the forces in the members of a truss.

   (*b*)   Show an example of this step for the truss in Fig. 6.8*a*.

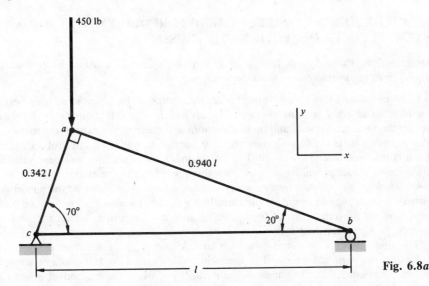

Fig. 6.8*a*

❚   (*a*)   The preliminary step in solving for the forces in truss members is to draw a free-body diagram of the entire truss and solve for the unknown reaction forces of the supporting foundation on the truss.

   From the description of a plane truss as an assemblage of straight members which are pin-connected at their ends, with the applied loads acting at these junctions, it may be concluded that the truss can be physically connected to the foundation *only by a hinge or a roller-type support*. It also follows that *no applied moment may act on the truss*. Thus, the system of forces which acts on the free-body diagram of the truss may be only a two-dimensional force system with no applied moments.

   (*b*)   Figure 6.8*a* shows a basic truss consisting of three members. The three joints are designated *a*, *b*, and *c*. The truss is supported by a hinge at joint *c* and a roller at joint *b*. The free-body diagram of the

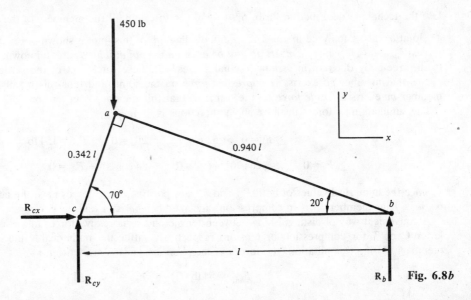

Fig. 6.8*b*

truss is shown in Fig. 6.8*b*. The span length of the truss is *l*, and the lengths of the remaining two sides are as shown in the figure. The requirement of moment equilibrium about joint *c* is

$$\sum M_c = 0 \qquad -450(0.342l)\cos 70° + R_b l = 0 \qquad R_b = 52.6\,\text{lb}$$

The forces are summed to zero in the *x* and *y* directions, with the result

$$\sum F_x = 0 \qquad R_{cx} = 0$$

$$\sum F_y = 0 \qquad R_{cy} - 450 + R_b = 0 \qquad R_{cy} - 450 + 52.6 = 0 \qquad R_{cy} = 397\,\text{lb}$$

The complete free-body diagram is now obtained, and this is a preliminary step in solving for the forces in the truss members.

6.2 METHOD OF JOINTS USING EQUILIBRIUM REQUIREMENTS FOR A CONCURRENT FORCE SYSTEM, PULLEY CONNECTIONS TO TRUSSES

6.9 Describe how the forces in truss members may be found by using the *method of joints*, together with the equilibrium requirements for a *concurrent* force system.

▌ The method of joints may be used when the values of the forces in *all* members are required. In using this method, a free-body diagram is drawn of each joint. The directions of the forces in the truss members which act on the joint are known, and these directions are those of the axes of the members. The set of forces which acts at each joint is thus a *concurrent* force system. As was shown in Prob. 5.10, the equilibrium requirements for a concurrent force system are that the summation of forces in two perpendicular directions be zero.

A double subscript will be used for each unknown force in a member. *This double subscript will identify the joints at either end of the member under consideration.* The order of the subscripts has no significance, since either F_{ab} or F_{ba} would represent the force in the member which has joints *a* and *b* at its ends. For consistency, however, the subscripts should be written in an ascending order. That is, the force in member *be* would be written as F_{be} rather than F_{eb}.

The joints may be analyzed in any order. However, *no more than two unknown member forces may be solved for at a particular joint.* This limitation follows from the fact that a concurrent force system requires two independent equations to define its equilibrium. In using the method of joints to solve for the forces in the truss members, some of the free-body diagrams of the joints will have only one unknown force acting at the joint. Two equations of equilibrium may be written for the forces which act at each joint. For the situation described above, one of these equations of equilibrium may be used to solve for the unknown member force. The second of these equations may then be used to verify the correctness of the numerical computations. These free-body diagrams which have only one unknown force will occur near the end of the solution, and this effect will be illustrated in the following problems.

The method of joints, together with the equilibrium requirements for a concurrent force system, is the fundamental method for solving for the forces in the members of a truss. An alternative method of solving for these member forces, which uses the closed force triangle or polygon law of force addition, is shown in Prob. 6.29.

6.10 Use the technique described in Prob. 6.9 to find the forces in all the members of the truss shown in Prob. 6.8.

▌ Joint *b* will be analyzed first. The free-body diagram of this joint is shown in Fig. 6.10*a*. The senses of the forces F_{ab} and F_{bc}, *which members ab and bc exert on the pin at joint b*, are unknown. These forces are drawn in the free-body diagram in *assumed* actual senses. If these forces act in the senses shown in the figure, it means that member *ab* exerts a *compressive force* on the joint, and thus this member is in compression, and member *bc* exerts a *tensile force* on the joint, so that this member is in tension.

The summation of forces in the *x* and *y* directions is then

$$\sum F_y = 0 \qquad 52.6 - F_{ab}\sin 20° = 0 \qquad F_{ab} = 154\,\text{lb}$$

$$\sum F_x = 0 \qquad F_{ab}\cos 20° - F_{bc} = 0 \qquad 154\cos 20° - F_{bc} = 0 \qquad F_{bc} = 145\,\text{lb}$$

Since the numerical values of both F_{ab} and F_{bc} are positive, these forces have the actual senses shown in Fig. 6.10*a*. Thus member *ab* is in compression, and member *bc* is in tension.

A convenient way of distinguishing between tensile and compressive forces in a truss member is to place the letter C or T for compression or tension, respectively, after the numerical value of the force in the truss member. Using this notation, the forces in members *ab* and *bc* are written as

$$F_{ab} = 154\,\text{lb (C)} \qquad F_{bc} = 145\,\text{lb (T)}$$

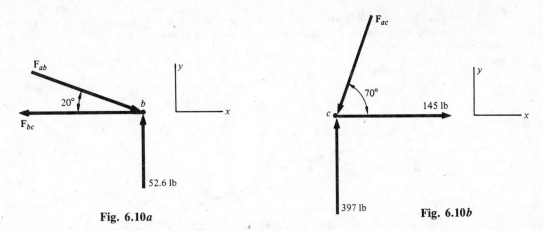

Fig. 6.10a   **Fig. 6.10b**

The free-body diagram of joint c is drawn next, as shown in Fig. 6.10b. Force F_{bc} is a known tensile force, in member bc, which was found above. This force must be shown in the free-body diagram of joint c as exerting a tensile force, or pulling, on the pin at joint c. The unknown force F_{ac} is assumed to be a compressive force, as shown in the figure.

It may be seen from the figure that F_{ac} is the only unknown force in the free-body diagram. Two equations of equilibrium can be written for this concurrent force system. One of these equations may be used to solve for F_{ac}, and the other equilibrium equation can be used as a check on the accuracy of the numerical calculations. For force equilibrium in the x direction,

$$\sum F_x = 0 \qquad 145 - F_{ac}\cos 70° = 0 \qquad F_{ac} = 424 \text{ lb (C)}$$

Since this value is positive, the assumed sense of F_{ac} in Fig. 6.10b was correct.

As a check on the numerical computations, the forces are summed to zero in the y direction. This requirement is

$$\sum F_y = 0$$
$$-F_{ac}\sin 70° + 397 = 0$$
$$424 \sin 70° \overset{?}{=} 397 \qquad 398 \approx 397$$

The above result is within the range of accuracy that is expected in working with three significant figures.

The values of the forces in the truss members may be recorded in tabular form, or they may be written directly on the truss diagram above the members. In using this latter technique, a complete statement of the solution would be as shown in Fig. 6.10c. It may be observed that the forces in the truss members *are independent of the magnitude of the length l.*

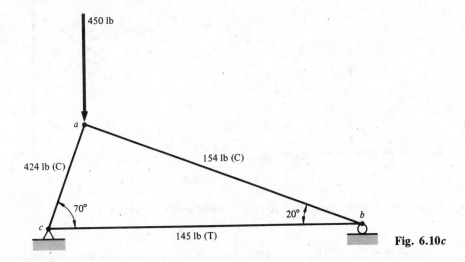

Fig. 6.10c

6.11 Show an exploded free-body diagram of the truss in Fig. 6.10c, including all of the forces that act on the members and on the pins.

■ An exploded free-body diagram of the entire truss is shown in Fig. 6.11. The member forces F_{ab} and F_{ac} exert compressive forces on the pins and on the members, and the member force F_{bc} exerts a tensile force on these elements.

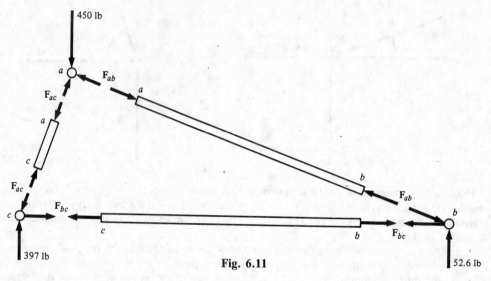

Fig. 6.11

6.12   Show how to identify *zero-force members* in a truss.

■ In certain truss configurations some of the members may be seen, by inspection, to have zero force in them. Two such arrangements are shown in Figs. 6.12a and b.

In Fig. 6.12a, joint a is assumed to have three member forces acting on it. If the forces are summed in the y direction, it may be immediately concluded that

$$\sum F_y = 0 \qquad F_{ab} = 0$$

If the forces are summed in the x direction, it follows that

$$\sum F_x = 0 \qquad F_{ac} = F_{ad}$$

In Fig. 6.12b, a summation of forces normal to the direction of F_{ab} or F_{bd} yields

$$F_{bc} = 0$$

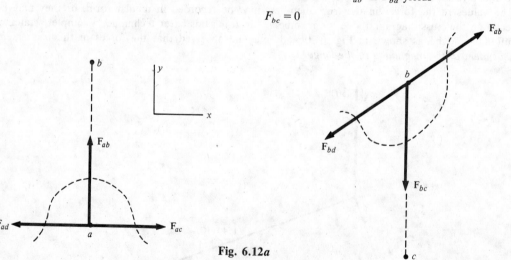

Fig. 6.12a

Fig. 6.12b

If the forces are summed along the above direction, the result is

$$F_{ab} = F_{bd}$$

Members ab, in Fig. 6.12a, and bc, in Fig. 6.12b, are referred to as zero-force members. The fact that a truss contains one or more zero-force members does not mean that these members serve no useful purpose. In certain applications, the original loading on a truss may be changed, and a zero-force member may become a load-bearing member. This effect will be illustrated in the following problems.

6.13 **(a)** Show that the truss in Fig. 6.13a is stable.

(b) Solve for the forces in all the truss members by using the method of joints.

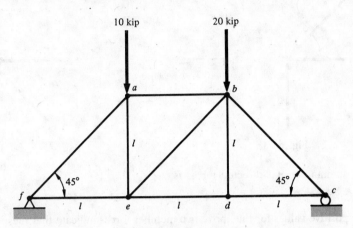

Fig. 6.13a

▌ **(a)** The stability of this truss may be verified by using the criterion in Prob. 6.3. For this case, $m' = 9$ and $n = 6$. Thus,

$$m = 2n - 3 \qquad 9 \overset{?}{=} 2(6) - 3 \qquad 9 \equiv 9$$

(b) The free-body diagram of the truss is shown in Fig. 6.13b. The reaction forces are found from

$$\sum M_f = 0$$

$$-10l - 20(2l) + R_c(3l) = 0 \qquad R_c = 16.7 \text{ kip}$$

$$\sum F_y = 0$$

$$R_{fy} - 10 - 20 + R_c = 0 \qquad R_{fy} - 10 - 20 + 16.7 = 0 \qquad R_{fy} = 13.3 \text{ kip}$$

$$\sum F_x = 0 \qquad R_{fx} = 0$$

Joint d has the configuration of the joint shown in Fig. 6.12a. It may be concluded immediately that member bd is a zero-force member.

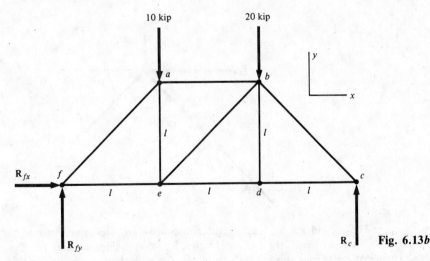

Fig. 6.13b

In using the method of joints, no more than two unknown forces may be solved for at any joint. Based on this, the analysis of this example must start with either joint c or joint f. Joint c is arbitrarily chosen, and the free-body diagram of this element is shown in Fig. 6.13c. The forces will be summed to zero along the x and y axes. If the forces were summed along the x axis first, the resulting equation would contain the two unknowns F_{bc} and F_{cd}. The forces in the y direction are thus summed first, with the result

$$\sum F_y = 0 \qquad -F_{bc}\sin 45° + 16.7 = 0 \qquad F_{bc} = 23.6 \text{ kip (C)}$$

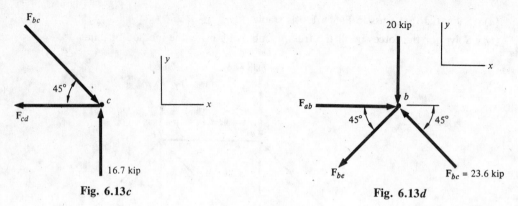

Fig. 6.13c   **Fig. 6.13d**

The summation of x-directed forces is

$$\sum F_x = 0 \qquad F_{bc} \cos 45° - F_{cd} = 0 \qquad 23.6 \cos 45° - F_{cd} = 0 \qquad F_{cd} = 16.7 \text{ kip (T)}$$

The positive values for the above two member forces indicate that these forces act in the senses shown in Fig. 6.13c. Member bc experiences a compressive force, while member cd is in tension. From consideration of joint d, it follows that

$$F_{de} = F_{cd} = 16.7 \text{ kip (T)}$$

The free-body diagram of joint b is drawn next, as shown in Fig. 6.13d. The member force F_{bc} in this diagram is known from the preceding solution. For equilibrium at joint b,

$$\sum F_y = 0 \qquad -20 - F_{be} \sin 45° + 23.6 \sin 45° = 0 \qquad F_{be} = -4.7 \text{ kip}$$

The minus sign in the above result indicates that the sense of force F_{be} in Fig. 6.13d was incorrectly assumed. This situation may be handled in two different ways. First, the free-body diagram is immediately corrected by drawing a wavy line through the vector with the incorrect sense and drawing in the magnitude and *correct* sense of this vector. Figure 6.13e illustrates this effect for the force F_{be} in this problem. In practice, the free-body diagram would not be redrawn. Rather, the change would be made on the original free-body diagram.

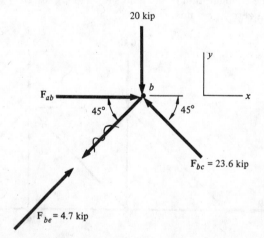

Fig. 6.13e

The second way of treating the situation where a computed force has a minus sign is to *keep* the original sense of the vector in the free-body diagram and let the *magnitude of this term be negative*. Using this method, the force F_{be}, with the sense shown in Fig. 6.13d, would be described as having the magnitude -4.7 kip. The first method of treating computed negative force components will be used consistently throughout this chapter on truss analysis. The second method will be used in the following chapter on framework analysis.

For force equilibrium in the x direction, for the joint in Fig. 6.13e,

$$\sum F_x = 0$$

$$F_{ab} + 4.7 \cos 45° - 23.6 \cos 45° = 0 \qquad F_{ab} = 13.4 \text{ kip (C)}$$

The free-body diagram of joint e is shown in Fig. 6.13f. The equilibrium requirements are

$$\sum F_x = 0$$

$$-F_{ef} - 4.7 \cos 45° + 16.7 = 0 \qquad F_{ef} = 13.4 \text{ kip (T)}$$

$$\sum F_y = 0$$

$$F_{ae} - 4.7 \sin 45° = 0 \qquad F_{ae} = 3.3 \text{ kip (T)}$$

The final joint to be analyzed is a. The free-body diagram of this joint is drawn in Fig. 6.13g. It may be observed that the single unknown member force F_{af} acts at this joint. For force equilibrium in the x direction,

$$\sum F_x = 0$$

$$F_{af} \cos 45° - 13.4 = 0 \qquad F_{af} = 19 \text{ kip (C)}$$

The remaining equation of equilibrium, in the y direction, will now be used to check the correctness of the numerical values for the forces. From Fig. 6.13g,

$$\sum F_y = 0$$

$$-10 + F_{af} \sin 45° - 3.3 \overset{?}{=} 0 \qquad 19 \sin 45° \overset{?}{=} 10 + 3.3 \qquad 13.4 \approx 13.3$$

Further analysis of the truss structure in Fig. 6.13a is given in Prob. 6.24.

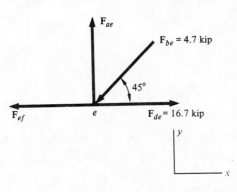

Fig. 6.13f

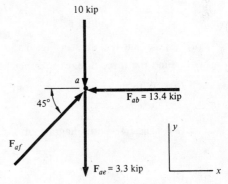

Fig. 6.13g

6.14 (*a*) Show that the truss in Fig. 6.14a is stable.

(*b*) Find the force in each member of the truss.

▌ (*a*) With $m = 3$, $n = 3$, the stability requirement is

$$m = 2n - 3 \qquad 3 \overset{?}{=} 2(3) - 3 \qquad 3 \equiv 3$$

(*b*) The free-body diagram of the truss is shown in Fig. 6.14b. The length l_{ab} of member ab is found from

$$\frac{l_{ab}}{\sin 40°} = \frac{6}{\sin 110°} \qquad l_{ab} = 4.10 \text{ ft}$$

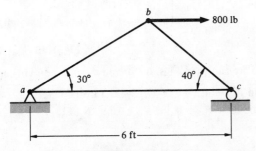

Fig. 6.14a

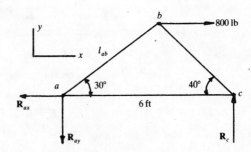

Fig. 6.14b

For equilibrium of the truss,

$$\Sigma M_a = 0 \qquad -800(4.10 \sin 30°) + R_c(6) = 0 \qquad R_c = 273 \text{ lb}$$
$$\Sigma F_y = 0 \qquad -R_{ay} + R_c = 0 \qquad R_{ay} = 273 \text{ lb}$$
$$\Sigma F_x = 0 \qquad R_{ax} = 800 \text{ lb}$$

The free-body diagram of joint c is shown in Fig. 6.14c. For equilibrium,

$$\Sigma F_y = 0 \qquad -F_{bc} \sin 40° + 273 = 0 \qquad F_{bc} = 425 \text{ lb (C)}$$
$$\Sigma F_x = 0 \qquad -F_{ac} + F_{bc} \cos 40° = 0 \qquad F_{ac} = 326 \text{ lb (T)}$$

The free-body diagram of joint b is seen in Fig. 6.14d. The equilibrium requirement is

$$\Sigma F_y = 0 \qquad -F_{ab} \sin 30° + 425 \sin 40° = 0 \qquad F_{ab} = 546 \text{ lb (T)}$$

As a check on the calculations, using Fig. 6.14d,

$$\Sigma F_x = 0$$
$$-546 \cos 30° - 425 \cos 40° + 800 \overset{?}{=} 0 \qquad 546 \cos 30° + 425 \cos 40° \overset{?}{=} 800 \qquad 798 \approx 800$$

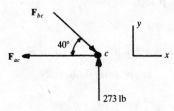

Fig. 6.14c

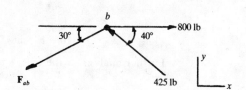

Fig. 6.14d

6.15 (*a*) Show that the truss in Fig. 6.15a is stable.

(*b*) Find the force in each member of the truss.

❚ (*a*) With five members and four joints,

$$m = 2n - 3 \qquad 5 \overset{?}{=} 2(4) - 3 \qquad 5 \equiv 5$$

(*b*) The free-body diagram of the truss is seen in Fig. 6.15b. For equilibrium of the truss,

$$\Sigma M_d = 0 \qquad -R_a(3) + 20(1.50) + 5(2.52) = 0 \qquad R_a = 14.2 \text{ kN}$$
$$\Sigma F_x = 0 \qquad R_{dx} - 5 = 0 \qquad R_{dx} = 5 \text{ kN}$$
$$\Sigma F_y = 0 \qquad R_a - 20 + R_{dy} = 0 \qquad R_{dy} = 5.8 \text{ kN}$$

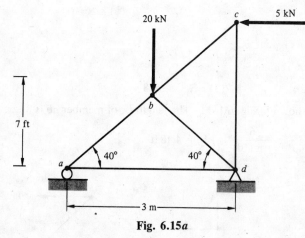

Fig. 6.15a

Fig. 6.15b

The free-body diagrams of joints a and c are shown in Figs. 6.15c and d. For equilibrium of joint a,

$$\Sigma F_y = 0 \qquad -F_{ab} \sin 40° + 14.2 = 0 \qquad F_{ab} = 22.1 \text{ kN (C)}$$
$$\Sigma F_x = 0 \qquad -F_{ab} \cos 40° + F_{ad} = 0 \qquad F_{ad} = 16.9 \text{ kN (T)}$$

For equilibrium of joint c,

$$\Sigma F_x = 0 \qquad F_{bc} \cos 40° - 5 = 0 \qquad F_{bc} = 6.53 \text{ kN (C)}$$
$$\Sigma F_y = 0 \qquad F_{bc} \sin 40° - F_{cd} = 0 \qquad F_{cd} = 4.20 \text{ kN (T)}$$

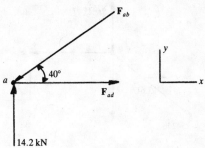

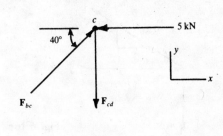

Fig. 6.15c   Fig. 6.15d

The free-body diagram of joint d is shown in Fig. 6.15e. For equilibrium in the y direction,

$$\Sigma F_y = 0 \qquad -F_{bd}\sin 40° + 4.2 + 5.8 = 0 \qquad F_{bd} = 15.6 \text{ kN (C)}$$

As a check on the calculations,

$$\Sigma F_x = 0$$
$$15.6\cos 40° + 5 - 16.9 \overset{?}{=} 0 \qquad 15.6\cos 40° + 5 \overset{?}{=} 16.9 \qquad 17.0 \approx 16.9$$

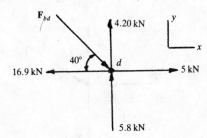

Fig. 6.15e

6.16   Find the forces in the members of the truss shown in Fig. 6.16a.

I   The free-body diagram of the truss is shown in Fig. 6.16b. For equilibrium of the truss,

$$\Sigma M_c = 0 \qquad 300(6) + 275(3) - R_b(3) = 0 \qquad R_b = 875 \text{ lb}$$
$$\Sigma F_y = 0 \qquad -300 - 275 + R_b - R_{cy} = 0 \qquad R_{cy} = 300 \text{ lb}$$
$$\Sigma F_x = 0 \qquad R_{cx} = 0$$

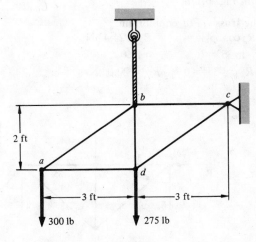

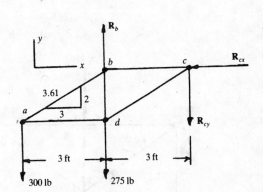

Fig. 6.16a   Fig. 6.16b

From the free-body diagram of joint a, shown in Fig. 6.16c,

$$\Sigma F_y = 0 \qquad \frac{2}{3.61}F_{ab} - 300 = 0 \qquad F_{ab} = 542 \text{ lb (T)}$$

$$\Sigma F_x = 0 \qquad \frac{3}{3.61}F_{ab} - F_{ad} = 0 \qquad F_{ad} = 450 \text{ lb (C)}$$

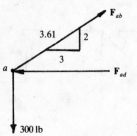

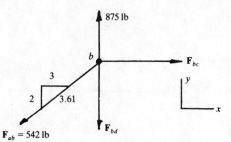

Fig. 6.16c

Fig. 6.16d

From Fig. 6.16d, the free-body diagram of joint b,

$$\Sigma F_y = 0 \qquad 875 - \frac{2}{3.61}(542) - F_{bd} = 0 \qquad F_{bd} = 575 \text{ lb (T)}$$

$$\Sigma F_x = 0 \qquad -\frac{3}{3.61}(542) + F_{bc} = 0 \qquad F_{bc} = 450 \text{ lb (T)}$$

The final free-body diagram, for joint d, is shown in Fig. 6.16e. For equilibrium,

$$\Sigma F_x = 0 \qquad 450 - \frac{3}{3.61}F_{cd} = 0 \qquad F_{cd} = 542 \text{ lb (C)}$$

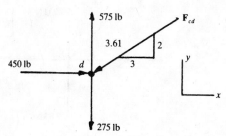

Fig. 6.16e

As a check on the calculations,

$$\Sigma F_y = 0$$

$$575 - \frac{2}{3.61}(542) - 275 \stackrel{?}{=} 0 \qquad \frac{2}{3.61}(542) + 275 \stackrel{?}{=} 575 \qquad 575 \equiv 575$$

6.17 Find the force in each member of the truss shown in Fig. 6.17a.

▌ Figure 6.17b shows the free-body diagram of the truss. For equilibrium,

$$\Sigma M_a = 0 \qquad -40(3) + (R_c \cos 35°)6 = 0 \qquad R_c = 24.4 \text{ kN}$$

$$\Sigma F_x = 0 \qquad R_{ax} - R_c \sin 35° = 0 \qquad R_{ax} = 14.0 \text{ kN}$$

$$\Sigma F_y = 0 \qquad R_{ay} - 40 + R_c \cos 35° = 0 \qquad R_{ay} = 20.0 \text{ kN}$$

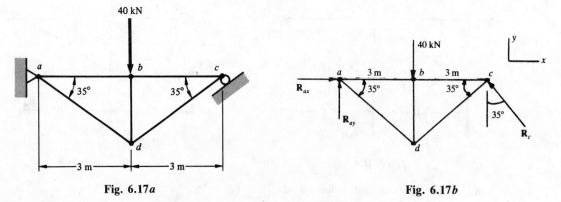

Fig. 6.17a

Fig. 6.17b

Figure 6.17c shows the free-body diagram of joint a. For equilibrium,

$$\Sigma F_y = 0 \qquad 20 - F_{ad} \sin 35° = 0 \qquad F_{ad} = 34.9 \text{ kN (T)}$$

$$\Sigma F_x = 0 \qquad 14 + F_{ad} \cos 35° - F_{ab} = 0 \qquad F_{ab} = 42.6 \text{ kN (C)}$$

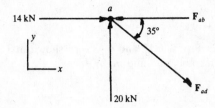

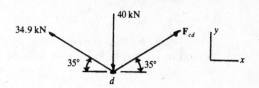

Fig. 6.17c **Fig. 6.17d**

By inspection,

$$F_{bc} = F_{ab} = 42.6 \text{ kN (C)} \qquad F_{bd} = 40 \text{ kN (C)}$$

The free-body diagram of joint d is shown in Fig. 6.17d. From symmetry,

$$F_{cd} = 34.9 \text{ kN (T)}$$

As a check on the calculations, from Fig. 6.17d,

$$\Sigma F_y = 0$$
$$2(34.9 \sin 35°) - 40 \overset{?}{=} 0 \qquad 2(34.9 \sin 35°) \overset{?}{=} 40 \qquad 40 \equiv 40$$

6.18 How would the forces in the members of the truss in Prob. 6.17 change if the method of support of the right end were changed to that shown in Fig. 6.18a?

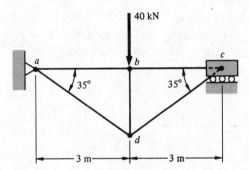

Fig. 6.18a

▌ The free-body diagram of the the truss now has the form shown in Fig. 6.18b. For equilibrium of the entire truss,

$$\Sigma M_a = 0 \qquad -40(3) + R_c(6) = 0 \qquad R_c = 20.0 \text{ kN}$$
$$\Sigma F_y = 0 \qquad R_{ay} - 40 + R_c = 0 \qquad R_{ay} = 20 \text{ kN}$$
$$\Sigma F_x = 0 \qquad R_{ax} = 0$$

For equilibrium of joint a, following Fig. 6.18c,

$$\Sigma F_y = 0 \qquad 20 - F_{ad} \sin 35° = 0 \qquad F_{ad} = 34.9 \text{ kN (T)}$$
$$\Sigma F_x = 0 \qquad F_{ad} \cos 35° - F_{ab} = 0 \qquad F_{ab} = 28.6 \text{ kN (C)}$$

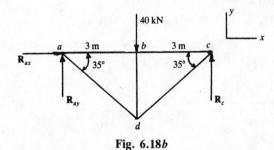

Fig. 6.18b **Fig. 6.18c**

By inspection of Fig. 6.18a,

$$F_{bc} = F_{ab} = 28.6 \text{ kN (C)} \qquad F_{bd} = 40 \text{ kN (C)}$$

From symmetry,

$$F_{cd} = F_{ad} = 34.9 \text{ kN (T)}$$

The results for the two types of support of joint c, in Probs. 6.17 and 6.18, are shown in Table 6.1. The method of support in Prob. 6.18 reduces the forces in members ab and bc by

$$\frac{42.6 - 28.6}{42.6} (100) = 33\%$$

The other member forces are unchanged.

TABLE 6.1

	Prob. 6.17	Prob. 6.18
F_{ab}	42.6 (C)	28.6 (C)
F_{ad}	34.9 (T)	34.9 (T)
F_{bd}	40.0 (C)	40.0 (C)
T_{bc}	42.6 (C)	28.6 (C)
F_{cd}	34.9 (T)	34.9 (T)
R_{ax}	14.0	0
R_{ay}	20.0	20.0
R_c	24.4	20.0

6.19 Find the forces in all of the members of the truss shown in Fig. 6.19a.

▌ From Fig. 6.19b, for equilibrium of the truss,

$$\Sigma M_b = 0 \qquad -(750 \cos 20°)280 - (450 \cos 30°)280 - (450 \sin 30°)150 + R_a(150) = 0 \qquad R_a = 2,270 \text{ N}$$
$$\Sigma F_x = 0 \qquad -R_{bx} + 750 \sin 20° - 450 \sin 30° + R_a = 0 \qquad R_{bx} = 2,300 \text{ N}$$
$$\Sigma F_y = 0 \qquad R_{by} - 750 \cos 20° - 450 \cos 30° = 0 \qquad R_{by} = 1,090 \text{ N}$$

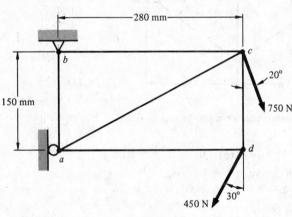

Fig. 6.19a

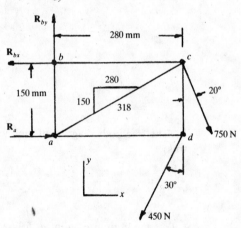

Fig. 6.19b

From Fig. 6.19c, the free-body diagram of joint d,

$$\Sigma F_y = 0 \qquad F_{cd} - 450 \cos 30° = 0 \qquad F_{cd} = 390 \text{ N (T)}$$
$$\Sigma F_x = 0 \qquad F_{ad} - 450 \sin 30° = 0 \qquad F_{ad} = 225 \text{ N (C)}$$

From inspection of Fig. 6.19b,

$$F_{bc} = R_{bx} = 2,300 \text{ N (T)} \qquad F_{ab} = R_{by} = 1,090 \text{ N (T)}$$

From the free-body diagram of joint a in Fig. 6.19d,

$$\Sigma F_y = 0 \qquad 1,090 - \frac{150}{318} F_{ac} = 0 \qquad F_{ac} = 2,310 \text{ N (C)}$$

As a check on the solutions above, using Fig. 6.19d,

$$\Sigma F_x = 0$$

$$2,270 - \frac{280}{318} (2,310) - 225 \overset{?}{=} 0 \qquad \frac{280}{318} (2,310) + 225 \overset{?}{=} 2,270 \qquad 2,260 \approx 2,270$$

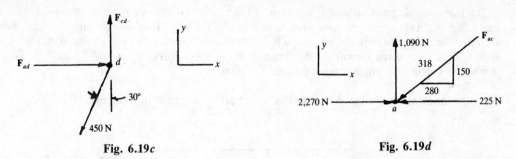

Fig. 6.19c   Fig. 6.19d

6.20   Show the force transmission through a *pulley connection* to the joint of a truss.

▎ *Pulleys*, or *sheaves*, are used to change the direction of a moving or stationary cable.   A pulley is a disk with a groove around its circumference.   It is attached to another structural element by a hinge pin through its center.

   Figure 6.20a shows a typical connection of a pulley to a plane truss.   The force in the cable is represented by the symbol P.   This cable tensile force is constant throughout the length of the cable, and the weight of the pulley is neglected.   The known angle θ in the figure defines the direction of the cable.   For the situation shown, the cable could be moving, or holding in position, the weight which is shown by the dashed outline.   The force effect of the pulley on joint b will be investigated next.

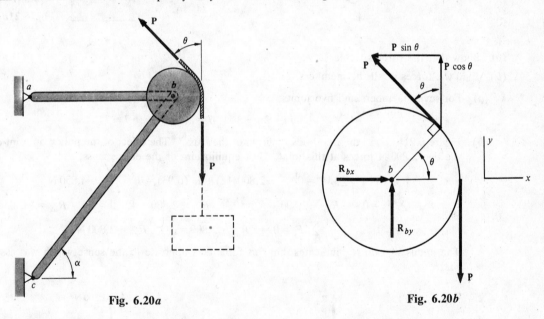

Fig. 6.20a   Fig. 6.20b

   Figure 6.20b shows the free-body diagram of the pulley.   The forces R_{bx} and R_{by} are the forces of joint b on the center of the pulley.   Moment equilibrium about the center of the pulley is identically satisfied, since the pulley is of constant radius and the cable tensile force P is constant.   For force equilibrium in the x and y directions,

$$\sum F_x = 0 \qquad R_{bx} - P\sin\theta = 0 \qquad R_{bx} = P\sin\theta$$

$$\sum F_y = 0 \qquad R_{by} + P\cos\theta - P = 0 \qquad R_{by} = P(1 - \cos\theta)$$

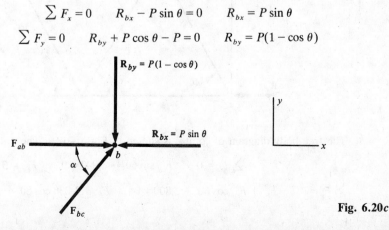

Fig. 6.20c

The free-body diagram of joint b is shown in Fig. 6.20c. By Newton's third law, the forces R_{bx} and R_{by} in Fig. 6.20b will have *opposite senses* when drawn in Fig. 6.20c. From consideration of these figures, it may be seen that the effect of the pulley on joint b is as if the cable forces, with their known magnitudes, directions, and senses, were transferred *directly* to this joint. The member forces F_{ab} and F_{bc} would be found in the usual manner by solving the equilibrium equations for joint b.

6.21   The plane truss shown in Fig. 6.21a acts as a supporting boom, with a pulley at joint c.

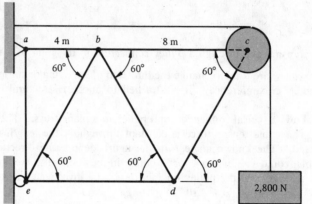

Fig. 6.21a

(*a*)   Show that the truss is stable.

(*b*)   Find the forces in all the members.

▌ (*a*)   For seven members and five joints,

$$m = 2n - 3 \qquad 7 \overset{?}{=} 2(5) - 3 \qquad 7 \equiv 7$$

(*b*)   Figure 6.21b shows the free-body diagram of the truss. The effect of the pulley on joint c is shown by the two 2,800-N forces at this joint. For equilibrium of the entire truss,

$$\sum M_a = 0 \qquad -2,800(12) + R_e(6.93) = 0 \qquad R_e = 4,850 \text{ N}$$

$$\sum F_x = 0 \qquad R_{ax} - 2,800 + R_e = 0 \qquad R_{ax} - 2,800 + 4,850 = 0 \qquad R_{ax} = -2,050 \text{ N}$$

$$\sum F_y = 0 \qquad R_{ay} - 2,800 = 0 \qquad R_{ay} = 2,800 \text{ N}$$

The minus sign on R_{ax} indicates that this force acts opposite to the sense which was assumed in Fig. 6.21b.

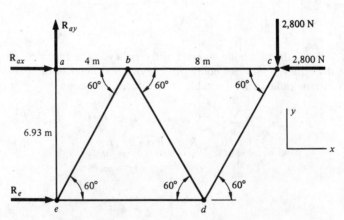

Fig. 6.21b

The free-body diagram of joint c is shown in Fig. 6.21c. The equilibrium requirements are

$$\sum F_y = 0 \qquad F_{cd} \sin 60° - 2,800 = 0 \qquad F_{cd} = 3,230 \text{ N (C)}$$

$$\sum F_x = 0 \qquad F_{bc} + F_{cd} \cos 60° - 2,800 = 0 \qquad F_{bc} + 3,230 \cos 60° - 2,800 = 0 \qquad F_{bc} = 1,190 \text{ N (C)}$$

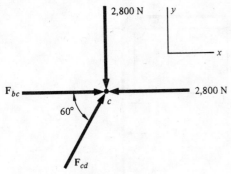

Fig. 6.21c

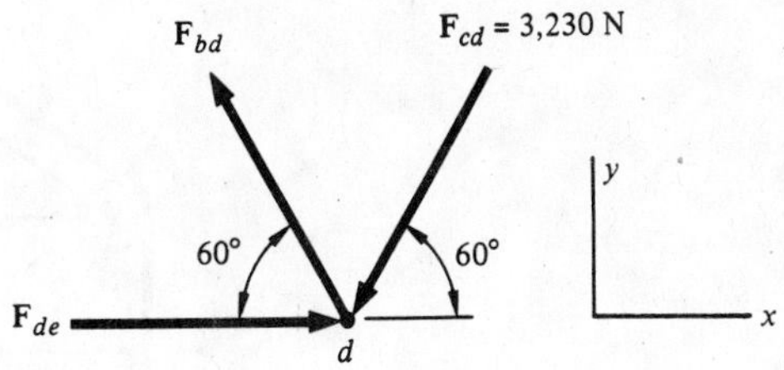

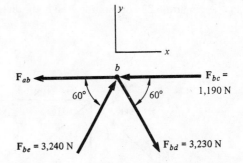

Fig. 6.21d

Joint d is analyzed next, as shown in Fig. 6.21d. For equilibrium,

$$\sum F_y = 0 \qquad F_{bd} \sin 60° - 3{,}230 \sin 60° = 0 \qquad F_{bd} = 3{,}230 \text{ N (T)}$$
$$\sum F_x = 0$$

$$F_{de} - F_{bd} \cos 60° - 3{,}230 \cos 60° = 0 \qquad F_{de} - 3{,}230 \cos 60° - 3{,}230 \cos 60° = 0 \qquad F_{de} = 3{,}230 \text{ N (C)}$$

The next joint to be analyzed is e, and this free-body diagram is shown in Fig. 6.21e. For equilibrium,

$$\sum F_x = 0 \qquad 4{,}850 - F_{be} \cos 60° - 3{,}230 = 0 \qquad F_{be} = 3{,}240 \text{ N (C)}$$

$$\sum F_y = 0 \qquad F_{ae} - F_{be} \sin 60° = 0 \qquad F_{ae} - 3{,}240 \sin 60° = 0 \qquad F_{ae} = 2{,}810 \text{ N (T)}$$

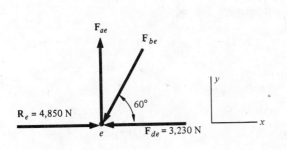

Fig. 6.21e

Fig. 6.21f

The final joint to be analyzed is joint b. The free-body diagram of this joint is shown in Fig. 6.21f. This joint has the one unknown force F_{ab} acting on it. Thus, one of the equilibrium equations may be used to check the calculations. For equilibrium in the x direction,

$$\sum F_x = 0 \qquad -F_{ab} + 3{,}240 \cos 60° + 3{,}230 \cos 60° - 1{,}190 = 0 \qquad F_{ab} = 2{,}050 \text{ N (T)}$$

The summation of forces in the y direction, within the accuracy of three-significant-figure computations, is identically zero. It is left as an exercise for the reader to verify that joint a is in equilibrium under the influence of the four forces F_{ab}, F_{ae}, R_{ax}, and R_{ay} which act on it.

6.22   The truss in Fig. 6.22a maintains a fixed separation distance between the cables. Find the forces in the members of this truss. The diameters of all pulleys are 300 mm.

❚ The lengths l of members ae and de are found as

$$l = 1.5 - \frac{0.3}{2} = 1.35 m$$

The cable forces acting on the pulleys are transferred directly to joints a, b, and c. The free-body diagram of the truss is shown in Fig. 6.22b. For equilibrium of the truss,

$$\sum M_e = 0 \qquad 15(\cancel{1.35}) - 15(\cancel{1.35}) + R_d(\cancel{1.35}) = 0 \qquad R_d = 0$$
$$\sum F_y = 0 \qquad -15 + R_{ey} - 15 = 0 \qquad R_{ey} = 30 \text{ kN}$$
$$\sum F_x = 0 \qquad R_{ex} = 0$$

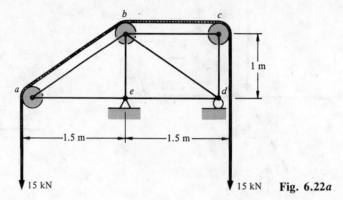

15 kN   15 kN   **Fig. 6.22a**

It may be observed that the pairs of 15-kN forces acting along the directions of members ab and bc are self-canceling and thus produce no moments about joint e or resultant forces in these member directions.

The free-body diagram for joint a is shown in Fig. 6.22c. For equilibrium,

$$\Sigma F_y = 0 \qquad -15 + \frac{1}{1.68}(15) + \frac{1}{1.68} F_{ab} = 0 \qquad F_{ab} = 10.2 \text{ kN (T)}$$

$$\Sigma F_x = 0 \qquad \frac{1.35}{1.68}(15) + \frac{1.35}{1.68} F_{ab} - F_{ae} = 0 \qquad F_{ae} = 20.3 \text{ kN (C)}$$

By inspection,

$$F_{de} = F_{ae} = 20.3 \text{ kN (C)} \qquad F_{be} = R_{ey} = 30.0 \text{ kN (C)}$$
$$F_{cd} = 15.0 \text{ kN (C)} \qquad F_{bc} = 15.0 \text{ kN (C)}$$

Figure 6.22d shows the free-body diagram for joint d. For equilibrium,

$$\Sigma F_x = 0 \qquad 20.3 - \frac{1.35}{1.68} F_{bd} = 0 \qquad F_{bd} = 25.3 \text{ kN (T)}$$

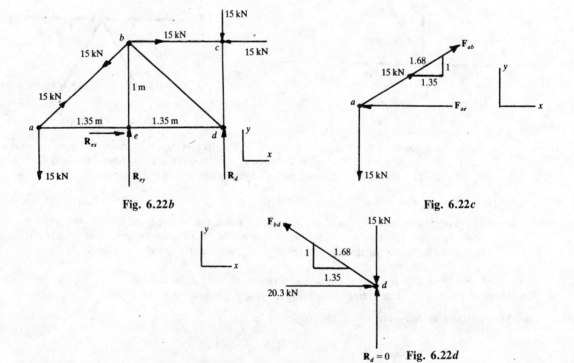

Fig. 6.22b   **Fig. 6.22c**

$R_d = 0$ **Fig. 6.22d**

As a check on the above calculations,

$$\Sigma F_y = 0$$

$$\frac{1}{1.68} F_{bd} - 15 \overset{?}{=} 0 \qquad \frac{1}{1.68}(25.3) \overset{?}{=} 15 \qquad 15.1 \approx 15$$

6.23 The truss of Prob. 6.22 is inverted, as shown in Fig. 6.23a. Find the forces in the truss members and compare these forces with the forces found in Prob. 6.22.

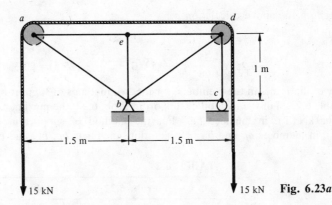

15 kN   15 kN   **Fig. 6.23a**

I The free-body diagram of the truss is shown in Fig. 6.23b. The pair of 15-kN forces acting along the direction *aed* is self-canceling and thus produces no moment about *b* or resultant force on the truss. For equilibrium,

$$\Sigma M_b = 0 \qquad 15(1.35) - 15(1.35) + R_c(1.35) = 0 \qquad R_c = 0$$
$$\Sigma F_y = 0 \qquad -15 + R_{by} - 15 = 0 \qquad R_{by} = 30 \text{ kN}$$
$$\Sigma F_x = 0 \qquad R_{bx} = 0$$

The free-body diagram for joint *a* is shown in Fig. 6.23c. For equilibrium,

$$\Sigma F_y = 0 \qquad -15 + \frac{1}{1.68} F_{ab} = 0 \qquad F_{ab} = 25.2 \text{ kN (C)}$$

$$\Sigma F_x = 0 \qquad 15 - \frac{1.35}{1.68} F_{ab} + F_{ae} = 0 \qquad F_{ae} = 5.25 \text{ kN (T)}$$

By inspection of the truss diagram in Fig. 6.23b,

$$F_{de} = F_{ae} = 5.25 \text{ kN (T)}$$
$$F_{be} = 0 \qquad F_{cd} = 0 \qquad F_{bc} = 0$$

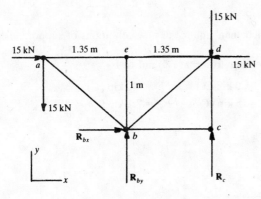

Fig. 6.23b

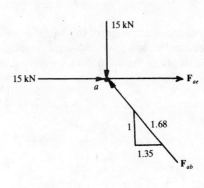

Fig. 6.23c

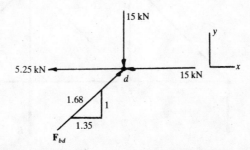

Fig. 6.23d

The free-body diagram of joint d is seen in Fig. 6.23d. The equilibrium requirement is

$$\Sigma F_y = 0 \qquad \frac{1}{1.68} F_{bd} - 15 = 0 \qquad F_{bd} = 25.2 \text{ kN (C)}$$

As a check on the above calculations,

$$\Sigma F_x = 0$$

$$-5.25 + \frac{1.35}{1.68} F_{bd} - 15 \overset{?}{=} 0 \qquad \frac{1.35}{1.68} (25.2) \overset{?}{=} 20.3 \qquad 20.3 \equiv 20.3$$

A summary of the forces in the members, for the two methods of support of the truss in Probs. 6.22 and 6.23, is given in Table 6.2. For the method of support in Prob. 6.22, the maximum value of force in any member of the truss, is 30.0 kN (T), in member be. For the method of support in Prob. 6.23, the maximum force is 25.2 kN (C), in members ab and bd. In addition, members bc, be, and cd are zero-force members.

TABLE 6.2

	Prob. 6.22	Prob. 6.23
F_{ab}	12.0 kN (T)	25.2[†] kN (C)
F_{ae}	20.3 kN (C)	5.25 kN (T)
F_{bc}	15.0 kN (C)	0
F_{bd}	25.3 kN (T)	25.2[†] kN (C)
F_{be}	30.0[†] kN (C)	0
F_{cd}	15.0 kN (C)	0
F_{de}	20.3 kN (C)	5.25 kN (T)

[†]Maximum value.

6.24   The following problem is posed for the plane truss shown in Prob. 6.13. The existing physical truss must support the applied loads of 10 and 20 kip at joints a and b, respectively, as shown in Fig. 6.13a. In the original design, the truss is supported by the ground at joints c and f.

Because of repairs which are to be made to the foundation, the truss must be supported temporarily first at joints d and f and then at joints c and e. Could these two different methods of support of the truss result in forces in the members which are in excess of those in the original design?

▌ Figure 6.24a shows the truss supported at joints d and f. The reaction forces of the ground on the truss are found to be

$$R_d = 25 \text{ kip} \qquad R_f = 5 \text{ kip}$$

From consideration of the figure it may be concluded that, under the new conditions of support, members bc, cd, and de are now zero-force members.

By using the method of joints, the remaining member forces are found to be

$$F_{af} = 7.1 \text{ kip (C)} \qquad F_{ef} = 5.0 \text{ kip (T)} \qquad F_{ab} = 5.0 \text{ kip (C)}$$
$$F_{ae} = 5.0 \text{ kip (C)} \qquad F_{bd} = 25 \text{ kip (C)} \qquad F_{be} = 7.1 \text{ kip (T)}$$

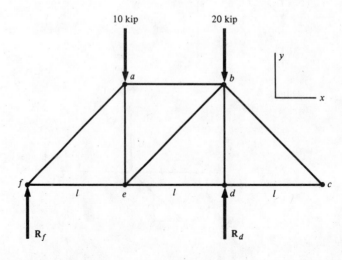

Fig. 6.24a

Figure 6.24b shows the free-body diagram for the case where the truss is supported at joints c and e. The reaction forces of the ground on the truss are

$$R_e = 20 \text{ kip} \qquad R_c = 10 \text{ kip}$$

By inspection, the members af, bd, and ef are zero-force members. Since member af has zero force in it, it follows that member ab is also a zero-force member.

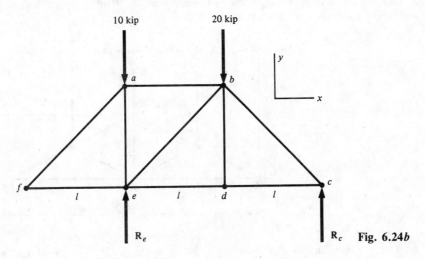

Fig. 6.24b

By using the method of joints, the remaining member forces are found to be

$$F_{bc} = 14.1 \text{ kip (C)}$$
$$F_{be} = 14.1 \text{ kip (C)}$$
$$F_{cd} = 10.0 \text{ kip (T)}$$
$$F_{ae} = 10.0 \text{ kip (C)}$$
$$F_{de} = 10.0 \text{ kip (T)}$$

The results for the three methods of support of the truss of Prob. 6.13 are shown in Table 6.3.

TABLE 6.3									
	ab	bc	cd	de	ef	af	ae	be	bd
10 kip, 20 kip (supported at f and c)	13.4 (C)	23.6† (C)	16.7 (T)	16.7 (T)	13.4 (T)	19.0 (C)	3.3 (T)	4.7 (C)	0
10 kip, 20 kip (supported at f and d)	5.0 (C)	0	0	0	5.0 (T)	7.1 (C)	5.0 (C)	7.1 (T)	25.0† (C)
10 kip, 20 kip (supported at e and c)	0	14.1† (C)	10.0 (T)	10.0 (T)	0	0	10.0 (C)	14.1† (C)	0

† Maximum value.

The maximum member force in the original method of support at joints c and f in Prob. 6.13 is $F_{bc} = 23.6$ kip. For support at joints d and f, the maximum member force is $F_{bd} = 25.0$ kip. For support at joints c and e, the maximum member force is 14.1 kN, in members bc and be. Thus, the temporary support of the truss *does not* produce any member forces which are *greater* than those in the original method of support, shown in Prob. 6.13.

6.25 A roof truss is shown in Fig. 6.25a. The amount of "slant" of the roof is given by the pitch, defined as h/l. Find the forces in the truss members for $h/l = \frac{1}{3}$.

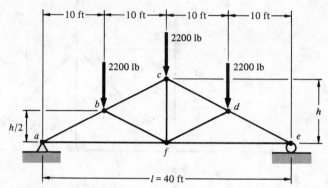

Fig. 6.25a

▌ The free-body diagram of the truss is shown in Fig. 6.25b. From symmetry, using $\Sigma F_y = 0$,

$$R_{ay} = R_e = \frac{3(2,200)}{2} = 3,300 \text{ lb} \qquad \Sigma F_x = 0 \qquad R_{ax} = 0$$

Height h is found from

$$\frac{h}{l} = \frac{1}{3} = \frac{h}{40} \qquad h = 13.3 \text{ ft}$$

The free-body diagrams of joints a and b are shown in Figs. 6.25c and d. For equilibrium, from Fig. 6.25c,

$$\Sigma F_y = 0 \qquad -\frac{13.3}{24} F_{ab} + 3,300 = 0 \qquad F_{ab} = 5,950 \text{ lb (C)}$$

$$\Sigma F_x = 0 \qquad F_{af} - \frac{20}{24} F_{ab} = 0 \qquad F_{af} = 4,960 \text{ lb (T)}$$

Using Fig. 6.25d,

$$\Sigma F_x = 0 \qquad \frac{20}{24}(5,950 - F_{bc} - F_{bf}) = 0 \qquad -F_{bc} - F_{bf} + 5,950 = 0 \qquad (1)$$

$$\Sigma F_y = 0 \qquad -2,200 + \frac{13.3}{24}(5,950 + F_{bf} - F_{bc}) = 0 \qquad -F_{bc} + F_{bf} + 5,950 - 3,970 = 0 \qquad (2)$$

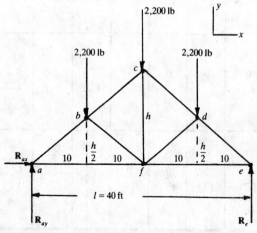

Fig. 6.25b

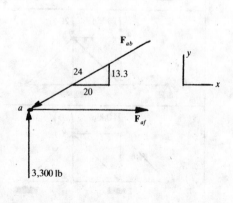

Fig. 6.25c

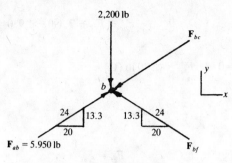

Fig. 6.25d

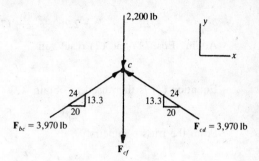

Fig. 6.25e

Equations (1) and (2) are added, with the result

$$-2F_{bc} + 7,930 = 0 \qquad F_{bc} = 3,970 \text{ lb (C)}$$

Using Eq. (1),

$$F_{bf} = 5,950 - F_{bc} = 1,980 \text{ lb (C)}$$

From the symmetry of the truss construction,

$$F_{de} = F_{ab} \qquad F_{cd} = F_{bc} \qquad F_{df} = F_{bf} \qquad F_{ef} = F_{af}$$

Figure 6.25e shows the free-body diagram of joint c. For equilibrium,

$$\Sigma F_y = 0 \qquad 2\left(\frac{13.3}{24}\right)3,970 - F_{cf} - 2,200 = 0 \qquad F_{cf} = 2,200 \text{ lb (T)}$$

6.26 Do the same as in Prob. 6.25, for $h/l = \frac{1}{4}$.

▮ The height h is given by

$$\frac{h}{l} = \frac{1}{4} = \frac{h}{40} \qquad h = 10 \text{ ft}$$

The truss reaction force components R_{ax}, R_{ay}, and R_e are the same as those found in Prob. 6.25. The free-body diagrams of joints a and b are seen in Figs. 6.26a and b. For equilibrium of these joints, using Fig. 6.26a first,

$$\Sigma F_y = 0 \qquad -\frac{10}{22.4} F_{ab} + 3,300 = 0 \qquad F_{ab} = 7,390 \text{ lb (C)}$$

$$\Sigma F_x = 0 \qquad -\frac{20}{22.4}(7,390) + F_{af} = 0 \qquad F_{af} = 6,600 \text{ lb (T)}$$

$$\Sigma F_x = 0 \qquad \frac{20}{22.4}(7,390 - F_{bc} - F_{bf}) = 0 \qquad -F_{bc} - F_{bf} + 7,390 = 0 \qquad (1)$$

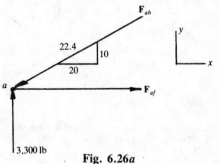

Fig. 6.26a

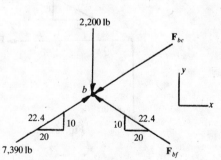

Fig. 6.26b

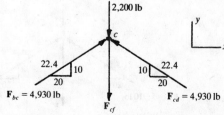

Fig. 6.26c

$$\Sigma F_y = 0 \qquad -2,200 + \frac{10}{22.4}(7,390 + F_{bf} - F_{bc}) = 0 \qquad -F_{bc} + F_{bf} + 7,390 - 4,930 = 0 \qquad (2)$$

Adding Eqs. (1) and (2) results in

$$-2F_{bc} + 9,850 = 0 \qquad F_{bc} = 4,930 \text{ lb (C)}$$

Equation (1) is then used to obtain

$$F_{bf} = 7,390 - F_{bc} = 2,460 \text{ lb (C)}$$

From the truss symmetry,

$$F_{de} = F_{ab} \qquad F_{cd} = F_{bc} \qquad F_{df} = F_{bf} \qquad F_{ef} = F_{af}$$

The free-body diagram of joint c is shown in Fig. 6.26c. The equilibrium requirement is

$$\Sigma F_y = 0 \qquad 2\left(\frac{10}{22.4}\right)4,930 - F_{cf} - 2,200 = 0 \qquad F_{cf} = 2,200 \text{ lb (T)}$$

6.27 (*a*) Find the forces in the members of the truss in Prob. 6.25, for $h/l = \frac{1}{5}$.

(*b*) Compare the truss member forces for the three truss shapes in Probs. 6.25, 6.26, and 6.27.

▌ (*a*) R_{ax}, R_{ay}, and R_e are the same as in Prob. 6.25.

$$\frac{h}{l} = \frac{1}{5} = \frac{h}{40} \qquad h = 8 \text{ ft}$$

For joint a, using Fig. 6.27a,

$$\Sigma F_y = 0 \qquad -\frac{8}{21.5}F_{ab} + 3,300 = 0 \qquad F_{ab} = 8,870 \text{ lb (C)}$$

$$\Sigma F_x = 0 \qquad -\frac{20}{21.5}F_{ab} + F_{af} = 0 \qquad F_{af} = 8,250 \text{ lb (T)}$$

For joint b, from Fig. 6.27b,

$$\Sigma F_x = 0 \qquad \frac{20}{21.5}(8,870 - F_{bc} - F_{bf}) = 0 \qquad -F_{bc} - F_{bf} + 8,870 = 0 \qquad (1)$$

$$\Sigma F_y = 0 \qquad -2,200 + \frac{8}{21.5}(8,870 + F_{bf} - F_{bc}) = 0 \qquad -F_{bc} + F_{bf} + 8,870 - 5,910 = 0 \qquad (2)$$

Equations (1) and (2) are added to obtain

$$-2F_{bc} + 11,800 = 0 \qquad F_{bc} = 5,910 \text{ lb (C)}$$

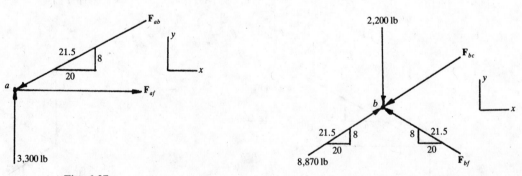

Fig. 6.27a

Fig. 6.27b

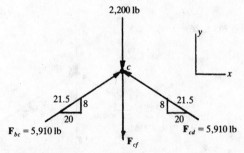

Fig. 6.27c

From Eq. (1),

$$F_{bf} = 8,870 - F_{bc} = 2,960 \text{ lb (C)}$$

From the symmetry of the truss structure,

$$F_{de} = F_{ab} \qquad F_{cd} = F_{bc} \qquad F_{df} = F_{bf} \qquad F_{ef} = F_{af}$$

For joint c, from Fig. 6.27c,

$$\Sigma F_y = 0 \qquad 2\left(\frac{8}{21.5}\right)5,910 - 2,200 - F_{cf} = 0 \qquad F_{cf} = 2,200 \text{ lb (T)}$$

(b) The truss member forces for the h/l values of Probs. 6.25, 6.26, and 6.27 are shown in Table 6.4. It may be seen from the table that all member forces, except those in member cf, increase with decreasing values of the pitch h/l. The interesting feature of this truss design is that member force F_{cf} has a *constant* value that is independent of the pitch.

TABLE 6.4

	Prob. 6.25 (a) $\frac{h}{l} = \frac{1}{3}$	Prob. 6.26 (b) $\frac{h}{l} = \frac{1}{4}$	Prob. 6.27 (c) $\frac{h}{l} = \frac{1}{5}$
F_{ab}, F_{de}	5,950 lb (C)	7,390 lb (C)	8,870 lb (C)
F_{af}, F_{ef}	4,960 lb (T)	6,600 lb (T)	8,250 lb (T)
F_{bc}, F_{cd}	3,970 lb (C)	4,930 lb (C)	5,910 lb (C)
F_{bf}, F_{df}	1,980 lb (C)	2,460 lb (C)	2,960 lb (C)
F_{cf}	2,200 lb (T)	2,200 lb (T)	2,200 lb (T)

6.28 (a) Show that the truss in Fig. 6.28a is stable.

(b) Find the forces in all of the members of the truss.

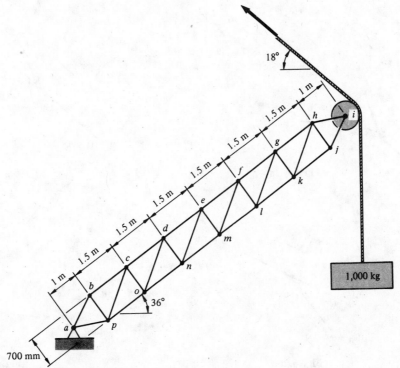

Fig. 6.28a

▮ (a) With 29 members and 16 joints,

$$m = 29 \qquad n = 16 \qquad m = 2n - 3$$
$$29 \overset{?}{=} 2(16) - 3 \qquad 29 \equiv 29$$

Since the truss has forces applied only at its ends (joints a and i), *the entire structure* behaves as a single truss link. Thus, the resultants of the applied forces at a and i must have directions that are collinear with the longitudinal axis of the truss.

The free-body diagram of the truss is shown in Fig. 6.28*b*. For equilibrium of the truss,

$$\Sigma F_x = 0 \qquad R_a - 2(9{,}810 \cos 54°) = 0 \qquad R_a = 11{,}500 \text{ N}$$

The free-body diagram of joint a is seen in Fig. 6.28*c*. From this figure,

$$\tan \theta = \frac{0.35}{1} \qquad \theta = 19.3°$$

From symmetry,

$$F_{ab} = F_{ap}$$

For equilibrium of joint a,

$$\Sigma F_x = 0 \qquad 11{,}500 - 2F_{ab} \cos 19.3° = 0 \qquad F_{ab} = F_{ap} = 6{,}090 \text{ N (C)}$$

From the symmetry of the ends of the structure,

$$F_{hi} = F_{ij} = F_{ab} = F_{ap} = 6{,}090 \text{ N (C)}$$

Figure 6.28*d* shows the free-body diagram of joint b. The equilibrium requirements for joint b are

$$\Sigma F_x = 0 \qquad 6{,}090 \cos 19.3° - F_{bc} = 0 \qquad F_{bc} = 5{,}750 \text{ N (C)}$$
$$\Sigma F_y = 0 \qquad 6{,}090 \sin 19.3° - F_{bp} = 0 \qquad F_{bp} = 2{,}010 \text{ N (T)}$$

From the symmetry of the ends of the truss, it follows that

$$F_{hj} = F_{bp} = 2{,}010 \text{ N (T)}$$

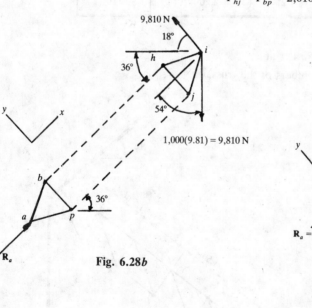

Fig. 6.28b

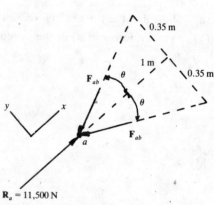

Fig. 6.28c

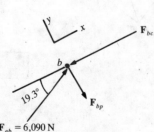

Fig. 6.28d

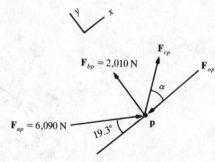

Fig. 6.28e

The free-body diagram of joint p is shown in Fig. 6.28e. For equilibrium of point p in the y direction,

$$\Sigma F_y = 0 \qquad -6{,}090 \sin 19.3° + 2{,}010 + F_{cp} \sin \alpha = 0 \qquad F_{cp} \sin \alpha = 0 \qquad F_{cp} = 0$$

It may be observed that the numerical value of α is not required in the above calculation.

For equilibrium of joint p in the x direction,

$$\Sigma F_x = 0 \qquad 6{,}090 \cos 19.3° - F_{op} = 0 \qquad F_{op} = 5{,}750 \text{ N (C)}$$

Since $F_{cp} = 0$, consideration of a free-body diagram of joint c would reveal that co is a zero-force member, so that $F_{co} = 0$. In view of the repeating pattern of the truss panels, it may be concluded that the forces in all of the diagonal members, and all of the cross members except bp and hj, are zero. The actual use of these diagonal and cross members is for lateral stability of the truss shape. It may further be concluded that the total compressive load on the truss, from the end joints a and i, is shared equally by all the outside members. Thus,

$$F_{cp} = F_{do} = F_{en} = F_{fm} = F_{gl} = F_{hk} = 0$$
$$F_{co} = F_{dn} = F_{em} = F_{fl} = F_{gk} = 0$$
$$F_{bc} = F_{cd} = F_{de} = F_{ef} = F_{fg} = F_{gh} = F_{jk} = F_{kl} = F_{lm} = F_{mn} = F_{no} = F_{op} = 5{,}750 \text{ N (C)}$$

6.3 METHOD OF JOINTS USING FORCE EQUILIBRIUM REQUIREMENT FOR A CLOSED FORCE TRIANGLE OR POLYGON

6.29 Describe how the forces in the members of a truss may be found by using the *method of joints*, *together with a closed force triangle or polygon*.

❚ Since the forces which act at each joint in a plane truss form a concurrent force system, the unknown forces in the members may also be found by using the triangle, or polygon, law of force addition. Although this is a graphical technique, it has a corresponding analytical interpretation.

Figure 6.29a shows the free-body diagram of a plane truss. The directions of the inclined members are given in terms of the slopes of these members. The free-body diagram of joint b is shown in Fig. 6.29b, and F_{ab} and F_{bc} are the unknown forces in members ab and bc, respectively. The forces which act on joint b form a concurrent force system which is in equilibrium. A necessary condition of this equilibrium is that the resultant of the forces which act at this joint be zero, or that the force triangle of the three forces close. This requirement was illustrated in Prob. 2.7. For the present problem, this effect is shown in Fig. 6.29c. The small triangle in the figure is defined by the slope of member ab, while the large triangle is the force triangle of the three forces which act at the joint. These two triangles are similar, so that their corresponding sides are in proportion. Thus,

$$\frac{F_{ab}}{5} = \frac{F_{bc}}{4} = \frac{2{,}000}{3} \qquad F_{ab} = 3{,}330 \text{ N (C)} \qquad F_{bc} = 2{,}670 \text{ N (T)}$$

When using the triangle law, the senses of the unknown forces are chosen in such a way as to make the triangle close. Member ab has a compressive force in it, and member bc experiences a tensile force. Figure 6.29c could have been solved graphically for the above values. The above equations are an analytical interpretation of this graphical solution.

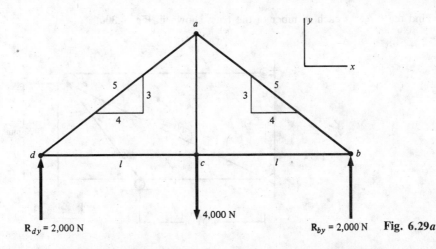

R_{dy} = 2,000 N R_{by} = 2,000 N   **Fig. 6.29a**

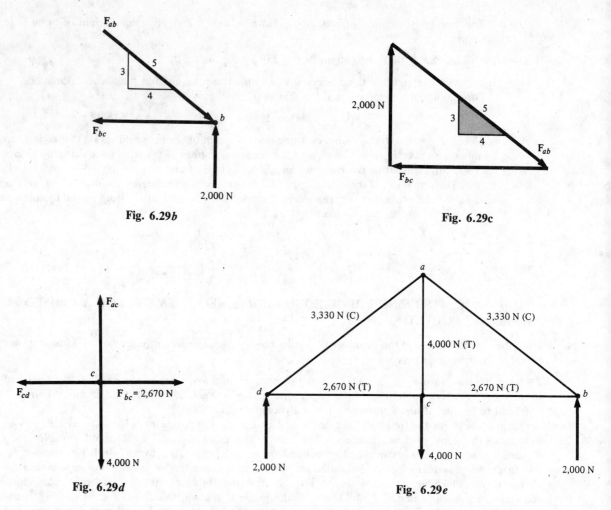

Fig. 6.29b

Fig. 6.29c

Fig. 6.29d

Fig. 6.29e

Figure 6.29d shows the free-body diagram of joint c. The forces F_{ac} and F_{cd} are found by inspection to be

$$F_{ac} = 4,000 \text{ N (T)} \qquad F_{cd} = 2,670 \text{ N (T)}$$

From symmetry considerations, it may be concluded that the force in member ad is the same as the force in member ab. The complete description of the forces in the members of the truss is shown in Fig. 6.29e.

It may be seen from the above solution that the use of the closed force triangle to find the member forces requires less computational effort than the use of the concurrent force system equilibrium equations, presented in Prob. 6.9. In certain problems involving complex truss geometry, however, using both techniques in the same problem, which minimizes the total computational effort, may be more efficient. This procedure will be seen in several of the following problems.

6.30 Find the force in each member of the truss shown in Fig. 6.30a.

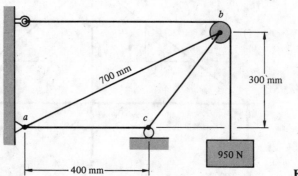

Fig. 6.30a

❚ The free-body diagram of the truss is shown in Fig. 6.30*b*. The dimension A is

$$A = \sqrt{700^2 - 300^2} = 632 \text{ mm}$$

For equilibrium of the truss,

$$\Sigma M_a = 0 \qquad R_c(400) + 950(300) - 950(632) = 0 \qquad R_c = 789 \text{ N}$$
$$\Sigma F_x = 0 \qquad R_{ax} - 950 = 0 \qquad R_{ax} = 950 \text{ N}$$
$$\Sigma F_y = 0 \qquad R_{ay} + R_c - 950 = 0 \qquad R_{ay} = 161 \text{ N}$$

The free-body diagram of joint a and the corresponding force triangle are shown in Figs. 6.30*c* and *d*. For equilibrium of this joint,

$$\frac{161}{300} = \frac{F_{ab}}{700} = \frac{950 - F_{ac}}{632} \qquad F_{ab} = 376 \text{ N (C)} \qquad F_{ac} = 611 \text{ N (C)}$$

The free-body diagram of joint c is shown in Fig. 6.30*e*. The force triangle is not drawn for this case, since there is only one unknown force, member force F_{bc}, acting at this joint. For equilibrium,

$$\Sigma F_x = 0 \qquad 611 - \frac{232}{379} F_{bc} = 0 \qquad F_{bc} = 998 \text{ N (C)}$$

As a check on the above calculations, using Fig. 6.30*e*,

$$\Sigma F_y = 0$$

$$789 - \frac{300}{379}(998) \overset{?}{=} 0 \qquad 789 \approx 790$$

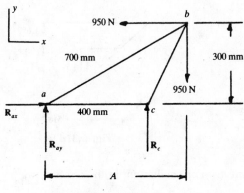

Fig. 6.30*b*

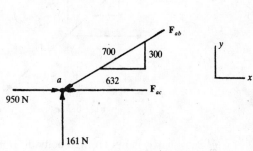

Fig. 6.30*c*

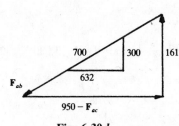

Fig. 6.30*d*

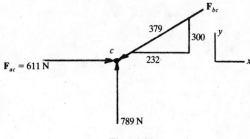

Fig. 6.30*e*

6.31 Find the member forces for the truss shown in Fig. 6.31*a*.

❚ The free-body diagram of the truss is seen in Fig. 6.31*b*. Dimension A is found from

$$\frac{A}{4} = \frac{7}{6.5} \qquad A = 4.31 \text{ ft}$$

For equilibrium of the entire truss,

$$\Sigma M_a = 0 \qquad \frac{6.5}{9.55}(F_{bc})4.31 + 1,650(4.31) - 1,650(7) = 0 \qquad F_{bc} = 1,510 \text{ lb (T)}$$

$$\Sigma F_x = 0 \qquad R_{ax} - \frac{7}{9.55} F_{bc} = 0 \qquad R_{ax} = 1,110 \text{ lb}$$

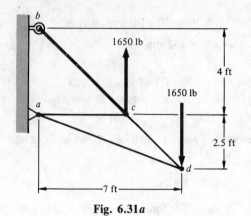

Fig. 6.31a

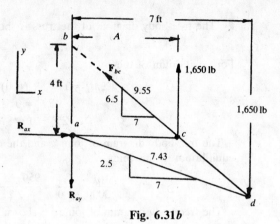

Fig. 6.31b

$$\Sigma F_y = 0 \qquad -R_{ay} + \frac{6.5}{9.55} F_{bc} + 1,650 - 1,650 = 0 \qquad R_{ay} = 1,030 \text{ lb}$$

The free-body diagram and the force triangle for joint c are shown in Figs. 6.31c and d. For equilibrium of joint c,

$$\frac{1,650}{6.5} = \frac{F_{ac}}{7} = \frac{F_{cd} - 1,510}{9.55} \qquad F_{ac} = 1,780 \text{ lb (T)} \qquad F_{cd} = 3,930 \text{ lb (T)}$$

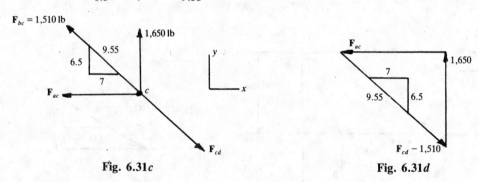

Fig. 6.31c   Fig. 6.31d

Figure 6.31e shows the free-body diagram of joint d. For equilibrium in the x direction,

$$\Sigma F_x = 0 \qquad \frac{7}{7.43} F_{ad} - \frac{7}{9.55} (3,930) = 0 \qquad F_{ad} = 3,060 \text{ lb (C)}$$

As a final check on the calculations,

$$\Sigma F_y = 0$$

$$-1,650 - \frac{2.5}{7.43}(3,060) + \frac{6.5}{9.55}(3,930) \overset{?}{=} 0 \qquad 1,650 + \frac{2.5}{7.43}(3,060) \overset{?}{=} \frac{6.5}{9.55}(3,930) \qquad 2,680 \approx 2,670$$

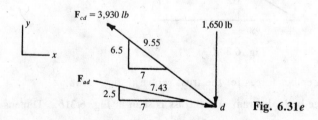

Fig. 6.31e

6.32   Find the forces in all of the members of the truss shown in Fig. 6.32a if the upward force at joint a is (a) 20 kip and (b) 40 kip.

▋ (a)   The free-body diagram of the truss is given in Fig. 6.32b. For equilibrium,

$$\Sigma M_b = 0 \qquad -20(4) + R_c(15) = 0 \qquad R_c = 5.33 \text{ kip}$$
$$\Sigma F_x = 0 \qquad R_{bx} = 0$$
$$\Sigma F_y = 0 \qquad 20 - R_{by} - 20 + R_c = 0 \qquad R_{by} = 5.33 \text{ kip}$$

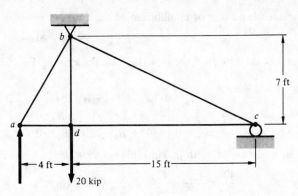

Fig. 6.32a

The free-body diagram and force triangle for joint a are shown in Figs. 6.32c and d. For equilibrium,

$$\frac{20}{7} = \frac{F_{ab}}{8.06} = \frac{F_{ad}}{4} \qquad F_{ab} = 23.0 \text{ kip (C)} \qquad F_{ad} = 11.4 \text{ kip (T)}$$

By inspection,

$$F_{bd} = 20.0 \text{ kip (T)} \qquad F_{cd} = F_{ad} = 11.4 \text{ kip (T)}$$

Figure 6.32e shows the free-body diagram of joint c. For equilibrium,

$$\Sigma F_y = 0 \qquad -\frac{7}{16.6} F_{bx} + 5.33 = 0 \qquad F_{bc} = 12.6 \text{ kip (C)}$$

As a check on the above calculations,

$$\Sigma F_x = 0$$

$$\frac{15}{16.6}(12.6) - 11.4 \stackrel{?}{=} 0 \qquad 11.4 \equiv 11.4$$

The pair of applied 20-kip forces acting on the truss form a couple of magnitude

$$M = -Fd = -20(4) = -80 \text{ kip} \cdot \text{ft}$$

The pair of reaction forces R_{by} and R_c also form a couple of magnitude

$$M_1 = F_1 d = 5.33(15) = 80 \text{ kip} \cdot \text{ft}$$

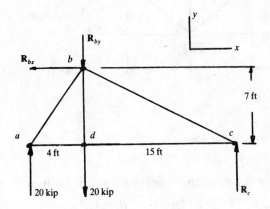

Fig. 6.32b

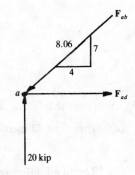

Fig. 6.32c

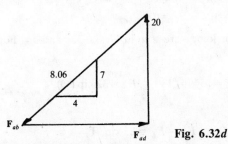

Fig. 6.32d

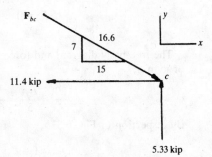

Fig. 6.32e

As an expected condition of equilibrium of the truss,

$$M + M_1 = -80 + 80 \equiv 0$$

(**b**) For the case of an upward force of 40 kip at joint a, using Fig. 6.32b with modification of the force at joint a,

$$\Sigma M_b = 0 \qquad -40(4) + R_c(15) = 0 \qquad R_c = 10.7 \text{ kip}$$
$$\Sigma F_x = 0 \qquad R_{bx} = 0$$
$$\Sigma F_y = 0 \qquad 40 - R_{by} - 20 + R_c = 0 \qquad R_{by} = 30.7 \text{ kip}$$

Using Figs. 6.32c and d, with 20 kip replaced by 40 kip,

$$\frac{40}{7} = \frac{F_{ab}}{8.06} = \frac{F_{ad}}{4} \qquad F_{ab} = 46.1 \text{ kip (C)} \qquad F_{ad} = 22.9 \text{ kip (T)}$$

By inspection,

$$F_{bd} = 20.0 \text{ kip (T)}$$
$$F_{cd} = F_{ad} = 22.9 \text{ kip (T)}$$

From Fig. 6.32e, with 11.4 kip replaced by 22.9 kip and 5.33 kip replaced by 10.7 kip,

$$\Sigma F_y = 0 \qquad -\frac{7}{16.6} F_{bc} + 10.7 = 0 \qquad F_{bc} = 25.4 \text{ kip (C)}$$

As a final check,

$$\Sigma F_x = 0$$
$$\frac{15}{16.6}(25.4) \stackrel{?}{=} 22.9 \qquad 23.0 \approx 22.9$$

6.33 (**a**) Show that the truss in Fig. 6.33a is stable.

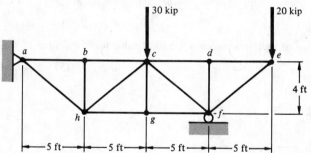

Fig. 6.33a

(**b**) Solve for the forces in all of the members.

(**a**) There are 13 members and 8 joints in the truss. For stability,

$$m = 2n - 3 \qquad 13 \stackrel{?}{=} 2(8) - 3 \qquad 13 \equiv 13$$

(**b**) The free-body diagram of the truss is shown in Fig. 6.33b. For equilibrium of the truss,

$$\Sigma M_a = 0 \qquad -30(10) + R_f(15) \qquad -20(20) = 0 \qquad R_f = 46.7 \text{ kip}$$
$$\Sigma F_y = 0 \qquad R_{ay} - 30 + R_f - 20 = 0 \qquad R_{ay} = 3.3 \text{ kip}$$
$$\Sigma F_x = 0 \qquad R_{ax} = 0$$

The free-body diagram and force triangle for joint e are seen in Figs. 6.33c and d. For equilibrium of joint e,

$$\frac{20}{4} = \frac{F_{de}}{5} = \frac{F_{ef}}{6.4} \qquad F_{de} = 25 \text{ kip (T)} \qquad F_{ef} = 32 \text{ kip (C)}$$

By inspection of Fig. 6.33b,

$$F_{bh} = F_{cg} = F_{df} = 0 \qquad F_{cd} = F_{de} = 25 \text{ kip (T)}$$

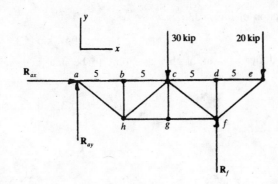

Fig. 6.33b

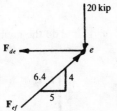

Fig. 6.33c

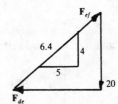

Fig. 6.33d

Figure 6.33e shows the free-body diagram of joint f. Since a simple force triangle cannot be drawn for the force system acting on joint f, the member forces F_{cf} and F_{fg} will be found by direct use of the equilibrium equations. The results are

$$\Sigma F_y = 0 \qquad -\frac{4}{6.4} F_{cf} - \frac{4}{6.4} (32) + 46.7 = 0 \qquad F_{cf} = 42.7 \text{ kip (C)}$$

$$\Sigma F_x = 0 \qquad -F_{fg} + \frac{5}{6.4} F_{cf} - \frac{5}{6.4} (32) = 0 \qquad F_{fg} = 8.4 \text{ kip (T)}$$

By inspection,

$$F_{gh} = F_{fg} = 8.4 \text{ kip (T)}$$

The free-body diagram of joint c is shown in Fig. 6.33f. For the reasons stated above, the equilibrium equations are written directly, with the results

$$\Sigma F_y = 0 \qquad \frac{4}{6.4} F_{ch} - 30 + \frac{4}{6.4} (42.7) = 0 \qquad F_{ch} = 5.3 \text{ kip (C)}$$

$$\Sigma F_x = 0 \qquad F_{bc} + \frac{5}{6.4} F_{ch} + 25 - \frac{5}{6.4} (42.7) = 0 \qquad F_{bc} = 4.2 \text{ kip (C)}$$

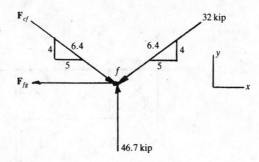

Fig. 6.33e

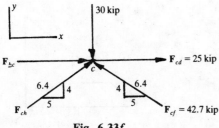

Fig. 6.33f

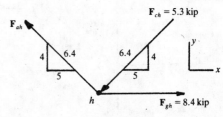

Fig. 6.33g

From inspection of Fig. 6.33b,

$$F_{ab} = F_{bc} = 4.2 \text{ kip (C)}$$

The free-body diagram of the last joint to be analyzed, joint h, is shown in Fig. 6.33g. For equilibrium,

$$\Sigma F_y = 0 \qquad \frac{4}{6.4} F_{ah} - \frac{4}{6.4} (5.3) = 0 \qquad F_{ah} = 5.3 \text{ kip (T)}$$

As a final check,

$$\Sigma F_x = 0$$

$$-\frac{5}{6.4} F_{ah} - \frac{5}{6.4} (5.3) + 8.4 \overset{?}{=} 0 \qquad \frac{5}{6.4} (5.3 + 5.3) \overset{?}{=} 8.4 \qquad 8.3 \approx 8.4$$

6.34 The truss in Prob. 6.33 is now modified to the form shown in Fig. 6.34a. How do the member forces in the new configuration compare with the original values?

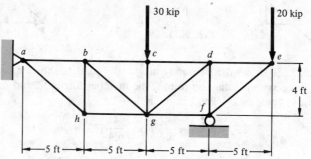

Fig. 6.34a

❚ The free-body diagram of the truss is shown in Fig. 6.34b. For equilibrium,

$$\Sigma M_a = 0 \qquad -30(10) - 20(20) + R_f(15) = 0 \qquad R_f = 46.7 \text{ kip}$$
$$\Sigma F_y = 0 \qquad R_{ay} - 30 - 20 + R_f = 0 \qquad R_{ay} = 3.3 \text{ kip}$$
$$\Sigma F_x = 0 \qquad R_{ax} = 0$$

The reaction forces are seen to be the same as those in Prob. 6.33.

The truss panel def has the same geometry and load as in Prob. 6.33. Using the results from Prob. 6.33,

$$F_{de} = 25 \text{ kip (T)} \qquad F_{ef} = 32 \text{ kip (C)}$$

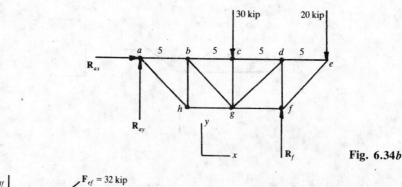

Fig. 6.34b

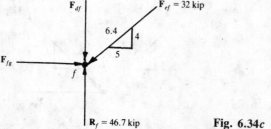

Fig. 6.34c

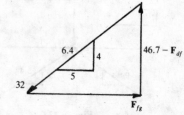

Fig. 6.34d

The free-body diagram and force triangle for joint f are shown in Figs. 6.34c and d. For equilibrium,

$$\frac{32}{6.4} = \frac{F_{fg}}{5} = \frac{46.7 - F_{df}}{4} \qquad F_{fg} = 25 \text{ kip (C)} \qquad F_{df} = 26.7 \text{ kip (C)}$$

By inspection of Fig. 6.34b,

$$F_{cg} = 30 \text{ kip (C)}$$

The free-body diagram and force triangle for joint d are shown in Figs. 6.34e and f. The equilibrium requirement is given by

$$\frac{26.7}{4} = \frac{F_{dg}}{6.4} = \frac{25 + F_{cd}}{5} \qquad F_{dg} = 42.7 \text{ kip (T)} \qquad F_{cd} = 8.4 \text{ kip (C)}$$

From inspection of Fig. 6.34b,

$$F_{bc} = F_{cd} = 8.4 \text{ kip (C)}$$

Figures 6.34g and h show the free-body diagram and force triangle for joint a. For equilibrium,

$$\frac{3.3}{4} = \frac{F_{ab}}{5} = \frac{F_{ah}}{6.4} \qquad F_{ab} = 4.1 \text{ kip (C)} \qquad F_{ah} = 5.3 \text{ kip (T)}$$

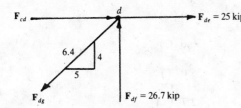

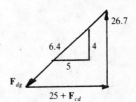

Fig. 6.34e Fig. 6.34f

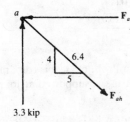

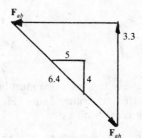

Fig. 6.34g Fig. 6.34h

For joint h, using Figs. 6.34i and j,

$$\frac{5.3}{6.4} = \frac{F_{gh}}{5} = \frac{F_{bh}}{4} \qquad F_{gh} = 4.1 \text{ kip (T)} \qquad F_{bh} = 3.3 \text{ kip (C)}$$

Figure 6.34k shows the free-body diagram and force triangle for the final joint to be analyzed, joint b. For equilibrium of this joint,

$$\Sigma F_y = 0 \qquad 3.3 - \frac{4}{6.4} F_{bg} = 0 \qquad F_{bg} = 5.3 \text{ kip (T)}$$

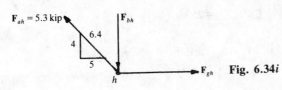

Fig. 6.34i

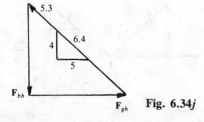

Fig. 6.34j

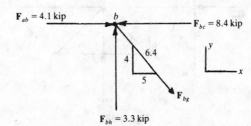

Fig. 6.34k

As a check on the above calculations,

$$\Sigma F_x = 0$$

$$4.1 + \frac{5}{6.4}(5.3) - 8.4 \overset{?}{=} 0 \qquad 8.2 \approx 8.4$$

The agreement of the above two numbers to less than 1 percent is due to the roundoff errors in the many computations in this problem.

A comparison of the member forces in Probs. 6.33 and 6.34 is shown in Table 6.5. It is interesting to note that the maximum member force of 42.7 kip is the same for both truss designs. In Prob. 6.33 this force is compressive, while in Prob. 6.34, it is tensile.

TABLE 6.5

	Prob. 6.33	Prob. 6.34
F_{ab}	4.2 (C)	4.1 (C)
F_{ah}	5.3 (T)	5.3 (T)
F_{bc}	4.2 (C)	8.4 (C)
F_{bg}		5.3 (T)
F_{bh}	0.0	3.3 (C)
F_{cd}	25.0 (T)	8.4 (C)
F_{cf}	42.7† (C)	
F_{cg}	0.0	30.0 (C)
F_{ch}	5.3 (C)	
F_{de}	25.0 (T)	25.0 (T)
F_{df}	0.0	26.7 (C)
F_{dg}		42.7† (T)
F_{ef}	32.0 (C)	32.0 (C)
F_{fg}	8.4 (T)	25.0 (C)
F_{gh}	8.4 (T)	4.1 (T)

†Maximum value.

6.35 Because of repairs which must be made to the support of the truss in Prob. 6.33, this truss must be temporarily supported at another joint. Find the forces in the members if the truss is supported at joint a and by a roller connection at joint g.

▮ The free-body diagram of the truss, for support at joint g, is shown in Fig. 6.35a. For equilibrium,

$$\Sigma M_a = 0 \qquad -30(10) - 20(20) + R_g(10) = 0 \qquad R_g = 70 \text{ kip}$$
$$\Sigma F_x = 0 \qquad R_{ax} = 0$$
$$\Sigma F_y = 0 \qquad -R_{ay} - 30 - 20 + R_g = 0 \qquad R_{ay} = 20 \text{ kip}$$

From Fig. 6.33c in Prob. 6.33,

$$F_{de} = 25 \text{ kip (T)} \qquad F_{ef} = 32 \text{ kip (C)}$$

By inspection, using Fig. 6.35a,

$$F_{cd} = F_{de} = 25 \text{ kip (T)} \qquad F_{df} = 0 \qquad F_{bh} = 0$$

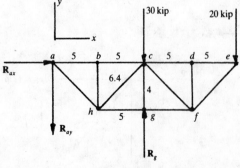

Fig. 6.35a

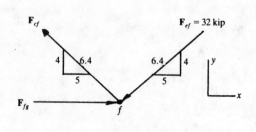

Fig. 6.35b

The free-body diagram of joint f is seen in Fig. 6.35b. For equilibrium,

$$\Sigma F_y = 0 \qquad \frac{4}{6.4} F_{cf} - \frac{4}{6.4} (32) = 0 \qquad F_{cf} = 32 \text{ kip (T)}$$

$$\Sigma F_x = 0 \qquad F_{fg} - 2\left(\frac{5}{6.4}\right)32 = 0 \qquad F_{fg} = 50 \text{ kip (C)}$$

By inspection, using Fig. 6.35a,

$$F_{gh} = F_{fg} = 50 \text{ kip (C)} \qquad F_{cg} = R_g = 70 \text{ kip (C)}$$

Figure 6.35c shows the free-body diagram of joint c. For equilibrium of this joint,

$$\Sigma F_y = 0 \qquad -30 - \frac{4}{6.4} F_{ch} + 70 - \frac{4}{6.4} (32) = 0 \qquad F_{ch} = 32 \text{ kip (T)}$$

$$\Sigma F_x = 0 \qquad -F_{bc} - \frac{5}{6.4} F_{ch} + \frac{5}{6.4} (32) + 25 = 0 \qquad F_{bc} = 25 \text{ kip (T)}$$

By inspection of Fig. 6.35a,

$$F_{ab} = F_{bc} = 25 \text{ kip (T)}$$

The last required free-body diagram, for joint h, is shown in Fig. 6.35d. For equilibrium,

$$\Sigma F_y = 0 \qquad -\frac{4}{6.4} F_{ah} + \frac{4}{6.4} (32) = 0 \qquad F_{ah} = 32 \text{ kip (C)}$$

As a final check on the calculations,

$$\Sigma F_x = 0$$

$$2\left[\frac{5}{6.4} (32)\right] - 50 \overset{?}{=} 0 \qquad 50 \equiv 50$$

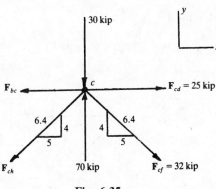

Fig. 6.35c

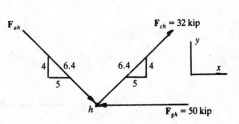

Fig. 6.35d

6.36   (a)   Do the same as in Prob. 6.35, for support of the truss in Prob. 6.33 at joint a and by a cable at joint d.

(b)   Compare the three methods of support in Probs. 6.33, 6.35, and 6.36.

$\blacksquare$ (a)   The free-body diagram of the truss, for support at joint d by cable tensile force T, is shown in Fig. 6.36a. For equilibrium of the truss,

$$\Sigma M_a = 0 \qquad -30(10) + T(15) - 20(20) = 0 \qquad T = 46.7 \text{ kip}$$
$$\Sigma F_y = 0 \qquad R_{ay} - 30 + T - 20 = 0 \qquad R_{ay} = 3.3 \text{ kip}$$
$$\Sigma F_x = 0 \qquad R_{ax} = 0$$

Using Fig. 6.33c in Prob. 6.33,

$$F_{de} = 25 \text{ kip (T)} \qquad F_{ef} = 32 \text{ kip (C)}$$

By inspection, using Fig. 6.36a,

$$F_{df} = T = 46.7 \text{ kip (T)} \qquad F_{cg} = 0 \qquad F_{bh} = 0 \qquad F_{cd} = F_{de} = 25 \text{ kip (T)}$$

Figure 6.36b shows the free-body diagram of joint f. For equilibrium of this joint,

$$\Sigma F_y = 0 \qquad -\frac{4}{6.4} F_{cf} + 46.7 - \frac{4}{6.4} (32) = 0 \qquad F_{cf} = 42.7 \text{ kip (C)}$$

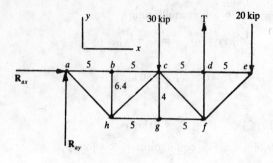

Fig. 6.36a

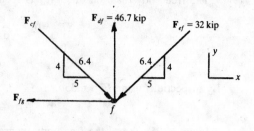

Fig. 6.36b

$$\Sigma F_x = 0 \qquad -F_{fg} + \frac{5}{6.4} F_{cf} - \frac{5}{6.4} (32) = 0 \qquad F_{fg} = 8.4 \text{ kip (T)}$$

By inspection of Fig. 6.36a,

$$F_{gh} = F_{fg} = 8.4 \text{ kip (T)}$$

The free-body diagram of joint c is shown in Fig. 6.36c. The equilibrium requirement is

$$\Sigma F_y = 0 \qquad -30 + \frac{4}{6.4} F_{ch} + \frac{4}{6.4} (42.7) = 0 \qquad F_{ch} = 5.3 \text{ kip (C)}$$

$$\Sigma F_x = 0 \qquad F_{bc} + \frac{5}{6.4} F_{ch} - \frac{5}{6.4} (42.7) + 25 = 0 \qquad F_{bc} = 4.2 \text{ kip (C)}$$

By inspection of Fig. 6.36a,

$$F_{ab} = F_{bc} = 4.2 \text{ kip (C)}$$

Using Fig. 6.36d, the free-body diagram of joint a,

$$\Sigma F_x = 0 \qquad -4.2 + \frac{5}{6.4} F_{ah} = 0 \qquad F_{ah} = 5.4 \text{ kip (T)}$$

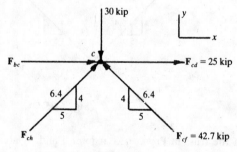

Fig. 6.36c

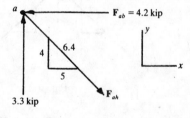

Fig. 6.36d

(b) The member forces for the three methods of support in Probs. 6.33, 6.35, and 6.36 are presented in Table 6.6. For the original method of support at joints a and f (Prob. 6.33), the maximum member force is 42.7 kip, in member cf. For support at joints a and g (Prob. 6.35), the maximum member force increases to 70 kip, in member cg, with a percent increase of

$$\frac{70 - 42.7}{42.7} (100) = 64\%$$

For the third case of support, at joints a and d (Prob. 6.3), the maximum member force is 46.7 kip. The percent increase for this case is

$$\frac{46.7 - 42.7}{42.7} (100) = 9.4\%$$

Further investigation of this problem would be necessary to determine whether the structure could fail if the truss were supported at joint g, due to the greatly increased member force F_{cg}. The analysis would require methods used in the study of strength, or mechanics, of materials.

TABLE 6.6

	Prob. 6.33 (support at a and f)	Prob. 6.35 (support at a and g)	Prob. 6.36 (support at a and d)
F_{ab}	4.2 (C)	25.0 (T)	4.2 (C)
F_{ah}	5.3 (T)	32.0 (C)	5.4 (T)
F_{bc}	4.2 (C)	25.0 (T)	4.2 (C)
F_{bh}	0.0	0.0	0.0
F_{cd}	25.0 (T)	25.0 (T)	25.0 (T)
F_{cf}	42.7[†] (C)	32.0 (T)	42.7 (C)
F_{cg}	0.0	70.0[†] (C)	0.0
F_{ch}	5.3 (C)	32.0 (T)	5.3 (C)
F_{de}	25.0 (T)	25.0 (T)	25.0 (T)
F_{df}	0.0	0.0	46.7[†] (T)
F_{ef}	32.0 (C)	32.0 (C)	32.0 (C)
F_{fg}	8.4 (T)	50.0 (C)	8.4 (T)
F_{gh}	8.4 (T)	50.0 (C)	8.4 (T)

[†]Maximum value.

6.37 The truss in Fig. 6.37a is to support a weight of 1,500 lb. The end of the cable may be connected to joint b, c, or d. Find the member forces if the cable is connected to joint b. Assume that the pulley diameter is small compared with the lengths of the members.

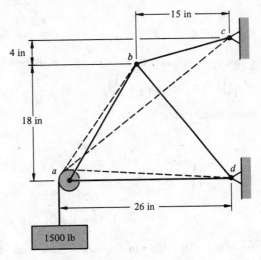

Fig. 6.37a

❚ Since the pulley diameter is assumed to be small compared to the lengths of the members, it follows that the direction of the cable between joints a and b is approximately equal to the direction of member ab. With this assumption, the free-body diagram of the truss has the form shown in Fig. 6.37b. The forces acting on the pulley are transferred directly to joint a. Also, the pair of 1,500-lb forces acting along the direction of member ab are self-canceling and do not affect the overall equilibrium of the truss. For equilibrium of the truss,

$$\Sigma M_d = 0 \qquad -\left(\frac{15}{15.5} F_{bc}\right)(18) - \left(\frac{4}{15.5} F_{bc}\right)15 + 1,500(26) = 0 \qquad F_{bc} = 1,830 \text{ lb (T)}$$

$$\Sigma F_x = 0 \qquad \frac{15}{15.5} F_{bc} - R_{dx} = 0 \qquad R_{dx} = 1,770 \text{ lb}$$

$$\Sigma F_y = 0 \qquad -1,500 + \frac{4}{15.5} F_{bc} + R_{dy} = 0 \qquad R_{dy} = 1,030 \text{ lb}$$

The free-body diagram and force triangle for joint d are shown in Figs. 6.37c and d. For equilibrium of joint d,

$$\frac{1,030}{18} = \frac{F_{bd}}{23.4} = \frac{1,770 - F_{ad}}{15} \qquad F_{bd} = 1,340 \text{ lb (C)} \qquad F_{ad} = 912 \text{ lb (C)}$$

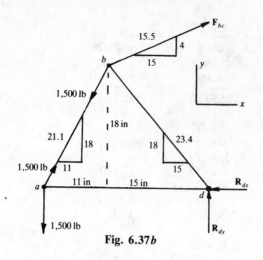

Fig. 6.37b

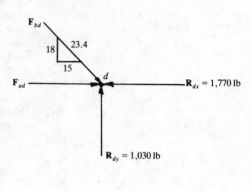

Fig. 6.37c

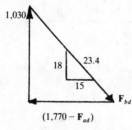

$(1,770 - \mathbf{F}_{ad})$

Fig. 6.37d

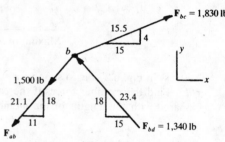

Fig. 6.37e

Figure 6.37e shows the free-body diagram of joint b. For equilibrium of this joint,

$$\Sigma F_x = 0 \qquad -\frac{11}{21.1}(F_{ab} + 1,500) + \frac{15}{15.5}(1,830) - \frac{15}{23.4}(1,340) = 0 \qquad F_{ab} = 249 \text{ lb (T)}$$

As a check on the above calculations,

$$\Sigma F_y = 0$$

$$-\frac{18}{21.1}(249 + 1,500) + \frac{18}{23.4}(1,340) + \frac{4}{15.5}(1,830) \overset{?}{=} 0$$

$$\frac{18}{23.4}(1,340) + \frac{4}{15.5}(1,830) \overset{?}{=} \frac{18}{21.1}(249 + 1,500)$$

$$1,500 \approx 1,490$$

6.38 Find the forces in the members of the truss in Prob. 6.37 if the end of the cable is connected to joint c.

❙ Since the diameter of the pulley is small compared to the member lengths, the direction of the cable is assumed to be the same as the direction of a line between joints a and c in Fig. 6.37a. Figure 6.38a shows the free-body diagram of the truss. For equilibrium of this structure,

$$\Sigma M_d = 0 \qquad 1,500(26) - \frac{22}{34.1}(1,500)(26) - \left(\frac{15}{15.5}F_{bc}\right)(18) - \left(\frac{4}{15.5}F_{bc}\right)(15) = 0 \qquad F_{bc} = 650 \text{ lb (T)}$$

$$\Sigma F_y = 0 \qquad -1,500 + \frac{22}{34.1}(1,500) + \frac{4}{15.5}F_{bc} + R_{dy} = 0 \qquad R_{dy} = 365 \text{ lb}$$

$$\Sigma F_x = 0 \qquad \frac{26}{34.1}(1,500) + \frac{15}{15.5}F_{bc} - R_{dx} = 0 \qquad R_{dx} = 1,770 \text{ lb}$$

Figures 6.38b and c show the free-body diagram and force triangle for joint d. The equilibrium requirements are

$$\frac{365}{18} = \frac{F_{bd}}{23.4} = \frac{1,770 - F_{ad}}{15} \qquad F_{bd} = 475 \text{ lb (C)} \qquad F_{ad} = 1,470 \text{ lb (C)}$$

The free-body diagram of joint b is seen in Fig. 6.38d. For equilibrium,

$$\Sigma F_x = 0 \qquad -\frac{11}{21.1}F_{ab} + \frac{15}{15.5}F_{bc} - \frac{15}{23.4}(475) = 0 \qquad F_{ab} = 623 \text{ lb (T)}$$

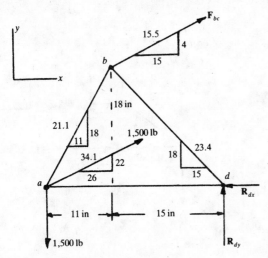

Fig. 6.38a

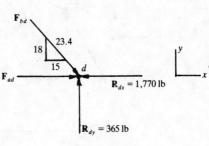

Fig. 6.38b

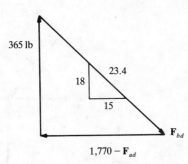

Fig. 6.38c

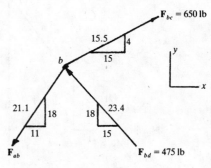

Fig. 6.38d

To check the above results, use

$$\Sigma F_y = 0 \qquad -\frac{18}{21.1}(623) + \frac{18}{23.4}(475) + \frac{4}{15.5}(650) \overset{?}{=} 0$$

$$\frac{18}{23.4}(475) + \frac{4}{15.5}(650) \overset{?}{=} \frac{18}{21.1}(623)$$

$$533 \approx 531$$

6.39 (a) Do the same as in Prob. 6.37 if the cable end is connected to joint d.

(b) Compare the member forces for the three types of cable end support in Probs. 6.37, 6.38, and 6.39.

▌ (a) The cable direction is assumed to be the same as that of member ad. The free-body diagram of the truss is shown in Fig. 6.39a. For equilibrium of the truss,

$$\Sigma M_d = 0 \qquad 1,500(26) - \left(\frac{4}{15.5}F_{bc}\right)15 - \left(\frac{15}{15.5}F_{bc}\right)18 = 0 \qquad F_{bc} = 1,830 \text{ lb (T)}$$

$$\Sigma F_x = 0 \qquad \frac{15}{15.5}F_{bc} - R_{dx} = 0 \qquad R_{dx} = 1,770 \text{ lb}$$

$$\Sigma F_y = 0 \qquad -1,500 + \frac{4}{15.5}F_{bc} + R_{dy} = 0 \qquad R_{dy} = 1,030 \text{ lb}$$

Figures 6.39b and c show the free-body diagram and force triangle for joint a. For equilibrium of this joint,

$$\frac{1,500}{18} = \frac{F_{ab}}{21.1} = \frac{F_{ad} - 1,500}{11} \qquad F_{ab} = 1,760 \text{ lb (T)} \qquad F_{ad} = 2,420 \text{ lb (C)}$$

The free-body diagram of joint d is given in Fig. 6.39d. For equilibrium of joint d,

$$\Sigma F_y = 0 \qquad -\frac{18}{23.4}F_{bd} + 1,030 = 0 \qquad F_{bd} = 1,340 \text{ lb (C)}$$

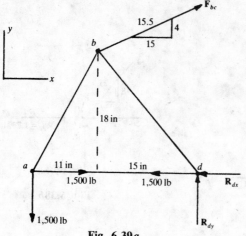

Fig. 6.39a

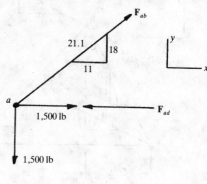

Fig. 6.39b

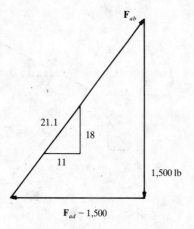

Fig. 6.39c

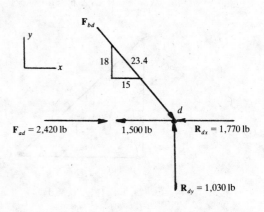

Fig. 6.39d

To check the above results use

$$\Sigma F_x = 0$$

$$2,420 + \frac{15}{23.4}(1,340) - 1,500 - 1,770 \stackrel{?}{=} 0$$

$$2,420 + \frac{15}{23.4}(1,340) \stackrel{?}{=} 1,550 + 1,770$$

$$3,280 \approx 3,270$$

(**b**)   A summary of the member forces for the three possible cable connection points in Probs. 6.37, 6.38, and 6.39 is given in Table 6.7.   It may be seen that connection of the cable end to joint c produces a maximum member force of 1,470 lb, while connection of the cable to joint d produces a maximum member force of 2,420 lb.

TABLE 6.7

	cable connection point		
	b	c	d
F_{ab}	249 (T)	623 (T)	1,760 (T)
F_{bc}	1,830[†] (T)	650 (T)	1,830 (T)
F_{bd}	1,340 (C)	475 (C)	1,340 (C)
F_{ad}	912 (C)	1,470[†] (C)	2,420[†] (C)

[†]Maximum value.

6.40 Find the forces in all of the members of the truss shown in Fig. 6.40a.

❚ The reaction forces of the ground on the truss are indicated by the dashed arrows in Fig. 6.40a. For equilibrium of the entire truss,

$$\sum M_c = 0 \qquad 30(5) + 50(2) + 80(5.5) - [(\tfrac{4}{5}R_d)(1.5) + (\tfrac{3}{5}R_d)(2)] = 0 \qquad R_d = 288 \text{ kN}$$

$$\sum F_y = 0 \qquad -30 - 50 + R_{cy} + \tfrac{3}{5}R_d = 0 \qquad -30 - 50 + R_{cy} + \tfrac{3}{5}(288) = 0 \qquad R_{cy} = -93 \text{ kN}$$

$$\sum F_x = 0 \qquad 80 - \tfrac{4}{5}R_d + R_{cx} = 0 \qquad 80 - \tfrac{4}{5}(288) + R_{cx} = 0 \qquad R_{cx} = 150 \text{ kN}$$

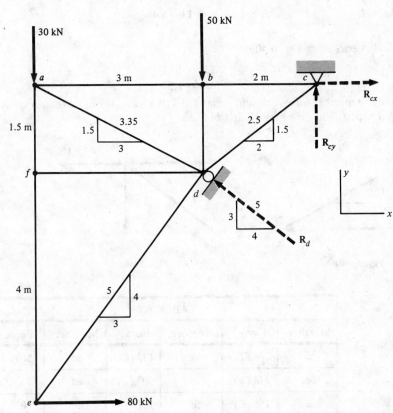

Fig. 6.40a

The minus sign on the component R_{cy} of reaction force at joint c indicates that this force was not assumed in the correct sense.

Figure 6.40b shows the free-body diagram of joint e, and the corresponding force triangle is depicted in Fig. 6.40c. The equilibrium requirement is

$$\frac{80}{3} = \frac{F_{de}}{5} = \frac{F_{ef}}{4} \qquad F_{de} = 133 \text{ kN (C)} \qquad F_{ef} = 107 \text{ kN (T)}$$

From the construction at joint f, Fig. 6.40a,

$$F_{df} = 0 \qquad F_{af} = F_{ef} = 107 \text{ kN (T)}$$

The free-body diagram and force triangle of joint a are shown in Figs. 6.40d and e. For equilibrium,

$$\frac{137}{1.5} = \frac{F_{ab}}{3} = \frac{F_{ad}}{3.35} \qquad F_{ab} = 274 \text{ kN (T)} \qquad F_{ad} = 306 \text{ kN (C)}$$

From the construction of joint b, Fig. 6.40a,

$$F_{bc} = F_{ab} = 274 \text{ kN (T)} \qquad F_{bd} = 50 \text{ kN (C)}$$

The final joint to be analyzed is joint c, which is shown in Fig. 6.40f. Since there is only one unknown member force acting at this joint, the second equation of equilibrium may be used to check the calculations. From Fig. 6.40g,

$$\frac{124}{2} = \frac{F_{cd}}{2.5} \qquad F_{cd} = 155 \text{ kN (C)}$$

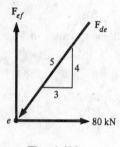

Fig. 6.40b

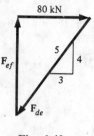

Fig. 6.40c

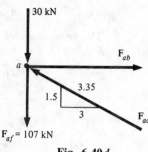

Fig. 6.40d

As the final check, using Fig. 6.40g,

$$\frac{124}{2} \overset{?}{=} \frac{93}{1.5} \qquad 62 \equiv 62$$

The values of the member forces are summarized in Table 6.8. It can be seen that the maximum truss member force of 274 kN occurs in members *ab* and *bc*.

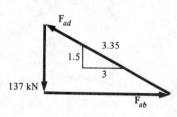

Fig. 6.40e

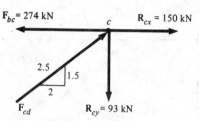

Fig. 6.40f

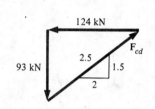

Fig. 6.40g

TABLE 6.8

Member	Force	Member	Force	Member	Force
ab	274 (T)	de	133 (C)	bd	50 (C)
bc	274 (T)	ef	107 (T)	ad	306 (C)
cd	155 (C)	af	107 (T)	df	0

6.4 METHOD OF SECTIONS

6.41 (*a*) Describe how the *method of sections* may be used to find the forces in selected members of a truss.

(*b*) Use this method to find the forces in members *ab*, *be*, and *de* in the truss of Prob. 6.13.

▌ (*a*) The first general method for solving for the forces in truss members is referred to as the method of joints. This method of solution was described in Probs. 6.9 and 6.29. The second general method of analysis for the solution of the forces in truss members is referred to as the method of sections. This technique may be used when the forces in only certain members of the truss are required. In using this method, a free-body diagram is drawn of a portion of the truss. The unknown member forces which are desired are treated as applied forces in the free-body diagram, and this free-body diagram is then analyzed as a general two-dimensional force system. This technique is illustrated in the following problem.

(*b*) The truss in Fig. 6.13*a* is redrawn in Fig. 6.41*a*. For this case, it is required to find the forces in members *ab*, *be*, and *de* only. An imagined cutting section, shown by the dashed line in Fig. 6.41*a*, is now passed through the truss in such a way as to "cut" these three members. Since the entire truss is in equilibrium, each of the two portions of the truss in Fig. 6.41*a* must be individually in equilibrium. The left-hand portion is arbitrarily chosen, and the free-body diagram of this structure is shown in Fig. 6.41*b*. The reaction force R_{fy} was originally found in Prob. 6.13.

The forces in members *ab*, *be*, and *de* are shown as applied forces in Fig. 6.41*b*, and the senses of these forces are arbitrarily chosen. The equilibrium requirements for this figure are that the sum of

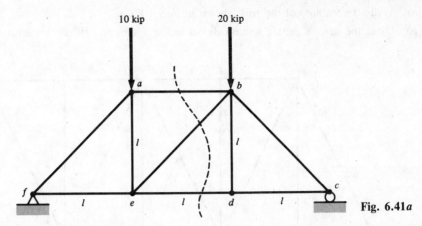

Fig. 6.41a

the forces in two perpendicular directions equals zero and that the sum of moments about any arbitrary point equals zero.

The moments will be summed now about point b. Since forces F_{ab} and F_{be} pass through this point, F_{de} may be found directly from

$$\sum M_b = 0 \qquad -13.3(2l) + 10l + F_{de}l = 0 \qquad F_{de} = 16.6 \text{ kip (T)}$$

Next the moments will be summed about point e. Since the forces F_{be} and F_{de} pass through this point, the force F_{ab} may be found directly from

$$\sum M_e = 0 \qquad -13.3l - F_{ab}l = 0 \qquad F_{ab} = -13.3 \text{ kip}$$

The minus sign indicates that F_{ab} has a sense which is opposite to that shown in Fig. 6.41b. The corrected free-body diagram is shown in Fig. 6.41c. Again, in working out a problem, this correction would be made on the original sketch. The force in member ab, then, may be formally stated as

$$F_{ab} = 13.3 \text{ kip (C)}$$

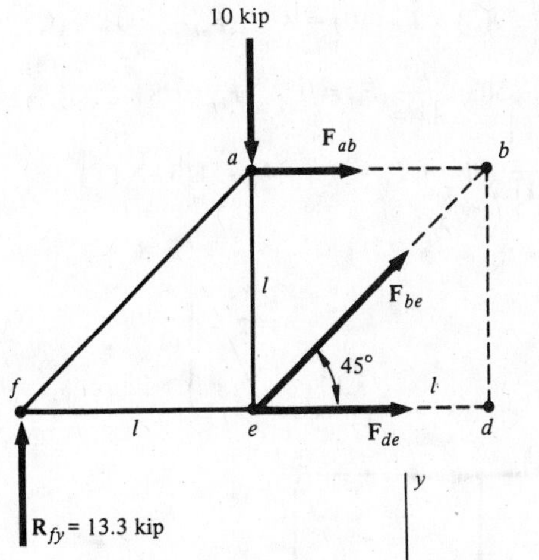

Fig. 6.41b

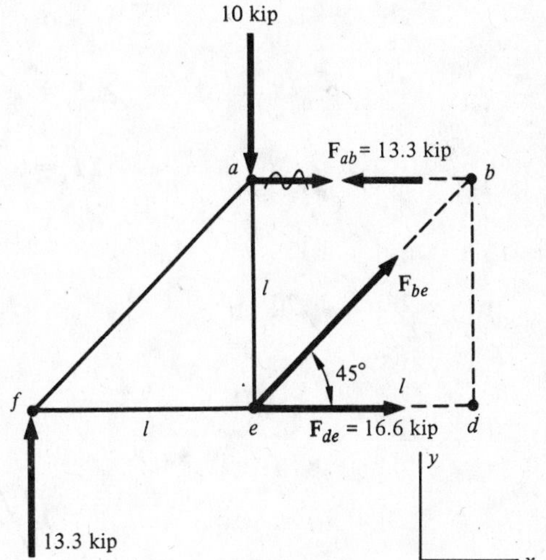

Fig. 6.41c

For equilibrium in the x direction,

$$\sum F_x = 0 \qquad -13.3 + F_{be} \cos 45° + 16.6 = 0 \qquad F_{be} = -4.7 \text{ kip}$$

The minus sign indicates that the actual sense of F_{be} is opposite to that shown in Fig. 6.41b. The sense of this force on the sketch of the free-body diagram would be changed, and the final statement of the force in member be would be

$$F_{be} = 4.7 \text{ kip (C)}$$

When the method of sections is used, *no more than three members with unknown forces may be imagined to be cut*. The reason is that only three equations of equilibrium may be written for the free-body diagram of any portion of the truss.

6.42 (*a*) Verify the stability of the truss shown in Fig. 6.42*a*.

(*b*) Using the dashed cutting section shown in the figure, find the forces in members *bc*, *cj*, and *ij*.

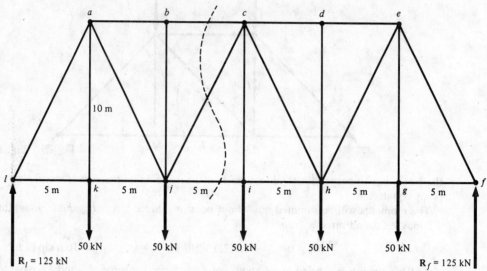

Fig. 6.42*a*

┃ (*a*) With $m = 21$ and $n = 12$,

$$m = 2n - 3 \qquad 21 \overset{?}{=} 2(12) - 3 \qquad 21 \equiv 21$$

(*b*) The left-hand portion of the sectioned truss is shown in Fig. 6.42*b*. The forces in the three truss members are then found from

$$\sum M_j = 0 \qquad -125(10) + 50(5) + F_{bc}(10) = 0 \qquad F_{bc} = 100 \text{ kN (C)}$$

$$\sum F_y = 0 \qquad 125 - 50 - 50 - \frac{10}{11.2} F_{cj} = 0 \qquad F_{cj} = 28 \text{ kN (C)}$$

$$\Sigma F_x = 0 \qquad -F_{bc} - \frac{5}{11.2} F_{cj} + F_{ij} = 0 \qquad F_{ij} = 113 \text{ kN (T)}$$

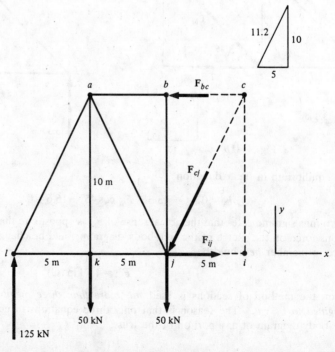

Fig. 6.42*b*

6.43 (*a*) Show that the truss in Fig. 6.43*a* is stable.

  (*b*) Find the forces in members *ab*, *bi*, and *hi* of the truss.

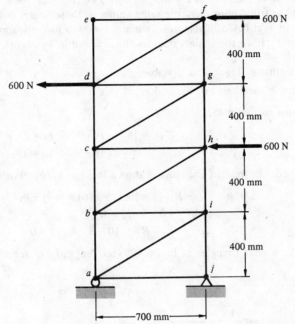

Fig. 6.43*a*

▌ (*a*) The truss has 17 members and 10 joints. Thus,

$$m = 17 \qquad n = 10 \qquad m = 2n - 3 \qquad 17 \overset{?}{=} 2(10) - 3 \qquad 17 \equiv 17$$

  (*b*) The free-body diagram of the truss is shown in Fig. 6.43*b*. For equilibrium,

$$\Sigma M_j = 0 \qquad -R_a(700) + 600(1,200) + 600(1,600) + 600(800) = 0 \qquad R_a = 3,090 \text{ N}$$
$$\Sigma F_x = 0 \qquad -600 - 600 - 600 + R_{jx} = 0 \qquad R_{jx} = 1,800 \text{ N}$$
$$\Sigma F_y = 0 \qquad R_a - R_{jy} = 0 \qquad R_{jy} = 3,090 \text{ N}$$

A cutting section is imagined to pass through members *ab*, *bi*, and *hi*, and the free-body diagram of the lower portion of the truss is shown in Fig. 6.43*c*. The member forces F_{ab}, F_{bi}, and F_{hi} are shown with assumed actual senses. For equilibrium,

$$\Sigma M_i = 0 \qquad F_{ab}(700) - 3,090(700) + 1,800(400) = 0 \qquad F_{ab} = 2,060 \text{ N (C)}$$
$$\Sigma F_x = 0 \qquad -F_{bi} + 1,800 = 0 \qquad F_{bi} = 1,800 \text{ N (T)}$$
$$\Sigma F_y = 0 \qquad -F_{ab} + 3,090 + F_{hi} - 3,090 = 0 \qquad F_{hi} = 2,060 \text{ N (T)}$$

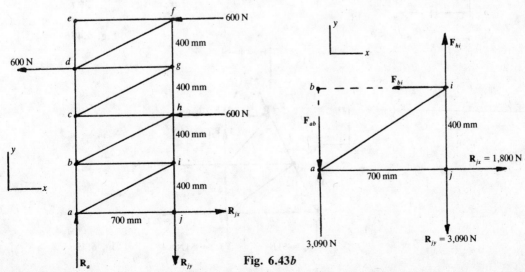

Fig. 6.43*b*   Fig. 6.43*c*

6.44 Do the same as in Prob. 6.43 if the horizontal force of 600 N at joint d is assumed to act on joint g. Compare the results with those found in Prob. 6.43.

❚ From consideration of the form of Fig. 6.43b, it may be seen that the ground reaction force components R_a, R_{jx}, R_{jy}, acting on the truss, must have the same values as those found in Prob. 6.43. It then follows that the free-body diagram in Fig. 6.43c is also valid when the 600-lb force is applied at joint g. Thus, the member forces F_{ab}, F_{bi}, and F_{hi} have the same values as those found in Prob. 6.43.

6.45 (a) Show that the truss in Fig. 6.45a is stable.

(b) Find the forces in members cd, cg, and gh of this truss.

❚ (a) For the truss in Fig. 6.45a,

$$m = 15 \qquad n = 9 \qquad m = 2n - 3$$
$$15 \overset{?}{=} 2(9) - 3 \qquad 15 \equiv 15$$

(b) The free-body diagram of the truss is shown in Fig. 6.45b. For equilibrium,

$$\Sigma M_i = 0 \qquad -R_a(4) + 10(3) + 8(6) + 6(9) + 4(12) = 0 \qquad R_a = 45 \text{ kN}$$
$$\Sigma F_y = 0 \qquad R_a - R_{iy} = 0 \qquad R_{iy} = R_a = 45 \text{ kN}$$
$$\Sigma F_x = 0 \qquad R_{ix} - 10 - 8 - 6 - 4 = 0 \qquad R_{ix} = 28 \text{ kN}$$

Using the similar triangles in Fig. 6.45b, the length l_{cg} of member cg is found to be

$$\frac{l_{cg}}{6} = \frac{4}{12} \qquad l_{cg} = 2 \text{ m}$$

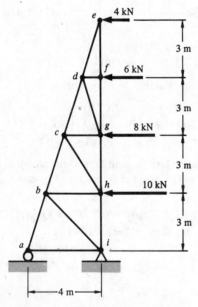

Fig. 6.45a

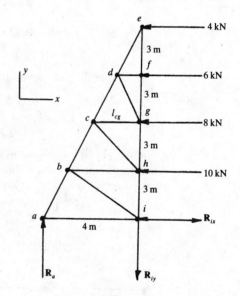

Fig. 6.45b

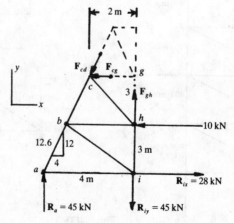

Fig. 6.45c

A cutting section is passed through members cd, cg, and gh, and the free-body diagram of the lower portion of the truss is shown in Fig. 6.45c. For equilibrium of the lower portion of the truss,

$$\Sigma M_c = 0 \qquad -45(2) + F_{gh}(2) - 10(3) + 28(6) - 45(2) = 0 \qquad F_{gh} = 21 \text{ kN (T)}$$

$$\Sigma F_y = 0 \qquad 45 - \frac{12}{12.6} F_{cd} + F_{gh} - 45 = 0 \qquad F_{cd} = 22.1 \text{ kN (C)}$$

$$\Sigma F_x = 0 \qquad -\frac{4}{12.6} F_{cd} - F_{cg} - 10 + 28 = 0 \qquad F_{cg} = 11 \text{ kN (C)}$$

It should be noted that the unknown member forces could also have been found by using a free-body diagram of the upper portion of the cut truss.

6.46 Figure 6.46a shows a roof truss. The term h/l is a ratio of the height of the truss to the span of the truss. Find the forces in members de, ei, and hi if $h/l = \frac{1}{3}$.

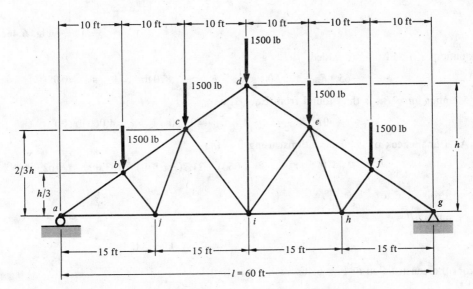

Fig. 6.46a

❚ Because of the symmetry of the structure and because of the applied loads, the reaction forces at a and g may be found by inspection to be

$$R_a = R_{ay} = R_g = R_{gy} = \frac{5(1,500)}{2} = 3,750 \text{ lb}$$

For $h/l = \frac{1}{3}$,

$$\frac{h}{l} = \frac{1}{3} = \frac{h}{60} \qquad h = 20 \text{ ft}$$

Members de, ei, and hi are imagined to be cut, and the free-body diagram of the right portion of the truss is shown in Fig. 6.46b. From the similar triangles dee' and dgi in Fig. 6.46b,

$$\frac{20 - A}{10} = \frac{20}{30} \qquad A = 13.3 \text{ ft}$$

For equilibrium of the structure shown in Fig. 6.46b,

$$\Sigma M_e = 0 \qquad -1,500(10) + 3,750(20) - F_{hi}(13.3) = 0 \qquad F_{hi} = 4,510 \text{ lb (T)}$$

$$\Sigma F_y = 0 \qquad -\frac{20}{36.1} F_{de} - \frac{13.3}{16.6} F_{ei} - 1,500 - 1,500 + 3,750 = 0 \tag{1}$$

$$\Sigma F_x = 0 \qquad \frac{30}{36.1} F_{de} - \frac{10}{16.6} F_{ei} - F_{hi} = 0 \tag{2}$$

Equation (1) is multiplied by $\frac{30}{20}(36.1)$, and Eq. (2) is multiplied by 36.1. The results are given as Eqs. (3) and (4):

$$-30 F_{de} - 43.4 F_{ei} + 40,600 = 0 \tag{3}$$

$$30 F_{de} - 21.7 F_{ei} - 163,000 = 0 \tag{4}$$

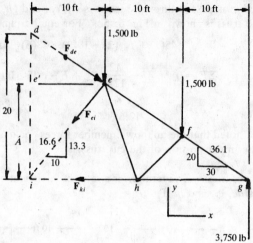

3,750 lb   **Fig. 6.46b**

Equations (3) and (4) are added, yielding

$$-65.1F_{ei} - 122,000 = 0 \qquad F_{ei} = -1,870 \text{ lb} \qquad F_{ei} = 1,870 \text{ lb (C)}$$

Member force F_{de} is then found from Eq. (3) as

$$30F_{de} = -43.4(-1,870) + 40,600 \qquad F_{de} = 4,060 \text{ lb (C)}$$

As a final check on the calculations, using Eq. (4),

$$30F_{de} - 21.7F_{ei} \overset{?}{=} 163,000 \qquad 30(4,060) - 21.7(-1,870) \overset{?}{=} 163,000 \qquad 162,000 \approx 163,000$$

6.47 Solve Prob. 6.46 for $h/l = \frac{1}{4}$.

❚ Using $h/l = \frac{1}{4}$,

$$\frac{h}{l} = \frac{1}{4} = \frac{h}{60} \qquad h = 15 \text{ ft}$$

From the construction of Fig. 6.46a, with $h = 15$ ft,

$$\frac{15 - A}{10} = \frac{15}{30} \qquad A = 10 \text{ ft}$$

The slopes of members ei and de are now as shown in Fig. 6.47. For equilibrium, using Fig. 6.46b, $A = 10$ ft, and the data in Fig. 6.47,

$$\Sigma M_e = 0 \qquad -1,500(10) + 3,750(20) - F_{hi}(10) = 0 \qquad F_{hi} = 6,000 \text{ lb (T)}$$

$$\Sigma F_y = 0 \qquad -\frac{15}{33.5} F_{de} - \frac{10}{14.1} F_{ei} - 1,500 - 1,500 + 3,750 = 0 \tag{1}$$

$$\Sigma F_x = 0 \qquad \frac{30}{33.5} F_{de} - \frac{10}{14.1} F_{ei} - F_{hi} = 0 \tag{2}$$

Equation (1) is multiplied by $2(33.5)/30$ and Eq. (2) by $33.5/30$, yielding

$$-F_{de} - 1.58F_{ei} + 1,680 = 0 \tag{3}$$

$$F_{de} - 0.792F_{ei} - 6,700 = 0 \tag{4}$$

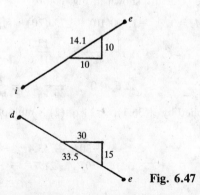

Fig. 6.47

Equations (3) and (4) are then added, with the result

$$-2.37F_{ei} - 5,020 = 0 \qquad F_{ei} = -2,120 \text{ lb} \qquad F_{ei} = 2,120 \text{ lb (C)}$$

Member force F_{de} is then found from Eq. (4) as

$$F_{de} = 0.792F_{ei} + 6,700 = 0.792(-2,120) + 6,700 = 5,020 \text{ lb (C)}$$

6.48   (*a*)   Solve Prob. 6.46 for $h/l = \frac{1}{5}$.

(*b*)   Compare the forces in members *de*, *ei*, and *hi* for the conditions of Probs. 6.46, 6.47, and 6.48.

▌ (*a*)   Using $h/l = \frac{1}{5}$,

$$\frac{h}{l} = \frac{1}{5} = \frac{h}{60} \qquad h = 12 \text{ ft}$$

From the construction of Fig. 6.46*a*, with $h = 12$ ft,

$$\frac{12 - A}{10} = \frac{12}{30} \qquad A = 8 \text{ ft}$$

The slopes of members *ei* and *de* are now as shown in Fig. 6.48. For equilibrium, using Fig. 6.46*b*, $A = 8$ ft,   and the data in Fig. 6.48,

$$\Sigma M_e = 0 \qquad -1,500(10) + 3,750(20) - F_{hi}(8) = 0 \qquad F_{hi} = 7,500 \text{ lb (T)}$$

$$\Sigma F_y = 0 \qquad -\frac{12}{32.3} F_{de} - \frac{8}{12.8} F_{ei} - 1,500 - 1,500 + 3,750 = 0 \tag{1}$$

$$\Sigma F_x = 0 \qquad \frac{30}{32.3} F_{de} - \frac{10}{12.8} F_{ei} - F_{hi} = 0 \tag{2}$$

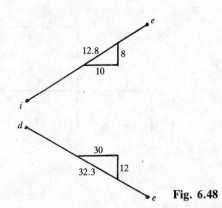

Fig. 6.48

Multiplying Eq. (1) by 32.3/12 and Eq. (2) by 32.3/30 and adding the two equations results in

$$-F_{de} - 1.68F_{ei} + 2,020 = 0 \tag{3}$$
$$F_{de} - 0.841F_{ei} - 8,080 = 0 \tag{4}$$

Equations (3) and (4) are added, to obtain

$$-2.52F_{ei} - 6,060 = 0 \qquad F_{ei} = -2,400 \text{ lb} \qquad F_{ei} = 2,400 \text{ lb (C)}$$

Member force F_{de} is then found from Eq. (4) as

$$F_{de} = 0.841(-2,400) + 8,080 \qquad F_{de} = 6,060 \text{ lb (C)}$$

(*b*)   The forces in members *de*, *ei*, and *hi*, for the three h/l values $\frac{1}{3}$, $\frac{1}{4}$, and $\frac{1}{5}$ are shown in Table 6.9. It may be seen that member forces F_{de}, F_{ei}, and F_{hi} all increase as the shape parameter h/l decreases.

TABLE 6.9

	Prob. 6.46, $h/l = \dfrac{1}{3}$	Prob. 6.47, $h/l = \dfrac{1}{4}$	Prob. 6.48, $h/l = \dfrac{1}{5}$
F_{de}	4,060 (C)	5,020 (C)	6,060 (C)
F_{ei}	1,870 (C)	2,120 (C)	2,400 (C)
F_{hi}	4,510 (T)	6,000 (T)	7,500 (T)

6.49 Find the forces in members *cd*, *cj*, *dj*, and *jk* of the Pratt truss shown in Fig. 6.49*a*.

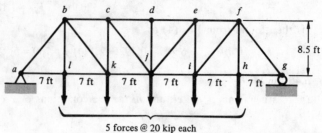

5 forces @ 20 kip each

Fig. **6.49***a*

I From the symmetry of the loading and from truss geometry, the vertical reaction forces at joints *a* and *g* are

$$R_a = R_{ay} = \frac{5(20)}{2} = 50 \text{ kip} \qquad R_g = R_{gy} = \frac{50(20)}{2} = 50 \text{ kip}$$

Members *cd*, *cj*, and *jk* are imagined to be cut, and the free-body diagram of the left portion of the truss is shown in Fig. 6.49*b*. For equilibrium.

$$\Sigma M_c = 0 \qquad -50(14) + 20(7) + F_{jk}(8.5) = 0 \qquad F_{jk} = 65.9 \text{ kip (T)}$$

$$\Sigma F_y = 0 \qquad 50 - 20 - 20 - \frac{8.5}{11.0} F_{cj} = 0 \qquad F_{cj} = 12.9 \text{ kip (T)}$$

$$\Sigma F_x = 0 \qquad -F_{cd} + \frac{7}{11.0} F_{cj} + F_{jk} = 0 \qquad F_{cd} = 74.1 \text{ kip (C)}$$

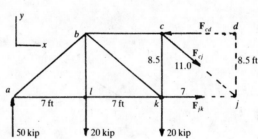

Fig. **6.49***b*

From inspection of Fig. 6.49*a*,

$$F_{dj} = 0$$

6.50 Figure 6.50*a* shows a roof truss.

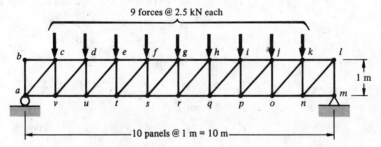

9 forces @ 2.5 kN each

10 panels @ 1 m = 10 m

Fig. **6.50***a*

(*a*) Show that the truss is stable.

(*b*) Find the forces in members *fg*, *gh*, *gr*, *gs*, *hr*, *rs*, and *qr*.

I (*a*) Using *m* = 41, *h* = 22,

$$m = 2n - 3 \qquad 41 \stackrel{?}{=} 2(22) - 3 \qquad 41 \equiv 41$$

(*b*) From the symmetry of the structure and from the loading,

$$R_a = R_{ay} = \frac{9(2.5)}{2} = 11.3 \text{ kN} \qquad R_m = R_{my} = \frac{9(2.5)}{2} = 11.3 \text{ kN}$$

Members fg, gs, and rs are imagined to be cut, and the free-body diagram of the left portion of the truss is seen in Fig. 6.50b. For equilibrium of the truss portion in Fig. 6.50b,

$$\Sigma M_s = 0 \qquad -11.3(4) + 2.5(1) + 2.5(2) + 2.5(3) + F_{fg}(1) = 0 \qquad F_{fg} = 30.2 \text{ kN (C)}$$

$$\Sigma F_y = 0 \qquad 11.3 - 4(2.5) \qquad -\frac{1}{1.41} F_{gs} = 0 \qquad F_{gs} = 1.83 \text{ kN (C)}$$

$$\Sigma F_x = 0 \qquad -F_{fg} - \frac{1}{1.41} F_{gs} + F_{rs} = 0 \qquad F_{rs} = 31.5 \text{ kN (T)}$$

The truss section required to find the forces in members gh, hr, and qr is seen in Fig. 6.50c. For equilibrium,

$$\Sigma M_r = 0 \qquad -11.3(5) + F_{gh}(1) + 2.5(1 + 2 + 3 + 4) = 0 \qquad F_{gh} = 31.5 \text{ kN (C)} \qquad (1)$$

$$\Sigma F_y = 0 \qquad 11.3 - 5(2.5) + \frac{1}{1.41} F_{hr} = 0 \qquad F_{hr} = 1.7 \text{ kN (T)}$$

$$\Sigma F_x = 0 \qquad -F_{gh} + \frac{1}{1.41} F_{hr} + F_{qr} = 0 \qquad F_{qr} = 30.3 \text{ kN (T)}$$

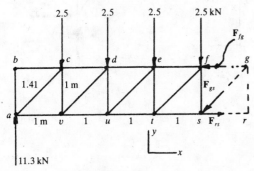

Fig. 6.50b Fig. 6.50c

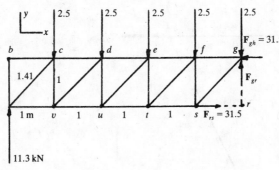

Fig. 6.50d

The construction required to find F_{gr}, using the known results $F_{gh} = 31.5 \text{ kN (C)}$ and $F_{rs} = 31.5 \text{ kN (T)}$, is shown in Fig. 6.50$d$. The equilibrium requirement is

$$\Sigma F_y = 0 \qquad 11.3 - 5(2.5) + F_{gr} = 0 \qquad F_{gr} = 1.2 \text{ kN}$$

As a final check on the calculations,
$$\Sigma M_s = 0$$
$$-11.3(4) + 2.5(1 + 2 + 3) - 2.5(1) + 31.5(1) + 1.2(1) \overset{?}{=} 0$$
$$2.5(1 + 2 + 3) + 31.5(1) + 1.2(1) \overset{?}{=} 11.3(4) + 2.5(1)$$
$$47.7 \equiv 47.7$$

6.5 CONNECTED TRUSSES

6.51   Describe the technique for solving for the reaction forces in the free-body diagrams of *connected trusses*.

❚ In certain truss configurations, the free-body diagram of the entire truss contains more than three unknown reaction forces of the foundation on the truss. Such a structure is shown in Fig. 6.51a. Only three equations

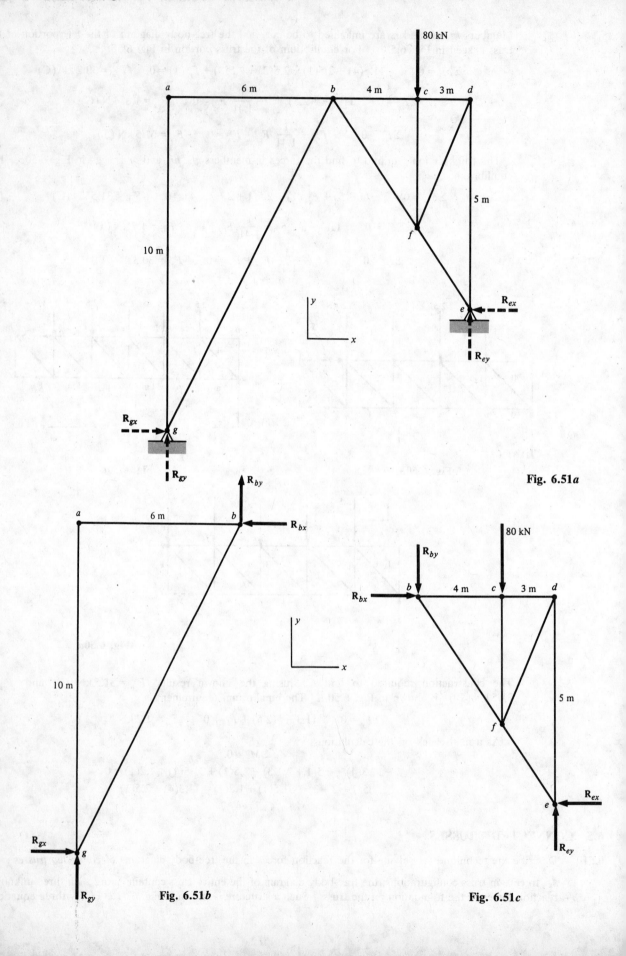

80 kN

a   6 m   b   4 m   c   3 m   d

10 m

5 m

f

y
x

$\mathbf{R}_{ex}$
e

$\mathbf{R}_{ey}$

$\mathbf{R}_{gx}$
g

$\mathbf{R}_{gy}$

Fig. 6.51a

$\mathbf{R}_{by}$

a   6 m   b $\mathbf{R}_{bx}$

10 m

y
x

$\mathbf{R}_{gx}$
g

$\mathbf{R}_{gy}$

Fig. 6.51b

80 kN

$\mathbf{R}_{by}$

$\mathbf{R}_{bx}$ b   4 m   c   3 m   d

y
x

5 m

f

$\mathbf{R}_{ex}$
e

$\mathbf{R}_{ey}$

Fig. 6.51c

of equilibrium, which can be used to solve for three unknown forces, can be written for the general two-dimensional force system. Thus the forces which act on the system in Fig. 6.51a cannot be found directly from this free-body diagram.

A careful inspection of this figure reveals that the structure shown is actually two separate trusses which share a common joint at b. The technique for the solution of this problem is to draw individual free-body diagrams for each of these trusses. This effect is shown in Fig. 6.51b. In Fig. 6.51b the forces R_{bx} and R_{by} are the forces on joint b of the left-side truss by the right-side truss. By Newton's third law, these forces must have opposite senses when drawn on the free-body diagram of the right-side truss, as shown in Fig. 6.51c. The two trusses shown in this figure have acting on them a total of six unknown forces. Three equations of equilibrium may be written for each truss, for a total of six equations. All of the unknown reaction forces may be found by using these six equations. The steps in this solution are shown in Prob. 6.52.

6.52 Find all of the reaction force components for the truss in Prob. 6.51.

❚ The moments are summed to zero about joint g, Fig. 6.51b, with the result

$$\sum M_g = 0 \qquad R_{bx}(10) + R_{by}(6) = 0 \qquad R_{bx} = -0.6R_{by} \tag{1}$$

The three equations of equilibrium for the truss in Fig. 6.51c are

$$\sum M_e = 0 \qquad -R_{bx}(5) + R_{by}(7) + 80(3) = 0 \tag{2}$$

$$\sum F_y = 0 \qquad -R_{by} - 80 + R_{ey} = 0 \tag{3}$$

$$\sum F_x = 0 \qquad R_{bx} - R_{ex} = 0 \tag{4}$$

Equation (1) is substituted into Eq. (2), with the result

$$-(-0.6R_{by})5 + R_{by}(7) + 80(3) = 0 \qquad 10R_{by} + 240 = 0 \qquad R_{by} = -24\text{ kN} \tag{5}$$

Equation (5) is substituted into Eq. (3), to get

$$-(-24) - 80 + R_{ey} = 0 \qquad R_{ey} = 56\text{ kN}$$

Equations (1) and (5) are combined, with the result

$$R_{bx} = -0.6(-24) = 14.4\text{ kN} \tag{6}$$

As the final step, Eq. (6) is combined with Eq. (4), to get

$$14.4 - R_{ex} = 0 \qquad R_{ex} = 14.4\text{ kN}$$

The remaining two equations of equilibrium for the truss in Fig. 6.51b are

$$\sum F_x = 0 \qquad R_{gx} - R_{bx} = 0 \qquad R_{gx} - 14.4 = 0 \qquad R_{gx} = 14.4\text{ kN}$$

$$\sum F_y = 0 \qquad R_{gy} + R_{by} = 0 \qquad R_{gy} + (-24) = 0 \qquad R_{gy} = 24\text{ kN}$$

With all the reaction forces known, the forces in the members can then be found by using either the method of joints or the method of sections.

6.53 Find the maximum value of the weight W which may be supported by the truss shown in Fig. 6.53a if no member is to experience a force of magnitude greater than 375 lb.

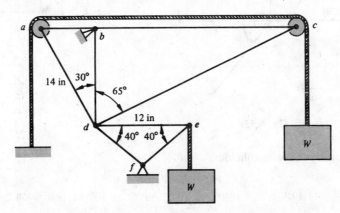

Fig. 6.53a

❚ The cable forces are transferred directly to truss *acd*, and the free-body diagrams of the two parts of the connected truss are shown in Figs. 6.53b and c. R_{dx} and R_{dy} are the reaction force components at the point of connection of the two trusses. Using Fig. 6.53b,

$$\Sigma M_b = 0 \qquad W(7) + R_{dx}(12.1) - W(25.9) = 0 \qquad R_{dx} = 1.56W$$
$$\Sigma F_x = 0 \qquad W - R_{bx} + R_{dx} - W = 0 \qquad R_{bx} = R_{dx} = 1.56W$$
$$\Sigma F_y = 0 \qquad -W + R_{by} - R_{dy} - W = 0 \tag{1}$$

From Fig. 6.53c,

$$\Sigma M_d = 0 \qquad -W(12) + R_{fx}(5.03) + R_{fy}(6) = 0 \tag{2}$$
$$\Sigma F_y = 0 \qquad R_{dy} + R_{fy} - W = 0 \tag{3}$$
$$\Sigma F_x = 0 \qquad -1.56W + R_{fx} = 0 \qquad R_{fx} = 1.56W \tag{4}$$

Using Eq. (4) in Eq. (2),

$$-0.420(1.56W) - 0.5R_{fy} + W = 0 \qquad R_{fy} = 0.690W \tag{5}$$

Using Eq. (5) in Eq. (3),

$$R_{dy} = W - 0.690W = 0.310W \tag{6}$$

Equation (6) is now substituted into Eq. (1) to obtain

$$R_{by} = R_{dy} + 2W = 2.31W$$

Using Fig. 6.53c, the free-body diagram of joint *d* has the form given in Fig. 6.53d. The equilibrium requirements for this joint are

$$\Sigma F_y = 0 \qquad 0.310W - F_{df}\sin 40° = 0 \qquad F_{df} = 0.482W \text{ (T)}$$
$$\Sigma F_x = 0 \qquad -1.56W + F_{de} + F_{df}\cos 40° = 0 \qquad F_{de} = 1.19W \text{ (T)}$$

Figure 6.53e shows the free-body diagram of joint *e*. For equilibrium,

$$\Sigma F_y = 0 \qquad F_{ef}\sin 40° - W = 0 \qquad F_{ef} = 1.56W \text{ (C)}$$

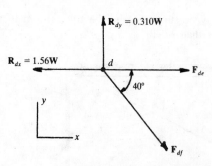

Fig. 6.53b

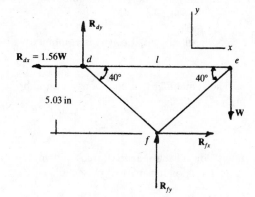

Fig. 6.53c

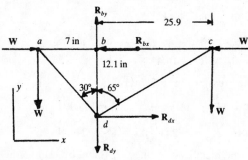

Fig. 6.53d

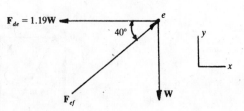

Fig. 6.53e

As a check on the above calculations,

$$\Sigma F_x = 0$$
$$-1.19W + F_{ef}\cos 40° \overset{?}{=} 0 \qquad (1.56W)\cos 40° \overset{?}{=} 1.19W \qquad 1.20W \approx 1.19W$$

FORCE ANALYSIS OF PLANE TRUSSES

At this stage of the solution all of the member forces, in terms of W, have been found for truss def. The forces in truss acd are found next. Figure 6.53f shows the free-body diagram of joint a. For equilibrium of this joint,

$$\Sigma F_y = 0 \qquad -W + F_{ad}\sin 60° = 0 \qquad F_{ad} = 1.15W\ (C)$$
$$\Sigma F_x = 0 \qquad W - F_{ab} - F_{ad}\cos 60° = 0 \qquad F_{ab} = 0.425W\ (C)$$

By inspection of Fig. 6.53b,

$$F_{bd} = R_{by} = 2.31W\ (T)$$

The free-body diagram of joint b is given in Fig. 6.53g. For equilibrium of joint b,

$$\Sigma F_x = 0 \qquad 0.425W - 1.56W + F_{bc} = 0 \qquad F_{bc} = 1.14W\ (T)$$

The free-body diagram of joint c is shown in Fig. 6.53h. For equilibrium,
$$\Sigma F_x = 0 \qquad F_{cd}\cos 25° - 1.14W - W = 0 \qquad F_{cd} = 2.36W\ (C)$$

As a final check on the above calculations,

$$\Sigma F_y = 0$$
$$F_{cd}\sin 25° - W \overset{?}{=} 0 \qquad (2.36W)\sin 25° \overset{?}{=} W \qquad 0.997W \approx 1.000W$$

The maximum force occurs in member cd, with $F_{cd} = 2.36W$. Using the limiting member force value of 375 lb,

$$F_{cd} = 375\ \text{lb} = 2.36W_{max} \qquad W_{max} = 159\ \text{lb}$$

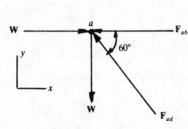

Fig. 6.53f

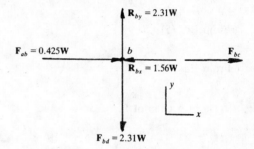

Fig. 6.53g

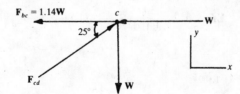

Fig. 6.53h

6.54   Find the forces in the members of the truss shown in Fig. 6.54a.

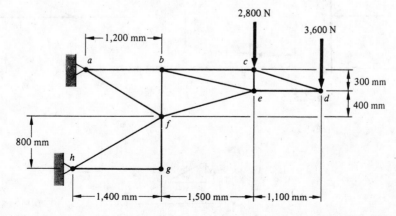

Fig. 6.54a

❚ The free-body diagrams of the two portions of the connected truss are shown in Figs. 6.54b and c. Using Fig. 6.54b, the equilibrium requirements are

$$\Sigma M_a = 0$$

$$R_{fy}(1,200) + R_{fx}(700) - 2,800(2,700) - 3,600(3,800) = 0 \qquad R_{fy} + 0.583 R_{fx} - 1.77 \times 10^4 = 0 \qquad (1)$$

$$\Sigma F_x = 0 \qquad -R_{ax} + R_{fx} = 0 \qquad (2)$$

$$\Sigma F_y = 0 \qquad -R_{ay} + R_{fy} - 2,800 - 3,600 = 0 \qquad -R_{ay} + R_{fy} - 6,400 = 0 \qquad (3)$$

From Fig. 6.54c, for equilibrium,

$$\Sigma M_h = 0 \qquad -R_{fy}(1,400) + R_{fx}(800) = 0 \qquad R_{fy} = 0.571 R_{fx} \qquad (4)$$

$$\Sigma F_x = 0 \qquad R_{hx} - R_{fx} = 0 \qquad (5)$$

$$\Sigma F_y = 0 \qquad R_{hy} - R_{fy} = 0 \qquad (6)$$

Equation (4) is substituted into Eq. (1), with the result

$$(0.571 R_{fx}) + 0.583 R_{fx} - 1.77 \times 10^4 = 0 \qquad R_{fx} = 15,300 \text{ N}$$

Using Eq. (4),

$$R_{fy} = 0.571 R_{fx} = 8,740 \text{ N}$$

From Eq. (2),

$$R_{ax} = R_{fx} = 15,300 \text{ N}$$

Using Eq. (3),

$$-R_{ay} + 8,740 - 6,400 = 0 \qquad R_{ay} = 2,340 \text{ N}$$

From Eq. (5),

$$R_{hx} = R_{fx} = 15,300 \text{ N}$$

From Eq. (6),

$$R_{hy} = R_{fy} = 8,740 \text{ N}$$

From inspection of Fig. 6.54c,

$$F_{fg} = 0 \qquad F_{gh} = 0$$

$$\therefore \ F_{fh} = R_h = \sqrt{R_{hx}^2 + R_{hy}^2} = \sqrt{15,300^2 + 8,740^2} \qquad F_{fh} = 17,600 \text{ N}$$

The free-body diagram of joint d is shown in Fig. 6.54d. For equilibrium of this joint,

$$\Sigma F_y = 0 \qquad \frac{300}{1,140} F_{cd} - 3,600 = 0 \qquad F_{cd} = 13,700 \text{ N (C)}$$

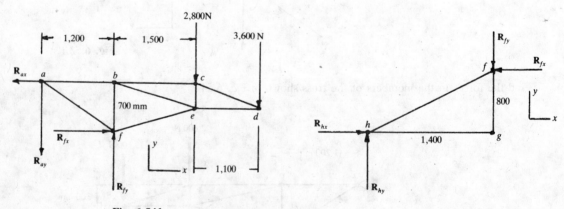

Fig. 6.54b   Fig. 6.54c

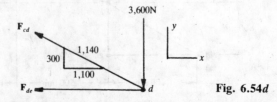

Fig. 6.54d

$$\Sigma F_x = 0 \qquad -\frac{1,100}{1,140} F_{cd} - F_{de} = 0 \qquad F_{de} = -13,200\,\text{N} \qquad F_{de} = 13,200\,\text{N (C)}$$

The free-body diagram and force triangle for joint c are shown in Figs. 6.54e and f. For equilibrium of joint c,

$$\frac{13,700}{1,140} = \frac{F_{bc}}{1,100} = \frac{F_{ce} - 2,800}{300} \qquad F_{bc} = 13,200\,\text{N (T)} \qquad F_{ce} = 6,410\,\text{N (C)}$$

The free-body diagram and force triangle for joint a are shown in Figs. 6.54g and h. The equilibrium requirement for joint a is

$$\frac{2,340}{700} = \frac{F_{af}}{1,390} = \frac{F_{ab} - 15,300}{1,200} \qquad F_{af} = 4,650\,\text{N (C)} \qquad F_{ab} = 19,300\,\text{N (T)}$$

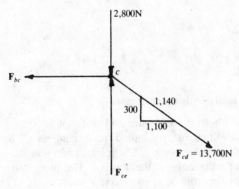

Fig. 6.54e

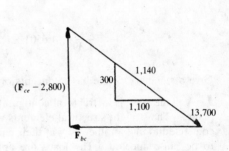

Fig. 6.54f

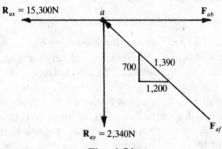

Fig. 6.54g

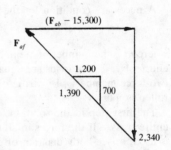

Fig. 6.54h

Figures 6.54i and j show the free-body diagram and force triangle for joint b. For equilibrium of joint b,

$$\frac{6,100}{1,500} = \frac{F_{bf}}{300} = \frac{F_{be}}{1,530} \qquad F_{bf} = 1,220\,\text{N (C)} \qquad F_{be} = 6,220\,\text{N (T)}$$

Figure 6.54k shows the free-body diagram of joint f. Member force F_{ef} is found from

$$\Sigma F_x = 0 \qquad 15,300 + \frac{1,200}{1,390}(4,650) - \frac{1,500}{1,550} F_{ef} = 0 \qquad F_{ef} = 20,000\,\text{N (C)}$$

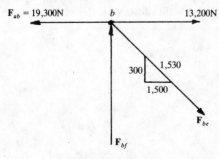

Fig. 6.54i

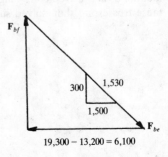

Fig. 6.54j

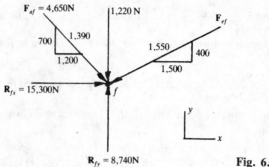

Fig. 6.54*k*

As a final check on the calculations above,

$$\Sigma F_y = 0$$

$$\frac{700}{1,390}(4,650) + 1,220 + \frac{400}{1,550}(20,000) \stackrel{?}{=} 8,740 \qquad 8,720 \approx 8,740$$

6.55 Give a summary of the basic concepts of the force analysis of plane trusses.

▮ A plane truss is a structural configuration in which both the structure and the applied loads lie in the same plane. The straight structural elements which comprise the truss are called truss members. The members are connected to one another only at their ends. These connections are called joints, and these joints are assumed to be pin-connected. The loads are applied to the truss only at the joints. The pins at the joints either distribute the applied loads to the members or distribute the load from one member to one or more other members.

A necessary condition for the stability of a plane truss is that

$$m = 2n - 3$$

where m is the number of members and n is the number of joints. A two-force member has loads applied at its ends only. These loads are equal in magnitude, opposite in sense, and collinear. Every member of a truss is a two-force member. By virtue of its construction, a truss may be connected to a foundation only by a hinge or roller support.

The first step in the solution for the forces in the members of a truss is to draw a free-body diagram of the entire truss. If the forces in all the members are required, the method of joints is used. When this technique is used, a free-body diagram of each joint is drawn. The member forces which act on each joint have known directions. The system of forces which acts at each joint forms a concurrent force system, and two equations of equilibrium may be written for each joint. No more than two unknown member forces may be solved for at a joint. With the method of joints, the forces in the members may also be found graphically.

When the forces in only certain members of the truss are required, the method of sections is used. In using this technique, a free-body diagram is drawn of a portion of the truss. This free-body diagram may contain no more than three unknown member forces. The set of forces which act on the isolated portion of the truss are a general two-dimensional force system, and three equations of equilibrium may be written for such a system.

A pulley may be connected to a truss at any joint. The cable forces which act on the rim of the pulley are transferred directly to the joint, and they are treated as applied forces at this joint.

A connected truss is a configuration in which the free-body diagram of the entire structure contains more than three unknown reaction forces. For this case, the stable trusses which comprise the original configuration are identified, and a free-body diagram of each of these trusses is drawn. The equilibrium equations of each of these trusses are then solved simultaneously, to obtain the reaction forces on the original structure.

CHAPTER 7

Force Analysis of Plane Frames and Machines

7.1 GEOMETRY AND LOADING OF FRAMES, COMPARISON OF TRUSS AND FRAME SOLUTIONS, MULTIFORCE AND TWO-FORCE MEMBERS IN A FRAME, GENERAL METHOD OF SOLUTION

7.1 (*a*) Describe the construction, and method of loading, of a *plane framework*, or *plane frame*.

(*b*) Show several examples of plane frames.

▮ (*a*) The basic characteristics of a plane frame are that all the members of the interconnected framework lie in a single plane and all the forces acting on this structure lie in the plane of the structure. These loads may be forces which are applied either at the connection points of the members or at any intermediate location along the length of the members. The applied loads may also be couples that are applied at any location on the member. Finally, the loading may be of a type which is distributed in some manner along the length of the frame member. This latter type of loading will be shown in Chap. 11.

The effect of the above methods of load application on the members of plane frames is that, for these members, the directions of the forces which act on these members are *not* known at the outset of the problem. In addition, these forces which act on the members will *not*, in general, have a single, uniform value along the length of the member, as was the case with a truss member.

(*b*) Figure 7.1 shows several examples of plane frames. The reader should study these diagrams carefully and determine that each has the framework characteristics described above and that none of these structures could be described as a truss.

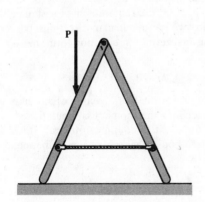

Fig. 7.1*a*

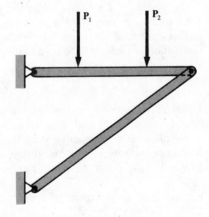

Fig. 7.1*b*

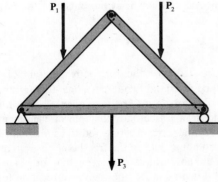

Fig. 7.1*c*

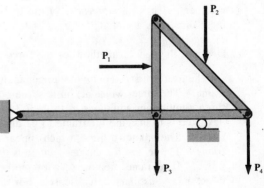

Fig. 7.1*d*

255

7.2   Compare the methods of application of the loads to a plane frame and to a plane truss.

▮   The plane truss, considered in Chap. 6, was seen to consist of members which are connected to one another only at their ends.   An additional requirement was that the loads acting on the truss lie in the plane of this structure, and that these loads be applied *only* at the joints where the truss members are joined to one another.   As a consequence, *every* member in the truss is a two-force member.   Thus, the *directions* of all the forces in the truss members are known at the outset of the problem, and these directions are along the axes of the members.

From this description, the plane frame configuration might appear to be the same as the plane truss configuration.   The major difference between the two, however, is that *no restriction is placed on how the loads may be applied to the members of the frame.*

7.3   Show the general form of loading on a plane frame member and state the requirements for equilibrium of plane frame members.

▮   The most general type of member found in the plane frame is referred to as a *multiforce* member.   Such a member, by definition, has three or more forces acting on it.   Figure 7.3 shows a typical multiforce member.   The forces in this figure all lie in the same plane, because of the earlier definition of loading on the members of a plane frame.

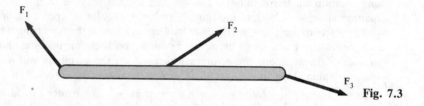

Fig. 7.3

The system of forces acting on the general multiforce member shown in Fig. 7.3 may be recognized as a general two-dimensional force system.   As was shown in Prob. 5.38, the requirements for static equilibrium of such a system of forces is that the summation of forces in two perpendicular directions be zero, and the summation of moments about any arbitrary reference point also be zero.   In equation form, these requirements are

$$\sum F_x = 0 \qquad \sum F_y = 0 \qquad \sum M = 0$$

The reference point for summing moments is usually chosen to be on the line of action of an unknown force, or force component, so that these latter force values do not enter into the moment computation.

The above three equations are the fundamental tools for the force analysis of plane frames.   It will be seen that this set of equations will be used repeatedly, with both the free-body diagrams of the frame members and the free-body diagram of the entire frame.

7.4   Explain how the directions of the forces in two-force members in a plane frame are determined.

▮   The plane frame may contain, in addition to the multiforce members described in Prob. 7.3, members which have no applied loads acting on them and which are pin-connected only at their ends to the remainder of the structure.   From this description, following Prob. 6.5, it may be seen that these members are two-force members.   Thus, the *directions* of the forces in these members are known, at the outset of the problem, and these directions are along the axes of the members.

7.5   Describe the general method for the force analysis of plane frames.

▮   The usual first step in the force analysis of a frame is to draw the free-body diagram of the entire frame.   The forces which act on this free-body diagram will be a general two-dimensional force system, and the three equilibrium equations in Prob. 7.3 may be used to solve for *up to three unknown reaction forces*.   The next step is to imagine the frame to be disassembled, and to draw a free-body diagram of each member of the frame.   The unknown forces which appear on these free-body diagrams are referred to as *internal forces in the frame*.   The reason for this designation is that these forces do not appear in the free-body diagram of the entire frame.   If a frame member is a two-force member, the unknown force acting on it is drawn with *known* direction and *assumed* actual sense.   For the case of a multiforce member, the unknown reaction forces acting on it are each represented by two perpendicular components with assumed actual senses.   The set of force components at the pin where one member joins another member will be a Newton's third law action-reaction pair.

A computed force, or force component, may have a negative value. That is, the force acts in a sense opposite to that which was assumed when the force was drawn on the free-body diagram. When this happened, for the cases of the plane trusses analyzed in Chap. 6, the sense of the force on the free-body diagram was immediately changed and the positive magnitude of this force was shown on the diagram. If, in the analysis of plane frames, the computed value of a force is negative, this term will *not* be altered on the free-body diagram. Rather, it will be carried through the solution as a negative quantity. The reasons for doing this are as follows. When drawing the free-body diagram of a particular member of a plane frame, any unknown internal reaction force that acts on the member has, by Newton's third law, an equal magnitude and *opposite* sense when acting on the member that joins the first member. Thus, if the technique of changing the sense of a reaction force when a negative value is obtained were used in frame analysis, this sense would always have to be changed on the free-body diagrams of *two* members. The usual mistake made when using this technique is that one of the forces is changed but not the other. To avoid the possibility of this, *all computed negative values of force will be treated as negative quantities throughout the remainder of the solution.* The effect of carrying a negative force component through a plane frame solution will be seen repeatedly in the following problems.

It is also possible to proceed with the solution of a frame problem by omitting the step of drawing a free-body diagram of the entire frame, and starting with the free-body diagrams of the frame members. For this case, the free-body diagram of the frame could be used in the final step to check the calculations.

7.6   (*a*)   Compare the solutions which are obtained in the force analyses of plane trusses and plane frames.

(*b*)   What is the central problem in plane frame force analysis?

▌(*a*)   A comparison may be made between the force analyses of plane trusses and plane frames. In the analysis of trusses, the results desired are the magnitude and sense of the forces in the members. For this case, the sense indicates whether the member experiences tensile or compressive forces. In frame analysis, the result usually sought is the magnitude of the total force acting on a member at a pin connection. This quantity is found as the square root of the sum of the squares of the components of the force. In this operation, the senses of the two force components have no effect on the final value of the resultant force.

(*b*)   *The central problem in plane frame force analysis is to solve for the forces which are exerted by the pins on the members.* This information is then used in a subsequent analysis to find the stresses in, or deflection of, the members of the frame. This latter analysis is treated in the subject strength, or mechanics, of materials.

7.7   (*a*)   Find the pin forces which act on the members of the plane frame shown in Fig. 7.7a.

(*b*)   Express the results in part (*a*) in formal vector notation.

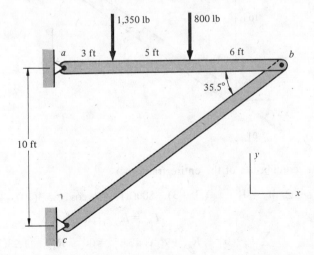

Fig. 7.7a

▌(*a*)   The free-body diagram of the frame is shown in Fig. 7.7b. Member bc is a two-force member, and thus the unknown reaction force R_c has the known direction 35.5°. The two components of the pin force at a are R_{ax} and R_{ay}, and these forces are drawn in assumed actual senses.

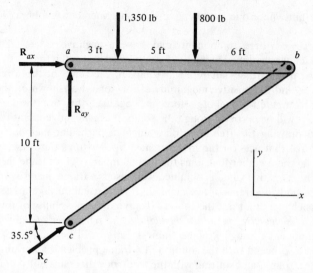

Fig. 7.7b

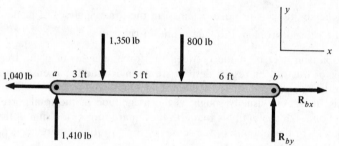

Fig. 7.7c

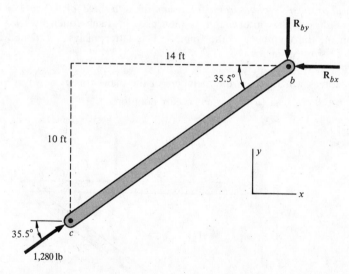

Fig. 7.7d

For static equilibrium of the entire frame,

$$\sum M_a = 0 \qquad -1{,}350(3) - 800(8) + (R_c \cos 35.5°)(10) = 0 \qquad R_c = 1{,}280 \text{ lb}$$

$$F_{bc} = R_c = 1{,}280 \text{ lb (C)}$$

$$\sum F_x = 0 \qquad R_{ax} + R_c \cos 35.5° = 0 \qquad R_{ax} + (1{,}280) \cos 35.5° = 0$$

$$R_{ax} = -1{,}040 \text{ lb}$$

$$\sum F_y = 0 \qquad R_c \sin 35.5° + R_{ay} - 1{,}350 - 800 = 0 \qquad 1{,}280 \sin 35.5° + R_{ay} - 1{,}350 - 800 = 0$$

$$R_{ay} = 1{,}410 \text{ lb}$$

The letter (C) or (T) accompanying the result for the force in a two-force member, as used with F_{bc} above, indicates whether this force is compressive or tensile. The minus sign on R_{ax} indicates that its actual sense is opposite to that which is assumed. The above computations complete the first step of the force analysis of the frame.

The free-body diagrams of the two members are shown in Figs. 7.7c and d. The known reaction forces at a and c, of the foundation on the frame, are drawn in these figures in their actual senses. The force exerted by member bc on member ab, through the pin connection at b, is shown in Fig. 7.7c, with the components R_{bx} and R_{by} drawn in assumed actual senses. These force components are transmitted through the pin at b, and the details of such a force transmission through a pin were presented in Prob. 6.6. By Newton's third law, the unknown force components R_{bx} and R_{by} of Fig. 7.7c *must have opposite senses when drawn in Fig. 7.7d.*

These two force components are examples of *internal forces* in a frame, since they do not appear in the free-body diagram of the entire frame shown in Fig. 7.7b. For force equilibrium of member ab,

$$\sum M_a = 0 \qquad -1{,}350(3) - 800(8) + R_{by}(14) = 0 \qquad R_{by} = 746 \text{ lb}$$

$$\sum F_x = 0 \qquad -1{,}040 + R_{bx} = 0 \qquad R_{bx} = 1{,}040 \text{ lb}$$

The correctness of the above results may be confirmed by using $\sum F_y = 0$. The magnitude of the total force R_b at joint b is

$$R_b = \sqrt{R_{bx}^2 + R_{by}^2} = \sqrt{1{,}040^2 + 746^2} = 1{,}280 \text{ lb}$$

An independent check of the above calculations may be made from consideration of Fig. 7.7d. Since member bc is a two-force member, the forces R_{bx} and R_{by} must be components of the known compressive force of 1,280 lb which acts on the member. Thus,

$$R_{bx} \overset{?}{=} 1{,}280 \cos 35.5° \qquad 1{,}040 \equiv 1{,}040$$
$$R_{by} \overset{?}{=} 1{,}280 \sin 35.5° \qquad 746 \approx 743$$

(*b*) The pin forces acting on member ab are

$$\mathbf{R}_a = -1{,}040\mathbf{i} + 1{,}410\mathbf{j} \quad \text{lb} \qquad \mathbf{R}_b = 1{,}040\mathbf{i} + 746\mathbf{j} \quad \text{lb}$$

The pin forces acting on member bc are

$$\mathbf{R}_b = -1{,}040\mathbf{i} - 746\mathbf{j} \quad \text{lb} \qquad \mathbf{R}_c = (1{,}280 \cos 35.5°)\mathbf{i} + (1{,}280 \sin 35.5°)\mathbf{j} = 1{,}040\mathbf{i} + 746\mathbf{j} \quad \text{lb}$$

7.8 The frame in Fig. 7.8a supports two 1-hp motors. The weight of each motor is 50 lb. When the motors are not running, there is a preload tensile force of 20 lb in each belt. Find the forces exerted by the pins on the frame members when the motors are not running.

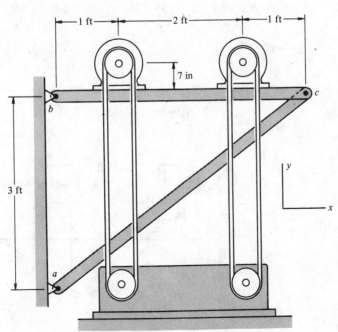

Fig. 7.8a

❚ Since both belt forces on each pulley are equal, the resultant of these forces may be shown as a single force acting at the center of the pulley. Member ac is a two-force member. The free-body diagram of member bc is shown in Fig. 7.8b. For equilibrium of member bc,

$$\Sigma M_b = 0 \qquad -90(1) - 90(3) + (\tfrac{3}{5}F_{ac})4 = 0 \qquad F_{ac} = 150 \text{ lb (C)}$$
$$\Sigma F_x = 0 \qquad -R_{bx} + \tfrac{4}{5}F_{ac} = 0 \qquad R_{bx} = 120 \text{ lb}$$
$$\Sigma F_y = 0 \qquad R_{by} - 90 - 90 + \tfrac{3}{5}F_{ac} = 0 \qquad R_{by} = 90 \text{ lb}$$
$$R_b = \sqrt{R_{bx}^2 + R_{by}^2} = \sqrt{120^2 + 90^2} = 150 \text{ lb}$$

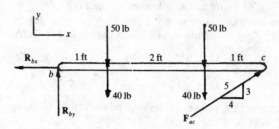

Fig. 7.8b

Pin forces R_a and R_c are given by

$$R_a = R_c = F_{ac} = 150 \text{ lb}$$

It may be observed that $R_b = R_c$. This is an expected result, because of the symmetry of the applied loading in Fig. 7.8b.

7.9 The sense of rotation of the motors in Prob. 7.8 is counterclockwise, and the pulley diameters are 4 in. When the motors are running, the tensile force in the right-side belt is 28.5 lb and that in the left-side belt is 16 lb. Find the forces exerted by the pins on the frame when both motors are running.

❚ Member ac, as before, is a two-force member. The combined free-body diagram of the motors and member bc is shown in Fig. 7.9. For equilibrium of the system in Fig. 7.9,

$$\Sigma M_b = 0 \qquad -16(10) - 50(12) - 28.5(14) - 16(34) - 50(36) - 28.5(38) + (\tfrac{3}{5}F_{ac})48 = 0$$
$$F_{ac} = 159 \text{ lb (C)}$$
$$\Sigma F_x = 0 \qquad -R_{bx} + \tfrac{4}{5}F_{ac} = 0 \qquad R_{bx} = 127 \text{ lb}$$
$$\Sigma F_y = 0 \qquad R_{by} - 16 - 28.5 - 16 - 28.5 + \tfrac{3}{5}F_{ac} - 50 - 50 = 0 \qquad R_{by} = 93.6 \text{ lb}$$
$$R_b = \sqrt{R_{bx}^2 + R_{by}^2} = \sqrt{127^2 + 93.6^2} = 158 \text{ lb}$$

Pin forces R_a and R_c are given by

$$R_a = R_c = F_{ac} = 159 \text{ lb}$$

As a comparison with the results $R_a = R_c = 150 \text{ lb}$ in Prob. 7.8,

$$R_b = 158 \text{ lb} \qquad R_c = 159 \text{ lb}$$

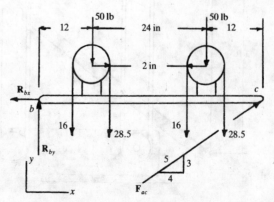

Fig. 7.9

7.10 Find the pin forces acting on the members of the frame in Fig. 7.10a.

▌ The cable forces acting on the pulley are transferred directly to member *ac*. Member *bc* is a two-force member, and the free-body diagram of the frame is shown in Fig. 7.10b. For equilibrium of the frame,

$$\Sigma M_a = 0 \qquad R_b(9) + 750(6) - 750(8) = 0 \qquad R_b = 167 \text{ lb}$$

$$F_{bc} = R_b = 167 \text{ lb (T)}$$

$$\Sigma F_x = 0 \qquad -R_b - 750 + R_{ax} = 0 \qquad R_{ax} = 917 \text{ lb}$$

$$\Sigma F_y = 0 \qquad R_{ay} - 750 = 0 \qquad R_{ay} = 750 \text{ lb}$$

$$R_a = \sqrt{R_{ax}^2 + R_{ay}^2} = \sqrt{917^2 + 750^2} = 1{,}180 \text{ lb} \qquad R_d = \sqrt{750^2 + 750^2} = 1{,}060 \text{ lb} \qquad R_c = R_b = F_{bc} = 167 \text{ lb}$$

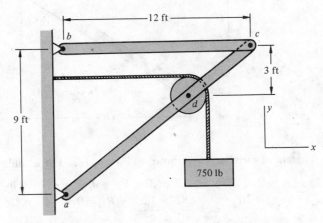

Fig. 7.10a

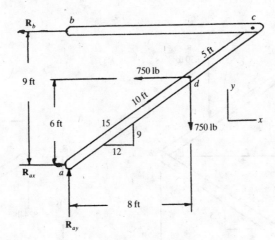

Fig. 7.10b

7.11 Find the pin forces acting on the members of the frame shown in Fig. 7.11a.

▌ Member *bc* is a two-force member, and Fig. 7.11b shows the free-body diagram of the frame. For equilibrium of the frame,

$$\Sigma M_a = 0 \qquad -1{,}250(3) + \frac{4}{4.72} R_c(5.5) = 0 \qquad R_c = 805 \text{ lb}$$

$$F_{bc} = R_c = 805 \text{ lb (C)}$$

$$\Sigma F_x = 0 \qquad -R_{ax} + \frac{4}{4.72} R_c = 0 \qquad R_{ax} = 682 \text{ lb}$$

$$\Sigma F_y = 0 \qquad R_{ay} + \frac{2.5}{4.72} R_c - 1{,}250 = 0 \qquad R_{ay} = 824 \text{ lb}$$

$$R_a = \sqrt{R_{ax}^2 + R_{ay}^2} = \sqrt{682^2 + 824^2} = 1{,}070 \text{ lb}$$

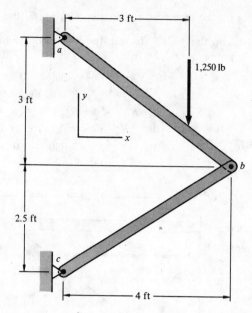

Fig. 7.11a

The free-body diagram of member ab is shown in Fig. 7.11c. For equilibrium of member ab,

$$\Sigma F_x = 0 \qquad -682 + R_{bx} = 0 \qquad R_{bx} = 682 \text{ lb}$$
$$\Sigma F_y = 0 \qquad 824 - 1{,}250 + R_{by} = 0 \qquad R_{by} = 426 \text{ lb}$$
$$R_b = \sqrt{R_{bx}^2 + R_{by}^2} = \sqrt{682^2 + 426^2} = 804 \text{ lb}$$
$$R_b = R_c = F_{bc} = 804 \text{ lb}$$

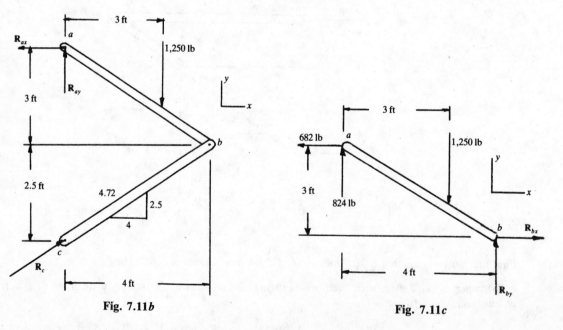

Fig. 7.11b

Fig. 7.11c

7.12 Find the pin forces acting on the members of the frame shown in Fig. 7.12a.

▮ Member bd is a two-force member, and Fig. 7.12b shows the free-body diagram of member ac. The equilibrium requirement is

$$\Sigma M_a = 0 \qquad \left(\frac{500}{583} F_{bd}\right)\left(\frac{500}{1{,}000}\right) + \left(\frac{300}{583} F_{bd}\right)\left(\frac{350}{1{,}000}\right) - 135\left(\frac{900}{1{,}000}\right) - 200 = 0 \qquad F_{bd} = 528 \text{ N (C)}$$

$$\Sigma F_x = 0 \qquad R_{ax} - \frac{300}{583} F_{bd} = 0 \qquad R_{ax} = 272 \text{ N}$$

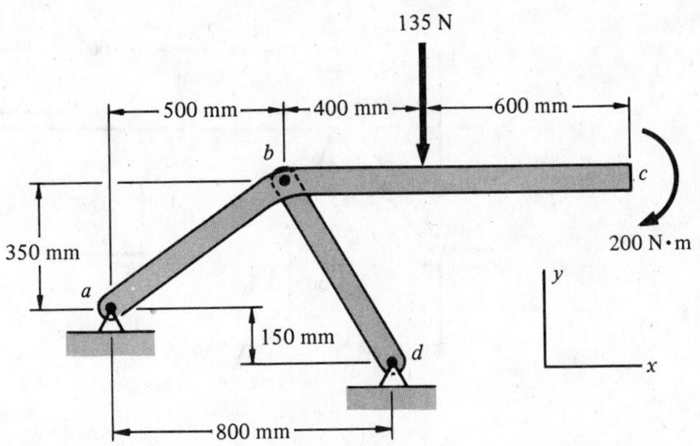

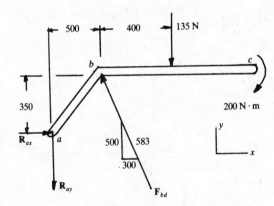

Fig. 7.12a

$$\Sigma F_y = 0 \qquad -R_{ay} + \frac{500}{583} F_{bd} - 135 = 0 \qquad R_{ay} = 318 \text{ N}$$

$$R_a = \sqrt{R_{ax}^2 + R_{ay}^2} = \sqrt{272^2 + 318^2} = 418 \text{ N} \qquad R_b = R_d = F_{bd} = 528 \text{ N}$$

As a check on the above results, using Fig. 7.12b,

$$\Sigma M_b = 0 \qquad -135\left(\frac{400}{1,000}\right) - 200 + 272\left(\frac{350}{1,000}\right) + 318\left(\frac{500}{1,000}\right) \overset{?}{=} 0$$

$$272\left(\frac{350}{1,000}\right) + 318\left(\frac{500}{1,000}\right) \overset{?}{=} 135\left(\frac{400}{1,000}\right) + 200 \qquad 254 \equiv 254$$

Fig. 7.12b

7.13 The applied couple is removed from the frame in Prob. 7.12

(a) What vertical force acting at c would result in the same force at pin d as that caused by the applied couple in the original problem?

(b) Find the corresponding value of the pin force at a.

▌ (a) From Prob. 7.12,

$$F_{bd} = 528 \text{ N (C)}$$

Figure 7.13 shows the free-body diagram of members ac, and P is the vertical force at c. For equilibrium,

$$\Sigma M_a = 0 \qquad \frac{500}{583}(528)500 + \frac{300}{583}(528)350 - 135(900) - P(1,500) = 0 \qquad P = 133 \text{ N}$$

$$\Sigma F_x = 0 \qquad R_{ax} - \frac{300}{583}(528) = 0 \qquad R_{ax} = 272 \text{ N}$$

$$\Sigma F_y = 0 \qquad -R_{ay} + \frac{500}{583}(528) - 135 - P = 0 \qquad R_{ay} = 185 \text{ N}$$

(b) The pin force at a is given by

$$R_a = \sqrt{R_{ax}^2 + R_{ay}^2} = \sqrt{272^2 + 185^2} = 329 \text{ N} \qquad R_b = R_d = F_{bd} = 528 \text{ N}$$

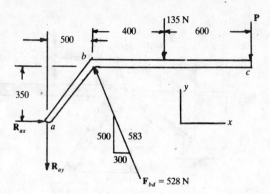

Fig. 7.13

7.14 Find the forces exerted by the pins on the members of the frame shown in Fig. 7.14a.

❚ Member df is a two-force member, and the free-body diagram of member ae is shown in Fig. 7.14b. For equilibrium,

$$\Sigma M_a = 0 \quad (1{,}500 \cos 22°)(1{,}500) + 800(150) - \left(\frac{600}{922} F_{df}\right)(1{,}250) = 0 \quad F_{df} = 2{,}710 \text{ N (T)}$$

$$\Sigma F_x = 0 \quad -1{,}500 \cos 22° + \frac{600}{922} F_{df} - R_{ax} = 0 \quad R_{ax} = 373 \text{ N}$$

$$\Sigma F_y = 0 \quad -1{,}500 \sin 22° - \frac{700}{922} F_{df} + 800 + R_{ay} = 0 \quad R_{ay} = 1{,}820 \text{ N}$$

$$R_a = \sqrt{R_{ax}^2 + R_{ay}^2} = \sqrt{373^2 + 1{,}820^2} - 1{,}860 \text{ N} \quad R_d = R_f = F_{df} = 2{,}710 \text{ N}$$

As a check,

$$\Sigma M_d = 0$$
$$(1{,}500 \cos 22°)(250) + 800(150) - 373(1{,}250) \overset{?}{=} 0 \quad 4.67 \times 10^5 \approx 4.66 \times 10^5$$

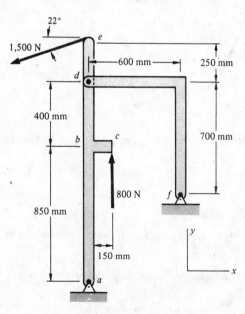

Fig. 7.14a

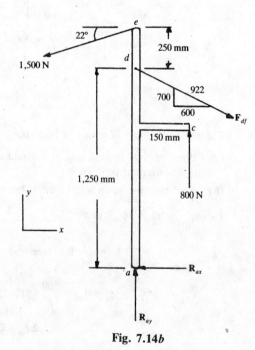

Fig. 7.14b

7.15 (a) Is it possible to replace the applied force at c in Prob. 7.14 with a single couple and obtain the same pin force at a as in the original problem?

(b) Do the same as in part (a) to obtain the same pin force at d as in the original problem.

(a) It is not possible to replace the applied force at c with a single couple and obtain the original pin force at a. The reason for this is that the resultant force of a couple is zero, and it is impossible to produce the required reaction force component R_{ay} with the couple. The reaction force component R_{ax} would be unaffected by the replacement of the 800-N applied force by a couple.

(b) Figure 7.15 shows the free-body diagram of the frame, and M is the applied couple. Member df is a two-force member, and, from Prob. 7.14,

$$F_{df} = 2,710\,\text{N (T)}$$

For equilibrium,

$$\Sigma M_a = 0 \qquad (1,500 \cos 22°)\left(\frac{1,500}{1,000}\right) - \frac{600}{922}(2,710)\left(\frac{1,250}{1,000}\right) + M = 0 \qquad M = 118\,\text{N·m}$$

$$\Sigma F_x = 0 \qquad -1,500 \cos 22° + \frac{600}{922}(2,710) - R_{ax} = 0 \qquad R_{ax} = 373\,\text{N}$$

$$\Sigma F_y = 0 \qquad -1,500 \sin 22° - \frac{700}{922}(2,710) + R_{ay} = 0 \qquad R_{ay} = 2,620\,\text{N}$$

$$R_a = \sqrt{R_{ax}^2 + R_{ay}^2} = \sqrt{373^2 + 2,620^2} = 2,650\,\text{N}$$

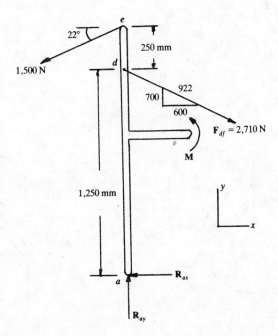

Fig. 7.15

7.16 **(a)** For the frame shown in Fig. 7.16a, find the pin force at b and the x and y components of the force exerted by the members on the ground. The cylinder weighs 300 lb.

(b) Express the results in part (a) in formal vector notation.

(a) From the geometry relationships in Fig. 7.16b,

$$\tan \theta = \frac{1.5}{2} \qquad \theta = 36.9° \qquad A = \frac{1.25}{\tan \theta} = 1.66\,\text{ft}$$

The free-body diagram of the cylinder is shown in Fig. 7.16c. From symmetry,

$$R_e = R_d$$
$$\Sigma F_y = 0 \qquad 2R_e \sin \theta - 300 = 0 \qquad R_e = R_d = 250\,\text{lb}$$

Figure 7.16d shows the free-body diagram of member ad of the symmetrical frame. From the symmetry of the entire structure,

$$R_{ay} = R_{cy} = \tfrac{1}{2}(300) = 150\,\text{lb}$$

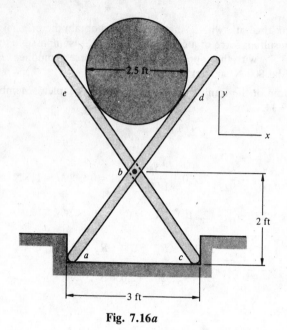

Fig. 7.16a

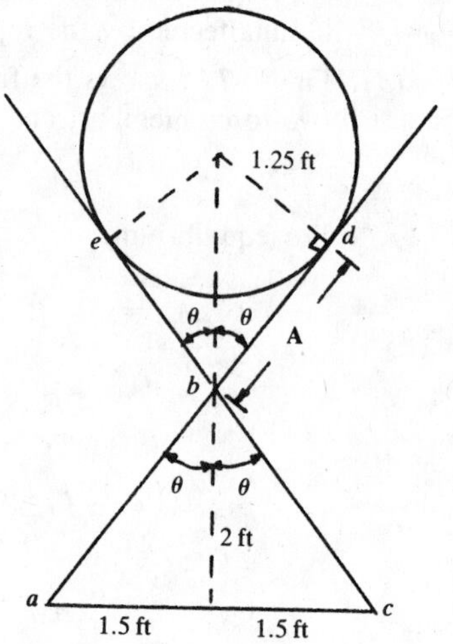

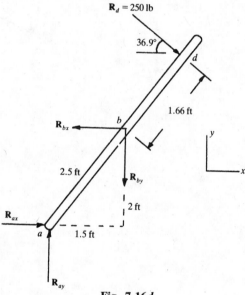

Fig. 7.16b

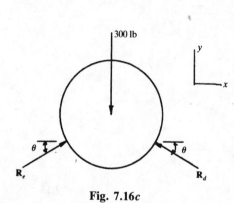

Fig. 7.16c

Fig. 7.16d

For equilibrium,

$$\Sigma F_y = 0 \qquad R_{ay} - R_{by} - 250 \sin 36.9° = 0 \qquad 150 - R_{by} - 250 \sin 36.9° = 0$$
$$R_{by} = 0$$
$$\Sigma M_a = 0 \qquad R_{bx}(2) - 250(2.5 + 1.66) = 0 \qquad R_{bx} = 520 \text{ lb}$$
$$R_b = R_{bx} = 520 \text{ lb}$$
$$\Sigma F_x = 0 \qquad R_{ax} - R_{bx} + 250 \cos 36.9° = 0 \qquad R_{ax} = 320 \text{ lb}$$

From symmetry, the reaction force components R_{cx} and R_{cy}, acting on member ce, act to the left and upward, and

$$R_{cx} = 320 \text{ lb} \qquad R_{cy} = 150 \text{ lb}$$

The senses of the forces exerted on the ground are opposite those shown in the free-body diagram of member ad in Fig. 7.16d.

(**b**) The pin force on member ab is

$$\mathbf{R}_b = -520\mathbf{i} \qquad \text{lb}$$

The pin force on member *ce* is

$$\mathbf{R}_b = 520\mathbf{i} \quad \text{lb}$$

The forces exerted by the members on the ground are

$$\mathbf{R}_a = -320\mathbf{i} - 150\mathbf{j} \quad \text{lb} \qquad \mathbf{R}_c = 320\mathbf{i} - 150\mathbf{j} \quad \text{lb}$$

7.17   The framework structure in Fig. 7.17*a* is to support an applied clockwise couple of 600 in · lb.   The couple may be applied to either member *ab* or member *bc*.   Which of these two methods of support of the applied couple will result in the smaller values of the forces exerted by the pins on the members?

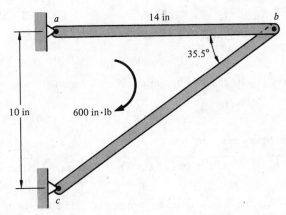

Fig. 7.17*a*

I   The couple is first assumed to be applied to member *ab*.   For this configuration, member *bc* behaves as a two-force member, and the free-body diagram of the structure is shown in Fig. 7.17*b*.   For equilibrium of the entire frame,

$$\sum M_a = 0 \qquad -600 + (R_c \cos 35.5°)10 = 0 \qquad R_c = 73.7 \text{ lb}$$
$$F_{bc} = R_c = 73.7 \text{ lb (C)}$$
$$\sum F_x = 0 \qquad R_{ax} + R_c \cos 35.5° = 0 \qquad R_{ax} + 73.7 \cos 35.5° = 0$$
$$R_{ax} = -60 \text{ lb}$$
$$\sum F_y = 0 \qquad R_{ay} + R_c \sin 35.5° = 0 \qquad R_{ay} + 73.7 \sin 35.5° = 0$$
$$R_{ay} = -42.8 \text{ lb}$$

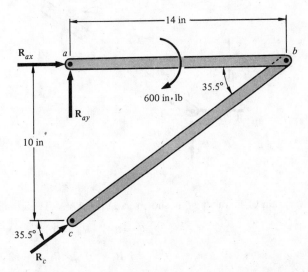

Fig. 7.17*b*

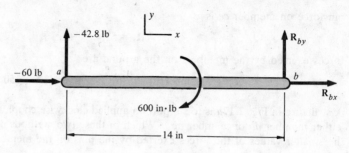

Fig. 7.17c

The free-body diagram of member ab is shown in Fig. 7.17c. For equilibrium of this member,

$$\sum M_a = 0 \qquad -600 + R_{by}(14) = 0 \qquad R_{by} = 42.9 \text{ lb}$$

$$\sum F_x = -60 + R_{bx} = 0 \qquad R_{bx} = 60 \text{ lb}$$

The above calculations may be checked by using $\sum F_y = 0$ in the free-body diagram of Fig. 7.17c. The result is

$$\sum F_y = -42.8 + R_{by} \overset{?}{=} 0 \qquad 42.9 \approx 42.8$$

The values of the pin forces are then

$$R_a = \sqrt{R_{ax}^2 + R_{ay}^2} = \sqrt{(-60)^2 + (-42.8)^2} = 73.7 \text{ lb} \qquad R_b = \sqrt{R_{bx}^2 + R_{by}^2} = \sqrt{60^2 + 42.9^2} = 73.8 \text{ lb}$$
$$R_b = 73.8 \text{ lb} \approx R_c = F_{bc} = 73.7 \text{ lb}$$

The couple is now assumed to be applied to member bc. The free-body diagram of the entire frame for this case is shown in Fig. 7.17d. Member bc is no longer a two-force member, since the load on this member is applied at a location other than the ends of the member. Member ab, however, is a two-force member for this condition of loading. For equilibrium of the entire frame,

$$\sum M_c = 0 \qquad -600 - R_a(10) = 0 \qquad R_a = -60 \text{ lb}$$

$$F_{ab} = R_a = 60 \text{ lb (T)}$$

$$\sum F_x = 0 \qquad R_a + R_{cx} = 0 \qquad -60 + R_{cx} = 0 \qquad R_{cx} = 60 \text{ lb}$$

$$\sum F_y = 0 \qquad R_{cy} = 0$$

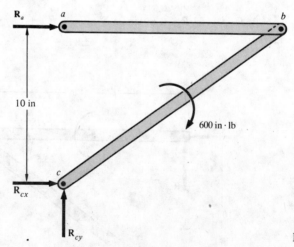

Fig. 7.17d

The free-body diagram of member bc is shown in Fig. 7.17e. The equilibrium requirements are

$$\sum F_y = 0 \qquad R_{by} = 0$$

$$\sum F_x = 0 \qquad 60 + R_{bx} = 0 \qquad R_{bx} = -60 \text{ lb}$$

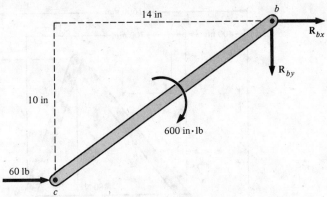

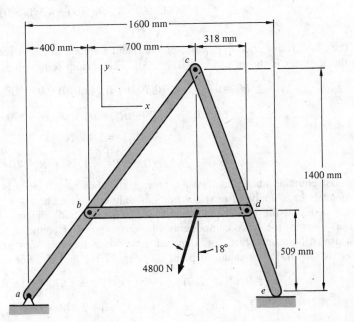

Fig. 7.17e

The above calculations may be checked from

$$\sum M_c = 0 \qquad -600 - R_{bx}(10) \stackrel{?}{=} 0 \qquad -600 - (-60)10 \equiv 0$$

The pin forces for this case are found to be

$$R_a = F_{ab} = 60 \text{ lb} \qquad R_b = R_{bx} = F_{ab} = 60 \text{ lb} \qquad R_c = R_{cx} = 60 \text{ lb}$$

The pin forces are seen to be smaller when the applied couple acts on member bc rather than on member ab.

7.18 (*a*) Find the forces exerted by the pins on the members of the frame shown in Fig. 7.18*a*.

(*b*) Express the forces of the pins on the members in formal vector notation.

Fig. 7.18a

▌ (*a*) The free-body diagram of the entire frame is shown in Fig. 7.18*b*. The equilibrium requirements are

$$\sum M_a = 0 \qquad 1,480(509) - 4,570(1,100) + R_e(1,600) = 0 \qquad R_e = 2,670 \text{ N}$$

$$\sum F_y = 0 \qquad R_{ay} - 4,570 + 2,670 = 0 \qquad R_{ay} = 1,900 \text{ N}$$

$$\sum F_x = 0 \qquad R_{ax} - 1,480 = 0 \qquad R_{ax} = 1,480 \text{ N}$$

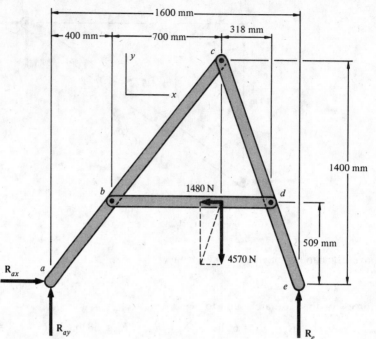

Fig. 7.18b

As a check on the above calculations, a moment summation is made about point e. The result is

$$\sum M_e = 0$$

$$-R_{ay}(1,600) + 1,480(509) + 4,570(500) \overset{?}{=} 0$$
$$1,900(1,600) \overset{?}{=} 1,480(509) + 4,570(500)$$
$$3.04 \times 10^6 \equiv 3.04 \times 10^6$$

The frame is imagined to be disassembled, and member bd is analyzed first. The free-body diagram of this member is shown in Fig. 7.18c. The equilibrium requirements are

$$\sum M_b = 0 \qquad -4,570(700) - R_{dy}(1,020) = 0 \qquad R_{dy} = -3,140 \text{ N}$$

$$\sum F_y = 0 \qquad R_{by} - 4,570 - R_{dy} = 0 \qquad R_{by} - 4,570 - (-3,140) = 0$$

$$R_{by} = 1,430 \text{ N}$$

$$\sum F_x = 0 \qquad R_{bx} - 1,480 - R_{dx} = 0 \qquad (1)$$

The last equation above has two unknowns, and thus cannot be solved at this stage of the solution.

Member ce is analyzed next since it has only one force, acting at the bottom, instead of the two forces acting at the bottom of member ac. The free-body diagram of member ce is shown in Fig. 7.18d. Again, the force components where one member connects to another have opposite senses, because of Newton's third law, depending on which member they are shown to be acting on. This effect is shown, for example, at point d in Figs. 7.18c and d. For equilibrium of member ce,

$$\sum M_c = 0 \qquad R_{dx}(891) + (-3,140)(318) + 2,670(500) = 0 \qquad R_{dx} = -378 \text{ N}$$

$$\sum F_y = 0 \qquad -R_{cy} + (-3,140) + 2,670 = 0 \qquad R_{cy} = -470 \text{ N}$$

$$\sum F_x = 0 \qquad -R_{cx} + (-378) = 0 \qquad R_{cx} = -378 \text{ N}$$

The force component R_{bx} is now found from Eq. (1) as

$$R_{bx} - 1,480 - R_{dx} = 0 \qquad R_{bx} - 1,480 - (-378) = 0 \qquad R_{bx} = 1,100 \text{ N}$$

The values computed above may be checked by considering the free-body diagram of member ac, which is shown Fig. 7.18e. The equilibrium requirements of this member are

$$\sum F_x = 0 \qquad 1,480 + R_{cx} - R_{bx} = 0$$

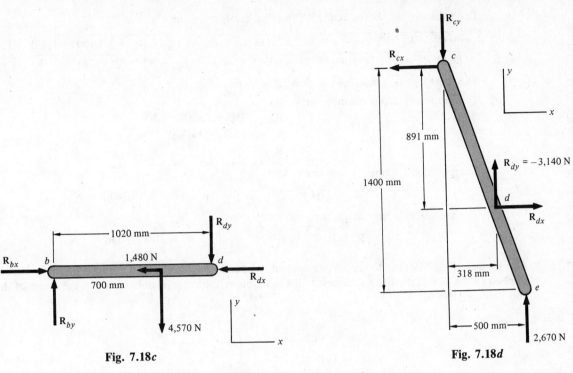

Fig. 7.18c

Fig. 7.18d

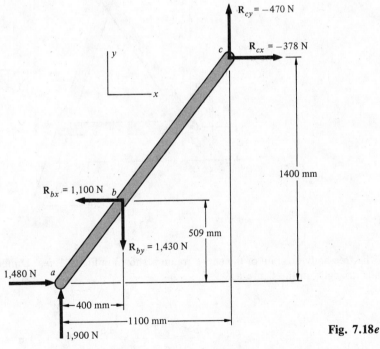

Fig. 7.18e

$$1,480 + (-378) - 1,100 \overset{?}{=} 0 \qquad 1,480 \overset{?}{=} 378 + 1,100 \qquad 1,480 \equiv 1,480$$

$$\sum F_y = 0 \qquad 1,900 - R_{by} + R_{cy} = 0$$

$$1,900 - 1,430 + (-470) \overset{?}{=} 0 \qquad 1,900 \overset{?}{=} 1,430 + 470 \qquad 1,900 \equiv 1,900$$

$$\sum M_a = 0$$

$$-R_{by}(400) + R_{bx}(509) - R_{cx}(1,400) + R_{cy}(1,100) = 0$$

$$-1,430(400) + 1,100(509) - (-378)(1,400) + (-470)(1,100) \overset{?}{=} 0$$

$$1,430(400) + 470(1,100) \overset{?}{=} 1,100(509) + 378(1,400)$$

$$1.09 \times 10^6 \equiv 1.09 \times 10^6$$

The values of the forces acting on the pins are found from

$$R_a = \sqrt{R_{ax}^2 + R_{ay}^2} = \sqrt{1{,}480^2 + 1{,}900^2} = 2{,}410 \text{ N} \qquad R_b = \sqrt{R_{bx}^2 + R_{by}^2} = \sqrt{1{,}100^2 + 1{,}430^2} = 1{,}800 \text{ N}$$

$$R_c = \sqrt{R_{cx}^2 + R_{cy}^2} = \sqrt{(-378)^2 + (-470)^2} = 603 \text{ N} \qquad R_d = \sqrt{R_{dx}^2 + R_{dy}^2} = \sqrt{(-378)^2 + (-3{,}140)^2} = 3{,}160 \text{ N}$$

The most heavily loaded pin is seen to be pin d.

(b) The pin forces acting on member ac are

$$\mathbf{R}_a = 1{,}480\mathbf{i} + 1{,}900\mathbf{j} \qquad \text{N}$$
$$\mathbf{R}_b = -1{,}100\mathbf{i} - 1{,}430\mathbf{j} \qquad \text{N}$$
$$\mathbf{R}_c = -378\mathbf{i} - 470\mathbf{j} \qquad \text{N}$$

The pin forces acting on member ce are

$$\mathbf{R}_c = 378\mathbf{i} + 470\mathbf{j} \qquad \text{N} \qquad \mathbf{R}_d = -378\mathbf{i} - 3{,}140\mathbf{j} \qquad \text{N}$$

The pin forces acting on member bd are

$$\mathbf{R}_b = 1{,}100\mathbf{i} + 1{,}430\mathbf{j} \qquad \text{N} \qquad \mathbf{R}_d = 378\mathbf{i} + 3{,}140\mathbf{j} \qquad \text{N}$$

7.19 The frame in Fig. 7.19a has to support an applied counterclockwise couple of 6,500 N · m. This couple may be applied to any one of the three members. If the couple is applied to member ac, find the forces of the pins on the frame members.

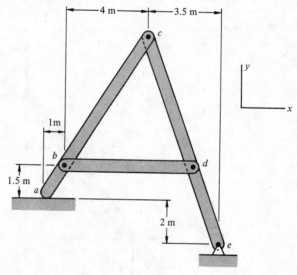

Fig. 7.19a

❚ The free-body diagram of the entire frame is shown in Fig. 7.19b. Dimensions A and B are found from the similar triangles in the figure as

$$\frac{1}{1.5} = \frac{4}{A} \qquad A = 6 \text{ m}$$

$$\frac{B}{3.5} = \frac{3.5}{A + 3.5} = \frac{3.5}{9.5} \qquad B = 1.29 \text{ m}$$

For equilibrium of the frame,

$$\Sigma M_e = 0 \qquad -R_a(8.5) + 6{,}500 = 0 \qquad R_a = 765 \text{ N}$$
$$\Sigma F_y = 0 \qquad R_a - R_{ey} = 0 \qquad R_{ey} = 765 \text{ N}$$
$$\Sigma F_x = 0 \qquad R_{ex} = 0$$

When the couple is applied to member ac, member bd acts as a two-force member. The free-body diagram of member ac is shown in Fig. 7.19c. For equilibrium of member ac,

$$\Sigma M_c = 0 \qquad -765(5) - F_{bd}(6) + 6{,}500 = 0 \qquad F_{bd} = 446 \text{ N (C)}$$
$$\Sigma F_x = 0 \qquad -F_{bd} + R_{cx} = 0 \qquad R_{cx} = 446 \text{ N}$$
$$\Sigma F_y = 0 \qquad 765 - R_{cy} = 0 \qquad R_{cy} = 765 \text{ N}$$

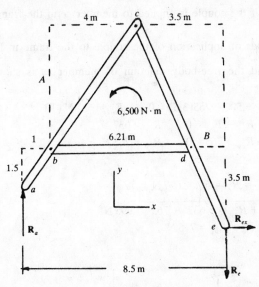

Fig. 7.19b

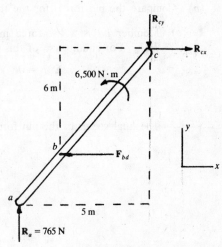

Fig. 7.19c

The final values of the pin forces are

$$R_b = R_d = F_{bd} = 446\,\text{N} \qquad R_c = \sqrt{R_{cx}^2 + R_{cy}^2} = \sqrt{446^2 + 765^2} = 886\,\text{N} \qquad R_e = R_{ey} = 765\,\text{N}$$

7.20 Find the pin forces if the couple is applied to member *bd* in the frame in Prob. 7.19.

❚ For this loading, the frame contains no two-force members. Figure 7.20*a* shows the free-body diagram of member *bd*. For equilibrium of this member,

$$\Sigma M_d = 0 \qquad -R_{by}(6.21) + 6,500 = 0 \qquad R_{by} = 1,050\,\text{N}$$
$$\Sigma F_y = 0 \qquad R_{by} - R_{dy} = 0 \qquad R_{dy} = 1,050\,\text{N} \qquad \Sigma F_x = 0 \qquad R_{bx} - R_{dx} = 0$$

The free-body diagram of member *ac* is shown in Fig. 7.20*b*. For equilibrium of member *ac*,

$$\Sigma M_c = 0 \qquad 1,050(4) - R_{bx}(6) - 765(5) = 0 \qquad R_{bx} = 62.5\,\text{N}$$

Using $R_{bx} - R_{dx} = 0$,

$$R_{dx} = R_{bx} = 62.5\,\text{N}$$

$$\Sigma F_x = 0 \qquad -R_{bx} + R_{cx} = 0 \qquad R_{cx} = 62.5\,\text{N}$$
$$\Sigma F_y = 0 \qquad 765 - 1,050 + R_{cy} = 0 \qquad R_{cy} = 285\,\text{N}$$

The pin forces have the final values

$$R_b = \sqrt{R_{bx}^2 + R_{by}^2} = \sqrt{62.5^2 + 1,050^2} = 1,050\,\text{N}$$
$$R_c = \sqrt{R_{cx}^2 + R_{cy}^2} = \sqrt{62.5^2 + 285^2} = 292\,\text{N}$$
$$R_d = \sqrt{R_{dx}^2 + R_{dy}^2} = \sqrt{62.5^2 + 1,050^2} = 1,050\,\text{N}$$
$$R_e = R_{ey} = 765\,\text{N}$$

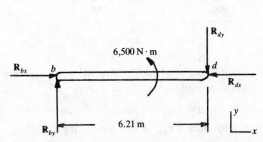

Fig. 7.20a

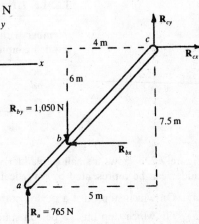

Fig. 7.20b

7.21   (a)   Find the forces of the pins on the members if the couple is applied to member *ce* in the frame in Prob. 7.19.

       (b)   Compare the pin forces for the three methods of application of the couple to the frame in Prob. 7.19.

▌ (a)   Member *bd* is a two-force member, and the free-body diagram of member *ce* is shown in Fig. 7.21.   For equilibrium of this member,

$$\Sigma M_c = 0 \qquad -F_{bd}(6) + 6{,}500 - 765(3.5) = 0 \qquad F_{bd} = 637 \text{ N (T)}$$
$$\Sigma F_x = 0 \qquad R_{cx} - F_{bd} = 0 \qquad R_{cx} = 637 \text{ N}$$
$$\Sigma F_y = 0 \qquad R_{cy} - 765 = 0 \qquad R_{cy} = 765 \text{ N}$$

The final values of the pin forces are

$$R_b = R_d = F_{bd} = 637 \text{ N}$$
$$R_c = \sqrt{R_{cx}^2 + R_{cy}^2} = \sqrt{637^2 + 765^2} = 995 \text{ N}$$
$$R_e = R_{ey} = 765 \text{ N}$$

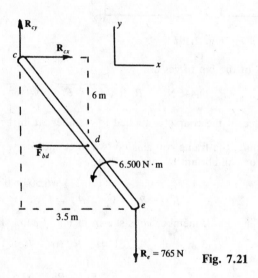

Fig. 7.21

    (b)   The pin forces for the three methods of application of the couple in Probs. 7.19, 7.20, and 7.21 are shown in Table 7.1.   The best design is application of couple to member *ac*, with a maximum pin force (pin *c*) of 886 N.   The worst design is application of couple to member *bd*, with a maximum pin force (pins *b* and *d*) of 1,050 N.   It may be observed that the pin force at *e* is the same for all three cases.

TABLE 7.1

member with applied couple	pin force, N			
	R_b	R_c	R_d	R_e
ac	446	886	446	765
bd	1,050	292	1,050	765
ce	637	995	637	765

7.22   Figure 7.22*a* shows a small stepladder which rests on a smooth plane surface.   The weight of a person on the ladder can be represented by a vertical force acting at the center of the width of a step.

       (a)   On which step must a person stand to produce the maximum force of pin *f* on the members?

       (b)   On which step must a person stand to produce the maximum force in the connecting cable?

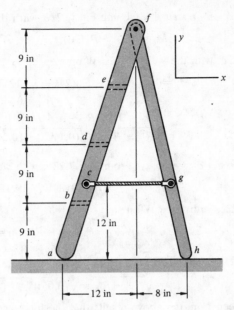

Fig. 7.22a

(a) *Weight of person on step e*   Figure 7.22b shows the free-body diagram of the stepladder, with the weight of the person on step *e*.   For moment equilibrium of the ladder,

$$\Sigma M_h = 0 \qquad -R_a(20) + W(8+3) = 0 \qquad R_a = 0.55W$$

The free-body diagram of member *af* is shown in Fig. 7.22c, and it may be observed that the connecting cable is a two-force member.   For equilibrium of this member,

$$\Sigma M_f = 0 \qquad W(3) + F_{cg}(24) - R_a(12) = 0 \qquad 3W + 24F_{cg} - 12(0.55W) = 0$$
$$F_{cg} = 0.15W \text{ (T)}$$
$$\Sigma F_x = 0 \qquad F_{cg} - R_{fx} = 0 \qquad R_{fx} = 0.15W$$
$$\Sigma F_y = 0 \qquad R_a - W + R_{fy} = 0 \qquad 0.55W - W + R_{fy} = 0$$
$$R_{fy} = 0.45W$$
$$R_f = \sqrt{R_{fx}^2 + R_{fy}^2} = \sqrt{(0.15W)^2 + (0.45W)^2} = 0.474W$$

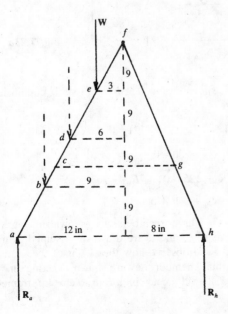

Fig. 7.22b

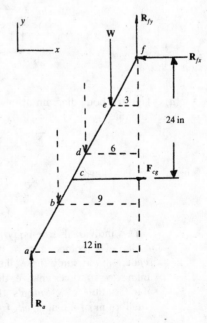

Fig. 7.22c

Weight of person on step d   Using Fig. 7.22b, with the force W acting on step d,

$$\Sigma M_h = 0 \quad -R_a(20) + W(8+6) = 0 \quad R_a = 0.7W$$

From Fig. 7.22c, with the weight force W on step d,

$$\Sigma M_f = 0 \quad W(6) + F_{cg}(24) - R_a(12) = 0 \quad F_{cg} = 0.1W \text{ (T)}$$
$$\Sigma F_x = 0 \quad F_{cg} - R_{fx} = 0 \quad R_{fx} = 0.1W$$
$$\Sigma F_y = 0 \quad R_a - W + R_{fy} = 0 \quad R_{fy} = 0.3W$$
$$R_f = \sqrt{R_{fx}^2 + R_{fy}^2} = \sqrt{(0.1W)^2 + (0.3W)^2} \quad R_f = 0.316W$$

Weight of person on step b   From Fig. 7.22b, with weight force W acting on step b,

$$\Sigma M_h = 0 \quad -R_a(20) + W(8+9) = 0 \quad R_a = 0.85W$$

Using Fig. 7.22c, with force W on step b,

$$\Sigma M_f = 0 \quad W(9) + F_{cg}(24) - R_a(12) = 0 \quad F_{cg} = 0.05W \text{ (T)}$$
$$\Sigma F_x = 0 \quad F_{cg} - R_{fx} = 0 \quad R_{fx} = 0.05W$$
$$\Sigma F_y = 0 \quad R_a - W + R_{fy} = 0 \quad R_{fy} = 0.15W$$
$$R_f = \sqrt{R_{fx}^2 + R_{fy}^2} = \sqrt{(0.05W)^2 + (0.15W)^2} = 0.158W$$

It may be seen from the above results that the force acting on pin f is maximum when the person stands on step e.

(b) The force F_{cg} in the connecting cable is maximum when the person stands on step e.

7.23   (a)   Find the forces of all the pins on the members of the frame shown in Fig. 7.23a.

(b)   Express the forces of the pins on the members in formal vector notation.

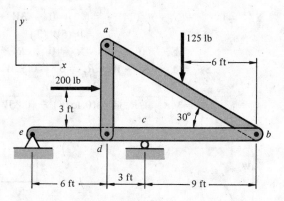

Fig. 7.23a

▌ (a)   The free-body diagram of the entire frame is shown in Fig. 7.23b.   For static equilibrium of the entire structure,

$$\Sigma M_e = 0 \quad -200(3) + R_c(9) - 125(12) = 0 \quad R_c = 233 \text{ lb}$$

$$\Sigma F_x = 0 \quad -R_{ex} + 200 = 0 \quad R_{ex} = 200 \text{ lb}$$

$$\Sigma F_y = 0 \quad R_{ey} + R_c - 125 = 0 \quad R_{ey} + 233 - 125 = 0$$

$$R_{ey} = -108 \text{ lb}$$

The individual free-body diagrams of the three members are shown in Figs. 7.23c through 7.23e.   The components of the pin forces at a, b, and d are drawn in assumed actual senses.   The reader should study these diagrams carefully and observe how the *senses* of the same components of unknown pin forces change, depending on which member they are shown as acting on.   This effect is a manifestation of Newton's third law, and it will occur in *all* problems in frame analysis.   For equilibrium of member ad, from Fig. 7.23c,

$$\Sigma M_d = 0 \quad -200(3) + R_{ax}(6.93) = 0 \quad R_{ax} = 87 \text{ lb}$$

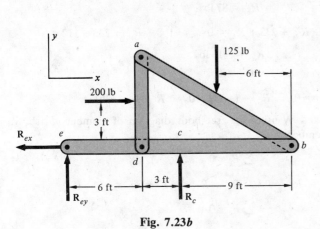

Fig. 7.23b

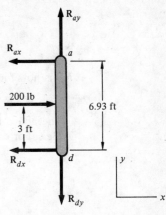

Fig. 7.23c

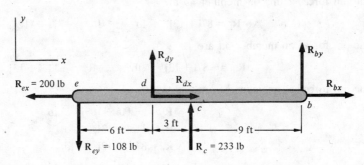

Fig. 7.23d

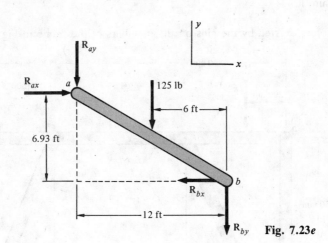

Fig. 7.23e

$$\sum F_x = 0 \qquad -R_{ax} + 200 - R_{dx} = 0 \qquad -87 + 200 - R_{dx} = 0$$

$$R_{dx} = 113 \text{ lb}$$

$$\sum F_y = 0 \qquad R_{ay} - R_{dy} = 0$$

It may be observed that the above computations do not yield all the numerical values of the unknown force components which act on member ad.

From Fig. 7.23d, the equilibrium requirements of member eb are

$$\sum M_d = 0 \qquad 108(6) + 233(3) + R_{by}(12) = 0 \qquad R_{by} = -112 \text{ lb}$$

$$\sum F_x = 0 \qquad -200 + R_{dx} + R_{bx} = 0 \qquad -200 + 113 + R_{bx} = 0$$

$$R_{bx} = 87 \text{ lb}$$

$$\sum F_y = 0 \qquad -108 + R_{dy} + 233 + R_{by} = 0 \qquad -108 + R_{dy} + 233 - 112 = 0$$

$$R_{dy} = -13 \text{ lb}$$

R_{ay} is now found as

$$R_{ay} - R_{dy} = 0 \qquad R_{ay} - (-13) = 0 \qquad R_{ay} = -13 \text{ lb}$$

The above results may be checked by using the free-body diagram of member ab, shown in Fig. 7.23e. For equilibrium of this member,

$$\sum F_x = 0$$

$$R_{ax} - R_{bx} = 0 \qquad R_{ax} \overset{?}{=} R_{bx} \qquad 87 \equiv 87$$

$$\sum F_y = 0$$

$$-R_{ay} - 125 - R_{by} \overset{?}{=} 0 \qquad -R_{ay} - R_{by} \overset{?}{=} 125 \qquad -(-13) - (-112) \overset{?}{=} 125 \qquad 125 \equiv 125$$

(*b*) The pin forces acting on member ab are

$$\mathbf{R}_a = 87\mathbf{i} + 13\mathbf{j} \quad \text{lb} \qquad \mathbf{R}_b = -87\mathbf{i} + 112\mathbf{j} \quad \text{lb}$$

The pin forces on member ad are

$$\mathbf{R}_a = -87\mathbf{i} - 13\mathbf{j} \quad \text{lb} \qquad \mathbf{R}_d = -113\mathbf{i} + 13\mathbf{j} \quad \text{lb}$$

The pin forces on member be are

$$\mathbf{R}_e = -200\mathbf{i} - 108\mathbf{j} \quad \text{lb}$$
$$\mathbf{R}_d = 113\mathbf{i} - 13\mathbf{j} \quad \text{lb}$$
$$\mathbf{R}_b = 87\mathbf{i} - 112\mathbf{j} \quad \text{lb}$$

As seen in the above equations, the senses of the components of the pin forces $\mathbf{R}_a$, $\mathbf{R}_b$, and $\mathbf{R}_d$ are opposite when shown as forces acting on different members. This is an expected result from Newton's third law.

7.24 Find the forces exerted by the pins on the members of the frame in Fig. 7.24a.

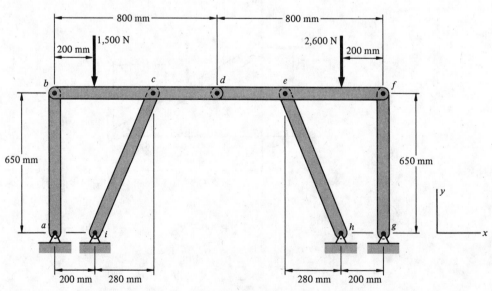

Fig. 7.24a

▌ Members ab, ci, eh, and fg are two-force members. Figure 7.24b shows the free-body diagram of member bd. For equilibrium of this member,

$$\Sigma M_b = 0 \qquad -1{,}500(200) + \left(\frac{650}{708} F_{ci}\right)(480) - R_{dy}(800) = 0 \qquad (1)$$

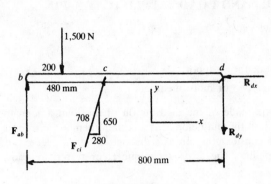

Fig. 7.24b

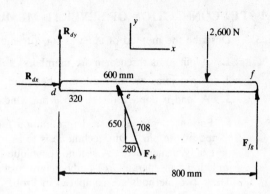

Fig. 7.24c

$$\Sigma F_x = 0 \qquad \frac{280}{708} F_{ci} - R_{dx} = 0 \qquad (2)$$

$$\Sigma F_y = 0 \qquad F_{ab} - 1{,}500 + \frac{650}{708} F_{ci} - R_{dy} = 0 \qquad (3)$$

The free-body diagram of member df is shown in Fig. 7.24c. For equilibrium of member df,

$$\Sigma M_d = 0 \qquad \frac{650}{708} F_{eh}(320) - 2{,}600(600) + F_{fg}(800) = 0 \qquad (4)$$

$$\Sigma F_x = 0 \qquad R_{dx} - \frac{280}{708} F_{eh} = 0 \qquad (5)$$

$$\Sigma F_y = 0 \qquad R_{dy} + \frac{650}{708} F_{eh} - 2{,}600 + F_{fg} = 0 \qquad (6)$$

F_{eh} and F_{fg} from Eqs. (5) and (6) are substituted into Eq. (4), with the result

$$1.39 R_{dx} + R_{dy} - 650 = 0 \qquad (7)$$

Using Eq. (2) with Eq. (1), to eliminate F_{ci},

$$1.39 R_{dx} - R_{dy} - 375 = 0 \qquad (8)$$

R_{dy} is eliminated between Eqs. (7) and (8) to obtain

$$R_{dx} = 369 \text{ N}$$

Using Eq. (2),

$$F_{ci} = \frac{708}{280} R_{dx} = \frac{708}{280} (369) = 933 \text{ N (C)}$$

From Eq. (7),

$$R_{dy} = -1.39 R_{dx} + 650 = -1.39(369) + 650 = 137 \text{ N}$$

$$R_d = \sqrt{R_{dx}^2 + R_{dy}^2} = \sqrt{369^2 + 137^2} = 394 \text{ N}$$

From Eq. (3),

$$F_{ab} = 1{,}500 - \frac{650}{708} F_{ci} + R_{dy} = 1{,}500 - \frac{650}{708} (933) + 137 = 780 \text{ N}$$

Using Eq. (5),

$$F_{eh} = \frac{708}{280} R_{dx} = \frac{708}{280} (369) = 933 \text{ N (C)}$$

From Eq. (6),

$$F_{fg} = -R_{dy} - \frac{650}{708} F_{eh} + 2{,}600 = -137 - \frac{650}{708} (933) + 2{,}600 = 1{,}610 \text{ N (C)}$$

The pin forces acting on the members are

$$R_a = R_b = F_{ab} = 780 \text{ N} \qquad R_c = R_i = F_{ci} = 933 \text{ N}$$
$$R_d = 394 \text{ N} \qquad R_e = R_h = F_{eh} = 933 \text{ N} \qquad R_f = R_g = F_{fg} = 1{,}610 \text{ N}$$

7.2 PIN CONNECTION OF SEVERAL MEMBERS, AND LOADS APPLIED AT A PIN

7.25 Describe the method of analysis for a frame in which

1. A pin joins three or more members
2. A pin joins two or more members and a foundation support
3. A load is applied directly to a pin which joins two or more members

❚ For the above cases, a determination must be made of the distribution of the load among the members joined by the pin. One technique is to assume that the pin belongs to a certain member and forms a physical part of that member. A second technique is to consider the pin as a distinct structural element of the frame and draw a separate free-body diagram of the pin. The latter technique is preferred for most problems. These two methods are compared in Prob. 7.26.

7.26 The frame of Prob. 7.7 now has an additional 1,000-lb load applied on pin b, as shown in Fig. 7.26a. Find the forces exerted by the pins on the members in the following two cases:

(a) A separate free-body diagram of pin b is used.

(b) Pin b is considered to be an integral part of member ab.

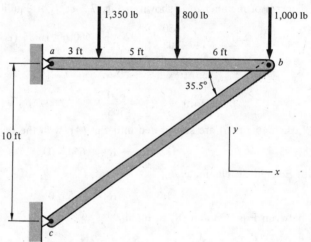

Fig. 7.26a

❚ (a) The earlier conclusion that member bc is a two-force member remains unchanged, and the free-body diagram of the entire frame is shown in Fig. 7.26b. For equilibrium of the frame,

$$\sum M_a = 0 \qquad -1,350(3) - 800(8) - 1,000(14) + (R_c \cos 35.5°)10 = 0$$

$$R_c = 3,000 \text{ lb} \qquad F_{bc} = R_c = 3,000 \text{ lb (C)}$$

$$\sum F_x = 0 \qquad R_{ax} + R_c \cos 35.5° = 0 \qquad R_{ax} + 3,000 \cos 35.5° = 0$$

$$R_{ax} = -2,440 \text{ lb}$$

$$\sum F_y = 0 \qquad R_{ay} - 1,350 - 800 - 1,000 + R_c \sin 35.5° = 0$$

$$R_{ay} - 1,350 - 800 - 1,000 + 3,000 \sin 35.5° = 0 \qquad R_{ay} = 1,410 \text{ lb}$$

In the first method of analysis, the pin will be considered to be a separate element of the structure. The free-body diagram of pin b is shown in Fig. 7.26c. The force components R_{bx} and R_{by} are, in this figure, forces exerted by member ab on the pin. For equilibrium of the pin,

$$\sum F_x = 0 \qquad -R_{bx} + 3,000 \cos 35.5° = 0 \qquad R_{bx} = 2,440 \text{ lb}$$

$$\sum F_y = 0 \qquad -1,000 + 3,000 \sin 35.5° + R_{by} = 0 \qquad R_{by} = -742 \text{ lb}$$

The free-body diagram of member ab is shown in Fig. 7.26d. As a check on the preceding computations, the equilibrium of this member will be verified.

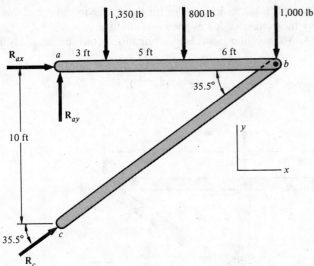

Fig. 7.26b

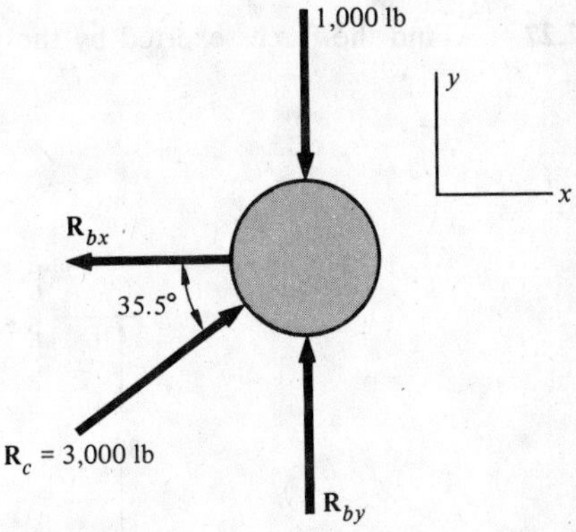

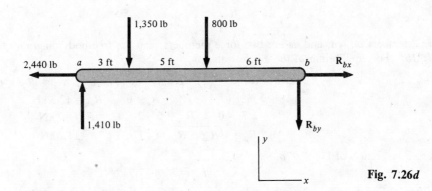

Fig. 7.26c

Fig. 7.26d

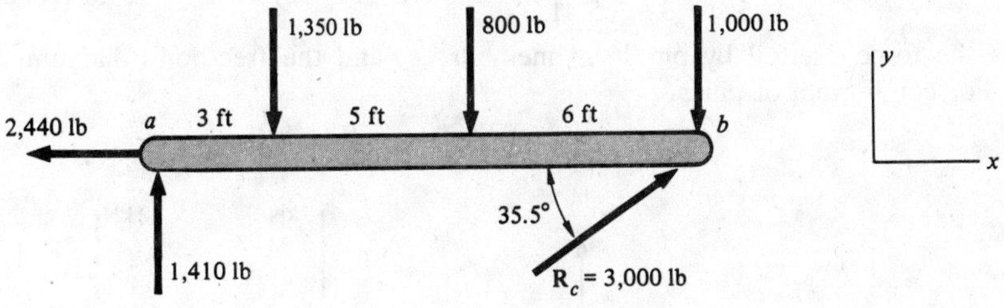

Fig. 7.26e

$$\sum F_x = 0$$

$$-2{,}440 + R_{bx} \overset{?}{=} 0 \qquad 2{,}440 \equiv 2{,}440$$

$$\sum F_y = 0$$

$$1{,}410 - 1{,}350 - 800 - R_{by} \overset{?}{=} 0 \qquad 1{,}410 - 1{,}350 - 800 - (-742) \overset{?}{=} 0$$

$$1{,}410 + 742 \overset{?}{=} 1{,}350 + 800 \qquad 2{,}150 \equiv 2{,}150$$

$$\sum M_a = 0$$

$$-1{,}350(3) - 800(8) - R_{by}(14) \overset{?}{=} 0 \qquad -1{,}350(3) - 800(8) - (-742)14 \overset{?}{=} 0$$

$$1{,}350(3) + 800(8) \overset{?}{=} 742(14) \qquad 10{,}500 \approx 10{,}400$$

(*b*) In the second method of solution, the pin is assumed to form an integral part of member *ab*. The free-body diagram reactions found from Fig. 7.26*b* will be the same as before. The free-body diagram of member *ab* is shown in Fig. 7.26*e*. It is left as an exercise for the reader to show that this member is in equilibrium when acted on by the system of forces shown in the figure.

It should be observed that the part (*b*) solution yields less information about the pin loads than does the part (*a*) solution.

7.27 Find the forces exerted by the pins on the members of the frame in Fig. 7.27*a*.

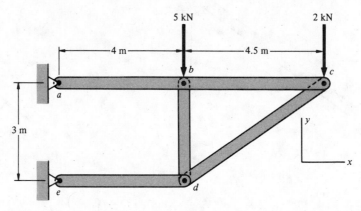

Fig. 7.27*a*

▌ Members *bd*, *cd*, and *de* are two-force members, and the free-body diagram of the frame is shown in Fig. 7.27*b*. For equilibrium of the entire frame,

$$\Sigma M_a = 0 \qquad -5(4) - 2(8.5) + R_e(3) = 0 \qquad R_e = 12.3 \text{ kN}$$
$$\Sigma F_x = 0 \qquad -R_{ax} + R_e = 0 \qquad R_{ax} = 12.3 \text{ kN}$$
$$\Sigma F_y = 0 \qquad R_{ay} - 5 - 2 = 0 \qquad R_{ay} = 7 \text{ kN}$$
$$R_a = \sqrt{R_{ax}^2 + R_{ay}^2} = \sqrt{12.3^2 + 7^2} = 14.2 \text{ kN}$$

From inspection of Fig. 7.27*b*,

$$F_{de} = R_e = 12.3 \text{ kN (C)}$$

The free-body diagram and force triangle for pin *d* are shown in Figs. 7.27*c* and *d*. For equilibrium of pin *d*,

$$\frac{12.3}{4.5} = \frac{F_{bd}}{3} = \frac{F_{cd}}{5.41} \qquad F_{bd} = 8.2 \text{ kN (T)} \qquad F_{cd} = 14.8 \text{ kN (C)}$$

R_{ab} is the force exerted by pin *b* on member *ac*, and the free-body diagram of pin *b* is shown in Fig. 7.27*e*. For equilibrium of pin *b*,

$$\Sigma F_x = 0 \qquad R_{ab,x} = 0$$

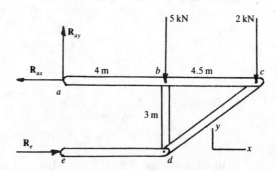

Fig. 7.27*b*

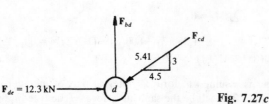

Fig. 7.27*c*

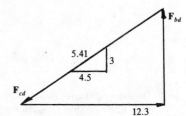

Fig. 7.27*d*

$$\Sigma F_y = 0 \qquad -5 + R_{ab,y} - 8.2 = 0 \qquad R_{ab,y} = 13.2 \text{ kN}$$
$$R_{ab} = R_{ab,y} = 13.2 \text{ kN}$$

R_{ac} is the force exerted by pin c on member ac, and the free-body diagram of pin c is seen in Fig. 7.27f. For equilibrium of pin c,

$$\Sigma F_x = 0 \qquad \frac{4.5}{5.41}(14.8) - R_{ac,x} = 0 \qquad R_{ac,x} = 12.3 \text{ kN}$$

$$\Sigma F_y = 0 \qquad -2 + \frac{3}{5.41}(14.8) - R_{ac,y} = 0 \qquad R_{ac,y} = 6.21 \text{ kN}$$

$$R_{ac} = \sqrt{R_{ac,x}^2 + R_{ac,y}^2} = \sqrt{12.3^2 + 6.21^2} = 13.8 \text{ kN}$$

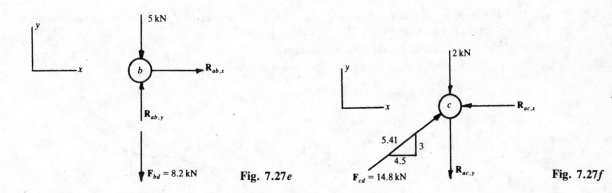

Fig. 7.27e **Fig. 7.27f**

7.28 Would the results in Prob. 7.27 change if the 5-kN load were applied at pin d instead of at pin b?

❚ The reaction forces R_{ax}, R_{ay}, and R_e are the same as in Prob. 7.27, and members bd, cd, and de are again two-force members. The free-body diagram of pin d is shown in Fig. 7.28a. For equilibrium of pin d,

$$\Sigma F_x = 0 \qquad 12.3 - \frac{4.5}{5.41} F_{cd} = 0 \qquad F_{cd} = 14.8 \text{ kN (C)}$$

$$\Sigma F_y = 0 \qquad F_{bd} - \frac{3}{5.41} F_{cd} - 5 = 0 \qquad F_{bd} = 13.2 \text{ kN (T)}$$

Member force F_{cd} is the same as in Prob. 7.27. It therefore follows, from Fig. 7.27f, that pin force R_{ac} is also the same, found in that problem as

$$R_{ac} = 13.8 \text{ kN}$$

The free-body diagram of pin b is shown in Fig. 7.28b. For equilibrium of this pin,

$$\Sigma F_x = 0 \qquad R_{ab,x} = 0$$
$$\Sigma F_y = 0 \qquad R_{ab,y} - 13.2 = 0 \qquad R_{ab} = R_{ab,y} = 13.2 \text{ kN}$$

A comparison of the pin forces for the two different points of application of the 5-kN load is shown in Table 7.2. The net effect of applying the 5-kN force at pin d instead of at pin b is to increase the force in member bd from 8.2 to 13.2 kN.

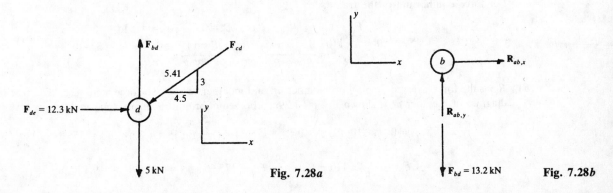

Fig. 7.28a **Fig. 7.28b**

TABLE 7.2

pin force	5-kN force applied at b, kN	5-kN force applied at d, kN
F_{bd}	8.2	13.2
F_{cd}	14.8	14.8
F_{de}	12.3	12.3
R_{ab}	13.2	13.2
R_{ac}	13.8	13.8
R_a	14.2	14.2

7.29 (*a*) Find the required rod force P to keep the crank arrangement in Fig. 7.29*a* in the equilibrium position shown, and the force exerted by the foundation on pin f.

(*b*) Find the forces exerted by the pins on members af and df.

(*c*) Express the results in parts (*a*) and (*b*) in formal vector notation.

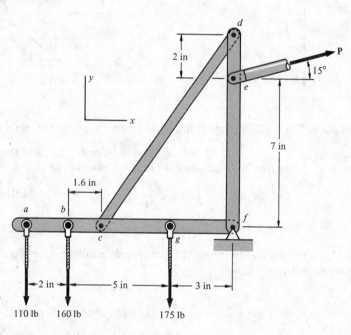

110 lb   160 lb   175 lb

Fig. 7.29a

▌ (*a*) R_f is the force exerted by the foundation on pin f. The free-body diagram of the frame is shown in Fig. 7.29*b*. For moment equilibrium of the frame,

$$\Sigma M_f = 0 \qquad -(P \cos 15°)(7) + 175(3) + 160(8) + 110(10) = 0 \qquad P = 430 \text{ lb}$$

For force equilibrium of the frame,

$$\Sigma F_x = 0 \qquad -R_{fx} + P \cos 15° = 0 \qquad R_{fx} = 415 \text{ lb}$$
$$\Sigma F_y = 0 \qquad -110 - 160 - 175 + R_{fy} + P \sin 15° = 0 \qquad R_{fy} = 334 \text{ lb}$$
$$R_f = \sqrt{R_{fx}^2 + R_{fy}^2} = \sqrt{415^2 + 334^2} = 533 \text{ lb}$$

(*b*) R_{af} is the force exerted by pin f on member af, and member cd is a two-force member. The free-body diagram of member af is shown in Fig. 7.29*c*. For equilibrium of member af,

$$\Sigma M_f = 0 \qquad 110(10) + 160(8) + 175(3) - \left(\frac{9}{11} F_{cd}\right)6.4 = 0 \qquad F_{cd} = 555 \text{ lb}$$

$$\Sigma F_y = 0 \qquad -110 - 160 + \frac{9}{11} F_{cd} - 175 - R_{af,y} = 0 \qquad R_{af,y} = 9.1 \text{ lb}$$

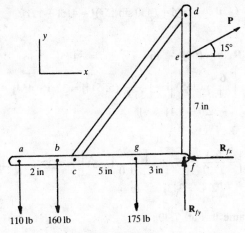

Fig. 7.29b

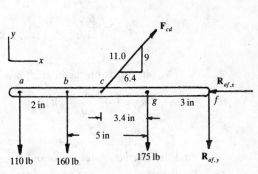

Fig. 7.29c

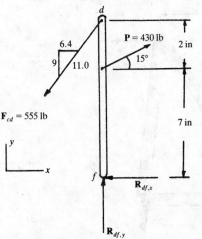

Fig. 7.29d

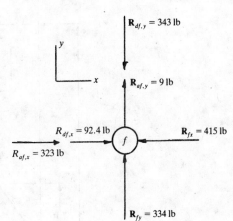

Fig. 7.29e

$$\Sigma F_x = 0 \qquad \frac{6.4}{11} F_{cd} - R_{af,x} = 0 \qquad R_{af,x} = 323 \text{ lb}$$

$$R_{af} = \sqrt{R_{af,x}^2 + R_{af,y}^2} = \sqrt{323^2 + 9.1^2} = 323 \text{ lb}$$

R_{df} is the force exerted by pin f on member df, and the free-body diagram of this member is shown in Fig. 7.29d. For equilibrium of member df,

$$\Sigma F_x = 0 \qquad -\frac{6.4}{11}(555) + 430 \cos 15° - R_{df,x} = 0 \qquad R_{df,x} = 92.4 \text{ lb}$$

$$\Sigma F_y = 0 \qquad -\frac{9}{11}(555) + 430 \sin 15° + R_{df,y} = 0 \qquad R_{df,y} = 343 \text{ lb}$$

$$R_{df} = \sqrt{R_{df,x}^2 + R_{df,y}^2} = \sqrt{92.4^2 + 343^2} = 355 \text{ lb}$$

As a final check on the calculations, the equilibrium of pin f will be confirmed. The free-body diagram of this pin is seen in Fig. 7.29e. For equilibrium of pin f,

$$\Sigma F_x = 0 \qquad 92 + 323 \overset{?}{=} 415 \qquad 415 \equiv 415$$
$$\Sigma F_y = 0 \qquad 334 + 9 \overset{?}{=} 343 \qquad 343 \equiv 343$$

The pin forces on member af are

$$R_a = 110 \text{ lb} \qquad R_b = 160 \text{ lb} \qquad R_c = F_{cd} = 555 \text{ lb}$$
$$R_{af} = 323 \text{ lb} \qquad R_g = 175 \text{ lb}$$

The pin forces on member df are

$$R_d = F_{cd} = 555 \text{ lb} \qquad R_e = P = 430 \text{ lb} \qquad R_{df} = 355 \text{ lb}$$

(c) $$\mathbf{P} = (P\cos 15°)\mathbf{i} + (P\sin 15°)\mathbf{j} = (430\cos 15°)\mathbf{i} + (430\sin 15°)\mathbf{j} = 415\mathbf{i} + 111\mathbf{j} \quad \text{lb}$$

The pin forces on member af are

$$\mathbf{R}_a = -110\mathbf{j} \quad \text{lb} \qquad \mathbf{R}_b = -160\mathbf{j} \quad \text{lb} \qquad \mathbf{R}_g = -175\mathbf{j} \quad \text{lb}$$

$$\mathbf{R}_c = \left(\frac{6.4}{11.0}F_{cd}\right)\mathbf{i} + \left(\frac{9}{11.0}F_{cd}\right)\mathbf{j} = \left[\frac{6.4}{11.0}(555)\right]\mathbf{i} + \left[\frac{9}{11.0}(555)\right]\mathbf{j} = 323\mathbf{i} + 454\mathbf{j} \quad \text{lb}$$

$$\mathbf{R}_{af} = -323\mathbf{i} - 9.1\mathbf{j} \quad \text{lb}$$

The pin forces on member df are

$$\mathbf{R}_d = \left(-\frac{6.4}{11.0}F_{cd}\right)\mathbf{i} - \left(\frac{9}{11.0}F_{cd}\right)\mathbf{j} = \left[-\frac{6.4}{11.0}(555)\right]\mathbf{i} - \left[\frac{9}{11.0}(555)\right]\mathbf{j} = -323\mathbf{i} - 454\mathbf{j} \quad \text{lb}$$

$$\mathbf{R}_{df} = -92.4\mathbf{i} + 343\mathbf{j} \quad \text{lb}$$

7.30 (*a*) Find the reaction forces at *a* and *e* for the frame in Fig. 7.30*a*.

(*b*) Find the forces exerted by the pins on members *ac* and *ce*.

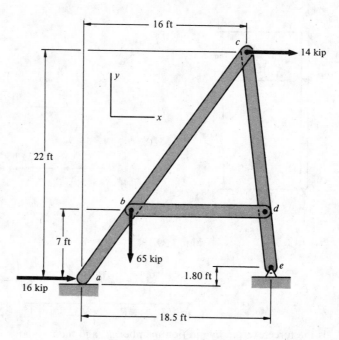

Fig. 7.30*a*

(*a*) Figure 7.30*b* shows the free-body diagram of the frame. From the similar triangles in this figure,

$$\frac{A}{7} = \frac{16}{22} \qquad A = 5.09 \text{ ft}$$

For equilibrium of the frame,

$$\Sigma M_e = 0 \qquad -R_a(18.5) + 16(1.8) + 65(13.4) - 14(20.2) = 0 \qquad R_a = 33.4 \text{ kip}$$

$$\Sigma F_y = 0 \qquad R_a - 65 + R_{ey} = 0 \qquad R_{ey} = 31.6 \text{ kip}$$

$$\Sigma F_x = 0 \qquad 16 + 14 - R_{ex} = 0 \qquad R_{ex} = 30 \text{ kip}$$

$$R_e = \sqrt{R_{ex}^2 + R_{ey}^2} = \sqrt{30^2 + 31.6^2} = 43.6 \text{ kip}$$

(*b*) Member *bd* is a two-force member, and R_{ab} is the force exerted on member *ac* by pin *b*. R_{ac} is the force exerted on member *ac* by pin *c*.

The free-body diagrams of pin *b* and member *ac* are shown in Figs. 7.30*c* and *d*. For equilibrium, using Fig. 7.30*c*,

$$\Sigma F_y = 0 \qquad R_{ab,y} - 65 = 0 \qquad R_{ab,y} = 65 \text{ kip}$$

Using Fig. 7.30*d*,

$$\Sigma M_c = 0 \qquad -33.4(16) + 16(22) + R_{ab,y}(10.9) - R_{ab,x}(15) = 0 \qquad R_{ab,x} = 35.1 \text{ kip}$$

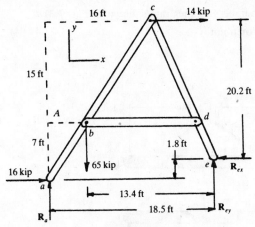

Fig. 7.30b

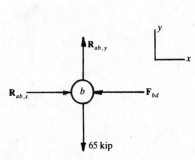

Fig. 7.30c

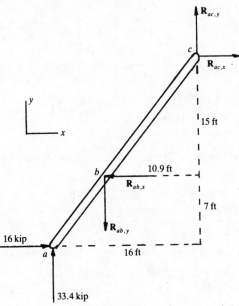

Fig. 7.30d

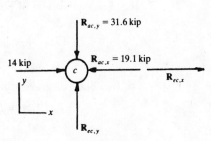

Fig. 7.30e

$$R_{ab} = \sqrt{R_{ab,x}^2 + R_{ab,y}^2} = \sqrt{35.1^2 + 65^2} = 73.9 \text{ kip}$$
$$\Sigma F_x = 0 \qquad 16 - R_{ab,x} + R_{ac,x} = 0 \qquad R_{ac,x} = 19.1 \text{ kip}$$
$$\Sigma F_y = 0 \qquad 33.4 - R_{ab,y} + R_{ac,y} = 0 \qquad R_{ac,y} = 31.6 \text{ kip}$$
$$R_{ac} = \sqrt{R_{ac,x}^2 + R_{ac,y}^2} = \sqrt{19.1^2 + 31.6^2} = 36.9 \text{ kip}$$

Using Fig. 7.30c,

$$\Sigma F_x = 0 \qquad R_{ab,x} - F_{bd} = 0 \qquad F_{bd} = 35.1 \text{ kip (C)}$$

R_{ec} is the force exerted on member ce by pin c. The free-body diagram of pin c is shown in Fig. 7.30e. For equilibrium of pin c,

$$\Sigma F_x = 0 \qquad 14 - 19.1 + R_{ec,x} = 0 \qquad R_{ec,x} = 5.1 \text{ kip}$$
$$\Sigma F_y = 0 \qquad R_{ec,y} - 31.6 = 0 \qquad R_{ec,y} = 31.6 \text{ kip}$$
$$R_{ec} = \sqrt{R_{ec,x}^2 + R_{ec,y}^2} = \sqrt{5.1^2 + 31.6^2} = 32.0 \text{ kip}$$

The pin forces on member ac are

$$R_{ab} = 73.9 \text{ kip} \qquad R_{ac} = 36.9 \text{ kip}$$

The forces on member ce are

$$R_{ec} = 32.0 \text{ kip} \qquad R_b = F_{bd} = 35.1 \text{ kip}$$

7.31 (*a*) Find the forces exerted by the pins on the members of the frame shown in Fig. 7.31*a*.

    (*b*) Express the results in part (*a*) in formal vector notation.

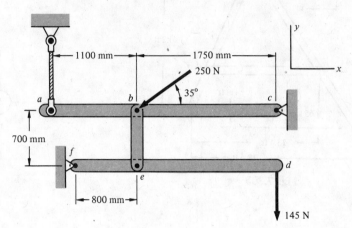

Fig. 7.31*a*

  (*a*) Member *be* is a two-force member, and the free-body diagram of member *df* is shown in Fig. 7.31*b*. For equilibrium of member *df*,

$$\Sigma M_f = 0 \qquad F_{be}(800) - 145(2{,}550) = 0 \qquad F_{be} = 462 \text{ N (T)}$$
$$\Sigma F_y = 0 \qquad -R_{fy} + F_{be} - 145 = 0 \qquad R_{fy} = 317 \text{ N}$$
$$\Sigma F_x = 0 \qquad R_{fx} = 0$$

R_b is the force exerted by pin *b* on member *ac*. The free-body diagram of pin *b* is seen in Fig. 7.31*c*. For equilibrium of pin *b*,

$$\Sigma F_x = 0 \qquad R_{bx} - 250 \cos 35° = 0 \qquad R_{bx} = 205 \text{ N}$$
$$\Sigma F_y = 0 \qquad R_{by} - 250 \sin 35° - F_{be} = 0 \qquad R_{by} = 605 \text{ N}$$
$$R_b = \sqrt{R_{bx}^2 + R_{by}^2} = \sqrt{205^2 + 605^2} = 639 \text{ N}$$

The free-body diagram of member *ac* is shown in Fig. 7.31*d*. For equilibrium of this member,

$$\Sigma M_c = 0 \qquad -R_a(2{,}850) + 605(1{,}750) = 0 \qquad R_a = 371 \text{ N}$$
$$\Sigma F_y = 0 \qquad R_a - 605 + R_{cy} = 0 \qquad R_{cy} = 234 \text{ N}$$
$$\Sigma F_x = 0 \qquad -205 + R_{cx} = 0 \qquad R_{cx} = 205 \text{ N}$$
$$R_c = \sqrt{R_{cx}^2 + R_{cy}^2} = \sqrt{205^2 + 234^2} = 311 \text{ N}$$

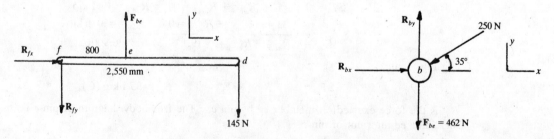

Fig. 7.31*b*   Fig. 7.31*c*

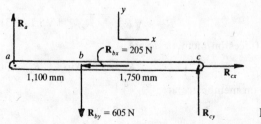

Fig. 7.31*d*

The pin forces on member *ac* are

$$R_a = 371 \text{ N} \qquad R_b = 639 \text{ N} \qquad R_c = 311 \text{ N}$$

The pin forces on member *df* are

$$R_e = F_{be} = 462 \text{ N} \qquad R_f = R_{fy} = 317 \text{ N}$$

The pin forces on member *be* are

$$R_b = F_{be} = 462 \text{ N} \qquad R_e = F_{be} = 462 \text{ N}$$

A free-body diagram of the entire frame was not used in this problem. It is left as an exercise for the reader to use such a construction to check that the overall equilibrium of the frame is satisfied.

(**b**) The forces of pins *a* and *c* on member *ac* are

$$\mathbf{R}_a = 371\mathbf{j} \quad \text{N} \qquad \mathbf{R}_c = 205\mathbf{i} + 234\mathbf{j} \quad \text{N}$$

The force exerted by pin *b* on member *ac* is

$$\mathbf{R}_b = -205\mathbf{i} - 605\mathbf{j} \quad \text{N}$$

The forces of pins *e* and *f* on member *df* are

$$\mathbf{R}_e = 462\mathbf{j} \quad \text{N} \qquad \mathbf{R}_f = -317\mathbf{j} \quad \text{N}$$

7.32 Find the forces exerted by the pins on the members of the frame in Fig. 7.32*a*.

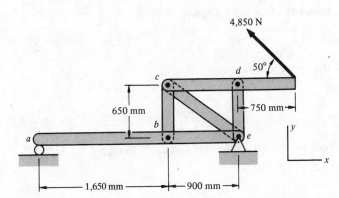

Fig. 7.32*a*

▌ R_e is the force exerted by the foundation on pin *e*, and the free-body diagram of the frame is shown in Fig. 7.32*b*. For equilibrium of the frame,

$$\Sigma M_e = 0 \qquad -R_a(2,550) + (4,850 \cos 50°)(650) + (4,850 \sin 50°)(750) = 0 \qquad R_a = 1,890 \text{ N}$$
$$\Sigma F_x = 0 \qquad -4,850 \cos 50° + R_{ex} = 0 \qquad R_{ex} = 3,120 \text{ N}$$
$$\Sigma F_y = 0 \qquad R_a - R_{ey} + 4,850 \sin 50° = 0 \qquad R_{ey} = 5,610 \text{ N}$$

Members *cb*, *ce*, and *de* are two-force members. R_{dc} is the force exerted by pin *c* on member *cd*. The free-body diagram of member *cd* is seen in Fig. 7.32*c*. For equilibrium of this member,

$$\Sigma M_c = 0 \qquad -F_{de}(900) + (4,850 \sin 50°)(1,650) = 0$$
$$F_{de} = 6,810 \text{ N (T)}$$
$$\Sigma F_y = 0 \qquad R_{dc,y} - F_{de} + 4,850 \sin 50° = 0 \qquad R_{dc,y} = 3,090 \text{ N}$$
$$\Sigma F_x = 0 \qquad R_{dc,x} - 4,850 \cos 50° = 0 \qquad R_{dc,x} = 3,120 \text{ N}$$
$$R_{dc} = \sqrt{R_{dc,x}^2 + R_{dc,y}^2} = \sqrt{3,120^2 + 3,090^2} = 4,390 \text{ N}$$

R_{ae} is the force exerted by pin *e* on member *ae*, and the free-body diagram of this member is shown in Fig. 7.32*d*. For equilibrium of member *ae*,

$$\Sigma M_e = 0 \qquad -1,890(2,550) + F_{bc}(900) = 0 \qquad F_{bc} = 5,360 \text{ N (C)}$$
$$\Sigma F_y = 0 \qquad 1,890 - F_{bc} + R_{ae,y} = 0 \qquad R_{ae,y} = 3,470 \text{ N}$$
$$\Sigma F_x = 0 \qquad R_{ae,x} = 0 \qquad R_{ae} = R_{ae,y} = 3,470 \text{ N}$$

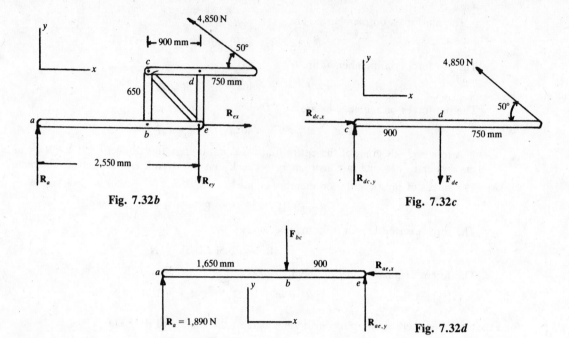

Fig. 7.32b   Fig. 7.32c

Fig. 7.32d

Figure 7.32e shows the free-body diagram of pin c. For equilibrium of this pin,

$$\Sigma F_x = 0 \qquad \frac{900}{1,110} F_{ce} - 3,120 = 0 \qquad F_{ce} = 3,850 \text{ N (T)}$$

As a check on the above value,

$$\Sigma F_y = 0$$

$$-3,090 - \frac{650}{1,110}(3,850) \overset{?}{=} 5,360 \qquad 5,340 \approx 5,360$$

A separate check on the calculations may be made using pin e. Figure 7.32f shows the free-body diagram of this pin. For equilibrium of pin e,

$$\Sigma F_x = 0$$

$$-\frac{900}{1,110}(3,850) + 3,120 \overset{?}{=} 0 \qquad 3,120 \equiv 3,120$$

$$\Sigma F_y = 0$$

$$\frac{650}{1,110}(3,850) + 6,810 \overset{?}{=} 3,470 + 5,610 \qquad 9,060 \approx 9,080$$

Pin e is the connection point of three frame members to the foundation. Readers are urged to study Fig. 7.32f carefully and establish clearly in their minds the details of the force transmission through this connection.

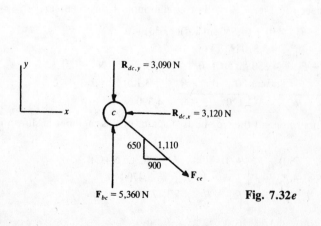

Fig. 7.32e

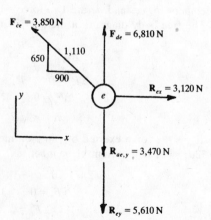

Fig. 7.32f

7.33 Find the maximum value of the load P which may be supported by the frame in Fig. 7.33a if the force exerted by a pin on any member may not exceed 1,450 lb.

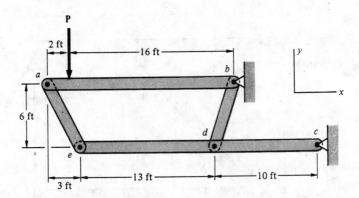

Fig. 7.33a

▎ R_{ab} is the force exerted by pin b on member ab. Members ae and bd are two-force members. Figure 7.33b shows the free-body diagram of member ab. For equilibrium of this member,

$$\Sigma M_b = 0 \qquad -\left(\frac{6}{6.71} F_{ae}\right)18 + 16P = 0 \qquad F_{ae} = 0.994P \text{ (C)}$$

$$\Sigma F_x = 0 \qquad -\frac{3}{6.71} F_{ae} + R_{ab,x} = 0 \qquad R_{ab,x} = 0.444P$$

$$\Sigma F_y = 0 \qquad \frac{6}{6.71} F_{ae} - P + R_{ab,y} = 0 \qquad R_{ab,y} = 0.111P$$

$$R_{ab} = \sqrt{R_{ab,x}^2 + R_{ab,y}^2} = \sqrt{(0.444P)^2 + (0.111P)^2} = 0.458P$$

The free-body diagram of member ce is shown in Fig. 7.33c. For equilibrium,

$$\Sigma M_c = 0 \qquad \left(\frac{6}{6.71} F_{ae}\right)23 - \left(\frac{6}{6.32} F_{bd}\right)10 = 0 \qquad F_{bd} = 2.15P$$

$$\Sigma F_x = 0 \qquad \frac{3}{6.71} F_{ae} + \frac{2}{6.32} F_{bd} - R_{cx} = 0 \qquad R_{cx} = 1.12P$$

$$\Sigma F_y = 0 \qquad -\frac{6}{6.71} F_{ae} + \frac{6}{6.32} F_{bd} - R_{cy} = 0 \qquad R_{cy} = 1.15P$$

$$R_c = \sqrt{R_{cx}^2 + R_{cy}^2} = \sqrt{(1.12P)^2 + (1.15P)^2} = 1.61P$$

The maximum pin force is

$$R_d = F_{bd} = 2.15P$$

Using the maximum allowable value of 1,450 lb for a pin force,

$$2.15P = 1,450 \text{ lb} \qquad P_{max} = 674 \text{ lb}$$

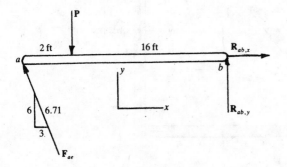

Fig. 7.33b

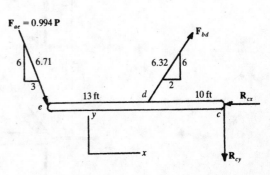

Fig. 7.33c

7.34 The plane frame shown in Fig. 7.34*a* is to support the 3,500-N tensile force shown. This load may be applied to either pin *c* or pin *d*.

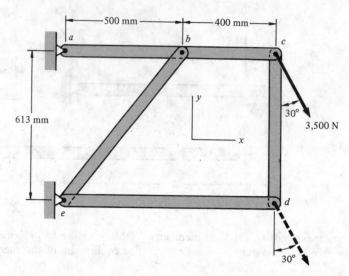

Fig. 7.34*a*

(*a*) Find the forces exerted by the pins at *a*, *b*, and *e* on the members if the 3,500-N tensile force is applied to pin *c*.

(*b*) Do the same as in part (*a*) if the 3,500-N force is applied to pin *d*.

▮ (*a*) The 3,500-N force is applied to pin *c*. The free-body diagram of the structure for this case is shown in Fig. 7.34*b*. For equilibrium of the entire frame,

$$\sum M_e = 0 \qquad R_{ax}(613) = 1{,}750(613) - 3{,}030(900) = 0 \qquad R_{ax} = 6{,}200 \text{ N}$$

$$\sum F_x = 0 \qquad -R_{ax} + R_{ex} + 1{,}750 = 0 \qquad -6{,}200 + R_{ex} + 1{,}750 = 0 \qquad R_{ex} = 4{,}450 \text{ N}$$

$$\sum F_y = 0 \qquad R_{ay} + R_{ey} - 3{,}030 = 0 \tag{1}$$

From consideration of Fig. 7.34*b*, it may be concluded that members *be*, *cd*, and *de* are two-force members. Members *cd* and *de* have the configuration of the zero-force members shown in Prob. 6.12. Thus, the forces in these two members are zero. The free-body diagram of member *ac* is shown in Fig. 7.34*c*. The equilibrium requirements for this member are

$$\sum M_a = 0 \qquad -R_{by}(500) - 3{,}030(900) = 0 \qquad R_{by} = -5{,}450 \text{ N}$$

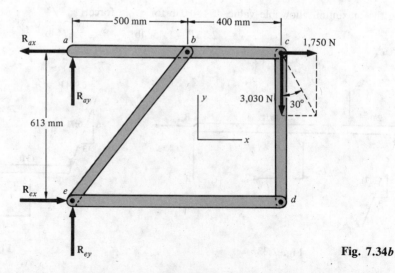

Fig. 7.34*b*

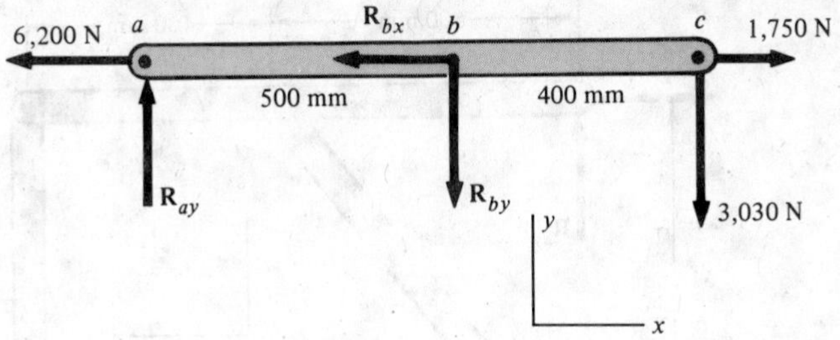

6,200 N a R_{bx} b   c   1,750 N

500 mm   400 mm

R_{ay} R_{by} 3,030 N

y

x

Fig. 7.34c

$$\sum F_x = 0 \qquad -6,200 - R_{bx} + 1,750 = 0 \qquad R_{bx} = -4,450\,\text{N}$$

$$\sum F_y = 0 \qquad R_{ay} - R_{by} - 3,030 = 0 \qquad R_{ay} - (-5,450) - 3,030 = 0 \qquad R_{ay} = -2,420\,\text{N}$$

Next the force component R_{ey} is found from Eq. (1) as

$$R_{ay} + R_{ey} - 3,030 = 0 \qquad -2,420 + R_{ey} - 3,030 = 0 \qquad R_{ey} = 5,450\,\text{N}$$

As a check on the above calculations, the equilibrium of member be will be confirmed. Figure 7.34d shows the free-body diagram of this member. It may be seen that the equilibrium requirements in the x and y directions are satisfied. For moment equilibrium about point e,

$$R_{by}(500) - R_{bx}(613) \overset{?}{=} 0 \qquad (-5,450)(500) - (-4,450)(613) \overset{?}{=} 0$$
$$5,450(500) \overset{?}{=} 4,450(613) \qquad 2.73 \times 10^6 \equiv 2.73 \times 10^6$$

It was observed earlier that member be is a two-force member. Thus, the resultant force at b and e in Fig. 7.34d must have the direction of member be. As a check,

$$\frac{4,450}{5,450} \overset{?}{=} \frac{500}{613} \qquad 0.817 \approx 0.816$$

As the final step in the solution of the problem, the total forces exerted by the pins on the members at a and b are found to be

$$R_a = \sqrt{R_{ax}^2 + R_{ay}^2} = \sqrt{6,200^2 + (-2,420)^2} = 6,660\,\text{N}$$
$$R_b = \sqrt{R_{bx}^2 + R_{by}^2} = \sqrt{(-4,450)^2 + (-5,450)^2} = 7,040\,\text{N}$$

Since member be is a two force member,

$$R_e = R_b = 7,040\,\text{N}$$

It may be seen that, for the load application of Fig. 7.34b, pins b and e are the most heavily loaded pins in the structure.

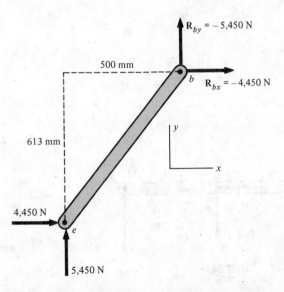

$R_{by} = -5,450\,\text{N}$

500 mm

b $R_{bx} = -4,450\,\text{N}$

y

x

613 mm

4,450 N

e

5,450 N

Fig. 7.34d

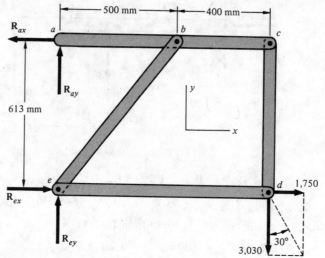

Fig. 7.34e

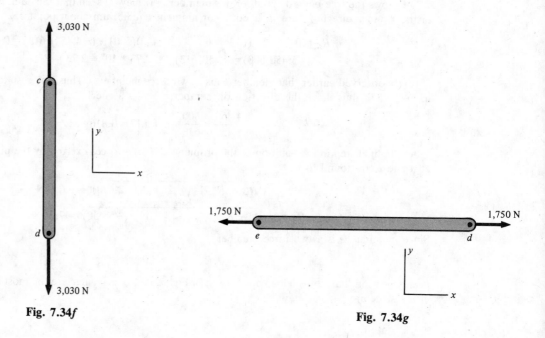

Fig. 7.34f

Fig. 7.34g

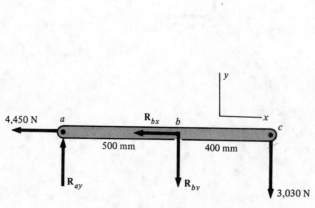

Fig. 7.34h

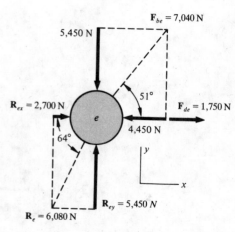

Fig. 7.34i

(*b*) For application of the 3,500-N load at pin *d*, the free-body diagram is as shown in Fig. 7.34*e*. For equilibrium of the entire structure,

$$\sum M_e = 0 \quad R_{ax}(613) - 3,030(900) = 0 \quad R_{ax} = 4,450 \text{ N}$$

$$\sum F_x = 0 \quad -R_{ax} + R_{ex} + 1,750 = 0 \quad -4,450 + R_{ex} + 1,750 = 0. \quad R_{ex} = 2,700 \text{ N}$$

$$\sum F_y = 0 \quad R_{ay} + R_{ey} - 3,030 = 0 \tag{2}$$

Members *cd* and *de* are two-force members, and the free-body diagrams of those members are shown in Figs. 7.34*f* and *g*. The free-body diagram of member *ac* is shown in Fig. 7.34*h*. The equilibrium requirements are given in the following equations:

$$\sum M_a = 0 \quad -R_{by}(500) - 3,030(900) = 0 \quad R_{by} = -5,450 \text{ N}$$

$$\sum F_y = 0 \quad R_{ay} - R_{by} - 3,030 = 0 \quad R_{ay} - (-5,450) - 3,030 = 0 \quad R_{ay} = -2,420 \text{ N}$$

$$\sum F_x = 0 \quad -4,450 - R_{bx} = 0 \quad R_{bx} = -4,450 \text{ N}$$

The force component R_{ey} is found from Eq. (2) as

$$R_{ay} + R_{ey} - 3,030 = 0 \quad -2,420 + R_{ey} - 3,030 = 0 \quad R_{ey} = 5,450 \text{ N}$$

The free-body diagram of the pin at *e* is shown in Fig. 7.34*i*. The force components R_{ex} and R_{ey} in this figure are the components of force of the foundation on the structure, as shown in the free-body diagram of Fig. 7.34*e*. The remaining forces in Fig. 7.34*i* are internal forces in the frame. The total forces acting on pins *a* and *b* are

$$R_a = \sqrt{R_{ax}^2 + R_{ay}^2} = \sqrt{4,450^2 + (-2,420)^2} = 5,070 \text{ N}$$
$$R_b = \sqrt{R_{bx}^2 + R_{by}^2} = \sqrt{(-4,450)^2 + (-5,450)^2} = 7,040 \text{ N}$$
and
$$R_e = \sqrt{R_{ex}^2 + R_{ey}^2} = \sqrt{2,700^2 + 5,450^2} = 6,080 \text{ N}$$

It may be seen that application of the load at point *d* rather than at point *c* reduces the force acting on pin *a* while leaving the force on pin *b* unchanged.

7.35 Find the forces exerted by the pins on the members of the frame in Fig. 7.35*a*.

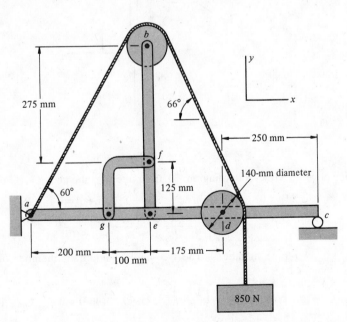

Fig. 7.35a

▌ R_a is the force exerted by the foundation on pin *a*, and the free-body diagram of the frame is shown in Fig. 7.35*b*. For equilibrium of the frame,

$$\sum M_a = 0 \quad -850(545) + R_c(725) = 0 \quad R_c = 639 \text{ N}$$
$$\sum F_x = 0 \quad R_{ax} = 0$$

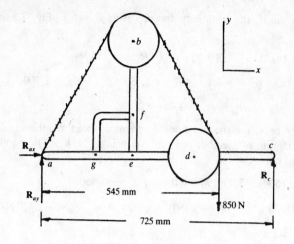

Fig. 7.35b

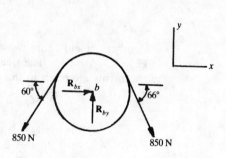

Fig. 7.35c

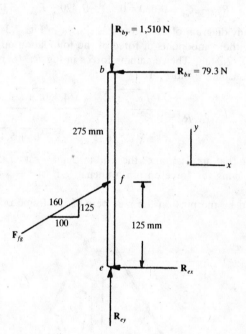

Fig. 7.35d

$$\Sigma F_y = 0 \qquad R_{ay} - 850 + R_c = 0 \qquad R_{ay} = 211 \text{ N}$$

Figure 7.35c shows the free-body diagram of the pulley on member *be*. For equilibrium of this element,

$$\Sigma F_y = 0 \qquad -850 \sin 60° + R_{by} - 850 \sin 66° = 0 \qquad R_{by} = 1{,}510 \text{ N}$$
$$\Sigma F_x = 0 \qquad R_{bx} - 850 \cos 60° + 850 \cos 66° = 0 \qquad R_{bx} = 79.3 \text{ N}$$
$$R_b = \sqrt{R_{bx}^2 + R_{by}^2} = \sqrt{79.3^2 + 1{,}510^2} = 1{,}510 \text{ N}$$

Member *fg* is a two-force member. The free-body diagram of member *be* is seen in Fig. 7.35d. For equilibrium of member *be*,

$$\Sigma M_e = 0 \qquad -\left(\frac{100}{160} F_{fg}\right)125 + 79.3(400) = 0 \qquad F_{fg} = 406 \text{ N}$$

$$\Sigma F_x = 0 \qquad \frac{100}{160} F_{fg} - 79.3 - R_{ex} = 0 \qquad R_{ex} = 174 \text{ N}$$

$$\Sigma F_y = 0 \qquad -1{,}510 + \frac{125}{160} F_{fg} + R_{ey} = 0 \qquad R_{ey} = 1{,}190 \text{ N}$$

$$R_e = \sqrt{R_{ex}^2 + R_{ey}^2} = \sqrt{174^2 + 1{,}190^2} = 1{,}200 \text{ N}$$

Figure 7.35e shows a free-body diagram of the pulley at d. For equilibrium of this pulley,

$$\Sigma F_x = 0 \quad -850 \cos 66° + R_{dx} = 0 \quad R_{dx} = 346 \text{ N}$$
$$\Sigma F_y = 0 \quad 850 \sin 66° + R_{dy} - 850 = 0 \quad R_{dy} = 73.5 \text{ N}$$
$$R_d = \sqrt{R_{dx}^2 + R_{dy}^2} = \sqrt{346^2 + 73.5^2} = 354 \text{ N}$$

R_{ca} is the force exerted by pin a on member ac, and the free-body diagram of this pin is shown in Fig. 7.35f. For equilibrium of this pin,

$$\Sigma F_x = 0 \quad 850 \cos 60° - R_{ca,x} = 0 \quad R_{ca,x} = 425 \text{ N}$$
$$\Sigma F_y = 0 \quad -R_{ca,y} + 850 \sin 60° + 211 = 0 \quad R_{ca,y} = 947 \text{ N}$$
$$R_{ca} = \sqrt{R_{ca,x}^2 + R_{ca,y}^2} = \sqrt{425^2 + 947^2} = 1{,}040 \text{ N}$$

The forces exerted by the pins on the members are

$$R_b = 1{,}510 \text{ N} \quad R_d = 354 \text{ N} \quad R_e = 1{,}200 \text{ N}$$
$$R_f = F_{fg} = 406 \text{ N} \quad R_g = F_{fg} = 406 \text{ N}$$

The force exerted by pin a on member ac is

$$R_{ca} = 1{,}040 \text{ N}$$

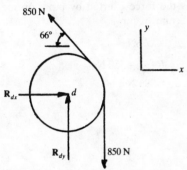

Fig. 7.35e

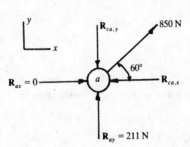

Fig. 7.35f

7.36 The frame in Prob. 7.35 must be temporarily supported at pin e instead of at location c. Will this method of support result in larger pin forces on the members than the original method of support did?

▮ The free-body diagram of the frame is shown in Fig. 7.36a, and R_e is the force exerted by the foundation on pin e. For equilibrium of the frame,

$$\Sigma M_a = 0 \quad R_e(300) - 850(545) = 0 \quad R_e = 1{,}540 \text{ N}$$
$$\Sigma F_y = 0 \quad -R_{ay} + R_e - 850 = 0 \quad R_{ay} = 690 \text{ N}$$
$$\Sigma F_x = 0 \quad R_{ax} = 0$$

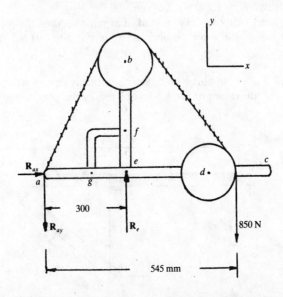

Fig. 7.36a

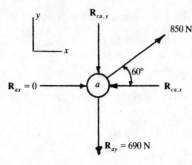

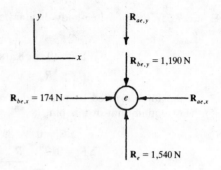

Fig. 7.36b

Fig. 7.36c

R_{ca} is the force exerted by pin a on member ac. The free-body diagram of pin a is seen in Fig. 7.36b. For equilibrium of pin a,

$$\Sigma F_x = 0 \qquad 850 \cos 60° - R_{ca,x} = 0 \qquad R_{ca,x} = 425\ N$$
$$\Sigma F_y = 0 \qquad -R_{ca,y} + 850 \sin 60° - 690 = 0 \qquad R_{ca,y} = 46.1\ N$$
$$R_{ca} = \sqrt{R_{ca,x}^2 + R_{ca,y}^2} = \sqrt{425^2 + 46.1^2} = 427\ N$$

The pin forces at b, d, f, and g are the same as in Prob. 7.35. R_{be} is the force exerted by pin e on member be. This is the same force as $R_e = 1,200\ N$ in Prob. 7.35. R_{ae} is the force exerted by pin e on member ac. Figure 7.36c shows the free-body diagram of pin e. For equilibrium of pin e,

$$\Sigma F_x = 0 \qquad 174 - R_{ae,x} = 0 \qquad R_{ae,x} = 174\ N$$
$$\Sigma F_y = 0 \qquad -R_{ae,y} - 1,190 + 1,540 = 0 \qquad R_{ae,y} = 350\ N$$
$$R_{ae} = \sqrt{R_{ae,x}^2 + R_{ae,y}^2} = \sqrt{174^2 + 350^2} = 391\ N$$

The forces exerted by the pins on the members are as follows.
The force of pin a on member ac is

$$R_{ca} = 427\ N$$

(Compare this result to $R_{ca} = 1,040\ N$ in Prob. 7.35.)
The pin forces at b, d, f, and g are the same as in Prob. 7.35:

$$R_b = 1,510\ N \qquad R_d = 354\ N \qquad R_f = F_{fg} = 406\ N \qquad R_g = F_{fg} = 406\ N$$

The force of pin e on member be is

$$R_{be} = 1,200\ N$$

(Compare this result to $R_e = 1,200\ N$ in Prob. 7.35.)
The force of pin e on member ac is

$$R_{ae} = 391\ N$$

(Compare this result to $R_e = 1,200\ N$ in Prob. 7.35.)
The method of support being at pin e instead of at pin c reduces the force of pin a on member ac from 1,040 to 427 N, and the force of pin e on member ac from 1,200 to 391 N. The force of pin e on member be is unchanged.

7.37 The frame in Fig. 7.37a must support a load, acting vertically downward, of 7,000 N. This load may be applied at pin b, c, or e. Find the values of the forces exerted by the pins on the members if the load is applied at pin b.

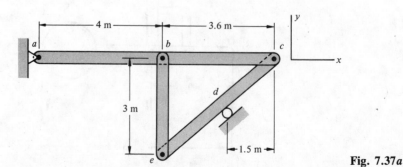

Fig. 7.37a

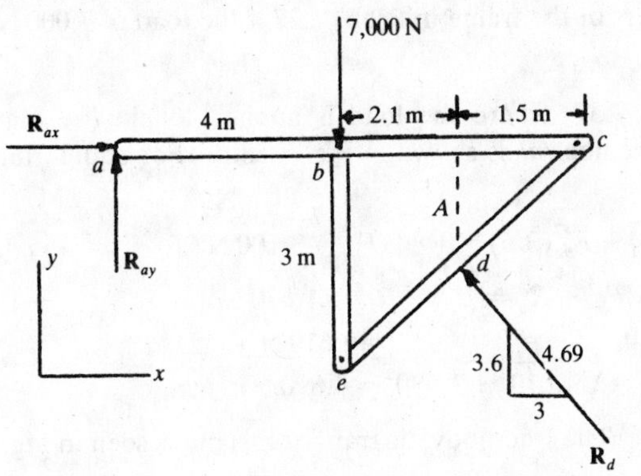

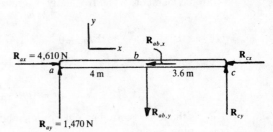

Fig. 7.37b

Fig. 7.37c

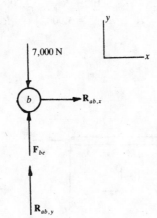

Fig. 7.37d

▌ Figure 7.37b shows the free-body diagram of the frame. From the similar triangles in the figure,

$$\frac{A}{1.5} = \frac{3}{3.6} \qquad A = 1.25 \text{ m}$$

For equilibrium of the frame,

$$\Sigma M_a = 0 \qquad -7,000(4) + \left(\frac{3.6}{4.69} R_d\right)6.1 - \left(\frac{3}{4.69} R_d\right)1.25 = 0 \qquad R_d = 7,210 \text{ N}$$

$$\Sigma F_x = 0 \qquad R_{ax} - \frac{3}{4.69} R_d = 0 \qquad R_{ax} = 4,610 \text{ N}$$

$$\Sigma F_y = 0 \qquad R_{ay} - 7,000 + \frac{3.6}{4.69} R_d = 0 \qquad R_{ay} = 1,470 \text{ N}$$

$$R_a = \sqrt{R_{ax}^2 + R_{ay}^2} = \sqrt{4,610^2 + 1,470^2} = 4,840 \text{ N}$$

R_{ab} is the force exerted by pin b on member ac. Member be is a two-force member. Figures 7.37c and d show the free-body diagrams of member ac and pin b. For equilibrium, using Fig. 7.37c,

$$\Sigma M_c = 0 \qquad -1,470(7.6) + R_{ab,y}(3.6) = 0 \qquad R_{ab,y} = 3,100 \text{ N}$$

Using Fig. 7.37d,

$$\Sigma F_x = 0 \qquad R_{ab,x} = 0 \qquad R_{ab} = R_{ab,y} = 3,100 \text{ N}$$
$$\Sigma F_y = 0 \qquad -7,000 + F_{be} + R_{ab,y} = 0 \qquad F_{be} = 3,900 \text{ N (C)}$$

$$R_e = F_{be} = 3,900 \text{ N}$$

Using Fig. 7.37c,

$$\Sigma F_x = 0 \qquad 4,610 - R_{cx} = 0 \qquad R_{cx} = 4.610 \text{ N}$$
$$\Sigma F_y = 0 \qquad 1,470 - R_{ab,y} + R_{cy} = 0 \qquad R_{cy} = 1,630 \text{ N}$$
$$R_c = \sqrt{R_{cx}^2 + R_{cy}^2} = \sqrt{4,610^2 + 1,630^2} = 4,890 \text{ N}$$

7.38 Find the forces exerted by the pins on the members of the frame in Prob. 7.37 if the load of 7,000 N is applied to pin e.

▮ R_{ax}, R_{ay}, and R_d are the same as in Prob. 7.37, where the load is applied to pin b. Member be is a two-force member. The free-body diagram of member ac is shown in Fig. 7.38a. For equilibrium of member ac,

$$\Sigma M_c = 0 \qquad -1,470(7.6) + F_{be}(3.6) = 0 \qquad F_{be} = 3,100 \text{ N (T)}$$

$$\Sigma F_y = 0 \qquad 1,470 - F_{be} + R_{cy} = 0 \qquad R_{cy} = 1,630 \text{ N}$$

$$\Sigma F_x = 0 \qquad 4,610 - R_{cx} = 0 \qquad R_{cx} = 4,610 \text{ N}$$

$$R_c = \sqrt{R_{cx}^2 + R_{cy}^2} = \sqrt{4,610^2 + 1,630^2} = 4,890 \text{ N}$$

R_{ce} is the force exerted by pin e on member ce. The free-body diagram of this pin is seen in Fig. 7.38b. For equilibrium of pin e,

$$\Sigma F_x = 0 \qquad R_{ce,x} = 0$$

$$\Sigma F_y = 0 \qquad 3,100 + R_{ce,y} - 7,000 = 0 \qquad R_{ce,y} = 3,900 \text{ N}$$

$$R_{ce} = R_{ce,y} = 3,900 \text{ N} \qquad R_b = F_{be} = 3,100 \text{ N}$$

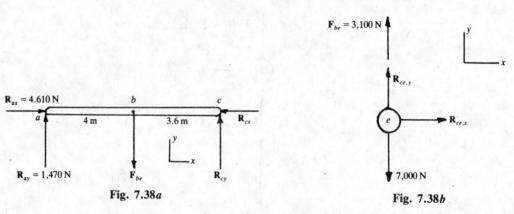

Fig. 7.38a   Fig. 7.38b

7.39 (a) Find the forces exerted by the pins on the members of the frame in Prob. 7.37 if the load of 7,000 N is applied to pin c.

(b) Compare the forces exerted by the pins on the frame members for the three different methods of loading in Probs. 7.37, 7.38 and 7.39.

▮ (a) The free-body diagram of the frame is shown in Fig. 7.39a. For equilibrium of the frame,

$$\Sigma M_a = 0 \qquad -7,000(7.6) - \left(\frac{3}{4.69} R_d\right)1.25 + \left(\frac{3.6}{4.69} R_d\right)6.1 = 0 \qquad R_d = 13,700 \text{ N}$$

$$\Sigma F_x = 0 \qquad R_{ax} - \frac{3}{4.69} R_d = 0 \qquad R_{ax} = 8,760 \text{ N}$$

$$\Sigma F_y = 0 \qquad -R_{ay} + \frac{3.6}{4.69} R_d - 7,000 = 0 \qquad R_{ay} = 3,520 \text{ N}$$

$$R_a = \sqrt{R_{ax}^2 + R_{ay}^2} = \sqrt{8,760^2 + 3,520^2} = 9,440 \text{ N}$$

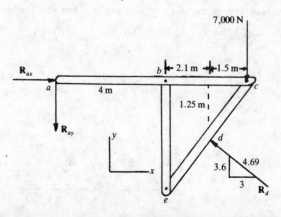

Fig. 7.39a

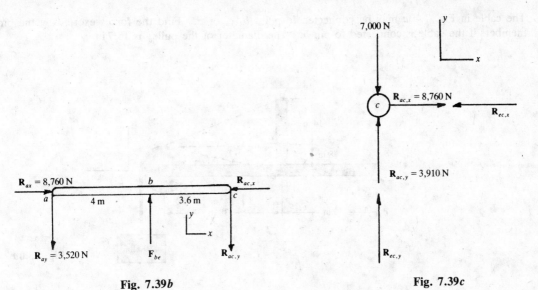

Fig. 7.39b **Fig. 7.39c**

R_{ac} is the force exerted by pin c on member ac. Member be is a two-force member. Figure 7.39b shows the free-body diagram of member ac. For equilibrium of member ac,

$$\Sigma M_c = 0 \qquad 3,520(7.6) - F_{be}(3.6) = 0 \qquad F_{be} = 7,430 \text{ N (C)}$$
$$\Sigma F_y = 0 \qquad -3,520 + F_{be} - R_{ac,y} = 0 \qquad R_{ac,y} = 3,910 \text{ N}$$

$$\Sigma F_x = 0 \qquad 8,760 - R_{ac,x} = 0 \qquad R_{ac,x} = 8,760 \text{ N}$$
$$R_{ac} = \sqrt{R_{ac,x}^2 + R_{ac,y}^2} = \sqrt{8,760^2 + 3,910^2} = 9,590 \text{ N}$$

R_{ec} is the force exerted by pin c on member ce, and the free-body diagram of this pin is shown in Fig. 7.39c. For equilibrium of pin c,

$$\Sigma F_x = 0 \qquad 8,760 - R_{ec,x} = 0 \qquad R_{ec,x} = 8,760 \text{ N}$$
$$\Sigma F_y = 0 \qquad -7,000 + 3,910 + R_{ec,y} = 0 \qquad R_{ec,y} = 3,090 \text{ N}$$
$$R_{ec} = \sqrt{R_{ec,x}^2 + R_{ec,y}^2} = \sqrt{8,760^2 + 3,090^2} = 9,290 \text{ N}$$

(**b**) A summary of the forces exerted by the pins on the members for the three methods of loading in Probs. 7.37, 7.38 and 7.39 is given in Table 7.3. It may be seen that the maximum force is the same for application of the 7,000-N force to pin b or pin e. The maximum force is significantly greater when this force is applied to pin c.

TABLE 7.3

Prob. 7.37	Prob. 7.38	Prob. 7.39
	applied force, N	
at pin b	at pin e	at pin c
	member ac	
$R_a = 4,840$	$R_a = 4,840$	$R_a = 9,440$
$R_b = 3,100$	$R_b = 3,100$	$R_b = 7,430$
$R_c = 4,890^\dagger$	$R_c = 4,890^\dagger$	$R_c = 9,590^\dagger$
	member ce	
$R_c = 4,890^\dagger$	$R_c = 4,890^\dagger$	$R_c = 9,290$
$R_e = 3,900$	$R_e = 3,900$	$R_e = 7,430$
	member be	
$R_b = 3,900$	$R_b = 3,100$	$R_b = 7,430$
$R_e = 3,900$	$R_e = 3,100$	$R_e = 7,430$

†Maximum value.

7.40 The cable in Fig. 7.40a may be connected to point b, c, or d. Find the forces exerted by the pins on the members if the cable is connected to pin b. The diameter of the pulley is 15.7 in.

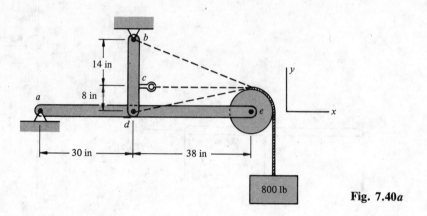

Fig. 7.40a

∎ From the geometry of Fig. 7.40b,

$$\tan \theta_1 = \frac{38}{22} \qquad \sin \theta_2 = \frac{8}{43.9}$$
$$\theta_1 = 59.9° \qquad \theta_2 = 10.3° \qquad \theta_3 = 90° - \theta_1 - \theta_2 = 19.8°$$

Figure 7.40c shows the free-body diagram of the pulley. For equilibrium of this element,

$$\Sigma F_x = 0 \qquad R_{ex} - 800 \cos 19.8° = 0 \qquad R_{ex} = 753 \text{ lb}$$
$$\Sigma F_y = 0 \qquad R_{ey} + 800 \sin 19.8° - 800 = 0 \qquad R_{ey} = 529 \text{ lb}$$
$$R_e = \sqrt{R_{ex}^2 + R_{ey}^2} = \sqrt{753^2 + 529^2} = 920 \text{ lb}$$

Member bd is a two-force member, and Fig. 7.40d shows the free-body diagram of member ae. For equilibrium of this member,

$$\Sigma M_a = 0 \qquad F_{bd}(30) - 529(68) = 0 \qquad F_{bd} = 1,200 \text{ lb (T)}$$
$$\Sigma F_y = 0 \qquad -R_{ay} + F_{bd} - 529 = 0 \qquad R_{ay} = 671 \text{ lb}$$
$$\Sigma F_x = 0 \qquad R_{ax} - 753 = 0 \qquad R_{ax} = 753 \text{ lb}$$
$$R_a = \sqrt{R_{ax}^2 + R_{ay}^2} = \sqrt{753^2 + 671^2} = 1,010 \text{ lb} \qquad R_b = R_d = F_{bd} = 1,200 \text{ lb}$$

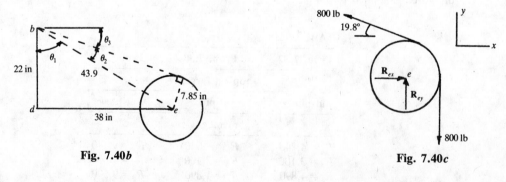

Fig. 7.40b

Fig. 7.40c

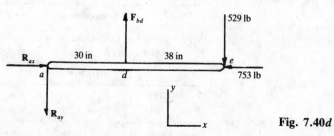

Fig. 7.40d

7.41 Find the pin forces acting on the members of the frame in Prob. 7.40 if the cable is connected to point c.

❚ For this case, the frame contains no two-force members. Figure 7.41a shows the free-body diagram of the pulley. For equilibrium of this element,

$$\Sigma F_x = 0 \qquad R_{ex} - 800 = 0 \qquad R_{ex} = 800 \text{ lb}$$
$$\Sigma F_y = 0 \qquad R_{ey} - 800 = 0 \qquad R_{ey} = 800 \text{ lb}$$
$$R_e = \sqrt{R_{ex}^2 + R_{ey}^2} = \sqrt{800^2 + 800^2} = 1,130 \text{ lb}$$

The free-body diagrams of members bd and ae are shown in Figs. 7.41b and c. For equilibrium of member bd, from Fig. 7.41b,

$$\Sigma M_b = 0 \qquad 800(14) - R_{dx}(22) = 0 \qquad R_{dx} = 509 \text{ lb}$$

For equilibrium of member ae, using Fig. 7.41c,

$$\Sigma M_a = 0 \qquad R_{dy}(30) - 800(68) = 0 \qquad R_{dy} = 1,810 \text{ lb}$$
$$R_d = \sqrt{R_{dx}^2 + R_{dy}^2} = \sqrt{509^2 + 1,810^2} = 1,880 \text{ lb}$$
$$\Sigma F_y = 0 \qquad -R_{ay} + R_{dy} - 800 = 0 \qquad R_{ay} = 1,010 \text{ lb}$$
$$\Sigma F_x = 0 \qquad R_{ax} + R_{dx} - 800 = 0 \qquad R_{ax} = 291 \text{ lb}$$
$$R_a = \sqrt{R_{ax}^2 + R_{ay}^2} = \sqrt{291^2 + 1,010^2} = 1,050 \text{ lb}$$

Using Fig. 7.41b,

$$\Sigma M_d = 0 \qquad R_{bx}(22) - 800(8) = 0 \qquad R_{bx} = 291 \text{ lb}$$
$$\Sigma F_y = 0 \qquad R_{by} - R_{dy} = 0 \qquad R_{by} = 1,810 \text{ lb}$$
$$R_b = \sqrt{R_{bx}^2 + R_{by}^2} = \sqrt{291^2 + 1,810^2} = 1,830 \text{ lb}$$

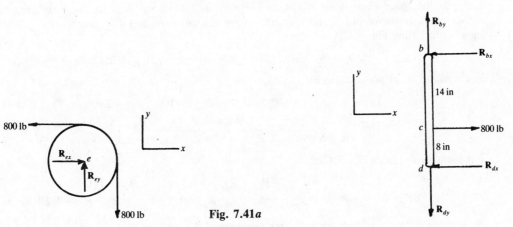

Fig. 7.41a

Fig. 7.41b

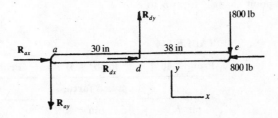

Fig. 7.41c

7.42 (a) Find the pin forces acting on the members of the frame in Prob. 7.40 if the cable is connected to pin d.
(b) Compare the pin forces for the three methods of connection of the cable in Probs. 7.40, 7.41, and 7.42.

❚ (a) Figure 7.42a shows the free-body diagram of the pulley. From the geometry of this figure,

$$\sin \theta_3 = \frac{7.85}{38} \qquad \theta_3 = 11.9°$$

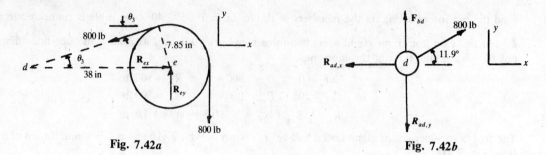

Fig. 7.42a

Fig. 7.42b

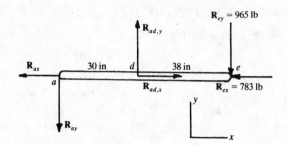

Fig. 7.42c

For equilibrium of the pulley,

$$\Sigma F_x = 0 \qquad -800 \cos 11.9° + R_{ex} = 0 \qquad R_{ex} = 783 \text{ lb}$$
$$\Sigma F_y = 0 \qquad -800 \sin 11.9° + R_{ey} - 800 = 0 \qquad R_{ey} = 965 \text{ lb}$$
$$R_e = \sqrt{R_{ex}^2 + R_{ey}^2} = \sqrt{783^2 + 965^2} = 1{,}240 \text{ lb}$$

R_{ad} is the force exerted by pin d on member ae. Member bd is a two-force member. The free-body diagrams of pin d and member ae are shown in Figs. 7.42b and c. For equilibrium of member ae, using Fig. 7.42c,

$$\Sigma M_a = 0 \qquad R_{ad,y}(30) - 965(68) = 0 \qquad R_{ad,y} = 2{,}190 \text{ lb}$$

For equilibrium of pin d, from Fig. 7.42b,

$$\Sigma F_y = 0 \qquad F_{bd} + 800 \sin 11.9° - R_{ad,y} = 0 \qquad F_{bd} = 2{,}030 \text{ lb (T)}$$
$$\Sigma F_x = 0 \qquad -R_{ad,x} + 800 \cos 11.9° = 0 \qquad R_{ad,x} = 783 \text{ lb}$$
$$R_{ad} = \sqrt{R_{ad,x}^2 + R_{ad,y}^2} = \sqrt{783^2 + 2{,}190^2} = 2{,}330 \text{ lb}$$

For member ae, from Fig. 7.42c,

$$\Sigma F_x = 0 \qquad -R_{ax} + R_{ad,x} - 783 = 0 \qquad R_{ax} = 0$$
$$\Sigma F_y = 0 \qquad -R_{ay} + R_{ad,y} - 965 = 0 \qquad R_{ay} = 1{,}230 \text{ lb} \qquad R_a = R_{ay} = 1{,}230 \text{ lb}$$

(*b*) A summary of the pin forces for the three positions of the cable in Probs. 7.40, 7.41, and 7.42 is given in Table 7.4. It may be seen that the forces of the pins on the members increase as the cable connection point changes from b to c to d.

TABLE 7.4

Prob. 7.40	Prob. 7.41	Prob. 7.42
	applied force, lb	
pin *b*	pin *c*	pin *d*
	member *ae*	
$R_a = 1{,}010$	$R_a = 1{,}050$	$R_a = 1{,}230$
$R_d = 1{,}200$	$R_d = 1{,}880$	$R_d = 2{,}330$
$R_e = 920$	$R_e = 1{,}130$	$R_e = 1{,}240$
	member *bd*	
$R_b = 1{,}200$	$R_b = 1{,}830$	$R_b = 2{,}030$
$R_d = 1{,}200$	$R_d = 1{,}880$	$R_d = 2{,}030$

7.43 Find the forces exerted by the pins on the members of the frame shown in Fig. 7.43a. Each of the pulleys has a mass of 75 kg.

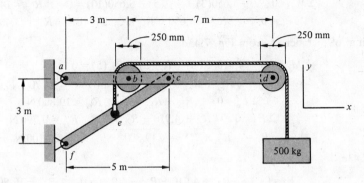

Fig. 7.43a

▮ The weight W of each pulley is

$$W = mg = 75(9.81) = 736 \text{ N}$$

The weight W_1 of the 500-kg mass is

$$W_1 = mg = 500(9.81) = 4{,}910 \text{ N}$$

The free-body diagrams of the two pulleys are shown in Fig. 7.43b. For equilibrium at pin b,

$$\Sigma F_x = 0 \qquad -R_{bx} + 4{,}910 = 0 \qquad R_{bx} = 4{,}910 \text{ N}$$
$$\Sigma F_y = 0 \qquad R_{by} - 736 - 4{,}910 = 0 \qquad R_{by} = 5{,}650 \text{ N}$$
$$R_b = \sqrt{R_{bx}^2 + R_{by}^2} = \sqrt{4{,}910^2 + 5{,}650^2} = 7{,}490 \text{ N}$$

For equilibrium at pin d,

$$\Sigma F_x = 0 \qquad R_{dx} - 4{,}910 = 0 \qquad R_{dx} = 4{,}910 \text{ N}$$
$$\Sigma F_y = 0 \qquad R_{dy} - 736 - 4{,}910 = 0 \qquad R_{dy} = 5{,}650 \text{ N}$$
$$R_d = \sqrt{R_{dx}^2 + R_{dy}^2} = \sqrt{4{,}910^2 + 5{,}650^2} = 7{,}490 \text{ N}$$

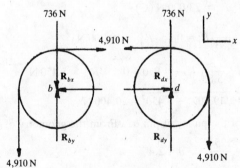

Fig. 7.43b

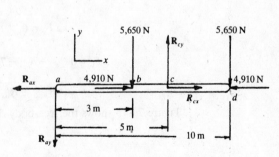

Fig. 7.43c

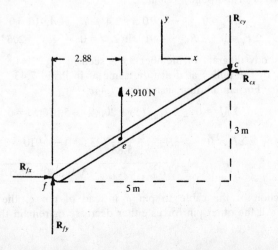

Fig. 7.43d

The free-body diagrams of members ad and cf are shown in Figs. 7.43c and d. For equilibrium, using Fig. 7.43c,

$$\Sigma M_a = 0 \qquad -5,650(3) + R_{cy}(5) - 5,650(10) = 0 \qquad R_{cy} = 14,700 \text{ N}$$
$$\Sigma F_y = 0 \qquad -R_{ay} - 5,650 + R_{cy} - 5,650 = 0 \qquad R_{ay} = 3,400 \text{ N}$$

For equilibrium of member cf, from Fig. 7.43d,

$$\Sigma M_f = 0 \qquad 4,910(2.88) + R_{cx}(3) - R_{cy}(5) = 0 \qquad R_{cx} = 19,800 \text{ N}$$
$$R_c = \sqrt{R_{cx}^2 + R_{cy}^2} = \sqrt{19,800^2 + 14,700^2} = 24,700 \text{ N}$$
$$\Sigma F_x = 0 \qquad R_{fx} - R_{cx} = 0 \qquad R_{fx} = 19,800 \text{ N}$$
$$\Sigma F_y = 0 \qquad R_{fy} + 4,910 - R_{cy} = 0 \qquad R_{fy} = 9,790 \text{ N}$$
$$R_f = \sqrt{R_{fx}^2 + R_{fy}^2} = \sqrt{19,800^2 + 9,790^2} = 22,100 \text{ N}$$

Using Fig. 7.43c,

$$\Sigma F_x = 0 \qquad -R_{ax} + 4,910 + R_{cx} - 4,910 = 0 \qquad R_{ax} = 19,800 \text{ N}$$
$$R_a = \sqrt{R_{ax}^2 + R_{ay}^2} = \sqrt{19,800^2 + 3,400^2} = 20,100 \text{ N}$$

The forces exerted by the pins on the members are

$$R_a = 20,100 \text{ N} \qquad R_b = 7,490 \text{ N} \qquad R_c = 24,700 \text{ N}$$
$$R_d = 7,490 \text{ N} \qquad R_e = 4,910 \text{ N} \qquad R_f = 22,100 \text{ N}$$

7.44 (*a*) Find the forces acting on the pins of the frame in Prob. 7.43 if the end of the cable is attached to pin a rather than pin b.

(*b*) Compare the values of the pin forces for the two different locations of cable attachment in Probs. 7.43 and 7.44.

▋ (*a*) The free-body diagram of the pulley and frame system is shown in Fig. 7.44a. Member cf is a two-force member. R_a is the force exerted by the foundation on pin a. For equilibrium of the frame and pulley system,

$$\Sigma M_a = 0 \qquad -736(3) - 736(10) - 4,910(10.1) + \left(\frac{5}{5.83} R_f\right)(3) = 0 \qquad R_f = 23,000 \text{ N}$$

$$R_c = R_f = F_{cf} = 23,000 \text{ N (C)}$$

$$\Sigma F_x = 0 \qquad -R_{ax} + \frac{5}{5.83} R_f = 0 \qquad R_{ax} = 19,700 \text{ N}$$

$$\Sigma F_y = 0 \qquad -R_{ay} - 736 - 736 + \frac{3}{5.83} R_f - 4,910 = 0 \qquad R_{ay} = 5,450 \text{ N}$$

$$R_a = \sqrt{R_{ax}^2 + R_{ay}^2} = \sqrt{19,700^2 + 5,450^2} = 20,400 \text{ N}$$

Figure 7.44b shows the free-body diagram of the pulley at pin b. From the geometry of this figure,

$$\sin \theta = \frac{125}{3,000} \qquad \theta = 2.4°$$

For equilibrium of the pulley at pin b,

$$\Sigma F_x = 0 \qquad -4,910 \cos 2.4° - R_{bx} + 4,910 = 0 \qquad R_{bx} = 4.31 \text{ N} \approx 0$$
$$\Sigma F_y = 0 \qquad R_{by} - 4,910 \sin 2.4° = 0 \qquad R_{by} = 206 \text{ N} \qquad R_b = R_{by} = 206 \text{ N}$$

The free-body diagram of member ad is seen in Fig. 7.44c. The forces R_{dx} and R_{dy} exerted on member ad by the pulley at d are the same as in Prob. 7.43. R_{da} is the force exerted by pin a on member ad. For equilibrium, using Fig. 7.44c,

$$\Sigma M_c = 0 \qquad R_{da,y}(5) + 206(2) - 5,650(5) = 0 \qquad R_{da,y} = 5,570 \text{ N}$$

$$\Sigma F_x = 0 \qquad -R_{da,x} + \frac{5}{5.83}(23,000) - 4,910 = 0 \qquad R_{da,x} = 14,800 \text{ N}$$

$$R_{da} = \sqrt{R_{da,x}^2 + R_{da,y}^2} = \sqrt{14,800^2 + 5,570^2} = 15,800 \text{ N}$$

(*b*) For connection of the cable to pin a instead of pin e, the force R_f increases from 22,100 to 23,000 N. All the other pin forces either decrease or remain the same.

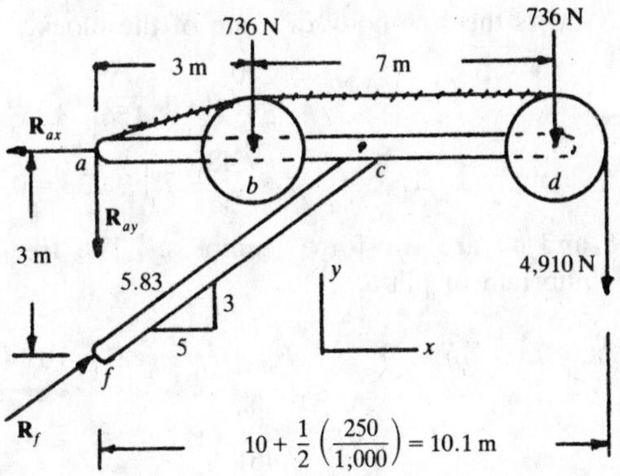

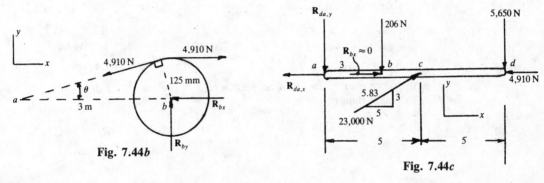

Fig. 7.44a

$$10 + \frac{1}{2}\left(\frac{250}{1,000}\right) = 10.1 \text{ m}$$

Fig. 7.44b

Fig. 7.44c

7.45 The block in Fig. 7.45a has a mass of 75 kg. The pulley diameter is small compared with other dimensions in the problem.

(*a*) Find the force in the cable.

(*b*) Find the forces exerted by the pins on the members of the frame.

(*c*) Express the results in part (*b*) in formal vector notation.

▌ (*a*) R_b is the force exerted by the foundation on pin b. The weight W of the block is

$$W = mg = 75(9.81) = 736 \text{ N}$$

The free-body diagram of the frame is shown in Fig. 7.45b. For equilibrium of the frame,

$$\Sigma M_b = 0 \qquad R_a(500) - 736(620) = 0 \qquad R_a = 913 \text{ N}$$
$$\Sigma F_x = 0 \qquad -R_{bx} + R_a = 0 \qquad R_{bx} = 913 \text{ N}$$
$$\Sigma F_y = 0 \qquad R_{by} - 736 = 0 \qquad R_{by} = 736 \text{ N}$$

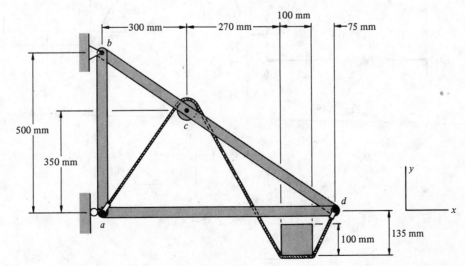

Fig. 7.45a

Figure 7.45c shows the free-body diagram of the block. For equilibrium of the block,

$$\Sigma F_x = 0 \qquad -\frac{270}{555} T_1 + \frac{75}{154} T_2 = 0 \qquad T_1 = T_2 = T$$

$$\Sigma F_y = 0 \qquad 2\left(\frac{485}{555} T\right) - 736 = 0 \qquad T = 421\ \text{N}$$

Members ab and ad are two-force members. The free-body diagram of pin a is shown in Fig. 7.45d. For equilibrium of pin a,

$$\Sigma F_y = 0 \qquad -F_{ab} + \frac{350}{461}(421) = 0 \qquad F_{ab} = 320\ \text{N (C)}$$

$$\Sigma F_x = 0 \qquad 913 + \frac{300}{461}(421) - F_{ad} = 0 \qquad F_{ad} = 1{,}190\ \text{N (C)}$$

R_{db} is the force exerted by pin b on member bd. Figure 7.45e shows the free-body diagram of pin b. For equilibrium of pin b,

$$\Sigma F_x = 0 \qquad R_{db,x} - 913 = 0 \qquad R_{db,x} = 913\ \text{N}$$
$$\Sigma F_y = 0 \qquad -R_{db,y} + 320 + 736 = 0 \qquad R_{db,y} = 1{,}060\ \text{N}$$
$$R_{db} = \sqrt{R_{db,x}^2 + R_{db,y}^2} = \sqrt{913^2 + 1{,}060^2} = 1{,}400\ \text{N}$$

R_{bd} is the force exerted by pin d on member bd. The free-body diagram of pin d is seen in Fig. 7.45f. For equilibrium of pin d,

$$\Sigma F_x = 0 \qquad -R_{bd,x} - \frac{75}{154}(421) + 1{,}190 = 0 \qquad R_{bd,x} = 985\ \text{N}$$

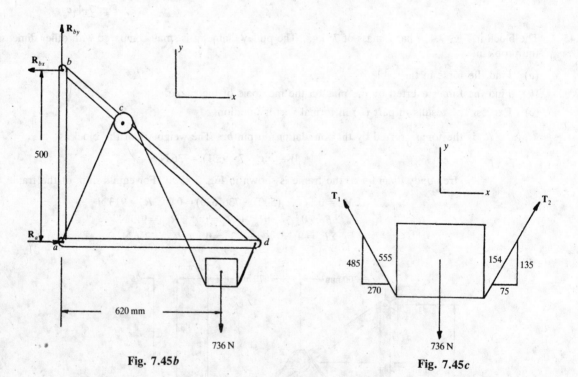

Fig. 7.45b

Fig. 7.45c

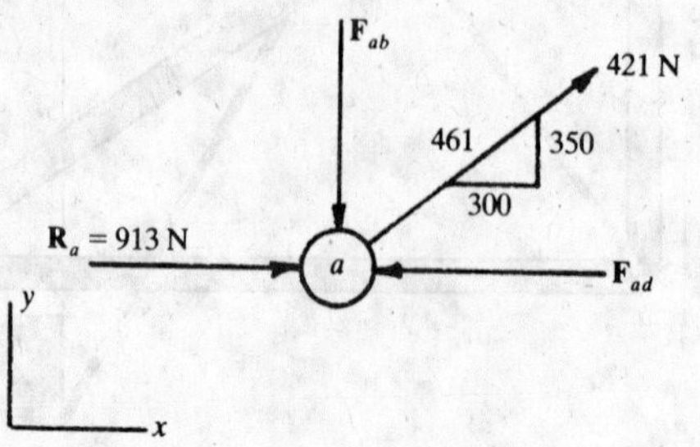

Fig. 7.45d

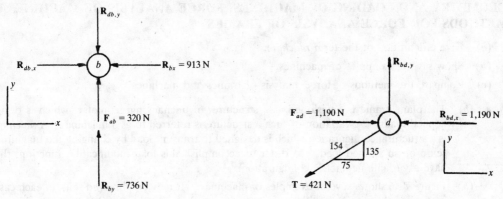

Fig. 7.45e   Fig. 7.45f

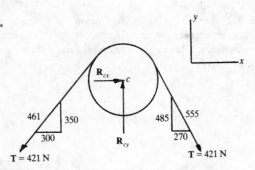

Fig. 7.45g

$$\Sigma F_y = 0 \qquad R_{bd,y} - \frac{135}{154}(421) = 0 \qquad R_{bd,y} = 369 \text{ N}$$

$$R_{bd} = \sqrt{R_{bd,x}^2 + R_{bd,y}^2} = \sqrt{985^2 + 369^2} = 1,050 \text{ N}$$

Figure 7.45g shows the free-body diagram of the pulley at pin c. For equilibrium of this element,

$$\Sigma F_x = 0 \qquad -\frac{300}{461}(421) + R_{cx} + \frac{270}{555}(421) = 0 \qquad R_{cx} = 69.2 \text{ N}$$

$$\Sigma F_y = 0 \qquad -\frac{350}{461}(421) - \frac{485}{555}(421) + R_{cy} = 0 \qquad R_{cy} = 688 \text{ N}$$

$$R_c = \sqrt{R_{cx}^2 + R_{cy}^2} = \sqrt{69.2^2 + 688^2} = 691 \text{ N}$$

The pin forces acting on member ab are

$$R_a = F_{ab} = 320 \text{ N} \qquad R_b = F_{ab} = 320 \text{ N}$$

The pin forces acting on member ad are

$$R_a = F_{ad} = 1,190 \text{ N} \qquad R_d = F_{ad} = 1,190 \text{ N}$$

The forces acting on member bd are

$$R_b = R_{db} = 1,400 \text{ N} \qquad R_c = 691 \text{ N} \qquad R_d = R_{bd} = 1,050 \text{ N}$$

(c) The pin forces acting on member ab are

$$\mathbf{R}_a = 320\mathbf{j} \quad \text{N} \qquad \mathbf{R}_b = -320\mathbf{j} \quad \text{N}$$

The pin forces acting on member ad are

$$\mathbf{R}_a = 1,190\mathbf{i} \quad \text{N} \qquad \mathbf{R}_d = -1,190\mathbf{i} \quad \text{N}$$

The pin forces acting on member bd are

$$\mathbf{R}_b = -913\mathbf{i} + 1,060\mathbf{j} \quad \text{N}$$
$$\mathbf{R}_c = -69.2\mathbf{i} - 688\mathbf{j} \quad \text{N}$$
$$\mathbf{R}_d = 985\mathbf{i} - 369\mathbf{j} \quad \text{N}$$

7.3 GEOMETRY AND LOADING OF MACHINES, FORCE ANALYSIS OF MACHINES USING THE METHODS FOR FORCE ANALYSIS OF FRAMES

7.46   (*a*)   Give a definition of the term *machine*.

(*b*)   Show several examples of machines.

(*c*)   Compare the methods of force analysis of frames and machines.

❚ (*a*)   The plane frame is envisioned as a structure, of unchanging geometry, which is designed to support applied loads.   A variation of such a structure is referred to as a machine.   A *machine*, by definition, is a structural configuration which is designed to transmit load by changing the magnitude, direction, or sense of an applied load.   In order to accomplish this load modification function, the machine must have a variable geometric configuration.

(*b*)   Figure 7.46 shows several examples of machines.   It may be observed that, in each case, some relative motion of the members with respect to one another must be possible if the machine is to perform its function.

(*c*)   The machine, for any particular configuration, is statically equivalent to the frame.   Thus, all the techniques which are presented in this chapter for the force analysis of frames may be also used to force-analyze machines.   It may be observed finally that the members of a frame or machine may be either straight or curved.

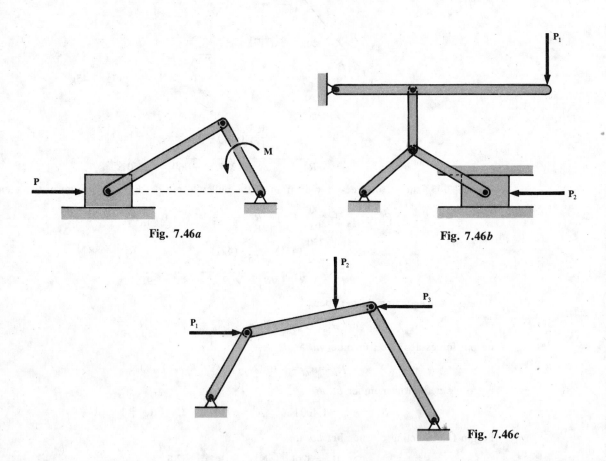

Fig. 7.46*a*   Fig. 7.46*b*

Fig. 7.46*c*

7.47   (*a*)   Find the force magnification ratio P_2/P_1 for the simple machine shown in Fig. 7.47*a*.   The slider block is smooth.

(*b*)   If $P_1 = 12$ kN,   find P_2 and the compressive force between the slider block and the wall of the guide.

(*c*)   Express the results in part (*b*) in formal vector notation.

❚ (*a*)   Both of the links in the machine are two-force members.   The free-body diagram of pin *b* is shown in Fig. 7.47*b*.   For equilibrium of this element,

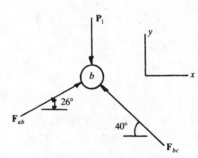

Fig. 7.47a

$$\Sigma F_x = 0 \qquad F_{ab}\cos 26° - F_{bc}\cos 40° = 0 \qquad F_{ab} = 0.852 F_{bc}$$
$$\Sigma F_y = 0 \qquad F_{ab}\sin 26° - P_1 + F_{bc}\sin 40° = 0$$
$$(0.852 F_{bc})\sin 26° - P_1 + F_{bc}\sin 40° = 0 \qquad F_{bc} = 0.984 P_1 \ (C)$$
$$F_{ab} = 0.852(0.984 P_1) = 0.838 P_1 \ (C)$$

Figure 7.47c shows the free-body diagram of pin *a*. R_a is the compressive force on the wall of the slider block. For equilibrium of this pin,

$$\Sigma F_y = 0 \qquad R_a - (0.838 P_1)\sin 26° = 0 \qquad R_a = 0.367 P_1$$

$$\Sigma F_x = 0 \qquad P_2 - (0.838 P_1)\cos 26° = 0 \qquad \frac{P_2}{P_1} = 0.753$$

(**b**) For $P_1 = 12$ kN,

$$\frac{P_2}{P_1} = 0.753 = \frac{P_2}{12} \qquad P_2 = 9.04 \text{ kN}$$

$$R_a = 0.367 P_1 = 0.367(12) = 4.40 \text{ kN}$$

(**c**)
$$\mathbf{P}_2 = 9.04\mathbf{i} \quad \text{kN} \qquad \mathbf{R}_a = 4.40\mathbf{j} \quad \text{kN}$$

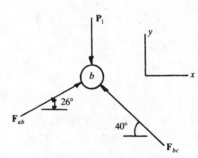

Fig. 7.47b

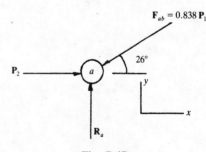

Fig. 7.47c

7.48 The slider block in Fig. 7.48a is smooth.

(**a**) Find the value of θ which will make $P_2/P_1 = 10$.

(**b**) If $P_1 = 200$ N, find the forces exerted by the pins on the members, and the compressive force on the slider block, for the value of θ found in part (a).

▌ (**a**) Member *bd* is a two-force member, and R_d is the compressive force on the slider block. The free-body diagrams of members *ac* and the slider block are shown in Figs. 7.48b and c. For equilibrium of member *ac*, using Fig. 7.48b,

$$\Sigma M_a = 0 \qquad (F_{bd}\sin\theta)(700) - P_1(1,200) = 0 \qquad F_{bd} = \frac{1.71 P_1}{\sin\theta}$$

For equilibrium of the slider block, from Fig. 7.48c,

$$\Sigma F_y = 0 \qquad R_d - F_{bd}\sin\theta = 0 \qquad R_d = \left(\frac{1.71 P_1}{\sin\theta}\right)\sin\theta = 1.71 P_1$$

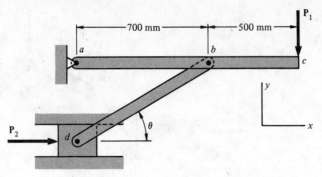

Fig. 7.48a

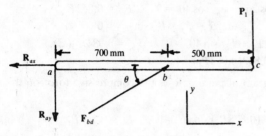

Fig. 7.48b

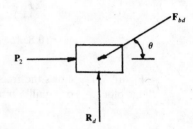

Fig. 7.48c

$$\Sigma F_x = 0 \qquad P_2 - F_{bd} \cos\theta = 0 \qquad P_2 - \left(\frac{1.71 P_1}{\sin\theta}\right)\cos\theta = 0$$

Using $P_2/P_1 = 10$,

$$\frac{P_2}{P_1} = \frac{1.71}{\tan\theta} = 10 \qquad \theta = 9.70°$$

(b) For equilibrium of member ac, from Fig. 7.48b,

$$\Sigma F_x = 0 \qquad -R_{ax} + F_{bd}\cos\theta = 0 \qquad R_{ax} = F_{bd}\cos\theta = \frac{1.71 P_1}{\sin\theta}\cos\theta = \frac{1.71 P_1}{\tan\theta} = P_2$$

$$\Sigma F_y = 0 \qquad -R_{ay} + F_{bd}\sin\theta - P_1 = 0$$

$$R_{ay} = -P_1 + \left(\frac{1.71 P_1}{\sin\theta}\right)\sin\theta = 0.71 P_1$$

Using $P_2/P_1 = 10$ and $P_1 = 200\,\text{N}$,

$$P_2 = 10 P_1 = 10(200) = 2,000\,\text{N} \qquad R_d = 1.71 P_1 = 1.71(200) = 342\,\text{N}$$
$$R_{ax} = P_2 = 2,000\,\text{N} \qquad R_{ay} = 0.71 P_1 = 0.71(200) = 142\,\text{N}$$
$$R_a = \sqrt{R_{ax}^2 + R_{ay}^2} = \sqrt{2,000^2 + 142^2} = 2,010\,\text{N}$$

$$F_{bd} = \frac{171 P_1}{\sin\theta} = \frac{171(200)}{\sin 9.70°} = 2,030\,\text{N (C)} \qquad R_b = R_d = F_{bd} = 2,030\,\text{N}$$

7.49 Figure 7.49a shows a model of a simple compacting device. Both members have the same direction with respect to the x axis. A force of 40 lb is applied to the handle.

(a) Find the forces exerted by the pins on the members.

(b) Find the force exerted by the smooth sliding element on the compacted material and the force of the wall on the element.

(c) Express the results in parts (a) and (b) in formal vector notation.

▮ (a) P is the force exerted by the sliding element on the material, and member ab is a two-force member. The free-body diagrams of member bc and the sliding element are shown in Figs. 7.49b and c. P is the force on the material and R_{cy} is the force of the wall on the sliding element. From the geometry of Fig. 7.49b,

$$12\cos\theta + 10\cos\theta = 18 \qquad \theta = 35.1°$$

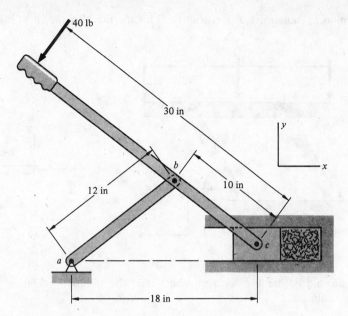

Fig. 7.49a

For equilibrium of member bc, using Fig. 7.49b,

$$\Sigma M_c = 0 \quad -(F_{ab} \cos 35.1°)(10 \sin 35.1°) - (F_{ab} \sin 35.1°)(10 \cos 35.1°) + 40(30) = 0$$
$$F_{ab} = 128 \text{ lb (C)}$$

$$\Sigma F_x = 0 \quad -40 \sin 35.1° + 128 \cos 35.1° - R_{cx} = 0 \quad R_{cx} = 81.7 \text{ lb}$$
$$\Sigma F_y = 0 \quad -40 \cos 35.1° + 128 \sin 35.1° - R_{cy} = 0 \quad R_{cy} = 40.9 \text{ lb}$$
$$R_c = \sqrt{R_{cx}^2 + R_{cy}^2} = \sqrt{81.7^2 + 40.9^2} = 91.4 \text{ lb}$$
$$R_a = R_b = F_{ab} = 128 \text{ lb}$$

(b) From the free-body diagram of the sliding element, using Fig. 7.49c,

$$\Sigma F_x = 0 \quad R_{cx} - P = 0 \quad P = R_{cx} = 81.7 \text{ lb}$$
$$\Sigma F_y = 0 \quad R_{cy} - R = 0 \quad R = R_{cy} = 40.9 \text{ lb}$$

(c) The pin forces which act on member ab are

$$\mathbf{R}_a = (F_{ab} \cos \theta)\mathbf{i} + (F_{ab} \sin \theta)\mathbf{j} = (128 \cos 35.1°)\mathbf{i} + (128 \sin 35.1°)\mathbf{j} = 105\mathbf{i} + 73.6\mathbf{j} \quad \text{lb}$$
$$\mathbf{R}_b = -105\mathbf{i} - 73.6\mathbf{j} \quad \text{lb}$$

The forces exerted by the pins on member bc are

$$\mathbf{R}_b = 105\mathbf{i} + 73.6\mathbf{j} \quad \text{lb} \qquad \mathbf{R}_c = -81.7\mathbf{i} - 40.9\mathbf{j} \quad \text{lb}$$

The force exerted by the slider on the material is

$$\mathbf{P} = 81.7\mathbf{i} \quad \text{lb}$$

The force of the wall on the slider is

$$\mathbf{R}_{cy} = 40.9\mathbf{j} \quad \text{lb}$$

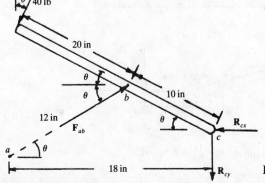

Fig. 7.49b

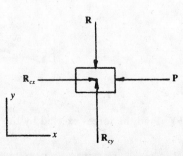

Fig. 7.49c

7.50 Find the variation in cable force P versus angle θ for the machine shown in Fig. 7.50a.

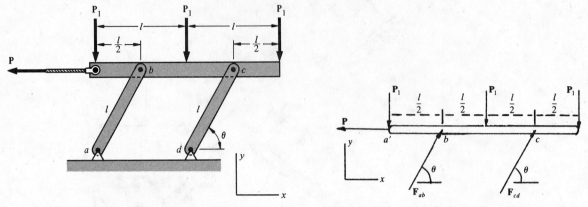

Fig. 7.50a **Fig. 7.50b**

❚ Members ab and cd are two-force members, and the free-body diagram of member bc is seen in Fig. 7.50b. For equilibrium,

$$\Sigma M_b = 0 \qquad P_1\left(\frac{l}{2}\right) - P_1\left(\frac{l}{2}\right) + (F_{cd}\sin\theta)l - P_1\left(\frac{3l}{2}\right) = 0 \tag{1}$$

$$\Sigma F_x = 0 \qquad -P + F_{ab}\cos\theta + F_{cd}\cos\theta = 0 \tag{2}$$

$$\Sigma F_y = 0 \qquad -3P_1 + F_{ab}\sin\theta + F_{cd}\sin\theta = 0 \tag{3}$$

From Eq. (1),

$$F_{cd} = \frac{3P_1}{2\sin\theta} \tag{4}$$

F_{cd} is eliminated between Eqs. (3) and (4), with the result

$$-3P_1 + F_{ab}\sin\theta + \left(\frac{3P_1}{2\sin\theta}\right)\sin\theta = 0 \qquad F_{ab} = \frac{3P_1}{2\sin\theta} \tag{5}$$

From comparison of Eqs. (4) and (5),

$$F_{ab} = F_{cd}$$

Using Eq. (2),

$$-P + \left(\frac{3P_1}{2\sin\theta}\right)\cos\theta + \left(\frac{3P_1}{2\sin\theta}\right)\cos\theta = 0 \qquad \frac{3\cos\theta}{\sin\theta}P_1 = P \qquad P = 3P_1\cot\theta$$

7.51 (a) Find the load magnification factor P_2/P_1 for the crushing device shown in Fig. 7.51a. The slider block is smooth.

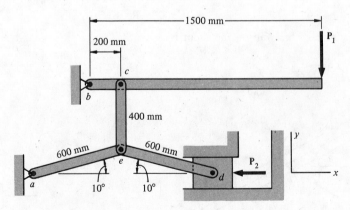

Fig. 7.51a

(b) Find the forces exerted by the pins on the members if $P_2 = 4{,}200\text{ N}$.

(a) Members ae, de, and ce are two-force members. The free-body diagram of the slider block is shown in Fig. 7.51b. For equilibrium,

$$\Sigma F_x = 0 \qquad F_{de} \cos 10° - P_2 = 0 \qquad F_{de} = \frac{P_2}{\cos 10°}$$

Figure 7.51c shows the free-body diagram of pin e. From the symmetry of the loading,

$$F_{de} = F_{ae}$$

For equilibrium of pin e,

$$\Sigma F_y = 0 \qquad -F_{ce} + 2F_{de} \sin 10° = 0 \qquad F_{ce} = 2\left(\frac{P_2}{\cos 10°}\right) \sin 10° = 2P_2 \tan 10°$$

The free-body diagram of member bc is shown in Fig. 7.51d. For equilibrium of this member,

$$\Sigma M_b = 0 \qquad F_{ce}(200) - P_1(1,500) = 0 \qquad 200(2P_2 \tan 10°) = 1,500P_1$$

$$\frac{P_2}{P_1} = \frac{1,500}{2(200)\tan 10°} = 21.3$$

(b) Using Fig. 7.51d,

$$\Sigma F_x = 0 \qquad R_{bx} = 0 \qquad \Sigma F_y = 0 \qquad -R_{by} + F_{ce} - P_1 = 0$$

$$R_{by} = F_{ce} - P_1 = 2P_2 \tan 10° - P_1 = 2(21.3P_1)\tan 10° - P_1 = 6.41P_1$$

For $P_2 = 4,200$ N,

$$F_{de} = F_{ae} = \frac{P_2}{\cos 10°} = \frac{4,200}{\cos 10°} = 4,260 \text{ N (C)} \qquad F_{ce} = 2P_2 \tan 10° = 2(4,200)\tan 10° = 1,480 \text{ N (C)}$$

$$R_b = R_{by} = 0.301P_2 = 0.301(4,200) = 1,260 \text{ N}$$

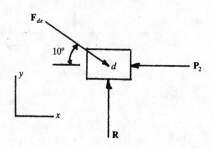

Fig. 7.51b

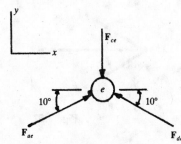

Fig. 7.51c

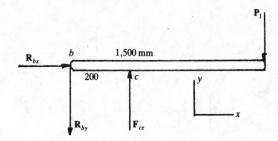

Fig. 7.51d

7.52 Figure 7.52a shows a model of an automobile "scissors" jack. A screw goes through blocks which are connected to pins b and e. Rotation of this screw causes these two points to approach each other. The forces P_1 in the figure represent the force of the screw threads on pins b and e. The force P_2 represents the load which is raised by the jack.

The function of this jack is to transform the screw-thread force P_1 into a lifting force P_2 to raise the automobile. As a measure of performance of the jack, a load ratio factor P^* will be defined as

$$P^* = \frac{P_2}{P_1}$$

This factor may be interpreted as the ratio of the output force P_2 of the machine to the input force P_1.

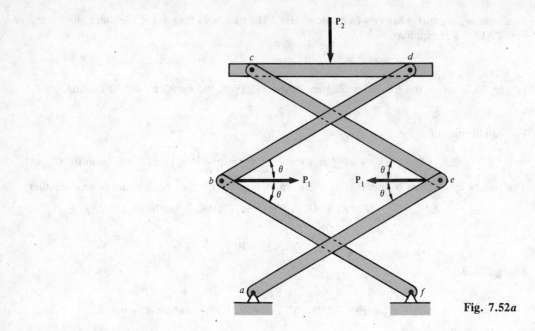

Fig. 7.52a

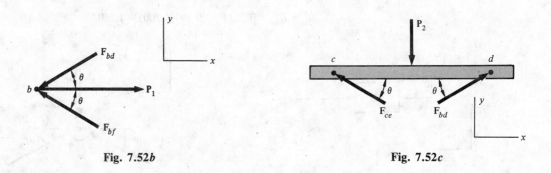

Fig. 7.52b

Fig. 7.52c

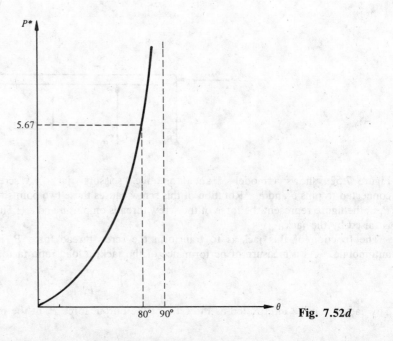

Fig. 7.52d

Compute and plot P^*, for the configuration shown in Fig. 7.52a, for a range of values of θ from $0°$ to $80°$.

▌ Every member in the jack, except cd, is a two-force member. Because of symmetry about the vertical center axis, only half of the structure will be analyzed. Figure 7.52b shows the free-body diagram of joint b. For equilibrium of this joint,

$$\sum F_y = 0 \qquad -F_{bd} \sin \theta + F_{bf} \sin \theta = 0 \qquad F_{bd} = F_{bf}$$

$$\sum F_x = 0 \qquad -F_{bd} \cos \theta + P_1 - F_{bf} \cos \theta = 0 \qquad 2F_{bd} \cos \theta = P_1 \qquad F_{bd} = \frac{P_1}{2 \cos \theta}$$

Figure 7.52c shows the free-body diagram of member cd. From symmetry considerations,

$$F_{ce} = F_{bd}$$

For equilibrium in the y direction,

$$\Sigma F_y = 0 \qquad F_{ce} \sin \theta + F_{bd} \sin \theta - P_2 = 0 \qquad 2F_{bd} \sin \theta = P_2$$

The above equations are combined, with the result

$$2 \left(\frac{P_1}{2 \cos \theta} \right) \sin \theta = P_2 \qquad P_1 \tan \theta = P_2 \qquad P^* = \frac{P_2}{P_1} = \tan \theta$$

The plot of this equation is shown in Fig. 7.52d. It may be seen that the load ratio factor P^* increases at a rapid rate for values of θ approaching $90°$.

7.53 A person pulls with a force of 200 N on the handle of the machine shown in Fig. 7.53a, which is in equilibrium.

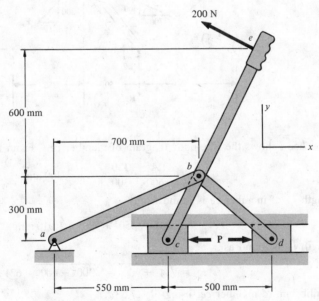

200 N

600 mm

700 mm

300 mm

a

b

c P d

550 mm

500 mm

e

Fig. 7.53a

(**a**) Find the value of the compressive forces P on the smooth slider blocks.

(**b**) Does each slider block experience the same compressive force from the wall of the guide?

▌ (**a**) Members ab and bd are two-force members. The free-body diagrams of slider blocks c and d are shown in Figs. 7.53b and c. N_c and N_d are the compressive forces exerted by the wall on the blocks, and R_c is the pin force at c. For equilibrium of slider c, from Fig. 7.53b,

$$\Sigma F_x = 0 \qquad R_{cx} - P = 0 \qquad R_{cx} = P \tag{1}$$
$$\Sigma F_y = 0 \qquad -R_{cy} + N_c = 0 \qquad N_c = R_{cy}$$

For equilibrium of slider d, using Fig. 7.53c,

$$\Sigma F_x = 0 \qquad -\frac{350}{461} F_{bd} + P = 0 \qquad F_{bd} = 1.32P \tag{2}$$

$$\Sigma F_y = 0 \qquad \frac{300}{461} F_{bd} - N_d = 0 \qquad N_d = \frac{300}{461} F_{bd} = 1.54F_{bd}$$

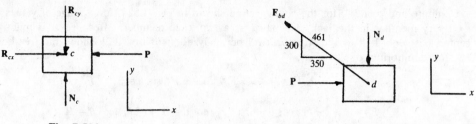

Fig. 7.53b   Fig. 7.53c

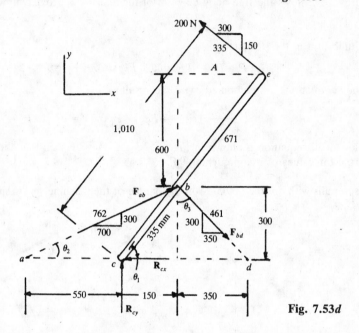

Fig. 7.53d

Figure 7.53d shows the free-body diagram of member ce. From the similar triangles in the figure,

$$\frac{A}{600} = \frac{150}{300} \qquad A = 300 \text{ mm}$$

The length l_{ce} of member ce is found from

$$l_{ce} = \sqrt{450^2 + 900^2} = 1{,}010 \text{ mm}$$

The length l_{be} is found as

$$l_{be} = \sqrt{A^2 + 600^2} = \sqrt{300^2 + 600^2} = 671 \text{ mm}$$

For equilibrium of member ce, using Fig. 7.53d,

$$\Sigma M_a = 0 \qquad R_{cy}(550) - \left(\frac{350}{461} F_{bd}\right)300 - \left(\frac{300}{461} F_{bd}\right)700 + \frac{300}{335}(200)900 + \frac{150}{335}(200)1{,}000 = 0 \tag{3}$$

F_{bd} is eliminated between Eqs. (2) and (3) to obtain

$$-R_{cy} + 1.64P - 456 = 0 \tag{4}$$

Instead of using a summation of forces in the x direction as the next equilibrium equation, a second moment summation will be used. For moment equilibrium about point b,

$$\Sigma M_b = 0 \qquad -R_{cy}(150) - R_{cx}(300) + 200(671) = 0$$

Using Eq. (1),

$$-R_{cy}(150) - P(300) + 200(671) = 0 \qquad R_{cy} + 2P - 895 = 0 \tag{5}$$

Equations (4) and (5) are added, with the result

$$P = 371 \text{ N}$$

(*b*) Using Eq. (4),

$$R_{cy} = 1.64P - 456 = 1.64(371) - 456 = 152 \text{ N}$$

Using Eq. (2),

$$F_{bd} = 1.32P = 1.32(371) = 490 \text{ N (T)}$$

For equilibrium of member *ce* in the *y* direction, using Fig. 7.53*d*,

$$\Sigma F_y = 0 \qquad R_{cy} + \frac{300}{762} F_{ab} - \frac{300}{461} F_{bd} + \frac{150}{335}(200) = 0$$

Using the above values for F_{bd} and R_{cy}, the final result for F_{ab} is

$$F_{ab} \doteq 196 \text{ N (C)}$$

Using Fig. 7.53*b*,

$$\Sigma F_y = 0 \qquad N_c - R_{cy} = 0 \qquad N_c = R_{cy} = 152 \text{ N}$$

From Fig. 7.53*c*,

$$\Sigma F_y = 0 \qquad \frac{300}{461} F_{bd} - N_d = 0 \qquad N_d = \frac{300}{461}(490) = 319 \text{ N}$$

It may be seen that slider block *d* experiences the larger compressive wall force, by a factor of

$$\frac{N_d}{N_c} = \frac{319}{152} = 2.10$$

7.54 A couple M_0 is applied to member *ab* in Fig. 7.54*a* to keep the system in the equilibrium position shown.   The slider block is smooth.

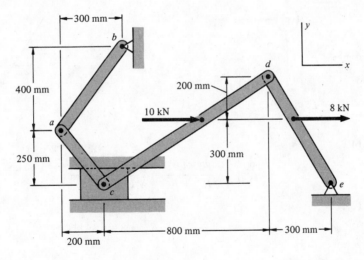

Fig. 7.54*a*

(*a*) Find the forces acting on pins *c*, *d*, and *e*.

(*b*) Find the force in member *ac* and the compressive force between the slider block and the wall of the guide.

(*c*) Find the force acting on pin *b* and find the magnitude and sense of M_0.

▌ (*a*) The free-body diagram of member *de* is shown in Fig. 7.54*b*.   For equilibrium of this member,

$$\Sigma M_e = 0 \qquad R_{dx}(500) + R_{dy}(300) - 8(300) = 0 \qquad (1)$$
$$\Sigma F_x = 0 \qquad -R_{dx} + 8 - R_{ex} = 0 \qquad (2)$$
$$\Sigma F_y = 0 \qquad -R_{dy} + R_{ey} = 0 \qquad (3)$$

Figure 7.54*c* shows the free-body diagram of member *cd*.   For moment equilibrium,

$$\Sigma M_c = 0 \qquad -10(300) - R_{dx}(500) + R_{dy}(800) = 0 \qquad (4)$$

R_{dx} is eliminated between Eqs. (1) and (4) to obtain

$$1,100 R_{dy} = 5,400 \qquad R_{dy} = 4.91 \text{ kN}$$

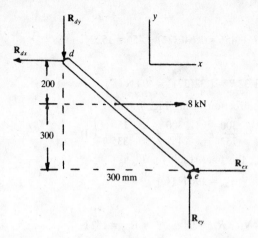

Fig. 7.54b

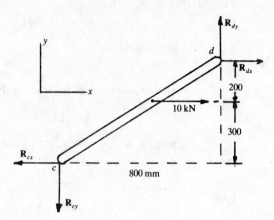

Fig. 7.54c

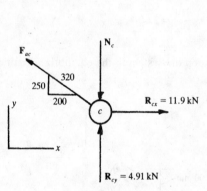

Fig. 7.54d

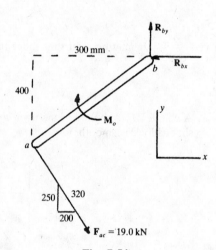

Fig. 7.54e

Using Eq. (4),

$$500R_{dx} - 800(4.91) + 3,000 = 0 \qquad R_{dx} = 1.86 \text{ kN}$$
$$R_d = \sqrt{R_{dx}^2 + R_{dy}^2} = \sqrt{1.86^2 + 4.91^2} = 5.25 \text{ kN}$$

Using Eq. (2),

$$R_{ex} = -R_{dx} + 8 = 6.14 \text{ kN}$$

Using Eq. (3),

$$R_{ey} = R_{dy} = 4.91 \text{ kN} \qquad R_e = \sqrt{R_{ex}^2 + R_{ey}^2} = \sqrt{6.14^2 + 4.91^2} = 7.86 \text{ kN}$$

Using Fig. 7.54c,

$$\Sigma F_x = 0 \qquad -R_{cx} + 10 + R_{dx} = 0 \qquad R_{cx} = 11.9 \text{ kN}$$
$$\Sigma F_y = 0 \qquad -R_{cy} + R_{dy} = 0 \qquad R_{cy} = 4.91 \text{ kN}$$
$$R_c = \sqrt{R_{cx}^2 + R_{cy}^2} = \sqrt{11.9^2 + 4.91^2} = 12.9 \text{ kN}$$

(b) Member ac is a two-force member. Figure 7.54d shows the free-body diagram of pin c. N_c is the compressive force exerted by the wall on the slider block, and this force is transmitted directly through the slider block to pin c. For equilibrium of pin c,

$$\Sigma F_x = 0 \qquad -\frac{200}{320} F_{ac} + 11.9 = 0 \qquad F_{ac} = 19.0 \text{ kN (T)}$$

$$\Sigma F_y = 0 \qquad \frac{250}{320} F_{ac} - N_c + 4.91 = 0 \qquad N_c = 19.8 \text{ kN}$$

(c) Figure 7.54e shows the free-body diagram of member ab. For equilibrium of this member,

$$\Sigma M_b = 0 \qquad -M_0 + \left(\frac{250}{320} F_{ac}\right)\frac{300}{1,000} + \left(\frac{200}{320} F_{ac}\right)\frac{400}{1,000} = 0$$

Using $F_{ac} = 19.0 \text{ kN}$ in the above equation,

$$M_0 = 9.20 \text{ kN} \cdot \text{m}$$

$$\Sigma F_x = 0 \qquad -R_{bx} + \frac{200}{320} F_{ac} = 0 \qquad R_{bx} = 11.9 \text{ kN}$$

$$\Sigma F_y = 0 \qquad R_{by} - \frac{250}{320} F_{ac} = 0 \qquad R_{by} = 14.8 \text{ kN}$$

$$R_b = \sqrt{R_{bx}^2 + R_{by}^2} = \sqrt{11.9^2 + 14.8^2} = 19.0 \text{ kN}$$

The two forces F_{ac} and R_b form a couple of magnitude $M_0 = 9.20 \text{ kN} \cdot \text{m}$. It is left as an exercise for the reader to show that the directions of these two forces are the same. The separation distance d between these two forces is

$$d = \frac{M_0}{F} = \frac{9,200}{19,000} = 0.484 \text{ m} = 484 \text{ mm}$$

7.55   (a) Find the magnitude of the force P which will keep the machine in Fig. 7.55a in the equilibrium position shown.

  (b) Find the force exerted by the foundation on pin a.

  (c) Find the forces exerted by pins b and c on member bd.

  (d) Express the results of parts (b) and (c) in formal vector notation.

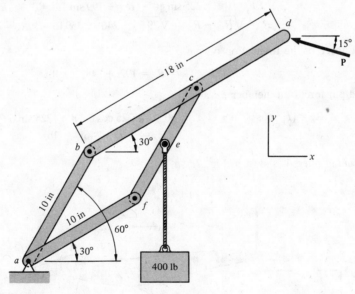

Fig. 7.55a

▌ (a) Figure 7.55b shows the free-body diagram of the system. R_a is the force exerted by the foundation on pin a. For equilibrium,

$$\Sigma M_a = 0$$
$$-400(10 \cos 30° + 5 \cos 60°) + (P \sin 15°)(10 \cos 60° + 18 \cos 30°) + (P \cos 15°)(10 \sin 60° + 18 \sin 30°) = 0$$
$$P = 199 \text{ lb}$$

(b) From Fig. 7.55b,

$$\Sigma F_x = 0 \qquad R_{ax} - P \cos 15° = 0 \qquad R_{ax} = 192 \text{ lb}$$
$$\Sigma F_y = 0 \qquad R_{ay} - 400 + P \sin 15° = 0 \qquad R_{ay} = 348 \text{ lb}$$
$$R_a = \sqrt{R_{ax}^2 + R_{ay}^2} = \sqrt{192^2 + 348^2} = 397 \text{ lb}$$

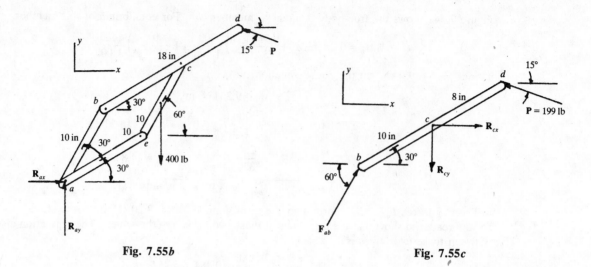

Fig. 7.55b

Fig. 7.55c

(c) Member ab is a two-force member. The free-body diagram of member bd is shown in Fig. 7.55c. For equilibrium of member bd,

$$\Sigma M_c = 0$$
$$-(F_{ab} \sin 60°)10 \cos 30° + (F_{ab} \cos 60°)10 \sin 30° + (199 \cos 15°)8 \sin 30° + (199 \sin 15°)8 \cos 30° = 0$$
$$F_{ab} = 225 \text{ lb (C)}$$

$$\Sigma F_x = 0 \qquad 225 \cos 60° + R_{cx} - 199 \cos 15° = 0 \qquad R_{cx} = 79.7 \text{ lb}$$
$$\Sigma F_y = 0 \qquad 225 \sin 60° - R_{cy} + 199 \sin 15° = 0 \qquad R_{cy} = 246 \text{ lb}$$
$$R_c = \sqrt{R_{cx}^2 + R_{cy}^2} = \sqrt{79.7^2 + 246^2} = 259 \text{ lb} \qquad R_b = F_{ab} = 225 \text{ lb}$$

(d) The force of the foundation on pin a is

$$\mathbf{R}_a = 192\mathbf{i} + 348\mathbf{j} \qquad \text{lb}$$

The pin forces on member bd are

$$\mathbf{R}_b = (F_{ab} \cos 60°)\mathbf{i} + (F_{ab} \sin 60°)\mathbf{j} = (225 \cos 60°)\mathbf{i} + (225 \sin 60°)\mathbf{j} = 113\mathbf{i} + 195\mathbf{j} \qquad \text{lb}$$
$$\mathbf{R}_c = 79.7\mathbf{i} - 246\mathbf{j} \qquad \text{lb}$$

7.56 A pair of locking pliers is shown in Fig. 7.56a.

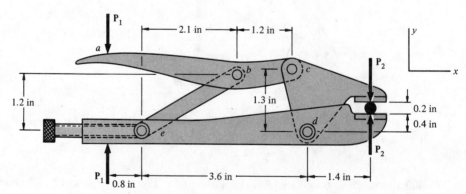

Fig. 7.56a

(a) If a person exerts a squeezing force P_1 on the handles, what compressive force P_2 is exerted by the jaws on the rod held in the jaws?

(b) What is the force magnification ratio P_2/P_1 of the pliers?

(c) Which is the most heavily loaded pin in the assembly?

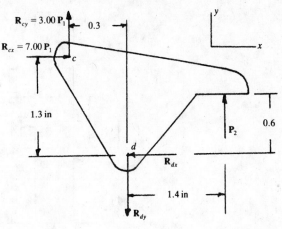

Fig. 7.56b   **Fig. 7.56c**

(a) Member be is a two-force member. The symbolic free-body diagram of the upper handle piece, member ac, is shown in Fig. 7.56b. For equilibrium of this member,

$$\Sigma M_c = 0 \qquad P_1(4.1) + \left(\frac{2.1}{2.42}\,F_{ab}\right)0.1 - \left(\frac{1.2}{2.42}\,F_{ab}\right)1.2 = 0 \qquad F_{ab} = 8.07P_1\ \text{(C)}$$

$$\Sigma F_x = 0 \qquad \frac{2.1}{2.42}\,(F_{ab}) - R_{cx} = 0 \qquad R_{cx} = 7.00P_1$$

$$\Sigma F_y = 0 \qquad -P_1 + \frac{1.2}{2.42}\,F_{ab} - R_{cy} = 0 \qquad R_{cy} = 3.00P_1$$

$$R_c = \sqrt{R_{cx}^2 + R_{cy}^2} = \sqrt{(7.00P_1)^2 + (3.00P_1)^2} = 7.62P_1$$

Figure 7.56c shows the free-body diagram of the upper jaw. For moment equilibrium of this element,

$$\Sigma M_d = 0 \qquad -(7.00P_1)1.3 - (3.00P_1)0.3 + P_2(1.4) = 0 \qquad P_2 = 7.14P_1$$

The compressive forces P_2 exerted by the jaws on the rod are 7.14 times greater than the forces P_1 exerted by the person on the handles.

(b) The force magnification ratio of the pliers is

$$\frac{P_2}{P_1} = 7.14$$

(c) Using Fig. 7.56c,

$$\Sigma F_x = 0 \qquad R_{cx} - R_{dx} = 0 \qquad R_{dx} = 7.00P_1$$
$$\Sigma F_y = 0 \qquad R_{cy} - R_{dy} + P_2 = 0 \qquad 3.00P_1 - R_{dy} + 7.14P_1 = 0$$
$$R_{dy} = 10.1P_1 \qquad R_d = \sqrt{R_{dx}^2 + R_{dy}^2} = \sqrt{(7.00P_1)^2 + (10.1P_1)^2} = 12.3P_1$$

The forces acting on the pins of the pliers are

$$R_a = F_{ab} = 8.07P_1 \qquad R_b = F_{ab} = 8.07P_1 \qquad R_c = 7.62P_1 \qquad R_d = 12.3P_1$$

Pin d is the most heavily loaded pin.

7.57 Figure 7.57a shows a model of half of the front end of an automobile.

(a) Find the value of the spring force which is required to maintain the equilibrium position shown.

(b) Find the forces acting on all the pins.

(a) Member bc is a two-force member. Figure 7.57b shows the free-body diagram of member ab. For equilibrium of this member,

$$\Sigma M_a = 0 \qquad -1,000(6) + F_{bc}(8) = 0 \qquad F_{bc} = 750\ \text{lb (C)}$$
$$\Sigma F_x = 0 \qquad -F_{bc} + R_{ax} = 0 \qquad R_{ax} = 750\ \text{lb}$$
$$\Sigma F_y = 0 \qquad 1,000 - R_{ay} = 0 \qquad R_{ay} = 1,000\ \text{lb}$$

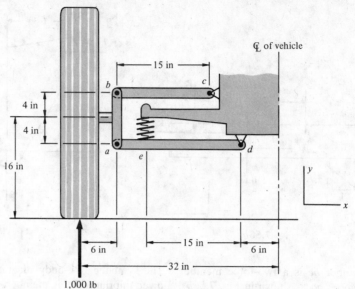

Fig. 7.57a

$$R_a = \sqrt{R_{ax}^2 + R_{ay}^2} = \sqrt{750^2 + 1,000^2} = 1,250 \text{ lb}$$

The free-body diagram of member ad is seen in Fig. 7.57c.

F_s is the force exerted by the spring on member ad. For moment equilibrium of member ad,

$$\Sigma M_d = 0 \qquad -1,000(20) + F_s(15) = 0 \qquad F_s = 1,330 \text{ lb}$$

(**b**) Using Fig. 7.57c,

$$\Sigma F_x = 0 \qquad -750 + R_{dx} = 0 \qquad R_{dx} = 750 \text{ lb}$$
$$\Sigma F_y = 0 \qquad 1,000 - F_s + R_{dy} = 0 \qquad R_{dy} = 330 \text{ lb}$$
$$R_d = \sqrt{R_{dx}^2 + R_{dy}^2} = \sqrt{750^2 + 330^2} = 819 \text{ lb}$$

The forces acting on the pins are

$$R_a = 1,250 \text{ lb} \qquad R_b = F_{bc} = 750 \text{ lb} \qquad R_c = F_{bc} = 750 \text{ lb} \qquad R_d = 819 \text{ lb}$$

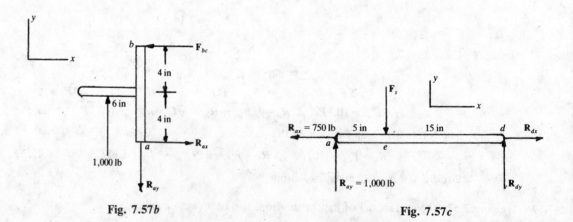

Fig. 7.57b   Fig. 7.57c

7.58 (**a**) Find the magnitude and sense of the couple M which will keep the machine in Fig. 7.58a in the equilibrium position shown.

(**b**) Find the forces acting on all the pins.

▌ (**a**) Members be and de are two-force members. Figure 7.58b shows the free-body diagram of member cd. For equilibrium,

$$\Sigma M_c = 0 \qquad -100(4) + (F_{de} \sin 68°)(10) = 0 \qquad F_{de} = 43.1 \text{ lb}$$

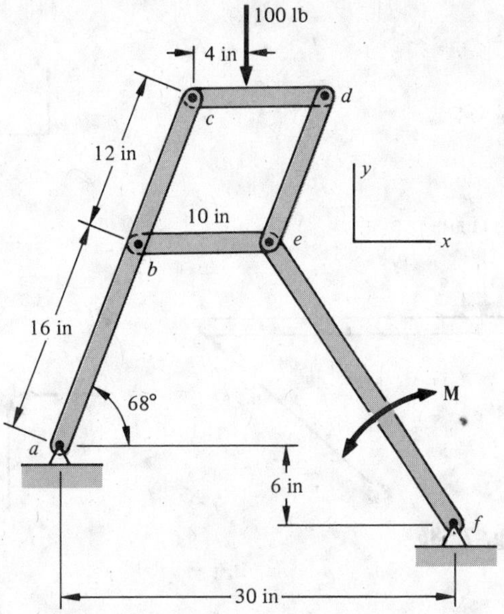

Fig. 7.58a

$$\Sigma F_y = 0 \quad R_{cy} - 100 + F_{de}\sin 68° = 0 \quad R_{cy} = 60\,\text{lb}$$
$$\Sigma F_x = 0 \quad -R_{cx} + F_{de}\cos 68° = 0 \quad R_{cx} = 16.1\,\text{lb}$$
$$R_c = \sqrt{R_{cx}^2 + R_{cy}^2} = \sqrt{16.1^2 + 60^2} = 62.1\,\text{lb}$$

The free-body diagram of member ac is shown in Fig. 7.58c. For equilibrium of this member,

$$\Sigma M_a = 0 \quad F_{be}(16\sin 68°) - 60(28\cos 68°) - 16.1(28\sin 68°) = 0 \quad F_{be} = 70.6\,\text{lb (C)}$$
$$\Sigma F_y = 0 \quad R_{ay} - 60 = 0 \quad R_{ay} = 60\,\text{lb}$$
$$\Sigma F_x = 0 \quad R_{ax} - F_{be} + 16.1 = 0 \quad R_{ax} = 54.5\,\text{lb}$$
$$R_a = \sqrt{R_{ax}^2 + R_{ay}^2} = \sqrt{54.5^2 + 60^2} = 81.1\,\text{lb}$$

R_e is the force exerted by pin e on member ef. The free-body diagram of pin e is shown in Fig. 7.58d. For equilibrium of this pin,

$$\Sigma F_x = 0 \quad 70.6 - 43.1\cos 68° - R_{ex} = 0 \quad R_{ex} = 54.5\,\text{lb}$$
$$\Sigma F_y = 0 \quad -43.1\sin 68° + R_{ey} = 0 \quad R_{ey} = 40\,\text{lb}$$
$$R_e = \sqrt{R_{ex}^2 + R_{ey}^2} = \sqrt{54.5^2 + 40^2} = 67.6\,\text{lb}$$

Figure 7.58e shows the free-body diagram of member ef. From Fig. 7.58a,

$$l_1 = 16\sin 68° + 6 = 20.8\,\text{in} \quad l_2 = 30 - (16\cos 68° + 10) = 14.0\,\text{in}$$

For equilibrium of member ef,

$$\Sigma M_f = 0 \quad -54.5(20.8) + 40(14) + M = 0 \quad M = 574\,\text{in·lb}$$

(b) From Fig. 7.58e,

$$\Sigma F_x = 0 \quad -R_{fx} + 54.5 = 0 \quad R_{fx} = 54.5\,\text{lb}$$
$$\Sigma F_y = 0 \quad -40 + R_{fy} = 0 \quad R_{fy} = 40\,\text{lb}$$
$$R_f = \sqrt{R_{fx}^2 + R_{fy}^2} = \sqrt{54.5^2 + 40^2} + 67.6\,\text{lb}$$

The forces acting on pins a, b, c, d, and f are

$$R_a = 81.1\,\text{lb} \quad R_b = F_{be} = 70.6\,\text{lb} \quad R_c = 62.1\,\text{lb}$$
$$R_d = F_{de} = 43.1\,\text{lb} \quad R_f = 67.6\,\text{lb}$$

The forces acting on pin e are 70.6 lb, by member be; 43.1 lb, by member de; and 67.6 lb, by member ef.

It may be observed that forces R_e (on member ef) and R_f form a couple of magnitude $M = 574$ in·lb. It is left as an exercise for the reader to show that these two forces have the same directions.

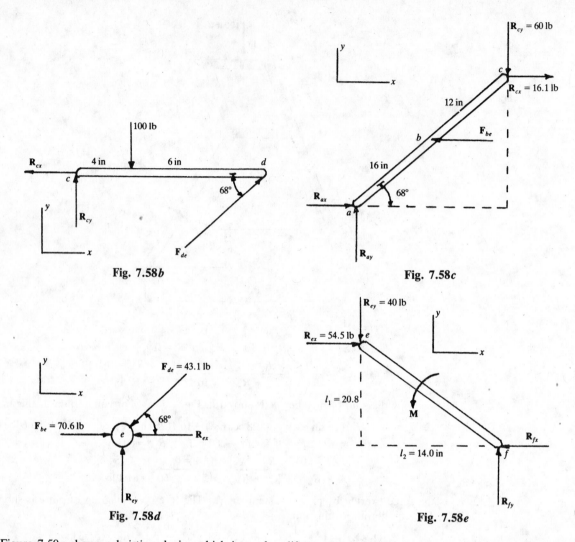

Fig. 7.58b

Fig. 7.58c

Fig. 7.58d

Fig. 7.58e

7.59 Figure 7.59a shows a hoisting device which is used to lift an automobile engine. The engine weighs 625 lb.

(a) For the position shown, find the force on the piston rod and the force of pin b on members ab and bc.

(b) For the position shown, find the reaction forces of the ground on the base of the hoisting device.

(c) What is the force on the piston rod, and the force of pin b on members ab and bc, when member bc is horizontal?

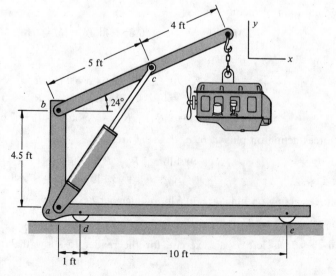

Fig. 7.59a

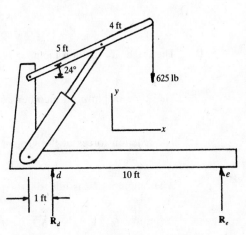

Fig. 7.59b   **Fig. 7.59c**

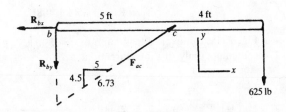

Fig. 7.59d

(a) Figure 7.59b shows the free-body diagram of member bc. The piston-rod assembly, member ac, is a two-force member. From this figure, using the law of cosines,

$$A^2 = 5^2 + 4.5^2 - 2(5)4.5 \cos 114° \qquad A = 7.97 \text{ ft}$$

Using the law of sines,

$$\frac{A}{\sin 114°} = \frac{7.97}{\sin 114°} = \frac{5}{\sin \theta_1} \qquad \theta_1 = 35°$$

For equilibrium of member bc,

$$\Sigma M_b = 0 \qquad (F_{ac} \sin 55°)(5 \cos 24°) - (F_{ac} \cos 55°)(5 \sin 24°) - 625(9 \cos 24°) = 0 \qquad F_{ac} = 2{,}000 \text{ lb}$$
$$\Sigma F_x = 0 \qquad -R_{bx} + F_{ac} \cos 55° = 0 \qquad R_{bx} = 1{,}150 \text{ lb}$$
$$\Sigma F_y = 0 \qquad -R_{by} + F_{ac} \sin 55° - 625 = 0 \qquad R_{by} = 1{,}010 \text{ lb}$$
$$R_b = \sqrt{R_{bx}^2 + R_{by}^2} = \sqrt{1{,}150^2 + 1{,}010^2} = 1{,}530 \text{ lb}$$

(b) Figure 7.59c shows the free-body diagram of the hoisting device. For equilibrium,

$$\Sigma M_d = 0 \qquad -625(9 \cos 24° - 1) + R_e(10) = 0 \qquad R_e = 451 \text{ lb}$$
$$\Sigma F_y = 0 \qquad R_d - 625 + R_e = 0 \qquad R_d = 174 \text{ lb}$$

(c) Figure 7.59d shows the free-body diagram of member bc in the horizontal position. For equilibrium,

$$\Sigma M_b = 0 \qquad \left(\frac{4.5}{6.73} F_{ac}\right)5 - 625(9) = 0 \qquad F_{ac} = 1{,}680 \text{ lb}$$

$$\Sigma F_x = 0 \qquad -R_{bx} + \frac{5}{6.73} F_{ac} = 0 \qquad R_{bx} = 1{,}250 \text{ lb}$$

$$\Sigma F_y = 0 \qquad -R_{by} + \frac{4.5}{6.73} F_{ac} - 625 = 0 \qquad R_{by} = 498 \text{ lb}$$

$$R_b = \sqrt{R_{bx}^2 + R_{by}^2} = \sqrt{1{,}250^2 + 498^2} = 1{,}350 \text{ lb}$$

(*d*) The force acting on pin *c* of the piston rod, for the position shown in Fig. 7.59*a*, is

$$\mathbf{R}_c = (-F_{ac}\cos 55°)\mathbf{i} - (F_{ac}\sin 55°)\mathbf{j}$$
$$= (-2,000\cos 55°)\mathbf{i} - (2,000\sin 55°)\mathbf{j} = -1,150\mathbf{i} - 1,640\mathbf{j} \quad \text{lb}$$

The force of pin *b* on member *bc* is

$$\mathbf{R}_b = -1,150\mathbf{i} - 1,010\mathbf{j} \quad \text{lb}$$

The force of pin *b* on member *ab* is

$$\mathbf{R}_b = 1,150\mathbf{i} + 1,010\mathbf{j} \quad \text{lb}$$

The reaction forces of the ground on the hoisting device are

$$\mathbf{R}_d = 174\mathbf{j} \quad \text{lb} \qquad \mathbf{R}_e = 451\mathbf{j} \quad \text{lb}$$

When member *bc* is horizontal, the force acting on pin *c* of the piston rod is

$$\mathbf{R}_c = \left(-\frac{5}{6.73}F_{ac}\right)\mathbf{i} - \left(\frac{4.5}{6.73}F_{ac}\right)\mathbf{j} = \left[-\frac{5}{6.73}(1.680)\right]\mathbf{i} - \left[\frac{4.5}{6.73}(1,680)\right]\mathbf{j} = -1,250\mathbf{i} - 1,120\mathbf{j} \quad \text{lb}$$

The corresponding value of the force of pin *b* on member *bc* is

$$\mathbf{R}_b = -1,250\mathbf{i} - 498 \quad \text{lb}$$

The force of pin *b* on member *ab*, for the case where member *bc* is horizontal, is

$$\mathbf{R}_b = 1,250\mathbf{i} + 498\mathbf{j}$$

7.60 The hydraulic cylinder in Fig. 7.60*a* can exert a maximum force of 40 kN. The pinhole at *d* is slotted, so that pin *d* can exert a force on the member only in the *x* direction.

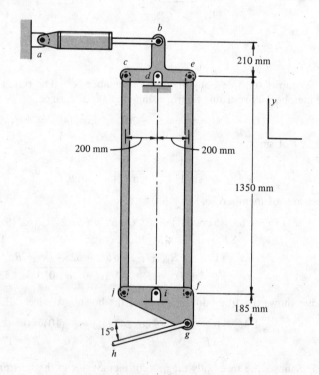

Fig. 7.60*a*

(*a*) What maximum force can be exerted by rod *gh*?

(*b*) Find the forces in members *cj* and *ef*.

(*c*) Find the forces acting on all of the pins.

▮ (*a*) Members *cj* and *ef* are two-force members. Figure 7.60*b* shows the free-body diagram of member *ce*. For equilibrium,

$$\Sigma M_d = 0 \qquad -40(210) + F_{cj}(200) + F_{ef}(200) = 0 \qquad (1)$$

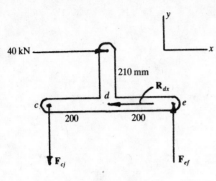

Fig. 7.60b

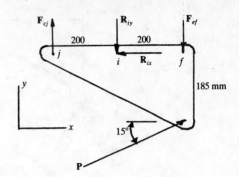

Fig. 7.60c

$$\Sigma F_x = 0 \qquad 40 - R_{dx} = 0 \qquad R_{dx} = 40 \text{ kN}$$
$$\Sigma F_y = 0 \qquad -F_{cj} + F_{ef} = 0 \qquad F_{cj} = F_{ef}$$

The free-body diagram of member jf is shown in Fig. 7.60c. For moment equilibrium,

$$\Sigma M_i = 0 \qquad (P\cos 15°)(185) + (P\sin 15°)(200) - F_{cj}(200) - F_{ef}(200) = 0 \qquad (2)$$

The quantity $F_{cj}(200) + F_{ef}(200)$ is eliminated between Eqs. (1) and (2), with the result

$$(P\cos 15°)(185) + (P\sin 15°)(200) = 40(210) \qquad P = 36.4 \text{ kN}$$

(b) Using Fig. 7.60c,

$$\Sigma F_x = 0 \qquad -R_{ix} + P\cos 15° = 0 \qquad R_{ix} = 36.4\cos 15° = 35.2 \text{ kN}$$
$$\Sigma F_y = 0 \qquad F_{cj} - R_{iy} - F_{ef} + P\sin 15° = 0$$

Using $F_{cj} = F_{ef} = F$,

$$R_{iy} = P\sin 15° = 36.4\sin 15° = 9.42 \text{ kN}$$

From Fig. 7.60c,

$$\Sigma M_j = 0 \qquad -R_{iy}(200) - F_{ef}(400) + (P\cos 15°)(185) + (P\sin 15°)(400) = 0$$
$$-9.42(200) - F(400) + (36.4\cos 15°)(185) + (36.4\sin 15°)(400) = 0 \qquad F = F_{ef} = F_{cj} = 21.0 \text{ kN}$$

F_{cj} is a tensile force, and F_{ef} is a compressive force

(c) The forces acting on the pins are

$$R_a = R_b = 40 \text{ kN} \qquad R_c = F_{cj} = F = 21.0 \text{ kN} \qquad R_d = R_{dx} = 40 \text{ kN}$$
$$R_e = F_{ef} = F = 21.0 \text{ kN} \qquad R_f = F_{ef} = F = 21.0 \text{ kN} \qquad R_g = P = 36.4 \text{ kN}$$
$$R_i = \sqrt{R_{ix}^2 + R_{iy}^2} = \sqrt{35.2^2 + 9.42^2} = 36.4 \text{ kN} \qquad R_j = F_{cj} = F = 21.0 \text{ kN}$$

It may be observed that $R_i = P$. This is an expected result, since the forces F_{cj} and F_{ef} form a couple which exerts no resultant force on member jf. Similarly, the equivalence of R_b and R_d is an expected result.

7.61 Figure 7.61a shows a hinge counterbalance arrangement for use on an automobile hood. Half of the hood weight W is carried by each hinge.

(a) Find the force which must be exerted by the spring to maintain the position shown in the figure, if the weight of the hood is 85 lb.

(b) Find the forces which act on the pins in the assembly.

▌ (a) Figure 7.61b shows the free-body diagram of the assembly consisting of the hood and bracket de. Dimension A is assumed to be small, and the slight curvature of the hood is neglected. Member ef is a two-force member. For equilibrium of the hood,

$$\Sigma M_d = 0 \qquad F_{ef}(3) + (42.5\sin 30°)(1.5) - (42.5\cos 30°)(37.6) = 0 \qquad F_{ef} = 451 \text{ lb}$$
$$\Sigma F_y = 0 \qquad -R_{dy} + F_{ef} - 42.5\cos 30° = 0 \qquad R_{dy} = 414 \text{ lb}$$
$$\Sigma F_x = 0 \qquad R_{dx} - 42.5\sin 30° = 0 \qquad R_{dx} = 21.3 \text{ lb}$$
$$R_d = \sqrt{R_{dx}^2 + R_{dy}^2} = \sqrt{21.3^2 + 414^2} = 415 \text{ lb}$$

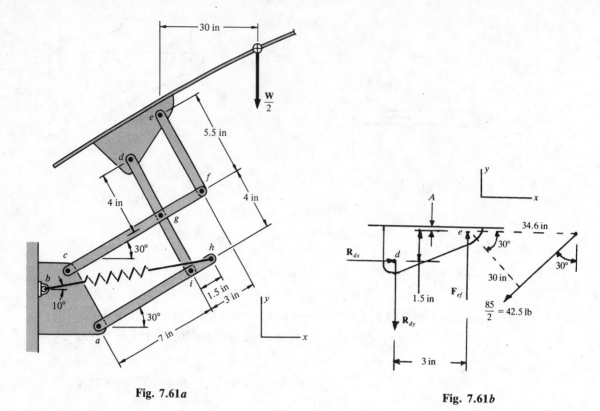

Fig. 7.61a   **Fig. 7.61b**

The free-body diagrams of members di and cf are shown in Figs. 7.61c and d. For equilibrium of member di, from Fig. 7.61c,

$$\Sigma M_i = 0 \qquad -R_{gx}(4) + 21.3(8) = 0 \qquad R_{gx} = 42.6 \text{ lb}$$
$$\Sigma F_x = 0 \qquad -R_{ix} + R_{gx} - 21.3 = 0 \qquad R_{ix} = 21.3 \text{ lb}$$
$$\Sigma F_y = 0 \qquad 414 - R_{gy} + R_{iy} = 0 \qquad (1)$$

From Fig. 7.61d,

$$\Sigma M_c = 0 \qquad R_{gy}(7) - 451(10) = 0 \qquad R_{gy} = 644 \text{ lb}$$

Using the above result in Eq. (1),

$$414 - 644 + R_{iy} = 0 \qquad R_{iy} = 230 \text{ lb}$$
$$R_i = \sqrt{R_{ix}^2 + R_{iy}^2} = \sqrt{21.3^2 + 230^2} = 231 \text{ lb}$$
$$R_g = \sqrt{R_{gx}^2 + R_{gy}^2} = \sqrt{42.6^2 + 644^2} = 645 \text{ lb}$$
$$\Sigma F_y = 0 \qquad -R_{cy} + R_{gy} - 451 = 0 \qquad R_{cy} = 193 \text{ lb}$$
$$\Sigma F_x = 0 \qquad R_{cx} - R_{gx} = 0 \qquad R_{cx} = 42.6 \text{ lb}$$
$$R_c = \sqrt{R_{cx}^2 + R_{cy}^2} = \sqrt{42.6^2 + 193^2} = 198 \text{ lb}$$

Figure 7.61e shows the free-body diagram of member ah. F is the force exerted by the spring on member ah. For equilibrium of member ah, using Fig. 7.61e,

$$\Sigma M_a = 0 \qquad -230(7) + (F \sin 20°)(8.5) = 0 \qquad F = 554 \text{ lb}$$
$$\Sigma F_x = 0 \qquad R_{ax} + 21.3 - 554 \cos 20° = 0 \qquad R_{ax} = 499 \text{ lb}$$
$$\Sigma F_y = 0 \qquad R_{ay} - 230 + 554 \sin 20° = 0 \qquad R_{ay} = 40.5 \text{ lb}$$
$$R_a = \sqrt{R_{ax}^2 + R_{ay}^2} = \sqrt{499^2 + 40.5^2} = 501 \text{ lb}$$

(*b*) The forces acting on the pins are

$$R_a = 501 \text{ lb} \qquad R_c = 198 \text{ lb} \qquad R_d = 415 \text{ lb} \qquad R_e = F_{ef} = 451 \text{ lb}$$
$$R_f = F_{ef} = 451 \text{ lb} \qquad R_g = 645 \text{ lb} \qquad R_h = F = 554 \text{ lb} \qquad R_i = 231 \text{ lb}$$

It may be seen that pin g is the most heavily loaded pin in the assembly.

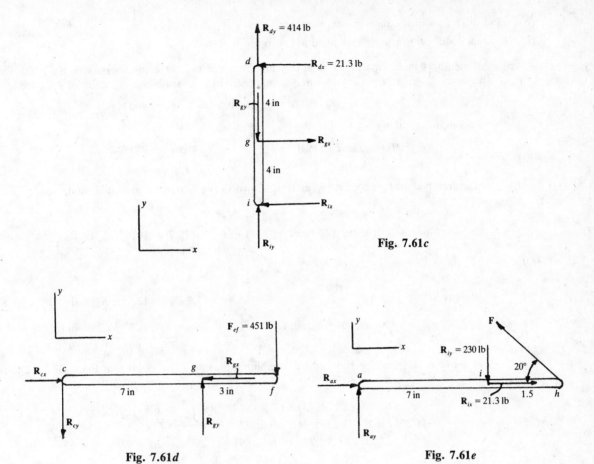

Fig. 7.61c

Fig. 7.61d

Fig. 7.61e

7.62 Figure 7.62a shows a clamping device for holding a workpiece between two jaws. A vertical force P_1 is applied to pin b, resulting in a compressive force P_2 at the slider block. This compressive force holds the workpiece. The guide in which the block moves is assumed to be smooth,

(*a*) If $P_1 = 800$ N for the configuration shown, what is the resulting force P_2?

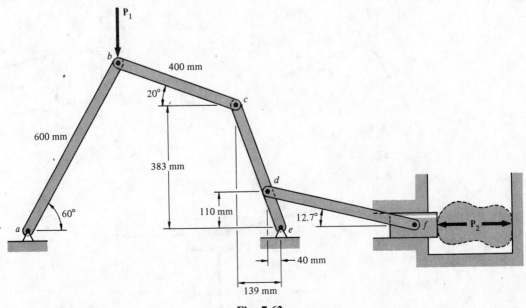

Fig. 7.62a

(b) Find the forces of the pins on the members.

(c) Find the load ratio factor P^*, defined as the ratio of P_2 to P_1.

▌ (a) All members in the machine, except member ce, are two-force members. Figure 7.62b shows the free-body diagram of pin b. A force summation normal to the line of action of F_{ab} yields

$$-800 \sin 30° + F_{bc} \cos 10° = 0 \qquad F_{bc} = 406 \text{ N (C)}$$

Summing forces along F_{ab} yields

$$-800 \cos 30° - F_{bc} \sin 10° + F_{ab} = 0 \qquad -800 \cos 30° - 406 \sin 10° + F_{ab} = 0$$
$$F_{ab} = 763 \text{ N (C)}$$

The free-body diagram of member ce is shown in Fig. 7.62c. For equilibrium,

$$\sum M_e = 0$$

$$-(406 \cos 20°)(383) + (406 \sin 20°)(139) + (F_{df} \cos 12.7°)(110) - (F_{df} \sin 12.7°)(40) = 0$$
$$F_{df} = 1{,}290 \text{ N (C)}$$

$$\sum F_x = 0$$

$$406 \cos 20° - F_{df} \cos 12.7° - R_{ex} = 0 \qquad 406 \cos 20° - (1{,}290) \cos 12.7° - R_{ex} = 0 \qquad R_{ex} = -877 \text{ N}$$

$$\sum F_y = 0 \qquad -406 \sin 20° + F_{df} \sin 12.7° + R_{ey} = 0 \qquad -406 \sin 20° + 1{,}290 \sin 12.7° + R_{ey} = 0$$
$$R_{ey} = -145 \text{ N} \qquad R_e = \sqrt{R_{ex}^2 + R_{ey}^2} = \sqrt{(-877)^2 + (-145)^2} = 889 \text{ N}$$

The above results may be checked by summing moments about point c in Fig. 7.62c. The result is

$$\sum M_c = 0$$
$$-(F_{df} \cos 12.7°)(273) + (F_{df} \sin 12.7°)(99) - R_{ex}(383) + R_{ey}(139) \overset{?}{=} 0$$
$$-(1{,}290 \cos 12.7°)(273) + (1{,}290 \sin 12.7°)(99) - (-877)(383) + (-145)(139) \overset{?}{=} 0$$
$$(1{,}290 \cos 12.7°)(273) + 145(139) \overset{?}{=} (1{,}290 \sin 12.7°)(99) + 877(383)$$
$$3.64 \times 10^5 \equiv 3.64 \times 10^5$$

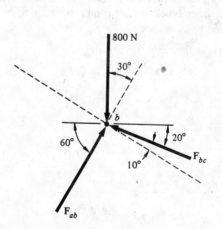

Fig. 7.62b

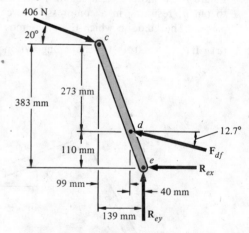

Fig. 7.62c

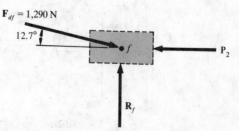

Fig. 7.62d

The free-body diagram of pin f is shown in Fig. 7.62d, and R_f is the normal force of the guide on the block.   For equilibrium of the block,

$$\sum F_x = 0 \qquad 1{,}290 \cos 12.7° - P_2 = 0 \qquad P_2 = 1{,}260 \text{ N}$$

The normal force R_f is found from

$$\sum F_y = 0 \qquad -1{,}290 \sin 12.7° + R_f = 0 \qquad R_f = 284 \text{ N}$$

(*b*)   The pin forces acting on member ab are

$$R_a = R_b = F_{ab} = 763 \text{ N}$$

The pin forces acting on member bc are

$$R_b = R_c = F_{bc} = 406 \text{ N}$$

The pin forces acting on member ce are

$$R_c = F_{bc} = 406 \text{ N} \qquad R_d = F_{df} = 1{,}290 \text{ N} \qquad R_e = 889 \text{ N}$$

The pin forces acting on member df are

$$R_d = R_f = F_{df} = 1{,}290 \text{ N}$$

7.63   Give a summary of the basic concepts of force analysis of plane frames and machines.

❙   In a plane framework both the structure and the applied forces lie in a common plane.   Unlike the plane truss, no restriction is placed on the manner in which the loads may be applied to the structure.   These loads may be applied at the connection points where members are joined or at any point along the member.   These loads may also be distributed along the length of the member, or the loading may be in the form of applied couples.

A structure which is similar to the plane framework is called a machine.   A machine is a structural configuration designed to transmit load by changing the magnitude, direction, or sense of an applied load.   A machine, for any particular configuration, is statically equivalent to a frame with the same geometry and loading as the machine.

The usual first step in the force analysis of a frame is to draw the free-body diagram of the entire frame.   The most general force system which may act on this free-body diagram is a general two-dimensional force system.   The unknown reaction forces of the ground on the frame may be found from the equilibrium conditions

$$\sum F_x = 0 \qquad \sum F_y = 0 \qquad \sum M = 0$$

The members of a frame may be either multiforce members or two-force members.   Free-body diagrams are drawn for each member.   The unknown forces at a pin connection of a multiforce member are represented by two reaction force components with assumed actual senses.   The pin force at connections where a two-force member joins another member is represented by a single force of known direction, unknown magnitude, and assumed actual sense.   In showing the same pin forces on members which are joined with one another, these forces will have opposite senses in the free-body diagrams of the two members.   This effect is due to Newton's third law.   The unknown forces which act on the members may be found by using the equilibrium equations above on each member.

A determination must be made of the distribution of the load among the members if any of the following three situations occurs:

1.   A pin joins three or more members.

2.   A pin joins two or more members and a foundation support.

3.   A load is applied directly to a pin which joins two or more members.

One technique of solution is to assume that the pin belongs to a particular member and forms an integral part of that member.   The second technique is to treat the pin as a separate structural element with its own free-body diagram.   This latter technique is the one usually used in the solution of plane frame problems.

CHAPTER 8
Analysis of Friction Forces

8.1 CHARACTERISTICS OF FRICTION FORCES, COEFFICIENT OF FRICTION, ANGLE OF FRICTION, ANGLE OF REPOSE, IMPENDING MOTION

8.1　　(*a*)　Describe the term *friction force*.

(*b*)　State the general characteristics of friction forces.

(*c*)　Give several examples of desirable and undesirable effects of friction forces.

❚　(*a*)　Friction forces are force effects which are produced when bodies whose surfaces are in contact have actual, or impending, sliding motion relative to one another. The lines of action of these friction forces lie in the plane which is tangent to the bodies at the point of contact.

(*b*)　All friction forces have two general characteristics:

1.　The direction of the friction force is the direction of the actual, or impending, relative sliding motion between the two contacting surfaces. The term direction, as before, is understood to be the orientation of the line of action of the friction force with respect to a known reference line.

2.　The sense of the friction force is *always to oppose* the actual, or impending, relative motion of the two contacting surfaces.

The above two effects are illustrated in Fig. 8.1. Object *A* is pressed against plane *B* by someone's hand. The surface of plane *B* is defined to be *rough*, and this term means that plane *B* will exert a force resistance to any object which tends to slide relative to this plane. Object *A* is made to traverse the dotted path *ab* shown in the figure. The friction forces which oppose the motion of object *A* are shown in Fig. 8.1 as short, straight arrows. These forces all lie in plane *B*. Their directions are at every point tangent to the curved path. Finally, the senses of these friction forces are such as to oppose the motion of object *A* as it traverses the path from *a* to *b*.

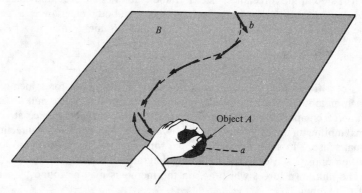

Fig. 8.1

(*c*)　Friction forces may be either useful or detrimental effects. In the case of brakes or clutches, of an automobile tire in contact with a pavement, or of a person walking along the ground, the friction effects are not only useful but also essential to the desired function in these situations. In the case of bearings, or where machine parts slide relative to one another, the effect of friction forces is detrimental since it results in wear of these parts.

8.2　　Figure 8.2*a* shows a block of weight *W* resting on a rough surface. As shown in the free-body diagram of Fig. 8.2*b*, the block is in equilibrium under the influence of the weight force *W* and the normal reaction force *N* of the surface on the block.

In Fig. 8.2*c* a horizontal force *P* is applied to the block. This force is chosen to be sufficiently small so that the block does not slide along the surface. Draw the free-body diagram of the block and find the friction force *F*, and the normal reaction force *N* acting on the block.

❚　The block is in a condition of static equilibrium, and the free-body diagram of this element is shown in Fig. 8.2*d*. The force *N*, as before, is the normal force of the surface on the block. The force *F* is the friction force

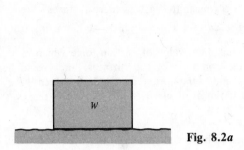

W

Fig. 8.2a

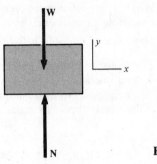

Fig. 8.2b

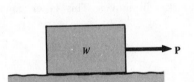

W P

Fig. 8.2c

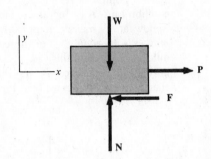

Fig. 8.2d

exerted by the surface on the block.   It lies in the surface, and its sense is such as to oppose the applied force P.   The statements of equilibrium of the block are

$$\sum F_x = 0 \qquad P - F = 0 \qquad F = P$$
$$\sum F_y = 0 \qquad -W + N = 0 \qquad N = W$$

8.3   Describe what happens to the block in Prob. 8.2 as the force P is slowly increased.

❙   If the force P in Fig. 8.2d is gradually increased, a condition will finally be reached where the friction force F of the surface on the block is no longer sufficient to prevent the onset of motion of the block.   The friction force for this limiting condition, *when motion is impending*, is the maximum value of friction force that the surface can exert on the block.   This maximum value of friction force will be designated F_{max}.

Figure 8.3 shows the free-body diagram for the case of impending motion of the block.   The normal force N is the same as in Prob. 8.2.   The equilibrium requirement in the x direction is

$$\sum F_x = 0 \qquad P - F_{max} = 0 \qquad P = F_{max}$$

If the force P is increased beyond the value $P = F_{max}$,   the block will have a resultant force of magnitude $P - F_{max}$ acting on it.   From Newton's second law, the block will accelerate in the positive x sense, and this problem is treated in dynamics.

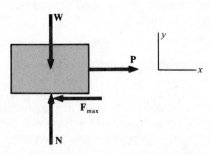

Fig. 8.3

8.4   (*a*)   What is the relationship between the maximum available friction force which may exist between two contacting surfaces and the normal force which is transmitted across these surfaces?

(*b*)   Show the relationship between *P* and *F* in Fig. 8.2*d* as *P* is slowly increased, starting from zero.

❚   (*a*)   It has been found experimentally that the maximum value of friction force which may exist between two contacting surfaces is a function of the condition of the surfaces, interpreted here to be the "roughness" of the materials of the surfaces, and of the normal contact force which these surfaces exert on each other. In equation form, the maximum available friction force is written as

$$F_{max} = \mu_s N$$

In the above equation, the term μ_s is defined to be the *coefficient of static friction*. This coefficient is experimentally determined, and it will be discussed in further detail in Prob. 8.6.

(*b*)   The relationships between the applied force *P* and the friction force *F*, for the block in Fig. 8.2*d*, are shown in Fig. 8.4. The applied force *P* is plotted on the abscissa, and the friction force *F* of the surface on the block is plotted on the ordinate. The situation of Fig. 8.2*d* is characterized by the general point *a* of regime *A*. For this case, $P = F$ and the block is at rest in a condition of static equilibrium. As *P* increases, the point *b* in Fig. 8.4 will be reached. The corresponding friction force is now at its maximum possible value, $F_{max} = \mu_s N$, and the description of this problem is that *motion is impending*. For this case, the block is again at rest and in a state of static equilibrium. The difference between this case and that of operation with the general condition *a* is that for any increase in the applied force *P*, no matter how slight, the block will move. This latter condition is characterized by regime *B* in Fig. 8.4.

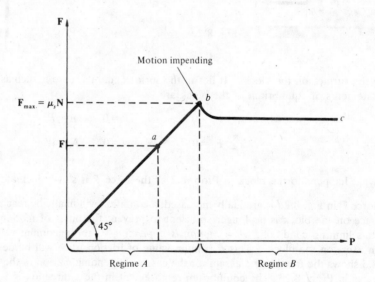

Fig. 8.4

8.5   What is the difference between the coefficients of friction for the case where the block in Prob. 8.2 is stationary and the case where it is moving?

❚   Experimental evidence indicates that once the sliding motion ensues, the coefficient of friction decreases somewhat. A distinction may thus be made between the coefficient of friction when sliding motion is impending and the coefficient of friction when the sliding motion actually occurs. For the former case, the coefficient is defined by $F_{max} = \mu_s N$. For the case of *actual* sliding motion, the friction force $F_{max, \text{sliding}}$ which opposes the sliding motion has the form

$$F_{max, \text{sliding}} = \mu_k N$$

where μ_k is defined to be the *coefficient of kinetic friction*, and

$$\mu_k < \mu_s$$

The effect of this slightly reduced coefficient of friction, and consequently the reduced value of friction force, during sliding motion of the block is illustrated by the typical curve shape *bc* in regime *B* of Fig. 8.4.

The friction force effects defined above are sometimes referred to as *Coulomb* friction.

8.6   (*a*)   What factors affect the value of the coefficient of friction?

        (*b*)   Show several typical values of the coefficient of friction.

   ▎ (*a*)   The value of a coefficient of friction is a function of the materials which comprise the two surfaces that are in contact, of the roughness of these surfaces, of the condition of lubrication which exists between these surfaces, and to some extent of the molecular attraction between these surfaces.

          In mechanical design and analysis, the value of the coefficient of friction is one of the most imprecise data of the problem. Designers may choose a value to work with, but circumstances beyond their control often can alter drastically the effective value of the available friction forces. In the case of an automobile, for example, the braking effect of brake shoes against brake drums, or of brake pads against disk rotors, is reasonably predictable under dry conditions. If, however, the vehicle drives through a deep puddle which wets the braking unit, experience shows that the braking capability may be drastically reduced. This results from the substantial decrease in the effective value of the coefficient of friction from the case of dry sliding surfaces to the case of the same surfaces when wet.

          Because of the general lack of knowledge of the exact value of the coefficient of friction, the values of this coefficient are rarely ever stated with more than two-significant-figure accuracy.

        (*b*)   Values of the coefficient of static friction are tabulated in engineering handbooks for a wide variety of combinations of mating materials with assumed conditions of lubrication. Table 8.1 contains a representative sample of values of coefficients of static friction.

TABLE 8.1
Representative Values of the Coefficient of Static Friction

	dry	lubricated
Steel on steel	0.8	0.16
Steel on brass	0.35	0.19
Steel on graphite	0.1	0.1
Steel on teflon	0.04	0.04
Aluminum on aluminum	1.35	0.3
Wood on wood (dry)	0.2–0.6	
Wood on wood (wet)	0.2	
Leather on wood	0.3–0.4	
Leather on metal (dry)	0.6	
Leather on metal (wet)	0.4	

8.7   (*a*)   Define the term *angle of friction*.

        (*b*)   How is the maximum value of the angle of friction related to the coefficient of friction?

   ▎ (*a*)   The block of Prob. 8.2 is shown in Fig. 8.7*a*. The force components N and F, shown in Fig. 8.7*b*, are the two components of the total reaction force of the surface on the block. An alternative description of this force is shown in Fig. 8.7*c*. R is the magnitude of the *total* reaction force exerted by the surface on the block and ϕ is the angle between this force and the normal to the surface. The system of forces shown in Fig. 8.7*c* is thus a concurrent two-dimensional force system, and the force triangle depicting the equilibrium of the block is shown in Fig. 8.7*d*. The angle ϕ in Figs. 8.7*c* and *d* is referred to as the angle of friction.

        (*b*)   The friction force F in Figs. 8.7*c* and *d* is not the maximum friction force which may exist between the block and the plane, since these forces correspond to the general point *a* in regime *A* of Fig. 8.4. The applied force P is now allowed to increase until the maximum static friction force $F_{max} = \mu_s N$ is reached. The force triangle for this case is shown in Fig. 8.7*e*. From this figure,

$$\tan \phi_s = \frac{\mu_s \cancel{N}}{\cancel{N}} = \mu_s$$

         ϕ_s is defined to be the *angle of static friction*. It is the *maximum* angle which may exist between the normal to the surface and the total reaction force R of the surface on the block. Thus, the angle ϕ_s corresponds to the *maximum* friction force F_{max} which may be exerted by the surface on the block. That is, $F_{max} = \mu_s N$ corresponds to $\phi = \phi_s$. It may also be seen from Fig. 8.7*e* that

$$\phi \le \phi_s$$

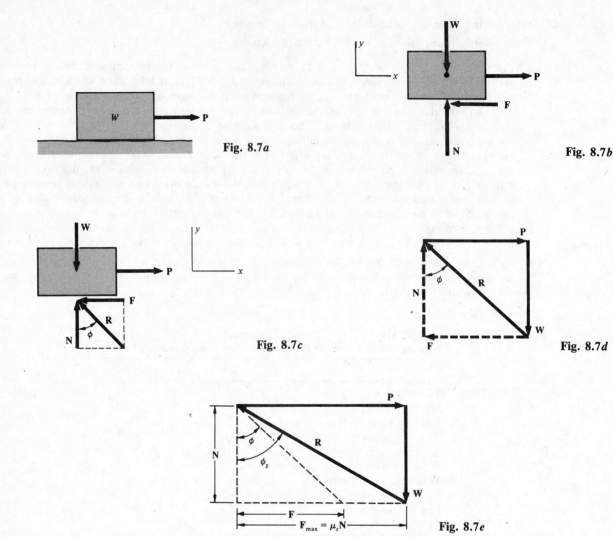

Fig. 8.7a

Fig. 8.7b

Fig. 8.7c

Fig. 8.7d

Fig. 8.7e

A similar derivation, for the case where sliding between the two surfaces occurs, results in

$$\tan \phi_k = \mu_k$$

where ϕ_k is the *angle of kinetic friction*. Since $\mu_k < \mu_s$, it follows that $\phi_k < \phi_s$. All the problems to be considered in this chapter will be of the "motion impending" type. Thus, only the coefficient of static friction μ_s will be used. (For simplicity, this coefficient will be written as $\mu_s = \mu$ in subsequent problems in this chapter.) The reader should be mindful, however, that for cases in which sliding motion actually occurs, the value of the coefficient of friction during the sliding motion will be somewhat less than that which exists when the motion is impending.

8.8 What is the mistake most frequently made in the solution of problems involving friction?

▌ The mistake most frequently made in the solution of problems involving friction is to automatically write the friction force F in the form

$$F = \mu N$$

It should be emphasized that, in the general case, the friction force acting on a body *is found by solving the equations of static equilibrium of the body*. It is *only* in the case of impending, or actual, sliding motion of bodies with respect to one another that the equation

$$F_{max} = \mu N$$

is valid.

8.9 Figure 8.9*a* shows a hinged plane element on which rests a block of weight *W*. For what value of θ will sliding motion of the block down the element be impending?

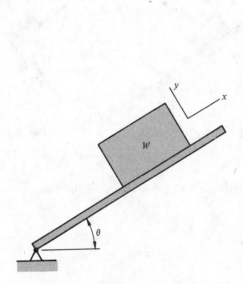

Fig. 8.9a

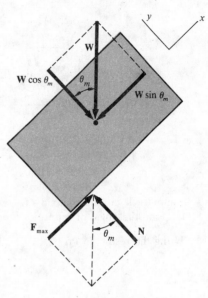

Fig. 8.9b

▮ The free-body diagram of the block is shown in Fig. 8.9*b*. Since the block tends to slide down the plane, the friction force on the block will oppose this motion, and thus this force must act in the positive *x*-coordinate sense. Since motion is assumed to be impending,

$$F_{max} = \mu N$$

The value of θ which corresponds to impending motion will be designated θ_m. The equilibrium requirements of the block are

$$\sum F_x = 0 \qquad -W \sin \theta_m + \mu_s N = 0$$
$$\sum F_y = 0 \qquad -W \cos \theta_m + N = 0$$

N is eliminated between the above two equations, with the result

$$\mu = \tan \theta_m \qquad \theta_m = \tan^{-1} \mu$$

Comparison of the above equations with the result in Prob. 8.7 reveals that, for this case, θ_m is equal to the angle of static friction. It may also be observed that the forces on the block in this problem are a collinear force system, since the block is in equilibrium under the effect of the weight force *W* and the total reaction force *R* of the plane on the block. This effect can be clearly seen in Fig. 8.9*b*.

8.10 Define the term *angle of repose*.

▮ If granular material is poured into a pile on a horizontal surface, it will form a mound which has the approximate shape of a right circular cylinder. Figure 8.10*a* shows such a formation, and Fig. 8.10*b* shows an idealization of one grain of the granular material. The sliding motion of this grain relative to the surface layer of the mound is assumed to be impending, so that the friction force acting on the grain has its maximum value. The free-body diagram of the grain is shown in Fig. 8.10*c*. Comparison of the free-body diagram in this figure with that shown in Fig. 8.9*b* reveals that these two-force systems are identical. It follows that

$$\mu = \tan \theta \qquad \theta = \tan^{-1} \mu$$

θ is referred to as the *angle of repose*. This angle may be used to compute the dimensions of a mound of loose material. Typical average values of the angle of repose are 27° for anthracite coal, 40° for coke of the same size, and 35 to 40° for screened materials. It may also be seen that the angle of repose is equal to ϕ_s, the angle of static friction.

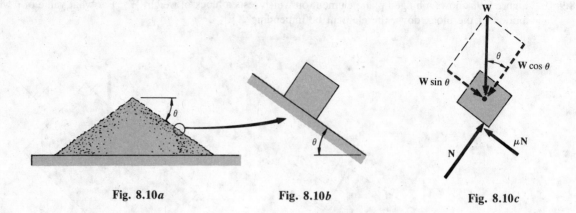

Fig. 8.10a   Fig. 8.10b   Fig. 8.10c

8.11 A dump truck empties a load of gravel onto the ground. The dimensions of the pile are shown in Fig. 8.11. Find the angle of repose of the material.

▮ Using the results in Prob. 8.10 we have

$$\mu = \tan \theta = \frac{2.5}{4} = 0.625 \qquad \theta = \tan^{-1} 0.625 = 32°$$

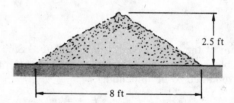

2.5 ft

8 ft

Fig. 8.11

8.12 The ladder of length l shown in Fig. 8.12a rests against a wall. The wall is assumed to be smooth, and μ is the coefficient of friction between the bottom of the ladder and the ground. The weight of the ladder is W. For what value of θ will sliding of the ladder be impending?

▮ The free-body diagram of the ladder is shown in Fig. 8.12b. Since the vertical wall is smooth, it can exert only the normal force N_a on the top of the ladder. Motion is assumed to be impending, so that the friction force of the ground on the bottom of the ladder has its maximum possible value μN_b. The corresponding value of θ is θ_m.

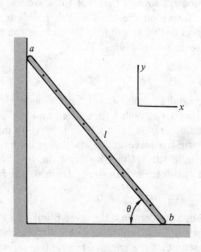

Fig. 8.12a

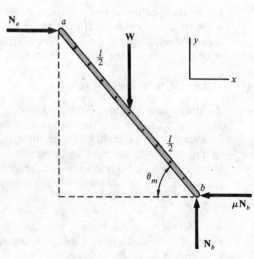

Fig. 8.12b

The equilibrium requirements are

$$\sum F_x = 0 \qquad N_a - \mu N_b = 0 \qquad\qquad (1)$$

$$\sum F_y = 0 \qquad -W + N_b = 0 \qquad\qquad (2)$$

$$\sum M_b = 0 \qquad -N_a(l \sin \theta_m) + W\left(\frac{l}{2} \cos \theta_m\right) = 0 \qquad\qquad (3)$$

The above system of equations contains the three unknowns N_a, N_b, and θ_m. The normal reaction forces are found from Eqs. (1) and (2) to be

$$N_b = W \qquad N_a = \mu N_b = \mu W$$

N_a is used in Eq. (3), and the limiting value of angle which corresponds to impending motion of the ladder is

$$\tan \theta_m = \frac{1}{2\mu_\phi}$$

θ_m given by the above equation is a limiting *minimum* value. For angles greater than this value, the ladder will be in static equilibrium when resting against the wall. The friction force F for this case is given by

$$F = N_\alpha$$

Using this result in Eq. (3), we get

$$F = \frac{W}{2 \tan \theta}$$

Since $\theta > \theta_m$, it follows that $\tan \theta > \tan \theta_m$ and therefore $F < F_{\max}$.

8.13 The ladder of weight W in Fig. 8.13a rests on an ice-coated ground on which the coefficient of friction may be assumed to be zero. The vertical wall is assumed to be smooth. For what range of values of θ will the ladder be in equilibrium?

❚ The free-body diagram of the ladder is shown in Fig. 8.13b. For equilibrium of this element,

$$\Sigma F_x = 0 \qquad N_a = 0$$
$$\Sigma F_y = 0 \qquad N_b - W = 0 \qquad N_b = W$$
$$\Sigma M_b = 0 \qquad -N_a(l \sin \theta) + W\left(\frac{l}{2} \cos \theta\right) = 0 \qquad \frac{Wl}{2} \cos \theta = 0 \qquad \cos \theta = 0 \qquad \theta = 90°$$

It may be seen that the only possible equilibrium position is that in which the ladder is vertical.

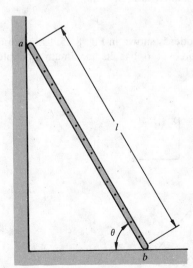

Fig. 8.13a

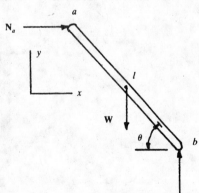

N_b Fig. 8.13b

8.14 A 23-ft ladder weighing 35 lb rests against the side of a building, as shown in Fig. 8.14a. The friction forces between the ladder and the building are assumed to be negligible. If sliding motion is impending in the position shown, find the coefficient of friction between the ladder and the ground.

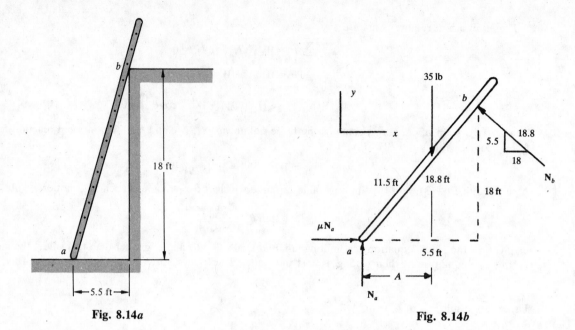

Fig. 8.14a   **Fig. 8.14b**

❚ Figure 8.14b shows the free-body diagram of the ladder. From the similar triangles in this figure,

$$\frac{A}{5.5} = \frac{11.5}{18.8} \qquad A = 3.36 \text{ ft}$$

For equilibrium of the ladder,

$$\Sigma M_a = 0 \qquad -35(3.36) + N_b(18.8) = 0 \qquad N_b = 6.26 \text{ lb}$$

$$\Sigma F_y = 0 \qquad N_a - 35 + \frac{5.5}{18.8} N_b = 0 \qquad N_a = 33.2 \text{ lb}$$

$$\Sigma F_x = 0 \qquad \mu N_a - \frac{18}{18.8} N_b = 0 \qquad \mu = 0.181$$

8.15   (a)   A person of weight P now ascends the ladder of Fig. 8.12a. The weight of the ladder is assumed to be negligible compared with the weight of the person. How far up the ladder, defined by the dimension s_m in Fig. 8.15, may the person climb before sliding motion of the ladder is impending?

   (b)   If $\theta = 55°$, $\mu = 0.35$, and $l = 15$ ft, find the value of s_m.

❚ (a)   The free-body diagram consisting of the person and the ladder is shown in Fig. 8.15. Since motion of the ladder is assumed to be impending, the friction force at b has its maximum possible value

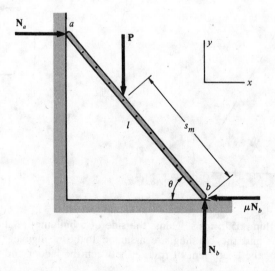

Fig. 8.15

μN_b. The angle θ is assumed to be known from the initial data of the problem. The equilibrium requirements are

$$\sum F_x = 0 \qquad N_a - \mu N_b = 0$$

$$\sum F_y = 0 \qquad -P + N_b = 0$$

$$\sum M_b = 0 \qquad -N_a(l \sin \theta) + P(s_m \cos \theta) = 0$$

The unknowns in the above three equations are N_a, N_b, and s_m. The above equations are solved, and these values are found to be

$$N_a = \mu P \qquad N_b = P \qquad s_m = \mu l \tan \theta$$

The value of s_m in the above equation is a limiting maximum value. For values of s less than s_m, the ladder will be in static equilibrium.

(b) For $\theta = 55°$, $l = 15$ ft, and $\mu = 0.35$,

$$s_m = \mu l \tan \theta = 0.35(15)(\tan 55°) = 7.50 \text{ ft}$$

For the particular numbers used in this case, the limiting position s_m is at the midpoint of the ladder length.

8.16 (a) Find the required coefficient of friction between the ladder in Prob. 8.15 and the ground if the ladder is to remain stationary for any position of the person on the ladder.

(b) Find the numerical value of the result in part (a) if $\theta = 55°$.

(c) The coefficient of friction between the ground and the ladder is now assumed to be 0.45. What is the minimum value θ_m of angle of the ladder with respect to the ground if the ladder must remain in equilibrium for any position of the person on the ladder.

▌ (a) If the ladder is to remain in static equilibrium for *any* position of the person on it, the limiting value s_m must satisfy the inequality

$$\frac{s_m}{l} > 1 \qquad s_m > l$$

The interpretation of this result is that the limiting position of the person is at a location which is greater than the actual length l of the ladder. Using the result $s_m = \mu l \tan \theta$ found in Prob. 8.15, this inequality appears as

$$\frac{s_m}{l} = \mu \tan \theta > 1$$

Thus, if the ladder is not to slide for any position of the person on it, the coefficient of friction between the ladder and the ground must satisfy the inequality

$$\mu > \frac{1}{\tan \theta}$$

(b) If $\theta = 55°$,

$$\mu > \frac{1}{\tan \theta} = \frac{1}{\tan 55°} = 0.700$$

(c) For equilibrium of the ladder, using $\mu = 0.45$,

$$\mu > \frac{1}{\tan \theta_m} \qquad 0.45 > \frac{1}{\tan \theta_m} \qquad \tan \theta_m > \frac{1}{0.45} = 2.22 \qquad \theta_m > 65.8°$$

If the ladder makes an angle with the ground which is equal to or greater than 65.8°, it will be stable for any position of the person on the ladder.

8.17 The automobile in Fig. 8.17a weighs 2,850 lb. The coefficient of friction between the rubber tires and the dry pavement is assumed to be 0.8. For the following three cases, find the maximum grade θ which the vehicle can drive up.

(a) Rear-wheel drive only.

(b) Front-wheel drive only.

(c) Four-wheel drive.

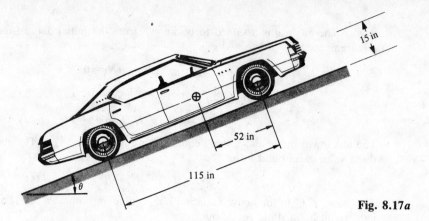

Fig. 8.17a

(a) The free-body diagram of the automobile for the general case is shown in Fig. 8.17b. The normal forces at the front and rear wheels are N_F and N_R, and the associated friction forces are F_F and F_R. The maximum value of angle θ will occur when sliding motion between the tires and the pavement is impending. For the case of rear-wheel drive only,

$$F_R = \mu N_R = 0.8 N_R \quad \text{and} \quad F_F = 0$$

For equilibrium of the automobile,

$$\Sigma M_b = 0 \quad (2{,}850 \cos \theta)(52) + (2{,}850 \sin \theta)(15) - N_R(115) = 0$$
$$\Sigma F_x = 0 \quad 0.8 N_R - 2{,}850 \sin \theta = 0$$

N_R is eliminated between the above two equations, resulting in

$$\tan \theta = 0.404 \quad \theta = 22.0°$$

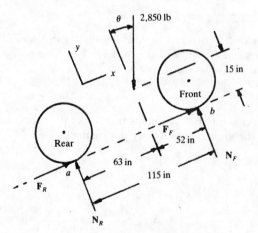

Fig. 8.17b

(b) For the case of front-wheel drive only,

$$F_F = \mu N_F = 0.8 N_F \quad \text{and} \quad F_R = 0$$

For equilibrium,

$$\Sigma M_a = 0 \quad -(2{,}850 \cos \theta)(63) + (2{,}850 \sin \theta)(15) + N_F(115) = 0$$
$$\Sigma F_x = 0 \quad 0.8 N_F - 2{,}850 \sin \theta = 0$$

Elimination of N_F between the above two equations results in

$$\tan \theta = 0.397 \quad \theta = 21.7°$$

(c) For the case where all four wheels drive the automobile,

$$F_R = \mu N_R = 0.8 N_R \quad \text{and} \quad F_F = \mu N_F = 0.8 N_F$$

For equilibrium,

$$\Sigma F_y = 0 \qquad N_R + N_F - 2{,}850 \cos \theta = 0$$
$$\Sigma F_x = 0 \qquad 0.8 N_R + 0.8 N_F - 2{,}850 \sin \theta = 0$$

The quantity $N_R + N_F$ is eliminated between the above two equations, with the final result

$$\tan \theta = 0.8 \qquad \theta = 38.7°$$

8.18   The cylinder in Fig. 8.18a weights 75 N. The coefficient of friction at all sliding surfaces is 0.15.

(a) For what value of P is motion of the cylinder impending?

(b) Express the forces exerted by the wall and ground on the cylinder in formal vector notation.

(c) Find the magnitudes of the forces in part (b).

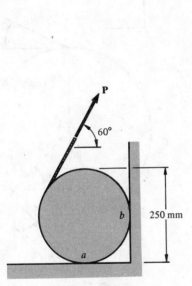

Fig. 8.18a

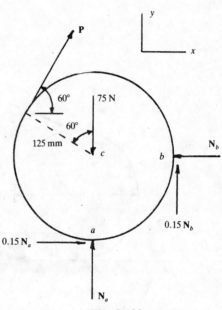

Fig. 8.18b

▌ (a) Figure 8.18b shows the free-body diagram of the cylinder. Since motion is impending, all friction forces have their maximum possible values. For equilibrium of the cylinder,

$$\Sigma M_c = 0 \qquad -P(125) + (0.15 N_b)(125) + (0.15 N_a)(125) = 0 \qquad P = 0.15(N_a + N_b) \qquad (1)$$
$$\Sigma F_y = 0 \qquad P \sin 60° - 75 + N_a + 0.15 N_b = 0 \qquad\qquad (2)$$
$$\Sigma F_x = 0 \qquad P \cos 60° + 0.15 N_a - N_b = 0 \qquad\qquad (3)$$

Eliminating P between Eqs. (1) and (2) results in

$$1.13 N_a + 0.280 N_b - 75 = 0 \qquad (4)$$

P is next eliminated between Eqs. (1) and (3) to obtain

$$0.225 N_a - 0.925 N_b = 0 \qquad N_a = 4.11 N_b \qquad (5)$$

N_a is eliminated between Eqs. (4) and (5), and N_b is found as

$$N_b = 15.2 \text{ N}$$

From Eq. (5),

$$N_a = 4.11 N_b = 4.11(15.2) = 62.5 \text{ N}$$

The value of P is found from Eq. (1) as

$$P = 0.15(N_a + N_b) = 0.15(62.5 + 15.2) = 11.7 \text{ N}$$

(b) The force exerted by the wall on the cylinder is

$$\mathbf{R}_b = -N_b \mathbf{i} + 0.15 N_b \mathbf{j} = -15.2 \mathbf{i} + 0.15(15.2) \mathbf{j} = -15.2 \mathbf{i} + 2.28 \mathbf{j} \qquad \text{N}$$

The force exerted by the ground on the cylinder is

$$\mathbf{R}_a = 0.15 N_a \mathbf{i} + N_a \mathbf{j} = 0.15(62.5)\mathbf{i} + 62.5\mathbf{j} = 9.38\mathbf{i} + 62.5\mathbf{j} \qquad \text{N}$$

(c)
$$R_a = \sqrt{R_{ax}^2 + R_{ay}^2} = \sqrt{9.38^2 + 62.5^2} = 63.2 \qquad \text{N}$$
$$R_b = \sqrt{R_{bx}^2 + R_{by}^2} = \sqrt{(-15.2)^2 + 2.28^2} = 15.4 \qquad \text{N}$$

8.19 The roller in Fig. 8.19a weighs 400 lb. The coefficient of friction is assumed to be the same at all sliding surfaces.

(a) Find the value of P which causes impending pivoting motion of the roller about the edge of the step.

(b) Find the minimum value of μ if the roller is to pivot about the edge rather than slide in the position shown.

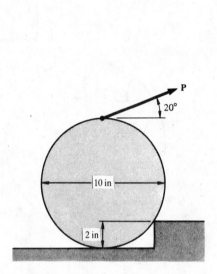

Fig. 8.19a

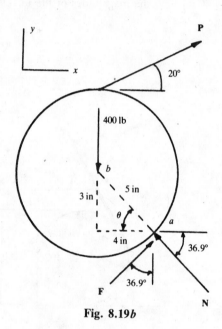

Fig. 8.19b

▌ (a) Figure 8.19b shows the free-body diagram of the roller. The friction force where the roller contacts the step is assumed to be sufficiently large to prevent sliding motion. From Fig. 8.19b,

$$\sin\theta = \frac{3}{5} \qquad \theta = 36.9°$$

For moment equilibrium of the roller,

$$\Sigma M_a = 0 \qquad (-P\sin 20°)(5)\cos 36.9° - (P\cos 20°)(5 + 5\sin 36.9°) + 400(5\cos 36.9°) = 0 \qquad P = 180 \text{ lb}$$

(b) For equilibrium of the roller,

$$\Sigma M_b = 0 \qquad (-P\cos 20°)5 + F(5) = 0 \qquad F = 180\cos 20° = 169 \text{ lb}$$
$$\Sigma F_y = 0 \qquad P\sin 20° - 400 + F\cos 36.9° + N\sin 36.9° = 0$$
$$180\sin 20° - 400 + 169\cos 36.9° + N\sin 36.9° = 0 \qquad N = 339 \text{ lb}$$

For impending sliding motion of the roller on the edge of the step, the friction force has its maximum value, given by

$$F_{\text{max}} = \mu N$$

Using the values of F and N found above,

$$\mu = \frac{F}{N} = \frac{169}{339} = 0.499$$

If $\mu \geq 0.499$, the roller will not slide on the edge of the step.

8.20 The roller in Fig. 8.20a weighs 18 lb. The coefficient of friction on all sliding surfaces is assumed to be the same. When $P = 10$ lb, motion of the roller is impending. Find μ.

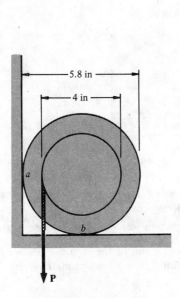

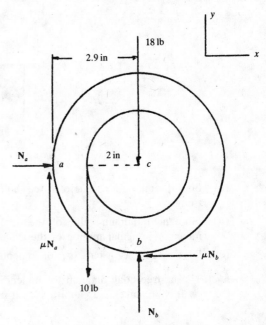

Fig. 8.20a   Fig. 8.20b

❚ The free-body diagram of the roller is shown in Fig. 8.20b. Since motion is impending, both friction forces have their maximum values. The senses of these friction forces are such as to oppose the impending counterclockwise motion of the roller. For equilibrium of the roller,

$$\Sigma M_c = 0 \qquad -(\mu N_a)2.9 + 10(2) - (\mu N_b)(2.9) = 0 \qquad (1)$$
$$\Sigma F_x = 0 \qquad N_a - \mu N_b = 0 \qquad (2)$$
$$\Sigma F_y = 0 \qquad \mu N_a - 10 - 18 + N_b = 0 \qquad (3)$$

Eliminating N_a between Eqs. (2) and (3) results in

$$N_b = \frac{28}{1 + \mu^2}$$

Using Eq. (2),

$$N_a = \mu N_b = \frac{28\mu}{1 + \mu^2}$$

The above values of N_a and N_b are now substituted into Eq. (1), with the result

$$61.2\mu^2 + 81.2\mu - 20 = 0 \qquad (4)$$

Equation (4) is a quadratic equation of the form

$$ax^2 + bx + c = 0$$

the solution to which is

$$x = \frac{-b \pm \sqrt{b^2 - 4ac}}{2a}$$

Thus the solution to Eq. (4) is

$$\mu = \frac{-81.2 \pm \sqrt{81.2^2 - 4(61.2)(-20)}}{2(61.2)} = \frac{-81.2 \pm 107}{2(61.2)}$$

The negative root of the above equation is discarded, and

$$\mu = \frac{-81.2 + 107}{2(61.2)} = 0.211$$

8.21   The cylinder in Fig. 8.21a weighs 5 lb. The coefficients of friction at surfaces a and b are 0.15 and 0.22, respectively.

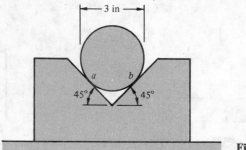

Fig. 8.21a

(*a*). Find the minimum value of force required to cause impending sliding motion of the cylinder out of the trough.

(*b*) Find the minimum value of the applied couple, acting in the plane of the figure, which will cause impending angular motion of the cylinder in a clockwise sense.

(*c*) Do the same as in part (*b*) for impending angular motion in a counterclockwise sense.

| (*a*) The free-body diagram for impending sliding motion out of the trough is shown in Fig. 8.21*b*. The friction forces $\mu_a N_a$ and $\mu_b N_b$ act in the z direction. From the symmetry of the loading,

$$N_a = N_b = N$$

For equilibrium,

$$\Sigma F_y = 0 \quad 2N \cos 45° - 5 = 0 \quad N = 3.54 \text{ lb}$$
$$\Sigma F_z = 0$$
$$P - \mu_a N_a - \mu_b N_b = 0 \quad P - 0.15(3.54) - 0.22(3.54) = 0 \quad P = 1.31 \text{ lb}$$

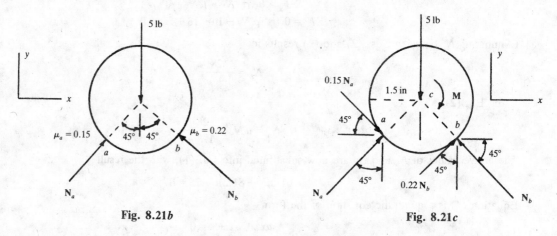

Fig. 8.21b   **Fig. 8.21c**

(*b*) Figure 8.21*c* shows the free-body diagram for impending clockwise motion of the cylinder. For equilibrium of the cylinder,

$$\Sigma F_x = 0 \quad 0.15N_a \cos 45° + N_a \sin 45° + (0.22N_b)(\sin 45°) - N_b \cos 45° = 0 \quad N_a = 0.678N_b \quad (1)$$
$$\Sigma F_y = 0 \quad -(0.15N_a)(\sin 45°) + N_a \cos 45° - 5 + (0.22N_b)(\cos 45°) + N_b \sin 45° = 0 \quad (2)$$

N_a from Eq. (1) is used in Eq. (2) with the result

$$N_a = 2.67 \text{ lb} \quad \text{and} \quad N_b = 3.94 \text{ lb}$$

M is the couple required to cause impending angular motion of the cylinder. For moment equilibrium,

$$\Sigma M_c = 0 \quad -M + (0.15N_a)(1.5) + (0.22N_b)(1.5) = 0$$
$$M = 0.15(2.67)(1.5) + 0.22(3.94)(1.5) = 1.90 \text{ in} \cdot \text{lb}$$

(*c*) For the case of impending counterclockwise motion, the friction forces $0.15N_a$ and $0.22N_b$, and the couple M, have senses opposite those shown in Fig. 8.21c. For equilibrium, using the above changes in Fig. 8.21c,

$$\Sigma F_x = 0 \quad -(0.15N_a)(\cos 45°) + N_a \sin 45° - (0.22N_b)(\sin 45°) - N_b \cos 45° = 0 \quad N_a = 1.44N_b$$
$$(3)$$

$$\Sigma F_y = 0 \quad (0.15N_a)(\sin 45°) + N_a \cos 45° - 5 - (0.22N_b)(\cos 45°) + N_b \sin 45° = 0 \quad (4)$$

N_a from Eq. (3) is used in Eq. (4) to obtain

$$N_a = 4.18 \text{ lb} \quad N_b = 2.90 \text{ lb}$$

For moment equilibrium,

$$\Sigma M_c = 0 \quad M - (0.15N_a)(1.5) - (0.22N_b)(1.5) = 0$$
$$M = 0.15(4.18)(1.5) + 0.22(2.90)(1.5) = 1.90 \text{ in} \cdot \text{lb}$$

It is interesting to note that the same magnitude of applied couple is required for both senses of rotation. The normal and friction forces, however, are significantly different in the two cases.

8.22   A block rests on a floor, as shown in Fig. 8.22a. The block has a mass of 300 kg, and the coefficient of friction between the block and the floor is assumed to be 0.35. The block is to be moved to the right by sliding it along the floor. Someone can either push on the block at corner a or pull on it at corner b, with the directions shown in Fig. 8.22a. Which point of load application would require less effort to cause impending motion of the block?

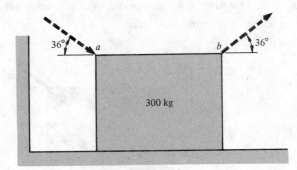

36° a b 36°

300 kg

Fig. 8.22a

❚ The free-body diagram of the block for the case of pushing on the block at corner a is shown in Fig. 8.22b. The weight W of the block is found from

$$W = mg = 300(9.81) = 2,940 \text{ N}$$

Since motion is assumed to be impending, the friction force is given by

$$F_{max} = \mu N = 0.35 N$$

For static equilibrium of the block,

$$\Sigma F_x = 0 \quad P_a \cos 36° - 0.35N = 0$$

$$\Sigma F_y = 0 \quad -P_a \sin 36° - 2,940 + N = 0$$

N is eliminated between the above two equations, with the result

$$P_a = 1,710 \text{ N}$$

The free-body diagram of the block, for application of the load at corner b, is shown in Fig. 8.22c. The equilibrium requirements are

$$\Sigma F_x = 0 \quad P_b \cos 36° - 0.35N = 0$$

$$\Sigma F_y = 0 \quad P_b \sin 36° - 2,940 + N = 0$$

N is eliminated, with the final result

$$P_b = 1,010 \text{ N}$$

It may be seen that a larger force is required if the applied load is a compressive force at corner a. The reason for this is that, in this latter case, the y component of the applied force increases the normal force N, and

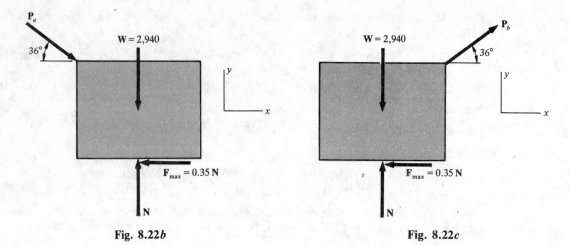

Fig. 8.22b   Fig. 8.22c

consequently the associated friction force. For the case of an applied tensile load at corner b, the y component of this force tends to reduce the normal force, and with it the friction force.

8.23 The block of Prob. 8.22 now has a tensile force P applied at corner b, as shown in Fig. 8.23a.

(a) For what value of angle θ will the force required to cause impending motion of the block be minimum?

(b) What is the magnitude of this minimum force?

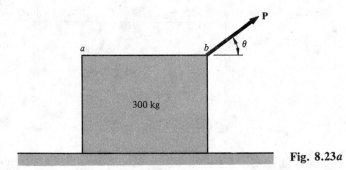

Fig. 8.23a

❚ (a) The free-body diagram of the block is shown in Fig. 8.23b. The body is in equilibrium when acted on by the applied force P, the weight force of 2,940 N, and the reaction force R of the floor on the block. ϕ_s is the angle of static friction. The force triangle representation of equilibrium is shown in Fig. 8.23c. P', P'', and P''' are three possible solutions for the force P. The minimum value of P will occur when the lines of action of P and R are perpendicular to each other, as shown by the force P'' in Fig. 8.23c. The force triangle will then be as shown in Fig. 8.23d.

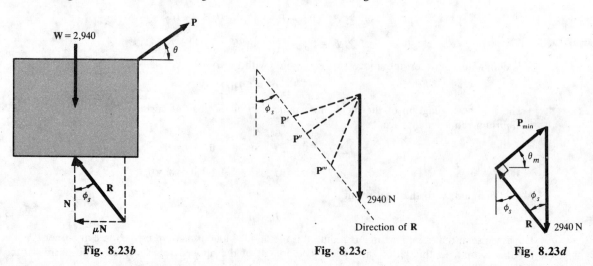

Fig. 8.23b   Fig. 8.23c   Fig. 8.23d

The angle of static friction ϕ_s, for the case of motion impending, is defined by

$$\tan \phi_s = \mu \qquad \tan \phi_s = 0.35 \qquad \phi_s = 19.3° = \theta_m$$

(b) The minimum force P_{min} is then found from Fig. 8.23d as

$$P_{min} = 2.940 \sin \phi_s = 2,940 \sin 19.3° = 972 \text{ N}$$

8.24 The crate shown in Fig. 8.24a has a mass of 580 kg.

(a) If $P = 6,000$ N, find the magnitude and sense of the friction force which acts on the crate.

(b) What value of P will cause the crate to have impending sliding motion up the plane?

(c) Find the minimum value of P required to keep the crate from sliding down the plane.

(d) For what range of values of P will the crate remain in the equilibrium position shown in Fig. 8.24a?

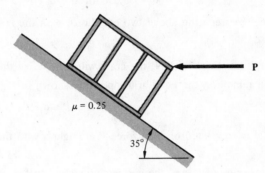

Fig. 8.24a

▮ (a) The weight W of the crate is

$$W = mg = 580(9.81) = 5,690 \text{ N}$$

The free-body diagram of the crate is shown in Fig. 8.24b. N is the normal force of the plane on the crate, and F is the friction force acting on the crate. For equilibrium of the crate,

$$\Sigma F_x = 0 \qquad F + 5,690 \sin 35° - 6,000 \cos 35° = 0 \qquad F = 1,650 \text{ N}$$
$$\Sigma F_y = 0 \qquad N - 5,690 \cos 35° - 6,000 \sin 35° = 0 \qquad N = 8,100 \text{ N}$$

The maximum possible value of the friction force is given by

$$F_{max} = \mu N = 0.25(8,100) = 2,030 \text{ N}$$

It may be seen that for this case the value of the friction force acting on the crate is *less than* its maximum possible value.

(b) The free-body diagram of the crate, for impending sliding motion up the plane, is seen in Fig. 8.24c. For this case the friction force acting on the crate has its maximum value, given by $F_{max} = \mu N$. This friction force has a sense in Fig. 8.24c which opposes the impending motion of the crate. For equilibrium,

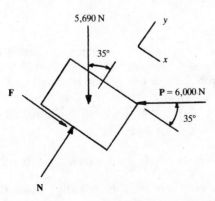

Fig. 8.24b

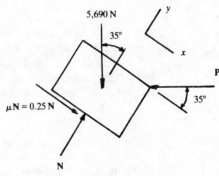

Fig. 8.24c

$$\Sigma F_y = 0 \qquad N - 5{,}690 \cos 35° - P \sin 35° = 0$$
$$\Sigma F_x = 0 \qquad 5{,}690 \sin 35° - P \cos 35° + 0.25N = 0$$

N is eliminated between the above two equations, with the result

$$P = 6{,}550 \, \text{N}$$

The above value of P is the maximum force that may be applied to the crate without causing this element to have motion up the plane.

(c) The free-body diagram in Fig. 8.24c represents the case of impending motion down the plane if the sense of the friction force is changed. The result for $\Sigma F_y = 0$ is the same as in part (b), given by

$$\Sigma F_y = 0 \qquad N - 5{,}690 \cos 35° - P \sin 35° = 0$$

For equilibrium in the x direction,

$$\Sigma F_x = 0 \qquad 5{,}690 \sin 35° - P \cos 35° - 0.25N = 0$$

N is eliminated between the above two equations, with the result

$$P = 2{,}180 \, \text{N}$$

For values of P less than 2,180 N, the crate will slide down the plane.

(d) The crate will remain in the equilibrium position shown in Fig. 8.24a if the applied force P satisfies

$$2{,}180 \, N \leq P \leq 6{,}550 \, \text{N}$$

8.25 The crate of Prob. 8.24 is now acted on by a force P at angle α with the horizontal direction as shown in Fig. 8.25a.

(a) Find the minimum value of P, and the corresponding value of α, required to keep the crate from sliding down the inclined surface.

(b) What is the minimum value of P, and the corresponding value of α, which will cause the crate to have impending sliding motion up the plane?

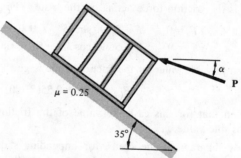

Fig. 8.25a

▌ (a) The free-body diagram of the crate is shown in Fig. 8.25b. R is the total reaction force of the plane on the crate, and motion is assumed to be impending. The angle ϕ_s of static friction is found using the definition

$$\tan \phi_s = \mu = 0.25 \qquad \phi_s = 14.0°$$

The crate is in equilibrium when acted on by the weight force W, the reaction force R (with the components N and μN), and the applied force P.

The triangle law of force addition (see Prob. 2.3) for this case is shown in Fig. 8.25c. The applied force P may have any of the typical values shown as P, P', and P'' in the figure. P will be minimum when its direction is normal to the direction of R. Thus, from Fig. 8.25c,

$$\alpha = 35° - \phi_s = 35° - 14° = 21°$$

$$\sin(35° - \phi_s) = \sin \alpha = \frac{P}{5{,}690} \qquad P = 5{,}690 \sin \alpha = 5{,}690 \sin 21° \qquad P = 2{,}040 \, \text{N}$$

The minimum required value of P to keep the crate from sliding down the plane is $P = 2{,}040 \, \text{N}$, and the corresponding value of the angle α is 21°.

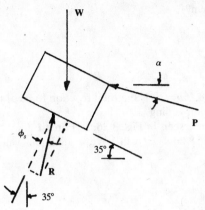

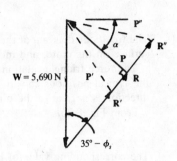

Fig. 8.25b

Fig. 8.25c

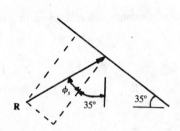

Fig. 8.25d

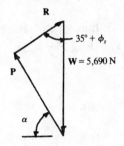

Fig. 8.25e

(b) For the case of impending motion up the plane, force R has the direction shown in Fig. 8.25d and forces W and P have the directions shown in Fig. 8.25b. The force triangle for this case is seen in Fig. 8.25e, where P is shown as a minimum value. As before, $\phi = 14.0°$. From Fig. 8.25e,

$$\alpha = 35° + \phi_s = 35° + 14° = 49°$$

$$\sin(35° + \phi_s) = \sin\alpha = \frac{P}{5,690} \qquad P = 5,690\sin\alpha = 5,690\sin 49° = 4,290 \text{ N}$$

The minimum required value of P to cause the crate to have impending sliding motion up the plane is $P = 4,290$ N, and the corresponding value of angle α is 49°.

8.26 Two people are to push the crate in Fig. 8.26a up the incline. One person exerts a horizontal force of 80 lb on the left corner. The second person pulls on the right corner with a tensile force P. Force P is at an angle α with the plane, and the coefficient of friction between the crate and the incline is 0.14.

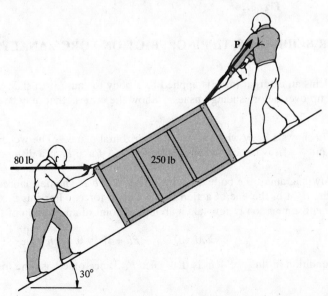

Fig. 8.26a

(*a*) What should angle α be if the force P is to be minimum?

(*b*) Find the value of this minimum force.

▌ (*a*) Using the definition of the angle of static friction,

$$\tan \phi_s = \mu = 0.14 \qquad \phi_s = 8°$$

The free-body diagram of the crate is shown in Fig. 8.26*b*. R is the total reaction force of the inclined surface on the crate, and motion is assumed to be impending.

The force triangle relationships for this problem are seen in Fig. 8.26*c*. Force P may have any of the typical positions shown as P, P', or P'' in Fig. 8.26*c*. P will have a minimum value when its direction is normal to the direction of R.

From Fig. 8.26*b*,

$$\beta = 30° + 8° = 38°$$

The corresponding value of α when P is minimum is then found from Fig. 8.26*c* as

$$30° + \alpha = \beta = 38° \qquad \alpha = 8°$$

It may be observed that the condition for force P to have a minimum value is given by $\alpha = \phi_s$.

(*b*) Using Fig. 8.26*c*,

$$A = \frac{80}{\tan 38°} = 102 \text{ lb} \qquad B = 250 - A = 250 - 102 = 148 \text{ lb}$$

$$P = B \sin \beta = 148 \sin 38° = 91.1 \text{ lb}$$

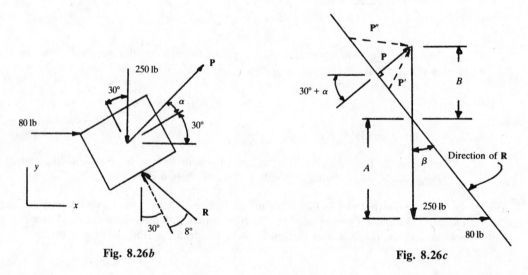

Fig. 8.26*b*   Fig. 8.26*c*

8.2 CRITERIA FOR SLIDING OR TIPPING, FRICTION FORCE ANALYSIS FOR CONNECTED BODIES

8.27   In many problems an external force is applied to a body to cause it to slide along a surface. In certain cases, the body may tip over before sliding ensues. Show the criteria that may be used to predict whether sliding or tipping occurs.

▌ Figure 8.27*a* shows a body which rests on a horizontal plane. The weight of the body is W, and it has the dimensions a and b. μ is the coefficient of friction. A force P, at a distance h above the plane, is applied to the body.

The free-body diagram of the body is shown in Fig. 8.27*b*. The normal reaction force N is assumed to act at distance c to the right of the line of action of the weight force. For this first case, the body is assumed not to slide. A moment summation is now taken about the point of application of the normal force N. The result is

$$\Sigma M = 0 \qquad -Ph + Wc = 0 \qquad h = \frac{Wc}{P}$$

A necessary condition in this problem is that $h \le b$, since otherwise the force P will not physically be acting on the body.

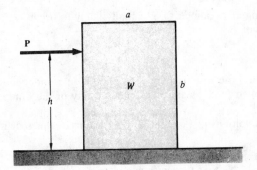

Fig. 8.27a

If P and W are assumed to be constants in a given problem, then the above equation shows that

$$c \sim h$$

That is, as the force P is applied at larger and larger values of h, the location of the line of action of the normal force N moves to the right. A limiting condition occurs when $c = a/2$. For this case, the normal force acts at the right-hand edge of the body, and the free-body diagram for this situation is shown in Fig. 8.27c. The body is in a condition of impending tipping motion about the right-hand corner, and the corresponding value of h is designated h_{max}.

For moment equilibrium about this corner, for the case of impending tipping,

$$\Sigma M = 0 \qquad -Ph_{max} + W\frac{a}{2} = 0$$

$$h_{max} = \frac{Wa}{2P} \qquad (1)$$

Again, a necessary condition is that $h_{max} \leq b$, since the force P must be acting on the body.

The condition for sliding will be determined next. If sliding motion of the body is assumed to be impending, the friction force F_{max} is given by

$$F_{max} = \mu N$$

The equilibrium requirements of the body are

$$\sum F_x = 0 \qquad P - \mu N = 0$$

$$\sum F_y = 0 \qquad -W + N = 0$$

The force P is then found to be

$$P = \mu W \qquad (2)$$

P in the above equation is the value of applied force which is required to cause impending sliding motion of the body.

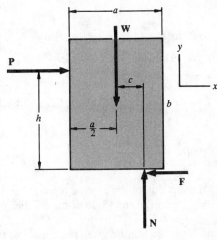

Fig. 8.27b

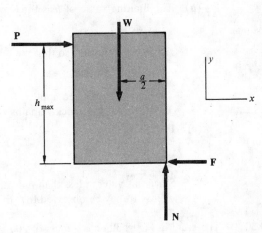

Fig. 8.27c

If sliding and tipping are assumed to be equally likely to occur, then W may be eliminated between Eqs. (1) and (2), with the result

$$h_{max} = \frac{a}{2\mu}$$

It may be observed that the result in the above equation is independent of the weight W and height b of the body, and of the applied force P, and is a function only of the dimension a and the coefficient of friction.

8.28 Show how, for the conditions in Prob. 8.27, five distinct regimes of motion may be identified.

> ❚ The five possible distinct regimes of motion are:

1. $h < h_{max}, P < \mu W$ Neither tipping nor sliding occurs, and the body remains at rest.
2. $h < h_{max}, P = \mu W$ The body does not tip, and sliding motion is impending.
3. $h < h_{max}, P > \mu W$ The body does not tip but sliding motion, with increasing velocity, occurs.
4. $h = h_{max}, P = \mu W$ Both sliding and tipping motion are impending, and the occurrence of either situation is equally likely. The results for this case are independent of the weight and height of the body.
5. $h > h_{max}$ The body will tip over for any value of μ, with one or the other of the following conditions. If $\mu \geq P/W$ or $P \leq \mu W$, the onset of tipping motion will occur with the tipping edge remaining stationary with respect to the surface. Otherwise, with $\mu < P/W$ or $P > \mu W$, the onset of tipping motion will be accompanied by the onset of sliding motion of the edge along the surface.

8.29 (a) For the body shown in Fig. 8.27a,

$$a = 12 \text{ in} \qquad b = 18 \text{ in} \qquad h = 15 \text{ in}$$
$$W = 100 \text{ lb} \qquad \mu = 0.24 \qquad P = 24 \text{ lb}$$

Will the body slide, tip, or remain at rest?

(b) The body described in part (a) now rests on a rougher surface, where the coefficient of friction is 0.35. Discuss the possible motions of the body.

(c) The block shown in Fig. 8.27a has the dimension $a = 6 \text{ in}$ and the coefficient of friction is 0.35. The remaining data are the same as in part (a) of this problem. Show that the block will tip over when acted on by the applied force P.

> ❚ (a) From Eq. (1) in Prob. 8.27,

$$h_{max} = \frac{Wa}{2P} = \frac{100(12)}{2(24)} = 25 \text{ in}$$

$h_{max} > b$, so that the force cannot act physically on the body. Thus, the body will not tip over. The maximum available friction force is found from

$$F_{max} = \mu W = 0.24(100) = 24 \text{ lb}$$

Comparison of this value with that of the applied force P reveals that these two forces are equal. Thus, the body is in a state of impending sliding motion with no tipping.

(b) The limiting value of h is the same as in part (a);

$$h_{max} = 25 \text{ in}$$

Again, $h_{max} > b$, so that the body will not tip over. The value of the available friction force F_{max} is

$$F_{max} = \mu W = 0.35(100) = 35 \text{ lb}$$

The magnitude of the applied force P is 24 lb. This value is less than the maximum friction force of 35 lb, so that the block remains at rest without tipping.

(c) From Eq. (1) in Prob. 8.27,

$$h_{max} = \frac{Wa}{2P} = \frac{100(6)}{2(24)} = 12.5 \text{ in}$$

Since this value is less than the value $h = 15 \text{ in}$, the location of the applied force P, the block will tip over. It may be observed that this conclusion is independent of the value of the coefficient of friction.

8.30 The crate in Fig. 8.30a has a mass of 200 kg, and the coefficient of friction between the crate and the ground is 0.35.

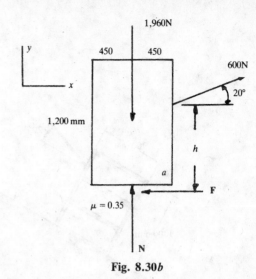

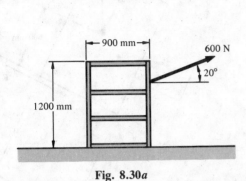

Fig. 8.30a

Fig. 8.30b

(a) Find the magnitude of the friction force which acts on the crate.

(b) Find the maximum height above the plane at which the 600-N force may be applied, if tipping is not to occur.

▌ (a) The weight W of the crate is

$$W = mg = 200(9.81) = 1,960 \text{ N}$$

The free-body diagram of the crate is shown in Fig. 8.30b. The height of the force above the plane is given by dimension h. For force equilibrium of the crate,

$$\Sigma F_x = 0 \qquad -F + 600 \cos 20° = 0 \qquad F = 564 \text{ N}$$
$$\Sigma F_y = 0 \qquad N + 600 \sin 20° - 1,960 = 0 \qquad N = 1,750 \text{ N}$$

The maximum value of the friction force that may exist between the crate and the ground is found from

$$F_{\text{max}} = \mu N = 0.35(1,750) = 613 \text{ N}$$

Since $F < F_{\text{max}}$, the magnitude of the friction force which acts on the crate is 564 N.

(b) When tipping motion is impending, the reaction force of the ground on the crate is located at point a in Fig. 8.30b. For moment equilibrium,

$$\Sigma M_a = 0 \qquad 1,960(450) - (600 \cos 20°)h = 0 \qquad h = 1,560 \text{ mm}$$

At this height, the force of 600 N does not act on the crate. Thus, the crate will not tip.

8.31 The crate of Prob. 8.30 is now placed on a ramp, as shown in Fig. 8.31a, and $\mu = 0.35$.

(a) Find the value of P for impending sliding motion up the plane.

(b) For what range of heights above the plane may this force be applied if tipping of the crate is not to occur?

(c) If the force P is removed, will the crate tip over in a counterclockwise sense?

(d) If the force P is removed, will the crate slide down the ramp?

▌ (a) Figure 8.31b shows the free-body diagram of the crate. For impending sliding motion up the plane, with $F = F_{\text{max}} = \mu N = 0.35 N$,

$$\Sigma F_y = 0 \qquad -1,960 \cos 30° + P \sin 20° + N = 0$$
$$\Sigma F_x = 0 \qquad -1,960 \sin 30° - 0.35N + P \cos 20° = 0$$

Eliminating N between the two equations results in

$$P = 1,490 \text{ N}$$

(b) When clockwise tipping motion is impending, the reaction force of the ramp on the crate is at edge a in Fig. 8.31b. For equilibrium,

$$\Sigma M_a = 0 \qquad -(P \cos 20°)h + (1,960 \sin 30°)(600) + (1,960 \cos 30°)(450) = 0 \qquad h = 965 \text{ mm}$$

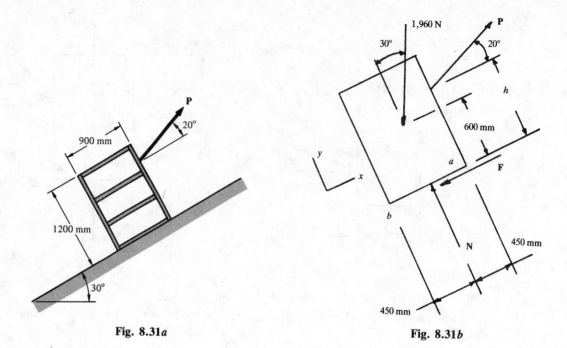

Fig. 8.31a   Fig. 8.31b

If $h < 965$ mm, the crate will not tip.

(c) For counterclockwise tipping, with force P removed from the crate, the reaction force of the ramp on the crate is located at edge b in Fig. 8.31b. The summation of moments about edge b is given by

$$\Sigma M_b = -(1{,}960 \cos 30°) \frac{450}{1{,}000} + (1{,}960 \sin 30°) \frac{600}{1{,}000} = -176 \text{ N} \cdot \text{m}$$

The negative value of resultant moment about edge b means that this moment has a clockwise sense. The interpretation of this result is that the reaction force of the ramp on the crate is to the right of edge b, and the crate will not tip in a counterclockwise sense.

(d) Using Fig. 8.31b, with force P removed and the friction force acting up and to the right,

$$\Sigma F_y = 0 \quad -1{,}960 \cos 30° + N = 0 \quad N = 1{,}700 \text{ N}$$
$$\Sigma F_x = 0 \quad -1{,}960 \sin 30° + F = 0 \quad F = 980 \text{ N}$$

F is the required friction force acting on the crate to keep this body in equilibrium. The maximum available value of friction force is given by

$$F_{\max} = \mu N = 0.35(1{,}700) = 595 \text{ N}$$

Since $F > F_{\max}$, the crate will slide down the ramp if the force P is removed.

8.32 (a) For what range of values of θ can the block in Fig. 8.32a slide without tipping?

(b) What is the range of values of coefficient of friction which corresponds to the answer to part (a)?

(c) Find the numerical values of these results if $a = 90$ min, $b = 135$ mm, $m = 6$ kg, and $P = 22$ N.

▌ (a) For the limiting case of tipping, the reaction force of the ground on the block acts at the right edge of the block. Figure 8.32b shows the free-body diagram for this case. F is the friction force exerted by the ground on the block. For moment equilibrium,

$$\Sigma M_c = 0 \quad mg \frac{a}{2} - (P \cos \theta)b = 0 \quad \cos \theta = \frac{mga}{2Pb}$$

If $\cos \theta \le mga/2Pb$, the block will not tip.

(b) When sliding motion is impending, or occurs, $F = F_{\max} = \mu N$. For force equilibrium,

$$\Sigma F_y = 0 \quad P \sin \theta - mg + N = 0$$
$$\Sigma F_x = 0 \quad P \cos \theta - \mu N = 0$$

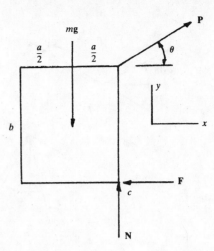

Fig. 8.32b

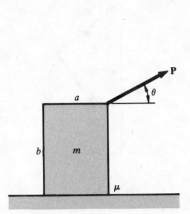

Fig. 8.32a

N is eliminated between the above two equations to obtain

$$\mu = \frac{P \cos \theta}{mg - P \sin \theta}$$

If $\mu \leq (P \cos \theta)/(mg - P \sin \theta)$, the block will be in a state of impending, or sliding, motion.

(c) From part (a), for the limiting case of tipping,

$$\cos \theta = \frac{mga}{2Pb} = \frac{6(9.81)(90)}{2(22)(135)} = 0.892 \qquad \theta = 26.9°$$

From part (b), for sliding motion, and using the limiting value of θ found above,

$$\mu = \frac{P \cos \theta}{mg - P \sin \theta} = \frac{22 \cos 26.9°}{6(9.81) - 22 \sin 26.9°} \qquad \mu = 0.401$$

If $\theta \leq 26.9°$, the block may slide without tipping. The corresponding range of values of μ is $\mu \leq 0.401$.

8.33 (a) Find the value of the angle α in Fig. 8.33a if the force P required to cause impending sliding motion of the block down the plane is to be minimum.

(b) Find the value of this minimum force.

(c) Is it possible that tipping motion of the block would occur before sliding occurred if the values of P and α found in parts (a) and (b) were used?

▌ (a) The free-body diagram of the block is shown in Fig. 8.33b, and R is the resultant reaction force of the plane on the block. The force triangle of the three forces acting on the block is shown in Fig. 8.33c. In order to make force P have a minimum value, the direction of P is chosen to be perpendicular to the line of action of R. From the definition of the angle of friction,

$$\tan \phi_s = \mu = 0.4 \qquad \phi_s = 21.8°$$

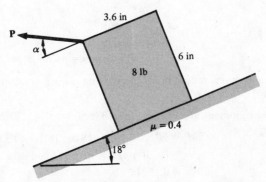

Fig. 8.33a

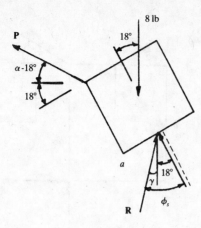

Fig. 8.33b

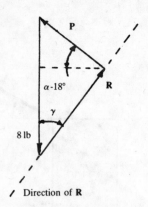

Fig. 8.33c

From Fig. 8.33b,

$$\gamma = \phi_s - 18° = 21.8° - 18° = 3.8°$$

Using Fig. 8.33c,

$$\gamma = \alpha - 18° \qquad \alpha = \gamma + 18° = 3.8° + 18° = 21.8°$$

(b) From Fig. 8.33c,

$$P = 8 \sin \gamma = 8 \sin 3.8° = 0.530 \, \text{lb}$$

The minimum force required for impending sliding motion of the block down the plane is $P = 0.530$ lb, with $\alpha = 21.8°$.

(c) For tipping to occur, the reaction force of the plane on the block must act at edge a in Fig. 8.33b. The *resultant* moment about a for this case, using the values of P and α found in part (b), is

$$\Sigma M_a = (P \cos \alpha)(6) - (8 \cos 18°)(1.8) + (8 \sin 18°)(3) = (0.530 \cos 21.8°)(6)$$
$$- (8 \cos 18°)(1.8) + (8 \sin 18°)(3) = -3.33 \, \text{in} \cdot \text{lb}$$

Since the resultant moment about a is negative, this moment acts in a clockwise sense. Thus, there is a net clockwise moment tending to keep the block in the position shown in Fig. 8.33a, and the block cannot tip for the values of P and α found in parts (a) and (b).

8.34   Solve Prob. 8.33 if the coefficient of friction between the block and the plane is 0.5.

▌ (a) Using the solution in Prob. 8.33,

$$\tan \phi = \mu = 0.5 \qquad \phi = 26.6°$$

From Fig. 8.33b,

$$\gamma = \phi_s - 18° = 8.6°$$

Using Fig. 8.33c,

$$\gamma = \alpha - 18° \qquad \alpha = \gamma + 18° = 26.6°$$

(b) From Fig. 8.33c,

$$P = 8 \sin \gamma = 8 \sin 8.6° = 1.20 \, \text{lb}$$

The minimum force required for impending sliding motion of the block down the plane is $P = 1.20$ lb, with $\alpha = 26.6°$.

(c) The resultant moment about a in Fig. 8.33c, for $P = 1.20$ lb and $\alpha = 26.6°$, is

$$\Sigma M_a = (P \cos \alpha)(6) - (8 \cos 18°)(1.8) + (8 \sin 18°)(3)$$
$$= (1.20 \cos 26.6°)(6) - (8 \cos 18°)(18) + (8 \sin 18°)(3)$$
$$= 0.159 \, \text{in} \cdot \text{lb}$$

Since the resultant moment about a is positive, the sense of this moment is counterclockwise, and there is a net counterclockwise moment tending to tip the block. Thus, the block would tip before sliding motion—with the values of P and α found in parts (a) and (b)—could occur.

8.35   (a)   A right circular cone rests on a rough inclined surface, as shown in Fig. 8.35a. The weight force acts through point a. For what value of θ would tipping of the cone be impending?

     (b)   What would be the required value of the coefficient of friction if the cone were to tip before it slid?

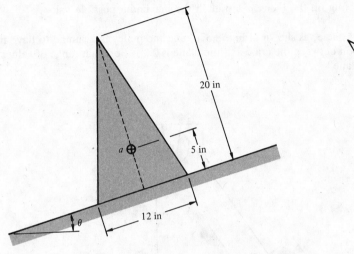

Fig. 8.35a

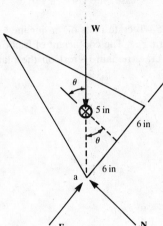

Fig. 8.35b

▎(a)   Figure 8.35b shows the free-body diagram of the cone. Tipping is impending when the line of action of the weight force passes through point a in Fig. 8.35b. For this condition,

$$\tan \theta = \tfrac{6}{5} \qquad \theta = 50.2°$$

An alternative solution to this problem is to sum the moments about point a, with the form

$$\Sigma M_a = 0 \qquad -(W \cos \theta)(6) + (W \sin \theta)(5) = 0 \qquad \tan \theta = \tfrac{6}{5} \qquad \theta = 50.2°$$

   (b)   For impending sliding motion, a cone on an inclined plane is equivalent to a block on an inclined plane (Prob. 8.9). Thus,

$$\tan \theta = \mu \qquad \tfrac{6}{5} = \mu \qquad \mu = 1.2$$

The required value of μ is $\mu \geq 1.2$ if the cone is to tip before it slides.

8.36   The block in Fig. 8.36a weights 50 lb and is in equilibrium in the position shown. The coefficient of friction between the block and the plane is 0.4. Find the value of the friction force acting on the block.

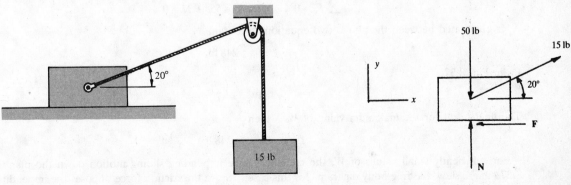

Fig. 8.36a Fig. 8.36b

▌ The free-body diagram of the block is seen in Fig. 8.36b. F is the friction force exerted by the ground on the block. For equilibrium,

$$\Sigma F_x = 0 \quad -F + 15 \cos 20° = 0 \quad F = 14.1 \text{ lb}$$

The above value of friction force will now be compared with the maximum available friction force. For equilibrium in the y direction,

$$\Sigma F_y = 0 \quad N + 15 \sin 20° - 50 = 0 \quad N = 44.9 \text{ lb}$$
$$F_{max} = \mu N = 0.4(44.9) = 18.0 \text{ lb}$$

For this case, the friction force acting on the block is less than the maximum possible value.

8.37 A 250-lb crate rests on an inclined plane, as shown in Fig. 8.37a, and the pulley is assumed to have frictionless bearings. The coefficient of friction between the crate and the plane is 0.3. For what range of values of W will the crate remain at rest on the plane?

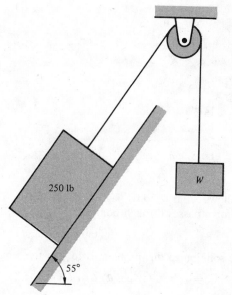

250 lb

55°

Fig. 8.37a

▌ If the magnitude of W is sufficiently large, the crate will slide up the plane. At the other extreme, for sufficiently small values of W, the crate will tend to slide down the plane. The case of maximum W will be considered first. For this case, the impending motion of the crate is up the plane. The free-body diagrams of the crate and the weight are shown in Figs. 8.37b and c, and T_1 is the cable tensile force. For equilibrium of the crate,

$$\Sigma F_x = 0 \quad T_1 - 250 \sin 55° - 0.3N = 0$$
$$\Sigma F_y = 0 \quad -250 \cos 55° + N = 0$$

N_1 is eliminated between the above two equations to obtain

$$T_1 = 248 \text{ lb}$$

From Fig. 8.37c,

$$W_{max} = T_1$$

The final result for the maximum value of W is then

$$W_{max} = 248 \text{ lb}$$

For sufficiently small values of W, the crate will have impending sliding motion down the plane. Figures 8.37d and e show the free-body diagrams for this case. Since the friction force opposes the impending motion, the sense of this force is in the positive x-coordinate sense.

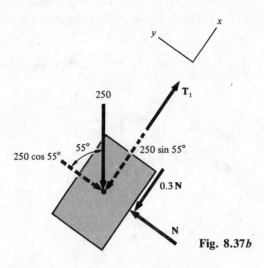

Fig. 8.37b

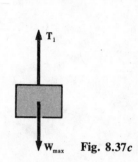

Fig. 8.37c

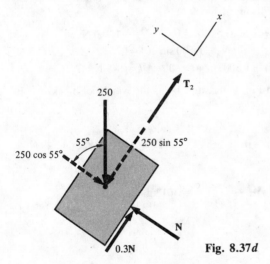

Fig. 8.37d

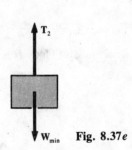

Fig. 8.37e

The equilibrium requirements of the crate are

$$\sum F_x = 0 \qquad T_2 - 250 \sin 55° + 0.3N = 0$$
$$\sum F_y = 0 \qquad -250 \cos 55° + N = 0$$

The cable tensile force is found from the above equations to be

$$T_2 = 162 \text{ lb}$$

The free-body diagram of Fig. 8.37e is used, and the minimum value of W is

$$W_{\min} = T_2 = 162 \text{ lb}$$

If the supported weight W is in the range

$$162 \text{ lb} \le W \le 248 \text{ lb}$$

the crate will remain at rest on the plane. It should be emphasized that the friction force which acts on the crate will be given by an equation of the form

$$F_{\max} = \mu N$$

only if $W \le 162$ lb or $W \ge 248$ lb. For any other values of W, the friction force may be found by solving the equilibrium equations of the crate.

8.38 What is the minimum value of the mass of block B required to maintain the equilibrium configurations shown in Fig. 8.38a?

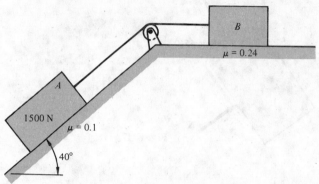

Fig. 8.38a

▮ The minimum value of the mass of block B corresponds to impending downward sliding motion of block A. The free-body diagram of block A for this condition is shown in Fig. 8.38b, and T is the tensile force exerted by the cable on block A. For equilibrium of block A,

$$\Sigma F_y = 0 \qquad N_A - 1{,}500 \cos 40° = 0 \qquad N_A = 1{,}150 \text{ N}$$
$$\Sigma F_x = 0 \qquad T + 0.1 N_A - 1{,}500 \sin 40° = 0$$
$$T + 0.1(1{,}150) - 1{,}500 \sin 40° = 0 \qquad T = 849 \text{ N}$$

The free-body diagram of block B is seen in Fig. 8.38c, and W_B is the weight of block B. For equilibrium of this element,

$$\Sigma F_y = 0 \qquad N_B - W_B = 0 \qquad N_B = W_B$$
$$\Sigma F_x = 0 \qquad -849 + 0.24 N_B = 0 \qquad -849 + 0.24 W_B = 0 \qquad W_B = 3{,}540 \text{ N}$$

The mass of block B is then found as

$$m_B = \frac{W_B}{g} = \frac{3{,}540}{9.81} = 361 \text{ kg}$$

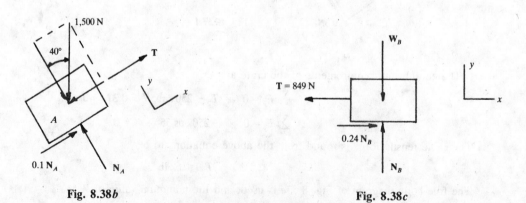

Fig. 8.38b   **Fig. 8.38c**

8.39 For what range of values of weight of block A will the system in Fig. 8.39a remain in equilibrium?

▮ Figure 8.39b shows the free-body diagram for the case of impending downward motion of block A, and T is the cable tensile force acting on the blocks. For equilibrium of block A,

$$\Sigma F_y = 0 \qquad N_A - W_A \cos 55° = 0 \qquad N_A = 0.574 W_A$$
$$\Sigma F_x = 0 \qquad T + 0.2 N_A - W_A \sin 55° = 0 \qquad T = 0.704 W_A$$

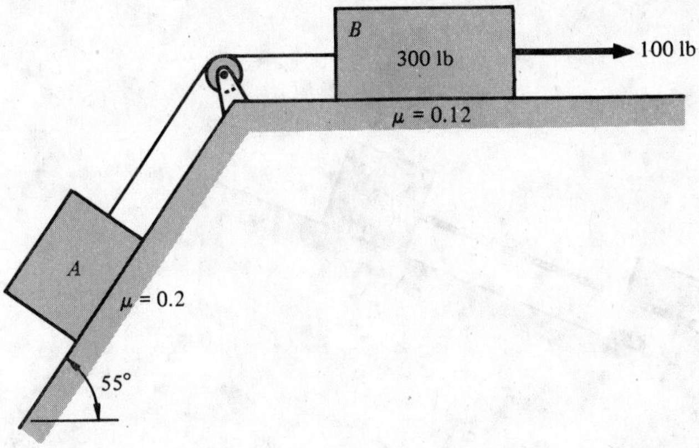

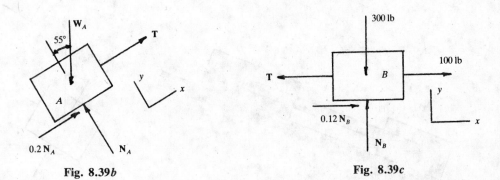

Fig. 8.39a

The free-body diagram of block B is shown in Fig. 8.39c. For equilibrium of this block,

$$\Sigma F_y = 0 \qquad N_B - 300 = 0 \qquad N_B = 300 \text{ lb}$$
$$\Sigma F_x = 0 \qquad 100 + 0.12 N_B - T = 0 \qquad 100 + 0.12(300) - 0.704 W_A = 0 \qquad W_A = 193 \text{ lb}$$

The free-body diagrams of the two blocks, for the case of impending motion of block A up the plane, are the same as in Figs. 8.39b and c, with the senses of the two friction forces changed. The normal forces N_A and N_B are the same as before. The equilibrium requirements for block A are

$$\Sigma F_x = 0 \qquad T - 0.2 N_A - W_A \sin 55° = 0$$
$$T - 0.2(0.574 W_A) - W_A \sin 55° = 0 \qquad T = 0.934 W_A$$

The equilibrium requirements for block B are

$$\Sigma F_x = 0 \qquad 100 - 0.12 N_B - T = 0$$
$$100 - 0.12(300) - 0.934 W_A = 0 \qquad W_A = 68.5 \text{ lb}$$

If W_A is between 68.5 and 193 lb, the system will remain in the equilibrium position shown in Fig. 8.39a.

Fig. 8.39b

Fig. 8.39c

8.40 What must be the relationship between the coefficients of friction of the two blocks on the inclined surface shown in Fig. 8.40a if the cable force is to be always tensile when motion is impending?

❚ The free-body diagrams of the two blocks are shown in Fig. 8.40b.
The equilibrium requirements for block A are

$$\Sigma F_y = 0 \qquad -W_A \cos \theta + N_A = 0 \qquad N_A = W_A \cos \theta$$
$$\Sigma F_x = 0 \qquad T - W_A \sin \theta + \mu_A N_A = 0 \qquad T = W_A (\sin \theta - \mu_A \cos \theta) \qquad (1)$$

The equilibrium requirements for block B are

$$\Sigma F_y = 0 \qquad -W_B \cos \theta + N_B = 0 \qquad N_B = W_B \cos \theta$$
$$\Sigma F_x = 0 \qquad -T - W_B \sin \theta + \mu_B N_B = 0 \qquad T = W_B (\mu_B \cos \theta - \sin \theta) \qquad (2)$$

The cable tensile force T must always be positive.

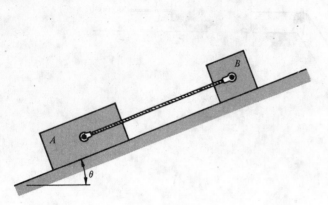

Fig. 8.40a

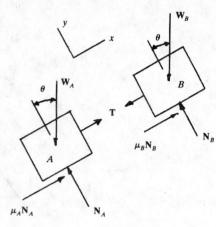

Fig. 8.40b

For $T > 0$, from Eq. (1)

$$\sin \theta > \mu_A \cos \theta \qquad \mu_A < \tan \theta \qquad (3)$$

For $T > 0$, from Eq. (2)

$$\mu_B \cos \theta > \sin \theta \qquad \mu_B > \tan \theta \qquad (4)$$

From Eqs. (3) and (4),

$$\mu_B > \mu_A$$

is the necessary condition for the cable force to be always tensile.

8.41 Block A in Fig. 8.41a weighs 10 lb, and block B weighs 40 lb. The coefficients of friction between blocks A and B are 0.1 and 0.45, respectively. The link may be assumed to be weightless. Find the resultant friction force which acts on the system of blocks.

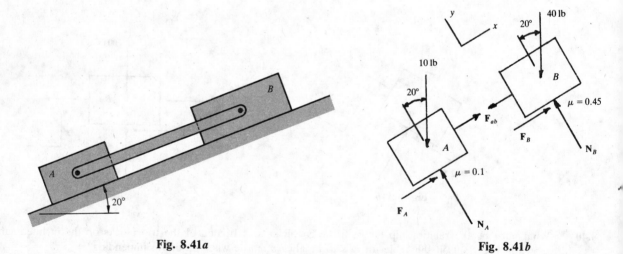

Fig. 8.41a

Fig. 8.41b

▌ Figure 8.41b shows the free-body diagrams of the two blocks. F_A and F_B are the friction forces which act on blocks A and B, respectively, and F_{ab} is the force exerted by the link on the blocks. For equilibrium of block A,

$$\Sigma F_y = 0 \qquad -10 \cos 20° + N_A = 0 \qquad N_A = 9.40 \text{ lb}$$
$$\Sigma F_x = 0 \qquad -10 \sin 20° + F_A + F_{ab} = 0 \qquad (1)$$

For equilibrium of block B,

$$\Sigma F_y = 0 \qquad -40 \cos 20° + N_B = 0 \qquad N_B = 37.6 \text{ lb}$$
$$\Sigma F_x = 0 \qquad -40 \sin 20° - F_{ab} + F_B = 0 \qquad (2)$$

Equations (1) and (2) are added, with the result

$$F_A + F_B = 50 \sin 20° = 17.1 \text{ lb}$$

The term $F_A + F_B$ is the resultant friction force acting on the system of two blocks. The maximum value of this resultant friction force is given by

$$F_{max} = \mu_A N_A + \mu_B N_B = 0.1(9.40) + 0.45(37.6) = 17.9 \text{ lb}$$

Since the actual resultant friction force $(F_A + F_B)$ is less than F_{max}, the final answer to the problem is

$$F_A + F_B = 17.1 \text{ lb}$$

8.42 The link in Fig. 8.42a is weightless. Block A has a mass of 185 kg, and block B has a mass of 65 kg. Block C rests on block B, as shown in the figure. What is the minimum value of the mass of block C needed to maintain the equilibrium configuration shown in the figure?

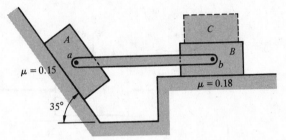

Fig. 8.42a

❚ The minimum value of the mass of block C corresponds to impending sliding motion of block A down the plane. The link is a two force member, and the force F_{ab} exerted by the link on block A has the direction of the link. The free-body diagram of block A is shown in Fig. 8.42b. For equilibrium of block A,

$$\Sigma F_y = 0 \qquad -1{,}810 + N_A \cos 35° + 0.15 N_A \sin 35° = 0 \qquad N_A = 2{,}000 \text{ N}$$
$$\Sigma F_x = 0 \qquad N_A \sin 35° - 0.15 N_A \cos 35° - F_{ab} = 0 \qquad F_{ab} = 901 \text{ N (C)}$$

(As previously, the letter C after the value of a force acting on a two-force member indicates that the force is compressive; the letter T, that it is tensile.)

The free-body diagram of block B is shown in Fig. 8.42c. W_C is the weight force of block C, and the friction force $0.18 N_B$ acts in a sense which opposes impending rightward motion of the block. For equilibrium of block B,

$$\Sigma F_x = 0 \qquad F_{ab} - 0.18 N_B = 0 \qquad 901 - 0.18 N_B = 0 \qquad N_B = 5{,}010 \text{ N}$$
$$\Sigma F_y = 0 \qquad -W_C - 638 + N_B = 0$$
$$W_C = 4{,}370 \text{ N} \qquad m_C = \frac{W_C}{g} = \frac{4{,}370}{9.81} = 446 \text{ kg}$$

If $m_C \geq 446 \text{ kg}$, the system will remain in the equilibrium position shown in Fig. 8.42a.

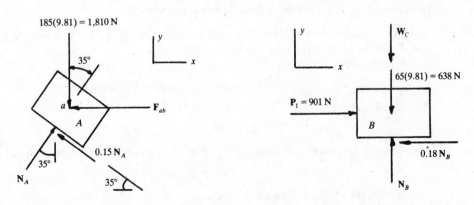

Fig. 8.42b   Fig. 8.42c

8.43 Block C is removed from the system shown in Prob. 8.42, and a horizontal force P is applied to block B. For what range of values of P will the system remain in the equilibrium configuration shown in the figure?

▌ For downward impending motion of block A, P is minimum. For this case, F_{ab} has the same value found in Prob. 8.42, given as $F_{ab} = 901\,\mathrm{N}\,(C)$. Figure 8.43 shows the free-body diagram of block B. For equilibrium of this element,

$$\Sigma F_y = 0 \qquad -638 + N_B = 0 \qquad N_B = 638\,\mathrm{N}$$
$$\Sigma F_x = 0 \qquad F_{ab} - 0.18N_B - P = 0 \qquad 901 - 0.18(638) - P = 0 \qquad P = 786\,\mathrm{N}$$

For impending upward sliding motion of block A, P has a maximum value. The free-body diagrams of blocks A and B are the same as those shown in Figs. 8.42b and 8.43, with the senses of the two friction forces $0.15N_A$ and $0.18N_B$ changed.

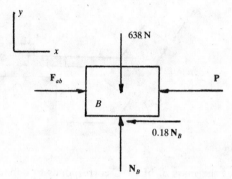

Fig. 8.43

For equilibrium of block A,

$$\Sigma F_y = 0 \qquad -1,810 + N_A \cos 35° - 0.15N_A \sin 35° = 0 \qquad N_A = 2,470\,\mathrm{N}$$
$$\Sigma F_x = 0 \qquad -F_{ab} + N_A \sin 35° + 0.15N_A \cos 35° = 0 \qquad F_{ab} = 1,720\,\mathrm{N}\,(C)$$

For equilibrium of block B,

$$\Sigma F_y = 0 \qquad -638 + N_B = 0 \qquad N_B = 638\,\mathrm{N}$$
$$\Sigma F_x = 0 \qquad F_{ab} + 0.18N_B - P = 0 \qquad 1,720 + 0.18(638) - P = 0 \qquad P = 1,830\,\mathrm{N}$$

The system will remain in the equilibrium configuration shown in Fig. 8.42a if the applied force P is in the range

$$786\,\mathrm{N} \le P \le 1,830\,\mathrm{N}$$

8.44 For the system shown in Fig. 8.44a, $W_A = W_B = 5\,\mathrm{lb}$ and the weight of the link is assumed to be negligible.

(*a*) For what value of θ is motion of the system impending?

(*b*) Express the forces acting on the link in formal vector notation.

▌ (*a*) The free-body diagrams of blocks A and B are shown in Fig. 8.44b. The link is a two-force member, and F_{ab} is the force exerted by the link on the blocks.

For equilibrium of block A,

$$\Sigma F_y = 0 \qquad -5 + F_{ab} \sin \theta = 0 \qquad F_{ab} \sin \theta = 5$$

For equilibrium of block B,

$$\Sigma F_y = 0 \qquad -5 - F_{ab} \sin \theta + N_B = 0$$

The quantity $(F_{ab} \sin \theta)$ is eliminated between the above two equations, with the result

$$-5 - 5 + N_B = 0 \qquad N_B = 10\,\mathrm{lb}$$

For equilibrium of block B in the x direction,

$$\Sigma F_x = 0 \qquad F_{ab} \cos \theta - 0.15N_B = 0$$

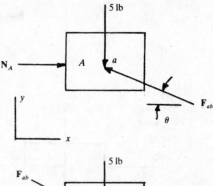

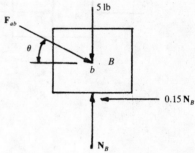

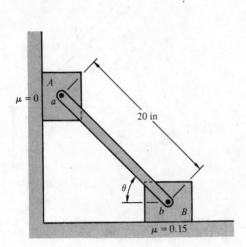

Fig. 8.44a

Fig. 8.44b

Using $F_{ab} = 5/(\sin \theta)$ in the above equation,

$$\frac{5}{\sin \theta} \cos \theta - 0.15(10) = 0 \qquad \tan \theta = 3.33 \qquad \theta = 73.3°$$

(b) The force acting on end a of the link is

$$R_a = (F_{ab} \cos \theta)i - (F_{ab} \sin \theta)j \qquad F_{ab} = \frac{5}{\sin \theta} = \frac{5}{\sin 73.3°} = 5.22 \text{ lb (C)}$$

$$R_a = (5.22 \cos 73.3°)i - (5.22 \sin 73.3°)j = 1.5i - 5j \qquad \text{lb}$$

Using the above result, the force acting on end b of the link is

$$R_b = -1.5i + 5j \qquad \text{lb}$$

8.45 (a) Solve Prob. 8.44 if a clockwise couple of 36 in · lb, in the plane of the figure, acts on the link.

(b) Express the forces and the couple acting on the link in formal vector notation.

(c) Compare the solutions for the two methods of loading in Probs. 8.44 and 8.45.

▌ (a) For this case, the link is no longer a two-force member. The free-body diagrams of block A and the link are shown in Figs. 8.45a and b. For equilibrium of block A,

$$\Sigma F_x = 0 \qquad N_A - R_{ax} = 0 \qquad R_{ax} = N_A$$
$$\Sigma F_y = 0 \qquad -5 + R_{ay} = 0 \qquad R_{ay} = 5 \text{ lb}$$

For equilibrium of the link,

$$\Sigma M_a = 0 \qquad -36 - R_{bx}(20 \sin \theta) + R_{by}(20 \cos \theta) = 0 \qquad (1)$$
$$\Sigma F_x = 0 \qquad R_{ax} - R_{bx} = 0 \qquad R_{bx} = R_{ax}$$
$$\Sigma F_y = 0 \qquad -R_{ay} + R_{by} = 0 \qquad R_{by} = R_{ay} = 5 \text{ lb} \qquad (2)$$

The free-body diagram of block B is given in Fig. 8.45c. For equilibrium,

$$\Sigma F_y = 0 \qquad N_B - R_{by} - 5 = 0 \qquad N_B = 10 \text{ lb}$$
$$\Sigma F_x = 0 \qquad R_{bx} - 0.15 N_B = 0 \qquad R_{bx} = 1.5 \text{ lb} \qquad (3)$$

R_{bx} and R_{by}, from Eqs. (2) and (3), are substituted into Eq. (1), with the result

$$-36 - 1.5(20 \sin \theta) + 5(20 \cos \theta) = 0 \qquad 30 \sin \theta + 36 = 100 \cos \theta$$

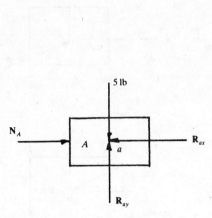

Fig. 8.45a

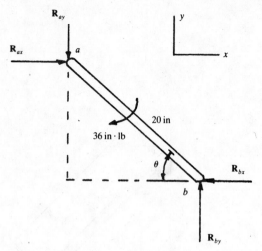

Fig. 8.45b

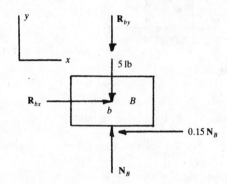

Fig. 8.45c

Both sides of the above equation are squared, with the result

$$900 \sin^2 \theta + 2(30)(36) \sin \theta + 36^2 = 100^2 \cos^2 \theta$$

Using $\cos^2 \theta = 1 - \sin^2 \theta$ in the above equation,

$$10,900 \sin^2 \theta + 2,160 \sin \theta - 8,700 = 0$$

The solution to the above quadratic equation is

$$\sin \theta = \frac{-2,160 \pm \sqrt{(2,160)^2 - 4(10,900)(-8,700)}}{2(10,900)} = \frac{-2,160 \pm 19,600}{2(10,900)}$$

where the positive root is

$$\sin \theta = 0.8 \qquad \theta = 53.1°$$

(b) The force acting on end a of the link is

$$\mathbf{R}_a = R_{ax}\mathbf{i} - R_{ay}\mathbf{j} = 1.5\mathbf{i} - 5\mathbf{j} \qquad \text{lb}$$

The force acting on end b of the link is

$$\mathbf{R}_b = -R_{bx}\mathbf{i} + R_{by}\mathbf{j} = -1.5\mathbf{i} + 5\mathbf{j} \qquad \text{lb}$$

The couple acting on the link is

$$\mathbf{M} = -36\mathbf{k} \text{ in} \cdot \text{lb}$$

(c) Application of the clockwise couple to the link reduces the equilibrium angle θ from 73.3° to 53.1°. The interesting observation in this problem is that the forces acting on ends a and b of the link are the *same* for both conditions of loading.

8.46   The coefficient of friction on all sliding surfaces in Fig. 8.46a is 0.1. Block A weighs 100 lb, and the link is assumed to be weightless.

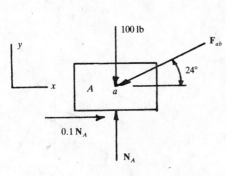

Fig. 8.46a

(a) Find the weight of block B which will cause impending sliding motion of the system, and the corresponding value of the force in the link.

(b) Express the forces exerted by the pins on the link in formal vector notation.

(c) Express the forces exerted by the plane surfaces on the blocks in formal vector notation.

(d) Use the results from part (c) to show that the system of two blocks and the link is in equilibrium.

▌ (a) Figure 8.46b shows the free-body diagram of block A, and F_{ab} is the force in the link. For equilibrium,

$$\Sigma F_y = 0 \qquad -100 - F_{ab} \sin 24° + N_A = 0$$
$$\Sigma F_x = 0 \qquad -F_{ab} \cos 24° + 0.1 N_A = 0$$

N_A is eliminated between the above two equations, with the result

$$F_{ab} = 11.5 \, \text{lb (C)}$$

N_A is then found from

$$-F_{ab} \cos 24° + 0.1 N_A = 0 \qquad -11.5 \cos 24° + 0.1 N_A = 0 \qquad N_A = 105 \, \text{lb}$$

The free-body diagram of block B is given in Fig. 8.46c. For equilibrium,

$$\Sigma F_y = 0 \qquad -W_B + F_{ab} \sin 24° + (0.1 N_B) \sin 40° + N_B \cos 40° = 0 \qquad (1)$$
$$\Sigma F_x = 0 \qquad F_{ab} \cos 24° + (0.1 N_B) \cos 40° - N_B \sin 40° = 0 \qquad (2)$$

Using $F_{ab} = 11.5 \, \text{lb}$ in Eq. (2) results in

$$N_B = 18.6 \, \text{lb}$$

This result is then used in Eq. (1) to obtain

$$W_B = 20.1 \, \text{lb}$$

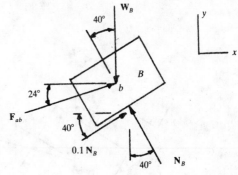

Fig. 8.46b   Fig. 8.46c

(b) The force exerted by pin a on the link is

$$\mathbf{R}_a = (F_{ab} \cos 24°)\mathbf{i} + (F_{ab} \sin 24°)\mathbf{j} = (11.5 \cos 24°)\mathbf{i} + (11.5 \sin 24°)\mathbf{j} = 10.5\mathbf{i} + 4.68\mathbf{j} \qquad \text{lb}$$

The force exerted by pin b on the link is

$$\mathbf{R}_b = -10.5\mathbf{i} - 4.68\mathbf{j} \qquad \text{lb}$$

(c) The force exerted by the surface on block A is

$$\mathbf{R}_A = 0.1 N_A \mathbf{i} + N_A \mathbf{j} = 0.1(105)\mathbf{i} + 105\mathbf{j} = 10.5\mathbf{i} + 105\mathbf{j} \quad \text{lb}$$

The force exerted by the surface on block B is

$$
\begin{aligned}
\mathbf{R}_B &= (0.1 N_B \cos 40° - N_B \sin 40°)\mathbf{i} + (0.1 N_B \sin 40° + N_B \cos 40°)\mathbf{j} \\
&= [0.1(18.6)\cos 40° - 18.6 \sin 40°]\mathbf{i} + [0.1(18.6)\sin 40° + 18.6 \cos 40°]\mathbf{j} \\
&= -10.5\mathbf{i} + 15.4\mathbf{j} \quad \text{lb}
\end{aligned}
$$

(d) The system shown in Fig. 8.46a is acted on by four external forces: the two reaction forces $\mathbf{R}_A$ and $\mathbf{R}_B$ and the two weight forces

$$\mathbf{W}_A = -100\mathbf{j} \quad \text{lb} \quad \text{and} \quad \mathbf{W}_B = -20.1\mathbf{j} \quad \text{lb}$$

For equilibrium of the system,

$$\mathbf{R}_A + \mathbf{R}_B + \mathbf{W}_A + \mathbf{W}_B = 0$$

As a check on the above equation,

$$(10.5\mathbf{i} + 105\mathbf{j}) + (-10.5\mathbf{i} + 15.4\mathbf{j}) - 100\mathbf{j} - 20.1\mathbf{j} \overset{?}{=} 0 \qquad (10.5 - 10.5)\mathbf{i} + (105 + 15.4 - 100 - 20.1)\mathbf{j} \overset{?}{=} 0$$

$$10.5 \equiv 10.5 \qquad 105 + 15.4 \overset{?}{=} 100 + 20.1 \qquad 120 \equiv 120$$

8.47 Block A has a weight of 50 N, and block B weighs 100 N. Sliding motion is impending in the configuration shown in Fig. 8.47a. Find the value of the coefficient of friction between block B and the inclined surface if the contacting surfaces of block A and the plane are assumed to be smooth.

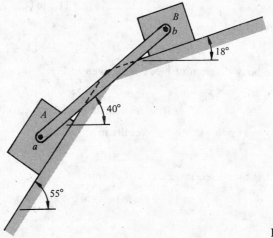

Fig. 8.47a

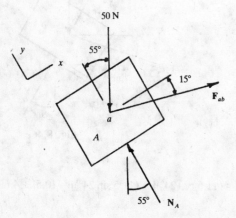

Fig. 8.47b

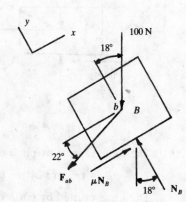

Fig. 8.47c

▌F_{ab} is the force in the link, and the free-body diagram of block A is seen in Fig. 8.47b. For equilibrium,

$$\Sigma F_x = 0 \qquad -50 \sin 55° + F_{ab} \cos 15° = 0 \qquad F_{ab} = 42.4\,\text{N (T)}$$

Figure 8.47c shows the free-body diagram of block B. For equilibrium of this block,

$$\Sigma F_y = 0 \qquad -100 \cos 18° - F_{ab} \sin 22° + N_B = 0 \qquad N_B = 111\,\text{N}$$
$$\Sigma F_x = 0 \qquad -100 \sin 18° - F_{ab} \cos 22° + \mu N_B = 0$$

Using the values of F_{ab} and N_B found above,

$$\mu = 0.633$$

8.48 The arm shown in Fig. 8.48a fits loosely on the vertical rod. The angle of static friction between the arm and rod is 8.5°.

(a) Find the minimum value of the dimension x_1 if the arm is not to slide down the rod.

(b) Do the same as in part (a) if, in addition to the force, a clockwise couple, in the plane of the figure, of 50 in · lb acts on the arm.

(c) Express the reaction forces of the rod on the arm, for the condition of part (a), in formal vector notation.

(d) Use the result in part (c) to verify the force equilibrium of the arm.

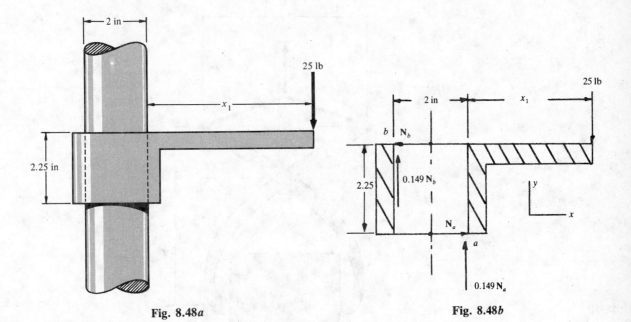

Fig. 8.48a Fig. 8.48b

▌(a) From the definition of the angle of static friction,

$$\tan \phi_s = \mu \qquad \mu = \tan 8.5° = 0.149$$

Impending sliding motion of the arm on the rod is assumed, and the free-body diagram of the arm is shown in Fig. 8.48b. For equilibrium of the arm,

$$\Sigma M_b = 0 \qquad -25(x_1 + 2) + N_a(2.25) + (0.149N_a)2 = 0 \qquad (1)$$
$$\Sigma F_x = 0 \qquad -N_b + N_a = 0 \qquad N_a = N_b$$
$$\Sigma F_y = 0 \qquad 0.149N_b + 0.149N_a - 25 = 0 \qquad N_a = N_b = 83.9\,\text{lb}$$

From Eq. (1),

$$-25(x_1 + 2) + 83.9(2.25) + 0.149(83.9)(2) = 0 \qquad x_1 = 6.55\,\text{in}$$

(b) The free-body diagram will be the same as in Fig. 8.48b, with the addition of a 50-in · lb clockwise couple to the figure. The normal forces N_a and N_b are the same as in part (a), given by

$$N_a = N_b = 83.9\,\text{lb}$$

For moment equilibrium,

$$\Sigma M_b = 0 \qquad -25(x_1 + 2) - 50 + N_a(2.25) + (0.149N_a)(2) = 0 \qquad x_1 = 4.55 \text{ in}$$

It may be seen that the addition of the couple to the problem decreases the value of x_1.

(c) For the conditions of part (a), the reaction forces of the rod on the arm are

$$\mathbf{R}_a = N_a\mathbf{i} + 0.149N_a\mathbf{j} = 83.9\mathbf{i} + 0.149(83.9)\mathbf{j} = 83.9\mathbf{i} + 12.5\mathbf{j} \qquad \text{lb}$$
$$\mathbf{R}_b = -N_B\mathbf{i} + 0.149N_b\mathbf{j} = -83.9\mathbf{i} + 0.149(83.9)\mathbf{j} = -83.9\mathbf{i} + 12.5\mathbf{j} \qquad \text{lb}$$

(d) The arm is acted on by the two external reaction forces $\mathbf{R}_a$ and $\mathbf{R}_b$, and the applied force $\mathbf{P} = -25\mathbf{j}$. For force equilibrium of the arm,

$$\mathbf{R}_a + \mathbf{R}_b + \mathbf{P} = 0$$

As a check on the above equation,

$$(83.9\mathbf{i} + 12.5\mathbf{j}) + (-83.9\mathbf{i} + 12.5\mathbf{j}) - 25\mathbf{j} \overset{?}{=} 0 \qquad (83.9 - 83.9)\mathbf{i} + (12.5 + 12.5 - 25)\mathbf{j} \overset{?}{=} 0$$

The above equation is seen to be identically satisfied.

8.49 The block in Fig. 8.49a weighs 20 lb, and the coefficient of friction between the block and the jaws of the lifting device is 0.26. The weight of the lifting device may be neglected.

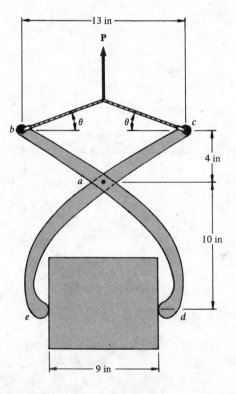

Fig. 8.49a

(a) Find the maximum value of θ, if the block is not to slip, and the corresponding value of the force on the connecting pin.

(b) Express all forces acting on member bd in formal vector notation.

❙ (a) Figure 8.49b shows the free-body diagram of the block, and motion is assumed to be impending. For equilibrium,

$$\Sigma F_y = 0 \qquad 2(0.26N) - 20 = 0 \qquad N = 38.5 \text{ lb}$$

The relationship between the cable forces T and the applied force P is shown in Fig. 8.49c. From consideration of the equilibrium of the entire system, $P = 20$ lb. Using Fig. 8.49c,

$$\Sigma F_y = 0 \qquad -2T \sin \theta + P = 0 \qquad -2T \sin \theta + 20 = 0 \qquad T = \frac{10}{\sin \theta}$$

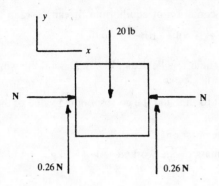

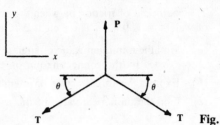

Fig. 8.49b

Fig. 8.49c

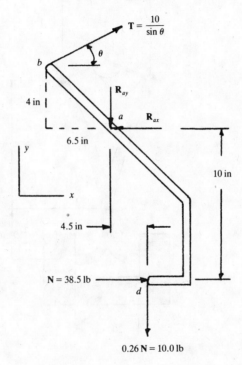

Fig. 8.49d

0.26 N = 10.0 lb

The free-body diagram of member bd of the lifting device is shown in Fig. 8.49d. For equilibrium,

$$\Sigma M_a = 0 \qquad -\left(\frac{10}{\sin\theta}\right)(\sin\theta)(6.5) - \left(\frac{10}{\sin\theta}\right)(\cos\theta)(4) - 10(4.5) + 38.5(10) = 0$$

$$\tan\theta = \frac{40}{275} \qquad \theta = 8.28°$$

If $\theta \le 8.28°$, the block will not slip.

(b) From Fig. 8.49d,

$$\Sigma F_x = 0 \qquad \frac{10}{\sin\theta}\cos\theta - R_{ax} + 38.5 = 0 \qquad R_{ax} = 107\ \text{lb}$$

$$\Sigma F_y = 0 \qquad \frac{10}{\sin\theta}\sin\theta - R_{ay} - 10 = 0 \qquad R_{ay} = 0 \qquad R_a = R_{ax} = 107\ \text{lb}$$

The forces acting on member bd are shown below in formal vector notation:

$$\mathbf{R}_b = (T\cos\theta)\mathbf{i} + (T\sin\theta)\mathbf{j}$$

Using $T = 10/(\sin\theta) = 10/(\sin 8.28°) = 69.4\ \text{lb}$,

$$\mathbf{R}_b = (69.4\cos 8.28°)\mathbf{i} + (69.4\sin 8.28°)\mathbf{j} = 68.7\mathbf{i} + 10\mathbf{j} \qquad \text{lb}$$
$$\mathbf{R}_a = -R_{ax}\mathbf{i} - R_{ay}\mathbf{j} = -107\mathbf{i} \qquad \text{lb}$$
$$\mathbf{R}_d = 38.5\mathbf{i} - 10\mathbf{j} \qquad \text{lb}$$

As a check on the above results, and as a necessary condition of equilibrium, it can be seen that

$$\mathbf{R}_a + \mathbf{R}_b + \mathbf{R}_d = (-107 + 68.7 + 38.5)\mathbf{i} + (10 - 10)\mathbf{j} \equiv 0$$

8.50 The coefficient of friction between the ends of the members of the lifting device in Fig. 8.50a and the workpiece of weight W is 0.43. The weight of the lifting device may be neglected

(a) Find the minimum value of angle θ, if the workpiece is not to slip, and the corresponding value of the force exerted by the connecting pin on the members.

(b) Express the forces acting on member cd in formal vector notation.

(c) Find the numerical value of the force in part (a) if the mass of the workpiece is 17.2 kg.

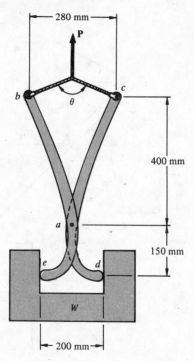

Fig. 8.50a

▌(a) Figure 8.50b shows the free-body diagram of the workpiece, and motion is assumed to be impending. For equilibrium,

$$\Sigma F_y = 0 \qquad 2(0.43N) - W = 0 \qquad N = 1.16W$$

The relationship between the cable forces T and the applied force P is shown in Fig. 8.50c. For equilibrium of the entire system, $P = W$. Using Fig. 8.50c,

$$\Sigma F_y = 0 \qquad -2T \sin \beta + P = 0 \qquad -2T \sin \beta + W = 0 \qquad T = \frac{W}{2 \sin \beta}$$

The free-body diagram of member cd of the lifting device is shown in Fig. 8.50d. For equilibrium,

$$\Sigma M_a = 0 \qquad \frac{W}{2 \sin \beta} \cos \beta \, 400 + \frac{W}{2 \sin \beta} \sin \beta \, 140 - (1.16W)150 - (0.5W)100 = 0$$

$$\tan \beta = 1.30 \qquad \beta = 52.4°$$

Using Fig. 8.50c,

$$\theta + 2\beta = 180° \qquad \theta = 75.2°$$

$$\Sigma F_x = 0 \qquad -\left(\frac{W}{2 \sin \beta}\right) \cos \beta + R_{ax} - 1.16W = 0 \qquad R_{ax} = 1.55W$$

$$\Sigma F_y = 0 \qquad \frac{W}{2 \sin \beta} \sin \beta + R_{ay} - 0.5W = 0$$

$$R_{ay} = 0 \qquad R_a = R_{ax} = 1.55W$$

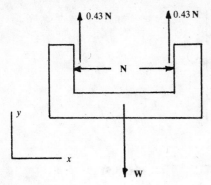

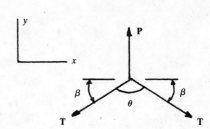

Fig. 8.50b

Fig. 8.50c

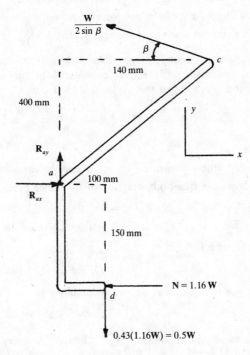

Fig. 8.50d

The workpiece will not slip if

$$\theta \geq 75.2°$$

The connecting pin exerts a force of 1.55W, in the x direction, on each member.

(b) The forces which act on member *cd* are

$$\mathbf{R}_a = R_{ax}\mathbf{i} + R_{ay}\mathbf{j} = 1.55W\mathbf{i}$$

$$\mathbf{R}_c = \left[\left(-\frac{W}{2\sin\beta}\right)\cos\beta\right]\mathbf{i} + \left[\left(\frac{W}{2\sin\beta}\right)\sin\beta\right]\mathbf{j} = \left(-\frac{W}{2}\cot 52.4°\right)\mathbf{i} + \frac{W}{2}\mathbf{j} = -0.385W\mathbf{i} + 0.5W\mathbf{j}$$

$$\mathbf{R}_d = -1.16W\mathbf{i} - 0.5W\mathbf{j}$$

(c) The weight of the workpiece is

$$W = mg = 17.2(9.81) = 169 \text{ N}$$

The force of the connecting pin on the members is

$$R_a = 1.55W = 1.55(169) = 262 \text{ N}$$

8.51 The slider block in Fig. 8.51a weighs 12 lb. The slider guide walls may be assumed to be frictionless, and the weight of the rod is neglected. The coefficient of friction between the rod and the ground is 0.38.

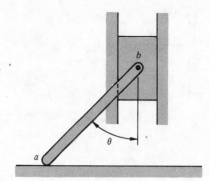

Fig. 8.51a

(*a*)   For what maximum value of θ may the equilibrium configuration shown in the figure be maintained?
(*b*)   Find the corresponding values of the pin force at *b* and the reaction force of the wall on the block.
(*c*)   Do the same as in parts (*a*) and (*b*) if the rod weighs 1.5 lb.

▋ (*a*)   Motion is assumed to be impending.   The rod is a two-force member.   Thus, the total force R_a exerted by the ground on the rod is collinear with the rod, and $F_{ab} = R_a$,   as shown in Fig. 8.51*b*.   Using the definition of the angle of static friction,

$$\tan\theta = \tan\phi_s = \frac{F}{N} = \frac{\mu N_a}{N_a} = \mu$$

$$\tan\theta = 0.38 \qquad \theta = 20.8°$$

The equilibrium position shown in Fig. 8.51*a* may be maintained if $\theta \le 20.8°$.

(*b*)   Figure 8.51*c* shows the free-body diagram of the slider block.   For equilibrium,

$$\Sigma F_y = 0 \quad F_{ab}\cos\theta - 12 = 0 \quad F_{ab}\cos 20.8° - 12 = 0 \quad F_{ab} = 12.8 \text{ lb (C)}$$
$$\Sigma F_x = 0 \quad F_{ab}\sin\theta - N = 0 \quad 12.8\sin 20.8° - N = 0 \quad N = 4.55 \text{ lb}$$

The pin force R_b at *b* is

$$R_b = F_{ab} = 12.8 \text{ lb}$$

The reaction force *N* of the wall on the slider block is

$$N = 4.55 \text{ lb}$$

(*c*)   Motion is again assumed to be impending.   The free-body diagrams of the slider block and the rod are shown in Figs. 8.51*d* and *e*, and the length of the rod is *l*.   For this case, the rod is no longer a two-force member.   Using Fig. 8.51*d*,

$$\Sigma F_y = 0 \quad R_{by} - 12 = 0 \quad R_{by} = 12 \text{ lb}$$

From Fig. 8.51*e*,

$$\Sigma F_y = 0 \quad N_a - 1.5 - R_{by} = 0 \quad N_a = 13.5 \text{ lb}$$

$$\Sigma M_b = 0 \quad 1.5\left(\frac{l}{2}\right)\sin\theta + (0.38N_a)l\cos\theta - N_a l\sin\theta = 0 \qquad \theta = 21.8°$$

$$\Sigma F_x = 0 \quad 0.38N_a - R_{bx} = 0 \quad R_{bx} = 5.13 \text{ lb}$$

$$R_b = \sqrt{R_{bx}^2 + R_{by}^2} = \sqrt{5.13^2 + 12^2} = 13.1 \text{ lb}$$

From Fig. 8.51*d*,

$$\Sigma F_x = 0 \quad R_{bx} - N = 0 \quad N = R_{bx} = 5.13 \text{ lb}$$

A summary of the results in this part of the problem is

$$\theta = 21.9° \qquad R_b = 13.1 \text{ lb} \qquad N = 5.13 \text{ lb}$$

The results from parts (*a*) and (*b*) are

$$\theta = 20.8° \qquad R_b = 12.8 \text{ lb} \qquad N = 4.55 \text{ lb}$$

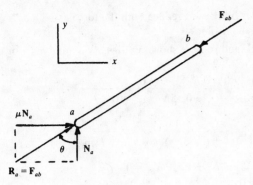

Fig. 8.51b

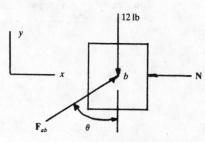

Fig. 8.51c

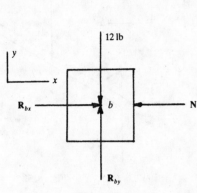

Fig. 8.51d

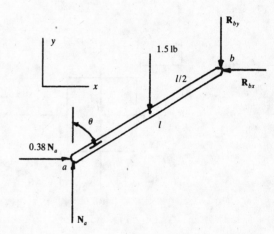

Fig. 8.51e

It may be seen that the inclusion of the rod weight increases angle θ, the pin force at b, and the value of the normal force of the wall on the block.

8.52 Solve Prob. 8.51 if the coefficient of friction between the slider guide walls and the block is 0.15.

▌ (a) The maximum value of θ is the same as in Prob. 8.51, found as $\theta = 20.8°$, since the equilibrium requirement of the rod is independent of the friction force acting on the slider block.

(b) The free-body diagram of the slider block is shown in Fig. 8.52a, and motion of this element is assumed to be impending. For equilibrium,

$$\Sigma F_y = 0 \qquad F_{ab} \cos 20.8° - 12 + 0.15N = 0$$
$$\Sigma F_x = 0 \qquad F_{ab} \sin 20.8° - N = 0 \qquad F_{ab} = 12.1 \text{ lb (C)} \qquad R_b = F_{ab} = 12.1 \text{ lb} \qquad N = 4.30 \text{ lb}$$

As a comparison, from Prob. 8.51, part (b),

$$F_{ab} = 12.8 \text{ lb (C)} \qquad N = 4.55 \text{ lb}$$

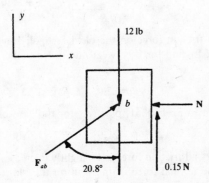

Fig. 8.52a

The effect of the friction force on the slider block is to reduce both the force acting on the rod and the normal reaction force of the wall on the slider block.

(c)   The free-body diagrams of the slider block and rod are shown in Figs. 8.52b and c.   For equilibrium, using Fig. 8.52b,

$$\Sigma F_x = 0 \qquad R_{bx} - N = 0 \tag{1}$$
$$\Sigma F_y = 0 \qquad R_{by} + 0.15N - 12 = 0 \tag{2}$$

From Fig. 8.52c,

$$\Sigma M_b = 0 \qquad 1.5\left(\frac{l}{2}\right)\sin\theta + (0.38N_a)l\cos\theta - N_a l\sin\theta = 0 \tag{3}$$

$$\Sigma F_y = 0 \qquad N_a - 1.5 - R_{by} = 0 \tag{4}$$
$$\Sigma F_x = 0 \qquad 0.38N_a - R_{bx} = 0 \tag{5}$$

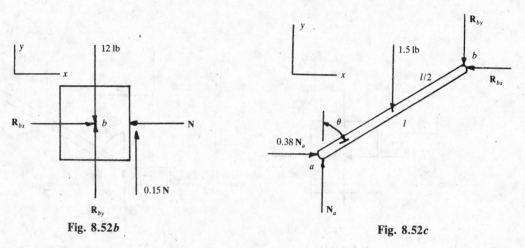

Fig. 8.52b   Fig. 8.52c

N is eliminated between Eqs. (1) and (2), with the result

$$R_{by} + 0.15R_{bx} = 12 \tag{6}$$

Using Eqs. (4) and (5) in Eq. (6),

$$(N_a - 1.5) + 0.15(0.38N_a) = 12 \qquad N_a = 12.8\ \text{lb}$$

Using Eq. (3),

$$\frac{1.5}{2}\sin\theta - 0.38(12.8)\cos\theta - 12.8\sin\theta = 0 \qquad \tan\theta = 0.404 \qquad \theta = 22.0°$$

Using Eq. (5),

$$R_{bx} = 0.38N_a = 0.38(12.8) = 4.86\ \text{lb}$$

Using Eq. (4),

$$R_{by} = N_a - 1.5 = 12.8 - 1.5 = 11.3\ \text{lb}$$

Using Eq. (1),

$$N = R_{bx} = 4.86\ \text{lb}$$
$$R_b = \sqrt{R_{bx}^2 + R_{by}^2} = \sqrt{4.86^2 + 11.3^2} = 12.3\ \text{lb}$$

The effect of the friction force on the block is to slightly increase θ and to reduce the pin force at b.
   For the case of friction force between the slider wall and the block, with the rod having a weight of 1.5 lb,

$$\theta = 22.0° \qquad R_b = 12.3\ \text{lb} \qquad N = 4.86\ \text{lb}$$

From the solution to Prob. 8.51, part (c), for the case of rod weight but no friction force,

$$\theta = 21.9° \qquad R_b = 13.1\ \text{lb} \qquad N = 5.13\ \text{lb}$$

The effect of the wall friction force is to slightly decrease the angle and to reduce the values of the pin force at b and the normal force of the wall on the block.

8.53 The device shown in Fig. 8.53*a* is used as a locking vise to hold a workpiece of weight 20 lb, and each link weighs 6.5 lb.

(*a*) Find the minimum required value of the coefficient of friction between the workpiece and the links if the workpiece is not to slip.

(*b*) Find the forces exerted by pins *a* and *d* on the links.

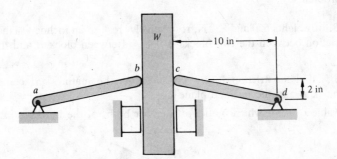

Fig. 8.53*a*

▌ (*a*) Figure 8.53*b* shows the free-body diagram of one link. For equilibrium,

$$\Sigma M_a = 0 \quad -6.5(5) - 10(10) + N(2) = 0 \quad N = 66.3 \text{ lb}$$

The free-body diagram of the workpiece is shown in Fig. 8.53*c*. Motion is assumed to be impending, so that the friction forces acting on the block have their maximum values

$$F_{max} = \mu N$$

For equilibrium of the workpiece,

$$\Sigma F_y = 0 \quad 2\mu N - 20 = 0 \quad 2\mu(66.3) - 20 = 0 \quad \mu = 0.151$$

The workpiece will not slip if $\mu \geq 0.151$.

(*b*) Using Fig. 8.53*b*,

$$\Sigma F_x = 0 \quad R_{ax} - N = 0 \quad R_{ax} = N = 66.3 \text{ lb}$$
$$\Sigma F_y = 0 \quad R_{ay} - 6.5 - 10 = 0 \quad R_{ay} = 16.5 \text{ lb}$$
$$R_a = \sqrt{R_{ax}^2 + R_{ay}^2} = \sqrt{66.3^2 + 16.5^2} = 68.3 \text{ lb}$$

From symmetry,

$$R_d = R_a = 68.3 \text{ lb}$$

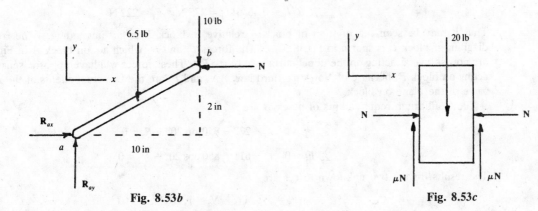

Fig. 8.53*b*　　　　　　　　　　　**Fig. 8.53*c***

8.3 MULTIPLE SLIDING SURFACES, THE WEDGE

8.54 Describe how the friction forces are analyzed in problems in which there are multiple sliding surfaces.

▌ In certain problems there may be more than one surface which may slide relative to a second surface. The friction forces on each pair of surfaces which may experience relative motion are analyzed exactly as for the case of a single pair of surfaces. That is, if motion is not impending, the friction forces which act between adjacent contacting surfaces may be found by using the equations of static equilibrium. The maximum friction force F_{max} which may exist between the two surfaces is given by

$$F_{max} = \mu N$$

where N is the compressive normal contact force between the two surfaces. The above effects are illustrated in the following problems.

8.55   Blocks A and B, of weights 650 and 850 N, respectively, rest on an inclined plane, as shown in Fig. 8.55a. The coefficient of friction between the two blocks is 0.45. Between block B and the plane this coefficient has the value 0.6.

(a)   If $\theta = 20°$, find the friction forces which act on the mating surfaces between blocks A and B, and between block B and the inclined plane C.

(b)   Find the value of θ for which motion of one, or both, of the blocks is impending.

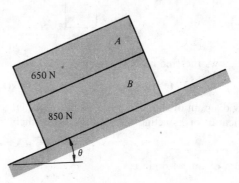

Fig. 8.55a

▌ (a)   The free-body diagram of block A is as shown in Fig. 8.55b. The equilibrium requirements are

$$\sum F_x = 0 \qquad -650 \sin 20° + F_1 = 0$$
$$\sum F_y = 0 \qquad -650 \cos 20° + N_1 = 0$$

from which

$$F_1 = 222 \text{ N} \qquad N_1 = 611 \text{ N}$$

The maximum value of the friction force which may exist on the contacting surfaces between blocks A and B is given by

$$F_{max} = \mu N_1 = 0.45(611) = 275 \text{ N} > F_1 = 222 \text{ N}$$

Thus it may be seen that motion of block A relative to block B is *not* impending. The free-body diagram of block B is shown in Fig. 8.55c. The forces F_1 and N_1 which act on block A in Fig. 8.55b are now shown as acting on the upper surface of block B. These forces will have opposite senses when acting on block B, because of Newton's third law. N_2 and F_2 are the two components of the reaction force of the plane on block B.

The equilibrium requirements of block B are

$$\sum F_x = 0 \qquad -222 - 850 \sin 20° + F_2 = 0$$
$$\sum F_y = 0 \qquad -611 - 850 \cos 20° + N_2 = 0$$

The results for the two unknown forces are

$$F_2 = 513 \text{ N} \qquad N_2 = 1,410 \text{ N}$$

The maximum value of friction force which may exist on the bottom surface of block B is given by

$$F_{max} = \mu N_2 = 0.6(1,410) = 846 \text{ N} > F_2 = 513 \text{ N}$$

Thus, motion of block B is *not* impending.

(b)   If angle θ is now allowed to increase, a limiting condition, given by θ_m, will be reached where the friction force between blocks A and B will be at its maximum value. The corresponding friction force

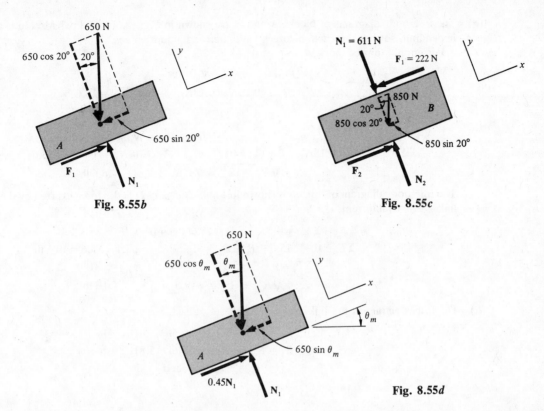

650 N

650 cos 20° 20°

650 sin 20°

A

F_1

N_1

Fig. 8.55b

$N_1 = 611$ N

$F_1 = 222$ N

850 N

20°

B

850 cos 20°

850 sin 20°

F_2

N_2

Fig. 8.55c

650 N

650 cos θ_m θ_m

θ_m

650 sin θ_m

A

0.45N_1

N_1

Fig. 8.55d

between block B and the plane, for this angle θ_m, will not have reached its maximum value, since $\mu = 0.45$ between blocks A and B is less than $\mu = 0.6$ between block B and the plane.

The free-body diagram for block A for this limiting situation is shown in Fig. 8.55d.

The equilibrium requirements are

$$\sum F_x = 0 \qquad -650 \sin \theta_m + 0.45N_1 = 0$$

$$\sum F_y = 0 \qquad -650 \cos \theta_m + N_1 = 0$$

N_1 is eliminated between the above two equations, and the angle θ_m is then found from

$$\tan \theta = 0.45 \qquad \theta_m = 24.2°$$

The above angle corresponds to impending sliding motion of block A along block B.

8.56 Block A weighs 5 lb and block B weighs 3 lb. The link is assumed to be weightless.

(**a**) Find the minimum value of the spring force needed to hold block A in the position shown in Fig. 8.56a, and the corresponding values of the pin forces at a and b.

(**b**) Express the forces acting on the link in formal vector notation.

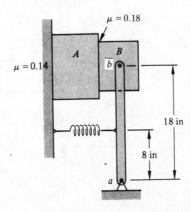

$\mu = 0.18$

A   B

$\mu = 0.14$ b

18 in

8 in

a

Fig. 8.56a

▌(a) The free-body diagrams of blocks A and B are shown in Figs. 8.56b and c. Motion is assumed to be impending, so that the friction forces have their maximum values.

For equilibrium of block A,

$$\Sigma F_x = 0 \qquad N_A - N_B = 0 \qquad N_A = N_B = N$$
$$\Sigma F_y = 0 \qquad 0.14N + 0.18N - 5 = 0 \qquad N = 15.6\,\text{lb}$$

For equilibrium of block B,

$$\Sigma F_x = 0 \qquad N_B - R_{bx} = 0 \qquad R_{bx} = N_B = N = 15.6\,\text{lb}$$
$$\Sigma F_y = 0 \qquad -0.18N_B - 3 + R_{by} = 0 \qquad R_{by} = 3 + 0.18N_B = 3 + 0.18(15.6) = 5.81\,\text{lb}$$
$$R_b = \sqrt{R_{bx}^2 + R_{by}^2} = \sqrt{15.6^2 + 5.81^2} = 16.6\,\text{lb}$$

The free-body diagram of link ab is shown in Fig. 8.56d, and F_s is the force exerted by the spring on the link. For equilibrium,

$$\Sigma M_a = 0 \qquad F_s(8) - 15.6(18) = 0 \qquad F_s = 35.1\,\text{lb}$$
$$\Sigma F_x = 0 \qquad 15.6 - F_s + R_{ax} = 0 \qquad R_{ax} = 35.1 - 15.6 = 19.5\,\text{lb}$$
$$\Sigma F_y = 0 \qquad -5.81 + R_{ay} = 0 \qquad R_{ay} = 5.81\,\text{lb}$$
$$R_a = \sqrt{R_{ax}^2 + R_{ay}^2} = \sqrt{19.5^2 + 5.81^2} = 20.3\,\text{lb}$$

(b) The forces acting on the link are

$$\mathbf{R}_a = R_{ax}\mathbf{i} + R_{ay}\mathbf{j} = 19.5\mathbf{i} + 5.81\mathbf{j} \qquad \text{lb}$$
$$\mathbf{R}_b = R_{bx}\mathbf{i} - R_{by}\mathbf{j} = 15.6\mathbf{i} - 5.81\mathbf{j} \qquad \text{lb}$$
$$\mathbf{F}_s = -F_s\mathbf{i} = -35.1\mathbf{i} \qquad \text{lb}$$

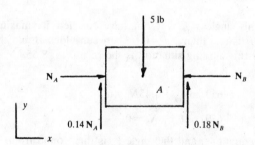

Fig. 8.56b

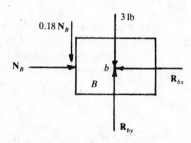

Fig. 8.56c

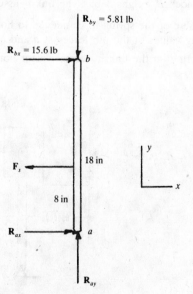

Fig. 8.56d

8.57 The spring in Prob. 8.56 is replaced by a counterclockwise couple, in the plane of the figure, applied to the link.

(a) For what value of this couple will motion of block A be impending?

(b) Find the corresponding values of the pin forces at a and b.

(c) State the results in part (b) in formal vector notation.

▌ (a) R_{bx} and R_{by} must have the same values as those found in Prob. 8.56, since the two forces are requirements for the equilibrium of blocks A and B. Figure 8.57 shows the free-body diagram of link ab. For equilibrium,

$$\Sigma M_a = 0 \qquad M_0 - 15.6(18) = 0 \qquad M_0 = 281 \text{ in} \cdot \text{lb}$$

(b)

$$\Sigma F_x = 0 \qquad 15.6 - R_{ax} = 0 \qquad R_{ax} = 15.6 \text{ lb}$$
$$\Sigma F_y = 0 \qquad -5.81 + R_{ay} = 0 \qquad R_{ay} = 5.81 \text{ lb}$$
$$R_a = \sqrt{R_{ax}^2 + R_{ay}^2} = \sqrt{15.6^2 + 5.81^2} = 16.6 \text{ lb}$$

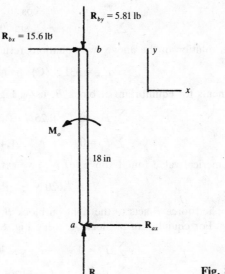

Fig. 8.57

The pin force at a is reduced when the spring is replaced by the applied couple.

(c) The forces acting on the link are

$$\mathbf{R}_a = -R_{ax}\mathbf{i} + R_{ay}\mathbf{j} = -15.6\mathbf{i} + 5.81\mathbf{j} \qquad \text{lb}$$
$$\mathbf{R}_b = R_{bx}\mathbf{i} - R_{by}\mathbf{j} = 15.6\mathbf{i} - 5.81\mathbf{j} \qquad \text{lb}$$

The applied couple acting on the link may be expressed as

$$\mathbf{M}_0 = M_0\mathbf{k} = 281\mathbf{k} \qquad \text{in} \cdot \text{lb}$$

8.58 In Fig. 8.58a block A is connected to the ground by the link ab. The masses of blocks A and B are 100 and 150 kg, respectively, and the coefficients of friction between the mating surfaces are $\mu_{AB} = 0.25$ and $\mu_{BC} = 0.35$. It is desired to slide block B out from beneath block A by applying a horizontal force P to block B.

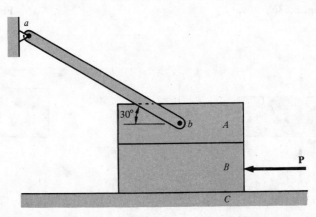

Fig. 8.58a

(a) If P is a compressive force to the left on block B, as shown in Fig. 8.58a, find the minimum value of P to cause impending leftward motion of block B.

(b) If the force P on block B acted to the right, would the magnitude of the force required to cause impending motion of block B be different from the result obtained in part (a)?

▎(a) Since member ab is a two-force member, the unknown force F_{ab} which acts in this member has the direction of the link. The weights of the two blocks are

$$W_A = m_A g = 100(9.81) = 981 \text{ N}$$
$$W_B = m_B g = 150(9.81) = 1{,}470 \text{ N}$$

The free-body diagrams of the two blocks, for the case where P is a leftward force on block B, are shown in Figs. 8.58b and c. N_A is the normal force between blocks A and B.

Motion is assumed to be impending, so that *all friction forces have their maximum values*. The equilibrium requirements of block A, from Fig. 8.58b, are

$$\sum F_x = 0. \quad F_{ab}\cos 30° - 0.25N_A = 0$$
$$\sum F_y = 0 \quad -F_{ab}\sin 30° - 981 + N_A = 0$$

Simultaneous solution of the above two equations results in

$$F_{ab} = 331 \text{ N (C)} \quad N_A = 1{,}150 \text{ N}$$

The requirements for equilibrium of block B, using Fig. 8.58c, are

$$\sum F_x = 0 \quad 0.25N_A + 0.35N_B - P = 0$$
$$\sum F_y = 0 \quad -N_A - 1{,}470 + N_B = 0$$

Using the numerical value found above for N_A, we get the final results

$$N_B = 2{,}620 \text{ N} \quad P = 1{,}200 \text{ N}$$

(b) When the applied force P acts to the right on block B, the free-body diagrams are as shown in Figs. 8.58d and e. For equilibrium of block A, using Fig. 8.58d,

$$\sum F_x = 0 \quad -F_{ab}\cos 30° + 0.25N_A = 0$$
$$\sum F_y = 0 \quad F_{ab}\sin 30° - 981 + N_A = 0$$

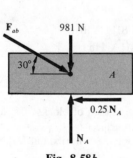
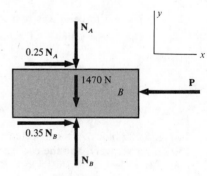

Fig. 8.58b

Fig. 8.58c

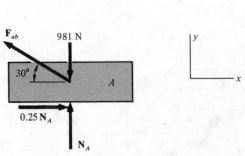
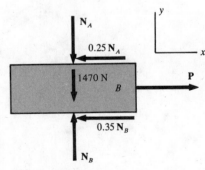

Fig. 8.58d

Fig. 8.58e

The solutions to the above equations are

$$F_{ab} = 248 \text{ N (T)} \qquad N_A = 857 \text{ N}$$

The equilibrium requirements for block B, using Fig. 8.58e, are

$$\Sigma F_x = 0 \qquad -0.25N_A - 0.35N_B + P = 0$$
$$\Sigma F_y = 0 \qquad -N_A - 1{,}470 + N_B = 0$$

Using $N_A = 857$ N, the final values for N_B and P are

$$N_B = 2{,}330 \text{ N} \qquad P = 1{,}030 \text{ N}$$

It may be seen that there is a significant difference between the required force $P = 1{,}030$ N for pulling rightward on block B and the required force $P = 1{,}200$ N for pushing leftward on this block. An explanation will now be presented for the difference in the magnitudes of these two forces.

For the case of the force tending to push the block leftward, the friction force of $0.25N_A$ shown in Fig. 8.58b causes a clockwise moment about end a of member ab. This effect requires an increased value of N_A to balance this moment, and this larger value of N_A results in larger values of N_B and of the two friction forces on block B.

For the case of the applied force P tending to cause rightward motion of block B, the friction force $0.25N_A$ in Fig. 8.58d causes a counterclockwise moment on the link with respect to end a. This effect is to reduce the value of the normal force N_A, and with it the values of N_B and the two friction forces on block B.

8.59 Block A in Fig. 8.59a weighs 40 lb, and block B weighs 70 lb. The angle of static friction at all sliding surfaces is 13.5°.

(a) Find the value of P for impending sliding motion of block B to the right.

(b) Do the same as in part (a) for leftward motion of block B.

(c) Which of the above two methods of loading requires the larger value of P, and which produces the larger force in the link?

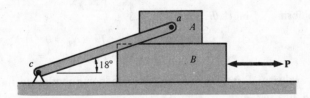

Fig. 8.59a

▮ (a) The coefficient of friction is found from

$$\tan \phi_s = \mu \qquad \mu = \tan 13.5° = 0.24$$

Figures 8.59b and c show the free-body diagrams for impending motion of block B to the right. For equilibrium of block A,

$$\Sigma F_x = 0 \qquad -F_{ac} \cos 18° + 0.24N_A = 0$$
$$\Sigma F_y = 0 \qquad -40 - F_{ac} \sin 18° + N_A = 0$$

N_A is eliminated between the above two equations, with the result

$$F_{ac} = 10.9 \text{ lb (T)} \qquad N_A = 43.4 \text{ lb}$$

For equilibrium of block B,

$$\Sigma F_y = 0 \qquad -N_A - 70 + N_B = 0 \qquad N_B = N_A + 70 = 43.4 + 70 = 113 \text{ lb}$$
$$\Sigma F_x = 0 \qquad -0.24N_A + P - 0.24N_B = 0 \qquad P = 0.24(N_A + N_B) = 0.24(43.4 + 113) = 37.5 \text{ lb}$$

(b) For impending motion of block B to the left, the free-body diagrams of the blocks are as shown in Figs. 8.59d and e. For equilibrium of block A, using Fig. 8.59d,

$$\Sigma F_x = 0 \qquad F_{ac} \cos 18° - 0.24N_A = 0$$
$$\Sigma F_y = 0 \qquad -40 + F_{ac} \sin 18° + N_A = 0$$

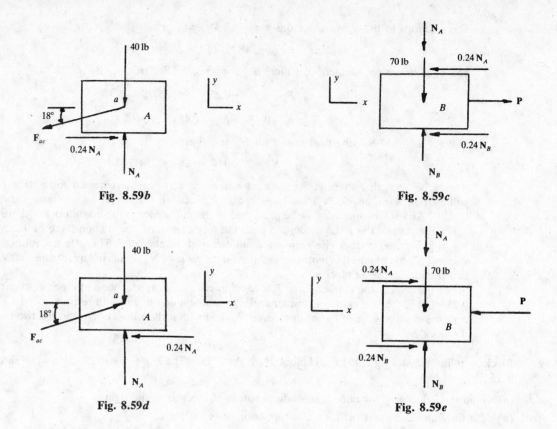

40 lb

18°

F_{ac}

a

A

$0.24\,N_A$

N_A

Fig. 8.59b

N_A

70 lb

$0.24\,N_A$

B

P

$0.24\,N_B$

N_B

Fig. 8.59c

40 lb

18°

F_{ac}

a

A

$0.24\,N_A$

N_A

Fig. 8.59d

N_A

$0.24\,N_A$

70 lb

B

P

$0.24\,N_B$

N_B

Fig. 8.59e

Simultaneous solution of the above two equations yields

$$F_{ac} = 9.36\,\text{lb (C)} \qquad N_A = 37.1\,\text{lb}$$

For equilibrium of block B, using Fig. 8.59e,

$$\Sigma F_y = 0 \qquad -N_A - 70 + N_B = 0 \qquad N_B = 107\,\text{lb}$$
$$\Sigma F_x = 0 \qquad 0.24 N_A - P + 0.24 N_B = 0 \qquad P = 34.6\,\text{lb}$$

(c) A larger value of P is required to cause impending motion to the right. The link force F_{ac} is also larger for this case.

8.60 The link in Fig. 8.60a is weightless. Block A weighs 100 N, and block B weighs 350 N. $\mu = 0.15$ between blocks A and B, and $\mu = 0.2$ between block B and the inclined surface.

(a) For what range of values of P will block B remain in the equilibrium position shown?

(b) Find the corresponding range of values of force in the link.

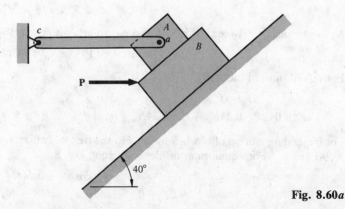

Fig. 8.60a

▌(a) Figure 8.60b shows the free-body diagram of block A for impending motion of block B up the plane. The link is a two-force member, and F_{ac} is the force exerted by the link on block A. For equilibrium of block A,

$$\Sigma F_y = 0 \qquad (0.15N_A)\sin 40° + N_A \cos 40° - 100 = 0 \qquad N_A = 116 \text{ N}$$
$$\Sigma F_x = 0 \qquad F_{ac} + (0.15N_A)\cos 40° - N_A \sin 40° = 0 \qquad F_{ac} = 61.2 \text{ N (C)}$$

The free-body diagram for block B, for impending motion up the plane, is given in Fig. 8.60c. For equilibrium,

$$\Sigma F_y = 0$$
$$N_B \cos 40° - (0.2N_B)\sin 40° - 350 - N_A \cos 40° - (0.15N_A)\sin 40° = 0 \qquad N_B = 706 \text{ N}$$
$$\Sigma F_{x'} = 0$$
$$P \cos 40° - 0.15N_A - 350 \sin 40° - 0.2N_B = 0 \qquad P = 501 \text{ N}$$

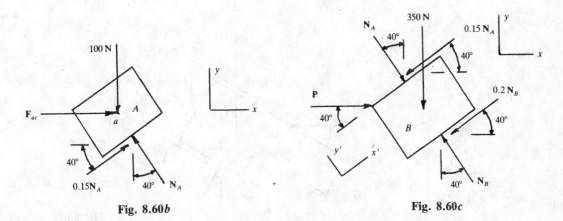

Fig. 8.60b   Fig. 8.60c

The solution for impending motion of block B down the plane is considered next. The free-body diagrams of blocks A and B are the same as those in Figs. 8.60b and c if the senses of the friction forces $0.15N_A$ and $0.2N_B$ are changed. For equilibrium of block A, using Fig. 8.60b,

$$\Sigma F_y = 0$$
$$-(0.15N_A)\sin 40° + N_A \cos 40° - 100 = 0 \qquad N_A = 149 \text{ N}$$
$$\Sigma F_x = 0$$
$$F_{ac} - (0.15N_A)\cos 40° - N_A \sin 40° = 0 \qquad F_{ac} = 113 \text{ N (C)}$$

For equilibrium of block B, with the senses of the forces $0.15N_A$ and $0.2N_B$ in Fig. 8.60c changed,

$$\Sigma F_y = 0$$
$$N_B \cos 40° + (0.2N_B)\sin 40° - 350 - N_A \cos 40° + (0.15N_A)\sin 40° = 0 \qquad N_B = 503 \text{ N}$$
$$\Sigma F_{x'} = 0$$
$$P \cos 40° + 0.15N_A - 350 \sin 40° + 0.2N_B = 0 \qquad P = 133 \text{ N}$$

If $133 \text{ N} \le P \le 501 \text{ N}$, block B will remain in the equilibrium position shown in Fig. 8.60a.

(b) The corresponding range of values of forces in the link is

$$113 \text{ N} \ge F_{ac} \ge 61.2 \text{ N}$$

8.61 **(a)** Solve Prob. 8.60 if a 120-N force acts vertically downward on the midpoint of the link.

(b) Find the forces exerted by the pins on the link.

▌(a) Figures. 8.61a and b show the free-body diagrams of block A and the link for the case of impending motion of block B up the plane. From symmetry, using Fig. 8.61a,

$$R_{ay} = R_{cy} = \frac{120}{2} = 60 \text{ N}$$

For equilibrium of block A,

$$\Sigma F_y = 0$$

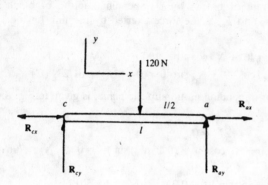

Fig. 8.61*a*

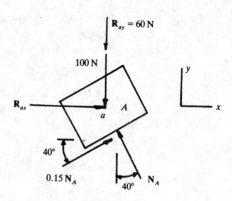

Fig. 8.61*b*

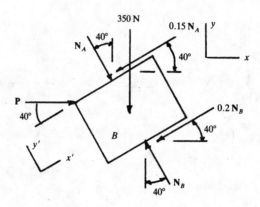

Fig. 8.61*c*

$$(0.15N_A) \sin 40° + N_A \cos 40° - 100 - 60 = 0 \qquad N_A = 186 \text{ N}$$
$$\Sigma F_x = 0$$
$$R_{ax} + (0.15N_A) \cos 40° - N_A \sin 40° = 0 \qquad R_{ax} = 98.2 \text{ N}$$

Figure 8.61*c* shows the free-body diagram of block *B*. For equilibrium of this element,

$$\Sigma F_y = 0$$
$$N_B \cos 40° - (0.2N_B) \sin 40° - 350 - N_A \cos 40° - (0.15N_A) \sin 40° = 0 \qquad N_B = 801 \text{ N}$$
$$\Sigma F_{x'} = 0$$
$$P \cos 40° - 0.15N_A - 350 \sin 40° - 0.2N_B = 0 \qquad P = 539 \text{ N}$$

The solution for impending motion of block *B* down the plane is the same as that shown above, with the senses of the friction forces $0.15N_A$ and $0.2N_B$ changed. The equilibrium requirements for block *A*, using Fig. 8.61*b*, are

$$\Sigma F_y = 0$$
$$-(0.15N_A) \sin 40° + N_A \cos 40° - 100 - 60 = 0 \qquad N_A = 239 \text{ N}$$
$$\Sigma F_x = 0$$
$$R_{ax} - (0.15N_A) \cos 40° - N_A \sin 40° = 0 \qquad R_{ax} = 181 \text{ N}$$

The equilibrium requirements for block *B*, from Fig. 8.61*c*, are

$$\Sigma F_y = 0$$
$$N_B \cos 40° + (0.2N_B) \sin 40° - 350 - N_A \cos 40° + (0.15N_A) \sin 40° = 0 \qquad N_B = 570 \text{ N}$$
$$\Sigma F_{x'} = 0$$
$$P \cos 40° + 0.15N_A - 350 \sin 40° + 0.2N_B = 0 \qquad P = 98.1 \text{ N}$$

If $98.1 \text{ N} \le P \le 539 \text{ N}$, block *B* will remain in the equilibrium position shown, with the 120-N force acting at the midpoint of the link. The corresponding range of values of *P* without the 120-N force on the link, from Prob. 8.60, is $133 \text{ N} \le P \le 501 \text{ N}$.

(b) From Fig. 8.61a,

$$\Sigma F_x = 0 \qquad -R_{cx} + R_{ax} = 0 \qquad R_{cx} = R_{ax}$$

For impending motion of block B up the plane,

$$R_{cx} = R_{ax} = 98.2\text{ N}$$
$$R_a = \sqrt{R_{ax}^2 + R_{ay}^2} = \sqrt{98.2^2 + 60^2} = 115\text{ N}$$
$$R_c = \sqrt{R_{cx}^2 + R_{cy}^2} = \sqrt{98.2^2 + 60^2} = 115\text{ N}$$

For impending motion of block B down the plane,

$$R_a = \sqrt{R_{ax}^2 + R_{ay}^2} = \sqrt{181^2 + 60^2} = 191\text{ N}$$
$$R_c = \sqrt{R_{cx}^2 + R_{cy}^2} = \sqrt{181^2 + 60^2} = 191\text{ N}$$

The range of values of $R_a = R_c$ that corresponds to the range of P found in the solution above is $191\text{ N} \geq R_a \geq 115\text{ N}$.

8.62   The two blocks in Fig. 8.62a are connected by a rod of negligible weight. Each block weighs 100 lb.

(a) Find the maximum value of the force P, acting to the right or left, which may be applied to block A without causing a force in the connecting link.

(b) Find the minimum value of the horizontal force P, acting to the left on block A, which will cause impending motion of the system. What is the force in the link for this case?

(c) Do the same as in part (b) if the force acts to the left on block B instead of on block A.

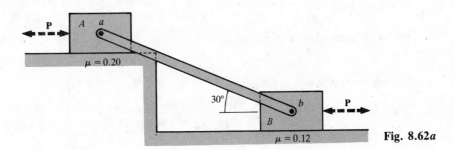

Fig. 8.62a

▌ (a) The free-body diagram of block A is shown in Fig. 8.62b, and force P is assumed to act to the left. Motion of block A is assumed to be impending. For equilibrium,

$$\Sigma F_y = 0 \qquad -100 + N_A = 0$$
$$\Sigma F_x = 0 \qquad -P + 0.2N_A = 0$$

Eliminate N_A to obtain

$$P = 20\text{ lb}$$

The same value for P would be obtained if this force were assumed to act to the right on block A.

(b) The free-body diagrams of blocks A and B, for the case where force P acts to the left on block A and motion is assumed to be impending, are shown in Figs. 8.62c and d. For equilibrium of block B,

$$\Sigma F_y = 0 \qquad F_{ab} \sin 30° - 100 + N_B = 0$$
$$\Sigma F_x = 0 \qquad -F_{ab} \cos 30° + 0.12N_B = 0$$

N_B is eliminated, with the result

$$F_{ab} = 13.0\text{ lb (T)}$$

For equilibrium of block A,

$$\Sigma F_y = 0 \qquad -100 - F_{ab} \sin 30° + N_A = 0 \qquad N_A = 107\text{ lb}$$
$$\Sigma F_x = 0 \qquad -P + 0.2N_A + F_{ab} \cos 30° = 0 \qquad P = 32.7\text{ lb}$$

(c) The free-body diagrams of blocks A and B, for the case where force P acts to the left on block B and motion is assumed to be impending, are shown in Figs. 8.62e and f. For equilibrium of block A,

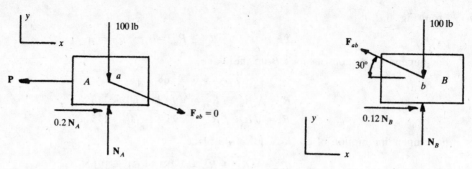

Fig. 8.62b   Fig. 8.62c

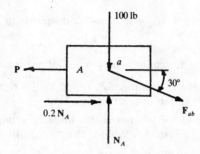

Fig. 8.62d

$$\Sigma F_y = 0 \qquad -100 + F_{ab} \sin 30° + N_A = 0$$
$$\Sigma F_x = 0 \qquad 0.2N_A - F_{ab} \cos 30° = 0$$

N_A is eliminated, with the result

$$F_{ab} = 20.7 \text{ lb (C)}$$

For equilibrium of block B,

$$\Sigma F_y = 0 \qquad -F_{ab} \sin 30° - 100 + N_B = 0 \qquad N_B = 110 \text{ lb}$$
$$\Sigma F_x = 0 \qquad F_{ab} \cos 30° + 0.12N_B - P = 0 \qquad P = 31.1 \text{ lb}$$

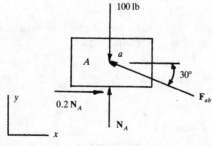

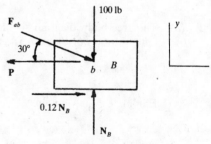

Fig. 8.62e   Fig. 8.62f

8.63 **(a)** Find the minimum value of the horizontal force P, acting to the right on block A, which will cause impending motion of the system in Prob. 8.62. What is the force in the link for this case?

(b) Do the same as in part (a) if the force P acts to the right on block B instead of on block A.

(c) Discuss the solutions for the four methods of loading the system in Fig. 8.62a.

▌ **(a)** The free-body diagrams for the case of force P acting to the right on block A are seen in Figs. 8.63a and b, and motion is assumed to be impending. For equilibrium of block B,

$$\Sigma F_y = 0 \qquad -F_{ab} \sin 30° - 100 + N_B = 0$$
$$\Sigma F_x = 0 \qquad F_{ab} \cos 30° - 0.12N_B = 0$$

N_B is eliminated between the above two equations, with the result

$$F_{ab} = 14.9 \text{ lb (C)}$$

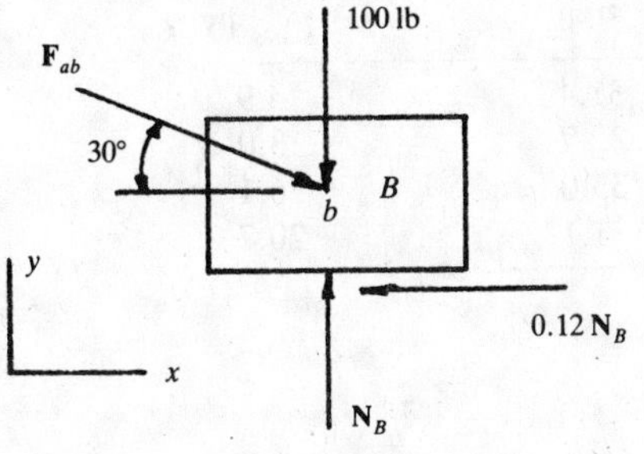

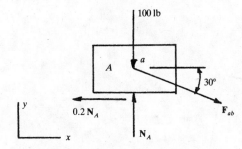

Fig. 8.63a

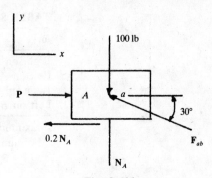

Fig. 8.63b

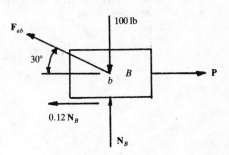

Fig. 8.63c

Fig. 8.63d

For equilibrium of block A,

$$\Sigma F_y = 0 \qquad -100 + F_{ab} \sin 30° + N_A = 0 \qquad N_A = 92.6 \text{ lb}$$
$$\Sigma F_x = 0 \qquad P - F_{ab} \cos 30° - 0.2 N_A = 0 \qquad P = 31.4 \text{ lb}$$

(b) For the case of force P acting to the right on block B, the free-body diagrams of blocks A and B have the forms shown in Figs. 8.63c and d. For equilibrium of block A,

$$\Sigma F_y = 0 \qquad -100 - F_{ab} \sin 30° + N_A = 0$$
$$\Sigma F_x = 0 \qquad -0.2 N_A + F_{ab} \cos 30° = 0$$

N_A is eliminated, with the result

$$F_{ab} = 26.1 \text{ lb (T)}$$

For equilibrium of block B,

$$\Sigma F_y = 0 \qquad F_{ab} \sin 30° - 100 + N_B = 0 \qquad N_B = 87.0 \text{ lb}$$
$$\Sigma F_x = 0 \qquad -F_{ab} \cos 30° - 0.12 N_B + P = 0 \qquad P = 33.0 \text{ lb}$$

(c) A summary of the values of P that cause impending motion, and the corresponding values of F_{ab}, for the four methods of loading in Probs. 8.62 and 8.63, is shown in Table 8.2. The maximum value of P is required when this force acts to the right on block B. The minimum value of this force is needed when the force acts to the left on block B. The percent difference between these two extreme values is

$$\%D = \frac{33.0 - 31.1}{33.0} \, 100 = 5.6\%$$

It may be seen that there is a relatively narrow range of values of P in this problem. By comparison, there is a significant difference between the extreme values of the link force F_{ab}. The minimum value of this force, 13.0 lb, occurs when P acts to the left on block A. The maximum value, 26.1 lb, occurs when P acts to the right on block B. The percent difference is

$$\%D = \frac{26.1 - 13.0}{26.1} \, 100 = 50\%$$

It may be concluded from the above results that the force in link ab is extremely sensitive to the point of application of force P, and less sensitive to the sense of this force.

TABLE 8.2

force	P, lb	F_{ab}, lb
Right on A	31.4	14.9
Left on A	32.7	13.0‡
Right on B	33.0†	26.1†
Left on B	31.1‡	20.7

†Maximum value.
‡Minimum value.

8.64 Two identical blocks rest on the inclined plane shown in Fig. 8.64a. Discuss what happens as angle θ is gradually increased from $\theta = 0°$.

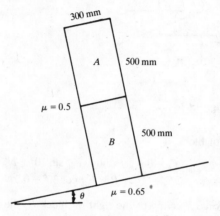

Fig. 8.64a

❚ There are four possible modes of sliding, or tipping, of the blocks.

Mode 1

Tipping of the combined system will occur when the line of action of the combined weight, $W_T = W_A + W_B$, passes through point a in Fig. 8.64b. For this case,

$$\tan \theta = \frac{150}{500} \qquad \theta = 16.7°$$

Mode 2

Sliding motion of block A with respect to block B is impending when

$$\tan \theta = \mu = 0.5 \qquad \theta = 26.6°$$

Mode 3

Sliding motion of the combined system with respect to the plane is impending when

$$\tan \theta = \mu = 0.65 \qquad \theta = 33.0°$$

Mode 4

Tipping of block A will occur when the weight force of block A acts through point b, as shown in Fig. 8.64c. For this case,

$$\tan \theta = \frac{150}{250} \qquad \theta = 31.0°$$

As θ is increased from 0°, the combined system will tip over when $\theta = 16.7°$.

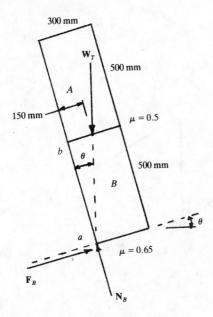

Fig. 8.64b

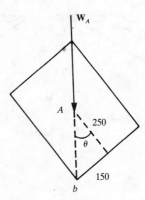

Fig. 8.64c

8.65 (*a*) Define the term *wedge*.

(*b*) Find the relationship between the force applied to a wedge and the force produced in the direction perpendicular to the applied force.

▌ (*a*) The wedge is a simple machine which is intended to transform an applied force into a force at approximately right angles to the direction of the applied force. A typical example of a wedge is shown in Fig. 8.65*a*. The wedge angle 2α is usually quite small. Typical values for this angle are 5 to 10°.

(*b*) P is the force applied to the wedge, and the weight of the wedge is assumed to be negligible compared with this force. Motion is assumed to be impending, so that the friction forces have their maximum possible values. R is the total force exerted on the wedge by the material which is in contact with the inclined faces, and ϕ_s is the angle of static friction. F_x is the x component of the force R. This force tends to separate material along the x direction.

The wedge is in equilibrium under the influence of the applied force P and the two reaction forces R. This system of three forces is a concurrent force system, and the force triangle for equilibrium is shown in Fig. 8.65*b*. From this figure,

$$R \sin(\phi_s + \alpha) = \frac{P}{2}$$

The above equation may be written in the form

$$R(\sin\phi_s \cos\alpha + \cos\phi_s \sin\alpha) = \frac{P}{2} \qquad (1)$$

The friction angle ϕ_s is related to the coefficient of friction by

$$\tan\phi_s = \mu$$

The above equation may be expressed graphically in the triangle of Fig. 8.65*c*. From this figure,

$$\sin\phi_s = \frac{\mu_s}{\sqrt{1 + \mu^2}} \qquad (2)$$

$$\cos\phi_s = \frac{1}{\sqrt{1 + \mu^2}} \qquad (3)$$

Equations (2) and (3) are used in Eq. (1) to obtain

$$R = \frac{P\sqrt{1 + \mu^2}}{2(\sin\alpha + \mu\cos\alpha)}$$

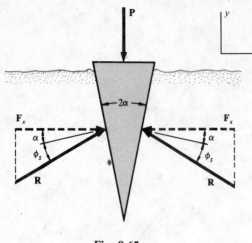

Fig. 8.65a

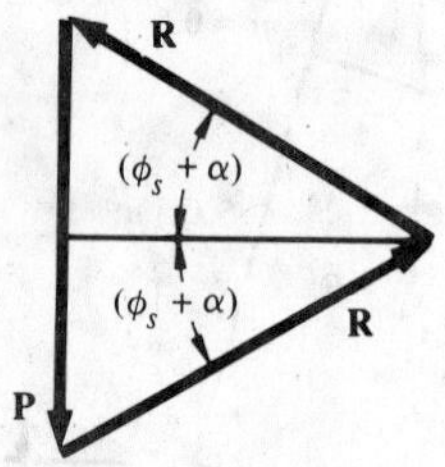

Fig. 8.65b

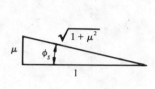

Fig. 8.65c

The force F_x in the x direction is given by

$$F_x = R \cos (\phi_s + \alpha) \tag{4}$$

Using the relationship

$$\cos (\phi_s + \alpha) = \cos \phi_s \cos \alpha - \sin \phi_s \sin \alpha$$

together with Eqs. (2) and (3), in Eq. (4), gives the final result

$$\frac{F_x}{P} = \frac{\cos \alpha - \mu \sin \alpha}{2(\sin \alpha + \mu \cos \alpha)}$$

F_x/P may now be interpreted as the ratio of the desired force effect F_x and the input force P. It is thus a representation of the *force-multiplying effect* of the wedge.

8.66 (a) Find the force-multiplying effect of a wedge if $\alpha = 3°$ and $\mu = 0.2$.
(b) Do the same as in part (a) if $\alpha = 3°$ and $\mu = 0$.
(c) Compare the results found in parts (a) and (b).

▎ (a) Using the result from Prob. 8.65,

$$\frac{F_x}{P} = \frac{\cos \alpha - \mu \sin \alpha}{2(\sin \alpha + \mu \cos \alpha)} = \frac{\cos 3° - 0.2 \sin 3°}{2(\sin 3° + 0.2 \cos 3°)} = 1.96$$

In the above case, the applied force P is approximately doubled in terms of the output force F_x of the wedge on the material which contacts the inclined faces.

(b) The force-multiplying effect of a wedge is very sensitive to the value of the coefficient of friction. A limiting case occurs if the inclined surfaces are assumed to be perfectly smooth, so that $\mu = 0$. For this case,

$$\frac{F_x}{P} = \frac{\cos \alpha}{2 \sin \alpha} = \frac{1}{2 \tan \alpha}$$

For $\alpha = 3°$, this limiting value of the force ratio would be

$$\frac{F_x}{P} = \frac{1}{2 \tan 3°} = 9.54$$

(c) Comparison of the results for F_x/P, with $\mu = 0.2$ and $\mu = 0$, reveals that there is almost a 5-to-1 difference factor in these values.

8.67 For a wedge of the type shown in Fig. 8.65a, plot F_x/P vs. μ for $\alpha = 2°$ and a range of values of μ from 0 to 0.50.

▌ For the given data,

$$\frac{F_x}{P} = \frac{\cos \alpha - \mu \sin \alpha}{2(\sin \alpha + \mu \cos \alpha)} = \frac{\cos 2° - \mu \sin 2°}{2(\sin 2° + \mu \cos 2°)}$$

The numerical results for this equation are contained in Table 8.3. These results are plotted in Fig. 8.67. It may be seen that the force ratio F_x/P drops off very rapidly for values of μ close to 0, and more slowly thereafter.

TABLE 8.3

μ	F_x/P	μ	F_x/P
0	14.3	0.30	1.49
0.10	3.69	0.40	1.13
0.20	2.11	0.50	0.92

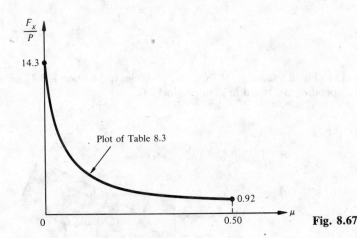

Fig. 8.67

8.68 (a) Find the general form for the force required to extract, or pull out, the wedge in Prob. 8.65.

(b) Find the value of this force if a wedge with $\alpha = 3°$ and $\mu = 0.2$ is inserted with a force of 500 N.

▌ (a) The free-body diagram of the wedge is shown in Fig. 8.68. P_1 is the force required to extract the wedge, and motion is assumed to be impending. For equilibrium of the wedge,

$$\Sigma F_y = 0 \qquad P_1 + 2N_1 \sin \alpha - 2\mu N_1 \cos \alpha = 0 \qquad P_1 = 2N_1(\mu \cos \alpha - \sin \alpha) \qquad (1)$$

In order to solve the above equation, the value of N_1 must be known. This normal force is produced by the original insertion force P.

Using the solution in Prob. 8.65, from Fig. 8.65a,

$$N_1 = R \cos \phi_s = \frac{P\sqrt{1 + \mu^2}}{2(\sin \alpha + \mu \cos \alpha)} \frac{1}{\sqrt{1 + \mu^2}} = \frac{P}{2(\sin \alpha + \mu \cos \alpha)} \qquad (2)$$

Eliminating N_1 between Eqs. (1) and (2) results in

$$P_1 = \frac{P(\mu \cos \alpha - \sin \alpha)}{\sin \alpha + \mu \cos \alpha}$$

P in the above equation is the force used to insert the wedge, and P_1 is the force required to extract the wedge.

(b) For $\alpha = 3°$, $\mu = 0.2$, and $P = 500$ N,

$$P_1 = \frac{P(\mu \cos \alpha - \sin \alpha)}{\sin \alpha + \mu \cos \alpha} = \frac{500(0.2 \cos 3° - \sin 3°)}{\sin 3° + 0.2 \cos 3°} = 292 \text{ N}$$

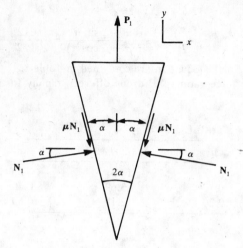

Fig. 8.68

8.69   Both blocks in Fig. 8.69a weigh 10 lb.   The coefficient of friction is 0.17 on all sliding surfaces.   Find the value of P for impending rightward motion of block A.

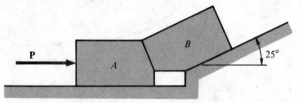

Fig. 8.69a

❙ Figure 8.69b shows the free-body diagram of block B.   For equilibrium,

$$\Sigma F_y = 0 \qquad -10 \cos 25° - 0.17N_2 + N_1 = 0$$
$$\Sigma F_x = 0 \qquad N_2 - 10 \sin 25° - 0.17N_1 = 0$$

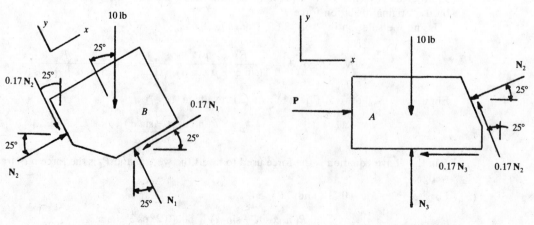

Fig. 8.69b

Fig. 8.69c

N_1 is eliminated, with the result

$$N_2 = 5.94\ \text{lb}$$

The free-body diagram of block A is given in Fig. 8.69c. For equilibrium of this element,

$$\Sigma F_y = 0 \quad -10 + N_3 - N_2 \sin 25° + (0.17N_2)\cos 25° = 0 \quad N_3 = 11.6\ \text{lb}$$
$$\Sigma F_x = 0 \quad P - 0.17N_3 - N_2 \cos 25° - (0.17N_2)\sin 25° = 0$$

Using the known values of N_2 and N_3 in the above equation,

$$P = 7.78\ \text{lb}$$

8.70 In Fig. 8.70a, block A weighs 25 N and block B weighs 18 N. The coefficient of friction at all surfaces is 0.11. For what range of values of P will the system be in equilibrium in the configuration shown?

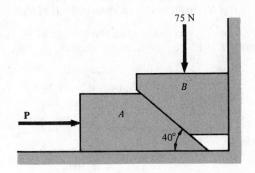

Fig. 8.70a

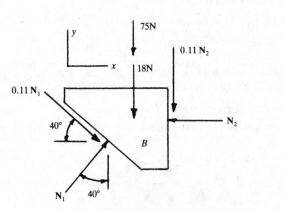

Fig. 8.70b

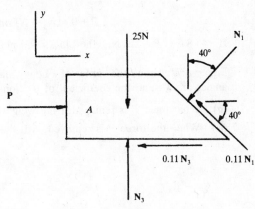

Fig. 8.70c

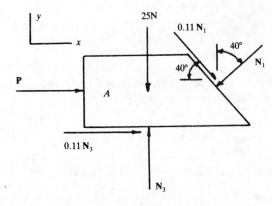

Fig. 8.70d

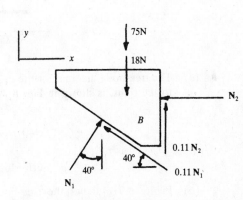

Fig. 8.70e

❙ The free-body diagram of block B, for impending upward motion of this element, is shown in Fig. 8.70b. For equilibrium of block B,

$$\Sigma F_x = 0$$
$$(0.11N_1)\cos 40° + N_1 \sin 40° - N_2 = 0 \qquad N_2 = 0.727N_1$$
$$\Sigma F_y = 0$$
$$-75 - 18 - (0.11N_1)\sin 40° + N_1 \cos 40° - 0.11N_2 = 0 \qquad N_1 = 151\text{ N}$$

Figure 8.70c shows the free-body diagram of block A for the assumption of impending upward motion of block B. For equilibrium of block A,

$$\Sigma F_y = 0$$
$$-25 - N_1 \cos 40° + (0.11N_1)\sin 40° + N_3 = 0 \qquad N_3 = 130\text{ N}$$
$$\Sigma F_x = 0$$
$$P - N_1 \sin 40° - (0.11N_1)\cos 40° - 0.11N_3 = 0 \qquad P = 124\text{ N}$$

Figures 8.70d and e show the free-body diagrams of blocks A and B, for the assumption of impending downward motion of block B. For equilibrium of block B,

$$\Sigma F_x = 0$$
$$N_1 \sin 40° - (0.11N_1)\cos 40° - N_2 = 0 \qquad N_2 = 0.559N_1$$
$$\Sigma F_y = 0$$
$$-75 - 18 + N_1 \cos 40° + (0.11N_1)\sin 40° + 0.11N_2 = 0 \qquad N_1 = 104\text{ N}$$

For equilibrium of block A,

$$\Sigma F_y = 0$$
$$-25 - N_1 \cos 40° - (0.11N_1)\sin 40° + N_3 = 0 \qquad N_3 = 112\text{ N}$$
$$\Sigma F_x = 0$$
$$P + 0.11N_3 - N_1 \sin 40° + (0.11N_1)\cos 40° = 0 \qquad P = 45.8\text{ N}$$

If $45.8\text{ N} \le P \le 124\text{ N}$, the blocks will not move.

8.71   Figure 8.71a shows the corner of a heavy machine which is to be leveled by inserting a wedge. The two wedge angles are 3°, and the coefficient of friction on all surfaces is 0.17.

(a) What force P is required if the weight to be raised is 40 kN?

(b) When the force P is removed, will the wedge remain in position?

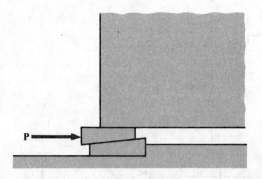

Fig. 8.71a

❙ (a) The free-body diagram for the case of raising the machine, with impending motion of the upper wedge to the right, is shown in Fig. 8.71b. For equilibrium of the wedge,

$$\Sigma F_y = 0$$
$$-40 + N \cos 3° - (0.17N)\sin 3° = 0 \qquad N = 40.4\text{ kN}$$
$$\Sigma F_x = 0$$
$$P - 0.17(40) - N \sin 3° - (0.17N)\cos 3° = 0 \qquad P = 15.8\text{ kN}$$

(b) Figure 8.71c shows the free-body diagram of the wedge with the force P removed. Impending outward motion of the wedge is assumed, and μ_m is the minimum required value of the coefficient of

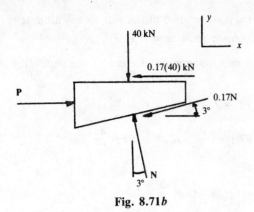

Fig. 8.71b

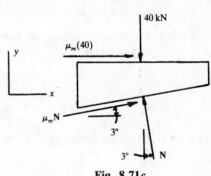

Fig. 8.71c

friction.   For equilibrium of the wedge,

$$\Sigma F_y = 0 \qquad -40 + (\mu_m N)\sin 3° + N\cos 3° = 0 \qquad N = \frac{40}{\mu_m \sin 3° + \cos 3°} \qquad (1)$$

$$\Sigma F_x = 0 \qquad \mu_m(40) + (\mu_m N)\cos 3° - N\sin 3° = 0 \qquad (2)$$

N from Eq. (1) is used in Eq. (2) to obtain

$$\mu_m^2 \sin 3° + 2\mu_m \cos 3° - \sin 3° = 0$$
$$\mu_m^2 + 2\mu_m \cot 3° - 1 = 0 \qquad \mu_m^2 + 38.2\mu_m - 1 = 0$$

The above quadratic equation may be used to find μ_m.   Six significant figures are used to avoid a large error due to subtracting two numbers of almost the same magnitude.   The result is

$$\mu_m^2 + 38.1623\mu_m - 1 = 0$$

$$\mu_m = \frac{-38.1623 \pm \sqrt{38.1623^2 - 4(1)(-1)}}{2(1)} = \frac{-38.1623 \pm 38.2147}{2} = 0.0262$$

Since $\mu = 0.17 \gg \mu_m = 0.0262$,   the wedge will not slide out.

   The solution presented in Prob. 8.68 could also have been used in this problem.   If the force P_1 in that solution were found to be a positive number, it would indicate that an externally applied force would be required to pull the wedge out.   Thus, in the absence of such a force, the wedge would remain in the position shown in Fig. 8.71a.

8.4   BELT FRICTION, FRICTION BRAKING, FRICTION FORCES IN PLANE MACHINES

8.72   Figure 8.72a shows a fixed circular drum over which passes a flexible element such as a belt, rope, or cable. β is the angle of contact of the belt on the drum.   The belt force T_2 is assumed to be constant.   As the belt tensile force T_1 is gradually increased, sliding motion of the belt with respect to the drum will be resisted by tangential friction forces between the belt and the drum.   If the magnitude of the force T_1 continues to increase, a limiting condition will be reached in which slipping motion of the belt relative to the drum is impending.

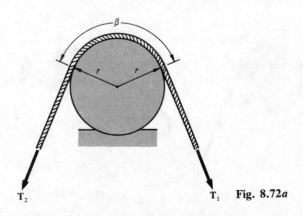

Fig. 8.72a

(a) Find the relationship between T_1 and T_2 when slipping motion of the belt on the drum is impending.

(b) Discuss the influence of the values of μ and β on the result in part (a).

▌ (a) Figure 8.72b shows the free-body diagram of a differential element of the belt. The equilibrium requirements are

$$\sum F_x = 0 \qquad dN - (T + dT)\sin\frac{d\theta}{2} - T\sin\frac{d\theta}{2} = 0$$

$$\sum F_y = 0 \qquad (T + dT)\cos\frac{d\theta}{2} + \mu\,dN - T\cos\frac{d\theta}{2} = 0$$

Using the small-angle approximations, we get

$$\sin\frac{d\theta}{2} \approx \frac{d\theta}{2} \qquad \cos\frac{d\theta}{2} \approx 1$$

and, by neglecting products of two differentials, the equilibrium equations appear as

$$\sum F_x = 0 \qquad dN - T\,d\theta = 0$$
$$\sum F_y = 0 \qquad dT + \mu\,dN = 0$$

The term dN is eliminated, with the result

$$dT + \mu T\,d\theta = 0 \qquad \frac{dT}{T} = -\mu\,d\theta \qquad \int_{T_1}^{T_2}\frac{dT}{T} = -\mu\int_0^\beta d\theta$$

$$\ln T\big|_{T_1}^{T_2} = \ln T_2 - \ln T_1 = -\mu\theta\big|_0^\beta = -\mu\beta$$

$$\ln\frac{T_2}{T_1} = -\mu\beta \qquad \ln\frac{T_1}{T_2} = \mu\beta$$

This result may be written in the more convenient form

$$\frac{T_1}{T_2} = e^{\mu\beta}$$

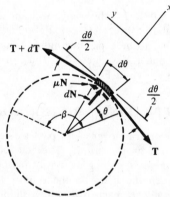

Fig. 8.72b

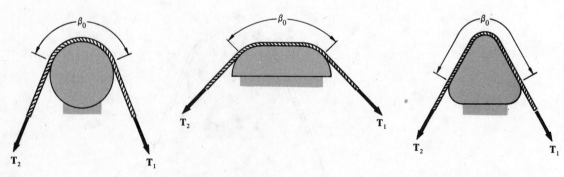

Fig. 8.72c

The above equation is the fundamental relationship between the belt tensile forces for *the case where slipping motion of the belt on the drum is impending.* μ is the coefficient of friction, and β is the angle of contact, expressed in radians. Since $e^{\mu\beta} > 1$, it follows that $T_1 > T_2$. The two terms T_1 and T_2 may be thought of as the larger tensile force and the smaller tensile force in the belt. The above result is often referred to as the belt friction equation.

It may be observed that the radius of the drum did not enter into the derivation of the above result. Thus, *this equation applies to a drum of any contour, and β is the total angle of contact.* Figure 8.72c shows three different drum profiles. The relationship between T_1 and T_2 for all three shapes, assuming the same value of μ, is

$$\frac{T_1}{T_2} = e^{\mu\beta_0} = \text{constant}$$

(b) It may be seen from consideration of the equation $T_1/T_2 = e^{\mu\beta_0}$ that the available friction force of the drum is very sensitive to the values of μ or β. In a given problem the value of μ is either known or assumed, based on the surface conditions, and materials, of the belt and the drum. Thus, μ *has a relatively narrow range of numerical values.* The value of β, by comparison, is a measure of how far around the circumference of the drum the belt is wrapped, and this value may be arbitrarily chosen.

8.73 (a) Find the minimum value of the rope force T_1 required to hold the block in the position shown in Fig. 8.73a.

(b) Do the same as in part (a) for Fig. 8.73b.

(c) Find the value of T_1 which will cause impending upward motion of the block in Fig. 8.73a.

(d) Do the same as in part (c) for Fig. 8.73b.

In all cases, find the magnitude of the resultant tangential friction force exerted by the rope on the drum.

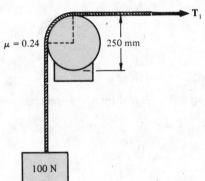

Fig. 8.73a

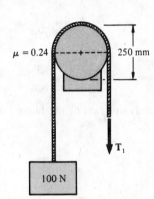

Fig. 8.73b

▌ (a) The larger of the two forces acting on the rope is 100 N. Thus,

$$\frac{100}{T_1} = e^{\mu\beta} = e^{0.24(\pi/2)} = 1.46 \qquad T_1 = \frac{100}{1.46} = 68.5 \text{ N}$$

The magnitude F of the resultant friction force exerted by the drum on the rope is $F = 100 - 68.5 = 31.5$ N.

(b) The 100-N force is again the larger force, and

$$\frac{100}{T_1} = e^{\mu\beta} = e^{0.24\pi} = 2.13 \qquad T_1 = \frac{100}{2.13} = 46.9 \text{ N}$$

The magnitude F of the friction force acting on the rope is

$$F = 100 - 46.9 = 53.1 \text{ N}$$

(c) T_1 is the larger force, so that

$$\frac{T_1}{100} = e^{\mu\beta} = e^{0.24(\pi/2)} = 1.46 \qquad T_1 = 146 \text{ N}$$

The magnitude of the friction force on the rope is

$$F = 146 - 100 = 46 \text{ N}$$

(d) T_1 is the larger force, and

$$\frac{T_1}{100} = e^{\mu\beta} = e^{0.24\pi} = 2.13 \qquad T_1 = 213 \text{ N}$$

The magnitude of the friction force is

$$F = 213 - 100 = 113 \text{ N}$$

From the above results, it may be concluded that the block in Fig. 8.73a will remain in the equilibrium position shown if $68.5 \text{ N} \leq T_1 \leq 146 \text{ N}$. The corresponding range of values of T_1 for the configuration in Fig. 8.73b is $46.9 \text{ N} \leq T_1 \leq 213 \text{ N}$.

8.74 The block in Fig. 8.74a is to be moved by applying a force T to a cable which slides over a fixed cylindrical surface. Find the value of T which will cause impending sliding motion of the block.

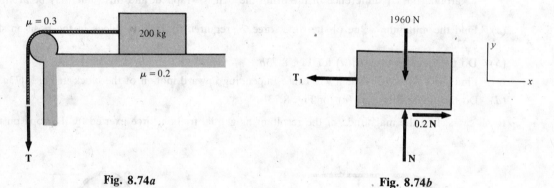

Fig. 8.74a   Fig. 8.74b

▮ The weight W of the block is

$$W = mg = 200(9.81) = 1{,}960 \text{ N}$$

The free-body diagram of the block is shown in Fig. 8.74b. The equilibrium requirements are

$$\sum F_x = 0 \qquad -T_1 + 0.2N = 0$$
$$\sum F_y = 0 \qquad -1{,}960 + N = 0$$

from which

$$T_1 = 0.2 \, N = 0.2(1{,}960) = 392 \text{ N}$$

The above value of T_1 is the lesser of the two forces acting on the cable wrapped around the curved surface. Using the belt friction equation,

$$\frac{T}{T_1} = e^{\mu\beta} \qquad \frac{T}{392} = e^{0.3(\pi/2)} = 1.60 \qquad T = 1.60(392) = 627 \text{ N}$$

8.75 (a) Find the magnitude of the total friction force exerted by the curved surface on the rope in Fig. 8.75.

(b) What is the magnitude of the weight which must be added to the 300-lb block to cause impending motion of the system?

(c) Do the same as in part (b) for the weight added to the 410-lb block.

▮ (a) The basic form of the belt friction equation for this problem is

$$\frac{T_1}{T_2} = e^{\mu\beta} = e^{0.18(117°)(\pi/180°)} = 1.44$$

For impending downward motion of the 300-lb block,

$$\frac{300}{T_2} = 1.44 \qquad T_2 = 208 \text{ lb}$$

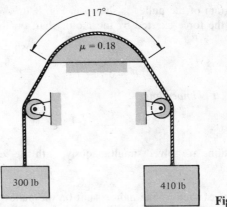

Fig. 8.75

Since   208 lb < 410 lb,   the 300-lb block does *not* move down.
   For impending downward motion of the 410-lb block,

$$\frac{410}{T_2} = 1.44 \qquad T_2 = 285 \text{ lb}$$

Since   285 lb < 300 lb,   the 410-lb block does *not* move down.
   It may therefore be concluded that, for the configuration of part (a), *motion is not impending.* It further follows that, *for this problem,*

$$\frac{T_1}{T_2} \neq e^{\mu\beta}$$

The magnitude of the friction force acting on the rope is then simply the difference of the two belt tensile forces, or

$$F = 410 - 300 = 110 \text{ lb}$$

(**b**)   For impending downward motion of the 300-lb block, the lesser belt tensile force is 410 lb.   Thus,

$$\frac{T_1}{410} = 1.44 \qquad T_1 = 590 \text{ lb}$$

The weight ΔW which must be added to the 300-lb weight is

$$\Delta W = 590 - 300 = 290 \text{ lb}$$

(**c**)   For impending downward motion of the 410-lb block, the lesser belt tensile force is 300 lb.   Therefore,

$$\frac{T_1}{300} = 1.44 \qquad T_1 = 432 \text{ lb}$$

The weight which must be added to the 410-lb weight is

$$\Delta W = 432 - 410 = 22 \text{ lb}$$

8.76   Figure 8.76a shows the pulley of a motor which transmits power through a flat belt.   The output torque of the motor is 1,050 in · lb, and the coefficient of friction between the pulley and the belt is 0.28.   What must be the initial tensile forces in the belt, and the horizontal force of the motor shaft on the pulley, if the belt is not to slip on the pulley?   The belt may be assumed to be inextensible.

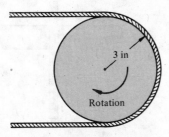

Fig. 8.76a

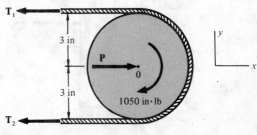

Fig. 8.76b

▌ The free-body diagram of the pulley is shown in Fig. 8.76b. The tensile force T_1 in the upper belt is the larger force, and P is the force exerted by the motor shaft on the pulley. The equations of equilibrium are

$$\sum M_0 = 0 \qquad T_1(3) - 1{,}050 - T_2(3) = 0$$

$$\sum F_x = 0 \qquad -T_1 + P - T_2 = 0$$

When slipping of the belt is impending,

$$\frac{T_1}{T_2} = e^{\mu\beta} = e^{0.28\pi} = 2.41$$

The above three equations are solved simultaneously, with the results

$$T_1 = 598 \text{ lb} \qquad T_2 = 248 \text{ lb} \qquad P = 846 \text{ lb}$$

The force P is the force exerted by the motor shaft on the pulley. It is assumed to be a constant force in the system, and it causes initial tensile forces in each belt when the motor is not running. For this latter case, the initial belt force T' in each belt is then

$$T' = \frac{846}{2} = 423 \text{ lb}$$

8.77   (a)   State how friction forces may be used in a braking system.

       (b)   Show two common examples of brake construction.

▌   (a)   A very common method of braking is to force a stationary element into contact with a rotating cylindrical drum or disk. The friction forces exerted by the braking element on the drum or disk will then either change the angular velocity of this body or hold it in a given position.

        In the following problems, several simple models of braking systems will be considered. Emphasis will be placed on the strong dependence of the braking effect on the *actual, or impending, sense of rotation of the drum*. For certain geometries, it will be seen that the brake is self-locking. This term is defined in Prob. 8.78.

    (b)   Figures 8.77a and b show two types of brake construction. The braking element may be either external or internal to a drum, as shown by the elements A and B in Fig. 8.77a. A common example of an external braking element is a contracting band about the outside of the brake drum. An example of an internal braking element is a brake shoe. The analysis of the braking effect caused by this latter element is beyond the scope of this book, and this subject is treated in texts on machine design. Figure 8.77b shows a disk brake construction. In this case, the forces exerted on the braking elements are equal and opposite, so that the resultant force normal to the disk is zero.

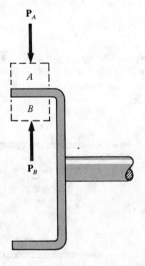

Fig. 8.77a

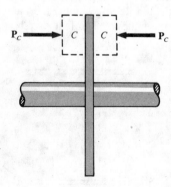

Fig. 8.77b

8.78   Figure 8.78a shows an elementary model of a braking system.   The force P applied to the lever causes a normal contact force between the braking element and the drum.   The friction force which is induced by this normal force is then used to brake the drum.

(*a*)   Find the relationship between the force P and the couple M_0 that may be resisted by the brake drum if the sense of the couple applied to the drum is clockwise.

(*b*)   Solve part (*a*) if the sense of the couple is counterclockwise.

(*c*)   Compare the result found in part (*a*) with the result found in part (*b*).

(*d*)   Under what conditions is the brake self-locking?

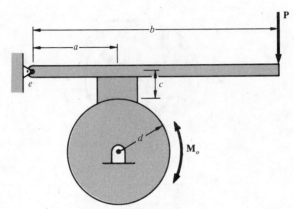

Fig. 8.78a

▍ (*a*)   The couple acting on the drum is assumed to have a clockwise sense, and the minimum value of P required for equilibrium is to be obtained.   Figure 8.78b shows the free-body diagram of the lever and the drum.   Impending slipping motion between the drum and the braking element is assumed, so that the friction force has its maximum value, μN.   For moment equilibrium of the lever,

$$\sum M_e = 0 \qquad (\mu N)(c) + Na - Pb = 0$$

and, for moment equilibrium of the drum,

$$\sum M_f = 0 \qquad (\mu N)d - M_0 = 0$$

N is eliminated between the above two equations, with the result

$$P = \frac{M_0(a + \mu c)}{\mu bd}$$

It may be seen that the required force P is *directly proportional* to the couple M_0 applied to the drum.

(*b*)   The couple M_0 acting on the drum is now assumed to have a counterclockwise sense, and the free-body diagrams for this case are shown in Fig. 8.78c.   The equilibrium requirements are

$$\sum M_e = 0 \qquad -(\mu N)c + Na - Pb = 0$$

$$\sum M_f = 0 \qquad -(\mu N)d + M_0 = 0$$

The final result for F is

$$P = \frac{M_0(a - \mu c)}{\mu bd}$$

Comparison of the above two results for P indicates that, for a given value of M_0, a smaller force P on the lever is required when M_0 acts in a counterclockwise sense.   It may also be seen from the above result for part (*b*) that P will be negative if $\mu c > a$.   A brake with this characteristic is called *self-locking*.   From consideration of Fig. 8.78c, it may be seen that the friction force acting on the lever tends to increase the normal force on the drum.   This, in turn, further increases the friction force.   In order to *release* the brake in a self-locking brake design, an *upward* force P must be applied to the lever.

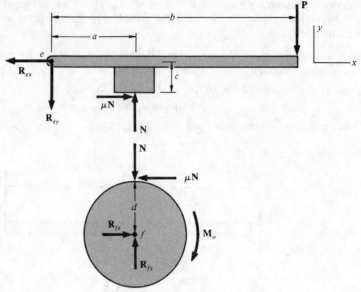

Fig. 8.78b

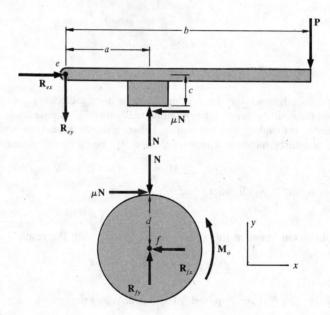

Fig. 8.78c

8.79 Figure 8.79a shows a brake and drum arrangement. The drum may be assumed to be weightless, and the vertical distance from the drum surface to the lever may be neglected.

(a) Find the minimum value of P required to hold the drum in the position shown and the corresponding pin forces at a and b.

(b) Express the forces acting on the lever and on the drum, for the condition of part (a), in formal vector notation.

▌ (a) The free-body diagrams are shown in Fig. 8.79b. The moment equilibrium requirements are

$$\sum M_a = 0 \qquad N(250) - P(700) = 0$$

$$\sum M_b = 0 \qquad (0.26N)90 - 80(150) = 0$$

The above two equations are solved, with the results

$$N = 512 \text{ N} \qquad P = 183 \text{ N}$$

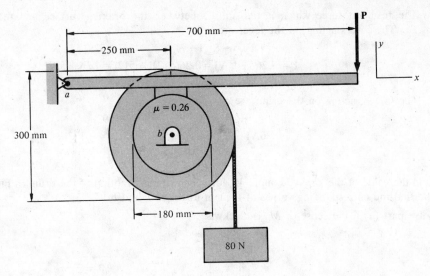

Fig. 8.79a

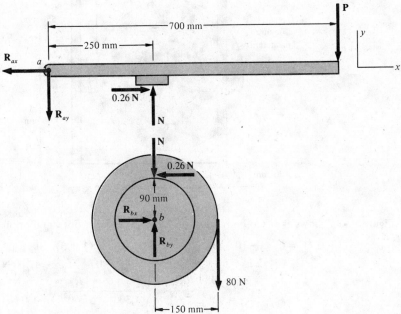

Fig. 8.79b

For force equilibrium of the brake lever,

$$\Sigma F_x = 0$$
$$-R_{ax} + 0.26N = 0 \qquad -R_{ax} + 0.26(512) = 0 \qquad R_{ax} = 133 \text{ N}$$
$$\Sigma F_y = 0$$
$$-R_{ay} + N - P = 0 \qquad -R_{ay} + 512 - 183 = 0 \qquad R_{ay} = 329 \text{ N}$$

For force equilibrium of the drum,

$$\Sigma F_x = 0$$
$$R_{bx} - 0.26N = 0 \qquad R_{bx} - 0.26(512) = 0 \qquad R_{bx} = 133 \text{ N}$$
$$\Sigma F_y = 0$$
$$R_{by} - N - 80 = 0 \qquad R_{by} - 512 - 80 = 0 \qquad R_{by} = 592 \text{ N}$$

The two hinge pin forces are

$$R_a = \sqrt{R_{ax}^2 + R_{ay}^2} = \sqrt{133^2 + 329^2} = 355 \text{ N} \qquad R_b = \sqrt{R_{bx}^2 + R_{by}^2} = \sqrt{133^2 + 592^2} = 607 \text{ N}$$

(b) The resultant force which is transmitted between the braking surface and the drum is designated **F**. The forces acting on the lever are

$$\mathbf{R}_a = -R_{ax}\mathbf{i} - R_{ay}\mathbf{j} = -133\mathbf{i} - 329\mathbf{j} \quad \text{N}$$
$$\mathbf{F} = 0.26N\mathbf{i} + N\mathbf{j} = 0.26(512)\mathbf{i} + 512\mathbf{j} = 133\mathbf{i} + 512\mathbf{j} \quad \text{N}$$
$$\mathbf{P} = -P\mathbf{j} = -183\mathbf{j} \quad \text{N}$$

The forces, acting on the drum are

$$\mathbf{R}_b = R_{bx}\mathbf{i} + R_{by}\mathbf{j} = 133\mathbf{i} + 592\mathbf{j} \quad \text{N}$$
$$\mathbf{F} = -0.26N\mathbf{i} - N\mathbf{j} = -0.26(512)\mathbf{i} - 512\mathbf{j} = -133\mathbf{i} - 512\mathbf{j} \quad \text{N}$$
$$\mathbf{W} = -80\mathbf{j} \quad \text{N}$$

8.80 (a) Find the value of the applied couple M_0 for which slipping motion of the drum in Fig. 8.80a is impending and find the corresponding values of the pin forces at a and b.

(b) Solve part (a) if the sense of M_0 is clockwise.

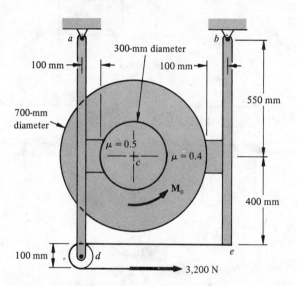

Fig. 8.80a

▌ (a) Figure 8.80b shows the free-body diagram of member ad. For equilibrium,

$$\Sigma M_a = 0 \qquad -(0.5N_1)100 - N_1(550) + 6,400(1,000) = 0 \qquad N_1 = 10,700 \text{ N}$$
$$\Sigma F_x = 0 \qquad R_{ax} - N + 6,400 = 0 \qquad R_{ax} = 4,300 \text{ N}$$
$$\Sigma F_y = 0 \qquad R_{ay} - 0.5N_1 = 0 \qquad R_{ay} = 5,350 \text{ N}$$
$$R_a = \sqrt{R_{ax}^2 + R_{ay}^2} = \sqrt{4,300^2 + 5,350^2} = 6,860 \text{ N}$$

The free-body diagram of member be is shown in Fig. 8.80c. For equilibrium,

$$\Sigma M_b = 0 \qquad -(0.4N_2)100 + N_2(550) - 3,200(950) = 0 \qquad N_2 = 5,960 \text{ N}$$
$$\Sigma F_x = 0 \qquad -R_{bx} + N_2 - 3,200 = 0 \qquad R_{bx} = 2,760 \text{ N}$$
$$\Sigma F_y = 0 \qquad -R_{by} + 0.4N_2 = 0 \qquad R_{by} = 2,380 \text{ N}$$
$$R_b = \sqrt{R_{bx}^2 + R_{by}^2} = \sqrt{2,760^2 + 2,380^2} = 3,640 \text{ N}$$

Figure 8.80d shows the free-body diagram of the drum. For equilibrium of this element,

$$\Sigma M_c = 0 \qquad M_0 - 5,350\left(\frac{150}{1,000}\right) - 2,380\left(\frac{350}{1,000}\right) = 0 \qquad M_0 = 1,640 \text{ N} \cdot \text{m}$$

(b) For the case of clockwise sense of M_0, the senses of all friction forces in Figs. 8.80b, c, and d change. For equilibrium of member ad,

$$\Sigma M_a = 0 \qquad -(0.5N_1)100 - N_1(550) + 6,400(1,000) = 0 \qquad N_1 = 12,800 \text{ N}$$
$$\Sigma F_x = 0 \qquad -N_1 + 6400 + R_{ax} = 0 \qquad R_{ax} = 6,400 \text{ N}$$
$$\Sigma F_y = 0 \qquad R_{ay} + 0.5N_1 = 0 \qquad R_{ay} = -6,400 \text{ N}$$
$$R_a = \sqrt{R_{ax}^2 + R_{ay}^2} = \sqrt{6,400^2 + (-6,400)^2} = 9,050 \text{ N}$$

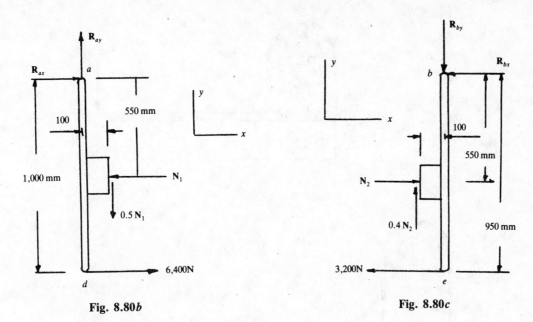

Fig. 8.80b

Fig. 8.80c

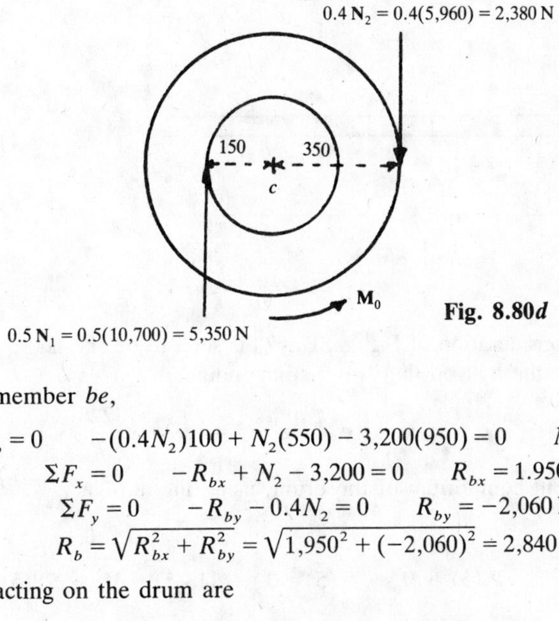

Fig. 8.80d

For equilibrium of member be,

$$\Sigma M_b = 0 \qquad -(0.4N_2)100 + N_2(550) - 3,200(950) = 0 \qquad N_2 = 5,150N$$
$$\Sigma F_x = 0 \qquad -R_{bx} + N_2 - 3,200 = 0 \qquad R_{bx} = 1.950\,\text{N}$$
$$\Sigma F_y = 0 \qquad -R_{by} - 0.4N_2 = 0 \qquad R_{by} = -2,060\,\text{N}$$
$$R_b = \sqrt{R_{bx}^2 + R_{by}^2} = \sqrt{1,950^2 + (-2,060)^2} = 2,840\,\text{N}$$

The friction forces acting on the drum are

Downward: $\qquad\qquad\qquad 0.5N_1 = 0.5(12,800) = 6,400\,\text{N}$

Upward: $\qquad\qquad\qquad 0.4N_2 = 0.4(5,150) = 2,060\,\text{N}$

For equilibrium of the drum, with clockwise sense of M_0,

$$\Sigma M_c = 0 \qquad -M_0 + 6,400\left(\frac{150}{1,000}\right) + 2,060\left(\frac{350}{1,000}\right) = 0 \qquad M_0 = 1,680\,\text{N} \cdot \text{m}$$

The braking capacity of the system is slightly larger when the drum resists a clockwise couple.

8.81 Figure 8.81a shows a model of an external contracting brake band assembly.

(**a**) Find the value of the couple M_0 which will cause impending counterclockwise motion of the drum.

(**b**) Do the same as in part (a) for impending clockwise motion.

❚ (**a**) The free-body diagrams for the condition of part (a) are shown in Fig. 8.81b. For moment equilibrium of the lever,

$$\sum M_b = 0 \qquad T_2(10) - 60(38) = 0 \qquad T_2 = 228\,\text{lb}$$

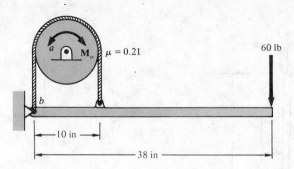

Fig. 8.81a

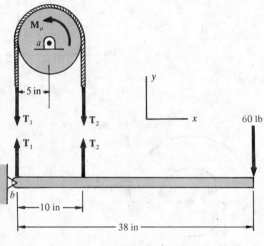

Fig. 8.81b

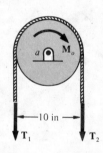

Fig. 8.81c

From consideration of Fig. 8.81b, T_2 is seen to be the larger of the two belt tensile forces. When slipping of the belt on the drum is impending,

$$\frac{T_2}{T_1} = e^{\mu\beta} = e^{0.21\pi} = 1.93 \qquad \frac{228}{T_1} = 1.93 \qquad T_1 = 118 \text{ lb}$$

For moment equilibrium of the drum, using Fig. 8.81b,

$$\sum M_a = 0$$

$$T_1(5) + M_0 - T_2(5) = 0 \qquad 118(5) + M_0 - 228(5) = 0 \qquad M_0 = 550 \text{ in} \cdot \text{lb}$$

(b) The free-body diagram of the drum for impending clockwise motion is shown in Fig. 8.81c. T_1 is now the larger of the two belt tensile forces. T_2 still has the value 228 lb, since this is a requirement of equilibrium of the lever. The belt friction equation now appears as

$$\frac{T_1}{T_2} = e^{\mu\beta} = e^{0.21\pi} = 1.93 \qquad \frac{T_1}{228} = 1.93 \qquad T_1 = 440 \text{ lb}$$

For equilibrium of the drum, using Fig. 8.81c,

$$\sum M_a = 0$$

$$T_1(5) - M_0 - T_2(5) = 0 \qquad 440(5) - M_0 - 228(5) = 0 \qquad M_0 = 1{,}060 \text{ in} \cdot \text{lb}$$

It may be seen that this brake design is much more efficient when resisting clockwise couples on the brake drum.

8.82 A weight W is to be raised by applying a couple M_0 to pulley A in Fig. 8.82a. It is assumed that the belt will not slip on the pulleys. Find the mass of the maximum weight that may be raised. Angle θ is given by $\sin\theta = (d_2 - d_1)/2a$, and the coefficient of friction between the belt and the pulleys is 0.22.

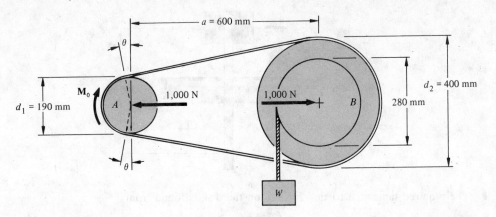

Fig. 8.82a

| Angle θ is found from

$$\sin \theta = \frac{d_2 - d_1}{2a} = \frac{400 - 190}{2(600)} \qquad \theta = 10.1°$$

The limiting condition for slipping of the belt on pulley A is found next.

$$\beta = 180° - 2\theta = 180° - 2(10.1°) = 160° \qquad \beta = 160°\left(\frac{\pi}{180°}\right) = 2.79 \text{ rad}$$

$$\frac{T_1}{T_2} = e^{\mu}\beta = e^{0.22(2.79)} = 1.85 \qquad (1)$$

Figure 8.82b shows the free-body diagram of pulley A. For equilibrium in the x direction,

$$\Sigma F_x = 0 \qquad (T_1 + T_2) \cos 10.1° - 1,000 = 0 \qquad T_1 + T_2 = 1,020 \text{ lb} \qquad (2)$$

Using Eq. (1) in Eq. (2),

$$1.85 T_2 + T_2 = 1,020 \qquad T_2 = 358 \text{ N}$$
$$T_1 = 1.85 T_2 = 1.85(358) = 662 \text{ N}$$

The free-body diagram of pulley B is seen in Fig. 8.82c. For equilibrium of this element,

$$\Sigma M_a = 0 \qquad (358 - 662)(200) + W(140) = 0 \qquad W = 434 \text{ N}$$

$$m = \frac{W}{g} = \frac{434}{9.81} = 44.2 \text{ kg}$$

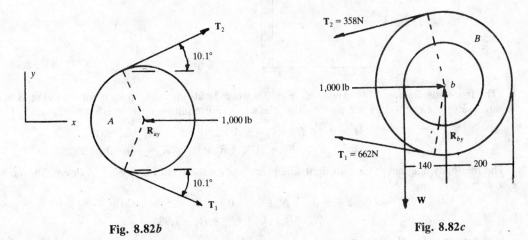

Fig. 8.82b   **Fig. 8.82c**

8.83   Figure 8.83 shows a model of a disk brake. What minimum value P of the force on the disk brake pads is required if the disk is not to slip when acted on by a couple of 1,070 N · m? The coefficient of friction between the pads and the disk is assumed to be 0.19.

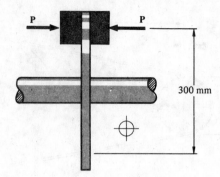

Fig. 8.83

▮ The required tangential friction force F on the disk is found from

$$F(150 \text{ mm})\left(\frac{1 \text{ m}}{1,000 \text{ mm}}\right) = 1,070 \text{ N} \cdot \text{m} \qquad F = 7,130 \text{ N}$$

The maximum value of the friction force exerted by each pad is $F_1 = \mu N = 0.19P$. Thus, since both pads exert equal friction forces on the disk,

$$F_{\text{total}} = 2(0.19P) = F = 7,130 \text{ N} \qquad P = 18,800 \text{ N} = 18.8 \text{ kN}$$

It may be observed that the disk brake operation is insensitive to the sense of the applied couple.

8.84   Fig. 8.84a shows a proposed design for a heavy-duty industrial disc brake. Find the maximum value of the torque which may be resisted by the disc brake.

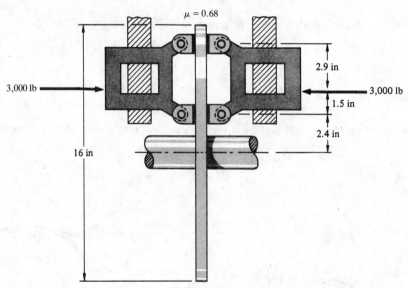

Fig. 8.84a

▮ The free-body diagram of the right-side element which holds the brake pads is shown in Fig. 8.84b. R_a and R_b are the forces exerted by the pins on this element. For equilibrium of this element,

$$\Sigma M_a = 0 \qquad R_b(4.4) - 3,000(2.9) = 0 \qquad R_b = 1,980 \text{ lb}$$
$$\Sigma F_x = 0 \qquad R_a - 3,000 + R_b = 0 \qquad R_a = 1,020 \text{ lb}$$

The free-body diagram of the two right-side brake pads is shown in Fig. 8.84c. For equilibrium of these two elements,

$$\Sigma F_x = 0 \qquad N_a - R_a = 0 \qquad N_a = R_a = 1,020 \text{ lb}$$
$$N_b - R_b = 0 \qquad N_b = R_b = 1,980 \text{ lb}$$

The maximum value of torque which may be resisted by the brake corresponds to impending sliding motion of the disc relative to the brake pads. From symmetry, the left-side pads exert the same normal forces as those exerted by the right-side pads. The total torque M that may be resisted by the brake disc is then

$$M = 2[0.68(1,020)6.8 + 0.68(1,980)2.4] = 15,900 \text{ in} \cdot \text{lb} = 1,330 \text{ ft} \cdot \text{lb}$$

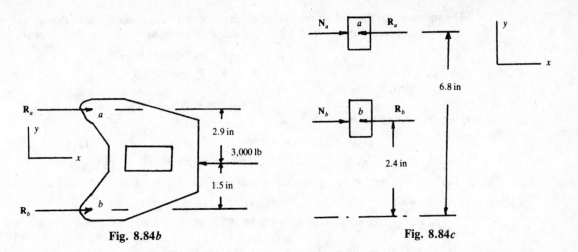

Fig. 8.84b   Fig. 8.84c

8.85 (a) For what range of values of P_2 will the simple machine shown in Fig. 8.85a remain in equilibrium in the position shown if $\mu = 0.15$ between the slider and the guide.

  (b) Find the numerical results for the solution in part (a) if $P_1 = 12$ kN.

  (c) Compare the results in part (b) with the solution to Prob. 7.47.

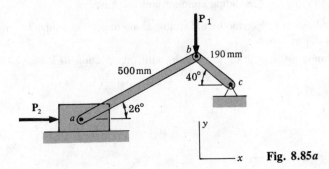

Fig. 8.85a

▌ (a) Both of the links are two-force members. Figure 8.85b shows the free-body diagram of the slider block for impending motion of this element to the right. For equilibrium of this element,

$$\Sigma F_y = 0 \qquad N - F_{ab} \sin 26° = 0$$
$$\Sigma F_x = 0 \qquad P_2 - F_{ab} \cos 26° - 0.15N = 0$$

N_2 is eliminated between the above two equations, with the result

$$F_{ab} = 1.04 P_2 \text{ (C)}$$

The free-body diagram of pin b is shown in Fig. 8.85c. For equilibrium,

$$\Sigma F_x = 0 \qquad F_{ab} \cos 26° - F_{bc} \cos 40° = 0 \qquad F_{bc} = 1.22 P_2 \text{ (C)}$$
$$\Sigma F_y = 0 \qquad -P_1 + F_{ab} \sin 26° + F_{bc} \sin 40° = 0 \qquad P_2 = 0.806 P_1$$

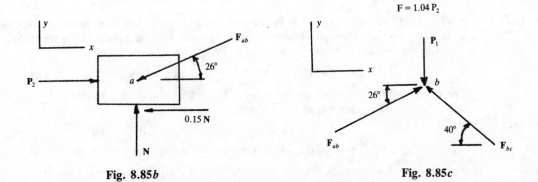

Fig. 8.85b   Fig. 8.85c

The solutions for the case of impending sliding motion of the slider block to the left are the same as those given above, with the sense of the friction force changed. For equilibrium of the block,

$$\Sigma F_x = 0 \qquad P_2 - F_{ab} \cos 26° + 0.15N = 0$$
$$\Sigma F_y = 0 \qquad N - F_{ab} \sin 26° = 0$$

N_2 is eliminated, to obtain

$$F_{ab} = 1.20P_2 \text{ (C)}$$

For equilibrium of the pin,

$$\Sigma F_x = 0 \qquad 1.20P_2 \cos 26° - F_1 \cos 40° = 0 \qquad F_{bc} = 1.41P_2 \text{ (C)}$$
$$\Sigma F_y = 0 \qquad -P_1 + F_{ab} \sin 26° + F_{bc} \sin 40° = 0 \qquad P_2 = 0.699P_1$$

If $0.699P_1 \le P_2 \le 0.806P_1$, the machine will remain in the equilibrium position shown in Fig. 8.85a.

(b) For $P_1 = 12$ kN,

$$P_{2,\min} = 0.699P_1 = 0.699(12) = 8.39 \text{ kN} \qquad P_{2,\max} = 0.806P_1 = 0.806(12) = 9.67 \text{ kN}$$

(c) The force analysis of the machine in Fig. 8.85a, for the case of no friction, was presented in Prob. 7.47. The result found in that problem was the single force $P_2 = 9.04$ kN required to keep the machine in equilibrium. It can be easily shown that this force is the mean of the two forces $P_{2,\max}$ and $P_{2,\min}$ found in part (b).

8.86 The system of Prob. 7.49 is repeated in Fig. 8.86a. A force of 40 lb is applied to the handle, and the coefficient of friction between the sliding element and the guide is 0.25.

(a) Find the force exerted by the sliding element on the compacted material and the forces exerted by the pins on the members.

(b) Compare the solution in part (a) with the solution to Prob. 7.49.

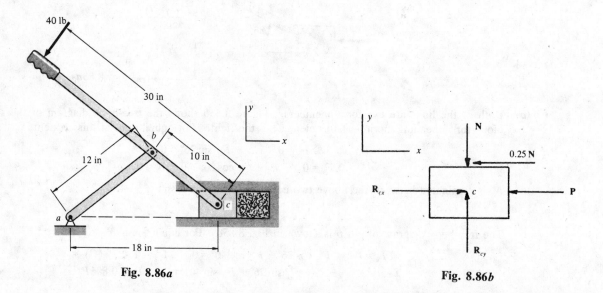

Fig. 8.86a   Fig. 8.86b

▌ (a) See Fig. 7.49b in Prob. 7.49. P is the force exerted by the sliding element on the material. Member ab is a two-force member. The forces acting on member bc are unaffected by the inclusion of friction forces in the problem. From the solution to Prob. 7.49,

$$F_{ab} = 128 \text{ lb (C)} \qquad R_{cx} = 81.7 \text{ lb} \qquad R_{cy} = 40.9 \text{ lb} \qquad R_c = 91.4 \text{ lb}$$

The free-body diagram of the slider is seen in Fig. 8.86b, and motion of this element is assumed to be impending. For equilibrium of this element,

$$\Sigma F_y = 0 \qquad -N + R_{cy} = 0 \qquad N = R_{cy} = 40.9 \text{ lb}$$
$$\Sigma F_x = 0 \qquad R_{cx} - 0.25N - P = 0 \qquad P = 81.7 - 0.25(40.9) = 71.5 \text{ lb}$$

For the case of no friction, from Prob. 7.49, $P = 81.7$ lb. The percent difference is

$$\%D = \frac{71.5 - 81.7}{81.7}(100) = 12.5\%$$

The effect of friction is to reduce the value of the compacting force P by 12.5 percent.

8.87 The system of Prob. 7.53 is repeated in Fig. 8.87a. Someone pulls with a force of 200 N on the handle of the machine, and the coefficient of friction on the sliding surfaces is 0.18.

(a) Find the value of the compressive forces P on the slider blocks, the forces in members ab and bd, and the pin force at c.

(b) Compare the solution in part (a) with the solution to part (a) of Prob. 7.53.

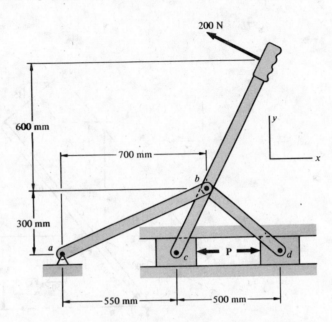

Fig. 8.87a

▌ (a) Motion of the slider blocks is assumed to be impending. The free-body diagrams of these two elements are shown in Figs. 8.87b and c. For equilibrium of slider c,

$$\Sigma F_y = 0 \qquad -R_{cy} + N_c = 0$$
$$\Sigma F_x = 0 \qquad R_{cx} - P - 0.18N_c = 0$$

N_c is eliminated, with the result

$$R_{cx} - P - 0.18R_{cy} = 0 \qquad (1)$$

For equilibrium of slider d,

$$\Sigma F_y = 0 \qquad -N_d + \frac{300}{461}F_{bd} = 0$$

$$\Sigma F_x = 0 \qquad P + 0.18N_d - \frac{350}{461}F_{bd} = 0$$

N_d is eliminated, with the result

$$P + 0.18\left(\frac{300}{461}F_{bd}\right) - \frac{350}{461}F_{bd} = 0$$
$$F_{bd} = 1.56P \qquad (2)$$

The next step is to obtain R_{cx} and R_{cy} in terms of force P. Figure 8.87d shows the free-body diagram of member ce. For moment equilibrium,

$$\Sigma M_b = 0 \qquad -300R_{cx} - 150R_{cy} + 200(671) = 0 \qquad (3)$$

R_{cx} is eliminated between Eqs. (1) and (3), with the result

$$R_{cy} = -1.47P + 658 \qquad (4)$$

Using Eq. (4) in Eq. (1),

$$R_{cx} = 0.735P + 118 \qquad (5)$$

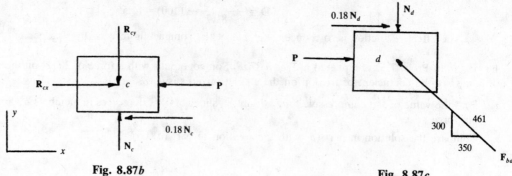

Fig. 8.87b

Fig. 8.87c

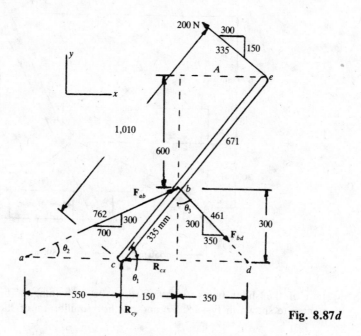

Fig. 8.87d

For force equilibrium of member ce,

$$\Sigma F_x = 0 \qquad -R_{cx} + \frac{700}{762} F_{ab} + \frac{350}{461} F_{bd} - \frac{300}{335}(200) = 0 \qquad (6)$$

$$\Sigma F_y = 0 \qquad R_{cy} + \frac{300}{762} F_{ab} - \frac{300}{461} F_{bd} + \frac{150}{335}(200) = 0 \qquad (7)$$

Eq. (5) is substituted into Eq. (6), and Eq. (2) is used, with the result

$$-(0.735P + 118) + \frac{700}{762} F_{ab} + \frac{350}{461}(1.56P) - \frac{300}{335}(200) = 0 \qquad 0.449P + 0.919F_{ab} - 297 = 0$$

$$(8)$$

Eq. (4) is substituted into Eq. (7), and Eq. (2) is used, with the result

$$(-1.47P + 658) + \frac{300}{762} F_{ab} - \frac{300}{461}(1.56P) + \frac{150}{335}(200) = 0 \qquad -2.49P + 0.394F_{ab} + 748 = 0 \quad (9)$$

F_{ab} is eliminated between Eqs. (8) and (9), with the result

$$P = 326 \text{ N}$$

Using Eq. (8),

$$0.449(326) + 0.919F_{ab} - 297 = 0 \qquad F_{ab} = 164 \text{ N (C)}$$

Using Eq. (2),

$$F_{bd} = 1.56(326) = 509 \text{ N (C)}$$

Using Eq. (4),

$$R_{cy} = -1.47(326) + 658 = 179 \text{ N}$$

Using Eq. (5),

$$R_{cx} = 0.735(326) + 118 = 358 \text{ N}$$
$$R_c = \sqrt{R_{cx}^2 + R_{cy}^2} = \sqrt{358^2 + 178^2} = 400 \text{ N}$$

(b) A comparison of the solutions with and without friction is shown in Table 8.4. It may be seen that friction reduces the compacting force by 12 percent.

Table 8.4

	$\mu = 0$	$\mu = 0.18$	%D[†]
P	371	326	−12
F_{ab}	196	164	−15
F_{bd}	490	509	4
R_{cx}	371	358	−4
R_{cy}	152	179	17
R_c	400	400	0

[†]Compared with $\mu = 0$.

8.88 Give a summary of the basic concepts of analysis of friction forces.

❚ Impending, or actual, relative sliding motion of one body on another body is resisted by forces called friction forces. These forces lie in the plane of contact between the two bodies. The sense of these friction forces is always such as to oppose impending or actual sliding motion.

The friction force that one surface can exert on a second surface reaches a maximum value when relative sliding motion of the surfaces is impending. This maximum force is given by

$$F_{max} = \mu_s N$$

where μ_s is the coefficient of static friction and N is the normal contact force between the surfaces. If sliding motion actually occurs, the retarding friction force has the magnitude $\mu_k N$, where μ_k is the coefficient of kinetic friction. The coefficient of kinetic friction is always less than the coefficient of static friction. Throughout this text, it is assumed that $\mu_s \approx \mu_k = \mu$, so that the maximum possible friction force is expressed by $F_{max} = \mu N$.

ϕ_s is the angle of static friction. It is the angle between the resultant force acting on the surface and the normal to this surface, when sliding motion of these surfaces is impending. This angle is related to the coefficient of friction by

$$\tan \phi_s = \mu$$

The angle of repose is the angle between the horizontal direction and the inclined surface of a pile of dry, granular material. This angle is equal to ϕ_s, the angle of static friction.

If the mating surfaces do not have impending sliding motion with respect to each other, the friction forces must be found by using the equations of static equilibrium.

When sliding motion of a belt or cable over a curved surface is impending, the two belt or cable forces are related by

$$\frac{T_1}{T_2} = e^{\mu\beta}$$

where T_1 is the larger of the two forces, β is the angle of contact, and μ is the coefficient of friction between the belt and the curved surface.

CHAPTER 9
Centroids of Plane Areas and Curves

9.1 DEFINITION OF THE CENTROID, CENTROIDS OF COMPOSITE AREAS, FIRST MOMENT OF AREA, CENTROIDS OF PATTERNS OF HOLE AREAS

9.1   Figure 9.1 shows a thin plate of homogeneous material and constant thickness.   The weight force of the plate is W.   The xy coordinate axes lie in the top surface plane of the plate.   Show how this model may be used to develop a physical interpretation of the concept of a *centroid* of a plane area.

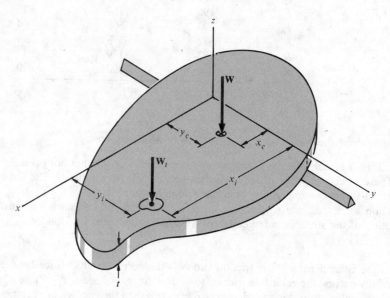

Fig. 9.1

▌   As a motivation toward ultimately obtaining the definition of a term which is referred to as a centroid, a knife edge is imagined to be positioned beneath the y axis, as shown in Fig. 9.1.   The plate is next imagined to be subdivided into n small elements.   The weight force of each of these elements will be designated W_i, where $i = 1, 2, 3, \ldots, n$.   The coordinates of the weight elements are x_i and y_i.   The system consisting of the weight forces of all the individual elements is a three-dimensional parallel force system.   The resultant force of this system of forces is the weight W of the plate.   The magnitude of this weight is given by

$$W = \sum_i W_i$$

It is now desired to find the location on the plate of the line of action of the weight force of the plate.   The coordinates of this point are defined by the coordinates x_c and y_c shown in Fig. 9.1.   Each of the elemental weight forces W_i has a moment about the y axis of magnitude $W_i x_i$.   The total moment of all of these elemental weight forces about the y axis is then

$$M_y = \sum_i W_i x_i$$

It may be observed from Fig. 9.1 that certain of the x_i terms will have negative values.   Thus, the above equation implicitly gives the *net* moment of the weight forces of the elements about the y axis.   The moment M_y must be equal to the moment of the total weight force W of the plate about the y axis, so that

$$Wx_c = \left(\sum_i W_i\right)x_c = \sum_i W_i x_i \qquad x_c = \frac{\sum_i W_i x_i}{\sum_i W_i} = \frac{\sum_i W_i x_i}{W}$$

A knife edge is now imagined to be placed beneath the x axis. A computation similar to the one above, for the resultant moment about the x axis, results in

$$y_c = \frac{\sum_i W_i y_i}{\sum_i W_i} = \frac{\sum_i W_i y_i}{W}$$

The location of the line of action of the resultant weight force W of the plate is given by the two coordinates x_c and y_c. The plane surface area of the plate is now designated A, with the elements A_i. The thickness of the plate is t, and γ is the specific weight of the plate material, with the units lb/ft^3 or N/m^3. With these definitions, the above equations may be written as

$$x_c = \frac{\sum_i (A_i t \gamma) x_i}{At\gamma} = \frac{t\gamma \sum_i x_i A_i}{t\gamma A} = \frac{\sum_i x_i A_i}{A} \qquad y_c = \frac{\sum_i (A_i t \gamma) y_i}{At\gamma} = \frac{t\gamma \sum_i y_i A_i}{t\gamma A} = \frac{\sum_i y_i A_i}{A}$$

The point on the plane area A located by the coordinates x_c and y_c is defined to be the *centroid* of this area.

9.2 (a) State the formal definition of the centroidal coordinates of a plane area.

(b) How is the location of the centroid affected by the placement of the xy axes with respect to the area?

I (a) If the subdivision of the plate area in Fig. 9.1 into the elemental areas A_i is continued, a limiting condition is reached in which the summation operations in Prob. 9.1 become integral operations, so that

$$x_c = \lim_{A_i \to 0} \frac{\sum_i x_i A_i}{A} = \frac{\int_A x\, dA}{A} \qquad \text{and} \qquad y_c = \lim_{A_i \to 0} \frac{\sum_i y_i A_i}{A} = \frac{\int_A y\, dA}{A}$$

The above two equations are the formal definition of the centroidal coordinates x_c and y_c.

Many authors use the symbols $\bar{x}$ and $\bar{y}$ to define the coordinates of the centroid. The use of these terms is intentionally avoided in this text, to avoid confusion with the use of a bar when writing a vector quantity.

The centroidal coordinates for several elementary plane area shapes are given in Table 9.9 at the end of this chapter. The values of the centroidal coordinates x_c and y_c in the table are found by direct integration of the above two equations.

(b) The location of the centroid is a function only of the *shape* of the area, and this location is *independent* of the placement of the coordinate axes with respect to the area.

9.3 Show that the centroidal coordinates of the triangular area in Fig. 9.3a are

$$x_c = \tfrac{2}{3}a \qquad y_c = \tfrac{1}{3}b$$

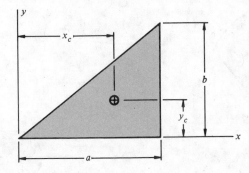

Fig. 9.3a

I The centroidal coordinate x_c will be found first. The area element dA for this case, shown in Fig. 9.3b, is

$$dA = \frac{b}{a} x\, dx$$

$$x_c = \frac{\int_A x\, dA}{A} = \frac{\int_0^a x\left(\dfrac{b}{a} x\right) dx}{\tfrac{1}{2} ab} = \frac{\dfrac{b}{a}\displaystyle\int_0^a x^2\, dx}{\tfrac{1}{2} ab} = \frac{\dfrac{b}{a}\left[\dfrac{x^3}{3}\right]_0^a}{\tfrac{1}{2} ab} = \frac{\dfrac{b}{a}\left(\dfrac{a^3}{3}\right)}{\tfrac{1}{2} ab} = \frac{2}{3}a$$

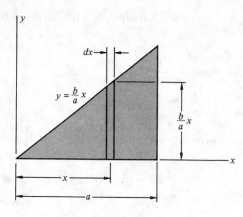

Fig. 9.3b

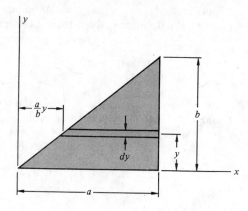

Fig. 9.3c

When solving for y_c, the area element in Fig. 9.3c is used.

$$dA = \left(a - \frac{a}{b} y\right) dy$$

$$y_c = \frac{\int y\, dA}{A} = \frac{\int_0^b y\left(a - \frac{a}{b} y\right) dy}{\frac{1}{2}ab} = \frac{\int_0^b \left(ay - \frac{a}{b} y^2\right) dy}{\frac{1}{2}ab} = \frac{\left[\frac{ay^2}{2} - \frac{a}{b}\frac{y^3}{3}\right]_0^b}{\frac{1}{2}ab} = \frac{\frac{ab^2}{2} - \frac{a}{b}\frac{b^3}{3}}{\frac{1}{2}ab} = \frac{\frac{ab^2}{6}}{\frac{1}{2}ab} = \frac{1}{3} b$$

9.4 (a) If a plane area has an axis of symmetry, show that the centroid of the area must lie on this axis.

(b) Where is the centroid of a plane area located if the area has two axes of symmetry?

▌ (a) Figure 9.4a shows a plane area which has an axis of symmetry. The moments about the y axis of the equally spaced, typical area elements A_i cancel one another. Thus it may be concluded that the centroid of an area which has an axis of symmetry must lie on this axis.

(b) If the area has two axes of symmetry, following the reasoning presented in part (a) above, it may be concluded that the centroid of this area lies on the intersection of the two axes of symmetry. Figure 9.4b shows an area with two axes of symmetry. The centroid for such a case is at the "center" of the area.

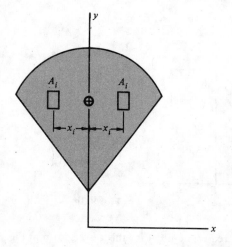

Fig. 9.4a

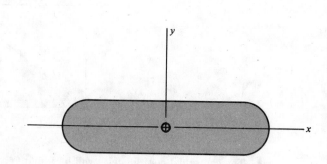

Fig. 9.4b

9.5 Show how the coordinates of the centroid of a composite area may be found.

▌ In many problems it is necessary to find the centroid of an area which is not one of the simple shapes shown in Table 9.9. A typical example of such a composite area is shown in Fig. 9.5a.

The technique of solution for this type of problem is to divide the original composite area into areas of elementary shapes, each of whose centroidal coordinates are known. The equations for the centroidal coordinates may then be written as

$$x_c = \sum_i \frac{x_i A_i}{A} = \frac{x_1 A_1 + x_2 A_2 + \cdots}{A_1 + A_2 + \cdots} \qquad \text{and} \qquad y_c = \sum_i \frac{y_i A_i}{A} = \frac{y_1 A_1 + y_2 A_2 + \cdots}{A_1 + A_2 + \cdots}$$

The quantities $A_1, A_2, \ldots$ are the elementary areas into which the original area has been subdivided, and the terms $x_1, x_2, \ldots, y_1, y_2, \ldots$ are the corresponding centroidal coordinates of these elementary areas.

The original composite area may be subdivided into elementary areas in different ways. Figures 9.5b and c show two methods of subdividing the area of Fig. 9.5a. This subdivision is done in such a way as to minimize the computational effort required in the subsequent calculations.

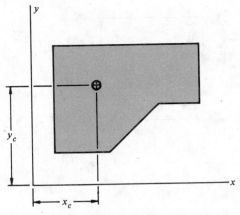

Fig. 9.5a

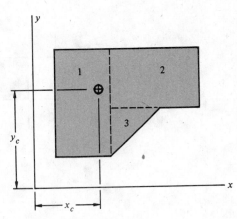

Fig. 9.5b

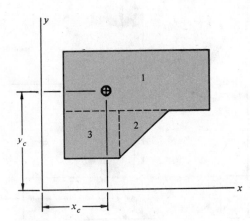

Fig. 9.5c

9.6 The area shown in Fig. 9.6a represents the cross section of a beam. Find the centroidal coordinates of this area.

❚ The given area is divided into the two rectangular areas shown in Fig. 9.6b.

From symmetry considerations,

$$x_c = 0$$

The centroidal coordinate y_c is found by using the equation presented in Prob. 9.5. A systematic technique will now be shown for using this equation and the similar equation form for x_c.

The centroidal coordinate y_1 and the area A_1 of element 1 are entered into the equation, and this particular construction appears as

$$y_c = \frac{y_1 A_1 +}{A_1 +} = \frac{(40 + \frac{175}{2})40(175) +}{40(175) +}$$

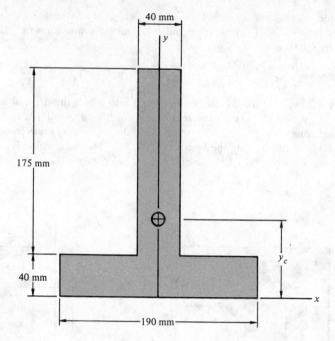

Fig. 9.6a

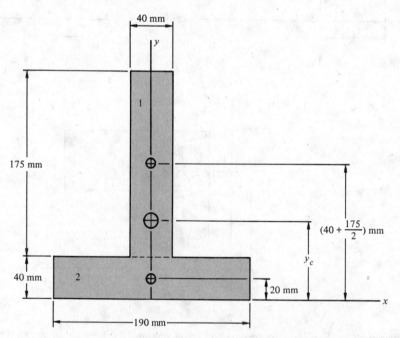

Fig. 9.6b

The area term is entered simultaneously in the numerator and the denominator of the centroidal coordinate equation for y_c. The effects of the terms y_2 and A_2 are then added into the numerator and denominator of the equation to obtain the final form

$$y_c = \frac{y_1 A_1 + y_2 A_2}{A_1 + A_2} = \frac{(40 + \frac{175}{2})40(175) + 20(190)40}{40(175) + 190(40)} = 71.5 \text{ mm}$$

The final statement of the solution is then

$$x_c = 0 \qquad y_c = 71.5 \text{ mm}$$

9.7 Find the centroidal coordinates of the area shown in Fig. 9.7a.

The given area is divided into the two simple area shapes shown in Fig. 9.7b. The centroidal coordinates are then found to be

$$x_c = \frac{x_1 A_1 + x_2 A_2}{A_1 + A_2} = \frac{(a/2)(a)2a + \frac{3}{2}a(a)^2}{a(2a) + a^2} = \frac{5}{6}a$$

Since the area has an axis of symmetry at 45° with the x and y axes, $y_c = x_c$.

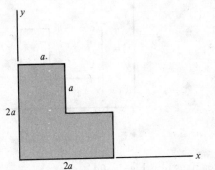

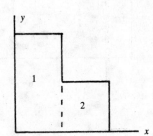

Fig. 9.7a   Fig. 9.7b

9.8 Find x_c and y_c for the area shown in Fig. 9.8a.

Figure 9.8b shows the division of the area into simple shapes. x_c and y_c are then found as

$$x_c = \frac{x_1 A_1 + x_2 A_2}{A_1 + A_2} = \frac{\frac{25}{2}(25)125 + [25 + \frac{1}{2}(65)]65(25)}{25(125) + 65(25)} = 27.9 \text{ mm}$$

$$A = 4,750 \text{ mm}^2$$

$$y_c = \frac{y_1 A_1 + y_2 A_2}{A_1 + A_2} = \frac{\frac{125}{2}(25)125 + \frac{25}{2}(65)25}{4,750} = 45.4 \text{ mm}$$

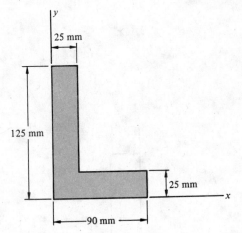

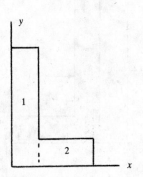

Fig. 9.8a   Fig. 9.8b

9.9 Find the location of the centroid of the area shown in Fig. 9.9a.

The division of the original area is seen in Fig. 9.9b. The values of the centroidal coordinates are then

$$x_c = \frac{x_1 A_1 + x_2 A_2 + x_3 A_3}{A_1 + A_2 + A_3} = \frac{\frac{25}{2}(25)125 + (25 + \frac{65}{2})65(25) + (25 + 65 + \frac{25}{2})25(60)}{25(125) + 65(25) + 25(60)} = 45.8 \text{ mm}$$

$$A = 6,250 \text{ mm}^2$$

$$y_c = \frac{y_1 A_1 + y_2 A_2 + y_3 A_3}{A_1 + A_2 + A_3} = \frac{\frac{125}{2}(25)125 + \frac{25}{2}(65)25 + \frac{60}{2}(25)60}{6,250} = 41.7 \text{ mm}$$

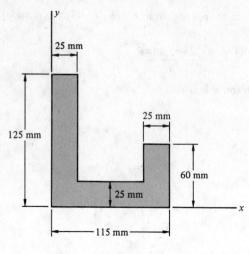

Fig. 9.9a

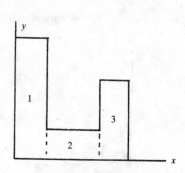

Fig. 9.9b

9.10 Find the values of the centroidal coordinates of the area shown in Fig. 9.10a.

▌ Figure 9.10b shows the division of the original area into elementary shapes. From symmetry,

$$x_c = \frac{90 + 25}{2} = 57.5 \text{ mm}$$

y_c is found from

$$y_c = \frac{y_1 A_1 + y_2 A_2 + y_3 A_3}{A_1 + A_2 + A_3} = \frac{\frac{125}{2}(25)125 + \frac{25}{2}(90)25 - \frac{100}{2}(25)100}{25(125) + 90(25) + 25(100)} = 12.5 \text{ mm}$$

It should be noted that y_3, the centroidal coordinate of area 3, is entered into the above equation as a negative quantity.

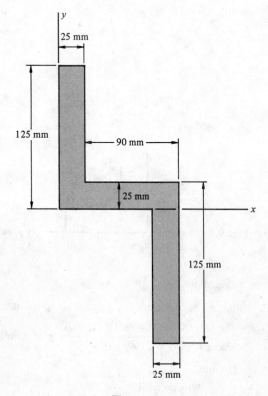

Fig. 9.10a

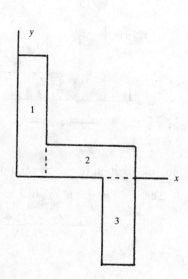

Fig. 9.10b

9.11 Find x_c and y_c for the area shown in Fig. 9.11a.

❙ The division of the area into elementary shapes is seen in Fig. 9.11b. The centroidal coordinates are then found to be

$$x_c = \frac{x_1A_1 + x_2A_2 + x_3A_3}{A_1 + A_2 + A_3} = \frac{\frac{2}{3}(713)\frac{1}{2}(713)850 + (713 + \frac{928}{2})928(850) + (713 + 928 + \frac{309}{3})\frac{1}{2}(309)850}{\frac{1}{2}(713)850 + 928(850) + \frac{1}{2}(309)850} = 1{,}060 \text{ mm}$$

$$A = 1.223 \times 10^6 \text{ mm}^2$$

$$y_c = \frac{y_1A_1 + y_2A_2 + y_3A_3}{A_1 + A_2 + A_3} = \frac{\frac{1}{3}(850)\frac{1}{2}(713)850 + \frac{850}{2}(928)850 + \frac{1}{3}(850)\frac{1}{2}(309)850}{1.223 \times 10^6} = 375 \text{ mm}$$

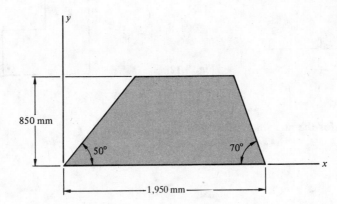

Fig. 9.11a

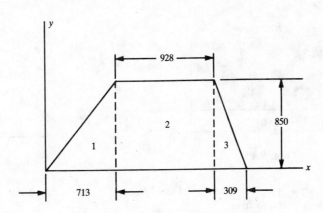

Fig. 9.11b

9.12 Find the centroidal coordinates of the area shown in Fig. 9.12a.

❙ The division of the given area is shown in Fig. 9.12b. The centroidal coordinates of area 1 are found by using Case 7 in Table 9.9, with the results

$$x_1 = 3 - \frac{4a}{3\pi} = 3 - \frac{4(3)}{3\pi} \qquad y_1 = 13 + \frac{4a}{3\pi} = 13 + \frac{4(3)}{3\pi}$$

x_c and y_c are then found to be

$$x_c = \frac{x_1A_1 + x_2A_2 + x_3A_3}{A_1 + A_2 + A_3} = \frac{[3 - 4(3)/3\pi]\frac{1}{4}\pi(3)^2 + \frac{3}{2}(3)13 + (3 + \frac{3}{2})3(4)}{\frac{1}{4}\pi(3)^2 + 3(13) + 3(4)} = 2.15 \text{ in}$$

$$A = 58.07 \text{ in}^2$$

$$y_c = \frac{y_1A_1 + y_2A_2 + y_3A_3}{A_1 + A_2 + A_3} = \frac{[13 + 4(3)/3\pi]\frac{1}{4}\pi(3)^2 + \frac{13}{2}(3)13 + \frac{4}{2}(3)4}{58.07} = 6.52 \text{ in}$$

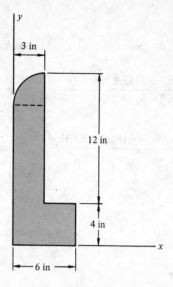

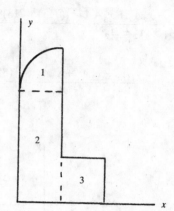

Fig. 9.12a

Fig. 9.12b

9.13 Find x_c and y_c for the area shown in Fig. 9.13a.

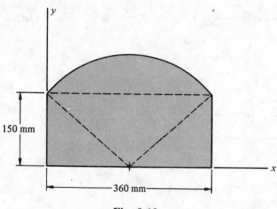

Fig. 9.13a

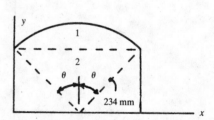

Fig. 9.13b

❚ The divided area is shown in Fig. 9.13b. Area 2 is the rectangular shape, and area 1 is the circular segment. Angle θ is found from

$$\tan \theta = \frac{180}{150} \qquad \theta = 50.2° = 0.876 \text{ rad}$$

From Case 9, Table 9.9, with $x_c \to y_1$ and $A \to A_1$,

$$A_1 = a^2\left(\theta - \frac{\sin 2\theta}{2}\right) = 234^2\left(0.876 - \frac{\sin[2(50.2°)]}{2}\right) = 2.10 \times 10^4 \text{ mm}^2$$

$$y_1 = \frac{2a \sin^3 \theta}{3\{\theta - [(\sin 2\theta)/2]\}} = \frac{2(234) \sin^3 50.2°}{3[0.876 - (\{\sin[2(50.2°)]\}/2)]} = 184 \text{ mm}$$

Using the above results,

$$y_c = \frac{y_1 A_1 + y_2 A_2}{A_1 + A_2} = \frac{184(2.10 \times 10^4) + \frac{150}{2}(360)150}{2.10 \times 10^4 + 360(150)} = 106 \text{ mm}$$

From the symmetry of the area in Fig. 9.13b,

$$x_c = \frac{360}{2} = 180 \text{ mm}$$

9.14   Define the *first moment of a plane area*.

❚   The final equations for x_c and y_c in Prob. 9.1 may be written in the form

$$\sum_i x_i A_i = x_c A = Q_y \qquad \sum_i y_i A_i = y_c A = Q_x$$

The integral forms of these equations are

$$\int_A x \, dA = x_c A = Q_y \qquad \int_A y \, dA = y_c A = Q_x$$

The two equations above define the terms Q_x and Q_y, which are referred to as the *first moments of the area A with respect to the x and y axes*.   The subscript on the Q terms indicate the axis to which the moment of area is referred.   A common application of the Q terms defined above is in the computation of the shearing stress in beams, and this topic is presented in strength, or mechanics, of materials.

It may be observed that *the units of the first moment of area are length raised to the third power*.   These units have no particular physical significance and merely result from the definitions given by the above equations.   The first moments of area of a composite area, using the forms in Prob. 9.5, are

$$Q_y = x_c A = x_1 A_1 + x_2 A_2 + \cdots \qquad and \qquad Q_x = y_c A = y_1 A_1 + y_2 A_2 + \cdots$$

9.15   The beam in Prob. 9.6 is now imagined to be made of two wood strips which are nailed together, with the cross section shown in Fig. 9.15*a*.   The theory used to determine the spacing of the nails, to withstand a given loading on the beam, is presented in strength, or mechanics, of materials.   A term required in this calculation is the first moment of area, with respect to the centroidal axis x_0, of area element 2 shown in the figure.   Find this area moment.   The cross-sectional area of the nail hole may be neglected.

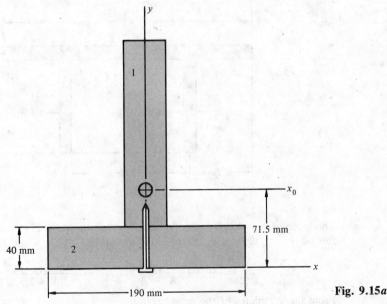

Fig. 9.15*a*

❚   Figure 9.15*b* shows the area A_2, and the x_0 axis passes through the centroid of the beam cross section.   The first moment Q_{x_0} of this area with respect to the x_0 axis is then

$$Q_{x_0} = y_2 A_2 = [71.5 - 0.5(40)][190(40)] = 3.91 \times 10^5 \text{ mm}^3$$

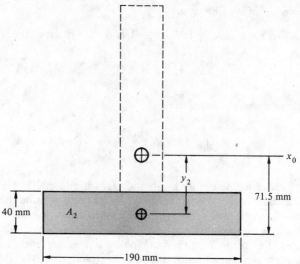

Fig. 9.15b

9.16 A cross section of a precast concrete floor beam is shown in Fig. 9.16a.

(*a*) Find the centroidal coordinates of the cross-sectional area.

(*b*) Find the first area moment, with respect to the horizontal centroidal axis, of the area bounded by this axis and the top edge of the cross section.

(*c*) Do the same as in part (*b*) for the area bounded by the horizontal centroidal axis and a horizontal line through the bottom of the area.

(*d*) Compare the results for parts (*b*) and (*c*).

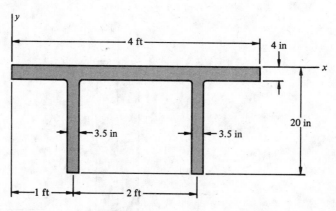

Fig. 9.16a

▌ (*a*) The given area is divided into the simple areas shown in Fig. 9.16b. Since the entire area lies below the *x* axis, y_1, y_2, and y_3 are negative quantities. y_c is found as

$$y_c = \frac{y_1 A_1 + y_2 A_2 + y_3 A_3}{A_1 + A_2 + A_3} = \frac{-\frac{4}{2}(48)4 - 2(4 + \frac{16}{2})3.5(16)}{48(4) + 2[3.5(16)]} = -5.68 \text{ in}$$

From the symmetry of the cross section,

$$x_c = 24 \text{ in}$$

(*b*) Figure 9.16c shows the area bounded by the centroidal x_0 axis and the top edge of the cross section. This area is divided into the simple areas shown in the figure. The first moment of the area is

$$Q_{x_0} = y_1 A_1 + 2y_4 A_4 = (1.68 + \frac{4}{2})48(4) + 2\left[\frac{1.68}{2}(3.5)1.68\right] = 716 \text{ in}^3$$

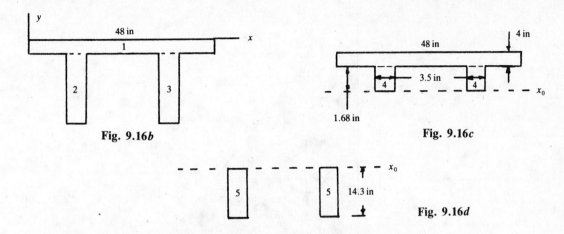

Fig. 9.16b

Fig. 9.16c

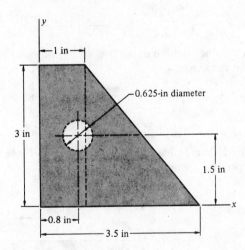

Fig. 9.16d

(c) The area bounded by the centroidal x_0 axis and the bottom of the cross-sectional area is shown in Fig. 9.16d. Using this figure,

$$Q_{x_0} = 2 y_5 A_5 = 2\left(\frac{14.3}{2}\right)3.5(14.3) = 716 \text{ in}^3$$

(d) The equality of the results in parts (b) and (c) follows from the definition of the centroidal y coordinate of the cross-sectional area.

9.17 How is the computation of the centroidal coordinates x_c and y_c of a plane area modified if the composite area contains holes or cutouts?

❙ If the composite area contains holes or cutouts, the areas of these elements are considered to be *negative* quantities when used in equations of the forms

$$x_c = \frac{x_1 A_1 + x_2 A_2 + \cdots}{A_1 + A_2 + \cdots} \quad \text{and} \quad y_c = \frac{y_1 A_1 + y_2 A_2 + \cdots}{A_1 + A_2 + \cdots}$$

9.18 A machine part in the form of a thin plate has the dimensions shown in Fig. 9.18a.

Fig. 9.18a

(a) Find the centroidal coordinates of the part.

(b) Find the centroidal coordinates if a hole, shown as the dashed circle in the figure, is drilled in the part.

❙ (a) The plate area is divided into the two elementary areas shown in Fig. 9.18b. For the plate *with no hole*,

$$x_c = \frac{x_1 A_1 + x_2 A_2}{A_1 + A_2} = \frac{0.5(1)(3) + [1 + 0.333(2.5)]0.5(2.5)(3)}{1(3) + 0.5(2.5)(3)} = 1.24 \text{ in}$$

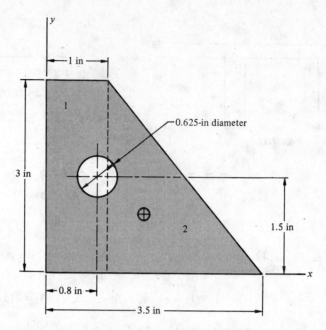

Fig. 9.18b

$$A = 6.75 \text{ in}^2$$

$$y_c = \frac{y_1 A_1 + y_2 A_2}{A_1 + A_2} = \frac{1.5(1)(3) + 1(0.5)(2.5)(3)}{6.75} = 1.22 \text{ in}$$

(b) For the case of the plate *with a hole*, the above two equations are used again, with the addition of terms to reflect the first moment, and area, of the hole.   The centroidal coordinates for this case are then

$$x_c = \frac{x_1 A_1 + x_2 A_2 + x_3(-A_3)}{A_1 + A_2 - A_3} = \frac{0.5(1)(3) + [1 + 0.333(2.5)]0.5(2.5)(3) - 0.8[\pi(0.625)^2/4]}{1(3) + 0.5(2.5)(3) - \pi(0.625)^2/4} = 1.26 \text{ in}$$

$$A = 6.443 \text{ in}^2$$

$$y_c = \frac{y_1 A_1 + y_2 A_2 + y_3(-A_3)}{A_1 + A_2 - A_3} = \frac{1.5(1)(3) + 1(0.5)(2.5)(3) - 1.5[\pi(0.625)^2/4]}{6.443} = 1.21 \text{ in}$$

It may be seen that the addition of the hole in this problem produces only small changes in the values of the centroidal coordinates.

9.19   Figure 9.19a shows a steel bearing plate.   The two holes are to be spaced so that the centroid of the plate is at its midheight.

(a)   What is the required value of the dimension y_1?

(b)   If the boundary of a hole is to have a minimum spacing of 2 in from the other hole, or the edge of the plate, what is the permissible range of values of x_1?

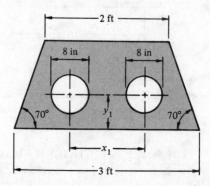

Fig. 9.19a

(a) Because of symmetry, only half of the area is required, and this area is shown in Fig. 9.19b. The area is divided into the three simple areas shown. Area 2 is a 12 by 16.5 in rectangle. The centroidal coordinate y_c has the form

$$y_c = \frac{y_1 A_1 + y_2 A_2 + y_3(-A_3)}{A_1 + A_2 - A_3} = \frac{\frac{1}{3}(16.5)\frac{1}{2}(6)16.5 + (16.5/2)(12)16.5 + y_3[-\pi(8)^2/4]}{\frac{1}{2}(6)16.5 + 12(16.5) - \pi(8)^2/4}$$

$$= \frac{1{,}906 - 50.27y_1}{197.2}$$

Since y_c must be the mid-height,

$$y_c = \frac{16.5}{2} = \frac{1{,}906 - 50.27y_1}{197.2} \qquad y_1 = 5.55 \text{ in}$$

(b) Figure 9.19c shows the minimum spacing of the holes with respect to each other. From this figure,

$$x_{1,min} = \tfrac{1}{2}(8) + 2 + \tfrac{1}{2}(8) = 10 \text{ in}$$

Figure 9.19d shows the minimum spacing of the hole from the edge of the plate. Dimension A is found, using Fig. 9.19e, from

$$\tan 70° = \frac{7.60}{A} \qquad A = 2.77 \text{ in}$$

From Fig. 9.19d,

$$B = 6 \sin 70° = 5.64 \text{ in}$$

$$A + B + \frac{x_{1,max}}{2} = 18 \qquad 2.77 + 5.64 + \frac{x_{1,max}}{2} = 18 \qquad x_{1,max} = 19.2 \text{ in}$$

The permissible range of spacing is

$$10 \text{ in} \le x_1 \le 19.2 \text{ in}$$

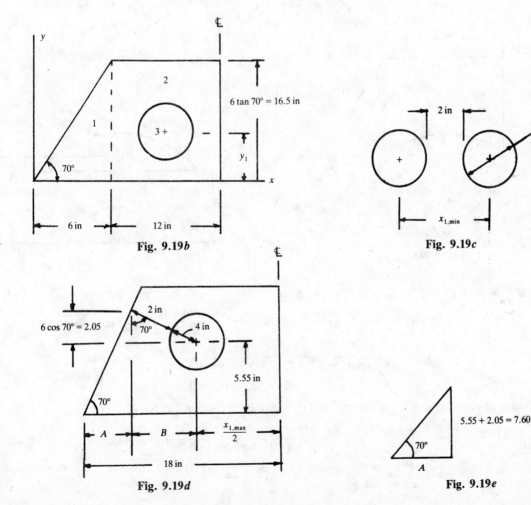

Fig. 9.19b

Fig. 9.19c

Fig. 9.19d

Fig. 9.19e

9.20 Find the centroidal coordinates of the area shown in Fig. 9.20a.

❙ The area is divided into the three elementary areas shown in Fig. 9.20b. Area 1 is an 18 by 15 in rectangle. Area 3 is shown in Fig. 9.20c, and angle θ is found from

$$\tan \theta = \frac{9}{11} \qquad \theta = 39.3° = 0.686 \text{ rad}$$

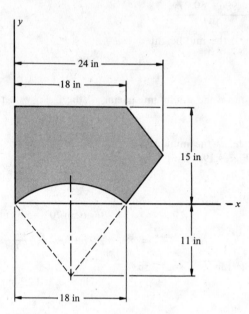

Fig. 9.20a

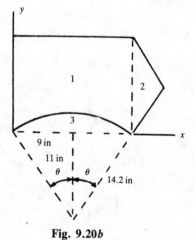

Fig. 9.20b

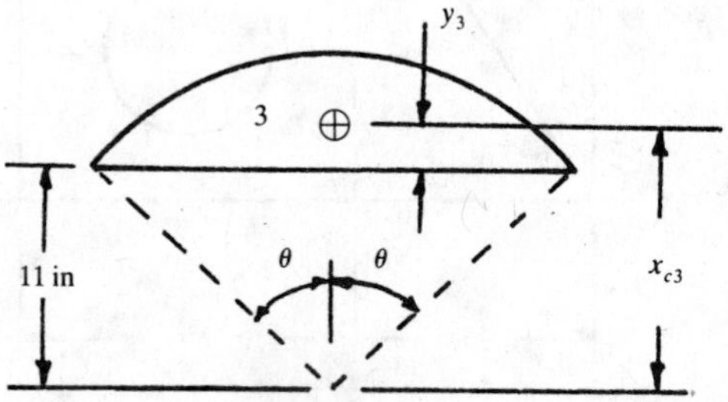

Fig. 9.20c

Using Case 9 in Table 9.9,

$$A_3 = a^2\left(\theta - \frac{\sin 2\theta}{2}\right) = 14.2^2\left\{0.686 - \frac{\sin\left[2(39.3°)\right]}{2}\right\} = 39.5 \text{ in}^2$$

$$x_{c3} = \frac{2a \sin^3 \theta}{3\{\theta - [(\sin 2\theta)/2]\}} = \frac{2(14.2) \sin^3 39.3°}{3(0.686 - (\{\sin [2(39.3°)]\}/2)} = 12.3 \text{ in}$$

$$y_3 = x_{c3} - 11 = 12.3 - 11 = 1.3 \text{ in}$$

The centroidal coordinates of the area in Fig. 9.20a are then found to be

$$x_c = \frac{x_1 A_1 + x_2 A_2 + x_3(-A_3)}{A_1 + A_2 - A_3} = \frac{\frac{18}{2}(18)15 + (18 + \frac{6}{3})0.5(15)6 + 9(-39.5)}{18(15) + 0.5(15)6 - 39.5} = 10.8 \text{ in} \qquad A = 276 \text{ in}^2$$

$$y_c = \frac{y_1 A_1 + y_2 A_2 + y_3(-A_3)}{A_1 + A_2 - A_3} = \frac{\frac{15}{2}(18)15 + \frac{15}{2}[\frac{1}{2}(15)6] - 1.3(39.5)}{276} = 8.37 \text{ in}$$

9.21 Find the location of the centroid of the area shown in Fig. 9.21a.

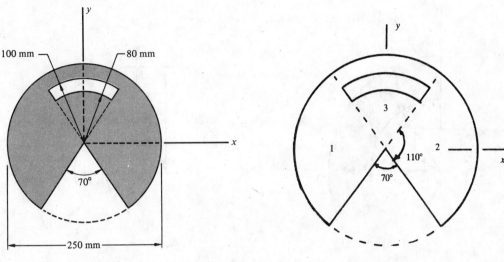

Fig. 9.21a   Fig. 9.21b

❙ Figure 9.21b shows the division of the area into three areas. A_1 and A_2 are both circular sectors with an included angle of 110°. A_3 is the remainder of the area in Fig. 9.21a. Area 3 is now divided further, as shown in Fig. 9.21c. Area 4 is a circular sector of radius 125 mm. Area 5 is a circular sector with radius 100 mm, and area 6 is a circular sector of radius 80 mm. The addition and subtraction operations for the area elements are indicated in Fig. 9.21c. Using Case 8 in Table 9.9, with r_c defined to be the radial centroidal coordinate (i.e., the coordinate from the origin of the xy axes in Case 8 to the centroid of the sector),

$$r_c = \frac{x_c}{\cos(\theta/2)} = \frac{2a \sin \theta}{3\theta \cos(\theta/2)} \qquad A = \frac{a^2\theta}{2} \qquad r_c A = \frac{a^3 \sin \theta}{3 \cos(\theta/2)}$$

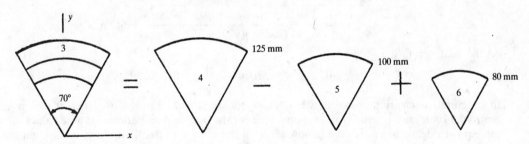

Fig. 9.21c

From Fig. 9.21c,

$$r_{c3} = \frac{r_{c4}A_4 - r_{c5}A_5 + r_{c6}A_6}{A_4 - A_5 + A_6}$$

Using $\theta = 70° = 1.22$ rad,

$$r_{c3} = \frac{[(\sin 70°)/(3 \cos 35°)](125^3 - 100^3 + 80^3)}{(1.22/2)(125^2 - 100^2 + 80^2)} = 76.3 \text{ mm}$$

$$A_3 = 7,340 \text{ mm}^2$$

$$A_1 = A_2 = \frac{125^2}{2}(110°)\frac{\pi}{180} = 1.50 \times 10^4 \text{ mm}^2$$

Using the above results, with Fig. 9.21b,

$$y_c = \frac{y_1A_1 + y_2A_2 + y_3A_3}{A_1 + A_2 + A_3} = \frac{0(1.50 \times 10^4) + 0(1.50 \times 10^4) + 76.4(7,340)}{1.50 \times 10^4 + 1.50 \times 10^4 + 7,340} = 15.0 \text{ mm}$$

From symmetry,

$$x_c = 0$$

9.22 Find the values of the centroidal coordinates of the area shown in Fig. 9.22*a*.

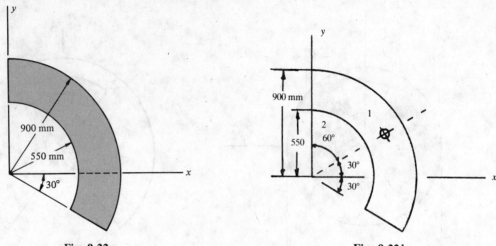

Fig. 9.22*a*   **Fig. 9.22*b***

▌ Figure 9.22*b* shows the two areas that represent the original area in Fig. 9.22*a*. Area 1 is a circular sector of radius 900 mm and included angle 120°. Area 2 has an included angle of 120° and a radius of 550 mm. Case 8 in Table 9.9 is used, and r_c is the radial centroidal coordinate. Using the results from Prob. 9.21,

$$r_c = \frac{x_c}{\cos(\theta/2)} = \frac{2a\sin\theta}{3\theta\cos(\theta/2)} \qquad A = \frac{a^2\theta}{2} \qquad r_c A = \frac{a^3\sin\theta}{3\cos(\theta/2)}$$

The net radial centroidal coordinate for the area in Fig. 9.22*a* is then

$$r_c = \frac{r_{c1}A_1 + r_{c2}(-A_2)}{A_1 - A_2}$$

Using $\theta = 120° = 2.09$ rad, $r_{c1} = 900$ mm, and $r_{c2} = 550$ mm in the above equation results in

$$r_c = \frac{\frac{1}{3}[(\sin 120°)/(\cos 60°)](900^3 - 550^3)}{(2.09/2)(900^2 - 550^2)} = 612 \text{ mm}$$

As the final step in the solution,

$$x_c = 612\cos 30° = 530 \text{ mm} \qquad y_c = 612\sin 30° = 306 \text{ mm}$$

9.23 In structural engineering problems, where elements are bolted or riveted together, it is necessary to find the centroidal coordinates of the cross-sectional areas of the bolt or rivet pattern. These values are then used in a subsequent calculation to determine how much of the total load that is transmitted through the connection is carried by each bolt or rivet. Figure 9.23*a* shows the pattern of rivet holes for the connection of two structural members.

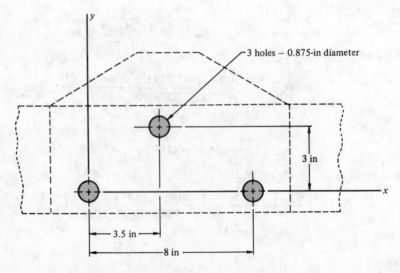

3 holes — 0.875-in diameter

3 in

3.5 in

8 in

Fig. 9.23*a*

(a) Find the centroidal coordinates x_c and y_c of this pattern of hole areas.

(b) What is the major difference between the problem in part (a) and the preceding centroid problems in this chapter?

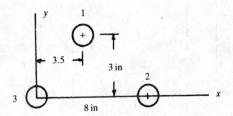

Fig. 9.23b

▌ (a) The hole areas are numbered as shown in Fig. 9.23b. A is the cross-sectional area of each hole, and the centroidal coordinates of the pattern of hole areas are given by

$$x_c = \frac{x_1 A_1 + x_2 A_2 + x_3 A_3}{A_1 + A_2 + A_3} = \frac{3.5A + 8A + 0(A)}{3A} = 3.83 \text{ in}$$

$$y_c = \frac{y_1 A_1 + y_2 A_2 + y_3 A_3}{A_1 + A_2 + A_3} = \frac{3A + 0(A) + 0(A)}{3A} = 1 \text{ in}$$

The above results are seen to be independent of the magnitudes of the hole areas since, in this case, all three holes have the same area.

(b) The major difference between this type of problem and the problems considered earlier is that the areas of the bolt or rivet cross sections *do not* physically contact each other.

9.24 Figure 9.24 shows the cross-sectional areas of four bolts that are used to connect two structural members. Find the centroidal coordinates of the array of cross-sectional areas of the bolts if all bolts have the same diameter.

▌ The bolts are all of equal diameter, and the cross-sectional areas of these elements are designated A.

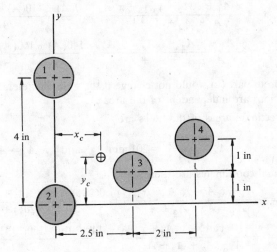

Fig. 9.24

The centroidal coordinates of the pattern of hole areas are

$$x_c = \frac{x_1 A_1 + x_2 A_2 + x_3 A_3 + x_4 A_4}{A_1 + A_2 + A_3 + A_4} = \frac{0(A) + 0(A) + 2.5A + 4.5A}{4A} = \frac{7A}{4A} = 1.75 \text{ in}$$

$$y_c = \frac{y_1 A_1 + y_2 A_2 + y_3 A_3 + y_4 A_4}{A_1 + A_2 + A_3 + A_4} = \frac{4A + 0(A) + (1)A + 2A}{4A} = \frac{7A}{4A} = 1.75 \text{ in}$$

9.25 Figure 9.25a shows the hole pattern in a steel structural member.

(a) Find the centroidal coordinates of the pattern of hole areas.

(b) Would the solution to part (a) change if all hole diameters were increased to 20 mm?

(c) Do the same as in part (a) if the diameters of holes b and c only are increased to 20 mm?

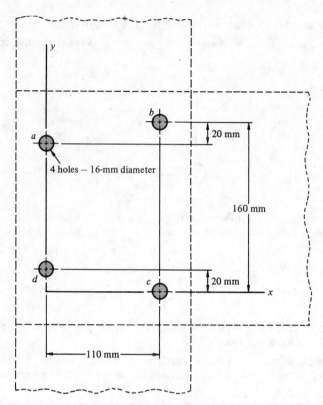

Fig. 9.25a

(a) The holes are numbered as shown in Fig. 9.25b. The hole area is designated A. The centroidal coordinates of the pattern of hole areas are then

$$x_c = \frac{x_1 A_1 + x_2 A_2 + x_3 A_3 + x_4 A_4}{A_1 + A_2 + A_3 + A_4} = \frac{0(A) + 110(A) + 110(A) + 0(A)}{4A} = 55 \text{ mm}$$

$$y_c = \frac{y_1 A_1 + y_2 A_2 + y_3 A_3 + y_4 A_4}{A_1 + A_2 + A_3 + A_4} = \frac{140(A) + 160(A) + 0(A) + 20(A)}{4A} = 80 \text{ mm}$$

(b) The solution to part (a) would not change if the hole diameters were increased, since x_c and y_c in the above solutions are independent of the area A.

(c) The cross-sectional areas of the holes are

$$A_1 = A_4 = \frac{\pi(16)^2}{4} = 201 \text{ mm}^2 \quad \text{and} \quad A_2 = A_3 = \frac{\pi(20)^2}{4} = 314 \text{ mm}^2$$

The centroidal coordinates are then found as

$$x_c = \frac{x_1 A_1 + x_2 A_2 + x_3 A_3 + x_4 A_4}{A_1 + A_2 + A_3 + A_4} = \frac{0(201) + 110(314) + 110(314) + 0(201)}{2(201) + 2(314)} = 67.1 \text{ mm}$$

$$y_c = \frac{y_1 A_1 + y_2 A_2 + y_3 A_3 + y_4 A_4}{A_1 + A_2 + A_3 + A_4} = \frac{140(201) + 160(314) + 0(314) + 20(201)}{1,030} = 80 \text{ mm}$$

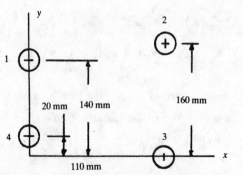

Fig. 9.25b

It may be seen that the effect of the two larger holes is to move the position of the centroid to the right, toward these holes. It may also be observed that the result $y_c = 80$ mm in parts (a) and (c) is expected, since the pattern of areas is symmetrical with respect to the line $y = 80$ mm.

9.2 TABULAR FORMS OF SOLUTION FOR CENTROIDAL COORDINATES, CENTROIDS OF SIMPLE AND COMPOSITE PLANE CURVES

9.26 Develop a tabular format to organize the calculations for the case in which the composite area is a complicated shape which requires the description of several elementary areas.

▮ Table 9.1 shows one such format. The identification numbers of the elementary areas are entered in column 1, and the values of these areas are shown in column 2. The centroidal coordinates of the elementary areas are listed in columns 3 and 5, and the first moments of these areas are shown in columns 4 and 6. Columns 2, 4, and 6 are then summed, and the final values for x_c and y_c are found by using the equations in the last row of the table.

TABLE 9.1					
(1)	(2)	(3)	(4)	(5)	(6)
Element	A_i	x_i	$x_i A_i$	y_i	$y_i A_i$
$A = \sum_i A_i$		$\sum_i x_i A_i$		$\sum_i y_i A_i$	
		$x_c = \dfrac{\sum_i x_i A_i}{A}$		$y_c = \dfrac{\sum_i y_i A_i}{A}$	

9.27 Find the centroidal coordinates of the flat plate shown in Fig. 9.27a. Organize the solution in a tabular form.

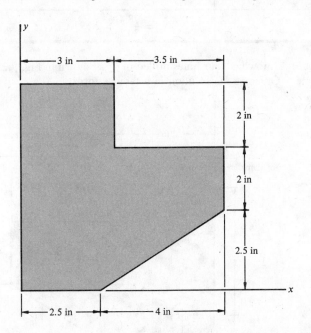

Fig. 9.27a

▮ The plate area is divided into four elements, as shown in Fig. 9.27b, and Table 9.2 is constructed. This table includes all the details of the calculations for a particular area, or for a centroidal coordinate of an elementary area, for possible subsequent use in checking the calculations.

From the last row of the table,

$$x_c = 2.59 \text{ in} \quad \text{and} \quad y_c = 3.13 \text{ in}$$

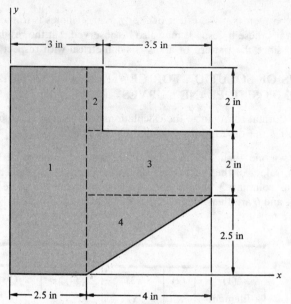

Fig. 9.27b

		TABLE 9.2			
Element	A_i	x_i	$x_i A_i$	y_i	$y_i A_i$
1	2.5(6.5) = 16.3	0.5(2.5) = 1.25	20.4	0.5(6.5) = 3.25	53.0
2	0.5(2) = 1	2.5 + 0.5(0.5) = 2.75	2.75	4.5 + 0.5(2) = 5.5	5.5
3	4(2) = 8	2.5 + 0.5(4) = 4.5	36.0	2.5 + 0.5(2) = 3.5	28.0
4	0.5(2.5)(4) = 5	2.5 + 0.333(4) = 3.83	19.2	0.667(2.5) = 1.67	8.35
$A = \sum_i A_i = 30.3$		$\sum_i x_i A_i = 78.4$		$\sum_i y_i A_i = 94.9$	
		$x_c = \dfrac{78.4}{30.3} = 2.59$ in		$y_c = \dfrac{94.9}{30.3} = 3.13$ in	

9.28 The plate of Prob. 9.27 now has four holes drilled in it, as shown in Fig. 9.28. Find the centroidal coordinates of the drilled plate. Organize the solution in a tabular form.

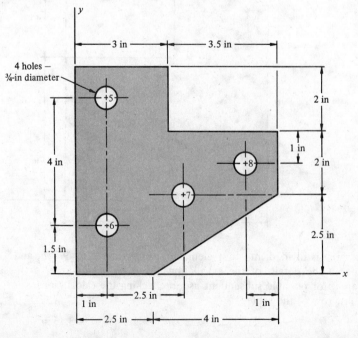

Fig. 9.28

▮ The table construction is shown in Table 9.3, and the first four rows are exactly the same as the corresponding rows in Table 9.2. The holes are designated as areas 5 through 8, and these circular areas are entered in the table as negative quantities. The values of the centroidal coordinates x_c and y_c are contained in the last row of the table. It may be observed that the addition of the holes in this example has virtually no effect on the location of the centroid.

				TABLE 9.3			
Element	A_i	x_i	x_iA_i		y_i	y_iA_i	
1	$2.5(6.5) = 16.3$	$0.5(2.5) = 1.25$	20.4		$0.5(6.5) = 3.25$	53.0	
2	$0.5(2) = 1$	$2.5 + 0.5(0.5) = 2.75$	2.75		$4.5 + 0.5(2) = 5.5$	5.5	
3	$4(2) = 8$	$2.5 + 0.5(4) = 4.5$	36.0		$2.5 + 0.5(2) = 3.5$	28.0	
4	$0.5(2.5)(4) = 5$	$2.5 + 0.333(4) = 3.83$	19.2		$0.667(2.5) = 1.67$	8.35	
5	$\dfrac{-\pi(0.75)^2}{4} = -0.44$	1	-0.44		$1.5 + 4 = 5.5$	-2.42	
6	-0.44	1	-0.44		1.5	-0.66	
7	-0.44	$1 + 2.5 = 3.5$	-1.54		2.5	-1.10	
8	-0.44	$2.5 + 4 - 1 = 5.5$	-2.42		$2.5 + 1 = 3.5$	-1.54	
$A = \sum\limits_i A_i = 28.5$		$\sum\limits_i x_iA_i = 73.5$			$\sum\limits_i y_iA_i = 89.1$		
		$x_c = \dfrac{73.5}{28.5} = 2.58$ in			$y_c = \dfrac{89.1}{28.5} = 3.13$ in		

9.29 Find the centroidal coordinates of the area shown in Fig. 9.29a. Organize the solution in a tabular form.

▮ The division of the area into elementary shapes is shown in Fig. 9.29b. Table 9.4 is constructed, and the values for x_c and y_c are shown in the last row of this table.

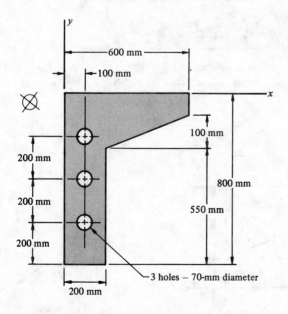

Fig. 9.29a

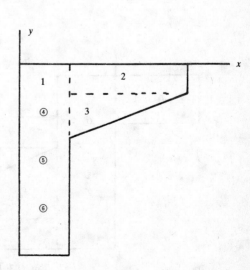

Fig. 9.29b

TABLE 9.4

element	A_i	x_i	x_iA_i	y_i	y_iA_i
1	$200(800) = 1.6 \times 10^5$	100	1.6×10^7	-400	-6.4×10^7
2	$400(150) = 6 \times 10^4$	$200 + \frac{400}{2} = 400$	2.4×10^7	$-\frac{150}{2} = -75$	-4.5×10^6
3	$\frac{1}{2}(400)100 = 2 \times 10^4$	$200 + \frac{1}{3}(400) = 333$	6.66×10^6	$-(150 + \frac{100}{3}) = -183$	-3.66×10^6
4	$-\frac{\pi(70)^2}{4} = -3,800$	100	-3.8×10^5	-200	7.6×10^5
5	$-3,800$	100	-3.8×10^5	-400	1.52×10^6
6	$-3,800$	100	-3.8×10^5	-600	2.28×10^6

$$A = \sum_i A_i = 2.29 \times 10^5 \qquad \sum_i x_iA_i = 4.55 \times 10^7 \qquad \sum_i y_iA_i = -6.76 \times 10^7$$

$$x_c = 199 \text{ mm} \qquad\qquad y_c = -295 \text{ mm}$$

9.30 Find x_c and y_c for the area shown in Fig. 9.30a. Use a tabular form to show the computations.

▌ The given area is divided into the simple area shapes shown in Fig. 9.30b. Figure 9.30c shows a view of area 5, the cutout. From this figure,

$$\frac{A}{4.5} = \frac{2}{9.25} \qquad A = 0.973 \text{ in}$$

Table 9.5 shows the details of the computations, and the values of x_c and y_c are shown in the bottom row of this table.

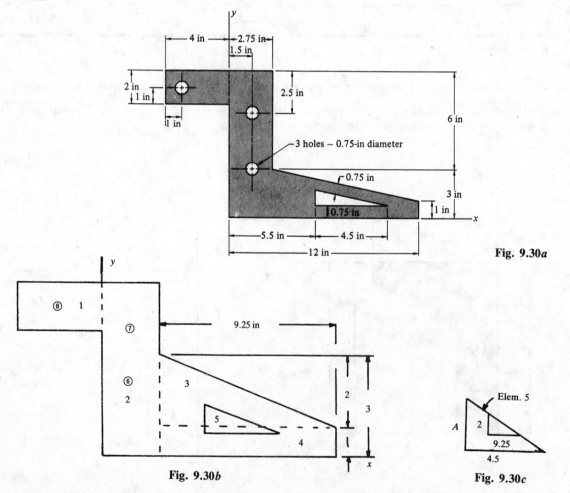

Fig. 9.30a

Fig. 9.30b

Fig. 9.30c

TABLE 9.5

element	A_i	x_i	x_iA_i	y_i	y_iA_i
1	$4(2) = 8$	-2	-16	$9 - 1 = 8$	64
2	$2.75(9) = 24.8$	$\dfrac{2.75}{2} = 1.38$	34.22	$\dfrac{9}{2} = 4.5$	111.6
3	$\frac{1}{2}(9.25)2 = 9.25$	$2.75 + \dfrac{9.25}{3} = 5.83$	53.93	$1 + \dfrac{2}{3} = 1.67$	15.45
4	$9.25(1) = 9.25$	$2.75 + \dfrac{9.25}{2} = 7.38$	68.27	0.5	4.63
5	$-\frac{1}{2}(4.5)0.973 = -2.19$	$5.5 + \dfrac{4.5}{3} = 7$	-15.33	$0.75 + \dfrac{0.973}{3} = 1.07$	-2.34
6	$-\dfrac{\pi(0.75)^2}{4} = -0.442$	1.5	-0.66	3	-1.33
7	-0.442	1.5	-0.66	6.5	-2.87
8	-0.442	-3	1.33	8	-3.54
$A = \sum\limits_i A_i = 47.8 \text{ in}^2$		$\sum\limits_i x_iA_i = 125$		$\sum\limits_i y_iA_i = 186$	
		$x_c = 2.62 \text{ in}$		$y_c = 3.89 \text{ in}$	

9.31   Find the location of the centroid of the area shown in Fig. 9.31a.   Give the details of the calculations in tabular form.

❚   Figure 9.31b shows the division into elementary areas.   Element 1 is a rectangular shape that has the outside dimensions of the area in Fig. 9.31a.   The details of the computation for the centroidal coordinates are given in Table 9.6.   The final results are seen in the last row of the table.

TABLE 9.6

element	A_i	x_i	x_iA_i	y_i	y_iA_i
1	$38(44) = 1,670$	19	$31,730$	-22	$-36,740$
2	$-\frac{1}{2}(10)9 = -45$	$\frac{10}{3} = 3.33$	-150	$-[15 + \frac{2}{3}(9)] = -21$	950
3	$-10(20) = -200$	5	$-1,000$	-34	$6,800$
4	$-\frac{1}{2}(8)9 = -36$	$30 + \frac{2}{3}(8) = 35.3$	$-1,270$	$-[15 + \frac{2}{3}(9)] = -21$	760
5	$-8(20) = -160$	34	$-5,440$	-34	$5,440$
6	$-\dfrac{\pi6^2}{4} = -28$	12	-340	-10	280
7	-28	26	-730	-10	280
8	-28	20	-560	$-(44 - 18) = -26$	730
9	-28	20	-560	$-(44 - 6) = -38$	$1,060$
$A = \sum\limits_i A_i = 1,120$		$\sum\limits_i x_iA_i = 21,700$		$\sum\limits_i y_iA_i = -20,400$	
		$x_c = 19.4 \text{ mm}$		$y_c = -18.3 \text{ mm}$	

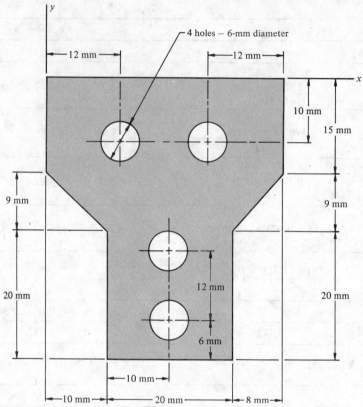

Fig. 9.31*a*

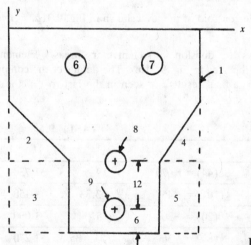

Fig. 9.31*b*

9.32 Find the values of the centroidal coordinates of the area shown in Fig. 9.32*a*. Organize the solution in tabular form.

▮ Figure 9.32*b* shows the division of the original area into six simple area shapes. Element 1 is a 12-in-diameter circular area, and element 2 is one quadrant of that area.

As a separate calculation, the centroidal coordinates of area 4 are found next. Figure 9.32*c* shows this area. Area 7 is a 4 by 4 in square. Case 7 of Table 9.9 is used to find the centroidal coordinates of area 8, and these values are shown in Fig. 9.32*c*. The centroidal coordinates of area 4 are then

$$x_{c4} = y_{c4} = \frac{x_{c7}A_7 + x_{c8}(-A_8)}{A_7 - A_8} = \frac{2(4)4 - 3.15(0.25)\pi(2)^2}{4(4) - 0.25\pi(2)^2} = 1.72 \text{ in}$$

$$A_4 = 12.9 \text{ in}^2$$

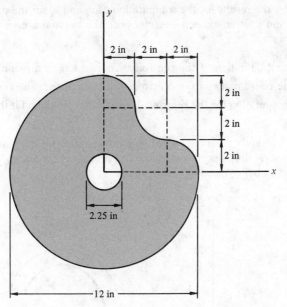

Fig. 9.32a

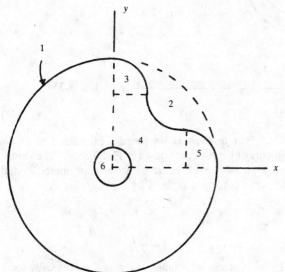

Fig. 9.32b

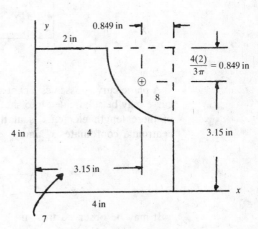

Fig. 9.32c

TABLE 9.7

element	A_i	x_i	x_iA_i	y_i	y_iA_i
1	$\pi(6)^2 = 113.1$	0	0	0	0
2	$-\frac{1}{4}\pi(6)^2 = -28.3$	$\frac{4(6)}{3\pi} = 2.55$	-72.17	$\frac{4(6)}{3\pi} = 2.55$	-72.17
3	$\frac{1}{4}\pi(2)^2 = 3.1$	0.849	2.63	$4 + 0.849 = 4.85$	15.04
4	12.9	1.72	22.19	1.72	22.19
5	$\frac{1}{4}\pi(2)^2 = 3.1$	$4 + 0.849 = 4.85$	15.04	0.849	2.63
6	$-\frac{\pi(2.25)^2}{4} = -4.0$	0	0	0	0
	$A = \sum_i A_i = 99.9$		$\sum_i x_iA_i = -32.3$		$\sum_i y_iA_i = -32.3$
		$x_c = -0.323$ in		$y_c = -0.323$ in	

Table 9.7 contains the results for the computation of x_c and y_c for the original area shown in Fig. 9.32b. The equivalence of x_c and y_c is an expected result, since the original area is symmetrical about a line which is 45° from the x and y axes.

9.33   (a)   State the formal definition of the centroidal coordinates of a plane curve.

         (b)   Show how the coordinates of the centroid of a composite plane curve may be determined.

         (c)   How is the computation for the centroidal coordinates simplified if the plane curve has one, or two, axes of symmetry?

▮   (a)   Figure 9.33a shows a curved line of length l which lies in the plane of the paper. This plane curve, as a geometric entity, has no thickness. For the purposes of this discussion, however, it may be envisioned as being made of thin wire.

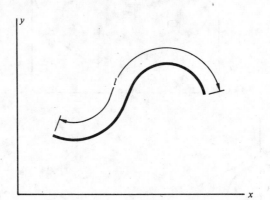

Fig. 9.33a

A plane curve possesses a centroid which is an inherent characteristic of the *shape* of the curve. The curve may be subdivided into short elements of length l_i, as shown in Fig. 9.33b. The centroid of each of these length elements is at the midpoint of the element, with the coordinates x_i and y_i. The centroidal coordinates x_c' and y_c' of the plane curve are defined by

$$x_c' = \frac{\sum_i x_i l_i}{\sum_i l_i} = \frac{\sum_i x_i l_i}{l} \quad \text{and} \quad y_c' = \frac{\sum_i y_i l_i}{\sum_i l_i} = \frac{\sum_i y_i l_i}{l}$$

It may be observed that, in general, the centroid of a plane curve is *not* physically located on the curve.

If the subdivision of the original length l into the elemental lengths l_i is continued, a limiting condition is reached in which the summation operations in the above equations become integral operations, with the forms

$$x_c' = \lim_{l_i \to 0} \frac{\sum_i x_i l_i}{l} = \frac{\int_l x \, dl}{l} \quad \text{and} \quad y_c' = \lim_{l_i \to 0} \frac{\sum_i y_i l_i}{l} = \frac{\int_l y \, dl}{l}$$

The above two equations are the formal definition of the centroidal coordinates of a plane curve.

The centroidal coordinates for several elementary plane curve shapes are given in Table 9.10 at the end of this chapter. The values of the centroidal coordinates x_c' and y_c' in this table are found by direct integration of the above two equations.

     (b)   Any composite curve shape may be divided into elementary curve shapes, each of whose centroidal coordinates are known. The centroidal coordinates of the original curve are then expressed as

$$x_c' = \frac{x_1 l_1 + x_2 l_2 + \cdots}{l_1 + l_2 + \cdots} \quad \text{and} \quad y_c' = \frac{y_1 l_1 + y_2 l_2 + \cdots}{l_1 + l_2 + \cdots}$$

     (c)   As in the case of plane areas, if a plane curve has an axis of symmetry, then the centroid must be located on this axis. If the plane curve possesses two axes of symmetry, the centroid is located at the intersection of these two axes. These effects are illustrated in Fig. 9.33c.

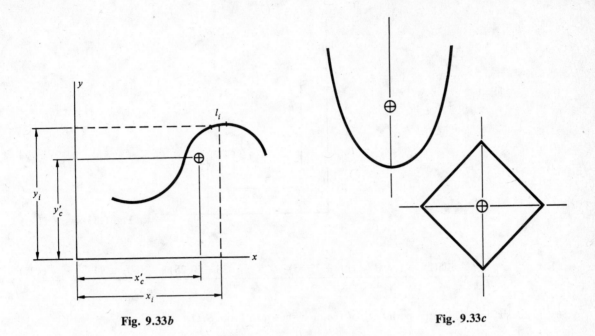

Fig. 9.33b   **Fig. 9.33c**

9.34   Where is the centroid of the boundary of the area shown in Prob. 9.7 located?

▮ The boundary of the area is divided into the six straight-line elements shown in Fig. 9.34. The centroidal coordinate x'_c of the plane curve that bounds the area is then

$$x'_c = \frac{\sum_i x_i l_i}{l}$$

$$x'_c = \frac{(a/2)(a) + a(a) + \frac{3}{2}a(a) + 2a(a) + a(2a) + 0(2a)}{a + a + a + a + 2a + 2a} = \frac{7}{8}a$$

Since the area shape is symmetrical with respect to a line which is at 45° with the x and y axes, it follows that

$$y'_c = x'_c = \tfrac{7}{8}a$$

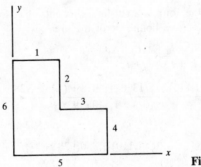

Fig. 9.34

9.35   Determine the values of the centroidal coordinates of the perimeter of the area shown in Prob. 9.9.

▮ The eight straight-line elements that bound the area are shown in Fig. 9.35. The coordinates of the plane curve that bounds the area are then found as

$$x'_c = \frac{\sum_i x_i l_i}{l} = \frac{\frac{25}{2}(25) + 25(100) + (25 + \frac{65}{2})65 + (25 + 65)35 + (25 + 65 + \frac{25}{2})25 + 115(60) + \frac{115}{2}(115) + 0(125)}{25 + 100 + 65 + 35 + 25 + 60 + 115 + 125}$$

$$= 46.9 \text{ mm}$$

$$l = 550 \text{ mm}$$

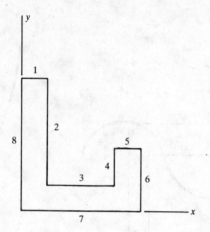

Fig. 9.35

$$y_c' = \frac{\sum_i y_i l_i}{l} = \frac{125(25) + (25 + \frac{100}{2})100 + 25(65) + (25 + \frac{35}{2})35 + 60(25) + \frac{60}{2}(60) + 0(115) + \frac{125}{2}(125)}{550} = 45.2 \text{ mm}$$

9.36 Find x_c' and y_c' for the perimeter of the area shown in Prob. 9.11.

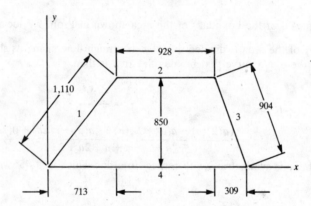

Fig. 9.36

I Figure 9.36 shows the four straight-line elements that form the perimeter of the area. The centroidal coordinates of the plane curve boundary of the area are

$$\dot{x}_c' = \frac{x_1 l_1 + x_2 l_2 + x_3 l_3 + x_4 l_4}{l_1 + l_2 + l_3 + l_4}$$

$$= \frac{\frac{713}{2}(1,110) + (713 + \frac{928}{2})928 + (713 + 928 + \frac{309}{2})904 + \frac{1,950}{2}(1,950)}{1,110 + 928 + 904 + 1,950} = 1,020 \text{ mm}$$

$$l = 4,892 \text{ mm}$$

$$y_c' = \frac{\frac{850}{2}(1,110) + 850(928) + \frac{850}{2}(904) + 0(1,950)}{4,892} = 336 \text{ mm}$$

9.37 Find the location of the centroid of the plane curve that bounds the area shown in Fig. 9.12a.

I The boundary of the area is divided into the six elementary curve shapes seen in Fig. 9.37. The centroidal coordinates of element 1, the circular arc, are found from the results for Case 6 in Table 9.10. For this case, with $\theta = 90° = \pi/2$,

$$x_c' = \frac{a \sin \theta}{\theta} = \frac{a(1)}{\pi/2} = \frac{2a}{\pi}$$

$$y_c' = \frac{2a \sin^2 (\theta/2)}{\theta} = \frac{2a \sin^2 45°}{\pi/2} = \frac{2a(\sqrt{2}/2)^2}{\pi/2} = \frac{2a}{\pi}$$

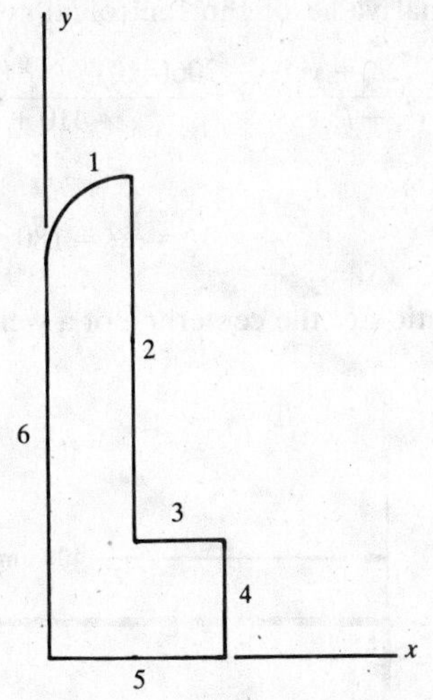

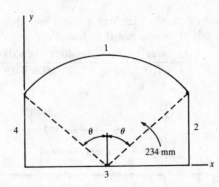

Fig. 9.37

The centroidal coordinates x_c' and y_c' for the plane curve which bounds the area are then

$$x_c' = \frac{\sum_i x_i l_i}{l}$$

$$x_c' = \frac{\{3 - [2(3)/\pi]\}(3\pi/2) + 3(12) + (3 + \frac{3}{2})3 + 6(4) + \frac{6}{2}(6) + 0(13)}{3\pi/2 + 12 + 3 + 4 + 6 + 13} = 2.26 \text{ in}$$

$$y_c' = \frac{\sum_i y_i l_i}{l} = \frac{[13 + 2(3)/\pi](3\pi/2) + [4 + \frac{12}{2}]12 + 4(3) + \frac{4}{2}(4) + 0(6) + \frac{13}{2}(13)}{42.71} = 6.90 \text{ in}$$

9.38 Find the centroidal coordinates of the boundary of the area shown in Prob. 9.13.

Fig. 9.38

▌ The boundary of the area is divided into the four elements shown in Fig. 9.38. From the solution to Prob. 9.13,

$$\theta = 50.2° = 0.876 \text{ rad}$$

The centroidal coordinates of element 1, the circular arc, may be found by using Case 6 in Table 9.10. From this table, using $\theta = 2(50.2°) = 100° = 1.75$ rad,

$$x_c' = \frac{a \sin \theta}{\theta} = \frac{234 \sin 100°}{1.75} = 132 \text{ mm}$$

The above value of x_c' is related to y_1', the centroidal y coordinate of curve 1 in Fig. 9.38, by

$$y_1' = \frac{x_c}{\cos(\theta/2)} = \frac{132}{\cos 50.2°} = 206 \text{ mm}$$

From Case 6 in Table 9.10,

$$l_1 = a\theta = 234(1.75) = 410 \text{ mm}$$

Using the above results, the final value of the centroidal coordinate y'_c of the boundary curve in Fig. 9.38 is

$$y'_c = \frac{y_1 l_1 + y_2 l_2 + y_3 l_3 + y_4 l_4}{l_1 + l_2 + l_3 + l_4} = \frac{206(410) + \frac{150}{2}(150) + 0(360) + \frac{150}{2}(150)}{410 + 150 + 360 + 150} = 100 \text{ mm}$$

From the symmetry in Fig. 9.38,

$$x'_c = \frac{360}{2} = 180 \text{ mm}$$

9.39 Figure 9.39a shows a representation of the centerline of a weld pattern on a flat plate. Find the location of the centroid of this centerline.

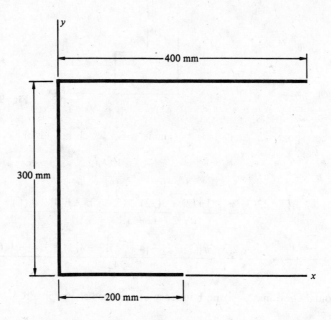

Fig. 9.39a

❚ The original shape is divided into the three straight-line elements 1, 2, and 3 shown in Fig. 9.39b. The centroidal coordinates of this centerline are then found as

$$x'_c = \frac{x_1 l_1 + x_2 l_2 + x_3 l_3}{l_1 + l_2 + l_3} = \frac{200(400) + 0(300) + 100(200)}{400 + 300 + 200} = 111 \text{ mm}$$

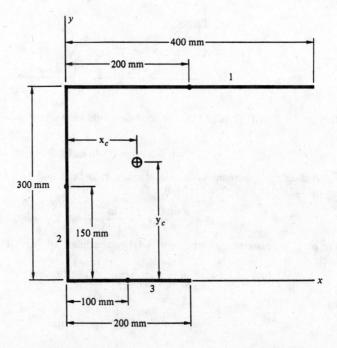

Fig. 9.39b

$$l = 900 \text{ mm}$$

$$y'_c = \frac{y_1 l_1 + y_2 l_2 + y_3 l_3}{l_1 + l_2 + l_3} = \frac{300(400) + 150(300) + 0(200)}{900} = 183 \text{ mm}$$

9.40 Find the location of the centroid of the centerline of the weld pattern shown in Fig. 9.40a.

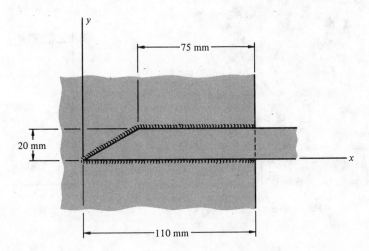

Fig. 9.40a

❚ The three line elements that form the centerline are shown in Fig. 9.40b. The centroidal coordinates of the centerline are then

$$x'_c = \frac{x_1 l_1 + x_2 l_2 + x_3 l_3}{l_1 + l_2 + l_3} = \frac{\frac{110}{2}(110) + \frac{35}{2}(40.3) + (35 + \frac{75}{2})75}{110 + 40.3 + 75} = 54.1 \text{ mm}$$

$$l = 225.3 \text{ mm}$$

$$y'_c = \frac{y_1 l_1 + y_2 l_2 + y_3 l_3}{l_1 + l_2 + l_3} = \frac{0(110) + \frac{20}{2}(40.3) + 20(75)}{225.3} = 8.45 \text{ mm}$$

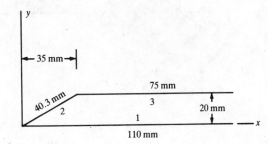

Fig. 9.40b

9.41 Find x'_c and y'_c for the centerline of the weld pattern shown in Fig. 9.41a,

❚ From the symmetry of the weld pattern,

$$x'_c = \frac{40 + 100 + 40}{2} = 90 \text{ mm}$$

Because of the above symmetry, only half of the weld pattern is needed to find y'_c, and Fig. 9.41b shows the line elements that comprise this plane curve. The centroidal coordinate y'_c is then found from

$$y'_c = \frac{y_1 l_1 + y_2 l_2 + y_3 l_3 + y_4 l_4}{l_1 + l_2 + l_3 + l_4} = \frac{\frac{240}{2}(240) + 240(40) + (240 - \frac{50}{2})50 + 190(50)}{240 + 40 + 50 + 50} = 154 \text{ mm}$$

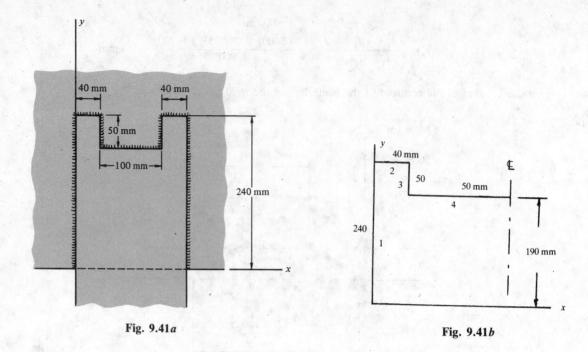

Fig. 9.41a

Fig. 9.41b

9.42   Figure 9.42a shows the centerline of a weld pattern. Find x'_c and y'_c for this plane curve.

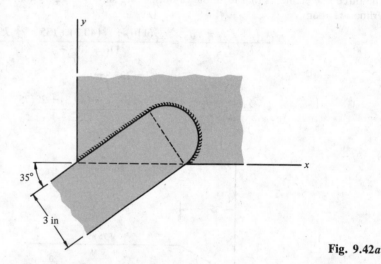

Fig. 9.42a

▮ The weld pattern in rotated and referenced to the $x_a y_a$ coordinate system shown in Fig. 9.42b, and the weld is divided into two simple curved shapes.   The length A of the straight portion of the weld is

$$A = \frac{3}{\tan 35°} = 4.28 \text{ in}$$

Case 5 of Table 9.10 is used, and

$$x'_{ca} = \frac{x_{1a}l_1 + x_{2a}l_2}{l_1 + l_2} = \frac{-(4.28/2)(4.28) + [2(1.5)/\pi][\pi(1.5)]}{4.28 + \pi(1.5)} = -0.518 \text{ in}$$

$$l = 8.99 \text{ in}$$

$$y'_{ca} = \frac{y_{1a}l_1 + y_{2a}l_2}{l_1 + l_2} = \frac{3(4.28) + 1.5[\pi(1.5)]}{8.99} = 2.21 \text{ in}$$

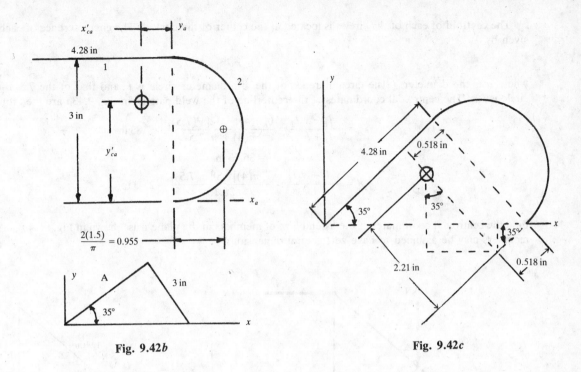

Fig. 9.42b

Fig. 9.42c

The weld pattern is now drawn in its original position, shown in Fig. 9.42c. The coordinates of the centroid are expressed in terms of dimensions along the xy axes. The results are

$$x'_c = \frac{4.28}{\cos 35°} - 0.518 \cos 35° - 2.21 \sin 35° = 3.53 \text{ in} \qquad \text{and} \qquad y'_c = 2.21 \cos 35° - 0.518 \sin 35° = 1.51 \text{ in}$$

9.43 Find the centroidal coordinates of the centerline of the weld pattern shown in Fig. 9.43.

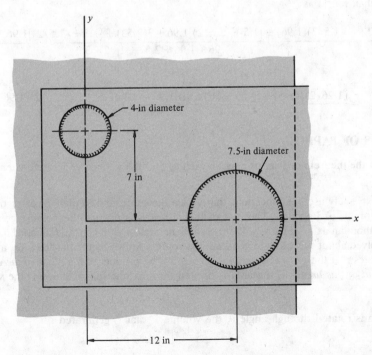

Fig. 9.43

▮ The centroid of each of the circles is located at the center of the circle. The circumference of each circle is given by

$$l = \pi d$$

where d is the diameter. The circumference of the 4-in-diameter circle is l_1 and that of the 7.5-in-diameter circle is l_2. The centroidal coordinates of the centerline of the weld pattern in Fig. 9.43a are then found to be

$$x'_c = \frac{x_1 l_1 + x_2 l_2}{l_1 + l_2} = \frac{0[\pi(4)] + 12[\pi(7.5)]}{\pi(4) + \pi(7.5)} = 7.83 \text{ in}$$

$$l = 36.13 \text{ in}$$

$$y'_c = \frac{y_1 l_1 + y_2 l_2}{l_1 + l_2} = \frac{7[\pi(4)] + 0[\pi(7.5)]}{36.13} = 2.43 \text{ in}$$

9.44 Find the centroidal coordinates of the assemblage of members in the plane truss shown in Fig. 9.44a. All of the members may be assumed to have zero lateral dimensions.

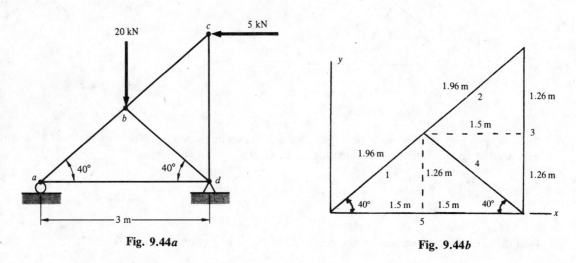

Fig. 9.44a   **Fig. 9.44b**

▮ The five straight-line elements of the truss are identified in Fig. 9.44b. The centroidal coordinates of the truss are then found as

$$x'_c = \frac{\sum_i x_i l_i}{l} = \frac{(1.5/2)(1.96) + [1.5 + (1.5/2)]1.96 + 3(2.52) + [1.5 + (1.5/2)]1.96 + \frac{3}{2}(3)}{1.96 + 1.96 + 2.52 + 1.96 + 3} = 1.96 \text{ m} = 1,960 \text{ mm}$$

$$l = 11.40 \text{ m}$$

$$y'_c = \frac{\sum_i x_i l_i}{l} = \frac{(1.26/2)(1.96) + [1.26 + (1.26/2)]1.96 + 1.26(2.52) + (1.26/2)(1.96) + 0(3)}{11.40} = 0.820 \text{ m} = 820 \text{ mm}$$

9.3 THEOREMS OF PAPPUS

9.45 Show how the theorem of Pappus may be used to find the volume generated when a plane area is rotated about an axis.

▮ Two extremely useful applications that utilize the centroid of a plane area or of a plane curve are contained in the theorems of Pappus. The first of these theorems defines the volume generated when a plane area is revolved about an axis. Figure 9.45 shows a plane area A with centroidal coordinate y_c. This area is imagined to be revolved about the x axis to generate the volume shown by the dashed lines in the figure. The theorem of Pappus states that the *generated volume* is equal to the product of *the plane area A* and the distance traveled by the *centroid of this area*. If the area is rotated through one full revolution, the volume V that is generated is

$$V = 2\pi y_c A$$

If the area is rotated through angle θ, the volume V that is generated is

$$V = y_c \theta A$$

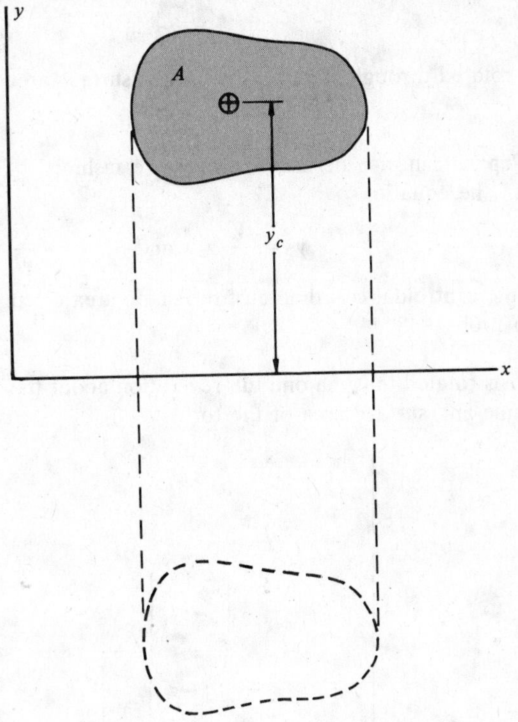

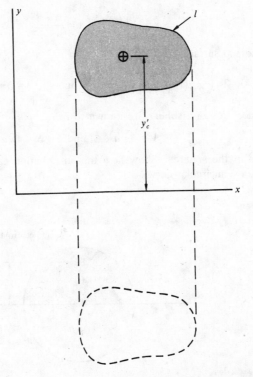

Fig. 9.45

9.46   How is the surface area of the volume described in Prob. 9.45 determined?

▮   The plane area of Prob. 9.45 is now redrawn in Fig. 9.46. l is the total length of the perimeter of this area, and y'_c is the centroidal coordinate of this *plane curve boundary*.   The perimeter of the area is now imagined to be revolved about the x axis.   The theorem of Pappus then states that the *surface area* of the generated volume is equal to the product of the length of the perimeter of the area and the distance traveled by the centroid of this plane curve.   If the perimeter in Fig. 9.46 is rotated through one full revolution, the surface area S which is

Fig. 9.46

generated is

$$S = 2\pi y'_c l$$

If the plane curve is rotated through angle θ, the curved surface area of the generated volume is

$$S = y'_c \theta l$$

The theorems of Pappus can also be used in a reverse fashion. That is, if the value of a volume or of a surface area is known, the equations

$$y_c = \frac{V}{\theta A} \quad \text{and} \quad y'_c = \frac{S}{\theta l}$$

may be used to find the centroidal coordinate of the plane area or curve used to generate this quantity. This effect is illustrated in Prob. 9.48.

9.47 The circle in Fig. 9.47 is rotated through one full revolution about the x axis. The resulting volume is called a *torus*. Find the volume and surface area of the torus.

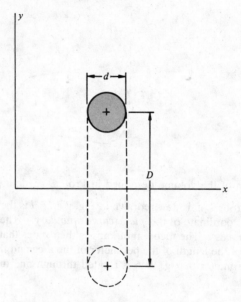

Fig. 9.47

▮ For this case,

$$y_c = y'_c = \frac{D}{2} \qquad A = \frac{\pi d^2}{4} \qquad l = \pi d$$

The generated volume and surface area are then

$$V = 2\pi \left(\frac{D}{2}\right)\left(\frac{\pi d^2}{4}\right) = \frac{\pi^2 D d^2}{4} \quad \text{and} \quad S = 2\pi \left(\frac{D}{2}\right)(\pi d) = \pi^2 D d$$

9.48 The volume and surface area of a sphere are known to be

$$V = \tfrac{4}{3}\pi a^3 \quad \text{and} \quad S = 4\pi a^2$$

where a is the radius of the sphere. Show how this information can be used to obtain the centroids of a semicircular area and a semicircular curve.

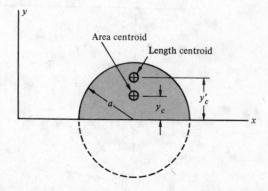

Fig. 9.48

▌ A sphere may be generated by revolving a semicircular area about an axis, as shown in Fig. 9.48. The volume generated is

$$2\pi y_c A = V \qquad 2\pi y_c \left(\frac{\pi a^2}{2}\right) = \frac{4}{3}\pi a^3 \qquad y_c = \frac{4a}{3\pi}$$

The generated surface area is expressed as

$$2\pi y'_c l = S \qquad 2\pi y'_c (\pi a) = 4\pi a^2 \qquad y'_c = \frac{2a}{\pi}$$

9.49 An ellipsoid is the volume that is formed by rotating an ellipse about its semiaxis. Find the volume of an ellipsoid that is 3 ft in diameter and 6 ft long.

▌ Figure 9.49 shows the semiellipse which is rotated about the x axis to generate the ellipsoid. From Case 11 in Table 9.9, for the semiellipse,

$$A = \frac{\pi ab}{8} \qquad y_c = \frac{2b}{3\pi}$$

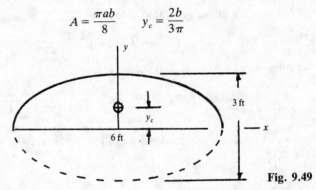

Fig. 9.49

For the present problem, $a = 6$ ft and $b = 3$ ft, so that

$$A = \frac{\pi ab}{8} = \frac{\pi(6)3}{8} = 7.07 \text{ ft}^2 \qquad \text{and} \qquad y_c = \frac{2b}{3\pi} = \frac{2(3)}{3\pi} = 0.637 \text{ ft}$$

The volume of the ellipsoid is then

$$V = 2\pi y_c A = 2\pi(0.637)(7.07) = 28.3 \text{ ft}^3$$

9.50 Figure 9.50 shows the preliminary dimensions of a fuel tank which is to be used on an experimental vehicle. The tank is to be formed as a body of revolution of the shaded area about the y axis.

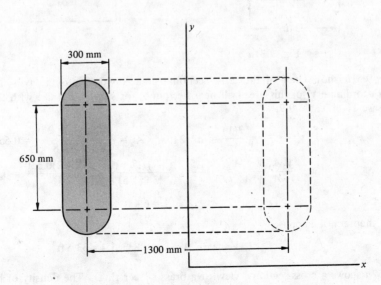

Fig. 9.50

(*a*) How many gallons (U.S.) of fuel will this tank hold?

(*b*) What is the total surface area, in square feet, of the tank? $1 \text{ ft}^3 = 7.48 \text{ gal (U.S.)}$.

■ (a) The cross-sectional area A of the shaded area is

$$A = \frac{\pi(300)^2}{4} + 300(650) = 2.66 \times 10^5 \text{ mm}^2$$

Using the theorem of Pappus, we get

$$V = 2\pi y_c A = (2\pi)\left(\frac{1,300}{2}\right)(2.66 \times 10^5) = 1.09 \times 10^9 \text{ mm}^3$$

This result is then converted to U.S. gallons, as

$$(1.09 \times 10^9 \text{ mm}^3)\left(\frac{1 \text{ m}}{1,000 \text{ mm}}\right)^3\left(\frac{3.28 \text{ ft}}{1 \text{ m}}\right)^3\left(\frac{7.48 \text{ gal}}{1 \text{ ft}^3}\right) = 288 \text{ gal (U.S.)}$$

(b) The perimeter of the shaded area is

$$l = \pi(300) + 2(650) = 2,240 \text{ mm}$$

The surface area of the generated volume is

$$S = 2\pi y_c' l = (2\pi)\left(\frac{1,300}{2}\right)2,240 = 9.15 \times 10^6 \text{ mm}^2$$

This value may be expressed in square feet as

$$(9.15 \times 10^6 \text{ mm}^2)\left(\frac{1 \text{ m}}{1,000 \text{ mm}}\right)^2\left(\frac{3.28 \text{ ft}}{1 \text{ m}}\right)^2 = 98.4 \text{ ft}^2$$

9.51 The cross section of the cylindrical tank shown in Fig. 9.51a includes two semielliptical shapes. Compute the volume of the tank.

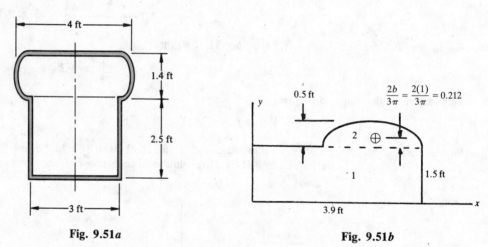

Fig. 9.51a   Fig. 9.51b

■ Figure 9.51b shows half of the cross-sectional area, divided into the two simple areas 1 and 2. The centroidal coordinate y_c of this area will next be found, for subsequent use with the theorem of Pappus. From Case 11 in Table 9.9,

$$y_{c2} = \frac{2b}{3\pi} = \frac{2(1)}{3\pi} = 0.212 \text{ ft} \qquad A_2 = \frac{\pi ab}{8} = \frac{\pi(1.4)1}{8} = 0.550 \text{ ft}$$

$$y_c = \frac{y_1 A_1 + y_2 A_2}{A_1 + A_2} = \frac{(1.5/2)(3.9)1.5 + (1.5 + 0.212)0.550}{3.9(1.5) + 0.550} = 0.833 \text{ ft}$$

$$A = 6.4 \text{ ft}^2$$

Using the theorem of Pappus,

$$V = 2\pi y_c A = 2\pi(0.833)6.4 = 33.5 \text{ ft}^3$$

9.52 Figure 9.52a shows a cross-sectional view of a brass spacer ring. The density of brass is $8,550 \text{ kg/m}^3$.

(a) Find the weight of the ring.

(b) Find the total surface area of the ring.

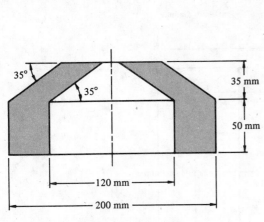

Fig. 9.52a   Fig. 9.52b

(a) Figure 9.52b shows one-half of the cross-sectional area. This area is divided into the five areas shown in the figure. Area 2 is a 35 by 100 mm rectangle. Dimension A is found from

$$A = \frac{35}{\tan 35°} = 50.0 \text{ mm}$$

The centroidal coordinate y_c is found from

$$y_c = \frac{y_1 A_1 + y_2 A_2 - y_3 A_3 - y_4 A_4 - y_5 A_5}{A_1 + A_2 - A_3 - A_4 - A_5}$$

$$= \frac{[(100 + 60)/2]50(40) + \frac{100}{2}(35)100 - [50 + \frac{2}{3}(50)]0.5(35)50 - [10 + \frac{1}{3}(50)]0.5(35)50 - \frac{10}{2}(35)10}{50(40) + 35(100) - 0.5(35)50 - 0.5(35)50 - 35(10)}$$

$$= 69.7 \text{ mm}$$

$$A = 3,400 \text{ mm}^2$$
$$V = 2\pi y_c A = 2\pi(69.7)3,400 = 1.49 \times 10^6 \text{ mm}^3$$

The weight of the ring is

$$W = 1.49 \times 10^6 \text{ mm}^3 \left(8,550 \frac{\text{kg}}{\text{m}^3}\right)\left(\frac{1 \text{ m}}{1,000 \text{ mm}}\right)^3\left(9.81 \frac{\text{m}}{\text{s}^2}\right) = 125 \text{ N}$$

(b) The boundary of the area shown in Fig. 9.52b is divided into the six elementary lengths shown in this figure. The centroidal coordinate y_c' of this boundary is

$$y_c' = \frac{y_1' l_1 + y_2' l_2 + y_3' l_3 + y_4' l_4 + y_5' l_5 + y_6' l_6}{l_1 + l_2 + l_3 + l_4 + l_5 + l_6}$$

$$= \frac{(60 + \frac{40}{2})40 + 100(50) + (50 + \frac{50}{2})61 + (10 + \frac{40}{2})40 + (10 + \frac{50}{2})61 + 60(50)}{40 + 50 + 61 + 40 + 61 + 50} = 63.3 \text{ mm}$$

$$l = 302 \text{ mm}$$

Using the theorem of Pappus, the curved surface area of the spacer ring is

$$S = 2\pi y_c' l = 2\pi(63.3)302 = 1.20 \times 10^5 \text{ mm}^2$$

The total surface area of the ring is found by adding the plane surface areas of the two end cross sections to the above value, with the result

$$A_{\text{total}} = 1.20 \times 10^5 + \pi(100^2 - 60^2) + \pi(50^2 - 10^2) = 1.48 \times 10^5 \text{ mm}^2\left(\frac{1 \text{ m}}{1,000 \text{ mm}}\right)^2 = 0.148 \text{ m}^2$$

9.53 The steel inner race of a ball bearing is shown in Fig. 9.53a. Find the weight of the race. $\rho = 7,830 \text{ kg/m}^3$.

Figure 9.53b shows the profile of the circular segment shaped area in the cross section. From this figure,

$$\cos\theta = \tfrac{3}{5} \qquad \theta = 53.1° = 0.927 \text{ rad}$$

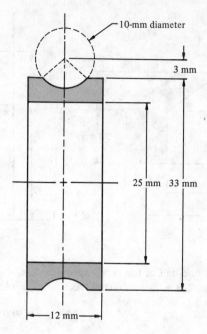

10-mm diameter

3 mm

25 mm 33 mm

├─12 mm─┤

Fig. 9.53a

Using Case 9 of Table 9.9,

$$A = a^2\left(\theta - \frac{\sin 2\theta}{2}\right) = 5^2\left(0.927 - \frac{\sin 2(53.1°)}{2}\right) = 11.2 \text{ mm}^2$$

Using the definition of x_c in the Case 9 illustration,

$$x_c = \frac{2a \sin^3 \theta}{3\{\theta - [(\sin 2\theta)/2]\}} = \frac{2(5) \sin^3 53.1°}{3(0.927 - \{[\sin 2(53.1°)]/2\})} = 3.81 \text{ mm}$$

The dimension Δx in Fig. 9.53b is

$$\Delta x = x_c - 3 = 0.81 \text{ mm}$$

Figure 9.53c shows the division of the cross-sectional area of the race into the two elementary shapes of a rectangle and a circular segment. The centroidal coordinate y_c of the area in Fig. 9.53c is

$$y_c = \frac{y_1 A_1 - y_2 A_2}{A_1 - A_2} = \frac{\frac{1}{2}(12.5 + 16.5)12(4) - (16.5 - 0.81)11.2}{12(4) - 11.2} = 14.1 \text{ mm} \qquad A = 36.8 \text{ mm}^2$$

Using the theorem of Pappus,

$$V = 2\pi y_c A = 2\pi(14.1)36.8 = 3,260 \text{ mm}^3$$

The weight of the race is

$$W = 3,260 \text{ mm}^3\left(7,830 \frac{\text{kg}}{\text{m}^3}\right)\left(\frac{1 \text{ m}}{1,000 \text{ mm}}\right)^3 9.81 \frac{\text{m}}{\text{s}^2} = 0.250 \text{ N}$$

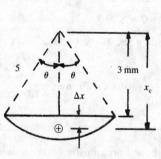

Fig. 9.53b

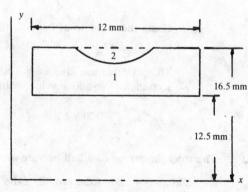

Fig. 9.53c

9.54   Figure 9.54*a* shows the cross section of a wheel that is used in a lightweight mechanism.   The hub is aluminium, with a specific weight of 0.1 lb/in³.   On the outer rim of the wheel is attached a plastic "tire" whose cross section is a semiellipse.   The tire material has a specific weight of 0.067 lb/in³.   Compute the weight, in ounces, of the wheel.

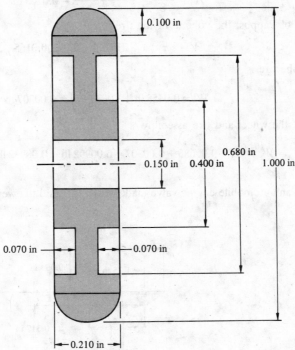

0.100 in

0.680 in

0.150 in   0.400 in   1.000 in

0.070 in ───   ─── 0.070 in

0.210 in

Fig. 9.54*a*

▎ The upper cross section of the hub is divided into the elementary shapes shown in Fig. 9.54*b*.   Area 1 is a 0.210 by 0.325 in rectangle.   The centroidal coordinate y_c of the area in Fig. 9.54*b* is

$$y_c = \frac{y_1 A_1 - 2y_2 A_2}{A_1 - 2A_2} = \frac{[(0.075+0.4)/2]0.210(0.325) - 2[(0.2+0.34)/2]0.070(0.14)}{0.210(0.325) - 2(0.070)0.14} = 0.224 \text{ in}$$

$$A = 0.0487 \text{ in}^2$$

Using the theorem of Pappus,

$$V = 2\pi y_c A = 2\pi(0.224)0.0487 = 0.0685 \text{ in}^3$$

The weight W_1 of the hub is then

$$W_1 = 0.0685 \text{ in}^3\left(0.1\ \frac{\text{lb}}{\text{in}^3}\right) = 0.00685 \text{ lb}$$

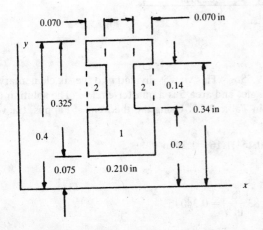

0.070 ───   ─── 0.070 in

y

2   2   0.14

0.34 in

0.325

0.2

0.4   1

0.075   0.210 in

x

Fig. 9.54*b*

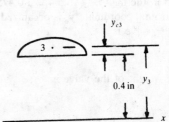

y_{c3}

3 · ──

0.4 in y_3

x   **Fig. 9.54*c***

The cross section of the "tire" is seen in Fig. 9.54c. Using Case 11 in Table 9.9, with $b = 0.2$ in and $a = 0.210$ in,

$$y_{c3} = \frac{2b}{3\pi} = \frac{2(0.2)}{3\pi} = 0.0424 \text{ in}$$

$$A_3 = \frac{\pi ab}{8} = \frac{\pi(0.210)0.2}{8} = 0.0165 \text{ in}^2$$

Using the theorem of Pappus, the volume of the tire is

$$V = 2\pi y_3 A = 2\pi(0.4 + 0.042)0.0165$$

The weight W_2 of the tire is

$$W_2 = 0.0458 \text{ in}^3\left(0.067 \frac{\text{lb}}{\text{in}^3}\right) = 0.00307 \text{ lb}$$

The total weight of the wheel and tire assembly is

$$W_{\text{total}} = W_1 + W_2 = 0.00685 + 0.00307 = 0.00992 \text{ lb} = 0.00992 \text{ lb}\left(\frac{16 \text{ oz}}{1 \text{ lb}}\right) = 0.159 \text{ oz}$$

9.55   Figure 9.55a shows an automobile engine valve made of steel. Find the weight of the valve. $\gamma = 0.283 \text{ lb/in}^3$.

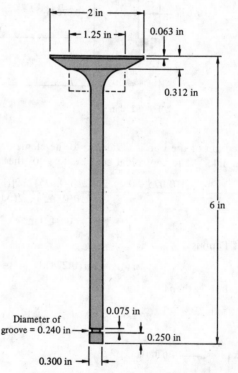

Fig. **9.55a**

▌ A cross-sectional view of the valve is shown in Fig. 9.55b. This view is divided into the six elementary areas seen in the figure. Area 4 is a 0.475 by 0.625 in rectangle, and area 5 is a quarter circle. The solution for the centroidal coordinate y_c is organized in the form seen in Table 9.8. Using the theorem of Pappus, the volume of the valve is

$$V = 2\pi y_c A = 2\pi(0.164)1.16 = 1.20 \text{ in}^3$$

The weight of the valve is

$$W = 1.20 \text{ in}^3\left(0.283 \frac{\text{lb}}{\text{in}^3}\right) = 0.340 \text{ lb}$$

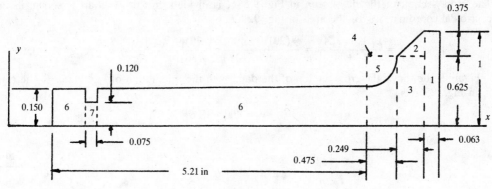

Fig. 9.55b

TABLE 9.8

element	A_i	y_i	y_iA_i
1	$0.063(1) = 0.063$	0.5	0.0315
2	$0.5(0.249)0.375 = 0.0467$	$0.625 + \frac{1}{3}(0.375) = 0.75$	0.0350
3	$0.249(0.625) = 0.1556$	$\frac{1}{2}(0.625) = 0.3125$	0.0486
4	$0.475(0.625) = 0.2969$	$\frac{1}{2}(0.625) = 0.3125$	0.0928
5	$-0.25\pi(0.475)^2 = -0.1772$	$0.625 - \dfrac{4(0.475)}{3\pi} = 0.4234$	-0.0750
6	$(5.21 - 0.075)0.150 = 0.7703$	$\frac{1}{2}(0.150) = 0.075$	0.0578
7	$0.075(0.120) = 0.0090$	$\frac{1}{2}(0.120) = 0.060$	0.0005
	$A = \sum_i A_i = 1.164 \text{ in}^2$	$\sum_i y_iA_i = 0.1912$	
		$y_c = 0.164 \text{ in}$	

9.56 Figures 9.56a and b show the dimensions of a proposed earth dam.

(a) How many cubic yards of earth are in this dam?

(b) Find the weight and mass of the dam material, if the dam is formed from packed, moist earth with a specific gravity of 1.6.

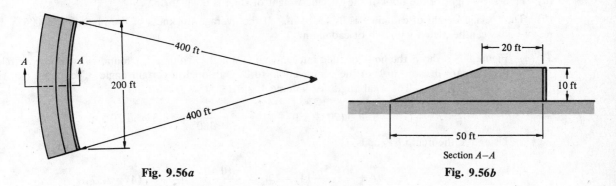

Fig. 9.56a

Section A–A

Fig. 9.56b

▌ (a) Figure 9.56c shows the top view, as in Fig. 9.56a, with the half angle θ. From this figure,

$$\sin \theta = \frac{100}{400} \qquad \theta = 14.5° = 0.253 \text{ rad}$$

The cross section of the dam, seen in Fig. 9.56b, is divided into the two areas shown in Fig. 9.56d. The centroidal coordinate x_c for the area in Fig. 9.56d is

$$x_c = \frac{x_1 A_1 + x_2 A_2}{A_1 + A_2} = \frac{\frac{20}{2}(20)10 + [20 + \frac{1}{3}(30)]0.5(30)10}{20(10) + 0.5(30)10} = 18.6 \text{ ft} \qquad A = 350 \text{ ft}^2$$

Figure 9.56e shows the cross section of the dam, with the orientation of Fig. 9.56a. The centroidal distance y_c in Fig. 9.56e is

$$y_c = 400 + 18.6 = 419 \text{ ft}$$

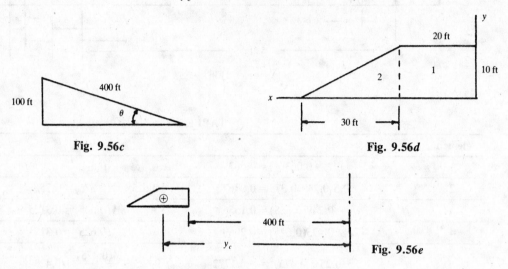

Fig. 9.56c   Fig. 9.56d

Fig. 9.56e

Using the theorem of Pappus, the volume of earth in the dam is

$$V = y_c \theta A = 2(419)0.253(350) = 74{,}200 \text{ ft}^3$$

$$V = 74{,}200 \text{ ft}^3 \left(\frac{1 \text{ yd}}{3 \text{ ft}}\right)^3 = 2{,}750 \text{ yd}^3$$

(b)   The weight of earth in the dam is

$$W = 74{,}200 \text{ ft}^3 (1.6)62.4 \frac{\text{lb}}{\text{ft}^3} = 7.41 \times 10^6 \text{ lb}\left(\frac{1 \text{ ton}}{2{,}000 \text{ lb}}\right) = 3{,}710 \text{ ton}$$

The mass of the earth is

$$m = \frac{W}{g} = \frac{7.41 \times 10^6}{32.2} = 2.30 \times 10^5 \text{ slug}$$

9.57   The hook shown in Fig. 9.57a is to be formed from 0.5-in-diameter steel rod and then plated with cadmium.

(a)   Find the weight of the hook if the specific weight of steel is 0.283 lb/in³.

(b)   The specific weight of cadmium is 0.313 lb/in³. If the average thickness of the plating is 0.0006 in, how many hooks can be plated with 1 lb of cadmium?

▮ (a)   Figure 9.57b shows the hook divided into volumes 1 and 2. Volume 1 is found by sweeping a circle of diameter 0.5 in, with a y_c of the center equal to 2 in, through a certain angle. Volume 2 is a right circular cylinder. The total angle for volume 1 is

$$\theta = (20° + 180° + 30° + 30°)\left(\frac{\pi \text{ rad}}{180°}\right) = 4.54 \text{ rad}$$

Using the theorem of Pappus,

$$V = y_c \theta A + \frac{\pi(0.5)^2}{4}(4) = 2(4.54)\frac{\pi(0.5)^2}{4} + \frac{\pi(0.5)^2}{4}(4) = 2.57 \text{ in}^3$$

The weight of the hook is given by

$$W = 2.57 \text{ in}^3\left(0.283 \frac{\text{lb}}{\text{in}^3}\right) = 0.727 \text{ lb}$$

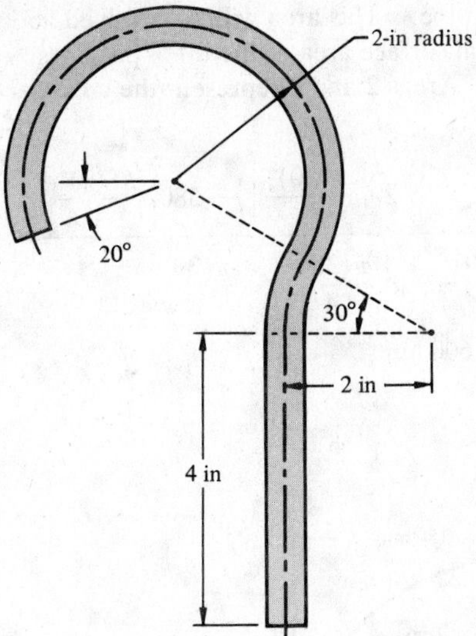

Fig. 9.57a

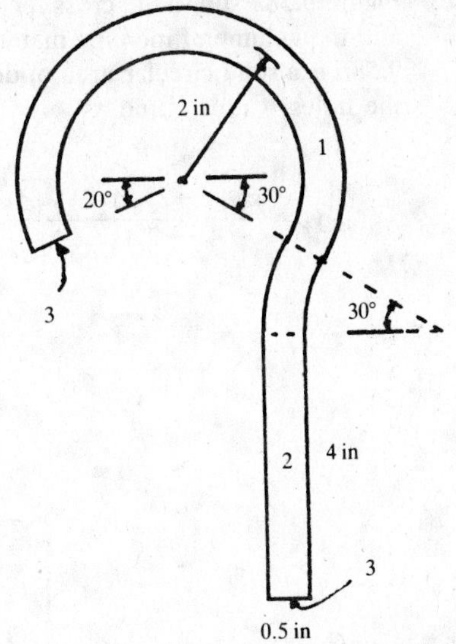

0.5 in   Fig. 9.57b

(b) The end cross sections in Fig. 9.57b are designated 3. The surface area of 1 is found by using the theorem of Pappus, and the surface area of 2 is that of a right circular cylinder. The centroidal coordinate y_c' for the circumference of the cross section is $y_c' = 2$ in, and the length of this circumference is given by $\pi d = \pi(0.5)$. The total surface area of the hook is then

$$S = y_c'\theta l + \pi(0.5)4 + 2\frac{\pi(0.5)^2}{4} = 2(4.54)\pi(0.5) + \pi(0.5)4 + 2\frac{\pi(0.5)^2}{4} = 20.9 \text{ in}^2$$

The weight W_1 of the plating on one hook is given by

$$W_1 = 20.9 \text{ in}^2(0.0006 \text{ in})0.313 \frac{\text{lb}}{\text{in}^3} = 0.00393 \text{ lb/hook}$$

The number of hooks N that can be plated with 1 lb of cadmium is then

$$N = \frac{1}{W_1} = \frac{1}{0.0039 \text{ lb/hook}} = 254 \text{ hooks}$$

9.58 A special steel fitting for a high-pressure system is shown in Fig. 9.58a.

(a) Find the weight of the fitting if $\rho = 7{,}830 \text{ kg/m}^3$.

(b) The combined internal surface area of the three passages in the curved pipe is required for use in a heat-transfer calculation. Find the value of this area.

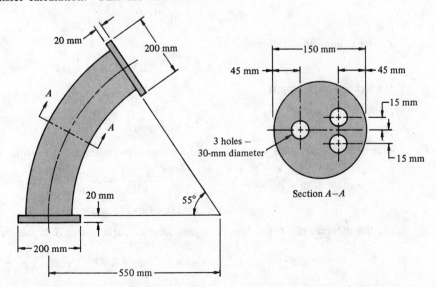

Fig. 9.58a

(a) Figure 9.58b shows the cross section of the curved pipe. This area will be revolved about the x axis to find the volume of the pipe material and the internal surface area of the three passages. Area 1 in Fig. 9.58b is a solid circular area of diameter 150 mm. Areas 2 and 3 represent the cross-sectional areas of the holes. y_c is found as

$$y_c = \frac{y_1 A_1 - 2y_2 A_2 - y_3 A_3}{A_1 - 2A_2 - A_3} = \frac{550\left[\dfrac{\pi(150)^2}{4}\right] - 2\left(520\left[\dfrac{\pi(30)^2}{4}\right]\right) - 580\left[\dfrac{\pi(30)^2}{4}\right]}{\dfrac{\pi(150)^2}{4} - 2\left[\dfrac{\pi(30)^2}{4}\right] - \dfrac{\pi(30)^2}{4}} = 550 \text{ mm}$$

$$A = 15,600 \text{ mm}^2$$

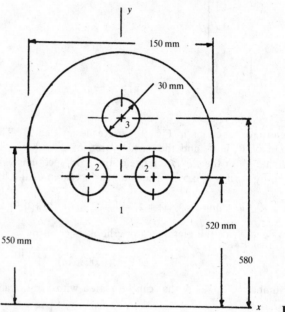

Fig. 9.58b

Using the theorem of Pappus,

$$V_1 = y_c \theta A = 550(55°)\left(\frac{\pi \text{ rad}}{180°}\right)15,600 = 8.24 \times 10^6 \text{ mm}^3$$

The volume V_2 of the two end plates is

$$V_2 = 2\left[\frac{\pi(200)^2}{4} - 3\left(\frac{\pi(30)^2}{4}\right)\right]20 = 1.17 \times 10^6 \text{ mm}^3$$

The total volume V of the fitting is

$$V = V_1 + V_2 = 8.24 \times 10^6 + 1.17 \times 10^6 = 9.41 \times 10^6 \text{ mm}^3$$

The weight of the fitting is

$$W_{\text{total}} = 9.41 \times 10^6 \text{ mm}^3\left(\frac{1 \text{ m}}{1,000 \text{ mm}}\right)^3 7,830 \frac{\text{kg}}{\text{m}^3}\left(9.81 \frac{\text{m}}{\text{s}^2}\right) = 723 \text{ N}$$

(b) The centroidal coordinate y_c' of the pattern of hole circumferences shown in Fig. 9.58b is

$$y_c' = \frac{2y_2' l_2 + y_3' l_3}{2l_2 + l_3} = \frac{2(520)\pi(30) + 580\pi(30)}{2\pi(30) + \pi(30)} = 540 \text{ mm}$$

$$l = 2l_2 + l_3 = 283 \text{ mm}$$

Using the theorem of Pappus, the combined internal surface area S of the three passages is given by

$$S = y_c' \theta l = 540(55°)\left(\frac{\pi \text{ rad}}{180°}\right)283 = 1,47 \times 10^5 \text{ mm}^2 = 1.47 \times 10^5 \text{ mm}^2\left(\frac{1 \text{ m}}{1,000 \text{ mm}}\right)^2 = 0.147 \text{ m}^2$$

9.4 SOLUTIONS USING THE INTEGRAL DEFINITIONS OF THE CENTROIDAL COORDINATES

9.59 Find the centroidal coordinates of the area bounded by the straight line and the parabolic curve in Fig. 9.59a. x and y are in millimeters.

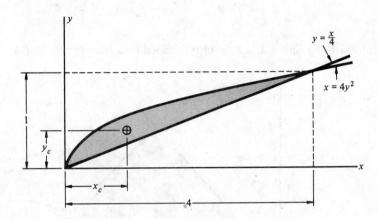

Fig. 9.59a

❚ The area element used to find x_c is shown in Fig. 9.59b.

$$dA = \left(\sqrt{\frac{x}{4}} - \frac{x}{4} \right) dx$$

$$x_c = \frac{\int_A x \, dA}{A} = \frac{\int_0^4 x \left(\sqrt{\frac{x}{4}} - \frac{x}{4} \right) dx}{\int_0^4 \left(\sqrt{\frac{x}{4}} - \frac{x}{4} \right) dx} = \frac{\int_0^4 \left(\frac{x^{3/2}}{2} - \frac{x^2}{4} \right) dx}{\int_0^4 \left(\frac{x^{1/2}}{2} - \frac{x}{4} \right) dx} = \frac{\left[\frac{2}{5} \left(\frac{x^{5/2}}{2} \right) - \frac{x^3}{3(4)} \right]_0^4}{\left[\frac{2}{3} \left(\frac{x^{3/2}}{2} \right) - \frac{x^2}{2(4)} \right]_0^4} = 1.60 \text{ mm}$$

The area element used to find y_c is shown in Fig. 9.59c.

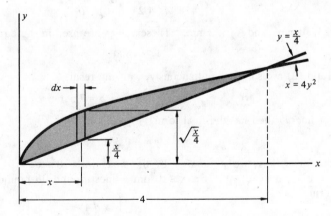

Fig. 9.59b

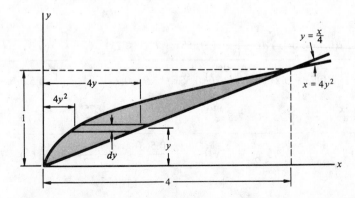

Fig. 9.59c

$$dA = (4y - 4y^2)\, dy$$

$$y_c = \frac{\int_A y\, dA}{A} = \frac{\int_0^1 4y(y - y^2)\, dy}{\int_0^1 4(y - y^2)\, dy} = \frac{\int_0^1 (y^2 - y^3)\, dy}{\int_0^1 (y - y^2)\, dy} = \frac{\left[\dfrac{y^3}{3} - \dfrac{y^4}{4}\right]_0^1}{\left[\dfrac{y^2}{2} - \dfrac{y^3}{3}\right]_0^1} = \frac{\dfrac{1}{3} - \dfrac{1}{4}}{\dfrac{1}{2} - \dfrac{1}{3}} = \frac{\dfrac{1}{12}}{\dfrac{1}{6}} = 0.5 \text{ mm}$$

9.60 Find the centroidal coordinates x_c and y_c of the shaded area shown in Fig. 9.60a. x and y are in millimeters.

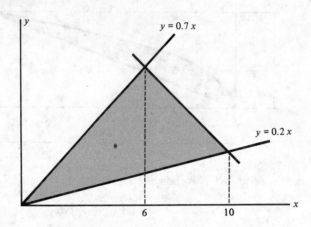

Fig. 9.60a

▌ The area is redrawn in Fig. 9.60b, with the differential area elements used for the computation of x_c. As a preliminary step, the equation of line ab is determined. The equation of this line has the form

$$y = mx + b$$

where m is the slope of the line and b is the y intercept. At point a, $x = 6$ mm and $y = 4.2$ mm. Using these values in the above equation,

$$4.2 = m(6) + b \tag{1}$$

At point b, $x = 10$ mm and $y = 2$ mm. These values are used in the equation of line $0a$ to obtain

$$2 = m(10) + b \tag{2}$$

Equations (1) and (2) are solved simultaneously, with the results

$$m = -0.55 \qquad b = 7.5 \text{ mm}$$

The equation of line ab then has the final form

$$y = -0.55x + 7.5$$

The integration for x_c has two intervals. In the first, with $0 \le x \le 6$ mm, lines $0a$ and $0b$ bound the area. In the second interval, with $6 \le x \le 10$ mm, lines ab and $0b$ bound the area. The equation for x_c then has the form

$$x_c = \frac{\int_A x\, dA}{\int_A dA}$$

$$x_c = \frac{\int_0^6 x(0.7x - 0.2x)\, dx + \int_6^{10} x(-0.55x + 7.5 - 0.2x)\, dx}{\int_0^6 (0.7x - 0.2x)\, dx + \int_6^{10} (-0.55x + 7.5 - 0.2x)} = \frac{\int_0^6 x(0.5x)\, dx + \int_6^{10} x(-0.75x + 7.5)\, dx}{\int_0^6 0.5x\, dx + \int_6^{10} (-0.75x + 7.5)\, dx}$$

$$= \frac{\left.\dfrac{0.5x^3}{3}\right|_0^6 + \left(-0.75\dfrac{x^3}{3} + 7.5\dfrac{x^2}{2}\right)\Big|_6^{10}}{0.5\dfrac{x^2}{2}\Big|_0^6 + \left(-0.75\dfrac{x^2}{2} + 7.5x\right)\Big|_6^{10}} = \frac{\dfrac{0.5(6)^3}{3} + \left[-\dfrac{0.75}{3}(10^3 - 6^3) + \dfrac{7.5}{2}(10^2 - 6^2)\right]}{\dfrac{0.5(6)^2}{2} + \left[-\dfrac{0.75}{2}(10^2 - 6^2) + 7.5(10 - 6)\right]} = 5.33 \text{ mm}$$

$$A = 15 \text{ mm}^2$$

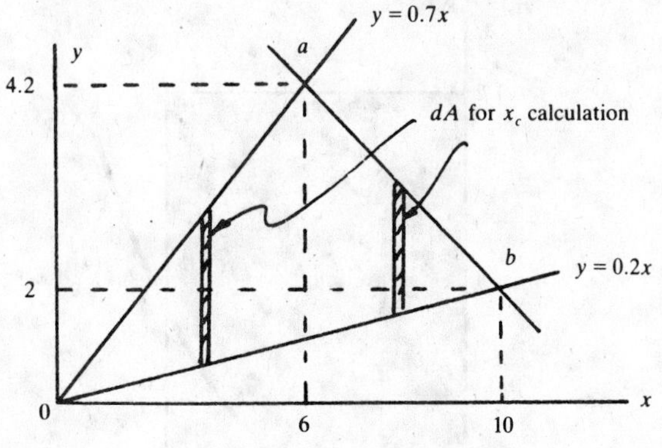

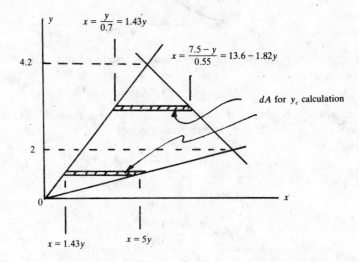

Fig. 9.60b

Fig. 9.60c

The differential area elements required for the computation of y_c are seen in Fig. 9.60c. y_c is then found as

$$y_c = \frac{\int_A y \, dA}{\int_A dA}$$

$$y_c = \frac{\int_0^2 y(5y - 1.43y) \, dy + \int_2^{4.2} y[(13.6 - 1.82y) - (1.43y)] \, dy}{15} = \frac{\int_0^2 3.57y^2 \, dy + \int_2^{4.2} (13.6y - 3.25^2) \, dy}{15}$$

$$= \frac{3.57 \left.\frac{y^3}{3}\right|_0^2 + \left[13.6 \frac{y^2}{2} - 3.25 \frac{y^3}{3}\right]_2^{4.2}}{15} = \frac{\frac{3.57}{3}(2)^3 + \left[\frac{13.6}{2}(4.2^2 - 2^2) - \frac{3.25}{3}(4.2^3 - 2^3)\right]}{15} = 2.05 \text{ mm}$$

9.61 (*a*) Find the location of the x coordinate of the centroid of the shaded area shown in Fig. 9.61*a*. The units of x and y are feet.

(*b*) Using the same area element as in part (*a*), find the y coordinate of the centroid of the area.

(*c*) Check the result in part (*b*) by using an area element which is parallel to the x axis.

▌ (*a*) The area element used to find x_c is shown in Fig. 9.61*b*.

$$x_c = \frac{\int_A x \, dA}{\int_A dA} = \frac{\int_0^{0.8} x(1.2 - x^2) \, dx}{\int_0^{0.8} (1.2 - x^2) \, dx} = \frac{\left[1.2\left(\frac{x^2}{2}\right) - \frac{x^4}{4}\right]_0^{0.8}}{\left[1.2x - \frac{x^3}{3}\right]_0^{0.8}} = 0.357 \text{ ft}$$

$$A = 0.789 \text{ ft}^2$$

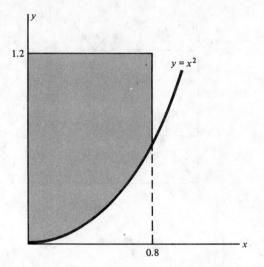

Fig. 9.61a

(b) The area element shown in Fig. 9.61b may be used to find y_c if use is made of the equation

$$\int y \, dA = \int y_{avg}(h \, dx)$$

where h is the height, in the y direction, of the area element, and y_{avg} is the y coordinate of the midpoint of this height. y_c is then found as

$$y_c = \frac{\int y_{avg}(h \, dx)}{\int_A dA} = \frac{\int_0^{0.8}\left(\frac{1.2+x^2}{2}\right)(1.2-x^2)\,dx}{0.789} = \frac{\int_0^{0.8}\frac{1}{2}(1.2^2-x^4)\,dx}{0.789}$$

$$= \frac{\left[\frac{1}{2}(1.2)^2 x - \frac{1}{2}\left(\frac{x^5}{5}\right)\right]\Big|_0^{0.8}}{0.789} = 0.689 \text{ ft}$$

(c) Figure 9.61c shows the standard area element used in the calculation for y_c. The area is divided into the two areas shown in the figure. y_c is then found from

$$y_c = \frac{\int_{A_1} y \, dA + y_2 A_2}{\int_{A_1} dA + A_2}$$

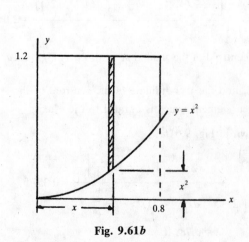

Fig. 9.61b

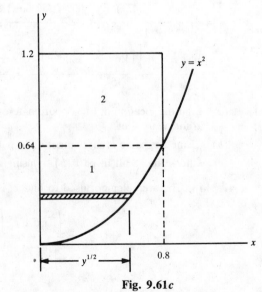

Fig. 9.61c

$$y_c = \frac{\displaystyle\int_0^{0.64} y(y^{1/2})\,dy + \frac{(1.2+0.64)}{2}0.8(1.2-0.64)}{\displaystyle\int_0^{0.64} y^{1/2}\,dy + 0.8(1.2-0.64)} = \frac{\left.\dfrac{2}{5}y^{5/2}\right|_0^{0.64} + 0.4122}{\left.\dfrac{2}{3}y^{3/2}\right|_0^{0.64} + 0.4480}$$

$$= \frac{\frac{2}{5}(0.64)^{5/2} + 0.4122}{\frac{2}{3}(0.64)^{3/2} + 0.4480} = 0.688\text{ ft}$$

It may be seen that the method used in part (b) requires less computational effort than that used in part (c).

9.62 Find the values of the centroidal coordinates of the shaded area shown in Fig. 9.62a. x and y are in meters.

I The equation of the straight line that bounds the area is

$$y = mx + b$$

Using $x = 0$ and $y = 2$, the value $b = 2$ is obtained. The area elements required for the computation of x_c and y_c are seen in Fig. 9.62b. The slope m of the curve is found from Fig. 9.62a as

$$m = \frac{3-2}{4} = 0.25 \qquad \text{and} \qquad y = mx + b = 0.25x + 2$$

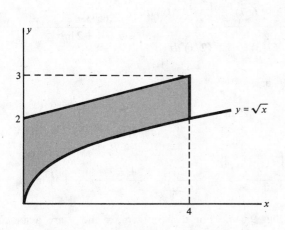

Fig. 9.62a

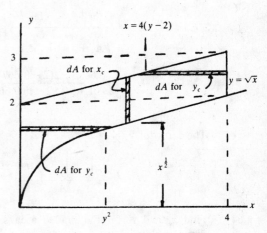

Fig. 9.62b

x_c and y_c are then found as

$$x_c = \frac{\displaystyle\int_A x\,dA}{\displaystyle\int_A dA}$$

$$x_c = \frac{\displaystyle\int_0^4 x(0.25x + 2 - x^{1/2})\,dx}{\displaystyle\int_0^4 (0.25x + 2 - x^{1/2})\,dx} = \frac{\left.\left(0.25\dfrac{x^3}{3} + \dfrac{2x^2}{2} - \dfrac{2x^{5/2}}{5}\right)\right|_0^4}{\left.\left(0.25\dfrac{x^2}{2} + 2x - \dfrac{2x^{3/2}}{3}\right)\right|_0^4} = 1.83\text{ m}$$

$$A = 4.67\text{ m}^2$$

$$y_c = \frac{\displaystyle\int_0^2 y(y^2\,dy) + \int_2^3 y[4 - 4(y-2)]\,dy}{4.67} = \frac{\displaystyle\int_0^2 y^3\,dy + 4\int_2^3 y(3-y)\,dy}{4.67} = \frac{\left.\dfrac{y^4}{4}\right|_0^2 + 4\left.\left(\dfrac{3y^2}{2} - \dfrac{y^3}{3}\right)\right|_2^3}{4.67}$$

$$= \frac{(2^4/4) + 4[\frac{3}{2}(3^2 - 2^2) - \frac{1}{3}(3^3 - 2^3)]}{4.67} = 1.86\text{ m}$$

9.63 Find the values of x_c and y_c for the shaded area shown in Fig. 9.63a. x and y are in inches.

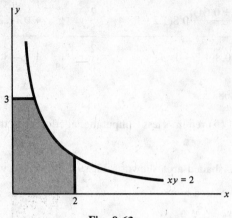

Fig. 9.63a

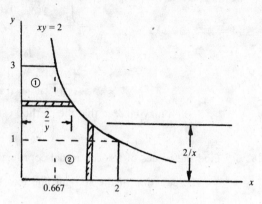

Fig. 9.63b

▮ Figure 9.63b shows the two area elements required for the computation of x_c and y_c. In this figure, area 1 is a rectangle bounded by the coordinate axes and $x = 0.667$, $y = 3$. The rectangular area 2 is bounded by the coordinate axes and $x = 2$, $y = 1$. x_c is found from

$$x_c = \frac{x_1 A_1 + \int_{x=0.667}^{x=2} x\, dA}{A_1 + \int_{x=0.667}^{x=2} dA} = \frac{(0.667/2)(0.667)3 + \int_{0.667}^{2} x(2/x)\, dx}{0.667(3) + \int_{0.667}^{2}(2/x)\, dx} = \frac{0.6673 + 2x\big|_{0.667}^{2}}{2.001 + 2\ln x\big|_{0.667}^{2}}$$

$$= \frac{0.6673 + 2(2 - 0.667)}{2.001 + 2(\ln 2 - \ln 0.667)} = 0.794 \text{ in}$$

$$A = 4.20 \text{ in}^2$$

The form for y_c is

$$y_c = \frac{y_2 A_2 + \int_{y=1}^{y=3} y\, dA}{A} = \frac{\frac{1}{2}(2)1 + \int_{1}^{3} y(2/y)\, dy}{4.20} = \frac{1 + 2y\big|_{1}^{3}}{4.20} = \frac{1 + 2(3 - 1)}{4.20} = 1.19 \text{ in}$$

9.64 Find x_c and y_c for the shaded area shown in Fig. 9.64a. For this problem, x and y are assumed to be dimensionless.

▮ Figure 9.64b shows the area element used to compute y_c. From symmetry,

$$x_c = \frac{\pi}{2}$$

y_c is now found using the form

$$\int y\, dA = \int y_{avg}(h\, dx)$$

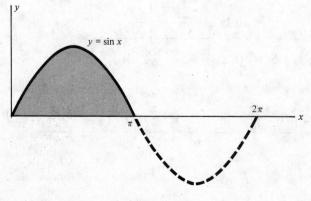

Fig. 9.64a

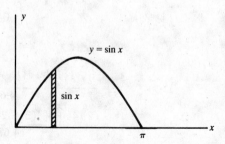

Fig. 9.64b

where h is the height of the area element.

$$y_c = \frac{\int_A y\, dA}{\int_A dA} = \frac{\int_0^\pi (\tfrac{1}{2}\sin x)\sin x\, dx}{\int_0^\pi \sin x\, dx}$$

Using the relationship

$$\int_0^\pi \sin^2 x\, dx = \left[\frac{x}{2} - \sin 2x\right]_0^\pi$$

$$y_c = \frac{\frac{1}{2}\left(\frac{x}{2} - \sin 2x\right)\Big|_0^\pi}{(-\cos x)\big|_0^\pi} = \frac{\frac{1}{2}\left[\left(\frac{\pi}{2} - 0\right) - (\sin \pi - 0)\right]}{-(-1-1)} = \frac{\pi}{8} = 0.393$$

9.65 Find the centroidal coordinates of the area bounded by the sine curve and the line in Fig. 9.65a. x and y are assumed to be dimensionless.

❚ The area is redrawn in Fig. 9.65b, and a vertical area element is chosen. The equation of the straight line in Fig. 9.65b is

$$y = mx + b$$

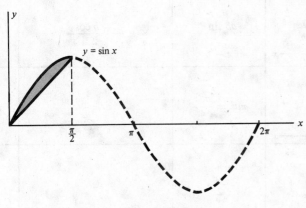

Fig. 9.65a

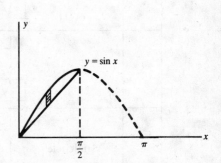

Fig. 9.65b

Using $x = 0$ and $y = 0$, it is found that $b = 0$. The slope m is found from Fig. 9.65a as

$$m = \frac{1}{\pi/2} = \frac{2}{\pi}$$

The final form of the equation of the line is

$$y = mx + b = \frac{2}{\pi}x$$

The centroidal coordinate x_c is given by

$$x_c = \frac{\int_0^{\pi/2} x(\sin x - (2/\pi)x)\, dx}{\int_0^{\pi/2} (\sin x - (2/\pi)x)\, dx} = \frac{\int_0^{\pi/2} (x\sin x - (2/\pi)x^2)\, dx}{\int_0^{\pi/2} (\sin x - (2/\pi)x)\, dx}$$

The term $\int x \sin x\, dx$ is now evaluated by using the method of integration by parts, with the form

$$\int u\, dv = uv - \int v\, du$$

Using $dv = \sin x\, dx$, $v = -\cos x$, $u = x$, and $du = dx$,

$$\int x \sin x\, dx = -x\cos x - \int (-\cos x)\, dx = -x\cos x + \sin x$$

TABLE 9.9 Centroids of Elementary Plane Areas

Case	Shape	A	x_c	y_c
1	Square	a^2	$\dfrac{a}{2}$	$\dfrac{a}{2}$
2	Rectangle	ab	$\dfrac{a}{2}$	$\dfrac{b}{2}$
3	Triangle†	$\dfrac{ab}{2}$	—	$\dfrac{b}{3}$
4	Parallelogram	$ab\sin\theta$	$\dfrac{a+b\cos\theta}{2}$	$\dfrac{b\sin\theta}{2}$
5	Circle	$\pi a^2 = \dfrac{\pi d^2}{4}$	$a = \dfrac{d}{2}$	$a = \dfrac{d}{2}$
6	Semicircle	$\dfrac{\pi a^2}{2}$	a	$\dfrac{4a}{3\pi}$

$$x_c = \frac{\left[(-x\cos x + \sin x) - \dfrac{2}{\pi}\left(\dfrac{x^3}{3}\right)\right]_0^{\pi/2}}{\left[-\cos x - \dfrac{2}{\pi}\left(\dfrac{x^2}{2}\right)\right]\Big|_0^{\pi/2}} = \frac{\left(-\dfrac{\pi}{2}\cos\dfrac{\pi}{2} + \sin\dfrac{\pi}{2}\right) - (0\cos 0 + \sin 0) - \dfrac{2}{3\pi}\left[\left(\dfrac{\pi}{2}\right)^3 - 0\right]}{-\left(\cos\dfrac{\pi}{2} - \cos 0\right) - \dfrac{1}{\pi}\left[\left(\dfrac{\pi}{2}\right)^2 - 0\right]} = 0.826$$

$$A = 0.215$$

For the y_c calculation, use is made of the form $\int y\, dA = \int y_{\text{avg}}(h\, dx)$.

$$y_c = \frac{\displaystyle\int_0^{\pi/2}\left(\frac{\sin x + (2/\pi)x}{2}\right)\left(\sin x - \frac{2}{\pi}x\right)dx}{0.215} = \frac{\dfrac{1}{2}\displaystyle\int_0^{\pi/2}\left(\sin^2 x - \frac{4}{\pi^2}x^2\right)dx}{0.215} = \frac{\dfrac{1}{2}\left[\left(\dfrac{x}{2} - \sin 2x\right) - \dfrac{4}{\pi^2}\left(\dfrac{x^3}{3}\right)\right]\Big|_0^{\pi/2}}{0.215}$$

$$= \frac{\dfrac{1}{2}\left\{\left(\dfrac{\pi}{4} - 0\right) - \left[\sin 2\left(\dfrac{\pi}{2}\right) - \sin 0\right] - \dfrac{4}{3\pi^2}\left[\left(\dfrac{\pi}{2}\right)^3 - 0\right]\right\}}{0.215} = 0.609$$

9.66 Give a summary of the basic concepts used to evaluate the centroids of plane areas and curves.

▌ The centroid of a plane area is a point which has a fixed location with respect to the area. If the plane area is positioned with respect to a set of coordinate axes, the coordinates of the centroid are

TABLE 9.9 (cont.)

Case	Shape	A	x_c	y_c
7	Quarter Circle	$\dfrac{\pi a^2}{4}$	$\dfrac{4a}{3\pi}$	$\dfrac{4a}{3\pi}$
8	Circular Sector‡	$\dfrac{a^2\theta}{2}$	$\dfrac{2a\sin\theta}{3\theta}$	$\dfrac{4a\sin^2\left(\frac{\theta}{2}\right)}{3\theta}$
9	Circular Segment‡	$a^2\left(\theta - \dfrac{\sin 2\theta}{2}\right)$	$\dfrac{2a\sin^3\theta}{3\left(\theta - \dfrac{\sin 2\theta}{2}\right)}$	0
10	Ellipse	$\dfrac{\pi ab}{4}$	$\dfrac{a}{2}$	$\dfrac{b}{2}$
11	Semiellipse	$\dfrac{\pi ab}{8}$	$\dfrac{a}{2}$	$\dfrac{2b}{3\pi}$

† The location of the area centroid of any triangle is at a height above any side which is equal to one-third of the altitude above that side.

‡ When θ appears directly as a factor, this quantity must be expressed in radians.

$$x_c = \frac{\sum_i x_i A_i}{A} \quad \text{and} \quad y_c = \frac{\sum_i y_i A_i}{A}$$

The integral forms of the definitions of the centroidal coordinates are

$$x_c = \frac{\int_A x\,dA}{A} \quad \text{and} \quad y_c = \frac{\int_A y\,dA}{A}$$

The location of the centroid on the plane area is independent of the location of the coordinate axes. If a plane area has an axis of symmetry, the centroid must lie on this axis. If the area has two axes of symmetry, the centroid lies on the intersection of these axes.

The centroid of a composite area may be found by subdividing this area into elementary area shapes, each of whose centroidal location is known. The centroidal coordinates of the composite area are then

$$x_c = \frac{x_1 A_1 + x_2 A_2 + \cdots}{A_1 + A_2 + \cdots} \quad \text{and} \quad y_c = \frac{y_1 A_1 + y_2 A_2 + \cdots}{A_1 + A_2 + \cdots}$$

where $A_1, A_2, \ldots$ are the elementary areas and the $x_1, x_2, \ldots$ and $y_1, y_2, \ldots$ are the centroidal coordinates of these areas. If a composite area has holes or cutouts, these areas are treated as negative quantities.

The first area moment of a plane area with respect to a reference axis is the product of the magnitude of the area and the distance between the reference axis and the centroid of the area. The units of a first area moment are length raised to the third power.

Case

Shape

TABLE 9.10 Centroids of Elementary Plane Curves

Case	Shape	l	x_c'	y_c'
1	Square 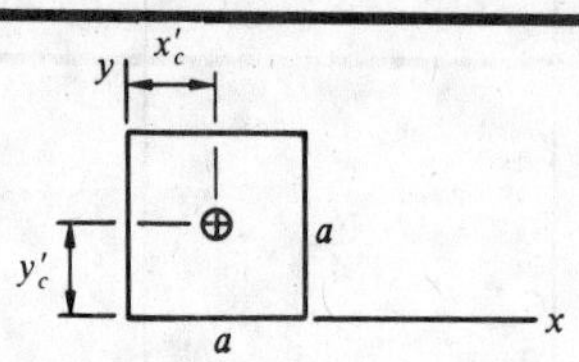	$4a$	$\dfrac{a}{2}$	$\dfrac{a}{2}$
2	Rectangle 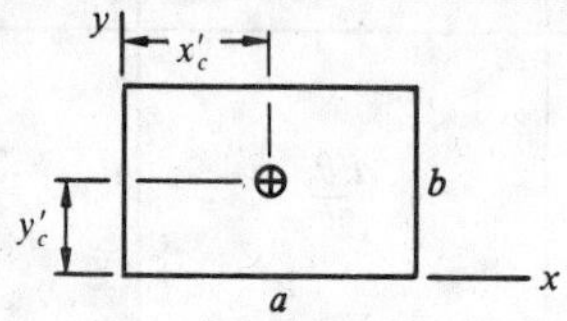	$2(a+b)$	$\dfrac{a}{2}$	$\dfrac{b}{2}$
3	Triangle 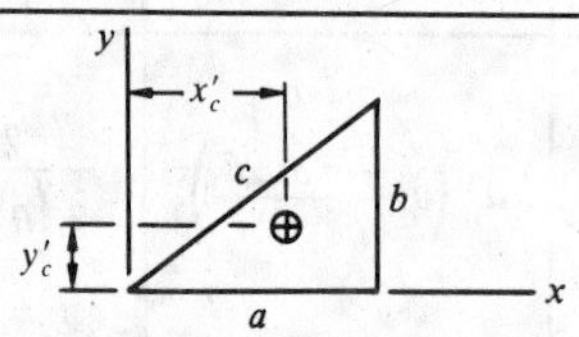	$a+b+c$	$\dfrac{a(a+2b+c)}{2(a+b+c)}$	$\dfrac{b(b+c)}{2(a+b+c)}$
4	Circle 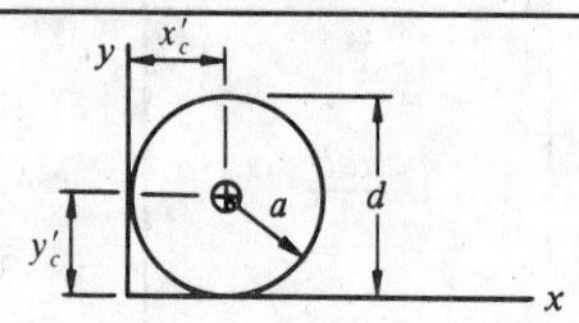	$2\pi a = \pi d$	$a = \dfrac{d}{2}$	$a = \dfrac{d}{2}$
5	Semicircle 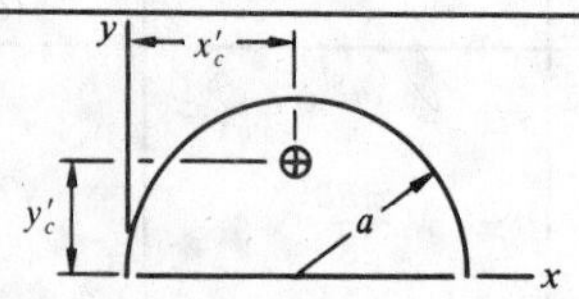	πa	a	$\dfrac{2a}{\pi}$
6	Circular arc† 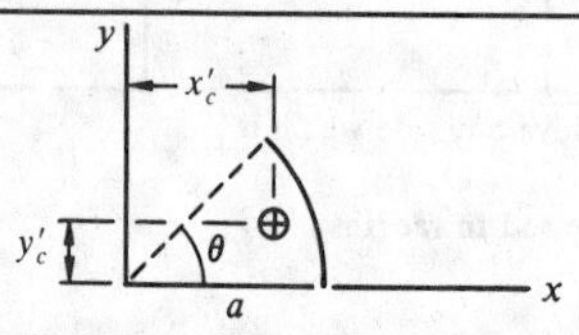	$a\theta$	$\dfrac{a\sin\theta}{\theta}$	$\dfrac{2a\sin^2\dfrac{\theta}{2}}{\theta}$

†When θ appears directly as a factor, this quantity must be expressed in radians.

A plane curve has a centroid which is located by the coordinates

$$x_c' = \frac{\sum_i x_i l_i}{l} \quad \text{and} \quad y_c' = \frac{\sum_i y_i l_i}{l}$$

The integral forms of the above two equations are

$$x_c' = \frac{\int_l x\,dl}{l} \quad \text{and} \quad y_c' = \frac{\int_l y\,dl}{l}$$

If the curve has an axis of symmetry, the centroid must lie on this axis. The centroid of a composite curve may be found by subdividing the curve into elementary curve shapes. The coordinates of the centroid of the composite curve are

$$x_c' = \frac{x_1 l_1 + x_2 l_2 + \cdots}{l_1 + l_2 + \cdots} \quad \text{and} \quad y_c' = \frac{y_1 l_1 + y_2 l_2 + \cdots}{l_1 + l_2 + \cdots}$$

where $l_1, l_2, \ldots$ are the elementary curve lengths and $x_1, x_2, \ldots y_1, y_2, \ldots$ are the centroidal coordinates of these lengths.

A plane area may be rotated about a reference axis to generate a volume. The theorem of Pappus states that the magnitude of this volume is equal to the product of the magnitude of the plane area and the distance traveled by the centroid of this area.

A second theorem of Pappus states that the magnitude of the surface area of the volume is equal to the product of the length of the plane curve that bounds the area and the distance traveled by the centroid of this plane curve.

CHAPTER 10
Moments and Products of Inertia of Plane Areas and Curves

10.1 MOMENTS OF INERTIA OF PLANE AREAS, POLAR MOMENT OF INERTIA, RADIUS OF GYRATION

10.1 Figure 10.1 shows a plane area A which is positioned with respect to a set of xy coordinate axes. The area is divided into the elementary areas A_i, with coordinates x_i and y_i.

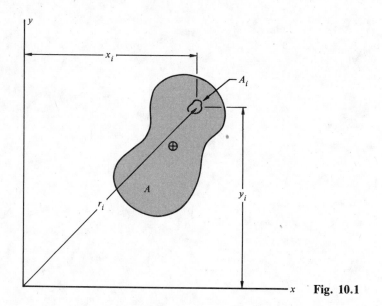

Fig. 10.1

(a) Show how this model may be used to develop a concept known as the *moment of inertia*.

(b) What information does the moment of inertia of a plane area provide about the area?

(c) What is an alternative term for the moment of inertia of a plane area?

▮ (a) A property of the *combined* system, consisting of the area and the position of this area with respect to the coordinate axes, is called the moment of inertia. For the area shown in Fig. 10.1, two moments of inertia are defined as

$$I_x = \sum_i y_i^2 A_i \qquad I_y = \sum_i x_i^2 A_i$$

The term I_x is referred to as the moment of inertia of the area A about the x axis, and the term I_y is the moment of inertia of this area about the y axis. It may be seen from the above equations that the units of an area moment of inertia are *length raised to the fourth power*.

(b) From consideration of the definitions of moment of inertia given by the above equations, two observations may be made. The first is that since A, x_i^2, and y_i^2 are inherently positive terms, *a moment of inertia is always positive*. The second observation is that each area element is multiplied by the *square* of its distance from the reference axis. Thus, elements of area which are more distant from this axis have a proportionately greater effect on the magnitude of the moment of inertia than do area elements which are closer to this axis. It follows that the moment of inertia is a measure of the *distribution* of the area elements within a plane area.

(c) The moment of inertia of a plane area is also referred to as the *second moment of area*.

10.2 (a) State the formal definitions of the moments of inertia of a plane area.

(b) What are some of the uses of the moment of inertia of a plane area?

❙ (*a*) As the subdivision of the area elements continues, a limiting condition is reached in which the summation operations for moments of inertia become integral operations, with the forms

$$I_x = \lim_{A_i \to 0} \sum_i y_i^2 A_i = \int_A y^2 \, dA \qquad I_y = \lim_{A_i \to 0} \sum_i x_i^2 A_i = \int_A x^2 \, dA$$

The above two equations are the formal definitions of the moment of inertia of a plane area.   The moments of inertia for several elementary plane area shapes are contained in Table 10.6 at the end of this chapter.   The values of these moments of inertia are found by direct integration of the above two equations.

(*b*) The area moment of inertia is used extensively in strength, or mechanics, of materials to find the stresses in, and deflection of, beams and shafts as well as the buckling loads of columns.   This quantity is also used in fluid mechanics to find the location of the resultant force on a submerged plane area.

10.3   State the fundamental difference between the concept of the *centroid* of a plane area and the concept of *moment of inertia* of a plane area.

❙ Although the centroid and moment of inertia are related terms, there is a fundamental difference between these two quantities.   Chapter 9 considered a property of a plane area or curve that is referred to as the centroid.   It was seen that the centroid is a point which has a fixed location with respect to the boundary of the plane area or curve, and that this point is *independent* of any coordinate axes which are used to locate the area or curve.   A complete definition of the moment of inertia, by comparison, *must* include a description of the *position of the coordinate axes* relative to the area under consideration.

10.4   Show that the moments of inertia of the plane area about the centroidal *xy* axes in Fig. 10.4*a* are $I_x = ab^3/12$ and $I_y = ba^3/12$.

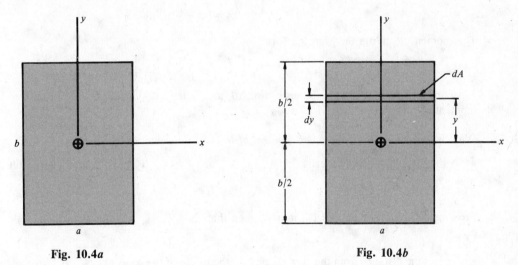

Fig. 10.4*a*	Fig. 10.4*b*

❙ I_x will be found first, and the required area element dA is shown in Fig. 10.4*b*.   Using

$$dA = a \, dy$$

the moment of inertia about the *x* axis has the form

$$I_x = \int_A y^2 \, dA = \int_{-b/2}^{b/2} y^2 (a \, dy) = a \int_{-b/2}^{b/2} y^2 \, dy = \frac{ay^3}{3} \Big|_{-b/2}^{b/2}$$

$$= \frac{a}{3} \left[\left(\frac{b}{2} \right)^3 - \left(-\frac{b}{2} \right)^3 \right] = \frac{a}{3} \left(\frac{b^3}{8} + \frac{b^3}{8} \right) = \frac{ab^3}{12}$$

From consideration of the above equation, it may be seen that the dimension *b*, which is perpendicular to the reference axis, has a much greater effect on the value of I_x than does the dimension *a*, which is parallel to this axis, because of the third-power effect.   This confirms the observation in Prob. 10.1, part (*b*), that area elements farther from the reference axis have greater effects on the moment of inertia than do area elements closer to this axis.

It follows from the above result that

$$I_y = \frac{ba^3}{12}$$

10.5 Show that the moment of inertia about the x axis of the triangular area in Fig. 10.5a is $ab^3/12$.

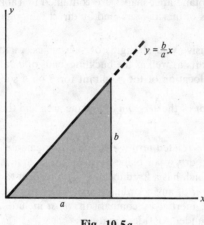

Fig. 10.5a

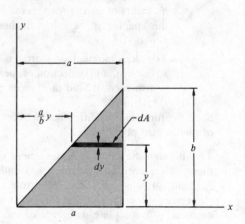

Fig. 10.5b

▮ The area element is shown in Fig. 10.5b, and has the form

$$dA = \left(a - \frac{a}{b}\, y\right) dy$$

The moment of inertia then has the form

$$I_x = \int_A y^2\, dA = \int_0^b y^2 \left(a - \frac{a}{b}\, y\right) dy = \int_0^b \left(ay^2 - \frac{a}{b}\, y^3\right) dy = \left[a\,\frac{y^3}{3} - \frac{a}{b}\,\frac{y^4}{4}\right]_0^b = \frac{ab^3}{3} - \frac{a}{b}\,\frac{b^4}{4} = \frac{ab^3}{12}$$

Section 10.5 gives several problems in which I_x and I_y are found by direct use of the integral definitions of the moments of inertia.

10.6 What is the definition of the *polar moment of inertia* of a plane area?

▮ If the area moments of inertia $I_x = \sum_i y_i^2 A_i$ and $I_y = \sum_i x_i^2 A_i$ are added, the result is

$$I_x + I_y = \sum_i (x_i^2 + y_i^2) A_i$$

From consideration of Fig. 10.1 it may be seen that

$$x_i^2 + y_i^2 = r_i^2$$

where r_i is the distance from the area element to the origin of the coordinates.   A term called the polar moment of inertia, with the symbol J, is now defined as

$$J = \sum_i r_i^2 A_i = I_x + I_y$$

The reference axis for J is an axis that is perpendicular to the xy plane and acts through the origin of these coordinates.   It may be seen that the value of J may be found readily if the values of I_x and I_y are known.   The integral form of the above equation is

$$J = \lim_{A_i \to 0} \sum_i r_i^2 A_i = \int_A r^2\, dA$$

The term polar moment of inertia finds widespread application in strength, or mechanics, of materials, where this term is used for the analysis of torsional stresses in, and deflection of, shafts.

10.7 The circular area shown in Fig. 10.7a has its center at the origin of the coordinate system.

(a) Show that the polar moment of inertia of this area, with respect to the origin of the coordinate system, is $\pi d^4/32$.

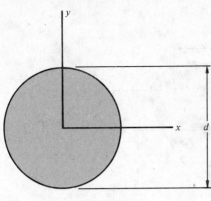

Fig. 10.7a

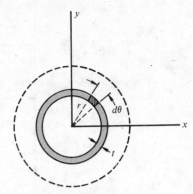

Fig. 10.7b

(b) Show how this result may be used to find I_x and I_y of the circular area.

▌ **(a)** As a preliminary step, the polar moment of inertia J_1 of the thin annular area shown in Fig. 10.7b will be found. The area element dA_1, shown as the shaded area in the figure, is

$$dA_1 = (r \, d\theta)t$$

$$J_1 = \int_A r^2 \, dA_1 = \int_0^{2\pi} r^2(r \, d\theta)t = r^3 t \int_0^{2\pi} d\theta = 2\pi r^3 t$$

It should be noted in the above equation that when an angle is used as a variable of integration, it must be expressed in radians.

The polar moment of inertia J of the circular area in Fig. 10.7a is the sum of the values of J_1 of all the annular areas shown in Fig. 10.7b. If

$$t \to dr$$

the final result is

$$J = \int_0^{d/2} 2\pi r^3 \, dr = 2\pi \int_0^{d/2} r^3 \, dr = 2\pi \left[\frac{r^4}{4} \right]_0^{d/2} = \frac{\pi d^4}{32}$$

(b) I_x, I_y, and J are related by

$$J = I_x + I_y$$

From the symmetry of the area of a circle,

$$I_x = I_y$$

Thus,

$$J = 2I_x \qquad 2I_x = \frac{\pi d^4}{32} \qquad I_x = I_y = \frac{\pi d^4}{64}$$

10.8 **(a)** What is meant by the *radius of gyration* of a plane area?

(b) State an important useful application of the concept of the radius of gyration.

▌ **(a)** The total area of a plane area is now imagined to be concentrated at the single point A shown in Fig. 10.8. The coordinates k_x and k_y of this point are the radii of gyration of the plane area with respect to the x and y axes. The subscripts on these terms denote the axes from which they are measured. These terms are defined by

$$k_x^2 A = I_x \qquad k_x = \sqrt{\frac{I_x}{A}}$$

$$k_y^2 A = I_y \qquad k_y = \sqrt{\frac{I_y}{A}}$$

It should be emphasized that the above equations are formal definitions of the radii of gyration and that *the point defined by these two terms bears no particular relationship to the location of the centroid of the area*. This latter conclusion follows from the fact that the centroid of an area is an invariant

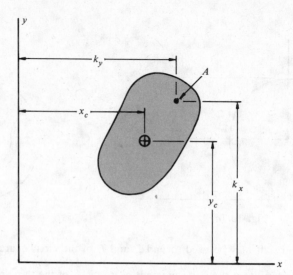

Fig. 10.8

property of that area, while the moments of inertia of the same area are functions of an arbitrary choice of reference axes. A radius of gyration which corresponds to the polar moment of inertia is designated k_p and defined by

$$k_p^2 A = J \qquad k_p = \sqrt{\frac{J}{A}}$$

From the above equations, it follows that

$$k_p^2 = k_x^2 + k_y^2$$

The value of a radius of gyration has no particular physical significance.

(b) An important use of the radius of gyration is in strength, or mechanics, of materials, in the analysis of buckling of columns.

10.9 Find the radii of gyration k_x, k_y, and k_p of the rectangular area in Prob. 10.4.

▮ From Prob. 10.4,

$$I_x = \frac{ab^3}{12} \qquad I_y = \frac{ba^3}{12} \qquad A = ab$$

Using the definitions of the radii of gyration,

$$k_x = \sqrt{\frac{I_x}{A}} = \sqrt{\frac{ab^3}{12ab}} = \frac{b}{\sqrt{12}} = 0.289b$$

$$k_y = \sqrt{\frac{I_y}{A}} = \sqrt{\frac{ba^3}{12ab}} = \frac{a}{\sqrt{12}} = 0.289a$$

$$k_p^2 = k_x^2 + k_y^2 = \frac{b^2 + a^2}{12} \qquad k_p = 0.289\sqrt{a^2 + b^2}$$

The half height of the rectangular area is $0.5b$. The ratio of k_x to this dimension is $0.289b/0.5b = 0.578$.

10.2 THE PARALLEL-AXIS, OR TRANSFER, THEOREM FOR AREA MOMENTS OF INERTIA, TABULAR FORMS OF SOLUTION

10.10 The moment of inertia of a plane area about one of its centroidal axes is known.

(a) Show how the moment of inertia of the area about an axis parallel to this centroidal axis may be found.

(b) What is the relationship between the magnitudes of the two moments of inertia?

▮ (a) Figure 10.10 shows a plane area A. d_x and d_y are the separation distances between the two parallel coordinate axis systems. The moment of inertia about the x axis is

$$I_x = \int_A y^2 \, dA$$

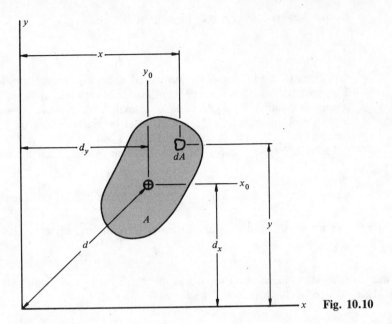

Fig. 10.10

Using

$$y = d_x + y_0$$

we can write the above result as

$$I_x = \int_A (d_x + y_0)^2 \, dA = \int_A d_x^2 \, dA + \int_A 2 d_x y_0 \, dA + \int_A y_0^2 \, dA$$

The x_0 and y_0 axes in Fig. 10.10 pass through the centroid of the area. The centroidal coordinate y_{0c}, measured in the x_0, y_0 coordinate system, is

$$y_{0c} = \frac{\int_A y_0 \, dA}{A} = 0 \qquad \int_A y_0 \, dA = 0$$

From this result, it follows that the second term on the right side of the above result for I_x is identically zero. This equation is now

$$I_x = d_x^2 \int_A dA + \int_A y_0^2 \, dA$$

The first term on the right side may be written as $A d_x^2$. The second term on the right side is the moment of inertia I_{0x} of the area about the centroidal x_0 axis, with the form

$$I_{0x} = \int_A y_0^2 \, dA$$

The final form for I_x is then

$$I_x = I_{0x} + A \, d_x^2 \qquad (1)$$

A similar analysis for the moment of inertia about the y axis results in

$$I_y = I_{0y} + A \, d_y^2 \qquad (2)$$

The above two equations define the *parallel-axis*, or *transfer*, *theorem*. If the moment of inertia about a centroidal axis is known, the moment of inertia about a parallel axis may be found by adding to the centroidal moment of inertia a term whose magnitude is the product of the area and the square of the separation distance between the two axes.

The parallel-axis theorem may be expressed in an alternative form. From consideration of Fig. 10.10, it may be seen that d_x and d_y are actually the centroidal coordinates of the plane area, measured in the xy coordinate system. Thus, with

$$d_x = y_c \qquad d_y = x_c$$

the parallel-axis theorem may be written as

$$I_x = I_{0x} + Ay_c^2 \qquad (3)$$
$$I_y = I_{0y} + Ax_c^2 \qquad (4)$$

Equations (1) and (2) may be considered to be the fundamental forms of the parallel-axis theorem. They may be used to find moments of inertia about axes parallel to the centroidal axes $x_0 y_0$, without a specific calculation of the centroidal coordinates. If, however, x_c and y_c are known, or to be found, in a given problem, then Eqs. (3) and (4) may be used to relate the moments of inertia about the sets of parallel axes. It is emphasized that Eqs. (1) and (2), and Eqs. (3) and (4), are equivalent statements.

The transfer theorem for the polar moment of inertia has the form

$$J = J_0 + Ad^2$$

where d is the separation distance between the centroidal axis and the axis of interest, as shown in Fig. 10.10.

(b) The terms $A\,d_x^2$ and $A\,d_y^2$ are always positive and never zero. It follows that

$$I_x > I_{0x} \qquad \text{and} \qquad I_y > I_{0y}$$

10.11 Find the moment of inertia with respect to the x axis of the rectangular area in Fig. 10.11.

❚ Using the parallel-axis theorem, we have

$$I_x = I_{0x} + Ay_c^2 = \frac{ab^3}{12} + ab\left(\frac{b}{2}\right)^2 = \frac{ab^3}{3}$$

It may be seen that the moment of inertia about the edge is 4 times greater than the moment of inertia about the centroidal axis.

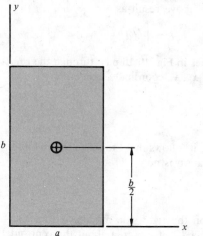

Fig. 10.11

10.12 How is the moment of inertia determined for a plane area that is not one of the simple shapes given in Table 10.6 at the end of this chapter?

❚ Very often the moment of inertia, with respect to a particular reference axis, is required for an area that is not an elementary shape. For this case, the given area or curve may be divided into elementary shapes and the moments of inertia of these shapes summed to obtain the final result. This division of the area is done in such a way as to minimize the computational effort. The moments of inertia of the elementary areas about the reference axis may be found directly from tabulations such as Table 10.6, or by using the parallel-axis theorem. The following problems illustrate the techniques used for finding the moments of inertia of composite areas.

10.13 The beam cross-sectional area of Prob. 9.6 is shown in Fig. 10.13a. Find the moments of inertia of this area about the centroidal x_0 axis and about the axes x and x' that lie along the two outer edges of the area.

❚ The original area is divided into the three elementary areas shown in Fig. 10.13b, and the moment of inertia about the centroidal x_0 axis is found first. The moments of inertia of areas 1 and 2 about the x_0 axis may be

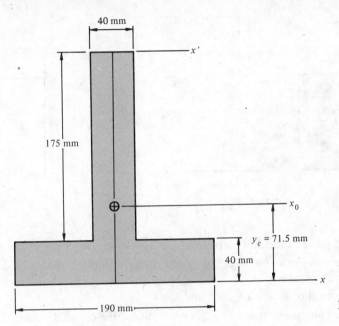

Fig. 10.13*a*

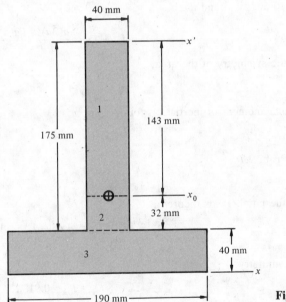

Fig. 10.13*b*

found directly, by using Case 2 of Table 10.6, as moments of inertia of a rectangle about its edge. The moment of inertia of area 3 about the x_0 axis is found by using the parallel-axis theorem. The result for I_{0x}, the moment of inertia of the cross-sectional area about the x_0 axis, is then

$$I_{0x} = I_{01} + I_{02} + I_{03} = \frac{40(143)^3}{3} + \frac{40(32)^3}{3} + \left[\frac{190(40)^3}{12} + (190)(40)(32+20)^2 \right] = 6.10 \times 10^7 \text{ mm}^4$$

The moment of inertia I_x may be found next by using the above result for the moment of inertia about the centroidal axis together with the parallel-axis theorem. This result is

$$I_x = I_{0x} + A d_x^2 = 6.10 \times 10^7 + [40(175) + 190(40)](71.5)^2 = 1.36 \times 10^8 \text{ mm}^4$$

In a similar fashion, the moment of inertia $I_{x'}$ about the x' axis is found to be

$$I_{x'} = I_{0x} + A d_x^2 = 6.10 \times 10^7 + [40(175) + 190(40)](143)^2 = 3.60 \times 10^8 \text{ mm}^4$$

10.14 (*a*) Find the moments of inertia I_x and I_y of the area shown in Fig. 10.14*a*.

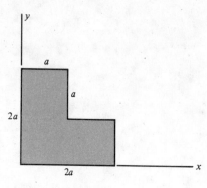

Fig. 10.14a

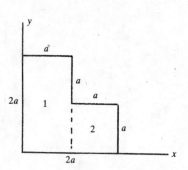

Fig. 10.14b

(b) Find the polar moment of inertia J of the area about the origin of the coordinate axes.

(c) Find the moments of inertia I_{0x}, I_{0y}, and J_0 of the area about the centroidal axes x_0 and y_0 that are parallel to the x and y axes.

(d) Find the radii of gyration k_x, k_y, and k_p.

(e) Find the radii of gyration k_{0x}, k_{0y}, and k_{0p} about the centroidal axes.

▌(a) Figure 10.14b shows the division of the area into two simple area shapes. Using Cases 1 and 2 of Table 10.6, I_x is found as

$$I_x = I_{x1} + I_{x2} = \frac{a(2a)^3}{3} + \frac{a(a)^3}{3} = 3a^4$$

From the symmetry of the area,

$$I_y = I_x = 3a^4$$

(b) The polar moment of inertia J is found from

$$J = I_x + I_y = 3a^4 + 3a^4 = 6a^4$$

(c) From Prob. 9.7,

$$x_c = y_c = \tfrac{5}{6}a = 0.833a \qquad A = 3a^2$$

Using the parallel-axis theorem,

$$I_x = I_{0x} + A\,d_x^2 = I_{0x} + Ay_c^2$$
$$I_{0x} = I_x - Ay_c^2 = 3a^4 - 3a^2(0.833a)^2 = 0.918a^4$$

From symmetry,

$$I_{0y} = I_{0x} = 0.918a^4$$

The polar moment of inertia about the origin of the x_0y_0 axes is given by

$$J_0 = I_{0x} + I_{0y} = 2(0.918a^4) = 1.84a^4$$

(d) The radii of gyration are

$$k_x = \sqrt{\frac{I_x}{A}} = \sqrt{\frac{3a^4}{3a^2}} = a \qquad k_y = k_x = a \qquad k_p = \sqrt{\frac{J}{A}} = \sqrt{\frac{6a^4}{3a^2}} = 1.41a$$

(e) The centroidal radii of gyration are

$$k_{0x} = \sqrt{\frac{I_{0x}}{A}} = \sqrt{\frac{0.918a^4}{3a^2}} = 0.553a \qquad k_{0y} = k_{0x}$$

$$k_{0p} = \sqrt{\frac{J_0}{A}} = \sqrt{\frac{1.84a^4}{3a^2}} = 0.783a$$

10.15 Find the quantities in Prob. 10.14 for the area shown in Fig. 10.15a.

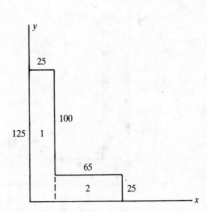

Fig. 10.15*a*   Fig. 10.15*b*

▌ (*a*)   The division of the area into two simple areas is shown in Fig. 10.15*b*.   Using Case 2 of Table 10.6,

$$I_x = I_{x1} + I_{x2} = \frac{25(125)^3}{3} + \frac{65(25)^3}{3} = 1.66 \times 10^7 \text{ mm}^4$$

$$I_y = I_{y1} + I_{y2} = \frac{125(25)^3}{3} + \left[\frac{25(65)^3}{12} + 65(25)\left(25 + \frac{65}{2}\right)^2\right] = 6.60 \times 10^6 \text{ mm}^4$$

(*b*)   The polar moment of inertia is found from

$$J = I_x + I_y = 1.66 \times 10^7 + 6.60 \times 10^6 = 2.32 \times 10^7 \text{ mm}^4$$

(*c*)   From Prob. 9.8,

$$x_c = 27.9 \text{ mm} \qquad y_c = 45.4 \text{ mm} \qquad A = 4,750 \text{ mm}^2$$

Using the parallel-axis theorem,

$$I_{0x} = I_x - Ay_c^2 = 1.66 \times 10^7 - 4,750(45.4)^2 = 6.81 \times 10^6 \text{ mm}^4$$
$$I_{0y} = I_y - Ax_c^2 = 6.60 \times 10^6 - 4,750(27.9)^2 = 2.90 \times 10^6 \text{ mm}^4$$

The polar moment of inertia is given by

$$J_0 = I_{0x} + I_{0y} = 6.81 \times 10^6 + 2.90 \times 10^6 = 9.71 \times 10^6 \text{ mm}^4$$

(*d*)   The radii of gyration are

$$k_x = \sqrt{\frac{I_x}{A}} = \sqrt{\frac{1.66 \times 10^7}{4,750}} = 59.1 \text{ mm} \qquad k_y = \sqrt{\frac{I_y}{A}} = \sqrt{\frac{6.60 \times 10^6}{4,750}} = 37.3 \text{ mm}$$

$$k_p = \sqrt{\frac{J}{A}} = \sqrt{\frac{2.32 \times 10^7}{4,750}} = 69.9 \text{ mm}$$

(*e*)   The centroidal radii of gyration are

$$k_{0x} = \sqrt{\frac{I_{0x}}{A}} = \sqrt{\frac{6.81 \times 10^6}{4,750}} = 37.9 \text{ mm} \qquad k_{0y} = \sqrt{\frac{I_{0y}}{A}} = \sqrt{\frac{2.90 \times 10^6}{4,750}} = 24.7 \text{ mm}$$

$$k_{0p} = \sqrt{\frac{J_0}{A}} = \sqrt{\frac{9.71 \times 10^6}{4,750}} = 45.2 \text{ mm}$$

10.16   (*a*)   Find the moments of inertia I_x, I_y, and J for the area shown in Fig. 10.16*a*.

   (*b*)   Find the moments of inertia I_{0x}, I_{0y}, and J_0 about the centroidal x_0y_0 axes, which are parallel to the *xy* axes.

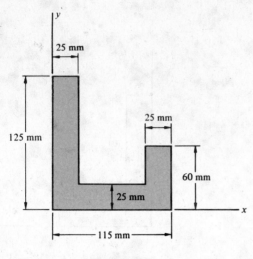

Fig. 10.16a

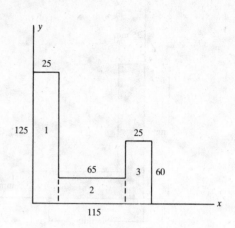

Fig. 10.16b

▌ (a) The area is divided into the three elementary areas shown in Fig. 10.16b.

$$I_x = I_{x1} + I_{x2} + I_{x3}$$

Using Case 2 of Table 10.6,

$$I_x = \frac{25(125)^3}{3} + \frac{65(25)^3}{3} + \frac{25(60)^3}{3} = 1.84 \times 10^7 \text{ mm}^4$$

$$I_y = I_{y1} + I_{y2} + I_{y3} = \frac{125(25)^3}{3} + \left[\frac{25(65)^3}{12} + 65(25)\left(25 + \frac{65}{2}\right)^2\right] + \left[\frac{60(25)^3}{12} + 25(60)\left(115 - \frac{25}{2}\right)^2\right]$$

$$= 2.24 \times 10^7 \text{ mm}^4$$

$$J = I_x + I_y = 1.84 \times 10^7 + 2.24 \times 10^7 = 4.05 \times 10^7 \text{ mm}^4$$

(b) From Prob. 9.9,

$$x_c = 45.8 \text{ mm} \qquad y_c = 41.7 \text{ mm} \qquad A = 6,250 \text{ mm}^2$$

Using the parallel-axis theorem,

$$I_{0x} = I_x - Ay_c^2 = 1.84 \times 10^7 - 6,250(41.7)^2 = 7.53 \times 10^6 \text{ mm}^4$$
$$I_{0y} = I_y - Ax_c^2 = 2.24 \times 10^7 - 6,250(45.8)^2 = 9.29 \times 10^6 \text{ mm}^4$$
$$J_0 = I_{0x} + I_{0y} = 1.68 \times 10^7 \text{ mm}^4$$

10.17 Find the quantities in Prob. 10.16 for the area shown in Fig. 10.17a.

▌ (a) Figure 10.17b shows the division of the area into three simple shapes and the values of the centroidal coordinates y_2 and y_3. The moment of inertia about the x axis is

$$I_x = I_{x1} + I_{x2} + I_{x3}$$

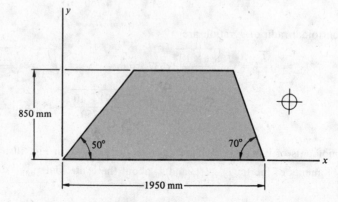

Fig. 10.17a

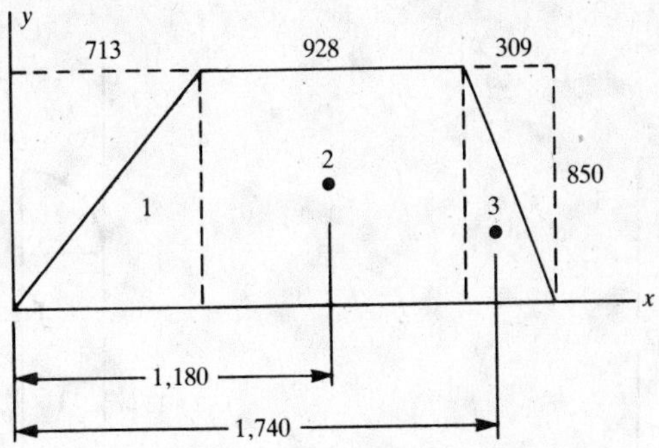

Fig. 10.17b

Using Cases 2 and 3 in Table 10.6,

$$I_x = \frac{713(850)^3}{12} + \frac{928(850)^3}{3} + \frac{309(850)^3}{12} = 2.42 \times 10^{11} \text{ mm}^4$$

The moment of inertia about the y axis is

$$I_y = I_{y1} + I_{y2} + I_{y3}$$

Using Case 3 of Table 10.6, and the parallel-axis theorem,

$$I_y = \frac{850(713)^3}{4} + \left[\frac{850(928)^3}{12} + (928)850(1,180)^2 \right] + \left[\frac{850(309)^3}{36} + \frac{1}{2}(309)850(1,740)^2 \right]$$

$$= 1.63 \times 10^{12} \text{ mm}^4$$

The polar moment of inertia has the value

$$J = I_x + I_y = 1.87 \times 10^{12} \text{ mm}^4$$

(*b*) From Prob. 9.11,

$$x_c = 1,060 \text{ mm} \qquad y_c = 375 \text{ mm} \qquad A = 1.22 \times 10^6 \text{ mm}^2$$

The moments of inertia about the centroidal axes are

$$I_{0x} = I_x - Ay_c^2 = 2.42 \times 10^{11} - 1.22 \times 10^6 (375)^2 = 7.04 \times 10^{10} \text{ mm}^4$$
$$I_{0y} = I_y - Ax_c^2 = 1.63 \times 10^{12} - 1.22 \times 10^6 (1,060)^2 = 2.59 \times 10^{11} \text{ mm}^4$$
$$J_0 = I_{0x} + I_{0y} = 3.29 \times 10^{11} \text{ mm}^4$$

10.18 Find the quantities in Prob. 10.16 for the area shown in Fig. 10.18*a*.

▌ (*a*) The area is divided into the three simple areas shown in Fig. 10.18*b*. From Case 7 in Table 9.9, for the case of a quarter circle,

$$x_c = y_c = \frac{4a}{3\pi}$$

Using Case 6 in Table 10.6,

$$I_{0x} = I_{0y} = \frac{a^4}{2} \left(\frac{\pi}{8} - \frac{8}{9\pi} \right)$$

$$I_x = I_{x1} + I_{x2} + I_{x3} = \frac{3(13)^3}{3} + \frac{3(4)^3}{3} + \left[\frac{3^4}{2} \left(\frac{\pi}{8} - \frac{8}{9\pi} \right) + \frac{\pi(3)^2}{4} \left(13 + \frac{4(3)}{3\pi} \right)^2 \right] = 3,710 \text{ in}^4$$

$$I_y = I_{y1} + I_{y2} + I_{y3} = \frac{13(3)^3}{3} + \left[\frac{4(3)^3}{12} + 3(4)4.5^2 \right] + \left[\frac{3^4}{2} \left(\frac{\pi}{8} - \frac{8}{9\pi} \right) + \frac{\pi(3)^2}{4} \left(3 - \frac{4(3)}{3\pi} \right)^2 \right] = 395 \text{ in}^4$$

$$J = I_x + I_y = 4,110 \text{ in}^4$$

(*b*) From Prob. 9.12,

$$x_c = 2.15 \text{ in} \qquad y_c = 6.52 \text{ in} \qquad A = 58.1 \text{ in}^2$$
$$I_{0x} = I_x - Ay_c^2 = 3,710 - 58.1(6.52)^2 = 1,240 \text{ in}^4$$
$$I_{0y} = I_y - Ax_c^2 = 395 - 58.1(2.15)^2 = 126 \text{ in}^4 \qquad J_0 = I_{0x} + I_{0y} = 1,370 \text{ in}^4$$

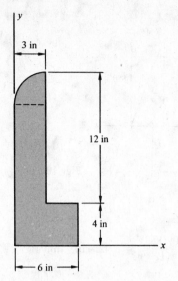

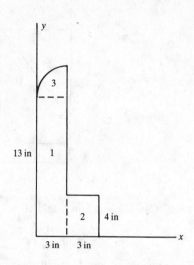

Fig. 10.18a

Fig. 10.18b

10.19 The polar moment of inertia J_a of the circular area in Fig. 10.19a, with respect to point a, is

$$J_a = \frac{\pi d^4}{32} + \frac{\pi d^2}{4} D^2$$

If the diameter d of the area is small with respect to D, the polar moment of inertia of the area about its centroid may be neglected. The approximate value of J_a is then

$$J_a \approx \frac{\pi d^2}{4} D^2$$

(a) For what range of values of the ratio d/D will the use of the approximate equation result in an error of 5 percent or less in the actual value of J_a?

(b) Sketch, to scale, the result computed in part (a).

▮ (a) Using the definition of the percent difference, given in Prob. 1.42,

$$\%D = \frac{J_{a,exact} - J_{a,approx}}{J_{a,exact}} = \frac{\left(\dfrac{\pi d^4}{32} + \dfrac{\pi d^2}{4} D^2\right) - \dfrac{\pi d^2}{4} D^2}{\dfrac{\pi d^4}{32} + \dfrac{\pi d^2}{4} D^2} = \frac{\dfrac{\pi d^4}{32}}{\dfrac{\pi d^4}{32} + \dfrac{\pi d^2}{4} D^2} = \frac{1}{1 + 8\left(\dfrac{D}{d}\right)^2}$$

Using $\%D = 5\% = 0.05$,

$$\frac{1}{1 + 8(D/d)^2} = 0.05 \qquad 1 + 8\left(\frac{D}{d}\right)^2 = \frac{1}{0.05} = 20 \qquad \frac{D}{d} = 1.54 \qquad \frac{d}{D} = 0.649$$

If $d \le 0.649D$, the error in using the approximate equation will be less than or equal to 5 percent.

(b) The limiting value of d is drawn to scale in Fig. 10.19b.

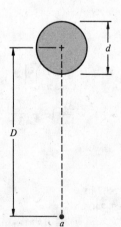

Fig. 10.19a

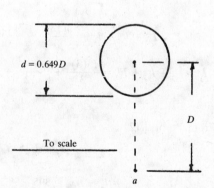

$d = 0.649D$

To scale

Fig. 10.19b

10.20 (*a*) Do the same as in Prob. 10.19 for the square area shown in Fig. 10.20*a*.

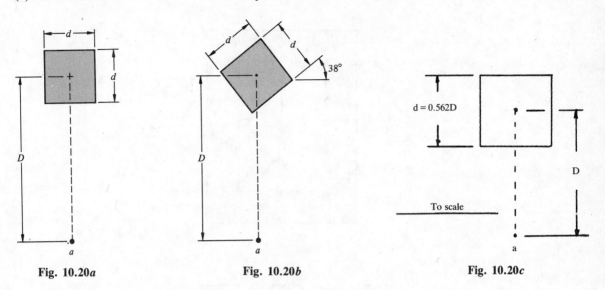

| Fig. 10.20*a* | Fig. 10.20*b* | Fig. 10.20*c* |

(*b*) Do the same as in part (*a*) if the square area is rotated as shown in Fig. 10.20*b*.

▌ (*a*) Using Case 1 of Table 10.6,

$$J = J_0 + A\,d^2 \qquad J_{a,\text{exact}} = \frac{d^4}{6} + d^2 D^2 \qquad J_{a,\text{approx}} = d^2 D^2$$

$$\%\text{D} = \frac{J_{a,\text{exact}} - J_{a,\text{approx.}}}{J_{a,\text{exact}}} = \frac{[(d^4/6) + d^2 D^2] - d^2 D^2}{(d^4/6) + d^2 D^2}$$

For %D = 5% = 0.05,

$$\frac{d^4/6}{(d^4/6) + d^2 D^2} = \frac{1}{1 + 6(D/d)^2} = 0.05$$

$$1 + 6\left(\frac{D}{d}\right)^2 = 20 \qquad \frac{D}{d} = 1.78 \qquad \frac{d}{D} = 0.562$$

If $d \le 0.562D$, the error in using the approximate equation will be less than or equal to 5 percent. The above value of d is shown to scale in Fig. 10.20*c*.

(*b*) The same result is obtained as in part (*a*), since J_0 is not a function of the angular orientation of the element.

10.21 What is the effect on the computation of the moment of inertia if the plane area has holes or cutouts?

▌ A composite area may have holes or cutouts in it. As a preliminary step toward treating this type of problem, the effect of subtracting moments of inertia will be considered.

Figure 10.21*a* shows a hollow, rectangular, cross-sectional area. It is desired to find the moment of inertia of this hollow area with respect to the centroidal x_0 axis.

The area shown in the figure may be envisioned to be the result of the operation shown in Fig. 10.21*b*. Here, the moment of inertia of the original cross-sectional area is found by subtracting the moment of inertia of the inner rectangle from that of the outer rectangle.

It was shown in Prob. 10.1 that an area element A_i is an inherently positive term and that, as a consequence, the moment-of-inertia terms are inherently positive. If the composite area being analyzed has holes or cutouts in it, the moments of inertia of these areas will be treated as negative quantities. The minus signs on these terms imply that the corresponding areas which result in these moments of inertia are *absent* in the original composite area, and *not* that these moments of inertia are computed negative quantities. An *algebraic summation* of the moments of inertia of all the elementary areas will then be the desired result for the *net* moment of inertia of the composite area.

The application of the above rule to the configuration of Fig. 10.21*b* results in

$$I_{0x} = I_{0x,1} + (-I_{0x,2}) = I_{0x,1} - I_{0x,2} = \frac{a_1 b_1^3}{12} - \frac{a_2 b_2^3}{12} = \frac{1}{12}\left(a_1 b_1^3 - a_2 b_2^3\right)$$

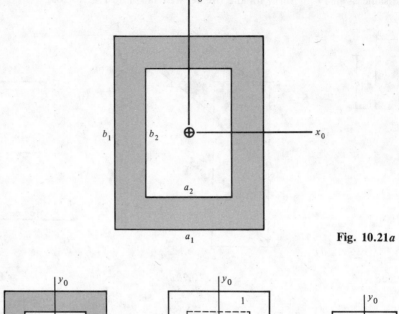

Fig. 10.21a

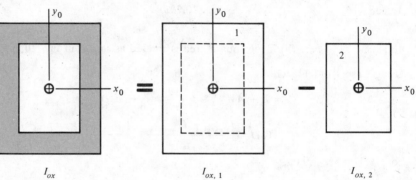

I_{ox} = $I_{ox,\,1}$ − $I_{ox,\,2}$

Fig. 10.21b

10.22 The area of Prob. 9.18 is redrawn in Fig. 10.22, together with its centroidal coordinates which were computed in that problem. Find the centroidal area moment of inertia I_{0x}.

❚ The plate is divided into the three elementary areas shown in the figure. The moment of inertia of the area with respect to the x axis will be found first. The parallel-axis theorem may be used then to find the final result for the moment of inertia about the centroidal x_0 axis. The moments of inertia of areas 1 and 2 about the x axis are found directly by using Cases 2 and 3 in Table 10.6 at the end of this chapter. The moment of inertia of the hole about the x axis is found with the aid of Case 5 of Table 10.6 and the parallel-axis theorem. Since the hole represents an area which is absent, the moment-of-inertia term for this case must be preceded by a minus

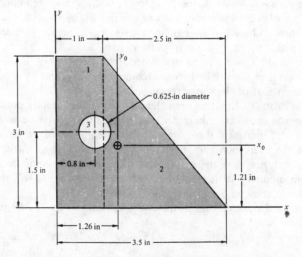

Fig. 10.22

sign. The result for I_x is then

$$I_x = I_{x1} + I_{x2} + (-I_{x3}) = \frac{1(3)^2}{3} + \frac{2.5(3)^3}{12} - \left[\frac{\pi(0.625)^4}{64} + \frac{\pi(0.625)^2}{4}(1.5)^2 \right] = 13.9 \text{ in}^4$$

The value of the area was found in Prob. 9.18 as $A = 6.44 \text{ in}^2$. Using the parallel-axis theorem, we have

$$I_x = I_{0x} + Ay_c^2 \qquad I_{0x} = I_x - Ay_c^2 = 13.9 - 6.44(1.21)^2 = 4.47 \text{ in}^4$$

10.23 Figure 10.23a shows the cross section of a precast concrete floor beam. Find the moment of inertia and radius of gyration about the horizontal centroidal axis of the cross section.

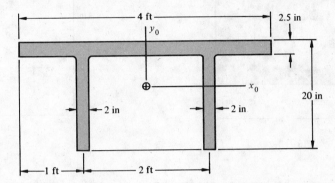

Fig. 10.23a

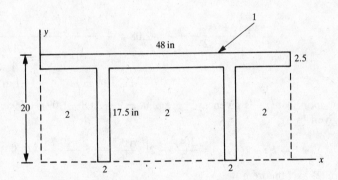

Fig. 10.23b

❙ The cross-sectional area is redrawn in Fig. 10.23b. Area 1 is a 48 in by 20 in rectangle. Areas 2 are rectangles of height 17.5 in. The centroidal coordinate y_c is found as

$$y_c = \frac{y_1 A_1 + y_2(-A_2)}{A_1 - A_2} = \frac{\frac{20}{2}(48)20 - (17.5/2)(48 - 2 - 2)17.5}{48(20) - (48 - 2 - 2)17.5} = 15.1 \text{ in}$$

$$A = 190 \text{ in}^2$$

The moment of inertia about the x axis has the form

$$I_x = I_{x1} + (-I_{x2}) = \frac{48(20)^3}{3} - \frac{(48 - 2 - 2)17.5^3}{3} = 4.94 \times 10^4 \text{ in}^4$$

Using the parallel-axis theorem,

$$I_{0x} = I_x - Ay_c^2 = 4.94 \times 10^4 - 190(15.1)^2 = 6,080 \text{ in}^4$$

The radius of gyration about the x_0 axis is

$$k_{0x} = \sqrt{\frac{I_{0x}}{A}} = \sqrt{\frac{6,080}{190}} = 5.66 \text{ in}$$

10.24 Find the centroidal moments of inertia I_{0x}, I_{0y}, and J_0 of the area shown in Fig. 10.24a. The centroidal x_0 axis is 2.08 in above the base of the area.

❙ The area is divided into the three areas in Fig. 10.24b. Area 1 is a 3 in by 4 in rectangle. Areas 2 are holes

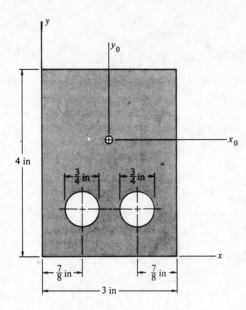

Fig. 10.24a

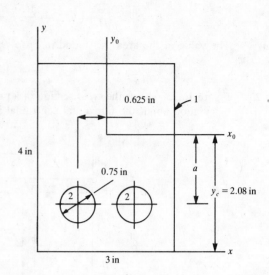

Fig. 10.24b

of 3/4 in diameter. The dimension a is found next. Using the definition of the centroidal coordinate y_c,

$$y_c = \frac{y_1 A_1 + 2y_2(-A_2)}{A_1 - 2A_2} = 2.08 = \frac{\frac{4}{2}(3)4 - 2\left[(2.08 - a)\frac{\pi(0.75)^2}{4}\right]}{3(4) - 2\left[\frac{\pi(0.75)^2}{4}\right]}$$

$$a = 1.09 \text{ in} \qquad A = 11.1 \text{ in}^2$$

The centroidal coordinate y_2 is given by $y_2 = 2.08 - a = 2.08 - 1.09 = 0.99$ in. The moment of inertia about the x axis is given by

$$I_x = I_{x1} + 2(-I_{x2}) = \frac{3(4)^3}{3} - 2\left[\frac{\pi(0.75)^4}{64} + \frac{\pi(0.75)^2}{4}(0.99)^2\right] = 63.1 \text{ in}^4$$

Using the parallel-axis theorem

$$I_{0x} = I_x - Ay_c^2 = 63.1 - 11.1(2.08)^2 = 15.1 \text{ in}^4$$

The area moment of inertia I_{0y} is found directly as

$$I_{0y} = I_{0y,1} - 2I_{0y,2} = \frac{4(3)^3}{12} - 2\left[\frac{\pi(0.75)^4}{64} + \frac{\pi(0.75)^2}{4}(0.625)^2\right] = 8.62 \text{ in}^4$$

The polar moment of inertia about the centroid is given by

$$J_0 = I_{0x} + I_{0y} = 23.7 \text{ in}^4$$

The polar radius of gyration is found as

$$k_{0p} = \sqrt{\frac{J_0}{A}} = \sqrt{\frac{23.7}{11.1}} = 1.46$$

10.25 Figure 10.25a shows a balance wheel which is to be used in a precision instrument. The wheel is punched from steel strip which is 0.0041 in thick.

(a) Find the polar moment of inertia J_0 of the area, shown in Fig. 10.25a, of the balance wheel.

(b) In order to achieve a desired final value of J_0, sets of small holes which are spaced 180° apart may be punched in the rim of the wheel. These 0.015-in-diameter holes are punched on a circle of 0.720 in mean diameter. Find the percent reduction in the original value of J_0 for each set of holes which is punched.

▮ (a) The area of the balance wheel is divided into the areas shown in Fig. 10.25b. The polar moment of inertia of the area is expressed by

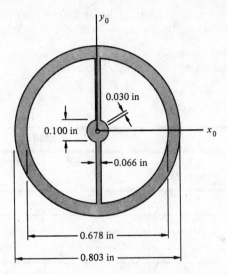

Fig. 10.25a

$$J_0 = J_{01} + 2J_{02} + J_{03}$$

For element 1, Case 5 of Table 10.6 is used, with the result

$$J_{01} = \frac{\pi}{32} [0.803^4 - 0.678^4] = 0.02007 \text{ in}^4$$

For element 2, using Case 2 of Table 10.1,

$$J_{02} = \frac{ab}{12}(a^2 + b^2) + Ad^2 = \frac{0.066(0.289)}{12}(0.066^2 + 0.289^2) + 0.066(0.289)\left(\frac{0.100}{2} + \frac{0.289}{2}\right)^2 = 0.00086 \text{ in}^4$$

Case 5 of Table 10.6 is used for element 3 to obtain

$$J_{03} = \frac{\pi}{32}[0.100^4 - 0.030^4] = 0.00001 \text{ in}^4$$

The polar moment of inertia then has the final form

$$J_0 = 0.02007 + 2(0.00086) + 0.00001 = 0.0218 \text{ in}^4$$

(b) Figure 10.25c shows the pair of hole areas. The polar moment of inertia ΔJ_0 of these two areas about the center of the original area is

$$\Delta J_0 = 2\left[\frac{\pi(0.015)^4}{32} + \frac{\pi(0.015)^2}{4}\left(\frac{0.720}{2}\right)^2\right] = 4.58 \times 10^{-5} \text{ in}^4$$

The percent reduction in the original value of J_0, per pair of holes, is then given by

$$\%D = \frac{\Delta J_0}{J_0} 100 = \frac{4.58 \times 10^{-5}}{0.0218} 100 = 0.21\% \approx \frac{1}{5}\%$$

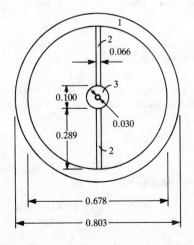

Fig. 10.25b

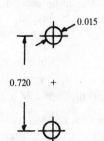

Fig. 10.25c

10.26 Develop a tabular format to organize the calculations for the moments of inertia when the composite area is a complicated shape that requires the description of several elementary areas.

❚ If the composite area comprises several elementary area shapes, the computations may be organized in a tabular form. One such format is shown in Table 10.1.

TABLE 10.1					
(1) Element	(2) I_{oi}	(3) d_i	(4) A_i	(5) $A_i d_i^2$	(6) I_i
1					
Reference Axis __		$A = \sum_i A_i$		I	
				$k = \sqrt{\dfrac{I}{A}}$	

The first step in finding the moment of inertia of the composite area is to divide this area into elementary areas, *each of whose moment of inertia about a particular reference axis is known*. These known values of moment of inertia typically would be found from a reference source such as Table 10.6. Each elementary area is numbered, and these numbers correspond to the rows in the table. The subscript i in the column headings corresponds to the number of the elementary area. The axis about which the moment of inertia of the composite area is to be found is also recorded.

Two possible situations may be identified now. In the first, the equation for the moment of inertia of the elementary area about the reference axis of the composite area is known. An example of this would be where the edge of a rectangular elementary area lies on the reference axis, with a value of moment of inertia known from Case 2 of Table 10.6. For this situation, the value of the moment of inertia would be entered directly in column 6 of Table 10.1. In the second, more general, case, the moment of inertia of the elementary area about its centroidal axis is known, and then the parallel-axis theorem is used to find the moment of inertia of this area about the reference axis. For this case, the centroidal moment of inertia I_{oi}, the separation distance d_i between the centroidal axis and the reference axis, and the elementary area A_i are entered in columns 2, 3, and 4, respectively. The transfer term $A_i d_i^2$ is then computed and entered in column 5. The final result for the moment of inertia of the elementary area, using the parallel-axis theorem, is then the sum of columns 2 and 5. This result is entered in column 6. The sum of the terms in column 6 is the final result for the moment of inertia of the composite area.

All the values required for the computation of any term are shown in the table, for possible subsequent use in checking the calculations. If the radius of gyration is desired, all the values of the elementary areas must be entered in column 4. This column is then summed, and the computation for k is contained in the last row of the table.

It should again be noted that, if an elementary area is a hole or cutout, the value in column 6 for its moment of inertia must be preceded by a minus sign. The tabular method described above may also be used for the computation of polar moments of inertia of plane areas.

10.27 The plate in Prob. 9.27 now has a rectangular hole cut in it, as shown in Fig. 10.27. Find the moment of inertia, and the radius of gyration, of the net plate area about the x axis.

❚ The plate is divided into the five elementary areas shown in the figure. Table 10.2 is constructed, and the appropriate values are entered. The final results are recorded in the last two rows of the table.

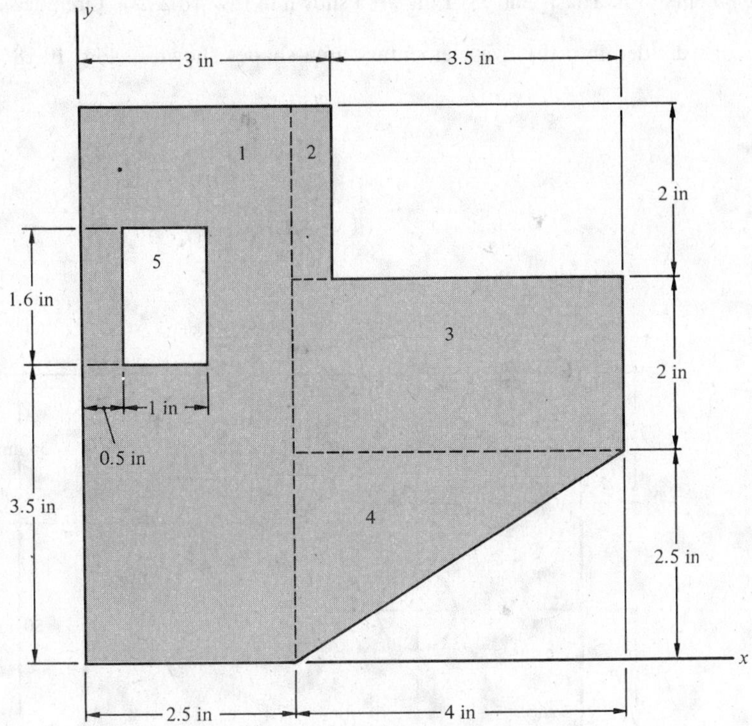

Fig. 10.27

		TABLE 10.2			
Element	I_{oi}	d_i	A_i	$A_i d_i^2$	I_i
1	—	—	$2.5(6.5) = 16.3$	—	$\dfrac{2.5(6.5)^3}{3} = 228.9$
2	$\dfrac{0.5(2)^3}{12} = 0.33$	$2.5 + 2 + 1 = 5.5$	$0.5(2) = 1$	30.25	30.6
3	$\dfrac{4(2)^3}{12} = 2.67$	$2.5 + 1 = 3.5$	$4(2) = 8$	98	100.7
4	$\dfrac{4(2.5)^3}{36} = 1.74$	$0.667(2.5) = 1.67$	$0.5(4)2.5 = 5$	13.94	15.7
5	$-\dfrac{1(1.6)^3}{12} = -0.34$	$3.5 + 0.8 = 4.3$	$-1(1.6) = -1.6$	-29.58	-29.9
Reference axis: x		$A = \sum_i A_i = 28.7 \text{ in}^2$		$I_x = 346 \text{ in}^4$	
		$k_x = \sqrt{\dfrac{I_x}{A}} = \sqrt{\dfrac{346}{28.7}} = 3.47 \text{ in}$			

10.28 Find the moments of inertia I_x and I_y of the area shown in Fig. 10.28a. Organize the solutions in tabular form.

❙ The area is divided into the five elementary area shapes shown in Fig. 10.28b. From this figure,

$$\tan 40° = \frac{35}{A} \qquad A = 41.7 \text{ mm}$$

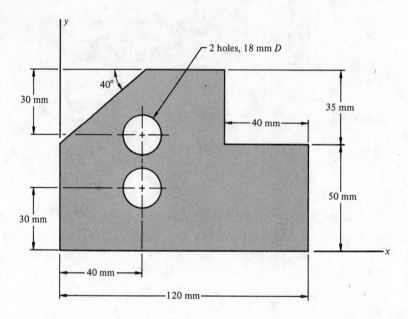

Fig. 10.28a

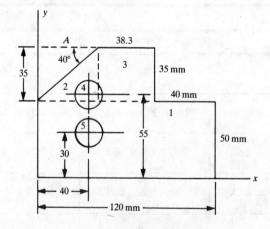

Fig. 10.28b

Table 10.3 is constructed, and the values for I_{0i}, d_i, A_i, and I_i, where appropriate, are entered in each row. The quantities $A_i d_i^2$ and I_i are then computed. The I_i are then summed to obtain the final values for I_x and I_y.

		TABLE 10.3			
element	I_{0i}	d_i	A_i	$A_i d_i^2$	I_i
1	—	—	—	—	$\dfrac{120(50)^3}{3} = 5.00 \times 10^6$
2	$\dfrac{41.7(35)^3}{36} = 4.97 \times 10^4$	$50 + \frac{35}{3} = 61.7$	$\frac{1}{2}(41.7)35 = 730$	2.78×10^6	2.83×10^6
3	$\dfrac{38.3(35)^3}{12} = 1.37 \times 10^5$	$50 + \frac{35}{2} = 67.5$	$38.3(35) = 1,340$	6.11×10^6	6.25×10^6
4	$-\dfrac{\pi(18)^4}{64} = -5,150$	55	$-\dfrac{\pi(18)^2}{4} = -254$	-7.68×10^5	-7.7×10^5
5	$-5,150$	30	-254	-2.29×10^5	-2.3×10^5
Axis: x					$I_x = 1.31 \times 10^7$ mm^4
1	—	—	—	—	$\dfrac{50(120)^3}{3} = 2.880 \times 10^7$
2	$\dfrac{35(41.7)^3}{36} = 7.05 \times 10^4$	$\frac{2}{3}(41.7) = 27.8$	$\frac{1}{2}(41.7)35 = 730$	5.64×10^5	6.3×10^5
3	$\dfrac{35(38.3)^3}{12} = 1.64 \times 10^5$	$41.7 + \frac{38.3}{2} = 60.9$	$38.3(35) = 1,340$	4.97×10^6	5.13×10^6
4	$-\dfrac{\pi(18)^4}{64} = -5,150$	40	$-\dfrac{\pi(18)^2}{4} = -254$	-4.06×10^5	-4.1×10^5
5	$-5,150$	40	-254	-4.06×10^5	-4.1×10^5
Axis: y					$I_y = 3.37 \times 10^7$ mm^4

10.29 Find the moments of inertia I_x and I_y for the area shown in Fig. 10.29a. Organize the solutions in tabular form.

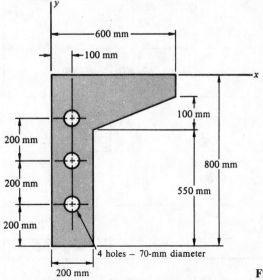

4 holes – 70-mm diameter

Fig. 10.29a

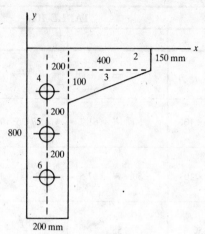

Fig. 10.29b

▎ Figure 10.29b shows the division of the given area into six simple area shapes.   Table 10.4 is constructed, and the final values for I_x and I_y are as shown in this table.

TABLE 10.4				
element	I_{0i}	d_i	A_i	$A_i d_i^2$	I_i
1	—	—	—	—	$\dfrac{200(800)^3}{3} = 3.413 \times 10^{10}$
2	—	—	—	—	$\dfrac{400(150)^3}{3} = 4.5 \times 10^8$
3	$\dfrac{400(100)^3}{36} = 1.11 \times 10^7$	$150 + \frac{100}{3} = 183$	$\frac{1}{2}(400)100 = 2 \times 10^4$	6.70×10^8	6.8×10^8
4	$-\dfrac{\pi(70)^4}{64} = -1.18 \times 10^6$	200	$-\dfrac{\pi(70)^2}{4} = -3{,}850$	-1.54×10^8	-1.6×10^8
5	-1.18×10^6	400	$-3{,}850$	-6.16×10^8	-6.2×10^8
6	-1.18×10^6	600	$-3{,}850$	-1.39×10^9	-1.39×10^9
Axis: x				$I_x = 3.31 \times 10^{10}$ mm^4	
1	—	—	—	—	$\dfrac{800(200)^3}{3} = 2.13 \times 10^9$
2	$\dfrac{150(400)^3}{12} = 8 \times 10^8$	$200 + \frac{400}{2} = 400$	$400(150) = 6 \times 10^4$	9.6×10^9	1.040×10^{10}
3	$\dfrac{100(400)^3}{36} = 1.78 \times 10^8$	$200 + \frac{400}{3} = 333$	$\frac{1}{2}(400)100 = 2 \times 10^4$	2.22×10^9	2.40×10^9
4	$-\dfrac{\pi(70)^4}{64} = -1.18 \times 10^6$	100	$-\dfrac{\pi(70)^2}{4} = -3{,}850$	-3.85×10^7	-4×10^7
5	-1.18×10^6	100	$-3{,}850$	-3.85×10^7	-4×10^7
6	-1.18×10^6	100	$-3{,}850$	-3.85×10^7	-4×10^7
Axis: y				$I_y = 1.48 \times 10^{10}$ mm^4	

10.3 MOMENTS OF INERTIA OF PATTERNS OF HOLE AREAS, PROPERTIES OF TYPICAL STRUCTURAL MEMBER CROSS SECTIONS

10.30 The pattern of hole areas in Prob. 9.23 is shown in Fig. 10.30a.

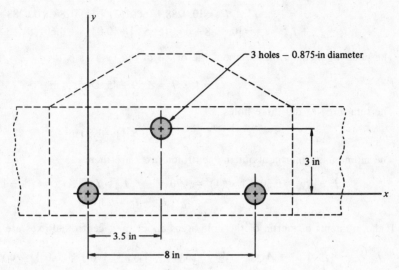

Fig. 10.30a

(a) Find the moments of inertia I_x and I_y of the pattern of hole areas. Then find the polar moment of inertia J, using $J = I_x + I_y$.

(b) Use the parallel-axis theorem to find the polar moment of inertia J_0 about the centroid of the pattern of hole areas.

(c) Do the same as in part (a), but neglect the moments of inertia of all hole areas about their centroidal axes. Compare these results with the results in part (a).

(d) Compute J_0 directly by using the distances between the hole centers and the centroid of the pattern of hole areas.

▌ (a) For Prob. 9.23,

$$x_c = 3.83 \text{ in} \qquad y_c = 1 \text{ in}$$

These values are used in Fig. 10.30b to compute the distances between each hole center and the centroid of the pattern of hole areas. The moment of inertia of the hole area about its centroidal axis, using Case 5 in Table 10.6, is

$$I_{0x} = I_{0y} = \frac{\pi(0.875)^4}{64} = 0.0288 \text{ in}^4$$

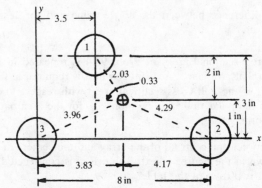

Fig. 10.30b

The area of the hole is

$$A = \frac{\pi(0.875)^2}{4} = 0.601 \text{ in}^2$$

Using the parallel-axis theorem, the moments of inertia about the x and y axes are

$$I_x = I_{x1} + I_{x2} + I_{x3} = [0.0288 + 0.601(3)^2] + 0.0288 + 0.0288 = 5.50 \text{ in}^4$$
$$I_y = I_{y1} + I_{y2} + I_{y3} = [0.0288 + 0.601(3.5)^2] + [0.0288 + 0.601(8)^2] + 0.0288 = 45.9 \text{ in}^4$$

The polar moment of inertia about the origin of the xy axes is then

$$J = I_x + I_y = 5.50 + 45.9 = 51.4 \text{ in}^4$$

(b)   The total area of the three holes is

$$A_{\text{total}} = 3(0.601) = 1.80 \text{ in}^2$$

The moments of inertia about the centroidal axes are then

$$I_{0x} = I_x - Ay_c^2 = 5.50 - 1.80(1)^2 = 3.70 \text{ in}^4 \qquad I_{0y} = I_y - Ax_c^2 = 45.9 - 1.80(3.83)^2 = 19.5 \text{ in}^4$$
$$J_0 = I_{0x} + I_{0y} = 23.2 \text{ in}^4$$

(c)   If the moments of inertia of the hole areas about their centroidal axes are neglected,

$$I_{x,\text{approx}} = \sum_i A_i d_i^2 = A_1 y_1^2 + A_2 y_2^2 + A_3 y_3^2 = A_1 y_1^2 + A_2(0) + A_3(0) = A_1 y_1^2 = 0.601(3)^2 = 5.41 \text{ in}^4$$

The percent difference between the exact value and the approximate value is

$$\%\text{D} = \frac{I_{x,\text{approx}} - I_{x,\text{exact}}}{I_{x,\text{exact}}} 100 = \frac{5.41 - 5.50}{5.50} 100 = -1.6\%$$

$$I_{y,\text{approx}} = \sum_i A_i d_i^2 = A_1 x_1^2 + A_2 x_2^2 + A_3 x_3^2 = A_1 x_1^2 + A_2 x_2^2 + A_3(0) = 0.601(3.5)^2 + 0.601(8)^2 = 45.8 \text{ in}^4$$

$$\%\text{D} = \frac{I_{y,\text{approx}} - I_{y,\text{exact}}}{I_{y,\text{exact}}} 100 = \frac{45.8 - 45.9}{45.9} 100 = -0.2\%$$

$$J_{\text{approx}} = I_{x,\text{approx}} + I_{y,\text{approx}} = 51.2 \text{ in}^4$$

$$\%\text{D} = \frac{J_{\text{approx}} - J_{\text{exact}}}{J_{\text{exact}}} 100 = \frac{51.2 - 51.4}{51.4} 100 = -0.4\%$$

(d)   For each hole area, using Case 5 in Table 10.6,

$$J_0 = \frac{\pi d^4}{32} = \frac{\pi(0.875)^4}{32} = 0.0575 \text{ in}^4$$

Using the parallel-axis theorem,

$$J_0 = \sum_i (J_{0i} + A_i d_i^2) = (J_{01} + A_1 d_1^2) + (J_{02} + A_2 d_2^2) + (J_{03} + A_3 d_3^2) = [0.0575 + 0.601(2.03)^2]$$

$$+ [0.0575 + 0.601(4.29)^2] + [0.0575 + 0.601(3.96)^2] = 23.1 \text{ in}^4$$

The 0.4% difference between the above value and the result in part (b) is due to computational roundoff error.

10.31   The array of bolt cross-sectional areas of Prob. 9.24 is repeated in Fig. 10.31, together with the centroidal coordinates found in that problem. A quantity which is required in the stress analysis of the bolts is the polar moment of inertia J_0 of the bolt cross-sectional areas with respect to an axis through the centroid of the bolt areas. Find J_0, and the polar radius of gyration k_{0p}, for the area pattern shown in the figure. Organize the solution in tabular form.

▮   The distances between the centroid of the array and the centers of the four circles are computed, and these dimensions are shown in the figure. Next, Table 10.5 is constructed, and the final results for J_0 and k_{0p} are contained in the last two rows of the table.

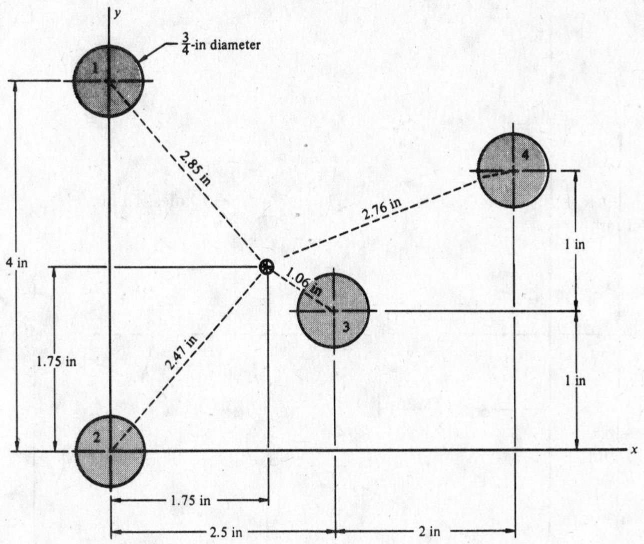

Fig. 10.31

		TABLE 10.5			
element	J_{0i}	d_i	A_i	$A_i d_i^2$	J_i
1	$\dfrac{\pi(0.75)^4}{32} = 0.03$	2.85	$\dfrac{\pi(0.75)^2}{4} = 0.44$	3.59	3.62
2	0.03	2.47	0.44	2.70	2.73
3	0.03	1.06	0.44	0.49	0.52
4	0.03	2.76	0.44	3.37	3.40
Reference axis: centroidal polar			$A = \sum_i A_i = 1.77 \text{ in}^2$	$J_0 = 10.2 \text{ in}^4$	
			$k_{0p} = \sqrt{\dfrac{J_0}{A}} = \sqrt{\dfrac{10.2}{1.77}} = 2.40 \text{ in}$		

10.32 Figure 10.32a shows the bolt hole pattern for a connection in an experimental vehicle design.

(a) Find I_x, I_y, and J for the pattern of hole areas.

(b) Find the moments of inertia I_{0x} and I_{0y}, and the polar moment of inertia J_0, about the centroidal axes of the pattern of hole areas.

▌ (a) The pattern of areas is repeated in Fig. 10.32b. Dimensions A and B are found from the similar triangles in the figure as

$$\frac{A}{90} = \frac{75}{165} \qquad A = 40.9 \text{ mm} \qquad \frac{B}{75} = \frac{65}{165} \qquad B = 29.5 \text{ mm}$$

For this problem, the form $I = \sum_i I_{0i} + \sum_i A d_i^2$ will be used to obtain the moments of inertia. For the 15-mm-diameter hole areas,

$$I_{0i} = \frac{\pi(15)^4}{64} = 2{,}490 \text{ mm}^4 \qquad A_i = \frac{\pi(15)^2}{4} = 177 \text{ mm}^2$$

For the 25-mm-diameter hole areas,

$$I_{0i} = \frac{\pi(25)^4}{64} = 19{,}200 \text{ mm}^4 \qquad A_i = \frac{\pi(25)^2}{4} = 491 \text{ mm}^2$$

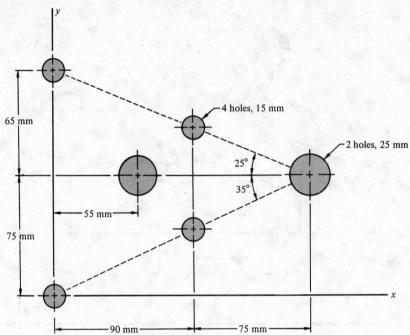

Fig. 10.32a

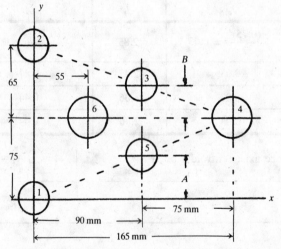

Fig. 10.32b

$$I_x = \sum_i I_{0x,i} + \sum_i A_i d_i^2 = 4(2,490) + 2(19,200) + 177(40.9^2 + 105^2 + 140^2) + 2(491)75^2 = 1.13 \times 10^7 \text{ mm}^4$$

$$I_y = \sum_i I_{0y,i} + \sum_i A_i d_i^2 = 4(2,490) + 2(19,200) + 2(177)90^2 + 491(55^2 + 165^2) = 1.78 \times 10^7 \text{ mm}^4$$

The polar moment of inertia is found from

$$J = I_x + I_y = 2.91 \times 10^7 \text{ mm}^4$$

(b) The centroidal coordinates of the pattern of hole areas are

$$x_c = \frac{2x_1 A_1 + 2x_3 A_3 + x_4 A_4 + x_6 A_6}{2A_1 + 2A_3 + A_4 + A_6} = \frac{2[0(177)] + 2[90(177)] + 165(491) + 55(491)}{2(177) + 2(177) + 2(491)} = 82.8 \text{ mm}$$

$$A = 1,690 \text{ mm}^2$$

$$y_c = \frac{y_1 A_1 + y_2 A_2 + y_3 A_3 + 2y_4 A_4 + y_5 A_5}{A}$$

$$= \frac{0(177) + 140(177) + 105(177) + 2[75(491)] + 40.9(177)}{1,690} = 73.5 \text{ mm}$$

The moments of inertia about the centroidal axes are then

$$I_{0x} = I_x - Ay_c^2 = 1.13 \times 10^7 - 1{,}690(73.5)^2 = 2.17 \times 10^6 \text{ mm}^4$$
$$I_{0y} = I_y - Ax_c^2 = 1.78 \times 10^7 - 1{,}690(82.8)^2 = 6.21 \times 10^6 \text{ mm}^4$$
$$J_0 = I_{0x} + I_{0y} = 8.38 \times 10^6 \text{ mm}^4$$

10.33   Two 10 [30 standard channels are riveted together to form a beam with the cross section shown in Fig. 10.33a.   In order to determine the spacing of the rivets along the length of the beam, the first area moment of one of the two channel cross-sectional areas with respect to the x axis is required.

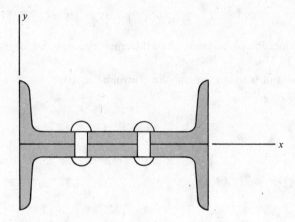

Fig. 10.33a

(a)   Find this first area moment.

(b)   Find the moment of inertia of the composite area about the x axis.   Neglect the effect of the rivet holes on the cross-sectional areas of the channels.

▮ (a)   The properties of the cross-sectional areas of structural members are tabulated for a wide variety of shapes and sizes.   These properties include the location of the centroid and the moments of inertia and radii of gyration about axes which pass through the centroid.   Also included in these tabulations are the dimensions of the cross section and the weight per foot of length of the member.   The weights given are for steel members.   The weights for structural members made of other materials may be found by multiplying the tabulated values of weight for steel by the ratio of the specific weight of the particular material to the specific weight of steel.

Tables 10.8 through 10.10 at the end of this chapter show typical values for wide-flange beams, standard channels, and standard angles with unequal legs.

The upper channel is shown in Fig. 10.33b.   The location of the centroid of the channel cross-sectional area is found from Table 10.9, and this dimension is shown in the figure.   The cross-sectional area is 8.80 in², and the first area moment with respect to the x axis is then

$$Q_x = y_c A = 0.65(8.80) = 5.72 \text{ in}^3$$

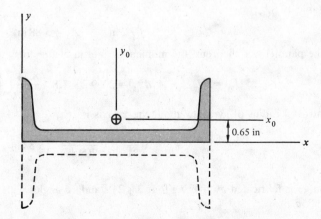

Fig. 10.33b

(**b**) The moment of inertia of one channel section about its centroidal x_0 axis is

$$I_{0x} = k_{0x}^2 A$$

The quantity k_{0x} in this problem is the term k_y in Table 10.9, with the value

$$k_{0x} = k_y = 0.67 \text{ in}$$

Thus,

$$I_{0x} = 0.67^2(8.80) = 3.95 \text{ in}^4$$

The moment of inertia I_x of the composite cross section, using the parallel-axis theorem, is

$$I_x = 2[I_{0x} + Ay_c^2] = 2[3.95 + 8.80(0.65)^2] = 15.3 \text{ in}^4$$

10.34 (**a**) Determine the location of the centroid of the cross-sectional area of the steel structural member assembly shown in Fig. 10.34a.

(**b**) Find the moment of inertia I_{0x} about the centroidal x_0 axis.

(**c**) Find the radius of gyration k_{0x}.

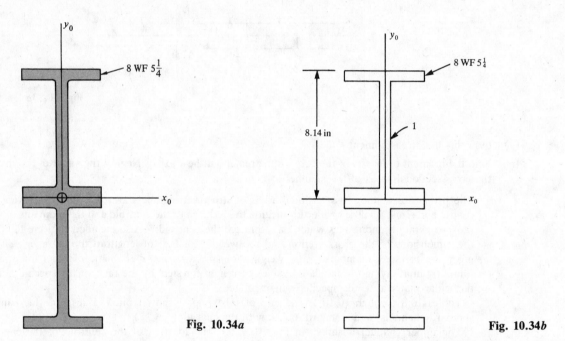

Fig. 10.34a   **Fig. 10.34b**

▮ (**a**) Figure 10.34b shows the cross section, with the beam height 8.14 in found in Table 10.8.   From the symmetry of the cross section, the centroid lies on the x_0 axis.

(**b**) From Table 10.8,

$$I_{0x,1} = 69.2 \text{ in}^4 \qquad A_1 = 5.88 \text{ in}^2$$

Using the parallel-axis theorem, the moment of inertia of the cross section about its centroidal axis is

$$I_{0x} = 2[I_{0x,1} + A_1 d_{x,1}^2] = 2\left[69.2 + 5.88\left(\frac{8.14}{2}\right)^2\right] = 333 \text{ in}^4$$

(**c**) The centroidal radius of gyration about the x_0 axis is

$$k_{0x} = \sqrt{\frac{I_{0x}}{A}} = \sqrt{\frac{333}{2(5.88)}} = 5.32 \text{ in}$$

10.35 A structural member is fabricated by riveting four $3 \times 2\frac{1}{2}$ standard angles with unequal legs to a web section, as shown in Fig. 10.35a.

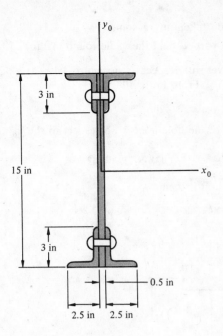

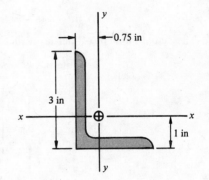

$$A = 2.50 \text{ in}^2$$
$$I_{xx} = 2.1 \text{ in}^4$$
$$I_{yy} = 1.3 \text{ in}^4$$

Fig. 10.35a

Fig. 10.35b

(a) Find the moments of inertia of the composite cross section about the centroidal axes.

(b) What part of each of the above results is contributed by the four angles?

(c) Find the radius of gyration about the x_0 and y_0 axes.

The effect of the rivet holes may be neglected in all of the above calculations.

▐ (a) Figure 10.35b shows the $3 \times 2\frac{1}{2}$ angle section, together with its properties from Table 10.10. The moment of inertia of the composite cross section about the centroidal x_0 axis is

$$I_{0x} = \frac{ab^3}{12} + 4(I_{xx} + A\, d_x^2) = \frac{0.5(15)^3}{12} + 4[2.1 + 2.5(6.5)^2] = 141 + 431 = 572 \text{ in}^4$$

The moment of inertia about the y_0 axis is

$$I_{0y} = \frac{ba^3}{12} + 4(I_{yy} + A\, d_y^2) = \frac{15(0.5)^3}{12} + 4\left[1.3 + 2.5\left(\frac{0.5}{2} + 0.75\right)^2\right] = 0.156 + 15.2 = 15.4 \text{ in}^4$$

The moment of inertia about the x_0 axis is significantly greater than the moment of inertia about the y_0 axis. The ratio is $572/15.4 = 37.1$.

(b) It may be seen from the above equations that the four angles contribute $(431/572)(100) = 75$ percent of the total value of the moment of inertia I_{0x}, while the web section contributes the remaining 25 percent. The four angles are seen to constitute $(15.2/15.4)(100) = 99$ percent of the moment of inertia I_{0y}.

(c) The area of the composite cross section is

$$A = 0.5(15) + 4(2.50) = 17.5 \text{ in}^2$$

The radius of gyration about the x_0 axis is

$$k_{0x} = \sqrt{\frac{I_{0x}}{A}} = \sqrt{\frac{572}{17.5}} = 5.72 \text{ in}$$

The radius of gyration about the y_0 axis is

$$k_{0y} = \sqrt{\frac{I_{0y}}{A}} = \sqrt{\frac{15.4}{17.5}} = 0.94 \text{ in}$$

10.36 The effect of the rivet holes in Prob. 10.35 was neglected in solving for the centroidal moment of inertia about the x_0 axis. The diameter of the rivets is $\frac{7}{16}$ in, and the spacing of the rivets from the centroidal x_0 axis is $5\frac{3}{4}$ in.

(*a*) Compute the moment of inertia I_{0x} if the holes are taken into account.

(*b*) Find the percent difference between the result in part (*a*) and the value found in Prob. 10.35.

▌ (*a*) Figure 10.36 shows the spacing of the upper rivet with respect to the centroidal x_0 axis.   The thickness dimension of the angle is found in Table 10.10.   From Prob. 10.35,

$$I_{0x,\text{no holes}} = 572 \text{ in}^4$$

Using $\frac{7}{16} = 0.438$ in,   the moment of inertia of the hole areas with respect to the x_0 axis is

$$I_{0x,\text{holes}} = 2\left[\frac{1.5(0.438)^3}{12} + 1.5(0.438)5.75^2 \right] = 43.5 \text{ in}^4$$

The actual moment of inertia of the cross section is then

$$I_{0x,\text{actual}} = I_{0x,\text{no holes}} - I_{0x,\text{holes}} = 572 - 43.5 = 529 \text{ in}^4$$

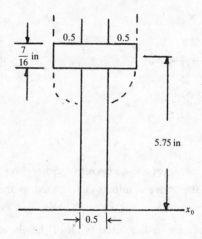

Fig. 10.36

(*b*) The percent difference between the above two results is

$$\%\text{D} = \frac{I_{0x,\text{actual}} - I_{0x,\text{no holes}}}{I_{0x,\text{no holes}}} 100 = \frac{529 - 572}{572} 100 = -7.5\%$$

10.37 A 14-in × 8-in-wide flange beam is to be reinforced by the addition of an 8 in × $\frac{3}{4}$ in strip along the top flange, as shown in Fig. 10.37*a*.

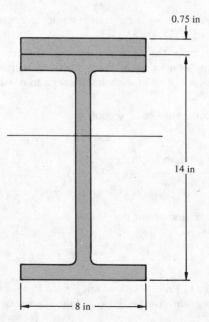

Fig. 10.37*a*

(a) Find the area moment of inertia about the horizontal axis which passes through the centroid of the composite area.

(b) Find the radius of gyration which corresponds to the moment of inertia found in part (a).

(c) By what percent does the addition of the strip increase the moment of inertia of the original cross section?

▮ (a) The combined system of the wide-flange beam and the strip is shown in Fig. 10.37b. The centroidal coordinate of the composite area is y_c. From Table 10.8, the area of the wide-flange section is 15.6 in². Then y_c is found as

$$y_c = \frac{7(15.6) + 14.4(8)0.75}{15.6 + 8(0.75)} = 9.06 \text{ in} \qquad A = 21.6 \text{ in}^2$$

Figure 10.37c shows the composite cross section and the distances between the centroidal locations. The moment of inertia of the wide-flange beam cross section about its centroidal x_0 axis, from Table 10.8, is 542 in⁴. The moment of inertia of the composite cross-sectional area about its centroidal axis, using the parallel-axis theorem, is

$$I_{0x} = 542 + 15.6(2.06)^2 + \frac{8(0.75)^3}{12} + 6(5.34)^2 = 780 \text{ in}^4$$

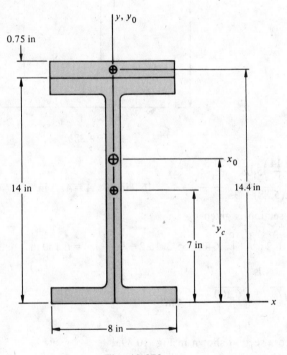

Fig. 10.37b

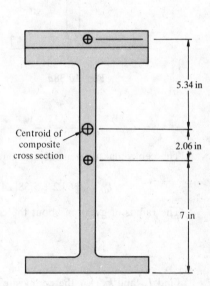

Fig. 10.37c

(b) The radius of gyration k_{0x} about the centroidal axis of the composite cross-sectional area is

$$k_{0x} = \sqrt{\frac{I_{0x}}{A}} = \sqrt{\frac{780}{21.6}} = 6.01 \text{ in}$$

(c) The addition of the strip to the beam increases the centroidal moment of inertia by [(780 − 542)/542]100 = 43.9 percent.

10.38 Find the moment of inertia I_{0x} and the radius of gyration k_{0x} of the structural cross section shown in Fig. 10.38a.

▮ The two beam cross-sectional areas are designated 1 and 2, as shown in Fig. 10.38b. From Table 10.8,

$$I_{0x,1} = 69.2 \text{ in}^4 \qquad A_1 = 5.88 \text{ in}^2$$
$$I_{0x,2} = 3,640 \text{ in}^4 \qquad A_2 = 35.3 \text{ in}^2$$

The centroidal coordinate y_c of the composite cross section is

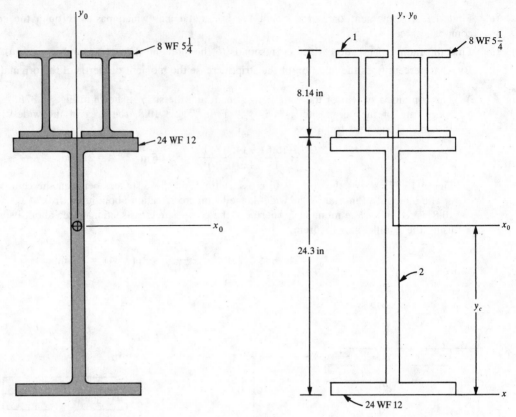

Fig. 10.38a   **Fig. 10.38b**

$$y_c = \frac{2y_1 A_1 + y_2 A_2}{2A_1 + A_2} = \frac{2\left(24.3 + \frac{8.14}{2}\right)5.88 + \frac{24.3}{2}(35.3)}{2(5.88) + 35.3} = 16.2 \text{ in} \quad A = 47.1 \text{ in}^2$$

The centroidal moment of the composite cross section is given by

$$I_{0x} = 2\left[69.2 + 5.88\left(24.3 + \frac{8.14}{2} - 16.2\right)^2\right] + \left[3.640 + 35.3\left(16.2 - \frac{24.3}{2}\right)^2\right] = 6,100 \text{ in}^4$$

The radius of gyration about the centroidal x_0 axis is

$$k_{0x} = \sqrt{\frac{I_{0x}}{A}} = \sqrt{\frac{6,100}{47.1}} = 11.4 \text{ in}$$

10.39 Find I_{0x} and k_{0x} for the composite structural cross section shown in Fig. 10.39a.

❚ Figure 10.39b shows the division of the given cross section into three area shapes. $I_{0x,2}$ is the term I_{yy} in Table 10.10, with the value

$$I_{0x,2} = 4.3 \text{ in}^4 \quad A_2 = 4.5 \text{ in}^2$$

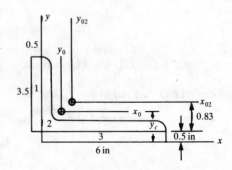

Fig. 10.39a   **Fig. 10.39b**

The y coordinate of the centroid of the composite cross section is

$$y_c = \frac{y_1 A_1 + y_2 A_2 + y_3 A_3}{A_1 + A_2 + A_3} = \frac{\left(0.5 + \frac{3.5}{2}\right)0.5(3.5) + (0.5 + 0.83)4.5 + \frac{0.5}{2}(6)0.5}{0.5(3.5) + 4.5 + 6(0.5)} = 1.15 \text{ in} \qquad A = 9.25 \text{ in}^2$$

The moment of inertia of the composite cross section about its centroidal axis is found as

$$I_{0x} = \left[\frac{0.5(3.5)^3}{12} + 0.5(3.5)\left(0.5 + \frac{3.5}{2} - 1.15\right)^2\right] + [4.3 + 4.5[(0.5 + 0.83) - 1.15]^2]$$
$$+ \left[\frac{6(0.5)^3}{12} + 6(0.5)\left(1.15 - \frac{0.5}{2}\right)^2\right] = 10.8 \text{ in}^4$$

The centroidal radius of gyration is

$$k_{0x} = \sqrt{\frac{I_{0x}}{A}} = \sqrt{\frac{10.8}{9.25}} = 1.08 \text{ in}$$

10.40 Find I_{0x} and k_{0x} for the structural cross section shown in Fig. 10.40. The height of the web section is 10 in.

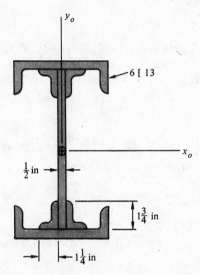

Fig. 10.40

▌ The 6[13 cross-sectional areas are designated areas 1, and the angle cross-sectional areas are areas 2. The rectangular web section is area 3. From symmetry, the centroid is at the mid-height of the composite area shown in Fig. 10.40. From Table 10.9, $A_1 = 3.81 \text{ in}^2$. The centroidal moment of inertia of the channel section about an axis parallel to the x axis is not given directly in Table 10.9. This quantity is found by using the values $k_{0y} = 0.53 \text{ in}$ and $A_1 = 3.81 \text{ in}^2$ in the equation

$$I_{0x,1} = k^2 A_1 = 0.53^2(3.81) = 1.07 \text{ in}^4$$

From Table 10.10,

$$A_2 = 0.69 \text{ in}^2 \qquad I_{0x,2} = 0.20 \text{ in}^4$$

The centroidal moment of inertia is expressed as

$$I_{0x} = 2I_1 + 4I_2 + I_3 = 2[1.07 + 3.81(5 + 0.437 - 0.52)^2] + 4[0.20 + 0.69(5 - 0.60)^2] + \frac{0.5(10)^3}{12} = 282 \text{ in}^4$$

$$A = 2(3.81) + 4(0.69) + 0.5(10) = 15.4 \text{ in}^2$$

The centroidal radius of gyration is

$$k_{0x} = \sqrt{\frac{I_{0x}}{A}} = \sqrt{\frac{282}{15.4}} = 4.28 \text{ in}$$

10.41 Find the percent decrease in the centroidal moment of inertia about the x_0 axis if the lower 6[13 section is removed from the area in Prob. 10.40, producing the cross section shown in Fig. 10.41a.

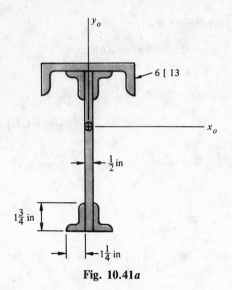

Fig. 10.41a

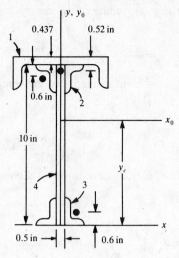

Fig. 10.41b

❚ Figure 10.41b shows the identification of the area shapes of the cross section. The centroidal coordinates of the channel and angle sections, and the web thickness of the channel, are found from Tables 10.9 and 10.10. These dimensions are shown in Fig. 10.41b. From Table 10.9,

$$A_1 = 3.81 \text{ in}^2 \qquad I_{0x,1} = k^2 A = 0.53^2(3.81) = 1.07 \text{ in}^4$$

From Table 10.10,

$$A_2 = A_3 = 0.69 \text{ in}^2 \qquad I_{0x,2} = I_{0x,3} = 0.20 \text{ in}^4$$

The centroidal coordinate y_c of the entire cross section is then found as

$$y_c = \frac{y_1 A_1 + 2y_2 A_2 + 2y_3 A_3 + y_4 A_4}{A_1 + 2A_2 + 2A_3 + A_4}$$

$$= \frac{(10 + 0.437 - 0.52)3.81 + 2(10 - 0.6)0.69 + 2(0.6)0.69 + \frac{10}{2}(0.5)10}{3.81 + 2(0.69) + 2(0.69) + 0.5(10)} = 6.62$$

$$A = 11.6 \text{ in}^2$$

The moment of inertia of the composite cross-sectional area about its centroidal axis is then found as

$$I_{0x} = I_{x1} + 2I_{x2} + 2I_{x3} + I_{x4} = [1.07 + 3.81[(10 + 0.437 - 0.52) - 6.62]^2]$$
$$+ 2[0.20 + 0.69[(10 - 0.6) - 6.62]^2] + 2[0.20 + 0.69(6.62 - 0.6)^2]$$

$$+ \left[\frac{0.5(10)^3}{12} + 0.5(10)(6.62 - 5)^2 \right] = 159 \text{ in}^4$$

The percent decrease in the moment of inertia, using the value of I_{0x} in Prob. 10.40 as the reference, is

$$\%D = \frac{159 - 282}{282} \; 100 = -44\%$$

It may be seen that the lower channel section in Fig. 10.40 makes a significant contribution to the centroidal moment of inertia of the cross section.

10.42 The cross section of a lightweight structural column is shown in Fig. 10.42a. The dashed lines indicate lattice members that do not contribute to the moment of inertia of the cross section. Find I_{0x} and I_{0y}.

❚ The centroidal coordinates of the angle sections, from Table 10.10, are shown in Fig. 10.42b. The area and moments of inertia of the angle sections, from Table 10.10, are

$$A = 2.50 \text{ in}^2 \qquad I_{xx} = 2.1 \text{ in}^4 \qquad I_{yy} = 1.3 \text{ in}^4$$

Using the parallel-axis theorem,

$$I_{0x} = 4[2.1 + 2.50(12 - 1)^2] = 1,220 \text{ in}^4 \qquad I_{0y} = 4[1.3 + 2.50(9 - 0.75)^2] = 686 \text{ in}^4$$

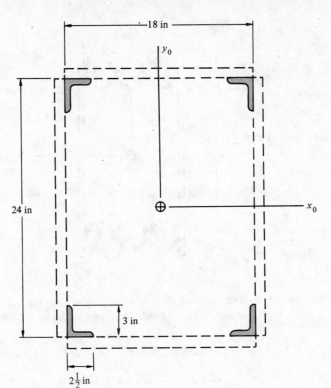

Fig. 10.42a

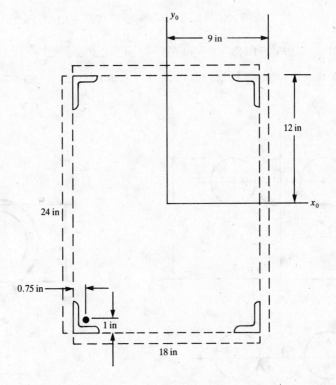

Fig. 10.42b

10.43 Figure 10.43a shows a cross-sectional view of a proposed lightweight beam design fabricated from six tubes with a supporting lattice structure. The tubes are seamless, cold-finished, carbon-steel mechanical tubing, with a mean diameter of 2 in and a wall thickness of 0.035 in. The lattice members shown as the dashed lines in the figure are assumed to not contribute to the moment of inertia. Find I_{0x} and I_{0y}.

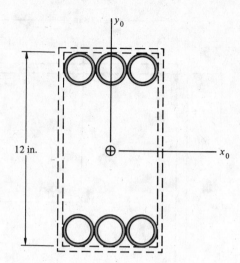

Fig. 10.43a

The tubes are located by the dimensions d_x and d_y shown in Fig. 10.43b, and Fig. 10.43c shows the dimensions of a tube cross section. The outside and inside diameters of the tube, given by d_o and d_i, respectively, are

$$d_o = 2 + 0.035 = 2.035 \text{ in} \qquad d_i = 2 - 0.035 = 1.965 \text{ in}$$

$$d_x = 6 - \frac{d_0}{2} = 6 - \frac{2.035}{2} = 4.98 \text{ in} \qquad d_y = d_o = 2.035 \text{ in}$$

Using Case 5 in Table 10.6, and the parallel-axis theorem, the centroidal moments of inertia of the beam design cross section are

$$I_{0x} = 6\left[\frac{\pi}{64}(2.035^4 - 1.965^4) + \frac{\pi}{4}(2.035^2 - 1.965^2)4.98^2\right] = 33.4 \text{ in}^4$$

$$I_{0y} = 4\left[\frac{\pi}{64}(2.035^4 - 1.965^4) + \frac{\pi}{4}(2.035^2 - 1.965^2)2.035^2\right] + 2\left[\frac{\pi}{64}(2.035^4 - 1.965^4)\right] = 4.30 \text{ in}^4$$

It may be seen that I_{0x} is much larger than I_{0y}, by a factor of $33.4/4.30 = 7.77$.

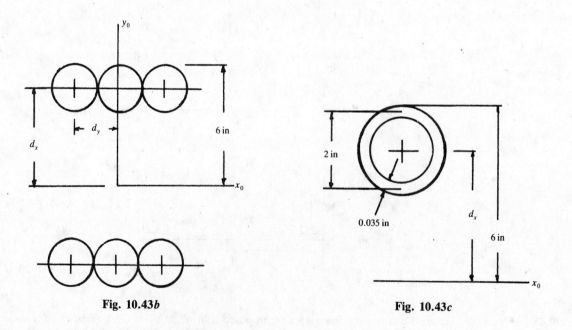

Fig. 10.43b

Fig. 10.43c

10.44 The beam cross section shown in Fig. 10.44a is of honeycomb construction. The thickness of the outer boundary is 4 mm, and the thickness of all internal elements is 2 mm. Find I_{0x}.

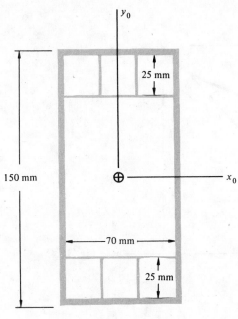

Fig. 10.44a

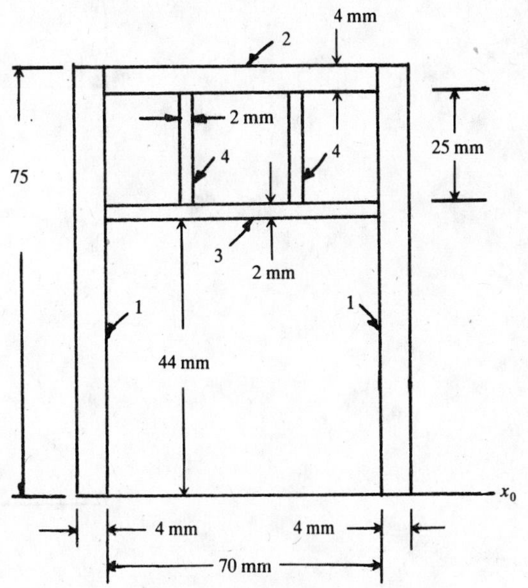

Fig. 10.44b

▌ The upper half of the cross section is shown in Fig. 10.44b. The centroidal moments of inertia of elements 2 and 3 are neglected. Using the parallel-axis theorem,

$$\tfrac{1}{2}I_{0x} = 2I_1 + I_2 + I_3 + 2I_4$$

$$= 2\left[\frac{4(75)^3}{3}\right] + 70(4)(75-2)^2 + 70(2)(44+1)^2 + 2\left[\frac{2(25)^3}{12} + 2(25)\left(44+2+\frac{25}{2}\right)^2\right]$$

$$= 6.50 \times 10^6 \text{ mm}^4$$

10.4 MOMENTS OF INERTIA OF PLANE CURVES, PRODUCTS OF INERTIA OF PLANE AREAS AND CURVES

10.45 Figure 10.45 shows a plane curve that is positioned with respect to a set of xy coordinate axes. The curve length is divided into the elementary lengths l_i, with coordinates x_i and y_i. Give the forms for the moments of inertia of the plane curve with respect to the x and y axes.

▌ The moments of inertia of the plane curve with respect to the coordinate axes are defined to be

$$I_x = \sum_i y_i^2 l_i \qquad I_y = \sum_i x_i^2 l_i$$

where the l_i are the length elements of the curve. The terms I_x and I_y given above are seen to be inherently positive terms. As is the case with moments of inertia of plane areas, elements more remote from the reference axis have a proportionately greater effect on the moments of inertia of the plane curve, because of the squaring effect. It may also be observed from the above equations that the units of the moment of inertia of a plane curve are length raised to the third power.

The integral forms of the above two equations are

$$I_x = \lim_{l_i \to 0} \sum_i y_i^2 l_i = \int_l y^2 \, dl \qquad I_y = \lim_{l_i \to 0} \sum_i x_i^2 l_i = \int_l x^2 \, dl$$

where l is the length of the curve.

For the plane curve shown in Fig. 10.45, the polar moment of inertia J is given by

$$J = \sum_i (x_i^2 + y_i^2)l_i = \sum_i r_i^2 l_i = I_x + I_y$$

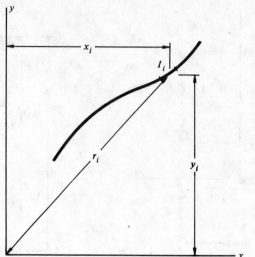

Fig. 10.45

and the integral form is

$$J = \int_l r^2 \, dl$$

The moments of inertia of several elementary plane curves are given in Table 10.7 at the end of this chapter. These values for the moments of inertia are found by direct integration of the above equations.

10.46 Figure 10.46a shows a straight line of length l positioned with respect to the xy coordinate axes. Show that the moments of inertia of this plane curve are

$$I_x = \frac{l^3}{12} \qquad I_y = a^2 l$$

❚ The length element used in the computation of I_x is shown in Fig. 10.46b. Using $I_x = \int_l y^2 \, dl$, with $dl = dy$,

$$I_x = \int_{-l/2}^{l/2} y^2 \, dy = \frac{y^3}{3}\bigg|_{-l/2}^{l/2} = \frac{1}{3}\left[\left(\frac{l}{2}\right)^3 - \left(-\frac{l}{2}\right)^3\right] = \frac{1}{3}\left(\frac{l^3}{8} + \frac{l^3}{8}\right) = \frac{l^3}{12}$$

When finding I_y,

$$x = a = \text{constant}$$

I_y then has the form

$$I_y = \int_l x^2 \, dl = \int_l a^2 \, dl = a^2 \int_l dl = a^2 l$$

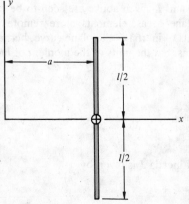

Fig. 10.46a

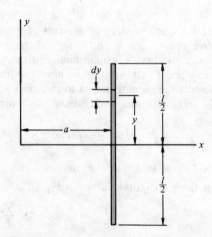

Fig. 10.46b

10.47   State the parallel-axis, or transfer, theorems for the plane curve shown in Fig. 10.47.

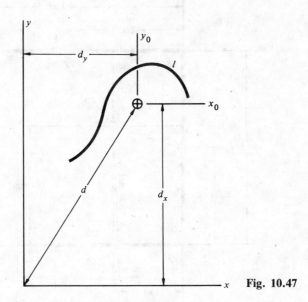

Fig. 10.47

▌ The parallel-axis, or transfer, theorems for the plane curve in Fig. 10.47 are

$$I_x = I_{0x} + l\,d_x^2 \tag{1}$$
$$I_y = I_{0y} + l\,d_y^2 \tag{2}$$

x_0 and y_0 are the centroidal axes of the plane curve, and these axes are parallel to the xy axes. I_{0x} and I_{0y} are the centroidal moments of inertia of the curve. J is the polar moment of inertia of the plane curve with respect to the origin of the xy coordinates, and J_0 is the polar moment of inertia about the centroid of the curve. The parallel-axis relationship between these two quantities is given by

$$J = J_0 + l\,d^2$$

   The parallel-axis theorem may be expressed in an alternative form. From consideration of Fig. 10.47, it may be seen that d_x and d_y are actually the centroidal coordinates of the plane curve, measured in the xy coordinate system. Thus, with

$$d_y = x_c' \quad \text{and} \quad d_x = y_c'$$

the parallel-axis theorem may be written as

$$I_x = I_{0x} + l y_c'^2 \tag{3}$$
$$I_y = I_{0y} + l x_c'^2 \tag{4}$$

   Equations (1) and (2) may be considered to be the fundamental forms of the parallel-axis therem. These two equations may be used to find moments of inertia about axes parallel to x_0 and y_0, without a specific calculation of the centroidal coordinates. If, however, x_c' and y_c' are known in a given problem, then Eqs. (3) and (4) may be used to relate the moments of inertia about parallel axes.
   It is emphasized that Eqs. (1) and (3), and Eqs. (2) and (4), are equivalent statements.

10.48   The weld pattern of Prob. 9.39 is redrawn in Fig. 10.48, together with its centroidal coordinates.   Find the polar moment of inertia of this plane curve with respect to its centroid.

▌ The given plane curve is divided into three elementary straight lines, as shown in the figure.   The distances between the centroid of the composite curve and the centroids of the three elementary lengths are computed and drawn in the figure.   The moment of inertia of a straight line about an axis normal to its centroid was shown in Prob. 10.46 to be $l^3/12$.   By using this result and the parallel-axis theorem, the polar moment of inertia J_0 is

$$J_0 = J_{01} + J_{02} + J_{03} = [\tfrac{1}{12}(400)^3 + 400(147)^2] + [\tfrac{1}{12}(300)^3 + 300(116)^2] + [\tfrac{1}{12}(200)^3 + 200(183)^2] = 2.76 \times 10^7 \text{ mm}^3$$

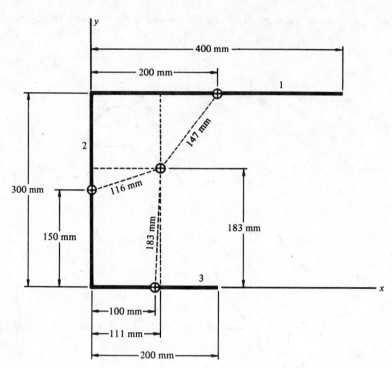

Fig. 10.48

10.49 (*a*) Find the moments of inertia I_x and I_y of the weld pattern shown in Fig. 10.49*a*.

(*b*) Find the polar moment of inertia J.

(*c*) Use the parallel-axis theorem to find the polar moment of inertia J_0 about the centroid of the weld pattern.

(*d*) Find the centroidal radius of gyration k_{0p}.

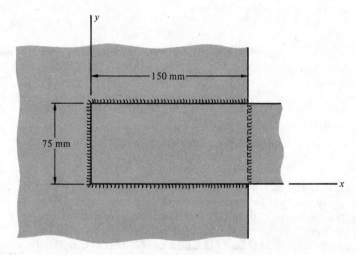

Fig. 10.49*a*

▌ (*a*) Figure 10.49*b* shows the numbering of the straight-line elements of the weld pattern. The moments of inertia are

$$I_x = I_{x1} + I_{x2} + I_{x3} + I_{x4}$$
$$I_{x1} = I_{x3} \qquad I_{x4} = 0 \qquad I_x = 2I_{x1} + I_{x2}$$

From Case 1, Table 10.7, with $\theta = 90°$,

$$I_{x1} = I_{x3} = \frac{l^3}{3} \qquad I_x = 2[\tfrac{1}{3}(75)^3] + 150(75)^2 = 1.13 \times 10^6 \text{ mm}^3$$

$$I_y = I_{y1} + I_{y2} + I_{y3} + I_{y4} \qquad I_{y2} = I_{y4} \qquad I_{y1} = 0 \qquad I_y = 2I_{y2} + I_{y3}$$

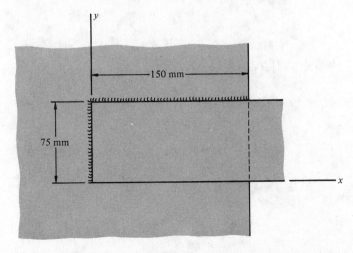

Fig. 10.49b

Using Case 1 in Table 10.7, with $\theta = 0°$,

$$I_{y2} = I_{y4} = \frac{l^3}{3} \qquad I_y = 2[\tfrac{1}{3}(150)^3] + 75(150)^2 = 3.94 \times 10^6 \text{ mm}^3$$

(b)
$$J = I_x + I_y = 5.07 \times 10^6 \text{ mm}^3$$

(c) From the symmetry of the weld pattern,

$$x'_c = 75 \text{ mm} \qquad y'_c = 37.5 \text{ mm}$$

The length of the curve is

$$l = 2(150) + 2(75) = 450 \text{ mm}$$

Using the parallel-axis theorem,

$$I_{0x} = I_x - ly'^2_c = 1.13 \times 10^6 - 450(37.5)^2 = 4.97 \times 10^5 \text{ mm}^3$$
$$I_{0y} = I_y - lx'^2_c = 3.94 \times 10^6 - 450(75)^2 = 1.41 \times 10^6 \text{ mm}^3$$

The centroidal polar moment of inertia is then found as

$$J_0 = I_{0x} + I_{0y} = 1.91 \times 10^6 \text{ mm}^3$$

(d) The centroidal radius of gyration k_{0p} is

$$k_{0p} = \sqrt{\frac{J_0}{l}} = \sqrt{\frac{1.91 \times 10^6}{450}} \qquad k_{0p} = 65.1 \text{ mm}$$

10.50 Solve Prob. 10.49 if the weld pattern is changed to that shown in Fig. 10.50a.

▮ (a) The numbering of the length elements is shown in Fig. 10.50b. The moments of inertia are found as

$$I_x = I_{x1} + I_{x2} = \tfrac{1}{3}(75)^3 + 150(75)^2 = 9.84 \times 10^5 \text{ mm}^3$$
$$I_y = I_{y2} = \tfrac{1}{3}(150)^3 = 1.13 \times 10^6 \text{ mm}^3$$

(b)
$$J = I_x + I_y = 2.11 \times 10^6 \text{ mm}^3$$

Fig. 10.50a

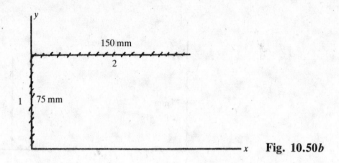

Fig. 10.50*b*

(*c*)   The centroidal coordinates are found as

$$x_c' = \frac{x_1 l_1 + x_2 l_2}{l_1 + l_2} = \frac{0(75) + \frac{150}{2}(150)}{75 + 150} = 50 \text{ mm} \qquad l = 225 \text{ mm}$$

$$y_c' = \frac{y_1 l_1 + y_2 l_2}{l_1 + l_2} = \frac{\frac{75}{2}(75) + 75(150)}{225} = 62.5 \text{ mm}$$

Using the parallel-axis theorem,

$$I_{0x} = I_x - l y_c'^2 = 9.84 \times 10^5 - 225(62.5)^2 = 1.05 \times 10^5 \text{ mm}^3$$
$$I_{0y} = I_y - l x_c'^2 = 1.13 \times 10^6 - 225(50)^2 = 5.68 \times 10^5 \text{ mm}^3$$

The polar moment of inertia about the centroid of the curve is

$$J_0 = I_{0x} + I_{0y} = 6.73 \times 10^5 \text{ mm}^3$$

(*d*)   The radius of gyration about the centroid of the curve is

$$k_{0p} = \sqrt{\frac{J_0}{l}} = \sqrt{\frac{6.73 \times 10^5}{225}} \qquad k_{0p} = 54.7 \text{ mm}$$

10.51   Find the polar moment of inertia of the weld pattern in Fig. 10.51 about the centroid of this weld pattern.

❚ The elementary curves are designated 1 for the circumference of the 4-in-diameter circle and 2 for the circumference of the 7.5-in-diameter circle. Using Case 3 in Table 10.7, with $\theta = 2\pi$, the centroidal

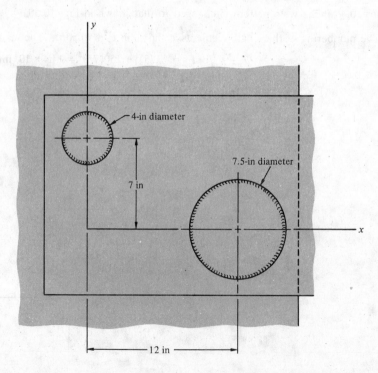

Fig. 10.51

moments of inertia of the circles have the form

$$I_{0x} = I_{0y} = \frac{a^3}{4}(2\theta - \sin 2\theta) = \frac{a^3}{4}[2(2\pi) - \sin 4\pi] = \pi a^3$$

The moments of inertia I_x and I_y of the weld pattern are then

$$I_x = I_{x1} + I_{x2} = [\pi(2)^3 + \pi(4)(7)^2] + \pi\left(\frac{7.5}{2}\right)^3 = 807 \text{ in}^3$$

$$I_y = I_{y1} + I_{y2} = \pi(2)^3 + \left[\pi\left(\frac{7.5}{2}\right)^3 + \pi(7.5)(12)^2\right] = 3,580 \text{ in}^3$$

From Prob. 9.43,

$$x'_c = 7.83 \text{ in} \qquad y'_c = 2.43 \text{ in} \qquad l = 36.1 \text{ in}$$

Using the parallel-axis theorem, in the form

$$I_{0x} = I_x - ly'^2_c$$

the moments of inertia I_{0x} and I_{0y} are found as

$$I_{0x} = 807 - 36.1(2.43)^2 = 594 \text{ in}^4$$
$$I_{0y} = I_y - lx'^2_c = 3,580 - 36.1(7.83)^2 = 1,370 \text{ in}^4$$

As the final step in the solution,

$$J_0 = I_{0x} + I_{0y} = 1,960 \text{ in}^3$$

10.52 (*a*) Show how the area in Fig. 10.1 may be used to develop a concept known as the *product of inertia* of a plane area.

(*b*) State the formal definition of the product of inertia of a plane area.

(*c*) What information does the product of inertia of a plane area provide about the area?

▮ (*a*) A property of the combined system, consisting of the area and the position of this area with respect to the coordinate axes, is called the product of inertia. For the area, and axes, shown in Fig. 10.1, this quantity is defined as

$$I_{xy} = \sum_i x_i y_i A_i$$

It may be seen from the above equation that the units of the area product of inertia are length raised to the fourth power. Unlike the moments of inertia of a plane area, the products of inertia may have both positive and negative values.

(*b*) As the subdivision of the area elements continues, a limiting condition is reached in which the summation operation in the equation in part (*a*) becomes an integral operation, with the form

$$I_{xy} = \lim_{A_i \to 0} \sum_i x_i y_i A_i = \int_A xy \, dA$$

The above equation is the formal definition of the product of inertia of a plane area. A direct solution of this equation is given in Prob. 10.69.

(*c*) Figure 10.52 shows a general plane area. The two equal area elements A_1 and A_2 are symmetrically located about the *y* axis. The contribution ΔI_{xy} of these two elements to the product of inertia I_{xy} is

$$\Delta I_{xy} = x_1 y_1 A_1 + x_2 y_2 A_2 = x_1 y_1 A_1 + (-x_1) y_2 A_2$$

Since $A_2 = A_1$ and $y_2 = y_1$, the value of the above expression is

$$\Delta I_{xy} = x_1 y_1 A_1 - x_1 y_1 A_1 \equiv 0$$

The very important conclusion may now be reached that the product of inertia I_{xy} of a plane area is *identically zero* if one of the axes *x* or *y* is an axis of symmetry of the area.

The reverse of the above statement is *not* true. That is, if the product of inertia of a plane area is zero, one of the *x* or *y* axes is not necessarily an axis of symmetry. The reader is urged to study Prob. 10.62, which provides a further discussion of this concept.

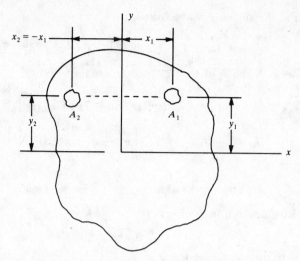

Fig. 10.52

10.53   Show that the product of inertia of the area shown in Fig. 10.53*a* is given by $I_{xy} = a^2b^2/8$.

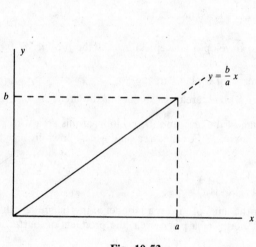

Fig. 10.53*a*

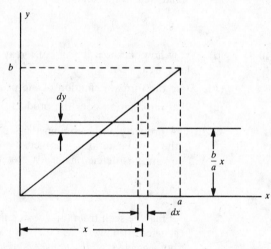

Fig. 10.53*b*

▌ Figure 10.53*b* shows the area element $dA = dx\,dy$.   The product of inertia of the plane area is given by

$$I_{xy} = \int_A xy\,dA = \int\int_A xy\,dx\,dy$$

The first integration is along the y direction in Fig. 10.53*b*, and the second integration is along the x direction.   The result is

$$I_{xy} = \int_0^a x\left[\int_0^{(b/a)x} y\,dy\right]dx = \int_0^a x\left[\frac{y^2}{2}\Big|_0^{(b/a)x}\right]dx = \int_0^a x\left[\frac{b^2}{2a^2}x^2\right]dx$$

$$= \frac{b^2}{2a^2}\int_0^a x^3\,dx = \frac{b^2}{2a^2}\left[\frac{x^4}{4}\Big|_0^a\right] = \frac{b^2}{2a^2}\frac{a^4}{4} = \frac{a^2b^2}{8}$$

10.54   The product of a plane area about its centroidal axes is known.   Show how the product of inertia of the area about a set of axes parallel to the centroidal axes may be found.

▌ From Fig. 10.10,

$$x = d_y + x_0$$
$$y = d_x + y_0$$

The product of inertia of the area about the xy axes is

$$I_{xy} = \int_A xy \, dA = \int_A (d_y + x_0)(d_x + y_0) \, dA = \int_A (d_x d_y + x_0 d_x + d_y y_0 + x_0 y_0) \, dA$$

$$= d_x d_y \int_A dA + d_x \int_A x_0 \, dA + d_y \int_A y_0 \, dA + \int_A x_0 y_0 \, dA$$

Since the $x_0 y_0$ axes pass through the centroid of the area,

$$x_{0c} = \frac{\int_A x_0 \, dA}{A} = 0 \qquad y_{0c} = \frac{\int_A y_0 \, dA}{A}$$

From the above two equations,

$$\int_A x_0 \, dA = 0 \qquad \int_A y_0 \, dA = 0$$

The equation for I_{xy} now has the form

$$I_{xy} = \int_A x_0 y_0 \, dA + d_x d_y \int_A dA$$

The first term on the right side of the above equation is the product of inertia of the area about the centroidal $x_0 y_0$ axes, given by

$$I_{0x0y} = \int_A x_0 y_0 \, dA$$

The final form for I_{xy} is

$$I_{xy} = I_{0x0y} + d_x d_y A$$

The above equation is referred to as the parallel-axis, or transfer, theorem, for products of inertia of a plane area.

The above result may be written in the alternative form

$$I_{xy} = I_{0x0y} + x_c y_c A$$

where x_c and y_c are the centroidal coordinates, measured in the xy coordinate system, of the area.

10.55 Find the product of inertia I_{xy} of the area shown in Fig. 10.55.

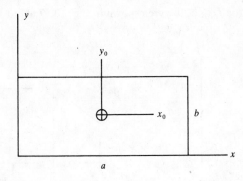

Fig. 10.55

▮ Using the parallel-axis theorem,

$$I_{xy} = I_{0x0y} + x_c y_c A$$

Since x_0 and y_0 are axes of symmetry of the area,

$$I_{0x0y} = 0$$

From the figure,

$$x_c = \frac{a}{2} \qquad y_c = \frac{b}{2}$$

The final form of I_{xy} is then

$$I_{xy} = x_c y_c A = \frac{a}{2}\left(\frac{b}{2}\right)ab = \frac{a^2 b^2}{4}.$$

10.56 The area in Prob. 10.55 is now positioned with respect to the xy axes as shown in Fig. 10.56.

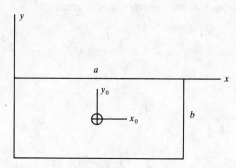

Fig. 10.56

(a) Find the product of inertia I_{xy}.

(b) Compare the result found in part (a) with the solution to Prob. 10.55.

▌ (a) From the symmetry of the area,

$$I_{0x0y} = 0$$

From the figure,

$$x_c = \frac{a}{2} \qquad y_c = -\frac{b}{2}$$

I_{xy} is then found as

$$I_{xy} = x_c y_c A = \frac{a}{2}\left(-\frac{b}{2}\right)ab = -\frac{a^2 b^2}{4}$$

(b) It may be seen from comparison of the results in Prob. 10.55 with those in part (a) above that the two products of inertia have the same magnitudes and opposite signs. This result illustrates that a product of inertia may have negative as well as positive values.

10.57 Find the product of inertia I_{0x0y} of the area shown in Fig. 10.57 about the centroidal axes x_0 and y_0 shown in the figure.

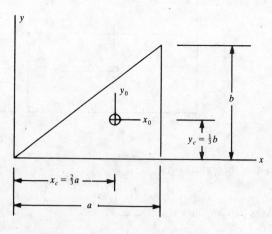

Fig. 10.57

▌ From Prob. 10.53,

$$I_{xy} = \frac{a^2 b^2}{8}$$

Using the parallel-axis theorem,

$$I_{xy} = I_{0x0y} + x_c y_c A \qquad I_{0x0y} = I_{xy} - x_c y_c A = \frac{a^2 b^2}{8} - \frac{2}{3} a (\frac{1}{3} b)(\frac{1}{2} ab) = \frac{a^2 b^2}{72}$$

10.58 Show how the product of inertia I_{xy} of the area in Prob. 10.53 may be found by using the parallel-axis theorem, together with a differential area element and a single integration.

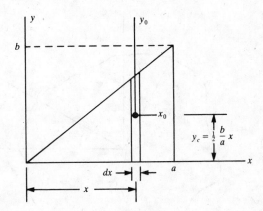

Fig. 10.58

I Figure 10.58 shows the differential area element, together with the centroidal axes x_0 and y_0 of this element. Using the parallel-axis theorem, the product of inertia of the differential area element with respect to the xy axes is

$$dI_{xy} = dI_{0x0y} + x_c y_c \, dA$$

From symmetry,

$$dI_{0x0y} = 0$$

and

$$x_c = x \qquad y_c = \frac{1}{2} \frac{b}{a} x \qquad dA = y \, dx = \frac{b}{a} x \, dx$$

The contribution of the area element to I_{xy} now has the form

$$dI_{xy} = x \left(\frac{1}{2} \frac{b}{a} x \right) \frac{b}{a} x \, dx = \frac{b^2}{2a^2} x^3 \, dx$$

The above equation is integrated, with the result

$$I_{xy} = \int dI_{xy} = \frac{b^2}{2a^2} \int_0^a x^3 \, dx = \frac{b^2}{2a^2} \left. \frac{x^4}{4} \right|_0^a = \frac{a^2 b^2}{8}$$

10.59 Find the product of inertia of the area shown in Prob. 10.15.

I Using Fig. 10.15b,

$$I_{xy} = (I_{0x0y,1} + x_1 y_1 A_1) + (I_{0x0y,2} + x_2 y_2 A_2)$$

From the symmetrical shapes of areas 1 and 2,

$$I_{0x0y,1} = I_{0x0y,2} = 0$$

$$I_{xy} = x_1 y_1 A_1 + x_2 y_2 A_2 = \frac{25}{2} (\frac{125}{2}) 25 (125) + (25 + \frac{65}{2}) \frac{25}{2} (65) 25 = 3.61 \times 10^6 \text{ mm}^4$$

10.60 Find I_{xy} for the pattern of hole areas in Prob. 10.30.

I From Fig. 10.30b,

$$I_{xy} = (I_{0x0y,1} + x_1 y_1 A_1) + (I_{0x0y,2} + x_2 y_2 A_2) + (I_{0x0y,3} + x_3 y_3 A_3)$$

The three centroidal products of inertia in the above equation are zero because of the symmetry of these areas with respect to their centroidal axes, and

$$A_1 = A_2 = A_3 = 0.601 \text{ in}^2$$

With the above results, and the dimensions from Fig. 10.30b, I_{xy} has the form

$$I_{xy} = [0(0)(0.601)] + [8(0)0.601] + [3.5(3)0.601] = 3.5(3)0.601 = 6.31 \text{ in}^4$$

10.61 Figure 10.61a shows an area located with respect to the xy axes. The quantities I_x, I_y, and I_{xy} are known. A second set of axes, $x'y'$, at direction θ with respect to xy, are placed on the area. Show how the terms $I_{x'}$, $I_{y'}$, and $I_{x'y'}$ may be expressed in terms of I_x, I_y, and I_{xy}.

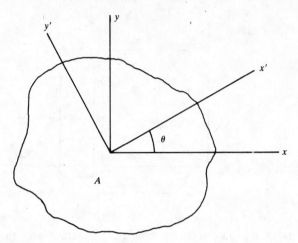

Fig. 10.61a

▌ The area is redrawn in Fig. 10.61b. The transformation equations between the two coordinate systems have the forms

$$x' = x \cos \theta + y \sin \theta \qquad y' = y \cos \theta - x \sin \theta$$

The moment of inertia $I_{x'}$ is defined by

$$I_{x'} = \int_A y'^2 \, dA = \int_A (y \cos \theta - x \sin \theta)^2 \, dA = \cos^2 \theta \int_A y^2 \, dA - 2 \sin \theta \cos \theta \int_A xy \, dA + \sin^2 \theta \int_A x^2 \, dA$$

Using the definitions

$$I_x = \int_A y^2 \, dA \qquad I_y = \int_A x^2 \, dA \qquad I_{xy} = \int_A xy \, dA \qquad \sin 2\theta = 2 \sin \theta \cos \theta$$

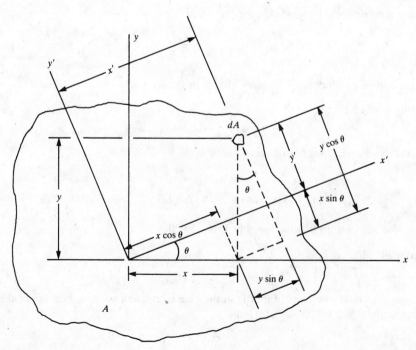

Fig. 10.61b

the final form of $I_{x'}$ is

$$I_{x'} = I_x \cos^2 \theta + I_y \sin^2 \theta - I_{xy} \sin 2\theta \qquad (1)$$

Similarly, the moment of inertia $I_{y'}$ is found to have the form

$$I_{y'} = I_x \sin^2 \theta + I_y \cos^2 \theta + I_{xy} \sin 2\theta \qquad (2)$$

The product of inertia of the area with respect to the $x'y'$ axes is

$$I_{x'y'} = \int_A x'y' \, dA = \int_A (x \cos \theta + y \sin \theta)(y \cos \theta - x \sin \theta) \, dA$$

$$= \cos^2 \theta \int_A xy \, dA + \sin \theta \cos \theta \int_A (y^2 - x^2) \, dA - \sin^2 \theta \int_A xy \, dA$$

Using the identities

$$\sin 2\theta = 2 \sin \theta \cos \theta \qquad \cos^2 \theta = \tfrac{1}{2}(1 + \cos 2\theta) \qquad \sin^2 \theta = \tfrac{1}{2}(1 - \cos 2\theta)$$

the final form for $I_{x'y'}$ is

$$I_{x'y'} = \frac{I_x - I_y}{2} \sin 2\theta + I_{xy} \cos 2\theta \qquad (3)$$

Equations (1) through (3) are the required transformation equations to find the moments and products of inertia with respect to the $x'y'$ axes when the values of these quantities with respect to the x and y axes are known.

10.62 (a) What is meant by the term *principal axes*?

(b) For what angle θ will the moments of inertia $I_{x'}$ and $I_{y'}$ in Prob. 10.61 have extreme values?

▌ (a) The principal axes, by definition, are those axes x' and y' with respect to which the product of inertia $I_{x'y'}$ is identically zero. The direction of these axes is found from the condition $I_{x'y'} = 0$. Using Eq. (3) in Prob. 10.61,

$$I_{x'y'} = \frac{I_x - I_y}{2} \sin 2\theta + I_{xy} \cos 2\theta = 0 \qquad \tan 2\theta = \frac{2I_{xy}}{I_y - I_x} \qquad (1)$$

It was shown in Prob. 10.52 that if the axes placed on an area are axes of symmetry, the product of inertia of the area with respect to these axes is identically zero. If, in a given problem, the product of inertia is zero, it does not necessarily mean that the area has an axis of symmetry. The result that I_{xy} is zero would simply mean that Eq. (1) above is satisfied. This effect is illustrated in Prob. 10.63.

(b) The extreme values of $I_{x'}$ and $I_{y'}$ in Prob. 10.61 may be found from the conditions

$$\frac{dI_{x'}}{d\theta} = 0 \qquad \frac{dI_{y'}}{d\theta} = 0$$

When these operations are carried out, it is found that the result is Eq. (1) in part (a) of this problem. Thus, *the maximum or minimum values of the moments of inertia of a plane area will be with respect to axes about which the product of inertia is zero.*

10.63 (a) Find the direction of the principal axes for the area in Prob. 10.15.

(b) Find the extreme values of the moments of inertia of this area.

▌ (a) The direction of the principal axes is found from

$$\tan 2\theta = \frac{2I_{xy}}{I_y - I_x}$$

From Prob. 10.15,

$$I_x = 1.66 \times 10^7 \text{ mm}^4 \qquad I_y = 6.60 \times 10^6 \text{ mm}^4$$

From Prob. 10.59,

$$I_{xy} = 3.61 \times 10^6 \text{ mm}^4$$

The above results are combined, to obtain

$$\tan 2\theta = \frac{2I_{xy}}{I_y - I_x} = \frac{2(3.61 \times 10^6)}{6.60 \times 10^6 - 1.66 \times 10^7} = -0.722 \qquad 2\theta = -35.8° \qquad \theta = -17.9°$$

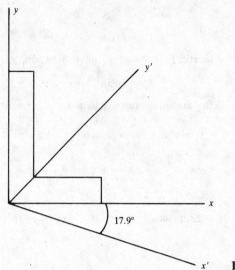

Fig. 10.63

The direction of the principal axes x' and y' are shown in Fig. 10.63.

(b) The moments of inertia of the area about the principal axes are

$$I_{x'} = I_x \cos^2 \theta + I_y \sin^2 \theta - I_{xy} \sin 2\theta = 1.66 \times 10^7 \cos^2(-17.9°) + 6.60 \times 10^6 \sin^2(-17.9°)$$
$$- 3.61 \times 10^6 \sin 2\,(-17.9°) = 1.78 \times 10^7 \text{ mm}^4$$
$$I_{y'} = I_x \sin^2 \theta + I_y \cos^2 \theta + I_{xy} \sin 2\theta = 1.66 \times 10^7 \sin^2(-17.9°) + 6.60 \times 10^6 \cos^2(-17.9°)$$
$$+ 3.61 \times 10^6 \sin 2\,(-17.9°) = 5.43 \times 10^6 \text{ mm}$$

It may be seen, as a necessary condition for the extreme values above, that

$$I_{x'} = 1.78 \times 10^7 \text{ mm}^4 > I_x = 1.66 \times 10^7 \text{ mm}^4 \quad \text{and} \quad I_{y'} = 5.43 \times 10^6 \text{ mm}^4 < I_y = 6.60 \times 10^6 \text{ mm}^4$$

10.64 (a) Find the maximum and minimum moments of inertia, of the pattern of hole areas in Prob. 10.30, about axes which lie in the plane of the areas and pass through the origin of the xy coordinates.

(b) Show that the product of inertia of the pattern of hole areas with respect to the axes in part (a) is zero.

▌ (a) From Prob. 10.30,

$$I_x = 5.50 \text{ in}^4 \qquad I_y = 45.9 \text{ in}^4 \qquad A = 0.601 \text{ in}^2$$

From Prob. 10.60,

$$I_{xy} = 6.31 \text{ in}^4$$

From the solution in Prob. 10.62, the moments of inertia will have their extreme values about axes x' and y' whose direction θ with respect to the x axis is found from

$$\tan 2\theta = \frac{2I_{xy}}{I_y - I_x} = \frac{2(6.31)}{45.9 - 5.50} \qquad \theta = 8.67°$$

The maximum and minimum values of the moments of inertia are

$$I_{x'} = I_x \cos^2 \theta + I_y \sin^2 \theta - I_{xy} \sin 2\theta = 5.50 \cos^2 8.67° + 45.9 \sin^2 8.67° - 6.31 \sin 2\,(8.67°) = 4.54 \text{ in}^4$$
$$I_{y'} = I_x \sin^2 \theta + I_y \cos^2 \theta + I_{xy} \sin 2\theta = 5.5 \sin^2 8.67° + 45.9 \cos^2 8.67° + 6.31 \sin 2\,(8.67°) = 46.9 \text{ in}^4$$

(b) Figure 10.64 shows the pattern of hole areas and the $x'y'$ axes. From this figure,

$$A = \frac{3.5}{\cos 8.67°} = 3.54 \text{ in} \qquad B = 3.5 \tan 8.67° = 0.534 \text{ in}$$
$$C = 3 - B = 3 - 0.534 = 2.47 \text{ in} \qquad D = C \sin 8.67° = 2.47 \sin 8.67° = 0.372 \text{ in}$$
$$E = C \cos 8.67° = 2.47 \cos 8.67° = 2.44 \text{ in}$$

The centroidal coordinates of the hole areas, measured in the $x'y'$ coordinate system, are

$$x_1' = 0 \qquad y_1' = c \qquad x_2' = 8 \cos 8.67° = 7.91 \text{ in} \qquad y_2' = -8 \sin 8.67° = -1.21 \text{ in}$$
$$x_3' = A + D = 3.54 + 0.372 = 3.91 \text{ in} \qquad y_3' = E = 2.44 \text{ in}$$

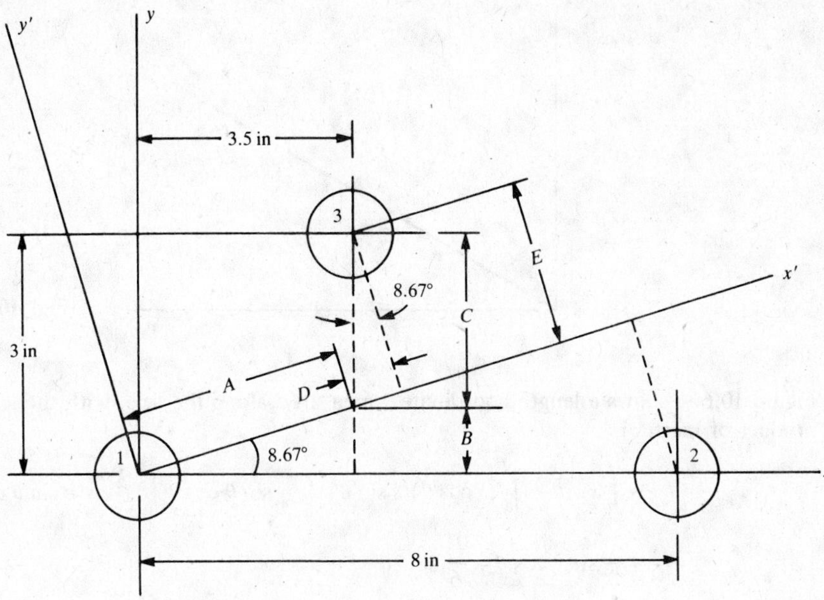

Fig. 10.64

The product of inertia about the $x'y'$ axes is

$$I_{x'y'} = x_1'y_1'A_1 + x_2'y_2'A_2 + x_3'y_3'A_3$$

Using $A_1 = A_2 = A_3 = 0.601$ in^2,

$$I_{x'y'} = 7.91(-1.21)0.601 + 3.91(2.44)0.601 \overset{?}{=} 0 \qquad 7.91(1.21)0.601 \overset{?}{=} 3.91(2.44)0.601 \qquad 5.75 \approx 5.73$$

10.65 (a) What is the definition of the product of inertia of a plane curve?

(b) State the parallel-axis, or transfer, theorem for the product of inertia of a plane curve.

▌ (a) A property of the *combined* system consisting of the plane curve and the position of the curve with respect to the coordinate axes is called the product of inertia. For the curve shown in Fig. 10.45, this quantity is defined as

$$I_{xy} = \sum_i x_i y_i l_i$$

The units of the product of inertia of a plane curve are length raised to the third power, and this quantity may have both positive and negative values. The integral form of the above equation is

$$I_{xy} = \int_l xy \, dl$$

(b) Using Fig. 10.47, and following the development in Prob. 10.54 for the case of a plane area, it can be shown that the parallel-axis, or transfer, theorem for the product of inertia of a plane curve has the form

$$I_{xy} = I_{0x0y} + d_x d_y l$$

In the above equation, I_{0x0y} is the product of inertia about the centroidal $x_0 y_0$ axes, I_{xy} is the product of inertia about the axes xy that are parallel to the centroidal axes, l is the length of the curve, d_x is the spacing between the x and x_0 axes, and d_y is the spacing between the y and y_0 axes. An alternative form of the above equation is

$$I_{xy} = I_{0x0y} + x_c'y_c'l$$

where x_c' and y_c' are the centroidal coordinates, measured in the xy coordinate system, of the curve.

10.66 (a) Show that the product of inertia I_{xy} of the line in Fig. 10.66a is $(l^3/6) \sin 2\theta$.

(b) For what angles θ does I_{xy} have extreme values?

(c) Find the product of inertia I_{0x0y} about the centroidal axes that are parallel to the xy axes.

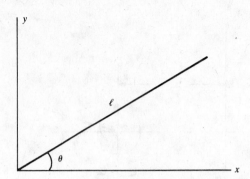

Fig. 10.66a

(a) Figure 10.66b shows a length coordinate s measured along the line, with the length element ds. The product of inertia is

$$I_{xy} = \int_l xy\, dl = \int_0^l (s\cos\theta)(s\sin\theta)\, ds = \sin\theta\cos\theta \int_0^l s^2\, ds = \sin\theta\cos\theta \left.\frac{s^3}{3}\right|_0^l$$

$$= \frac{l^3}{3}\sin\theta\cos\theta = \frac{l^3}{6}\sin 2\theta$$

(b) I_{xy} will have a minimum value $I_{xy,min} = 0$ when

$$\sin 2\theta = 0 \qquad \theta = 0°, 90°$$

The maximum value of I_{xy} will occur when

$$\sin 2\theta = 1 \qquad 2\theta = 90° \qquad \theta = 45° \qquad \text{and} \qquad I_{xy,max} = \frac{l^3}{6}$$

(c) Figure 10.66c shows the position of the centroidal axes x_0y_0 on the line. The parallel-axis equation is

$$I_{xy} = I_{0x0y} + d_x d_y l$$

$$I_{0x0y} = I_{xy} - d_x d_y l = \frac{l^3}{6}\sin 2\theta - \left(\frac{l}{2}\sin\theta\right)\left(\frac{l}{2}\cos\theta\right)l = \frac{l^3}{6}\sin 2\theta - \frac{l^3}{8}\sin 2\theta = \frac{l^3}{24}\sin 2\theta$$

The maximum and minimum values of I_{0x0y} occur at the values of angle θ found in part (b).

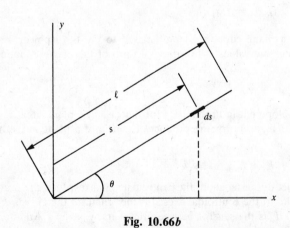

Fig. 10.66b

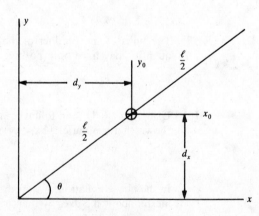

Fig. 10.66c

10.67 **(a)** Find the product of inertia I_{xy} of the weld pattern in Prob. 10.51.

(b) Find the product of inertia I_{0x0y} about the centroidal axes of the weld pattern. The x_0y_0 axes are parallel to the xy axes.

(a) Using the numbering system for the elementary shapes given in Prob. 10.51,

$$I_{xy} = I_{xy,1} + I_{xy,2} = (I_{0x0y,1} + x_1 y_1 l_1) + (I_{0x0y,2} + x_2 y_2 l_2)$$

From the symmetry of a circle,

$$I_{0x0y,1} = I_{0x0y,2} = 0$$

From Fig. 10.51a,

$$x_1 = 0 \qquad y_1 = 7 \text{ in} \qquad x_2 = 12 \text{ in} \qquad y_2 = 0$$
$$I_{xy} = x_1 y_1 l_1 + x_2 y_2 l_2 = 0(7)\pi(4) + 12(0)\pi(7.5) = 0$$

(b)
$$I_{xy} = I_{0x0y} + x_c' y_c' l$$

From Prob. 9.43 or Prob. 10.51,

$$x_c' = 7.83 \text{ in} \qquad y_c' = 2.43 \text{ in} \qquad l = 36.1 \text{ in}$$
$$I_{0x0y} = I_{xy} - x_c' y_c' l = 0 - 7.83(2.43)36.1 = -687 \text{ in}^3$$

10.5 SOLUTIONS USING THE INTEGRAL DEFINITIONS OF MOMENTS AND PRODUCTS OF INERTIA OF AREAS

10.68 The plane area in Fig. 10.68a is bounded by the curves $y = x/4$ and $x = 4y^2$. Find the moments of inertia of this area about the x and y axes, where x and y are in millimeters.

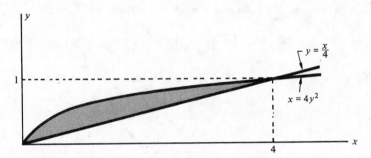

Fig. 10.68a

▮ The area element for the computation of I_x is shown in Fig. 10.68b.

$$dA = (4y - 4y^2)\, dy$$

$$I_x = \int_A y^2\, dA = \int_0^1 y^2(4y - 4y^2)\, dy = \int_0^1 (4y^3 - 4y^4)\, dy = \left[4\frac{y^4}{4} - 4\frac{y^5}{5}\right]_0^1 = 1 - \tfrac{4}{5} = \tfrac{1}{5} = 0.2 \text{ mm}^4$$

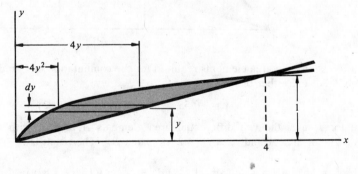

Fig. 10.68b

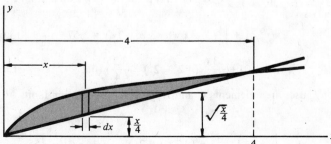

Fig. 10.68c

To compute I_y, the area element in Fig. 10.68c is used.

$$dA = \left(\sqrt{\frac{x}{4}} - \frac{x}{4}\right) dx$$

$$I_y = \int_A x^2\, dA = \int_0^4 x^2\left(\frac{x^{1/2}}{2} - \frac{x}{4}\right) dx = \int_0^4 \left(\frac{x^{5/2}}{2} - \frac{x^3}{4}\right) dx = \left[\frac{2}{7}\frac{x^{7/2}}{2} - \frac{x^4}{4(4)}\right]_0^4 = \frac{4^{7/2}}{7} - \frac{4^4}{4(4)} = 2.29 \text{ mm}^4$$

10.69 Find the product of inertia I_{xy} for the area in Prob. 10.68.

❚ The product of inertia is given by

$$I_{xy} = \int_A xy\, dA = \iint xy\, dx\, dy$$

The first integration is in the y direction along the area element in Fig. 10.68c. This operation appears as

$$I_{xy} = \int_0^4 x\left(\int_{x/4}^{\sqrt{x/4}} y\, dy\right) dx = \int_0^4 x\left(\frac{y^2}{2}\bigg|_{x/4}^{\sqrt{x/4}}\right) dx = \int_0^4 \frac{x}{2}\left[\left(\sqrt{\frac{x}{4}}\right)^2 - \left(\frac{x}{4}\right)^2\right] dx$$

$$= \int_0^4 \frac{x}{2}\left(\frac{x}{4} - \frac{x^2}{16}\right) dx = \frac{1}{32}\int_0^4 (4x^2 - x^3)\, dx$$

The above integral is now in a form to be integrated directly with respect to x.

$$I_{xy} = \frac{1}{32}\left[4\frac{x^3}{3} - \frac{x^4}{4}\right]_0^4 = \frac{1}{32}\left[\frac{4}{3}(4)^3 - \frac{4^4}{4}\right] = 0.667 \text{ mm}^4$$

10.70 Find I_x and I_y for the area shown in Fig. 10.70a. x and y are in millimeters.

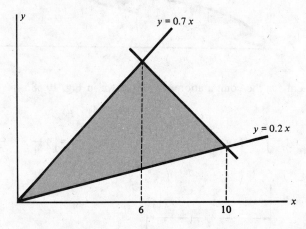

Fig. 10.70a

❚ Figure 10.70b shows the area elements required for the computation of I_x and I_y. The equation of line ab, found in Prob. 9.60, has either of the forms

$$y = -0.55x + 7.5 \qquad x = 13.6 - 1.82y$$

The solution for I_x uses the two differential area elements A and B, with the corresponding integration limits $0 \le y \le 2$ for area A and $2 \le y \le 4.2$ for area B. The result is

$$I_x = \int_A y^2\, dA = \int_0^2 y^2(5y - 1.43y)\, dy + \int_2^{4.2} y^2[(13.6 - 1.82y) - 1.43y]\, dy$$

$$= \int_0^2 3.57y^3\, dy + \int_2^{4.2}(13.6y^2 - 3.25y^3)\, dy = 3.57\frac{y^4}{4}\bigg|_0^2 + \left[13.6\frac{y^3}{3} - 3.25\frac{y^4}{4}\right]\bigg|_2^{4.2}$$

$$= \frac{3.57}{4}(2^4 - 0) + \frac{13.6}{3}(4.2^3 - 2^3) - \frac{3.25}{4}(4.2^4 - 2^4) = 74.1 \text{ mm}^4$$

The solution for I_y uses area elements C and D, with the integration limits $0 \le x \le 6$ and $6 \le x \le 10$. The result for I_y is

$$I_y = \int_A x^2\, dA = \int_0^6 x^2(0.7x - 0.2x)\, dx + \int_6^{10} x^2[(-0.55x + 7.5) - 0.2x]\, dx$$

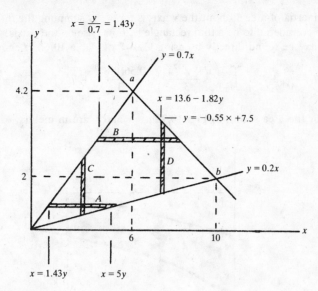

Fig. 10.70*b*

$$= 0.5 \left.\frac{x^4}{4}\right|_0^6 + \left(-0.75\,\frac{x^4}{4} + 7.5\,\frac{x^3}{3}\right)\Big|_6^{10}$$

$$= \frac{0.5}{4}(6^4 - 0) + \left[-\frac{0.75}{4}(10^4 - 6^4) + \frac{7.5}{3}(10^3 - 6^3)\right] = 490 \text{ mm}^4$$

10.71 Find the moments of inertia of the area in Fig. 10.71*a* about the *x* and *y* axes. *x* and *y* are in feet.

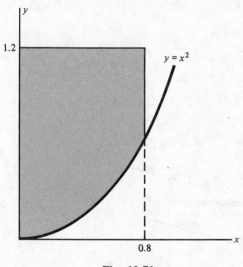

Fig. 10.71*a*

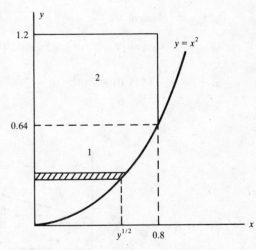

Fig. 10.71*b*

❚ The area is divided into the two areas shown in Fig. 10.71*b*. The contributions of area 1 to the moments of inertia will be found by using the integral definitions of these quantities. The moment of inertia about the *x* axis has the form

$$I_x = I_{x1} + I_{x2}$$

$$I_{x1} = \int_0^{0.64} y^2(y^{1/2}\,dy) = \int_0^{0.64} y^{5/2}\,dy = \tfrac{2}{7}y^{7/2}\big|_0^{0.64} = \tfrac{2}{7}(0.64^{7/2} - 0) = 0.0599 \text{ ft}^4$$

Using the parallel-axis theorem for area 2,

$$I_{x2} = \frac{0.8(0.56)^3}{12} + 0.8(0.56)\left(0.64 + \frac{0.56}{2}\right)^2 = 0.391 \text{ ft}^4 \qquad I_x = I_{x1} + I_{x2} = 0.451 \text{ ft}^4$$

The moment of inertia of area 1 about the y axis is found by summing the differential areas. Each of the area elements may be considered to be a thin rectangle with its base on the y axis. The contribution to I_y of each of these elements may be found directly by using Case 2 of Table 10.6. I_y now has the form

$$I_y = I_{y1} + I_{y2} = \int_0^{0.64} \frac{dy(y^{1/2})^3}{3} + \frac{0.56(0.8)^3}{3} + \frac{1}{3}(\frac{2}{5})y^{5/2}|_0^{0.64} + 0.0956 = 0.139 \text{ ft}^4$$

10.72 Find I_x and I_y for the area shown in Fig. 10.72a. x and y are in meters.

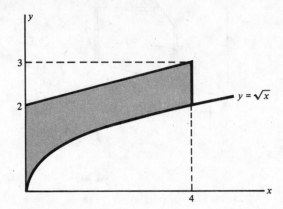

Fig. 10.72a

▮ The area elements required for the computation of x and y are seen in Fig. 10.72b. The equation of the straight line in the figure, from Prob. 9.62, is $y = 0.25x + 2$. Using area elements A and B,

$$I_x = \int_A y^2\, dA = \int_0^2 y^2(y^2\, dy) + \int_2^3 y^2[4 - (4y - 8)]\, dy = \int_0^2 y^4\, dy + \int_2^3 (12y^2 - 4y^3)\, dy$$

$$= \frac{y^5}{5}\Big|_0^2 + \left(12\frac{y^3}{3} - 4\frac{y^4}{4}\right)\Big|_2^3 = \frac{1}{5}(2^5 - 0) + [\frac{12}{3}(3)^3 - \frac{4}{4}(3)^4] - [\frac{12}{3}(2)^3 - \frac{4}{4}(2)^4] = 17.4 \text{ m}^4$$

Using area element C,

$$I_y = \int x^2\, dA = \int_0^4 x^2[(0.25x + 2) - x^{1/2}]\, dx = \int_0^4 (0.25x^3 + 2x^2 - x^{5/2})\, dx = \left(0.25\frac{x^4}{4} + 2\frac{x^3}{3} - \frac{2}{7}x^{7/2}\right)\Big|_0^4$$

$$= \frac{0.25}{4}(4^4 - 0) + \frac{2}{3}(4^3 - 0) - \frac{2}{7}(4^{7/2} - 0) = 22.1 \text{ m}^4$$

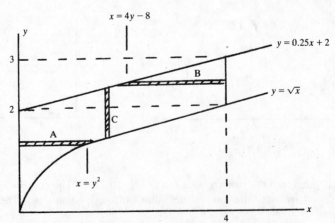

Fig. 10.72b

10.73 Find I_x and I_y for the area shown in Fig. 10.73a. x and y are in inches.

▮ Figure 10.73b shows the definitions of the differential areas. For the computation of I_x, element 1 is $gcdf$ and element 2 is $abcg$.

$$I_x = I_{x1} + I_{x2} = \frac{2(1)^3}{3} + \int_1^3 y^2\left(\frac{2}{y}\, dy\right) = \frac{2}{3} + \int_1^3 2y\, dy = \frac{2}{3} + 2\left(\frac{y^2}{2}\right)\Big|_1^3 = \frac{2}{3} + (3^2 - 1^2) = 8.67 \text{ in}^4$$

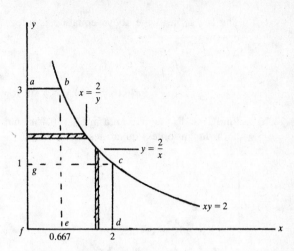

Fig. 10.73a   Fig. 10.73b

For the I_y computation, element 1 is *abef* and element 2 is *bcde*.

$$I_y = I_{y1} + I_{y2} = \frac{3(0.667)^3}{3} + \int_{0.667}^{2} x^2 \left(\frac{2}{x} \, dx \right) = 0.2967 + 2 \left(\frac{x^2}{2} \right) \Big|_{0.667}^{2} = 0.2967 + (2^2 - 0.667^2) = 3.85 \text{ in}^4$$

10.74 Find the moments of inertia I_x and I_y for the area under the half sine curve shown in Fig. 10.74a. For this problem, x and y are assumed to be dimensionless.

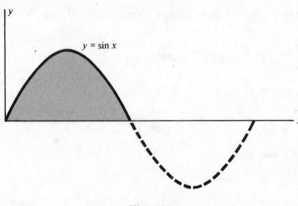

Fig. 10.74a

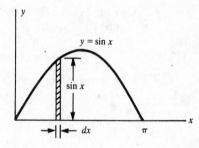

Fig. 10.74b

▌ Figure 10.74b shows the single differential area element which can be used to find both I_x and I_y. Using Case 2 of Table 10.6 for the moment of inertia of the differential element about its base on the x axis,

$$I_x = \int_0^\pi \frac{dx(\sin x)^3}{3} = \frac{1}{3} \int_0^\pi \sin^3 x \, dx = \frac{1}{3} \int_0^\pi \sin x (1 - \cos^2 x) \, dx = \frac{1}{3} \left[\int_0^\pi \sin x \, dx - \int_0^\pi \cos^2 x (\sin x \, dx) \right]$$

The last integral in the above equation is in the form $\int u^n \, du$, where $u = \cos x$, $n = 2$, and $du = -\sin x \, dx$. Using this result, I_x is

$$I_x = \frac{1}{3} \left[(-\cos x) - \left(-\frac{\cos^3 x}{3} \right) \right] \Big|_0^\pi = \frac{1}{3} \left[-(\cos \pi - \cos 0) + \frac{1}{3} (\cos^3 \pi - \cos^3 0) \right] = 0.444$$

The form for I_y is

$$I_y = \int x^2 \, dA = \int_0^\pi x^2 (\sin x \, dx)$$

The above equation is integrated by parts, with $u = x^2$, $du = 2x \, dx$, $dv = \sin x \, dx$, $v = -\cos x$, and $\int u \, dv = uv - \int v \, du$.

$$I_y = -x^2 \cos x - \int_0^\pi (-\cos x) 2x \, dx = -x^2 \cos x + 2 \int_0^\pi x \cos x \, dx$$

The last term in the above equation is integrated by parts, with $u = x$, $du = dx$, $dv = \cos x\, dx$, and $v = \sin x$.

$$I_y = -x^2 \cos x\big|_0^{\pi} + 2\left(x \sin x\big|_0^{\pi} - \int_0^{\pi} \sin x\, dx\right) = [-x^2 \cos x + 2(x \sin x + \cos x)]\big|_0^{\pi}$$

$$= -(\pi^2 \cos \pi - 0) + 2(\pi \sin \pi + \cos \pi) - 2(0 + \cos 0) = -\pi^2(-1) + 2(-1) - 2 = 5.87$$

10.75   Find I_x and I_y for the area in Fig. 10.75a that is bounded by the sine curve and the straight line. x and y are both dimensionless quantities.

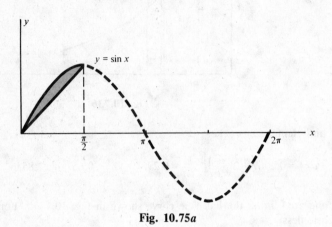

Fig. 10.75a   **Fig. 10.75b**

▎ Area 1 in Fig. 10.75b is the area under the quarter sine curve.   Area 2 is a triangle.   Using the result for I_x from Prob. 10.74, with $\pi \rightarrow \pi/2$,

$$I_{x1} = \frac{1}{3}\left[(-\cos x) - \left(-\frac{\cos 3x}{3}\right)\right]\Bigg|_0^{\pi/2} = \frac{1}{3}\left[-\left(\cos \frac{\pi}{2} - \cos 0\right) + \frac{1}{3}\left(\cos^3 \frac{\pi}{2} - \cos^3 0\right)\right]$$

$$= \frac{1}{3}\left[-(0-1) + \frac{1}{3}(0-1)\right] = 0.222$$

$$I_x = I_{x1} - I_{x2} = 0.222 - \frac{(\pi/2)(1)^3}{12} = 0.0911$$

Using the result for I_y from Prob. 10.74, with $\pi \rightarrow \pi/2$,

$$I_{y1} = [-x^2 \cos x + 2(x \sin x + \cos x)]\big|_0^{\pi/2}$$

$$= -\left[\left(\frac{\pi}{2}\right)^2 \cos \frac{\pi}{2} - 0\right] + 2\left(\frac{\pi}{2} \sin \frac{\pi}{2} + \cos \frac{\pi}{2}\right) - 2(0 \sin 0 + \cos 0) = 1.14 \qquad I_y = I_{y1} - I_{y2}$$

Using the parallel-axis theorem for I_{y2},

$$I_y = 1.14 - \left[\frac{1(\pi/2)^3}{36} + \frac{1}{2}\left(\frac{\pi}{2}\right)(1)\left(\frac{2}{3}\left[\frac{\pi}{2}\right]\right)^2\right] = 0.171$$

10.76   Give a summary of the basic concepts of moments and products of inertia of plane areas and curves.

▎ If a plane area is positioned with respect to a set of coordinate axes, quantities referred to as area moments of inertia may be defined as

$$I_x = \int_A y^2\, dA \qquad I_y = \int_A x^2\, dA$$

A complete definition of a moment of inertia must include a description of the position of the area with respect to the coordinate axes.

The polar moment of inertia J of a plane area, about a reference axis which is perpendicular to the plane of the area and which passes through the origin of the xy coordinate axes, is given by

$$J = \int_A r^2\, dA$$

Using
$$r^2 = x^2 + y^2$$

we get
$$J = I_x + I_y$$

Moments of inertia are inherently positive terms. Because of the squaring effect, area elements which are more distant from the reference axis have a greater effect on the magnitude of the moment of inertia than do area elements which are closer to this axis. The moment of inertia is also referred to as the second moment of area. The units of area moment of inertia are length raised to the fourth power.

The radii of gyration are quantities which are defined in terms of an area, and the moment of inertia of this area with respect to a reference axis, with the forms

$$k_x = \sqrt{\frac{I_x}{A}} \qquad k_y = \sqrt{\frac{I_y}{A}} \qquad k_p = \sqrt{\frac{J}{A}}$$

If the moment of inertia of a plane area about its centroidal axis is known, the moment of inertia about an axis which is parallel to this centroidal axis may be found by using the parallel-axis, or transfer, theorem. The relationship between these two moments of inertia has the form

$$I = I_0 + A\,d^2$$

I_0 is the moment of inertia of the area A about the centroidal axis. I is the moment of inertia about a parallel axis which is a distance d away from the centroidal axis.

The moment of inertia of a composite area may be found by dividing this area into elementary area shapes. The moments of inertia about the reference axis of these shapes are then summed to obtain the final value for the moment of inertia of the original area. The moments of inertia of holes or cutouts in the area are treated as negative quantities.

If a plane curve is positioned with respect to a set of coordinate axes, the moments of inertia of the curve are

$$I_x = \int_l y^2\,dl \qquad I_y = \int_l x^2\,dl$$

The polar moment of inertia of the curve is

$$J = \int_l r^2\,dl$$

Using
$$r^2 = x^2 + y^2$$

we get
$$J = I_x + I_y$$

The units of moment of inertia of a plane curve are length raised to the third power.

The moment of inertia of a composite plane curve may be found by subdividing this curve into elementary curve shapes. The moments of inertia of these shapes are then summed to obtain the final value for the moment of inertia of the original curve.

The product of inertia of a plane area that is positioned with respect to a set of coordinate axes is defined as

$$I_{xy} = \int_A xy\,dA$$

Products of inertia may have both positive and negative values, and the units of these quantities are length raised to the fourth power.

The parallel-axis theorem for products of inertia has the form

$$I_{xy} = I_{0x0y} + d_x\,d_y\,A$$

The centroidal x_0y_0 axes are parallel to the xy axes, and d_x and d_y are the separation distances between these sets of axes. I_{0x0y} and I_{xy} are the products of inertia with respect to the x_0y_0 and xy axes, respectively.

The axes placed with respect to an area may be rotated to new positions. The position of the coordinate axes for which $I_{xy} = 0$ are called principal axes. The moments of inertia of the area about the principal axes have extreme values.

The product of inertia of a plane curve is defined to be

$$I_{xy} = \int_l xy\,dl$$

The units of this quantity are length raised to the third power.

				TABLE 10.6. Moments of Inertia and Radii of Gyration of Several Elementary Plane Areas				
Case	Shape	I_{ox}	I_{oy}	J_o	I_x	I_y	k_{ox}	k_{oy}
1		$\dfrac{a^4}{12}$	$\dfrac{a^4}{12}$	$\dfrac{a^4}{6}$	$\dfrac{a^4}{3}$	$\dfrac{a^4}{3}$	$\dfrac{a}{\sqrt{12}}$	$\dfrac{a}{\sqrt{12}}$
2		$\dfrac{ab^3}{12}$	$\dfrac{ba^3}{12}$	$\dfrac{ab}{12}(a^2+b^2)$	$\dfrac{ab^3}{3}$	$\dfrac{ba^3}{3}$	$\dfrac{b}{\sqrt{12}}$	$\dfrac{a}{\sqrt{12}}$
3		$\dfrac{ab^3}{36}$	$\dfrac{ba^3}{36}$	$\dfrac{ab}{36}(a^2+b^2)$	$\dfrac{ab^3}{12}$	$\dfrac{ba^3}{4}$	$\dfrac{b}{\sqrt{18}}$	$\dfrac{a}{\sqrt{18}}$
4		$\dfrac{ab^3}{36}$	—	—	$\dfrac{ab^3}{12}$	—	$\dfrac{b}{\sqrt{18}}$	—
5		$\dfrac{\pi a^4}{4}=\dfrac{\pi d^4}{64}$	$\dfrac{\pi a^4}{4}=\dfrac{\pi d^4}{64}$	$\dfrac{\pi a^4}{2}=\dfrac{\pi d^4}{32}$	$\dfrac{5\pi a^4}{4}=\dfrac{5\pi d^4}{64}$	$\dfrac{5\pi a^4}{4}=\dfrac{5\pi d^4}{64}$	$\dfrac{a}{2}=\dfrac{d}{4}$	$\dfrac{a}{2}=\dfrac{d}{4}$
6		$a^4\left(\dfrac{\pi}{8}-\dfrac{8}{9\pi}\right)$	$\dfrac{\pi a^4}{8}$	$a^4\left(\dfrac{\pi}{4}-\dfrac{8}{9\pi}\right)$	$\dfrac{\pi a^4}{8}$	$\dfrac{5\pi a^4}{8}$	$a\sqrt{\dfrac{1}{4}-\dfrac{16}{9\pi^2}}$	$\dfrac{a}{2}$
7		$\dfrac{\pi ab^3}{64}$	$\dfrac{\pi ba^3}{64}$	$\dfrac{\pi ab}{64}(a^2+b^2)$	$\dfrac{5\pi ab^3}{64}$	$\dfrac{5\pi ba^3}{64}$	$\dfrac{b}{4}$	$\dfrac{a}{4}$

TABLE 10.7. Moments of Inertia of Several Elementary Plane Curves

Case	Shape	I_{0x}	I_{0y}	J_0	I_x	I_y
1	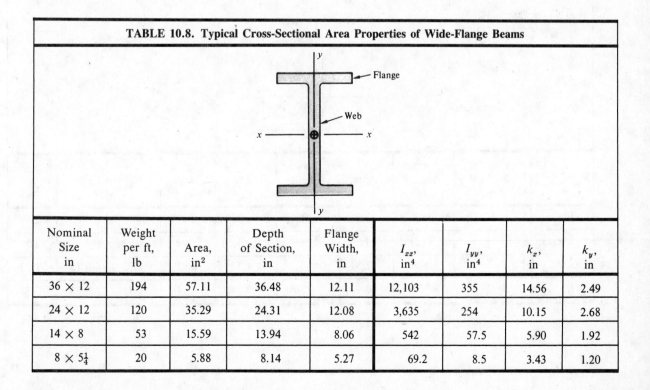	$\dfrac{l^3}{12}\sin^2\theta$	$\dfrac{l^3}{12}\cos^2\theta$	$\dfrac{l^3}{12}$	$\dfrac{l^3}{3}\sin^2\theta$	$\dfrac{l^3}{3}\cos^2\theta$
2		$\dfrac{a^3}{2\pi}(\pi^2-8)$	$\dfrac{\pi a^3}{2}$	$\dfrac{a^3}{\pi}(\pi^2-4)$	$\dfrac{\pi a^3}{2}$	$\dfrac{3\pi a^3}{2}$
3		$I_x = \dfrac{a^3}{4}(2\theta - \sin 2\theta)$ $I_y = \dfrac{a^3}{4}(2\theta + \sin 2\theta)$			$J = a^3\theta$	

TABLE 10.8. Typical Cross-Sectional Area Properties of Wide-Flange Beams

Nominal Size in	Weight per ft, lb	Area, in²	Depth of Section, in	Flange Width, in	I_{xx}, in⁴	I_{yy}, in⁴	k_x, in	k_y, in
36 × 12	194	57.11	36.48	12.11	12,103	355	14.56	2.49
24 × 12	120	35.29	24.31	12.08	3,635	254	10.15	2.68
14 × 8	53	15.59	13.94	8.06	542	57.5	5.90	1.92
8 × 5¼	20	5.88	8.14	5.27	69.2	8.5	3.43	1.20

TABLE 10.9. Typical Cross-Sectional Area Properties of American Standard Channels

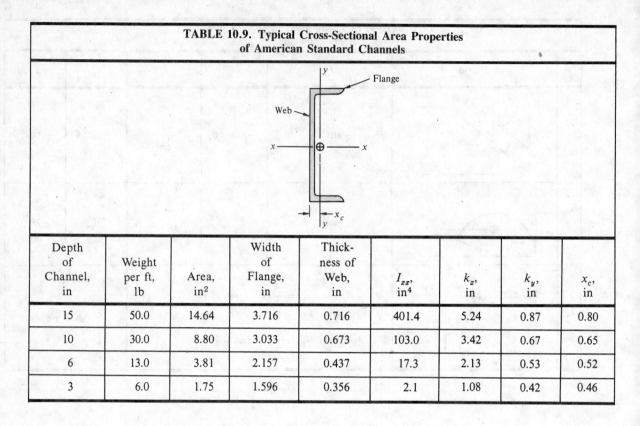

Depth of Channel, in	Weight per ft, lb	Area, in²	Width of Flange, in	Thickness of Web, in	I_{xx}, in⁴	k_x, in	k_y, in	x_c, in
15	50.0	14.64	3.716	0.716	401.4	5.24	0.87	0.80
10	30.0	8.80	3.033	0.673	103.0	3.42	0.67	0.65
6	13.0	3.81	2.157	0.437	17.3	2.13	0.53	0.52
3	6.0	1.75	1.596	0.356	2.1	1.08	0.42	0.46

TABLE 10.10. Typical Cross-Sectional Area Properties of Standard Angles with Unequal Legs

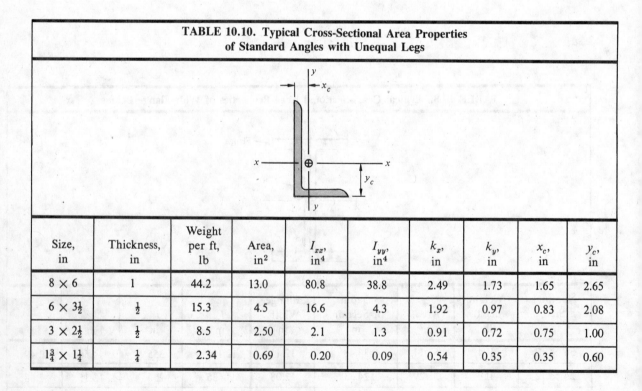

Size, in	Thickness, in	Weight per ft, lb	Area, in²	I_{xx}, in⁴	I_{yy}, in⁴	k_x, in	k_y, in	x_c, in	y_c, in
8 × 6	1	44.2	13.0	80.8	38.8	2.49	1.73	1.65	2.65
6 × 3½	½	15.3	4.5	16.6	4.3	1.92	0.97	0.83	2.08
3 × 2½	½	8.5	2.50	2.1	1.3	0.91	0.72	0.75	1.00
1¾ × 1¼	¼	2.34	0.69	0.20	0.09	0.54	0.35	0.35	0.60

CHAPTER 11
Distribution of Forces along Lengths and over Areas

11.1 FORCE DISTRIBUTION ALONG A LENGTH, UNIFORM, UNIFORMLY VARYING, AND GENERAL FORCE DISTRIBUTIONS

11.1 (*a*) What idealization of the physical problem is made when using a *concentrated force*?

(*b*) What is meant by a *distributed loading*?

(*c*) What is the significant difference between loadings that are represented by concentrated forces and loadings that are represented as distributed forces?

▮ (*a*) In all the problems involving forces which have been considered thus far in this text, the forces have been represented by arrows drawn on the body. A typical example of a physical loading on a beam is shown in Fig. 11.1*a*. The two weights are attached to the beam by thin cables. The idealization of this situation, in terms of the force loading on a beam, would be as shown in Fig. 11.1*b*. Although the cables are in contact with the beam over some measurable length, the forces are assumed to be acting at specific single locations on the beam. Thus, this representation of the loading does *not* take into account the width of the cables or the distribution of loading between the cables and the beam. The forces in this situation are referred to as concentrated forces.

(*b*) In this chapter another type of force application, referred to as distributed loading, is considered. For this case, the forces are distributed in some manner along a length of a member or over an area. This type of force loading would be representative of the weight of a structural member, of the effect of wind forces on a structure, or of the forces exerted by a liquid or gas on the walls of a pressure vessel. It would also represent the distribution, over a length or area, of the reaction force exerted by one body on another body.

In the following problems particular consideration will be given to three types of force distributions that occur very frequently. In the first case, the forces are *uniformly distributed* along a *length* of the body. This case is representative of the static weight of structural members. In the second case, the forces *vary uniformly* along the length of the body. This case will be identified subsequently as the force distribution, along the depth axis, that corresponds to the pressure in a stationary, incompressible liquid. It typifies the loading on the plane walls of tanks containing liquids, the loading on the walls of dams, or the loading on plane submerged areas. In the third case, the forces are *uniformly distributed* over a *plane area*. A typical example of this type of loading is the case of the force exerted by a compressed gas on the plane walls of a tank.

(*c*) There is a significant difference between loadings that are represented by concentrated forces and loadings that are distributed. In the case of a concentrated force, there is a clearly defined point of

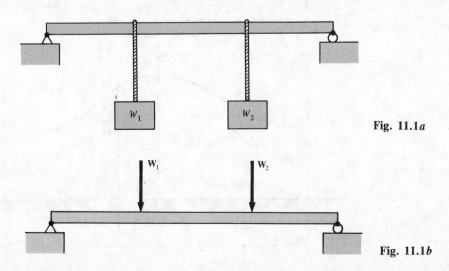

Fig. 11.1*a*

Fig. 11.1*b*

application of the force. In the case of distributed forces, there exists no *single* point of application of the forces. Rather, the total effect of the force distribution must be described in terms of the *resultant force* of the distribution of forces, together with the location of this resultant force.

11.2 (*a*) Give the general form for the magnitude of the resultant force of a force that is distributed along a straight length.

(*b*) What·is the interpretation of the result in part (*a*)?

(*c*) Where does the resultant force that was found in part (*a*) act?

I (*a*) Figure 11.2*a* shows a load distribution along a straight length *l* of a body. The magnitude of the loading is given by *w*, with the units of force per unit length. In the general case, *w* is a function of the location *x* along the member. It may be observed that the loading situation shown in Fig. 11.2*a* is a two-dimensional parallel force system.

The loading diagram is redrawn in Fig. 11.2*b*. The resultant force of this system of distributed loads is *F*, and the position of this force is given by x_0. Δx_i is a small element of length along the member, at position x_i, and the force which acts on this length element is $w\,\Delta x_i$. The sum of all the forces which act on the elementary lengths must be equal to the resultant force *F*, so that

$$F = \sum_i w\,\Delta x_i$$

If the elemental length Δx_i is allowed to decrease without limit, then

$$F = \lim_{\Delta x_i \to 0} \sum_i w\,\Delta x_i = \int_l w\,dx$$

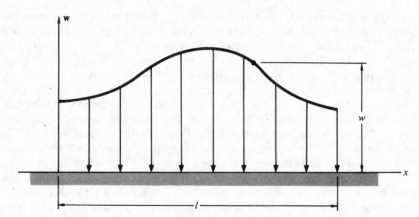

Fig. 11.2*a*

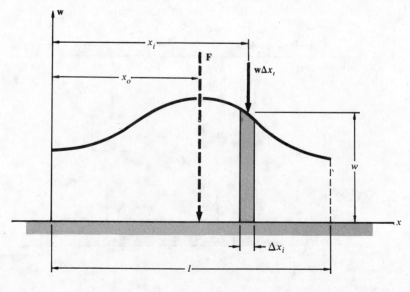

Fig. 11.2*b*

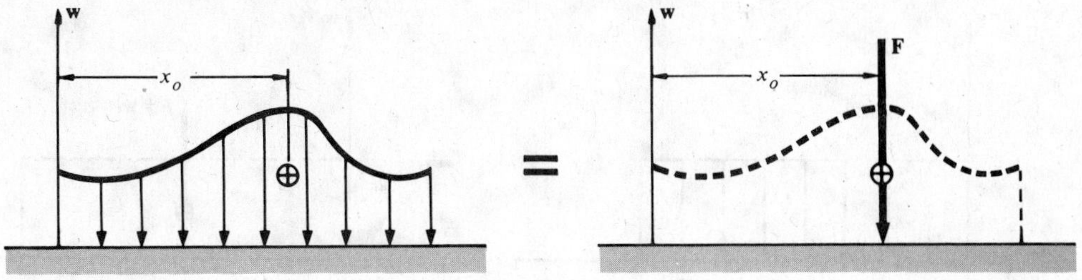

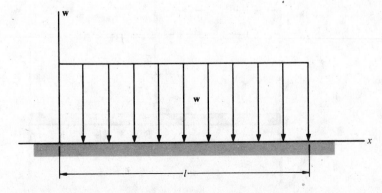

Fig. 11.2c

The above equation is the formal definition of the magnitude of the resultant force of the distributed loading.

(b) It may be seen from the result in part (a) that the magnitude of the resultant force is equal to the area under the load distribution curve.

(c) The moment about the origin in Fig. 11.2b, of the forces which act on the elementary lengths, must be equal to the moment of the resultant force about this point, so that

$$Fx_0 = \sum_i (w \, \Delta x_i)x_i \qquad x_0 = \frac{\sum\limits_i (w \, \Delta x_i)x_i}{F} = \frac{\sum\limits_i x_i(w \, \Delta x_i)}{\sum\limits_i w \, \Delta x_i}$$

The limiting form of the above equation is

$$x_0 = \frac{\lim\limits_{\Delta x_i \to 0} \sum\limits_i x_i(w \, \Delta x_i)}{\lim\limits_{\Delta x_i \to 0} \sum\limits_i w \, \Delta x_i} = \frac{\int_l x(w \, dx)}{\int_l w \, dx}$$

Inspection of this equation reveals that it has the basic form of the equation which defines a centroidal coordinate. The term $w \, \Delta x_i$ shown as the shaded area element in Fig. 11.2b, and the term $w \, dx$, play the role of the elementary area dA. The quantity $\int_l w \, dx$ is the total area under the load distribution curve. It may thus be concluded that *the line of action of the resultant force passes through the centroid of the load distribution area.* The force equivalence of the original load distribution and the resultant force of this distribution is illustrated in Fig. 11.2c.

11.3 (a) Find the magnitude and position of the resultant force of a force which is distributed *uniformly* along a length.

(b) Show a diagram of the replacement of the uniformly distributed force by the resultant force.

▎ (a) Figure 11.3a shows a uniform load distribution along a length of a member. Based on the results in Prob. 11.2, the resultant force F of the distributed loading is equal to the area under the load distribution curve, or

$$F = wl$$

Fig. 11.3a

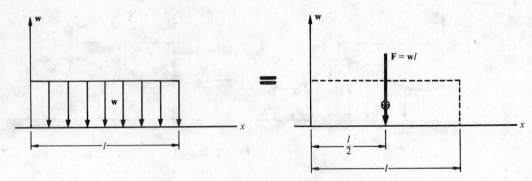

Fig. 11.3b

The line of action of this resultant force passes through the centroid of the load distribution area, at $x = l/2$.

(**b**) The replacement of the distributed loading by the statically equivalent resultant force is illustrated in Fig. 11.3b.

11.4 The 24 in × 12 in wide-flange beam shown in Fig. 11.4a has a span length of 30 ft and is supported at its ends.

(**a**) Find the magnitude and location of the resultant force of the weight of the beam.

(**b**) Find the reaction forces at the ends of the beam.

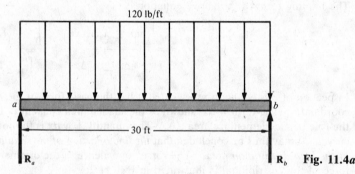

Fig. 11.4a

▮ (**a**) The free-body diagram of the beam is shown in Fig. 11.4b, and the weight per foot of this structural member is found from Table 10.8. The magnitude of the resultant force F, the weight of the beam, is equal to the area under the load distribution curve, or

$$F = 120\,\frac{\text{lb}}{\text{ft}}\,(30\,\text{ft}) = 3{,}600\,\text{lb}$$

The line of action of the resultant force passes through the centroid of the load distribution curve, and the final appearance of the resultant force loading on the beam is shown in Fig. 11.4b.

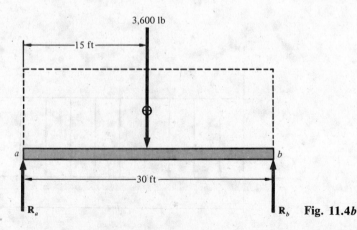

Fig. 11.4b

(*b*) From symmetry considerations, the support reaction forces at the ends of the beam are

$$R_a = R_b = 0.5(3,600) = 1,800 \text{ lb}$$

11.5 Find the force effects exerted by the wall on the beam in Fig. 11.5a.

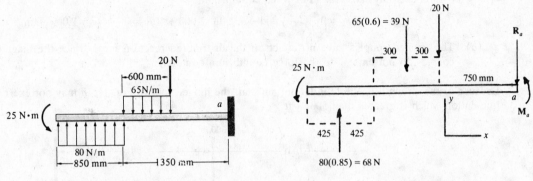

Fig. 11.5a   **Fig. 11.5b**

❙ The free-body diagram of the beam is shown in Fig. 11.5b, together with the resultant forces of the two uniform load distributions. Since the end of the beam is clamped, the wall exerts reaction force R_a and reaction moment M_a on the beam. For equilibrium of the beam,

$$\Sigma M_a = 0 \qquad M_a + 20\left(\frac{750}{1,000}\right) + 39\left(\frac{1,050}{1,000}\right) - 68\left(\frac{1,780}{1,000}\right) + 25 = 0 \qquad M_a = 40.1 \text{ N} \cdot \text{m}$$

$$\Sigma F_y = 0 \qquad 68 - 39 - 20 - R_a = 0 \qquad R_a = 9 \text{ N}$$

11.6 The reaction force of the ground on the beam in Fig. 11.6a is assumed to be uniformly distributed along the length of the beam.

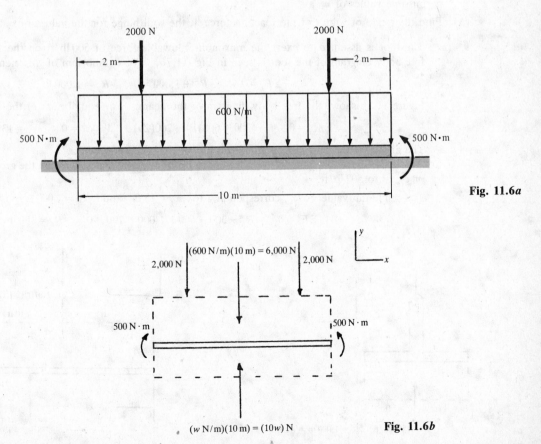

Fig. 11.6a

Fig. 11.6b

(*a*) Find the magnitude of this reaction force per unit length of the beam.

(*b*) How would the answer in part (*a*) change if the applied couples were removed from the beam?

▌ (*a*) Figure 11.6*b* shows the free-body diagram of the beam and the resultant force of each uniform force distribution. The distributed reaction force, w N/m, is to be determined. From symmetry, $\Sigma M = 0$ is identically satisfied. For force equilibrium,

$$\Sigma F_y = 0 \qquad -2{,}000 - 6{,}000 - 2{,}000 + 10w = 0 \qquad w = 1{,}000 \text{ N/m}$$

(*b*) The applied couples have no effect on the distributed reaction force, since the magnitudes of these couples do not enter into the force equilibrium equation.

11.7 Because of the strength of the foundation material, the hinged foot in Fig. 11.7*a* may not exert a force on the foundation which is greater than 500 lb/ft.

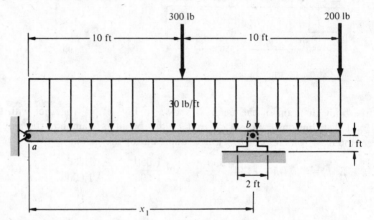

Fig. 11.7*a*

(*a*) If the force between the foot and the foundation is assumed to be uniformly distributed, find the range of permissible values of x_i.

(*b*) Find the range of values of the reaction force at the wall hinge for the values of x_i found in part (*a*).

▌ (*a*) The foot is assumed to exert the maximum allowable force of 500 lb/ft on the foundation, and the free-body diagram of the foot is seen in Fig. 11.7*b*. For equilibrium of this element,

$$\Sigma F_y = 0 \qquad -R_b + 1{,}000 = 0 \qquad R_b = 1{,}000 \text{ lb}$$

Figure 11.7*c* shows the free-body diagram of the beam. For equilibrium of the beam,

$$\Sigma M_a = 0 \qquad -(300 + 600)10 - 200(20) + 1{,}000x_1 = 0 \qquad x_1 = 13 \text{ ft}$$

The above result for x_1 is a minimum value.

If $x_1 \geq 13$ ft, the hinged foot will exert a uniformly distributed force on the ground that is less than or equal to 500 lb/ft.

(*b*) The minimum value of R_a, corresponding to $x_1 = 13$ ft and $R_b = 1{,}000$ lb, is found from

$$\Sigma F_y = 0 \qquad R_a - 300 - 600 + 1{,}000 - 200 = 0 \qquad R_a = 100 \text{ lb}$$

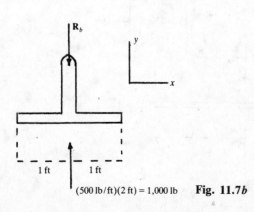

(500 lb/ft)(2 ft) = 1,000 lb   **Fig. 11.7***b*

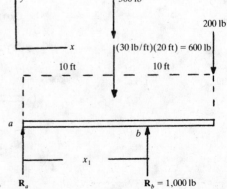

Fig. 11.7*c*

The maximum possible value of x_1 is 20 ft. For equilibrium of the beam, with reaction force R_b at the right end,

$$\Sigma M_b = 0 \qquad -R_a(20) + (300 + 600)10 = 0 \qquad R_a = 450 \text{ lb}$$

The range of values of R_a corresponding to the range of values of x_1 in part (a) is

$$100 \text{ lb} \le R_a \le 450 \text{ lb}$$

11.8 The system of Prob. 7.10 is repeated in Fig. 11.8a. The members are 6-in American standard channels, with a weight of 13 lb/ft.

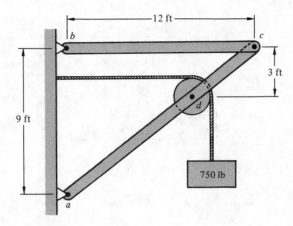

Fig. 11.8a

(a) Find the forces acting on the pins of the frame.

(b) Compare the solution in part (a) with the solution to Prob. 7.10.

▌ (a) The cable forces acting on the pulley are transferred directly to member ac. The weight forces of the two members are uniformly distributed along the lengths of the members. The resultant weight force of each member acts at the midpoint of the member length. The length of member ac is 15 ft, and the free-body diagram of the frame is shown in Fig. 11.8b. For equilibrium of the frame,

$$\Sigma M_a = 0 \qquad R_{bx}(9) - 156(6) + 750(6) - 195(6) - 750(8) = 0 \qquad R_{bx} = 401 \text{ lb}$$
$$\Sigma F_x = 0 \qquad -R_{bx} - 750 + R_{ax} = 0 \qquad R_{ax} = 1,150 \text{ lb}$$

Figure 11.8c shows the free-body diagram of member bc. For equilibrium,

$$\Sigma M_b = 0 \qquad -156(6) + R_{cy}(12) = 0 \qquad R_{cy} = 78 \text{ lb}$$
$$\Sigma F_x = 0 \qquad -401 + R_{cx} = 0 \qquad R_{cx} = 401 \text{ lb} \qquad R_c = \sqrt{R_{cx}^2 + R_{cy}^2} = \sqrt{401^2 + 78^2} = 409 \text{ lb}$$
$$\Sigma F_y = 0 \qquad R_{by} - 156 + R_{cy} = 0 \qquad R_{by} = 78 \text{ lb}$$
$$R_b = \sqrt{R_{bx}^2 + R_{by}^2} = \sqrt{401^2 + 78^2} = 409 \text{ lb}$$

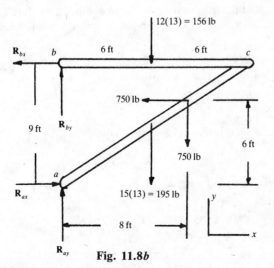

Fig. 11.8b

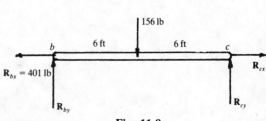

Fig. 11.8c

Using Fig. 11.8b,

$$\Sigma F_y = 0 \qquad R_{by} + R_{ay} - 156 - 195 - 750 = 0 \qquad R_{ay} = 1{,}020 \text{ lb}$$
$$R_a = \sqrt{R_{ax}^2 + R_{ay}^2} = \sqrt{1{,}150^2 + 1{,}020^2} = 1{,}540 \text{ lb}$$

The pin force R_d has the value

$$R_d = \sqrt{750^2 + 750^2} = 1{,}060 \text{ lb}$$

(*b*) A comparison of the pin forces for Problems 11.8 and 7.10 is shown in Table 11.1 below. It may be seen that the addition of the weight forces to the problem produces a significant increase in the pin forces at a, b, and c.

TABLE 11.1

	Prob. 11.8 pin force, lb	Prob. 7.10 pin force, lb
R_a	1,540	1,180
R_b	409	167
R_c	409	167
R_d	1,060	1,060

11.9 The system of Prob. 7.31 is repeated in Fig. 11.9a. The members all have a weight of 40 N/m.

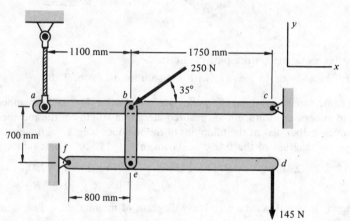

145 N   **Fig. 11.9a**

(*a*) Find the forces acting on pins a, c, e, and f of the frame, and the force exerted by pin b on member ac.
(*b*) Compare the solutions in parts (*a*) and (*b*) with the solution to Prob. 7.31.

▎ (*a*) The weight forces of the members are uniformly distributed along the lengths of the members. The resultant weight force of each member acts at the midpoint of the member length. Figure 11.9b shows the free-body diagram of member df, and R_e is the force exerted by pin e on this member. For equilibrium of member df,

$$\Sigma M_f = 0 \qquad R_e(800) - 102(1{,}280) - 145(2{,}550) = 0 \qquad R_e = 625 \text{ N}$$
$$\Sigma F_y = 0 \qquad -R_{fy} + R_e - 102 - 145 = 0 \qquad R_{fy} = 378 \text{ N}$$
$$\Sigma F_x = 0 \qquad R_{fx} = 0 \qquad R_f = R_{fy} = 378 \text{ N}$$

R_b is the force exerted by pin b on member be.
The free-body diagram of member be is seen in Fig. 11.9c. For equilibrium,

$$\Sigma F_y = 0 \quad_{\bullet} \quad R_b - 28 - 625 = 0 \qquad R_b = 653 \text{ N}$$

Figure 11.9d shows the free-body diagram of pin b, and R_{ab} is the force exerted by pin b on member ac. For equilibrium of pin b.

$$\Sigma F_x = 0 \qquad R_{ab,x} - 250 \cos 35° = 0 \qquad R_{ab,x} = 205 \text{ N}$$
$$\Sigma F_y = 0 \qquad R_{ab,y} - 250 \sin 35° - 653 = 0 \qquad R_{ab,y} = 796 \text{ N}$$
$$R_{ab} = \sqrt{R_{ab,x}^2 + R_{ab,y}^2} = \sqrt{205^2 + 796^2} = 822 \text{ N}$$

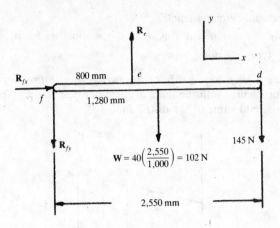

Fig. 11.9b

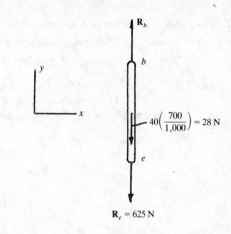

Fig. 11.9c

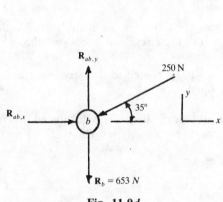

Fig. 11.9d

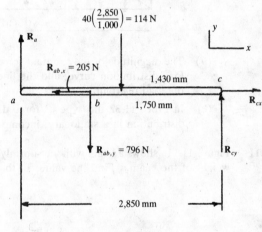

Fig. 11.9e

The free-body diagram of member ac is shown in Fig. 11.9e. For equilibrium of this member,

$$\Sigma M_c = 0 \qquad -R_a(2{,}850) + 796(1{,}750) + 114(1{,}430) = 0 \qquad R_a = 546\text{ N}$$
$$\Sigma F_x = 0 \qquad -205 + R_{cx} = 0 \qquad R_{cx} = 205\text{ N}$$
$$\Sigma F_y = 0 \qquad R_a - 796 - 114 + R_{cy} = 0 \qquad R_{cy} = 364\text{ N}$$
$$R_c = \sqrt{R_{cx}^2 + R_{cy}^2} = \sqrt{205^2 + 364^2} = 418\text{ N}$$

Table 11.2 shows a comparison of the results in this problem with those found in Prob. 7.31. The reference values for the percent difference are the values found in Prob. 7.31.

(b)

TABLE 11.2

	Prob. 11.9	Prob. 7.31	%D
member ac			
pin a	546	371	47%
pin b	822	639	29%
pin c	418	311	34%
member be			
pin b	653	462	41%
pin e	625	462	35%
member df			
pin e	625	462	35%
pin f	378	317	19%

11.10 (*a*) Show the form of a force which *varies uniformly* along a length.

(*b*) Explain how the magnitude and location of the resultant of this force distribution are found.

(*c*) What common physical problem does this type of loading represent?

▌ (*a*) A uniformly varying force along a length may have either of the two general forms shown in Fig. 11.10. The terms w_0, w_1, and w_2 in the figure, with the units of force per unit length, represent the magnitude of the distributed force at the endpoints of the distribution.

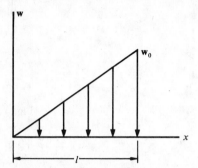 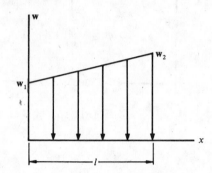

Fig. 11.10

(*b*) The magnitude of the resultant force in the two cases shown in the figure is equal to the area under the load distribution curve, and the lines of action of these resultant forces pass through the centroids of the load distribution areas.

(*c*) In Prob. 11.32 this type of force distribution will be related to the very important case of the force distribution in a stationary, incompressible liquid.

11.11 Figure 11.11*a* shows a beam with a uniformly varying load distribution. This distributed load includes the static weight of the beam. Find the values of the two support reactions R_a and R_b.

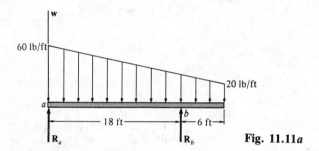

Fig. 11.11*a*

▌ The load distribution diagram is redrawn in Fig. 11.11*b*.

The magnitude of the resultant force F is equal to the area under the load distribution diagram, or

$$F = \left(\frac{60+20}{2} \ \frac{lb}{ft}\right)(24 \ ft) = 960 \ lb$$

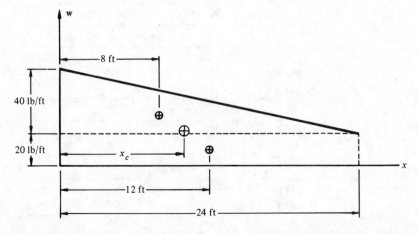

Fig. 11.11*b*

The centroidal coordinate x_c of the load distribution area is

$$x_c = \frac{8(0.5)24(40) + 12(24)20}{0.5(24)(40) + 24(20)} = 10.0 \text{ ft}$$

The free-body diagram of the beam is shown in Fig. 11.11c. The equilibrium requirements are

$$\sum M_a = 0 \qquad -960(10) + R_b(18) = 0 \qquad R_b = 533 \text{ lb}$$
$$\sum F_y = 0 \qquad R_a - 960 + R_b = 0 \qquad R_a = 427 \text{ lb}$$

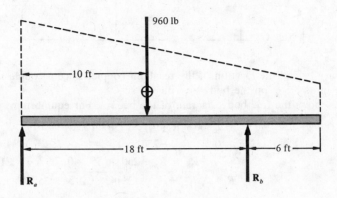

Fig. 11.11c

11.12 Find the forces exerted by the ground on the beam in Fig. 11.12a.

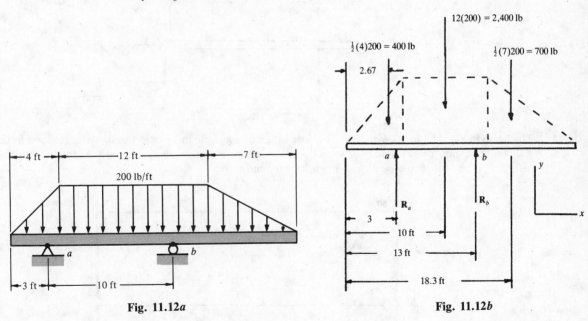

Fig. 11.12a   Fig. 11.12b

I Figure 11.12b shows the free-body diagram of the beam. The resultant forces of the two uniformly varying force distributions are shown acting through the centroids of their respective load diagrams. For equilibrium of the beam,

$$\sum M_a = 0 \qquad 400(3 - 2.67) - 2,400(7) + R_b(10) - 700(15.3) = 0 \qquad R_b = 2,740 \text{ lb}$$
$$\sum F_y = 0 \qquad -400 + R_a - 2,400 + R_b - 700 = 0$$

11.13 The beam shown in Fig. 11.13a is acted on by a uniformly varying distribution of couples and an applied force. Find the forces exerted by the ground on the beam.

I From the development in Prob. 11.10, it follows that the magnitude of the resultant couple M_R is equal to the area under the load distribution diagram. Thus,

$$M_R = \frac{300 + 100}{2} \cdot 12 = 2,400 \text{ in} \cdot \text{lb}$$

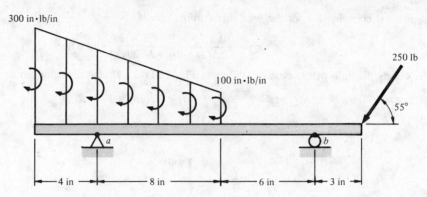

Fig. 11.13a

It is not necessary to find the location of the resultant couple, since a couple may be moved in its own plane without change in its effect on the body on which it acts.

Figure 11.13b shows the free-body diagram of the beam. For equilibrium,

$$\Sigma M_a = 0 \quad -2{,}400 + R_b(14) - (250 \sin 55°)17 = 0 \quad R_b = 420 \text{ lb}$$
$$\Sigma F_x = 0 \quad R_{ax} - 250 \cos 55° = 0 \quad R_{ax} = 143 \text{ lb}$$
$$\Sigma F_y = 0 \quad -R_{ay} + R_b - 250 \sin 55° = 0 \quad R_{ay} = 215 \text{ lb}$$
$$R_a = \sqrt{R_{ax}^2 + R_{ay}^2} = \sqrt{143^2 + 215^2} = 258 \text{ lb}$$

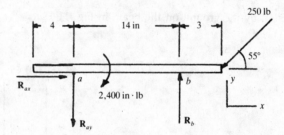

Fig. 11.13b

11.14 The beam in Fig. 11.14a fits with zero clearance into a mating hole in a vertical wall. The reaction forces of the wall on the beam are assumed to be uniformly varying, as shown in the figure. Find the magnitudes, in newtons per meter, of the peak load intensities w_1 and w_2.

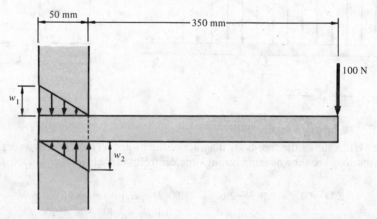

Fig. 11.14a

▌ Figure 11.14b shows the free-body diagram of the beam. The resultant forces of the two uniformly varying load distributions act through the centroids of the load distribution diagrams. For equilibrium of the beam,

$$\Sigma M_a = 0 \quad -100\left(\frac{350 + 33.3}{1{,}000}\right) + 0.5\left(\frac{50}{1{,}000}\right)w_2\left(\frac{16.7}{1{,}000}\right) = 0 \quad w_2 = 91{,}800 \text{ N/m} = 91.8 \text{ kN/m}$$

$$\Sigma F_y = 0 \quad -0.5\left(\frac{50}{1{,}000}\right)w_1 + 0.5\left(\frac{50}{1{,}000}\right)91{,}800 - 100 = 0 \quad w_1 = 87{,}800 \text{ N/m} = 87.8 \text{ kN/m}$$

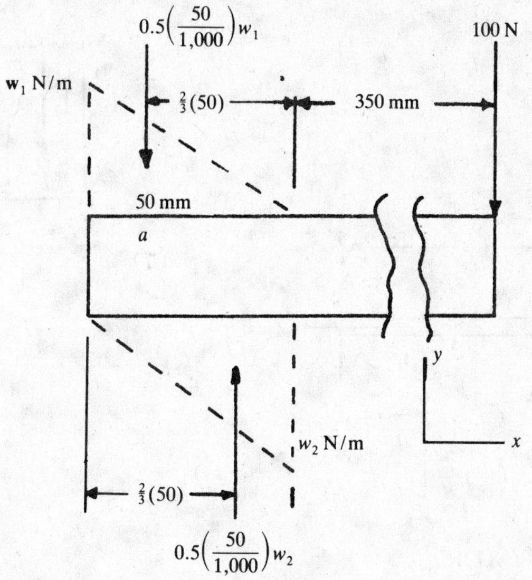

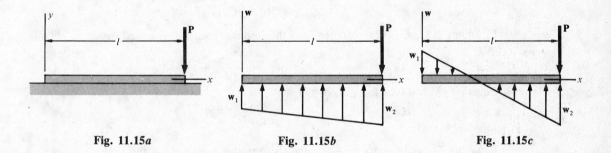

Fig. 11.14b

11.15   The rigid beam shown in Fig. 11.15a rests on an elastic foundation.   This foundation is assumed to be able to either "push" or "pull" on the beam and to exert a uniformly varying distributed reaction force on the beam.   Based on the above assumption, the reaction force distribution may be assumed to have either of the two general forms shown in Figs. 11.15b and c.

Fig. 11.15a   Fig. 11.15b   Fig. 11.15c

(a)   Show that the reaction force distribution in Fig. 11.15b will not satisfy the equilibrium equations of the beam.

(b)   Assume the reaction force distribution shown in Fig. 11.15c and find w_1 and w_2 in terms of P and l.

(c)   Find the numerical results for the solution in part (b), if $P = 6\,\text{lb}$ and $l = 10\,\text{in}$.

(d)   Check the solution found in part (c).

▌   (a)   Figure 11.15d shows the free-body diagram of the beam for the reaction force distribution given in Fig. 11.15b.   The resultant force R of the reaction force distribution acts through the centroid of the loading diagram, as shown in the figure.   The beam is acted on by only the two forces R and P.   Since these two forces are never concurrent, it may be concluded that it is not possible to satisfy moment equilibrium of the beam.

   (b)   Figure 11.15e shows the resultant forces, of the two uniformly varying load distributions of Fig. 11.15c, acting on the free-body diagram of the beam.   From the similar triangles in Fig. 11.15e,

$$\frac{w_1}{A} = \frac{w_2}{l - A} \qquad A = \frac{w_1 l}{w_1 + w_2} \tag{1}$$

For moment equilibrium of the beam,

$$\Sigma M_a = 0 \qquad -P[(l - A) + (A - \tfrac{1}{3}A)] + \tfrac{1}{2}(l - A)w_2[l - \tfrac{1}{3}A - \tfrac{1}{3}(l - A)] = 0$$

$$-P\left(l - \frac{A}{3}\right) + \tfrac{1}{2}(l - A)w_2\left(\frac{2l}{3}\right) = 0 \tag{2}$$

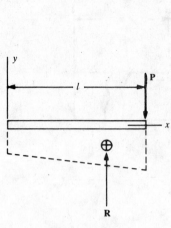

Fig. 11.15d   Fig. 11.15e

For force equilibrium of the beam,

$$\Sigma F_y = 0 \qquad -\tfrac{1}{2}Aw_1 + \tfrac{1}{2}(l-A)w_2 - P = 0 \qquad (3)$$

Equation (2) is solved for A, with the result

$$A = \frac{3Pl - w_2 l^2}{P - w_2 l} \qquad (4)$$

Equations (1) and (4) are equated, and the resulting equation is solved for w_1 to obtain

$$w_1 = \frac{w_2^2 l^2 - 3Plw_2}{2Pl} \qquad (5)$$

Equation (3) is solved for A to obtain

$$A = \frac{lw_2 - 2P}{w_1 + w_2} \qquad (6)$$

Equations (1) and (6) are equated, and the resulting equation has the form

$$Pw_1 + \frac{l}{2} w_1^2 = -Pw_2 + \frac{1}{2} lw_2^2 \qquad (7)$$

w_1 from Eq. (5) is now substituted into Eq. (7) to obtain the final result

$$w_2^3 - 6\left(\frac{P}{l}\right)w_2^2 + 9\left(\frac{P}{l}\right)^2 w_2 - 4\left(\frac{P}{l}\right)^3 = 0 \qquad (8)$$

With P and l known, w_2 may be found from the above cubic equation. With w_2 known, w_1 may be found from Eq. (5).

(c) Using $P = 6$ lb and $l = 10$ in, Eq. (8) has the form

$$w_2^3 - 3.6w_2^2 + 3.24w_2 - 0.864 = 0$$

The root w_2 which satisfies the above equation is

$$w_2 = 2.4 \text{ lb/in}$$

Using Eq. (5),

$$w_1 = \frac{w_2^2 l^2 - 3Plw_2}{2Pl} = \frac{2.4^2(10)^2 - 3(6)10(2.4)}{2(6)10} = 1.2 \text{ lb/in}$$

(d) The results found in part (c) are now used to check the original equilibrium equations. Using Eq. (1),

$$A = \frac{w_1 l}{w_1 + w_2} = \frac{1.2(10)}{1.2 + 2.4} = 3.33 \text{ in}$$

From Eq. (2),

$$-P\left(l - \frac{A}{3}\right) + \tfrac{1}{2}(l - A)w_2\left(\frac{2l}{3}\right) = 0 \qquad -6\left(10 - \frac{3.33}{3}\right) + \tfrac{1}{2}(10 - 3.33)2.4\left(\frac{2(10)}{3}\right) \overset{?}{=} 0 \qquad 53.3 \approx 53.4$$

Using Eq. (3),

$$-\tfrac{1}{2}Aw_1 + \tfrac{1}{2}(l - A)w_2 - P = 0 \qquad -\tfrac{1}{2}(3.33)1.2 + \tfrac{1}{2}(10 - 3.33)2.4 - 6 \overset{?}{=} 0 \qquad -2 + 8 - 6 \overset{?}{=} 0 \qquad 8 \equiv 8$$

11.16 Figure 11.16 shows a distribution of loading on a body. This loading is assumed to have a parabolic variation along the x axis.

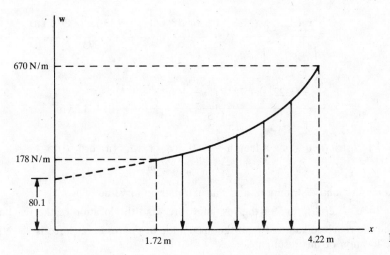

Fig. 11.16

(a) Find the functional form of $w(x)$.

(b) Find the magnitude and the location of the resultant of the force distribution over the length $x = 1.72\,\text{m}$ to $x = 4.22\,\text{m}$.

❙ (a) The parabolic load distribution has the general functional form

$$w = ax^2 + bx + c \tag{1}$$

This curve must pass through the points

$$x = 1.72\,\text{m} \qquad w = 178\,\text{N/m} \tag{2}$$
$$x = 4.22\,\text{m} \qquad w = 670\,\text{N/m} \tag{3}$$

An additional condition, from the symmetry of the curve with respect to the w axis, is

$$x = 0 \qquad \frac{dw}{dx} = 0 \tag{4}$$

From Eq. (1),

$$\frac{dw}{dx} = 2ax + b$$

Using Eq. (3),

$$0 = 0 + b \qquad b = 0$$

Using Eqs. (2) and (3) in Eq. (1),

$$178 = 1.72^2 a + c \tag{5}$$
$$670 = 4.22^2 a + c \tag{6}$$

Equation (5) is subtracted from Eq. (6), with the result

$$(4.22^2 - 1.72^2)a = 670 - 178 \qquad a = 33.1\,\text{N/m}^3$$

Using Eq. (2) in Eq. (1),

$$178 = 33.1(1.72)^2 + c \qquad c = 80.1\,\text{N/m}$$

The functional form of $w(x)$ is

$$w = 33.1x^2 + 80.1 \text{ N/m}$$

(**b**)   The magnitude of the resultant force is given by

$$F = \int w \, dx = \int_{1.72}^{4.22} (33.1x^2 + 80.1) \, dx = \left(33.1 \frac{x^3}{3} + 80.1x\right)\Big|_{1.72}^{4.22}$$

$$= \frac{33.1}{3} (4.22^3 - 1.72^3) + 80.1(4.22 - 1.72) = 973 \text{ N}$$

The coordinate x_c of the centroid of the load distribution area is found from

$$x_c = \frac{\int x \, dA}{\int dA} = \frac{\int xw \, dx}{\int w \, dx} = \frac{\int xw \, dx}{F} = \frac{\int_{1.72}^{4.22} x(33.1x^2 + 80.1) \, dx}{973} = \frac{\left(33.1 \frac{x^4}{4} + 80.1 \frac{x^2}{2}\right)\Big|_{1.72}^{4.22}}{973}$$

$$= \frac{\dfrac{33.1}{4} (4.22^4 - 1.72^4) + \dfrac{80.1}{2} (4.22^2 - 1.72^2)}{973} = 3.23 \text{ m}$$

11.17   A loading is distributed along a length $x = 0$ to $x = x_0$ according to the relationship

$$w = w_0(1 - e^{-4x/x_0})$$

(**a**)   Sketch the load distribution, and find the maximum value of w.

(**b**)   Find the general forms for the resultant force and the location of the resultant force.

(**c**)   Find the required value of w_0 if the magnitude of the resultant force is 950 N, when $x_0 = 3$ m, and find the location of this resultant force.

▮   (**a**)   The force distribution is shown in Fig. 11.17.   The maximum value of w occurs when $x = x_0$, and

$$w_{max} = w_0(1 - e^{-4}) = 0.982w_0$$

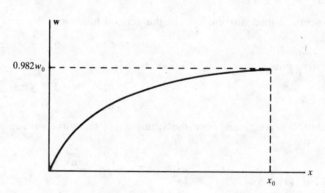

Fig. 11.17

(**b**)   The magnitude of the resultant force is given by

$$F = \int w \, dx = \int_0^{x_0} w_0(1 - e^{-4x/x_0}) \, dx = \int_0^{x_0} w_0 \, dx - w_0 \int_0^{x_0} e^{-4x/x_0} \, dx$$

$$= \int_0^{x_0} w_0 \, dx - w_0\left(-\frac{x_0}{4}\right)\int_0^{x_0} e^{-4x/x_0}\left(-\frac{4}{x_0} \, dx\right)$$

$$= w_0 x \Big|_0^{x_0} + w_0 \frac{x_0}{4} e^{-4x/x_0} \Big|_0^{x_0}$$

$$= w_0 x_0 + \frac{w_0 x_0}{4} (e^{-4} - 1)$$

$$= \tfrac{3}{4} w_0 x_0 + \frac{w_0 x_0}{4} e^{-4} = 0.755 w_0 x_0$$

The resultant force acts through the centroid of the load distribution area at the location given by

$$x_c = \frac{\int x\, dA}{\int dA} = \frac{\int xw\, dx}{\int w\, dx} = \int \frac{xw\, dx}{F} = \frac{\int_0^{x_0} xw_0(1 - e^{-4x/x_0})\, dx}{F} = \frac{w_0 \int_0^{x_0} x\, dx - w_0 \int_0^{x_0} x\, e^{-4x/x_0}\, dx}{F}$$

The second integral is evaluated by using integration by parts, with the general form

$$\int u\, dv = uv - \int v\, du \qquad u = x \qquad du = dx \qquad dv = e^{-4x/x_0}\, dx \qquad v = -\frac{x_0}{4} e^{-4x/x_0}$$

$$x_c = \frac{w_0 \left.\frac{x^2}{2}\right|_0^{x_0} - w_0\left(-\frac{x_0 x}{4} e^{-4x/x_0}\Big|_0^{x_0} - \int_0^{x_0} -\frac{x_0}{4} e^{-4x/x_0}\, dx\right)}{F}$$

$$= \frac{\frac{w_0 x_0^2}{2} + w_0\left(\frac{x_0 x}{4} e^{-4x/x_0}\Big|_0^{x_0} - \frac{x_0}{4}\int_0^{x_0} e^{-4x/x_0}\, dx\right)}{F}$$

$$= \frac{\frac{w_0 x_0^2}{2} + w_0\left[\frac{x_0 x}{4} e^{-4x/x_0} - \frac{x_0}{4}\left(-\frac{x_0}{4}\right) e^{-4x/x_0}\right]_0^{x_0}}{F}$$

$$= \frac{\frac{w_0 x_0^2}{2} + w_0\left[\frac{x_0}{4}(x_0 e^{-4}) + \frac{x_0^2}{16}(e^{-4} - e^0)\right]}{F}$$

$$= \frac{\frac{w_0 x^2}{2} + w_0 x_0^2\left[\left(\frac{1}{4} + \frac{1}{16}\right)e^{-4} - \frac{1}{16}\right]}{F} = \frac{0.4451 w_0 x_0^2}{0.755 w_0 x_0} = 0.590 x_0$$

It may be seen that the location $x_c = 0.590 x_0$ of the resultant force is independent of the value of w_0.

(c) For $F = 950\,\text{N}$ and $x_0 = 3\,\text{m}$,

$$F = 0.755 w_0 x_0 \qquad 950 = 0.755 w_0(3) \qquad w_0 \doteq 419\,\text{N/m}$$

The position of the resultant force is given by

$$x_c = 0.590 x_0 = 0.590(3) = 1.77\,\text{m}$$

11.2 FORCE DISTRIBUTION OVER AN AREA, UNIFORM AND UNIFORMLY VARYING FORCE DISTRIBUTIONS

11.18 Find the magnitude, direction, and location of the resultant force of a *uniform pressure* acting on a *plane area*.

▌ A pressure describes the intensity of force loading, in terms of force per unit area. As was shown in Prob. 1.13, typical units of pressure, in USCS units, are pounds per square inch, written as lb/in^2 or psi. In SI units, the pressure is expressed as newtons per square meter, written as N/m^2. A pressure of one newton per square meter is called a *pascal*, with the symbol Pa.

Figure 11.18 shows a plane area A on which acts a uniform force distribution, or pressure. The nature of this type of loading may be identified as a three-dimensional parallel force system. The resultant force F acting on

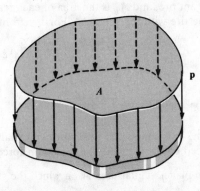

Fig. 11.18

plane A is the product of the magnitudes of the pressure and the area, or

$$F = pA$$

The resultant force of the pressure distribution is *perpendicular* to the plane of the area. The line of action of this resultant force passes through the *centroid* of the plane area.

11.19 Figure 11.19a shows a side view of a pressure vessel of rectangular cross section. The dimension of the tank normal to the plane of the paper is 800 mm. The tank contains a gas at a pressure of 415 kPa. Find the resultant forces on the tank walls shown in the figure.

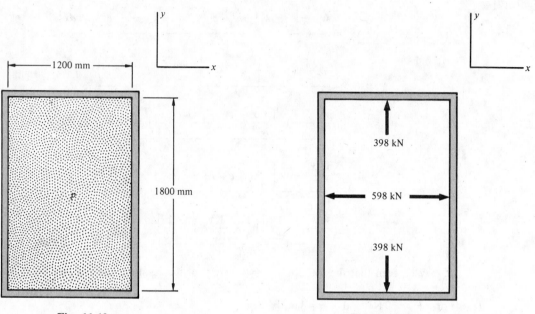

Fig. 11.19a   Fig. 11.19b

❚ Since the unit of the pascal is newtons per square meter, the resultant force F_x on the two vertical walls is

$$F_x = pA = \left(415 \times 10^3 \ \frac{N}{m^2}\right)(1,800)(800) \ mm^2 \left(\frac{1 \ m}{1,000 \ mm}\right)^2 = 5.98 \times 10^5 \ N = 598 \ N$$

The force F_y on the horizontal walls is

$$F_y = pA = \left(415 \times 10^3 \ \frac{N}{m^2}\right)(1,200)(800) \ mm^2 \left(\frac{1 \ m}{1,000 \ mm}\right)^2 = 3.98 \times 10^5 \ N = 398 \ kN$$

The resultant forces are shown in Fig. 11.19b.

11.20 The weight of an automobile is transmitted to the ground through the four "prints" that the tires make on the ground. A vehicle weighs 4,000 lb, and the tires are inflated to a pressure of 30 lb/in². If 60 percent of the vehicle weight is supported by the front wheels, find the required areas of the tire prints.

❚ A_F is the area of each front tire, and A_R is the area of each rear tire. The force is assumed to act uniformly over the contact area of the tire with the ground. For the front tires,

$$F = pA \qquad \frac{0.6(4,000)}{2} = 30 \ \frac{lb}{in^2} \ (A_F) \qquad A_F = 40 \ in^2$$

For the rear tires,

$$F = pA \qquad \frac{0.4(4,000)}{2} = 30(A_R) \qquad A_R = 26.7 \ in^2$$

11.21 The device shown in Fig. 11.21a is a model of a pressure relief valve for a steam supply system. If the pressure in the system may not exceed 650 lb/in², find the maximum force which the spring may exert on the valve.

❚ Figure 11.21b shows the plane annular area A on which the steam pressure acts. The value of this area is

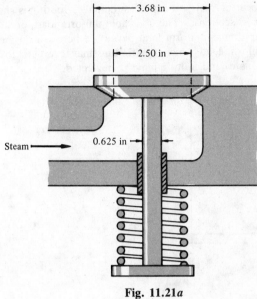

Fig. 11.21a　　　　　　　　　　　**Fig. 11.21b**

given by

$$A = \frac{\pi}{4}(2.5^2 - 0.625^2) = 4.60 \text{ in}^2$$

The resultant force F acting on the area, due to the steam pressure, is

$$F = pA = 650 \frac{\text{lb}}{\text{in}^2}(4.60 \text{ in}^2) \qquad F = 2,990 \text{ lb}$$

The maximum value of force which the spring may exert on the valve, to limit the pressure to 650 lb/in^2, is $2,990 \text{ lb}$.

11.22　The nut in Fig. 11.22 is tightened until there is a uniformly distributed load of $35,000 \text{ lb/in}^2$ over the cross section of the bolt. The flat washers are assumed to distribute load uniformly over the flat faces of the cylinder. Find the uniform load induced in the cross section of the walls of the cylinder by tightening the bolt.

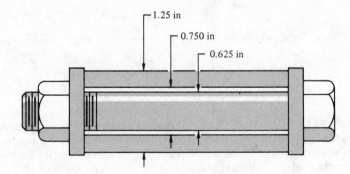

Fig. 11.22

❚　The cross-sectional area of the bolt is given by

$$A_{\text{bolt}} = \frac{\pi(0.625)^2}{4} = 0.307 \text{ in}^2$$

The bolt force produced by the distributed load of $35,000 \text{ lb/in}^2$ is found as

$$F_{\text{bolt}} = pA = 35,000(0.307) = 10,700 \text{ lb}$$

The tightening of the nut on the bolt produces the compressive force of $10,700 \text{ lb}$ on the cylinder wall cross section. The uniform load acting on this area is

$$p_{\text{cyl}} = \frac{F_{\text{bolt}}}{A_{\text{cyl}}} = \frac{10,700 \text{ lb}}{(\pi/4)(1.25^2 - 0.750^2) \text{ in}^2} = 13,600 \text{ lb/in}^2$$

11.23 The detail of a timber column resting on a steel bearing plate on a precast concrete footing is shown in Fig. 11.23. All three of these elements have square cross sections. The allowable uniform load, or stress, on the cross section of the timber is $1,450\ \text{lb/in}^2$. The maximum uniform load permitted between the bearing plate and the footing is $600\ \text{lb/in}^2$. The maximum uniform load which the footing may exert on the ground is $8,500\ \text{lb/ft}^2$. The weights of the column, bearing plate, and footing may be neglected.

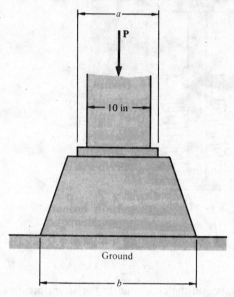

Fig. 11.23

(a) Find the maximum load P which may be supported by the column.

(b) For the load of part (a), find the minimum values of the dimensions a and b.

▌ (a) For the timber column,

$$P = pA = 1,450\ \frac{\text{lb}}{\text{in}^2}\ (10)^2\ \text{in}^2 \qquad P = 145,000\ \text{lb} = 72.5\ \text{ton}$$

(b) Force P is transmitted through the steel bearing plate, and it is assumed to produce a uniform loading on the footing. The dimension a of the bearing plate is then found from

$$P = pA \qquad 145,000 = 600a^2 \qquad a = 15.5\ \text{in}$$

The column force is transmitted through the footing to the ground. Assuming uniform load transfer to the ground,

$$P = pA \qquad 145,000 = \left(8,500\ \frac{\text{lb}}{\text{ft}^2}\right)\left(\frac{1\ \text{ft}}{12\ \text{in}}\right)^2 b^2 \qquad b = 49.6\ \text{in}$$

11.24 Figure 11.24a shows a right circular cylindrical, thin-wall pressure vessel with an internal pressure p. The mean radius is r, and the wall thickness t is assumed to be much less than r. As a consequence of this assumption, the forces in the wall material may be assumed to be uniformly distributed across the thickness. x is the longitudinal direction of the cylinder, and y has the direction of a line tangent to the circumference of the cylinder.

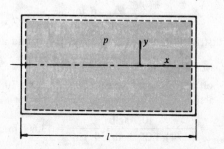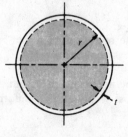

Fig. 11.24a

(a) Pass a cutting plane normal to the axis of the cylinder and draw the free-body diagram of one of the isolated portions. Show that the tensile load distribution w_x in the walls in the x direction is expressed by $w = pr/(2t)$.

(b) Pass a cutting plane through the center axis of the cylinder. From equilibrium considerations of the half cylinder, show that the tensile load distribution w_y in the walls in the y direction is $w_y = pr/t$.

▌ (a) The units of w_x and w_y are force per unit area. Figure 11.24b shows a view of a cutting plane passed normal to the axis of the cylinder. For force equilibrium in the x direction,

$$\Sigma F_x = 0 \qquad -p(\pi r^2) + w_x(2\pi rt) = 0 \qquad w_x = \frac{pr}{2t}$$

(b) The view of a cutting plane passed through the center axis of the cylinder is shown in Fig. 11.24c. l is the axial length of the cylinder. For force equilibrium in the y direction,

$$\Sigma F_y = 0 \qquad -p(2rl) + 2w_y(lt) = 0 \qquad w_y = \frac{pr}{t}$$

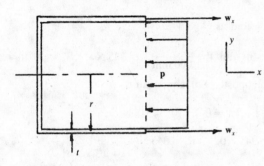

Fig. 11.24b

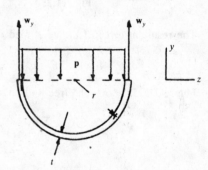

Fig. 11.24c

11.25 The stop sign shown in Fig. 11.25a experiences a uniformly distributed wind load of 720 Pa. Find the force and moment exerted by the ground on the signpost.

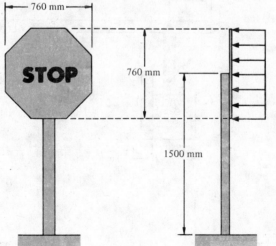

Fig. 11.25a

▌ Figure 11.25b shows the area of the sign. From this figure,

$$2l \cos 45° + l = 760 \qquad l = 315 \text{ mm}$$

The area of the sign is found from

$$A = 4[\tfrac{1}{2}(l \cos 45°)(l \sin 45°)] + l(760) + 2[l(l \cos 45°)]$$

Using the above value of l,

$$A = 4.79 \times 10^5 \text{ mm}^2$$

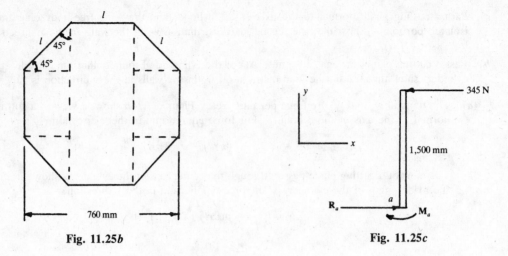

Fig. 11.25b

Fig. 11.25c

The resultant force of the wind load on the sign is found as

$$F = pA = 720 \, \frac{N}{m^2} \, (4.79 \times 10^5 \, mm^2) \left(\frac{1\,m}{1,000\,mm}\right)^2 = 345 \, N$$

Figure 11.25c shows the free-body diagram of the signpost. For equilibrium,

$$\Sigma M_a = 0 \qquad 345 \left(\frac{1,500}{1,000}\right) - M_a = 0 \qquad M_a = 518 \, N \cdot m$$

$$\Sigma F_x = 0 \qquad R_a - 345 = 0 \qquad R_a = 345 \, N$$

11.26 The device shown in Fig. 11.26a is used to compress material in a cylindrical well. For the loading shown on the handle, find the average pressure exerted on the compacted material.

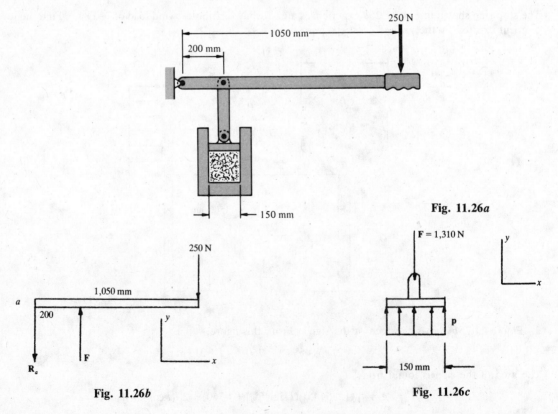

Fig. 11.26a

Fig. 11.26b

Fig. 11.26c

❚ Figure 11.26b shows the free-body diagram of the handle lever. The connecting link is a two-force member, and the force F in this line has the known direction of the link. For equilibrium of the handle lever,

$$\Sigma M_a = 0 \qquad F(200) - 250(1,050) = 0 \qquad F = 1,310\,\text{N}$$

The free-body diagram of the compacting element is seen in Fig. 11.26c. For equilibrium of this element, assuming that the compacting force is uniformly distributed over the area of the element,

$$\Sigma F_y = 0 \qquad -1,310 + \left(p\,\frac{\text{N}}{\text{m}^2} \right) \frac{\pi(150)^2}{4}\,\text{mm}^2 \left(\frac{1\,\text{m}}{1,000\,\text{mm}} \right)^2 \qquad p = 74,100\,\frac{\text{N}}{\text{m}^2} = 74,100\,\text{Pa} = 74.1\,\text{kPa}$$

The average pressure exerted on the compacted material is 74.1 kPa.

11.27   The pressure on the face of the cylindrical piston in Fig. 11.27a is 5.42 MPa.

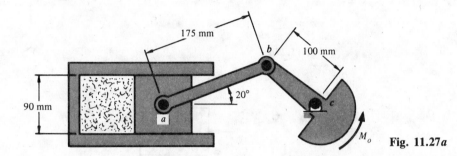

Fig. 11.27a

(a)   Find the range of values of moment M_0 for which the equilibrium configuration shown in the figure may be maintained, if the piston is assumed to be frictionless.   Also find the magnitude of the reaction force exerted by the ground on the crank.

(b)   Do the same as in part (a) if the coefficient of friction between the piston and the cylinder wall is 0.15.

▮ (a)   Figure 11.27b shows the free-body diagram of the piston.   Since the link is a two-force member, force F_{ab} has the known direction of 20° with respect to the x axis.   The resultant force on the face of the piston is expressed as

$$F = pA \qquad F = 5.42 \times 10^6\,\frac{\text{N}}{\text{m}^2}\,\frac{\pi(90)^2}{4}\,\text{mm}^2 \left(\frac{1\,\text{m}}{1,000\,\text{mm}} \right)^2 = 3.45 \times 10^4\,\text{N} = 34.5\,\text{kN}$$

For equilibrium of the piston,

$$\Sigma F_x = 0 \qquad 34.5 - F_{ab}\cos 20° = 0 \qquad F_{ab} = 36.7\,\text{kN (C)}$$

The symbol (C) following the magnitude of F_{ab} indicates that this force is compressive.

$$\Sigma F_y = 0 \qquad N - F_{ab}\sin 20° = 0$$
$$N = 12.6\,\text{kN}$$

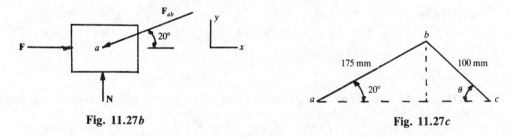

Fig. 11.27b   Fig. 11.27c

Using Fig. 11.27c, the direction θ of the crank is found by using the law of sines.   The result is

$$\frac{175}{\sin\theta} = \frac{100}{\sin 20°} \qquad \theta = 36.8°$$

Figure 11.27d shows a symbolic free-body diagram of the crank.   For equilibrium of this element,

$$\Sigma F_x = 0 \qquad 36.7\cos 20° - R_{cx} = 0 \qquad R_{cx} = 34.5\,\text{kN}$$
$$\Sigma F_y = 0 \qquad 36.7\sin 20° - R_{cy} = 0 \qquad R_{cy} = 12.6\,\text{kN}$$
$$R_c = \sqrt{R_{cx}^2 + R_{cy}^2} = \sqrt{34.5^2 + 12.6^2} = 36.7\,\text{kN}$$

It may be observed from the above result that $R_c = F_{ab} = 36.7\,\text{kN}$.

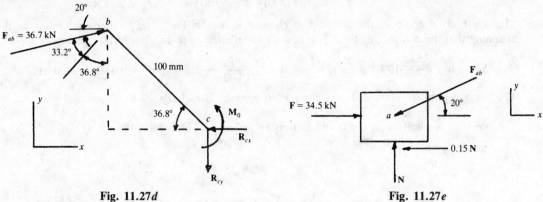

Fig. 11.27d   Fig. 11.27e

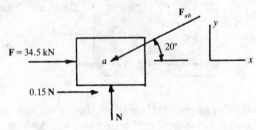

Fig. 11.27f

For moment equilibrium of the crank,

$$\Sigma M_c = 0 \qquad -36.7 \cos 33.2° \text{ (kN)}100 \text{ (mm)} + M_0 = 0 \qquad M_0 = 3,070 \text{ N} \cdot \text{m}$$

The system will remain in equilibrium for the single value

$$M_0 = 3,070 \text{ N} \cdot \text{m}$$

(b) Figure 11.27e shows the free-body diagram of the piston, for assumed impending motion of this element to the right.   For equilibrium,

$$\Sigma F_x = 0 \qquad 34.5 - F_{ab} \cos 20° - 0.15 \, N = 0$$
$$\Sigma F_y = 0 \qquad N - F_{ab} \sin 20° = 0$$

N is eliminated between the above two equations to obtain

$$F_{ab} = 34.8 \text{ kN (C)}$$

Using Fig. 11.27d,

$$\Sigma M_c = 0 \qquad -(34.8 \cos 33.2°)100 + M_0 = 0 \qquad M_0 = 2,910 \text{ N} \cdot \text{m}$$

The reaction force R_c is given by

$$R_c = F_{ab} = 34.8 \text{ kN}$$

The free-body diagram of the piston, for impending motion of this element to the left, is shown in Fig. 11.27f.   The equilibrium equations for the piston are

$$\Sigma F_x = 0 \qquad 34.5 - F_{ab} \cos 20° + 0.15 N = 0$$
$$\Sigma F_y = 0 \qquad N - F_{ab} \sin 20° = 0$$

Eliminating N between the above two equations results in

$$F_{ab} = 38.8 \text{ kN (C)}$$

From Fig. 11.27d,

$$\Sigma M_c = 0 \qquad -(38.8 \cos 33.2°)100 + M_0 = 0 \qquad M_0 = 3,250 \text{ N} \cdot \text{m}$$

The reaction force R_c is found as

$$R_c = F_{ab} = 38.8 \text{ kN}$$

The system will remain in equilibrium if

$$2{,}910 \, \text{N} \cdot \text{m} \le M_0 \le 3{,}250 \, \text{N} \cdot \text{m}$$

The corresponding range of values of the reaction force R_c is

$$34.8 \, \text{kN} \le R_c \le 38.8 \, \text{kN}$$

11.28 An outdoor billboard display is shown in Fig. 11.28a. The plane wall is connected to a strut in a vertical plane, at a, and is hinged to the ground at the two locations c. The maximum anticipated value of wind pressure which will act on the display is $15 \, \text{lb/ft}^2$, and this force distribution is shown in the figure. Find the force in the strut ab which supports the display and the force exerted by the ground on the plane wall. Neglect the weight of the wall.

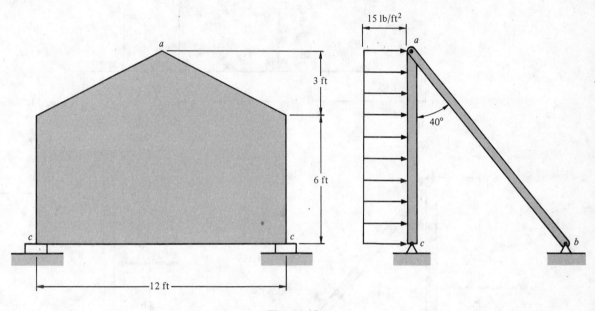

Fig. 11.28a

▌ The plane area is divided into the two elementary areas shown in Fig. 11.28b. The centroidal coordinate y_c of this area is

$$y_c = \frac{y_1 A_1 + y_2 A_2}{A_1 + A_2} = \frac{3(12)6 + 7(0.5)12(3)}{12(6) + 0.5(12)(3)} = 3.8 \, \text{ft} \qquad A = 90 \, \text{ft}^2$$

The resultant force F acting on the billboard, due to the wind loading, is

$$F = pA = 15 \, \frac{\text{lb}}{\text{ft}^2} \, (90 \, \text{ft}^2) = 1{,}350 \, \text{lb}$$

The location of this resultant force of the wind loading is shown in Fig. 11.28c. Hinge c in this figure represents the combined effect of the two hinges which attach the wall to the ground. Member ab is a two-force member, so that the direction of the force in this member is along the axis of the member. The free-body diagram of the wall is shown in Fig. 11.28d.

The equilibrium requirements are

$$\sum M_c = 0 \qquad -1{,}350(3.8) + (F_{ab} \sin 40°)(9) = 0 \qquad F_{ab} = 887 \, \text{lb}$$

$$\sum F_y = 0 \qquad -R_{cy} + F_{ab} \cos 40° = 0 \qquad R_{cy} = 679 \, \text{lb}$$

$$\sum F_x = 0 \qquad 1{,}350 + R_{cx} - F_{ab} \sin 40° = 0 \qquad R_{cx} = -780 \, \text{lb}$$

The minus sign in the above equation indicates that the actual sense of R_{cx} is to the left in Fig. 11.28d.

The resultant force R_c exerted by the ground on the vertical wall is then

$$R_c = \sqrt{R_{cx}^2 + R_{cy}^2} = \sqrt{(-780)^2 + 679^2} = 1{,}030 \, \text{lb}$$

Half of this force is transmitted through each of the two hinges.

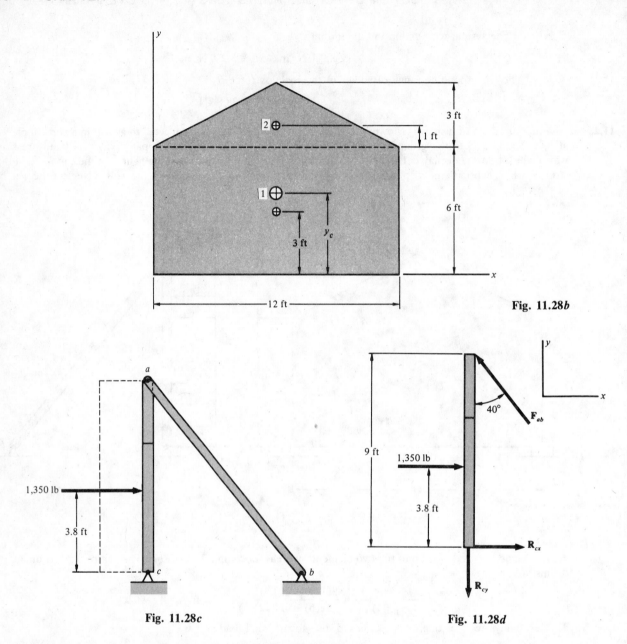

Fig. 11.28b

Fig. 11.28c

Fig. 11.28d

11.3 PRESSURE IN A STATIONARY, INCOMPRESSIBLE LIQUID, MAGNITUDE AND LOCATION OF RESULTANT FORCE ON A SUBMERGED PLANE

11.29 Figure 11.29a shows a tank filled with a stationary, incompressible liquid to a depth h. The coordinate x is measured from the surface of the liquid.

(a) Use the results from Prob. 11.18 to find the pressure at a point in the liquid.

(b) Find the magnitude and location of the maximum pressure in the liquid.

(c) What type of force distribution in the liquid is represented by the variation of pressure with depth?

(d) State several typical examples of the loading produced by a stationary, incompressible liquid on a body.

▮ (a) A column of liquid is shown by the dashed lines. The plane area of the bottom of this column is A. The weight W of this column of liquid is given by $W = Ax\gamma$, where the quantity Ax is the volume of the column of liquid and γ is the specific weight of the liquid, with the units force per unit volume. This weight of liquid is supported by the horizontal plane area A at the bottom of the column. Since the liquid is stationary, it may be concluded that the weight force of the column of liquid is *uniformly* distributed over the area A. The *pressure* which acts on plane A is then

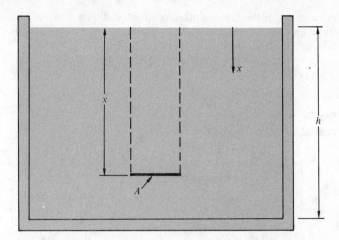

Fig. 11.29a

$$p = \frac{W}{A} = \frac{\cancel{A}x\gamma}{\cancel{A}} = x\gamma$$

The above result is a fundamental relationship of fluid mechanics. This equation is independent of A, and it states that the pressure at a point in a stationary, incompressible liquid is the product of the depth beneath the surface and the specific weight of the liquid. The height of the liquid column is also referred to as the *pressure head*.

(*b*) The maximum pressure p_{max} in the tank shown in Fig. 11.29a acts on the floor of the tank, where $x = h$, with the value

$$p_{max} = h\gamma$$

The pressure given by the equations above is referred to as a *gage* pressure since, according to this equation, $p = 0$ when $x = 0$. If a pressure p_0 exists above the liquid surface, the pressure has the form

$$p = p_0 + x\gamma$$

The above results are for the case of a horizontal plane area. It can be shown that, at a given point beneath the surface of a liquid, *the pressure is the same for all directions through the point, and this pressure is a function only of the depth*.

(*c*) A plot of pressure vs. depth along a vertical rectangular tank wall is shown in Fig. 11.29b. This distribution of the pressure forces along the wall may be recognized as the uniformly varying load distribution which was presented in Prob. 11.10.

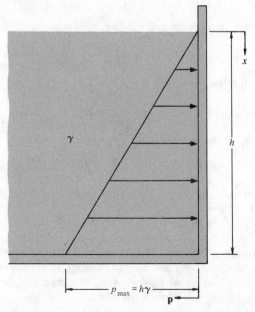

$$p_{max} = h\gamma$$

Fig. 11.29b

(d) Typical examples of the loading produced by a stationary, incompressible liquid on a body include the forces on dams, on the walls of tanks or pressure vessels, and on submerged surfaces. This subject is treated in detail in hydraulics and fluid mechanics.

11.30 A homogeneous body of square cross section floats in water, as shown in Fig. 11.30a. The density of water is 1,000 kg/m³.

(a) Find the weight of the body.

(b) Find the mass density of the body.

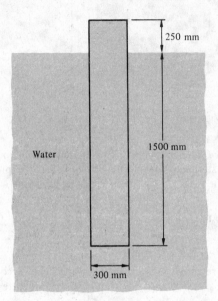

Fig. 11.30a   Fig. 11.30b

❚ (a) Figure 11.30b shows the free-body diagram of the body. W is the weight of the body and F is the resultant force of the liquid on the plane bottom area of the body, with the form

$$F = pA = x\gamma A = 1,500 \text{ mm}\left(1,000 \frac{\text{kg}}{\text{m}^3}\right)9.81 \frac{\text{m}}{\text{s}^2}(300 \text{ mm})^2\left(\frac{1 \text{ m}}{1,000 \text{ mm}}\right)^3 = 1,320 \text{ N}$$

For equilibrium of the body,

$$\Sigma F_y = 0 \qquad -W + F = 0 \qquad W = 1,320 \text{ N}$$

(b) The mass of the body is given by

$$m = \frac{W}{g} = \frac{1,320 \text{ N}}{9.81 \text{ m/s}^2} = 135 \text{ kg}$$

The volume of the body is

$$V = (300)^2 1,750 \text{ mm}^3\left(\frac{1 \text{ m}}{1,000 \text{ mm}}\right)^3 = 0.158 \text{ m}^3$$

The mass density of the body is then found as

$$\rho = \frac{m}{V} = \frac{135}{0.158} = 854 \text{ kg/m}^3$$

It may be observed that the density of the body is less than the density of the water, which is an expected result for a floating body.

11.31 Figure 11.31a shows a boat with a glass observation window in the floor. The boat floats in salt water, and the glass is a 15 in × 25 in rectangular plate. To ensure a watertight connection, the glass must exert a force of 1.5 lb/in on the sealing strip. The window is held down by 12 bolts, and the force of the glass on the seal may be assumed to be uniformly distributed. What is the minimum required bolt force if all bolts share the load equally? The specific weight of salt water is 64.0 lb/ft³.

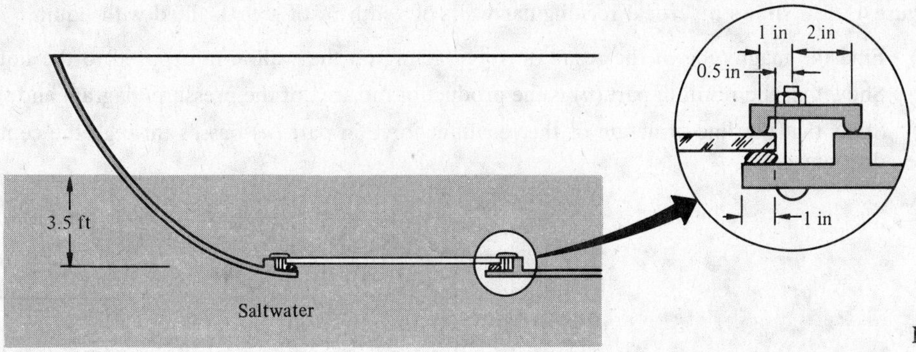

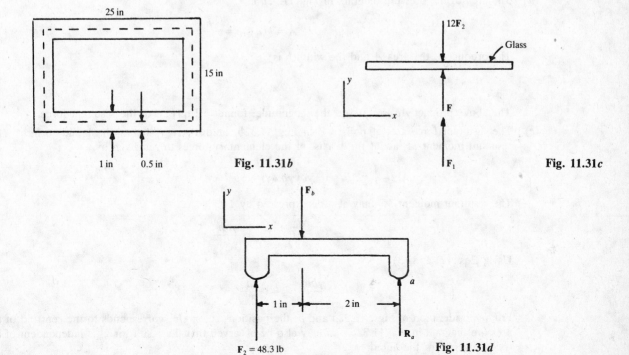

Fig. 11.31a

▮ A is the glass area exposed to the salt water. Subtracting the 1-in width of the sealing strip around the perimeter of the glass,

$$A = (15 - 2)(25 - 2) = 299 \text{ in}^2$$

F is the force exerted by the salt water on the window, given by

$$F = pA = x\gamma A = 3.5 \text{ ft} \left(6.40 \frac{\text{lb}}{\text{ft}^3} \right)(299 \text{ in}^2)\left(\frac{1 \text{ ft}}{12 \text{ in}} \right)^2 = 465 \text{ lb}$$

The dashed line in Fig. 11.31b shows the centerline of the sealing strip. The length l of this strip is

$$l = 2(15 - 1) + 2(25 - 1) = 76 \text{ in}$$

F_1 is the total required seal force, given by

$$F_1 = \left(1.5 \frac{\text{lb}}{\text{in}} \right)76 \text{ in} = 114 \text{ lb}$$

F_2 is the force exerted by each clamping lever on the glass. The glass window is subjected to the three forces F, F_1, and F_2, and the free-body diagram of the window is shown in Fig. 11.31c. For equilibrium of the window,

$$\Sigma F_y = 0 \qquad -12F_2 + F + F_1 = 0 \qquad F_2 = 48.3 \text{ lb}$$

F_b is the force exerted by the bolt on the clamping lever, and the free-body diagram of this element is shown in Fig. 11.31d. For equilibrium of the lever,

$$\Sigma M_a = 0 \qquad -48.3(3) + F_b(2) = 0 \qquad F_b = 72.5 \text{ lb}$$

Fig. 11.31b

Fig. 11.31c

Fig. 11.31d

11.32   Figure 11.32a shows a vertical rectangular wall, of width b, of a tank filled with liquid.

(a)   Find the magnitude of the resultant force acting on the wall area exposed to the liquid.

(b)   Show that the result in part (a) is the product of the area of the pressure diagram and the width of the wall.

(c)   Show that the line of action of the resultant force in part (a) passes through the centroid of the pressure diagram.

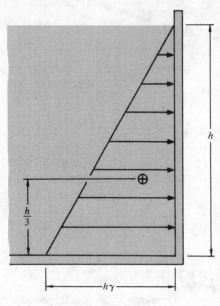

Fig. 11.32a

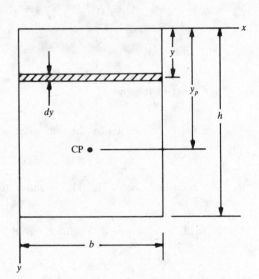

Fig. 11.32b

∎ (a)   Figure 11.32b shows the rectangular wall area exposed to the liquid.   The force on an elementary area is

$$dF = p \, dA = y\gamma b \, dy \qquad F = \int dF = \int_0^h y\gamma b \, dy = \gamma b \int_0^h y \, dy = \gamma b \left.\frac{y^2}{2}\right|_0^h = \frac{\gamma b h^2}{2} \qquad (1)$$

(b)   The area of the pressure diagram in Fig. 11.32a is

$$A = \tfrac{1}{2}(h\gamma)h = \frac{\gamma h^2}{2}$$

The product of the area A and the width b is

$$Ab = \frac{\gamma b h^2}{2}$$

The above product Ab is equal to the magnitude, found in Eq. (1), of the resultant force.

(c)   The resultant force acts at point CP in Fig. 11.32b, at distance y_p below the liquid surface.   The resultant moment M_x about the x axis, of the elementary forces dF, is given by

$$M_x = \int y \, dF = \int_0^h y(y\gamma b \, dy) = \gamma b \int_0^h y^2 \, dy = \gamma b \left.\frac{y^3}{3}\right|_0^h = \frac{\gamma b h^3}{3} \qquad (2)$$

The resultant moment M_x may also be expressed by

$$M_x = F y_p = \frac{\gamma b h^2}{2} \, y_p \qquad (3)$$

Using Eqs. (2) and (3),

$$\frac{\gamma b h^2}{2} \, y_p = \frac{\gamma b h^3}{3} \qquad y_p = \tfrac{2}{3}h$$

From consideration of Figs. 11.32a and b, the position $y_p = \tfrac{2}{3}h$ corresponds to the centroid of the pressure diagram in Fig. 11.32a.   It may also be observed that the result for y_p is independent of the specific weight of the liquid.

11.33 The tank of rectangular cross section shown in Fig. 11.33a is filled with water to the depth shown. The dimension of the tank normal to the plane of the figure is 600 mm.

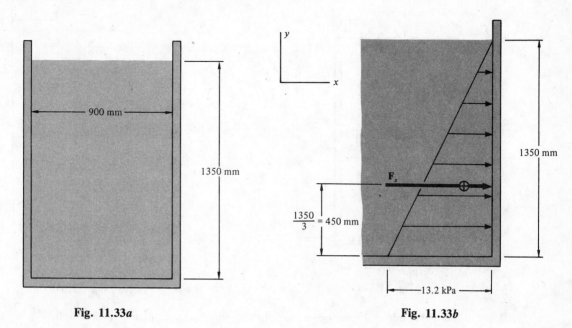

Fig. 11.33a   Fig. 11.33b

(a) Find the pressure, and the total force, acting on the bottom of the tank.

(b) Find the magnitude and location of the force which the water exerts on the vertical walls. The density of water is 10^3 kg/m^3.

▮ (a) The specific weight γ of the water is

$$\gamma = \rho g = 10^3 \, \frac{\text{kg}}{\text{m}^3}\left(9.81 \, \frac{\text{m}}{\text{s}^2}\right) = 9.81 \times 10^3 \, \frac{\text{kg} \cdot \text{m/s}^2}{\text{m}^3} = 9.81 \times 10^3 \, \frac{\text{N}}{\text{m}^3}$$

The pressure on the bottom of the tank is

$$p = h\gamma = (1{,}350 \, \text{mm})\left(9.81 \times 10^3 \, \frac{\text{N}}{\text{m}^3}\right)\left(\frac{1 \, \text{m}}{1{,}000 \, \text{mm}}\right) = 1.32 \times 10^4 \, \frac{\text{N}}{\text{m}^2} = 1.32 \times 10^4 \, \text{Pa} = 13.2 \, \text{kPa}$$

The area A of the tank floor is

$$A = 900(600) = 5.40 \times 10^5 \, \text{mm}^2$$

The total force on the bottom of the tank is

$$F = pA = \left(1.32 \times 10^4 \, \frac{\text{N}}{\text{m}^2}\right)(5.40 \times 10^5 \, \text{mm}^2)\left(\frac{1 \, \text{m}}{1{,}000 \, \text{mm}}\right)^2 = 7{,}130 \, \text{N}$$

It is left as an exercise for the reader to show that this force is equal to the weight of the water in the tank.

(b) The pressure distribution on the vertical wall is shown in Fig. 11.33b. The resultant force F_x is equal to the product of the area of the pressure diagram and the width of the wall, or

$$F_x = 0.5\left(1.32 \times 10^4 \, \frac{\text{N}}{\text{m}^2}\right)(1{,}350 \, \text{mm})(600 \, \text{mm})\left(\frac{1 \, \text{m}}{1{,}000 \, \text{mm}}\right)^2 = 5{,}350 \, \text{N}$$

The line of action of this force is at a distance $1{,}350/3 = 450$ mm above the bottom of the tank, as shown in Fig. 11.33b.

11.34 The dimension of the rectangular tank in Fig. 11.34a normal to the plane of the figure is 1,800 mm. The liquid in the tank has a specific gravity of 1.6, and the pressure in the space above the liquid is 75 kPa.

(a) Find the resultant force acting on the top of the tank.

(b) Find the resultant force acting on the bottom of the tank.

(c) Sketch the pressure distribution on the vertical walls of the tank.

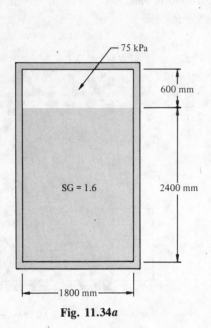

Fig. 11.34a

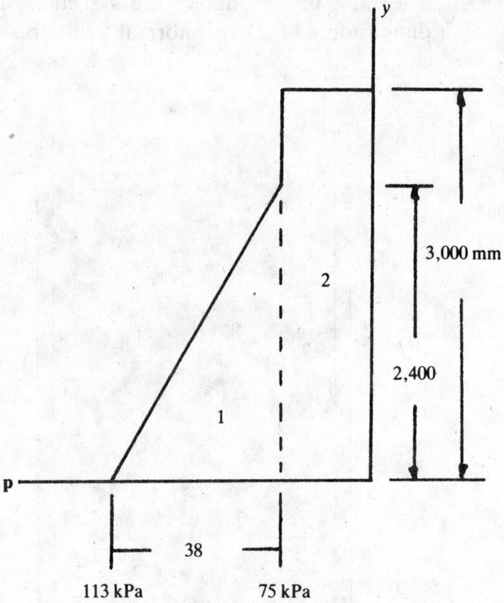

Fig. 11.34b

(*d*) Find the magnitude and location of the resultant force acting on each vertical wall.

▌ (*a*) The force acting on the top of the tank is given by

$$F = pA = 75,000 \frac{N}{m^2} (1,800)^2 \text{ mm}^2 \left(\frac{1 \text{ m}}{1,000 \text{ mm}}\right)^2 = 2.43 \times 10^5 \text{ N} = 243 \text{ kN}$$

(*b*) The pressure acting on the bottom of the tank is

$$p_{\text{bottom}} = p_0 + x\gamma = 75 \text{ kPa} + \left(\frac{2,400}{1,000} \text{ m}\right)1.6\left(1,000 \frac{\text{kg}}{\text{m}^3}\right)9.81 \frac{\text{m}}{\text{s}^2} \left(\frac{1 \text{ kPa}}{1,000 \text{ Pa}}\right) = 113 \text{ kPa}$$

The resultant force acting on the bottom of the tank is

$$F_{\text{bottom}} = p_{\text{bottom}}A = 113 \text{ kPa}(1,800)^2 \text{ mm}^2 \left(\frac{1 \text{ m}}{1,000 \text{ mm}}\right)^2 = 366 \text{ kN}$$

(*c*) The pressure distribution on the four walls of the tank is shown in Fig. 11.34*b*.

(*d*) The force on the wall is the product of the area of the pressure diagram and the width of the wall, given by

$$F = [\tfrac{1}{2}(38)2,400 + 75(3,000)](\text{kPa})(\text{mm})(1,800 \text{ mm})\left(\frac{1 \text{ m}}{1,000 \text{ mm}}\right)^2 = 487 \text{ kN}$$

The centroidal coordinate y_c of the pressure diagram in Fig. 11.34*b* is found as

$$y_c = \frac{y_1 A_1 + y_2 A_2}{A_1 + A_2} = \frac{\tfrac{1}{3}(2400)\tfrac{1}{2}(38)2400 + (3,000/2)(75)3,000}{\tfrac{1}{2}(38)2400 + 75(3,000)} = 1,380 \text{ mm}$$

The line of action of the resultant force is 1,380 mm above the bottom of the tank.

11.35 (*a*) Find the magnitude of the resultant force of a liquid on a submerged plane area at angle θ with the surface of the liquid.

(*b*) Find the location of the resultant force found in part (*a*).

(*c*) What term is used to describe the point where the resultant force acts on the plane?

▌ (*a*) The force on a vertical tank wall was found in Prob. 11.32 from direct consideration of the pressure distribution over the wall. That problem was a special case of the more general problem of finding the force on a submerged plane area. The present problem is devoted to this latter topic.

Figure 11.35 shows a plane area which is submerged in a stationary, incompressible liquid. The area of the plane is A, and the angle between the plane and the surface of the liquid is θ. The coordinate y

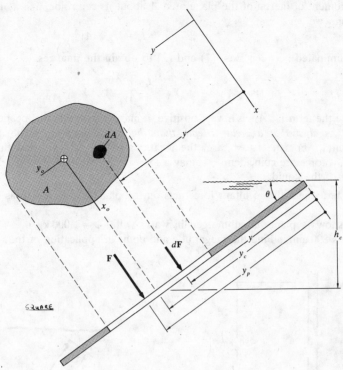

SQUARE

Fig. 11.35

is measured from the intersection line of the plane and the surface. y_c locates the centroid of the plane area, and y_p gives the location of the resultant force on the plane.

The force dF on a differential area dA of the plane is given by

$$dF = (y \sin \theta)\gamma \, dA$$

The total force F acting on the submerged plane is the summation of all the elementary forces dF, or

$$F = \int_A (y \sin \theta)\gamma \, dA = \gamma \sin \theta \int_A dA$$

From the definition of a centroidal coordinate,

$$\int_A y \, dA = y_c A \qquad \text{so that} \qquad F = (\gamma \sin \theta)y_c A$$

If h_c is defined to be the depth of the *centroid of the submerged plane area*, then

$$h_c = y_c \sin \theta$$

The final form for the resultant force is then

$$F = h_c \gamma A$$

The above equation states that the total force of the liquid on the submerged plane is *the product of the area of the plane and the pressure at the centroid of this area*.

(**b**) The location y_p of the resultant force is found next. The moment dM of the elementary force dF about the x axis in Fig. 11.35 is

$$dM = (dF)y = (y \sin \theta)(\gamma \, dA)y = (y^2 \sin \theta)(\gamma \, dA)$$

The moment of the resultant force F about this axis must be equal to the sum of the elementary moments given by the above equation, or

$$Fy_p = \int_A (y^2 \sin \theta)\gamma \, dA = \gamma \sin \theta \int_A y^2 \, dA$$

The integral term on the right-hand side of the equation above may be recognized as the moment of inertia of the plane area A about the x axis. This term is designated I_x, and the equation then appears as

$$Fy_p = (\gamma \sin \theta)I_x \qquad y_p = \frac{(\gamma \sin \theta)I_x}{F} = \frac{(\gamma \sin \theta)I_x}{(\gamma \sin \theta)Ay_c} = \frac{I_x}{Ay_c} \qquad (1)$$

The moment of inertia of the plane area A about its centroidal axis x_0 is I_{0x}. Using the parallel-axis theorem, we get

$$I_x = I_{0x} + Ay_c^2 \qquad (2)$$

I_x is eliminated between Eqs. (1) and (2) to obtain the final result

$$y_p = \frac{I_{0x} + Ay_c^2}{Ay_c} = y_c + \frac{I_{0x}}{Ay_c}$$

Since the term I_{0x}/Ay_c is always positive, it may be seen that the location of the resultant force, given by y_p, is *always at a greater depth than the centroid of the plane area* unless the plane area is horizontal. For the latter case, the locations of the centroid and the point of application of the resultant force are coincident. It may also be seen that the value of y_p is independent of the specific weight of the liquid.

(c) The point where the resultant force acts on the plane is referred to as the *center of pressure*.

11.36 Figure 11.36a shows a plane *ab* submerged in water with $\rho = 1,000 \text{ kg/m}^3$. The dimension normal to the plane of figure is 800 mm. Find the magnitude and point of application of the resultant force of the water on the plane.

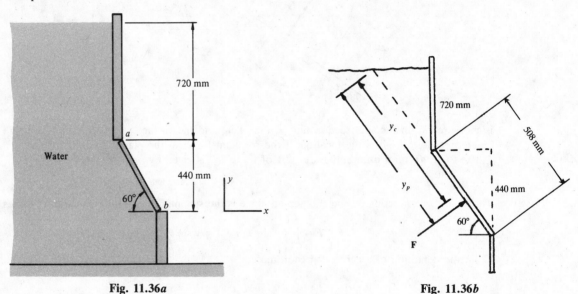

Fig. 11.36a   **Fig. 11.36b**

❙ Figure 11.36b shows the terms y_c and y_p, measured from the intersection line of the plane with the surface of the liquid. The resultant force of the liquid on the plane is given by

$$F = h_c \gamma A = \left(720 + \frac{440}{2}\right) \text{mm} \left(1,000 \ \frac{\text{kg}}{\text{m}^3}\right) 9.81 \ \frac{\text{m}}{\text{s}^2} \ 508(800) \ \text{mm}^2 \left(\frac{1 \text{ m}}{1,000 \text{ mm}}\right)^3 = 3,750 \text{ N}$$

The length y_c is found from

$$y_c = \frac{720 + 440/2}{\sin 60°} \doteq 1,090 \text{ mm}$$

The location of the center of pressure is then found as

$$y_p = y_c + \frac{I_{0x}}{Ay_c} = 1,090 + \frac{800(508)^3/12}{800(508)1,090} = 1,110 \text{ mm}$$

11.37 Figure 11.37a shows the back end of a flat-bottom rowboat. The rectangular rear transom board has a width of 1,200 mm. Find the magnitude and the point of application of the total force exerted by the liquid on the rear section of the boat. The density of water is 10^3 kg/m^3.

❙ The submerged plane *ab* is drawn in Fig. 11.37b. The specific weight of water is

$$\gamma = \rho g = 10^3 \ \frac{\text{kg}}{\text{m}^3} \left(9.81 \ \frac{\text{m}}{\text{s}^2}\right) = 9.81 \times 10^3 \ \frac{\text{N}}{\text{m}^3}$$

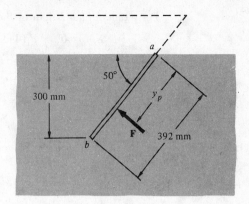

Fig. 11.37a   **Fig. 11.37b**

The total force of the liquid on the plane area is

$$F = h_c \gamma A = 150\text{ mm}\left(9.81 \times 10^3\ \frac{N}{m^3}\right)(392)(1{,}200)\text{ mm}^2\left(\frac{1\text{ m}}{1{,}000\text{ mm}}\right)^3 = 692\text{ N}$$

The location of the center of pressure, where the resultant force acts, is given by

$$y_p = y_c + \frac{I_{0x}}{Ay_c} = \frac{392}{2} + \frac{1{,}200(392)^3/12}{1{,}200(392)(392/2)} = 261\text{ mm}$$

11.38 The width of the gate in Fig. 11.38a is 1,600 mm, and the weight of the gate may be neglected.

(a) Find the force exerted by the gate on the stop.

(b) Find the reaction force of the ground on the hinge pin.

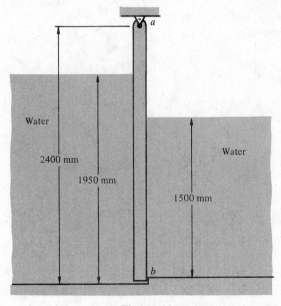

Fig. 11.38a

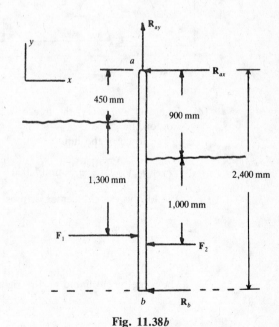

Fig. 11.38b

❚ (a) Figure 11.38b shows the free-body diagram of the gate. R_b is the force exerted by the stop on the gate. The force F_1 of the liquid which acts on the left side of the gate is given by

$$F_1 = h_{c1}\gamma A_1 = \frac{1{,}950}{2}\text{ mm}\left(1{,}000\ \frac{kg}{m^3}\right)\left(9.81\ \frac{m}{s^2}\right)1{,}950(1{,}600)\text{ mm}^2\left(\frac{1\text{ m}}{1{,}000\text{ mm}}\right)^3 = 29{,}800\text{ N} = 29.8\text{ kN}$$

The force F_2 of the liquid on the right side of the gate is found as

$$F_2 = h_{c2}\gamma A_2 = \frac{1{,}500}{2}(1{,}000)9.81(1{,}500)(1{,}600)\left(\frac{1}{1{,}000}\right)^3 = 17{,}700\text{ N} = 17.7\text{ kN}$$

For moment equilibrium of the gate,

$$\Sigma M_a = 0 \qquad F_1(450 + 1{,}300) - F_2(900 + 1{,}000) - R_b(2{,}400) = 0 \qquad R_b = 7.72 \text{ kN}$$

Using Newton's third law, the force of the gate on the stop is 7.72 kN, acting to the right in Fig. 11.38a.

(b) For force equilibrium of the gate,

$$\Sigma F_x = 0 \qquad -R_{ax} + F_1 - F_2 - R_b = 0 \qquad R_{ax} = 4.38 \text{ kN}$$
$$\Sigma F_y = 0 \qquad R_{ay} = 0 \qquad R_a = R_{ax} = 4.38 \text{ kN}$$

11.39 Do the same as in Prob. 11.38 for the case where the gate is inclined, as shown in Fig. 11.39a.

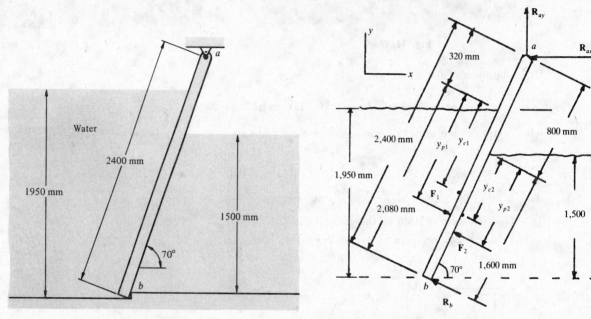

Fig. 11.39a   **Fig. 11.39b**

❚ (a) Figure 11.39b shows the free-body diagram of the inclined gate. R_b is again the force exerted by the stop on the gate.

The force of the liquid on the left side of the gate is

$$F_1 = h_{c1}\gamma A_1 = \frac{1{,}950}{2} \text{ mm}\left(1{,}000 \frac{\text{kg}}{\text{m}^3}\right)9.81 \frac{\text{m}}{\text{s}^2}(2{,}080)(1{,}600) \text{ mm}^2\left(\frac{1 \text{ m}}{1{,}000 \text{ mm}}\right)^3 = 31{,}800 \text{ N} = 31.8 \text{ kN}$$

The location of the center of pressure on the left side of the gate is found from

$$y_{c1} = \frac{2{,}080}{2} = 1{,}040 \text{ mm}$$

$$y_{p1} = y_{c1} + \frac{I_{0x,1}}{A_1 y_{c1}} = 1{,}040 + \frac{1{,}600(2{,}080)^3/12}{2{,}080(1{,}600)1{,}040} = 1{,}390 \text{ mm}$$

The force of the liquid on the right side of the gate is given by

$$F_2 = h_{c2}\gamma A_2 = \frac{1{,}500}{2}(1{,}000)9.81(1{,}600)1{,}600\left(\frac{1}{1{,}000}\right)^3 = 18{,}800 \text{ N} = 18.8 \text{ kN}$$

The location of the center of pressure on the right side of the gate is found from

$$y_{c2} = \frac{1{,}600}{2} = 800 \text{ mm} \qquad y_{p2} = y_{c2} + \frac{I_{0x}}{A_2 y_{c2}} = 800 + \frac{1{,}600(1{,}600)^3/12}{1{,}600(1{,}600)800} = 1{,}070 \text{ mm}$$

For moment equilibrium of the gate,

$$\Sigma M_a = 0 \qquad F_1(1{,}390 + 320) - F_2(1{,}070 + 800) - R_b(2{,}400) = 0 \qquad R_b = 8.01 \text{ kN}$$

From Newton's third law, the force of the gate on the stop is 8.01 kN, acting down and to the right in Fig. 11.39b.

(b) $\Sigma F_x = 0$ $-R_{ax} + F_1 \cos 20° - F_2 \cos 20° - R_b \cos 20° = 0$ $R_{ax} = 4.69\text{ kN}$

$\Sigma F_y = 0$ $R_{ay} - F_1 \sin 20° + F_2 \sin 20° = 0$ $R_{ay} = 4.45\text{ kN}$

$$R_a = \sqrt{R_{ax}^2 + R_{ay}^2} = \sqrt{4.69^2 + 4.45^2} = 6.47\text{ kN}$$

The percent increase in the pin force R_a for the case where the gate is inclined, using the above result and the value of R_a found in Prob. 11.38, is

$$\%D = \frac{6.47 - 4.38}{4.38}\, 100 = 48\%$$

11.40 Figure 11.40a shows a floodgate which is designed to open when the water level reaches a certain height h. For what value of h will the opening motion of the gate be impending? Neglect all friction effects in the seals and hinge pin.

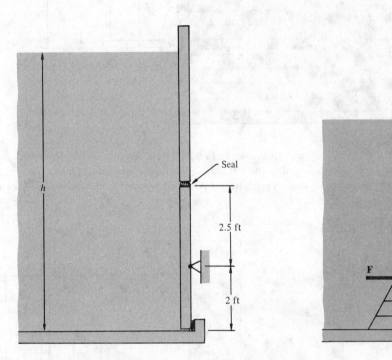

h

Seal

2.5 ft

2 ft

F

Fig. 11.40a   **Fig. 11.40b**

I The pressure distribution for a depth less than that required to open the gate is shown in Fig. 11.40b. For this case, the resultant force F acts on the gate at a location which is below the hinge pin, and this force tends to keep the gate closed against the stop. As the depth of the liquid is increased, the location of the resultant force on the gate moves upward. The limiting condition is reached when this resultant force acts through the hinge pin, as shown in Fig. 11.40c. For this position, opening motion of the gate is impending.

The location of the resultant force is given by

$$y_p = y_c + \frac{I_{0x}}{Ay_c} \qquad y_p - y_c = \frac{I_{0x}}{Ay_c}$$

From Fig. 11.40c,

$$y_p - y_c = 2.5 - 2.25 = 0.25\text{ ft}$$

The width of the gate is designated b, and

$$0.25 = \frac{I_{0x}}{Ay_c} = \frac{b(4.5)^3/12}{b(4.5)(y_c)} \qquad y_c = 6.75\text{ ft}$$

The final value for the depth h is, then,

$$h = y_c + 2.25 = 6.75 + 2.25 = 9\text{ ft}$$

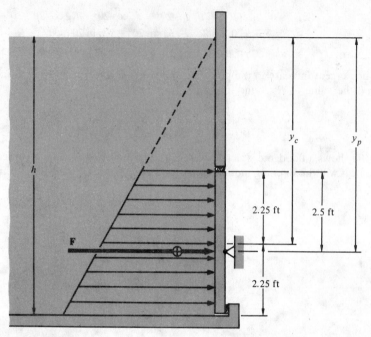

Fig. 11.40c

11.41   When the water level on the left side of the gate shown in Fig. 11.41a drops to a certain level, the gate is to open and allow water from the right side to flow into the left side.   Find the required weight W of the counterweight if opening motion of the gate is to be impending when the left-side water level reaches 750 mm.

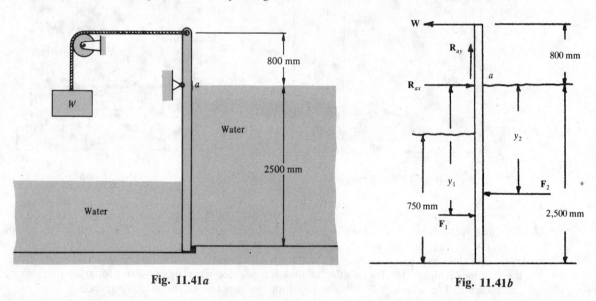

Fig. 11.41a   Fig. 11.41b

❚ The free-body diagram of the gate is shown in Fig. 11.41b.   The cable force W which acts on the gate is equal to the weight force of the counterweight.   The width of the gate is assumed to be 1 meter.   The resultant force of the liquid on the left side of the gate is given by

$$F_1 = h_{c1}\gamma A_1 = \frac{750}{2} \text{ mm}\left(1,000 \frac{\text{kg}}{\text{m}^3}\right)9.81 \frac{\text{m}}{\text{s}^2} (750 \text{ mm})(1 \text{ m})\left(\frac{1 \text{ m}}{1,000 \text{ mm}}\right)^2 = 2,760 \text{ N}$$

The location of F_1 is found as

$$y_1 = 2,500 - \tfrac{1}{3}(750) = 2,250 \text{ mm}$$

The resultant liquid force on the right side of the gate is

$$F_2 = h_{c2}\gamma A_2 = \frac{2,500}{2} (1,000)(9.81)2,500(1)\left(\frac{1}{1,000}\right)^2 = 30,700 \text{ N}$$

The position of force F_2 is found from

$$y_2 = \tfrac{2}{3}(2,500) = 1,670 \text{ mm}$$

For moment equilibrium of the gate,

$$\Sigma M_a = 0 \qquad W(800) - 30,700(1,670) + 2,760(2,250) = 0 \qquad W = 56,300 \text{ N} = 56.3 \text{ kN}$$

11.42 The gate in Fig. 11.42a is 5 ft long and weighs 250 lb. The sealing strip at the bottom of the gate is assumed to be frictionless. Find the minimum value of the force P required to keep the gate in the position shown.

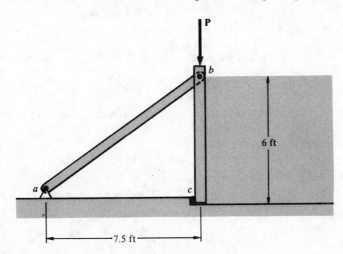

Fig. 11.42a

▌ The resultant force of the liquid on the gate is given by

$$F = h_c \gamma A = \frac{6}{2}(62.4)5(6) = 5,620 \text{ lb}$$

Figure 11.42b shows the free-body diagram of the gate. The link is a two-force member, so that F_{ab} has the known direction of the link. R_c is the normal force of the seal on the gate. If a friction force F_c were exerted by the sealing strip on the gate, it would have the direction shown by the dashed arrow in Fig. 11.42b. This situation is considered in Prob. 11.43.

For equilibrium of the gate,

$$\Sigma M_a = 0 \qquad -\left(\frac{7.5}{9.60} F_{ab}\right)6 + 5,620(2) = 0 \qquad F_{ab} = 2,400 \text{ lb}$$

$$\Sigma F_y = 0 \qquad P + \frac{6}{9.60}(2,400) - 250 = 0 \qquad P = 1,250 \text{ lb}$$

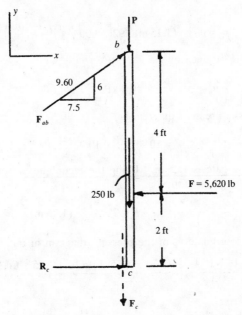

Fig. 11.42b

11.43 Do the same as in Prob. 11.42, if the coefficient of friction between the sealing strip and the gate is assumed to be 0.22.

❚ The sealing strip is now assumed to exert a friction force on the gate, as shown in Fig. 11.42b. Motion is assumed to be impending, so that the friction force may be written in the form

$$F_c = \mu R_c$$

The resultant force of the liquid on the gate has the same value as that found in Prob. 11.42. It follows that the force F_{ab} in the link is again

$$F_{ab} = 2,400 \text{ lb (C)}$$

For equilibrium of the gate,

$$\Sigma F_x = 0 \qquad \frac{7.5}{9.60} F_{ab} - 5,620 + R_c = 0 \qquad R_c = 3,750 \text{ lb}$$

$$\Sigma F_y = 0 \qquad -P + \frac{6}{9.60} F_{ab} - 250 - 0.22(3,750) = 0 \qquad P = 425 \text{ lb}$$

It may be seen that the friction force significantly reduces the value of force P required to keep the gate in the equilibrium position shown in Fig. 11.42a.

11.44 The width of the gate in Fig. 11.44a is 1.25 m.

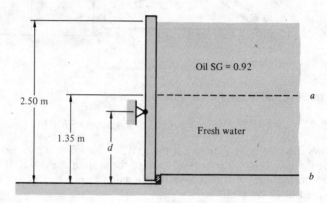

Fig. 11.44a

(*a*) Find the resultant force of the liquid on the gate in the closed position shown.

(*b*) Find the maximum permissible value of d if the gate is to remain in the closed position.

❚ (*a*) Figure 11.44b shows the variation of the pressure on the gate with depth. The pressure at a is found as

$$p_a = h\gamma = (1.15 \text{ m})0.92\left(1,000 \frac{\text{kg}}{\text{m}^3}\right)\left(9.81 \frac{\text{m}}{\text{s}^2}\right) = 10,400 \text{ Pa} = 10.4 \text{ kPa}$$

The pressure at b is given by

$$p_b = p_a + x\gamma = 10.4 \text{ kPa} + 1.35(1,000)(9.81)\left(\frac{1}{1,000}\right) = 23.6 \text{ kPa}$$

The area A of the pressure diagram is

$$A = \tfrac{1}{2}(10.4)1.15 + 10.4(1.35) + \tfrac{1}{2}(13.2)1.35 = 28.9 \frac{\text{kN}}{\text{m}}$$

The width of the gate is 1.25 m. Using the technique presented in Prob. 11.32, the resultant force which acts on the gate is found as

$$F = 28.9 \frac{\text{kN}}{\text{m}} (1.25 \text{ m}) = 36.1 \text{ kN}$$

(*b*) The centroidal coordinate y_c of the pressure diagram in Fig. 11.44b is found from

$$y_c = \frac{y_1 A_1 + y_2 A_2 + y_3 A_3}{A_1 + A_2 + A_3} = \frac{(1.35 + 1.15/3)\tfrac{1}{2}(10.4)1.15 + (1.35/2)(10.4)1.35 + (1.35/3)(\tfrac{1}{2})13.2(1.35)}{28.9}$$

$$= 0.825 \text{ m}$$

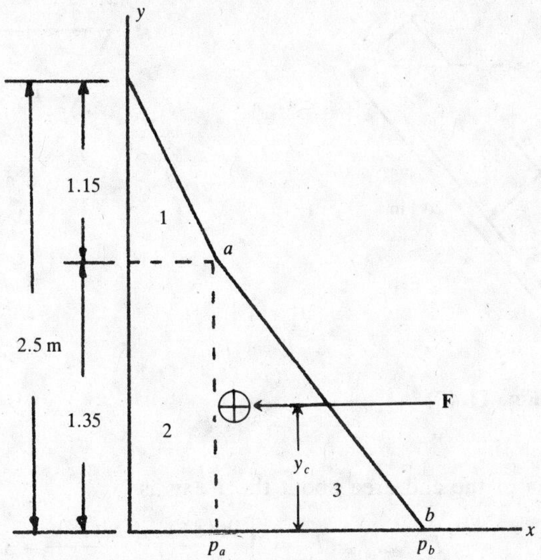

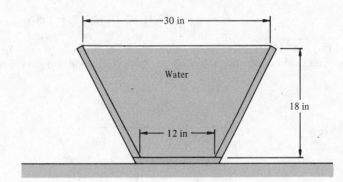

Fig. 11.44b

The gate will open if the resultant force of the liquid acts through a point which is below the hinge pin. Thus, d must be less than y_c, and $d_{max} = y_c = 0.825$ m.

11.45 Figure 11.45a shows a cross-sectional view of a watering trough for livestock. The end sections are vertical, and the length of the trough is 7 ft. Find the magnitude and location of the resultant forces of the water on the bottom, sides, and ends of the trough.

Fig. 11.45a

I Figure 11.45b shows the resultant force F_1 of the water on one inclined side of the trough. The magnitude of F_1 is given by

$$F_1 = h_c \gamma A = \frac{18}{2}\left(62.4\ \frac{\text{lb}}{\text{ft}^3}\right)\left(\frac{1\ \text{ft}}{12\ \text{in}}\right)^3 20.1(7)12 = 549\ \text{lb}$$

The location y_{p1} of the resultant force F_1 is found from

$$y_{p1} = y_c + \frac{I_{0x}}{Ay_c} \qquad I_{0x} = \frac{[7(12)]20.1^3}{12} = 5.68 \times 10^4\ \text{in}^4 \qquad y_{p1} = \frac{20.1}{2} + \frac{5.68 \times 10^4}{[20.1(7)12](20.1/2)} = 13.4\ \text{in}$$

The force F_2 of the water on the bottom of the trough is given by

$$F_2 = pA = 18(62.4)(\tfrac{1}{12})^3 12(7)12 = 655\ \text{lb}$$

F_2 acts through the centroid of the bottom area.

Figure 11.45c shows a vertical end of the trough. The area A and centroidal coordinate y_c are found as

$$A = 2[0.5(9)18] + 12(18) = 378\ \text{in}^2 \qquad y_c = \frac{2y_1A_1 + y_2A_2}{A} = \frac{2\left[\dfrac{18}{3}(0.5)9(18)\right] + \dfrac{18}{2}(12)18}{378} = 7.71\ \text{in}$$

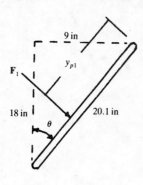

Fig. 11.45b

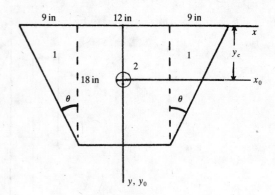

Fig. 11.45c

The moment of inertia of the end area about the x axis is

$$I_x = 2I_{x1} + I_{x2} = 2\left[\frac{9(18)^3}{12}\right] + \frac{12(18)^3}{3} = 32,100 \text{ in}^4$$

Using the parallel-axis theorem,

$$I_{0x} = I_x - Ay_c^2 = 32,100 - 378(7.71)^2 = 9,630 \text{ in}^4$$

The location of the resultant force is given by

$$y_p = y_c + \frac{I_{0x}}{Ay_c} = 7.71 + \frac{9,630}{378(7.71)} = 11.0 \text{ in}$$

The resultant force F_3 on the vertical ends is

$$F_3 = h_c\gamma A = 7.71(62.4)(\tfrac{1}{12})^3 378 = 105 \text{ lb}$$

11.46 Use the results in Prob. 11.45 to verify that the volume of water is in equilibrium when acted on by its own weight and the forces exerted on the water by the bottom and sides of the trough.

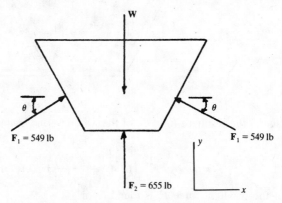

Fig. 11.46

❙ Figure 11.46 shows the free-body diagram of the volume of water. The weight of the water is the product of the specific weight and the volume, given by

$$w = \gamma V = 378(7)12(62.4)(\tfrac{1}{12})^3 = 1,150 \text{ lb}$$

For equilibrium of the volume of water

$$\Sigma F_y = 0 \qquad 2F_1 \sin\theta + F_2 \overset{?}{=} 1,150 \qquad 2(549)\left(\frac{9}{20.1}\right) + 655 \overset{?}{=} 1,150 \qquad 1,150 \equiv 1,150$$

It may be observed that forces F_3, on the vertical end walls, have no component in the y direction.

11.47 Figure 11.47a shows a masonry dam which rests on the ground. The average specific weight of the dam material is 140 lb/ft³.

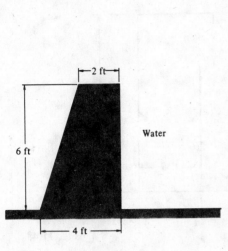

Fig. 11.47a

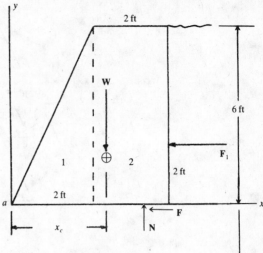

Fig. 11.47b

(a) Is there a possibility of the dam tipping over as the water level is increased until its maximum value of 6 ft is reached?

(b) For what value of the coefficient of friction between the base of the dam and the ground is sliding motion of the dam impending?

❚ (a) The cross-sectional area of the dam is divided into the two simple areas shown in Fig. 11.47b. The cross-sectional area A is given by

$$A = A_1 + A_2 = \tfrac{1}{2}(2)6 + 2(6) = 18 \text{ ft}^2$$

The centroidal coordinate x_c is given by

$$x_c = \frac{x_1 A_1 + x_2 A_2}{A_1 + A_2} = \frac{\tfrac{2}{3}(2)\tfrac{1}{2}(2)6 + (2 + \tfrac{2}{3})2(6)}{18} = 2.44 \text{ ft}$$

Figure 11.47b is now interpreted to be a partial free-body diagram of a 1-ft length of the dam, with the weight W given by

$$W = \gamma V = \left(140 \; \frac{\text{lb}}{\text{ft}^3}\right) 18(1) \text{ ft}^3 = 2{,}520 \text{ lb}$$

The resultant force F_1 of the liquid which acts on the vertical face is

$$F_1 = h_{c1} \gamma A_1 = 3 \text{ ft}\left(62.4 \; \frac{\text{lb}}{\text{ft}^3}\right) 6(1) \text{ ft}^2 = 1{,}120 \text{ lb}$$

The resultant moment, about point a, of the forces which act on the dam is found as

$$\Sigma M_a = -2{,}520(2.44) + 1{,}120(2) = -3{,}910 \text{ ft} \cdot \text{lb}$$

Since the resultant moment about a is negative, tipping motion of the dam is *not* impending:

(b) For impending sliding motion of the dam, the friction force F has its maximum possible value, given by

$$F_{\max} = \mu N \qquad \Sigma F_x = 0 \qquad \mu(2{,}520) - 1{,}120 = 0 \qquad \mu = 0.444$$

11.48 Do the same as in Prob. 11.47, if the inclined face of the dam is exposed to the 6-ft depth of water.

❚• (a) Figure 11.48 shows the dam with the inclined face exposed to the 6-ft depth of water. The magnitude and location of the resultant force F_2 of the liquid on this face are found as

$$F_2 = h_{c2} \gamma A_2 = \tfrac{6}{2}(62.4)6.32(1) = 1{,}180 \text{ lb}$$

$$y_p = y_c + \frac{I_{0x}}{A y_c} = \frac{6.32}{2} + \frac{1(6.32)^3/12}{6.32(1)(6.32/2)} = 4.21 \text{ ft}$$

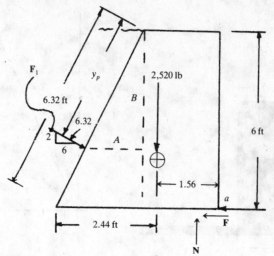

Fig. 11.48

From Fig. 11.48,

$$\frac{A}{2} = \frac{B}{6} = \frac{y_p}{6.32} = \frac{4.21}{6.32} \qquad A = 1.33 \text{ ft} \qquad B = 4.00 \text{ ft}$$

The resultant moment about point a is

$$\Sigma M_a = -\frac{6}{6.32}(1,180)2 + \frac{2}{6.32}(1,180)(1.33 + 2) + 2,520(1.56) = 2,930 \text{ ft} \cdot \text{lb}$$

Since the resultant moment about a is positive, tipping motion of the dam is not impending.

(**b**) N is the normal force exerted by the ground on the dam. For impending sliding motion of the dam

$$\Sigma F_y = 0 \qquad N - 2,520 - \frac{2}{6.32}(1,180) = 0 \qquad N = 2,890 \text{ lb}$$

$$\Sigma F_x = 0 \qquad \frac{6}{6.32}(1,180) - \mu(2,890) = 0 \qquad \mu = 0.388$$

Comparison of the above value with the result in Prob. 11.47 reveals that a smaller value of the coefficient of friction is required to prevent sliding motion when the face of the dam is inclined.

11.49 Figure 11.49a shows the board forms used in pouring a concrete barrier. Each section of the form is 4 ft long. The inclined board is staked to the ground, and the vertical form is held in position by the wires shown in the figure. There are two such wires, in a vertical plane, for each 1 ft of length of the boards. The specific weight of liquid concrete is 135 lb/ft³.

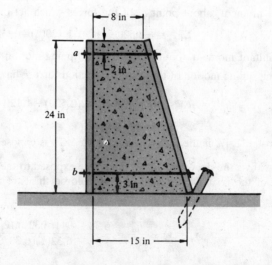

Fig. 11.49a

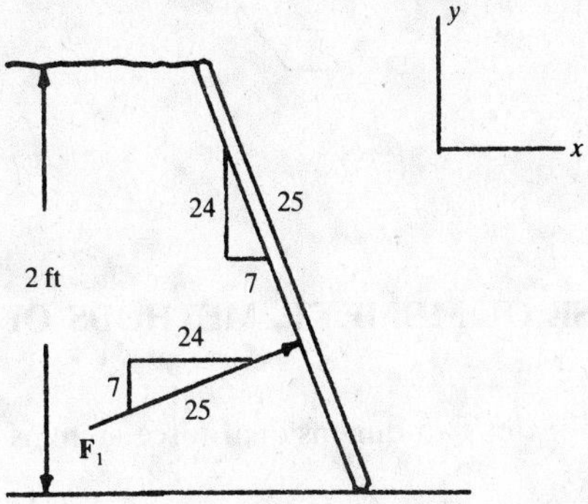

Fig. 11.49b 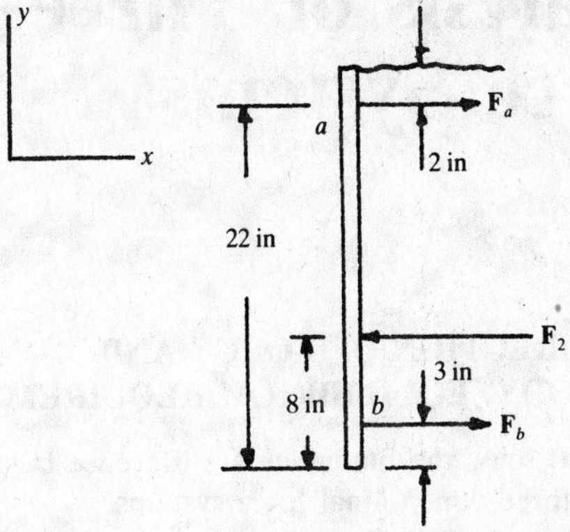   Fig. 11.49c

(a) Find the vertical force exerted by the liquid concrete on the inclined 4-ft length of form.

(b) Find the tensile forces in wires a and b.

❚ (a) Figure 11.49b shows the resultant force F_1 of the concrete on the inclined board. The magnitude of F_1 is given by

$$F_1 = h_{c1} \gamma A_1 = \frac{2}{2}(135)\,\frac{25}{12}\,(4) = 1{,}130\ \text{lb}$$

The component of the above force in the y direction is

$$F_y = \frac{7}{25}(1{,}130) = 316\ \text{lb}$$

(b) Figure 11.49c shows the free-body diagram of a 1-ft length of the vertical form. The resultant force F_2 of the concrete on the form is

$$F_2 = h_{c2} \gamma A_2 = \frac{2}{2}(135)2(1) = 270\ \text{lb}$$

For equilibrium of the form,

$$\Sigma M_b = 0 \qquad 270(8-3) - F_a(22-3) = 0 \qquad F_a = 71.1\ \text{lb}$$
$$\Sigma F_x = 0 \qquad F_a - 270 + F_b = 0 \qquad F_b = 199\ \text{lb}$$

11.50 Give a summary of the basic concepts of distribution of forces along lengths and over areas.

❚ In a distributed loading, the forces are distributed over some length or area of the structure. The total load effect of a distributed load must be described in terms of the magnitude, direction, sense, and location of the resultant force of the load distribution.

For the case of a load distribution along a length, the magnitude of the resultant force is equal to the area under the load distribution diagram. This force acts through the centroid of the area of this diagram. For a uniform load distribution over a plane area, the magnitude of the resultant force is equal to the product of the magnitudes of the pressure and of the plane area. This force acts through the centroid of the plane area.

The gage pressure at any point in a stationary, incompressible liquid is equal to the product of the depth below the surface of the liquid and the specific weight of the liquid. The pressure distribution on a plane wall submerged in a liquid is uniformly varying. The magnitude of the resultant force on this wall is equal to the product of the area of the pressure diagram and the width of the wall. This resultant force acts through the centroid of the pressure diagram.

For the general case of a submerged plane area, the magnitude of the resultant force acting on the area is equal to the product of this area and the pressure at the centroid of the area. The force acts at a point called the center of pressure. This point is always at a greater depth than the centroid of the plane area, except for the special case where the plane area is horizontal and parallel to the liquid surface. The distance between these two points, measured along the plane, is given by $I_{0x}/(Ay_c)$. I_{0x} is the moment of inertia of the plane area about a centroidal axis which is parallel to the liquid surface, and A is the area of the plane. y_c is the distance between the centroid of the area and the surface of the liquid, measured in the plane of the area.

CHAPTER 12
Analysis of Three-Dimensional Force Systems

12.1 DIRECTION OF FORCE AND MOMENT, SENSE OF MOMENT, METHODS OF SUPPORT OF BODY, EQUILIBRIUM REQUIREMENTS

12.1 Discuss the fundamental difference between the analysis of two-dimensional force systems and the analysis of three-dimensional force systems.

❚ Chapter 5 considered the problem of the two-dimensional force system. In this type of system, both the structure and the loads lie in a common plane. The general equilibrium requirements for such a system are that the force summations along two perpendicular directions in the plane be zero, and that the moment summation about any axis normal to the plane be zero. The plane truss was seen to be a special type of two-dimensional force system, in which the forces acting on the truss members have directions which are collinear with the axes of these members. The plane frame was seen to be another special type of two-dimensional force system which consists of an assemblage of multiforce, and possibly two-force, members. From consideration of the equilibrium requirements of the two-dimensional force system, it may be concluded that a maximum of *three* unknown reaction force effects may be found from each free-body diagram.

In this chapter, the problem of the three-dimensional force system is considered. In such a system, either the physical structure or the loads, or both, do not lie in a common plane. Such a system does not require new theory for its solution. Rather, the results previously obtained for the two-dimensional case are extended to include the effects of a third spatial coordinate.

In the following problems we review both the methods of describing the direction and sense of a force or moment in space and the types of foundation support in a three-dimensional system. Then there follows a formal definition of the equilibrium requirements of the three-dimensional force system, as well as an identification of special types of such force systems.

12.2 Show two methods which may be used to define the direction of the line of action of a force or moment in a three-dimensional coordinate system.

❚ There are two general ways to define the direction of the line of action of a force or moment in a three-dimensional space. One method is to specify the direction angles of the line. The concept of a direction angle was introduced in Chap. 1 and was used subsequently in Chaps. 2 and 3. A summary of the direction angle relationships is presented below.

Figure 12.2a shows a force which is referenced to an xyz coordinate system. The angles θ_x, θ_y, and θ_z between the line of action of the force and the coordinate axes are defined to be direction angles. The cosines of these angles are called direction cosines. The direction cosines must satisfy the relationship

$$\cos^2 \theta_x + \cos^2 \theta_y + \cos^2 \theta_z = 1$$

If any two of the three direction angles are known, the third angle may be found from this equation.

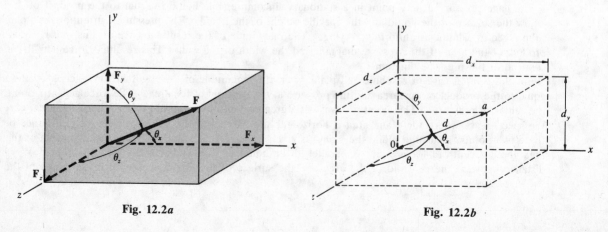

Fig. 12.2a   **Fig. 12.2b**

The components of the force in Fig. 12.2a are

$$F_x = F \cos \theta_x \qquad \cos \theta_x = \frac{F_x}{F}$$

$$F_y = F \cos \theta_y \qquad \cos \theta_y = \frac{F_y}{F}$$

$$F_z = F \cos \theta_z \qquad \cos \theta_z = \frac{F_z}{F}$$

and

$$F = \sqrt{F_x^2 + F_y^2 + F_z^2}$$

The above description of the direction angles is in terms of the magnitude F, and the components F_x, F_y, and F_z, of the force $\mathbf{F}$.

An alternative method of describing the direction angles is in terms of *spatial coordinates*. Figure 12.2b shows a line $0a$ referenced to the xyz axes. The length of this line is d, and the lengths of its components along the coordinate axes are d_x, d_y, d_z. The description of the direction angles in terms of *spatial quantities* is

$$\cos \theta_x = \frac{d_x}{d} \qquad \cos \theta_y = \frac{d_y}{d} \qquad \cos \theta_z = \frac{d_z}{d}$$

$$d = \sqrt{d_x^2 + d_y^2 + d_z^2}$$

A second way to define the direction of the line of action of a force or moment in space is by the use of angles with respect to fixed reference planes. Figure 12.2c shows the two angles α and β. α measures the position of the plane containing the force or moment and the y axis. β is the angle, measured in this plane, between the plane formed by the x and z axes and the force or moment. The positive senses of these two angles are as shown in the figure.

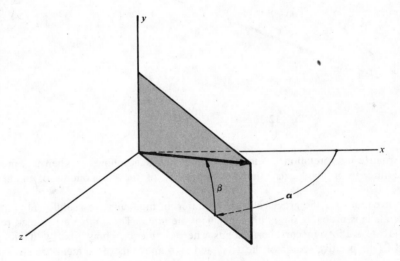

Fig. 12.2c

12.3   Show the sign convention for the sense of a moment or couple in a three-dimensional coordinate system.

▌ Two sign conventions for the sense of a moment or couple in a three-dimensional frame of reference were presented in Chap. 3. These are repeated below.

In the first convention, the moment is represented by a curved arrow which encircles the axis about which the moment acts. This effect is shown in Fig. 12.3a. The curved arrow emphasizes the fact that the moment is a turning, or rotation, effect. The tip is placed on the curved arrow so as to show the sense of this turning effect, and an arbitrary sense of rotation is chosen to represent positive moment. The curved arrow may be drawn in the plane of the force, or in a plane which is parallel to the plane of the force.

The second way of graphically depicting a moment is shown in Fig. 12.3b. In this case, the moment is represented by a straight arrow with a double arrowhead at the tip end. The direction of the double-headed arrow is along the axis about which the moment acts, and this direction is perpendicular to the plane which contains the force.

In order to establish the positive sense of the moment, the thumb of the right hand is imagined to be pointing in the positive sense of the double-headed arrow. The relaxed, curved fingers of the right hand will then be acting in the turning sense of the moment. This effect is shown in Fig. 12.3c.

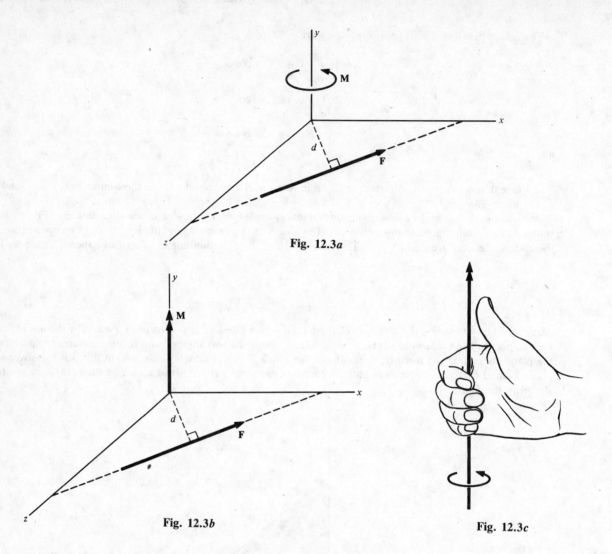

Fig. 12.3a

Fig. 12.3b

Fig. 12.3c

An alternative interpretation of the above statement is as follows. When looking outward from the origin along the three coordinate axes, the positive senses of the components of moment about these axes are clockwise.

Both conventions for the graphical representation of moments are useful. For simple systems, the curved-arrow technique is probably the more convenient method. The double-arrowhead representation of moments finds its greatest utility in complex systems where moments, about axes with different directions, must be combined. The positive senses of the x, y, and z components of moment are shown in Fig. 12.3d.

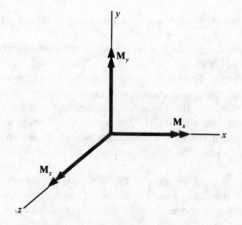

Fig. 12.3d

12.4 Summarize the reaction force and moment requirements in three-dimensional problems for support of a body by a cable, by a plane or curved surface, by a hinge pin or ball joint, and by a clamped connection.

I. The details of how a body may be physically connected to a foundation were presented in Chap. 4. The methods of support for the three-dimensional case, together with a description of the reaction force effects, are repeated below. The reaction force has the general components R_x, R_y, and R_z, and the reaction moment has the general components M_x, M_y, and M_z. These components are considered to be positive when acting in the positive coordinate senses.

Support by a Cable

Figure 12.4a shows the end of a body to which the cable is attached. The unknown reaction force R has the *known direction* of the cable and the *known sense* shown in the figure. The components of R are

$$R_x = R \cos \theta_x \qquad R_y = R \cos \theta_y \qquad R_z = R \cos \theta_z$$

The cable cannot transmit a moment. Thus, the components of the reaction moment are

$$M_x = M_y = M_z = 0$$

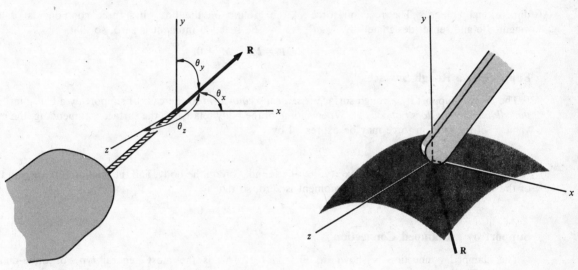

Fig. 12.4a   **Fig. 12.4b**

Support by a Plane, or Curved, Smooth Surface

Figure 12.4b shows the portion of a body which contacts the smooth surface. The reaction force R of the surface on the body has the known direction of the normal to the surface, and it is a compressive force effect on the body. The smooth surface cannot transmit a moment. Therefore, the components of the reaction moment are

$$M_x = M_y = M_z = 0$$

Support by a Hinge Pin

The hinge pin support is shown in Fig. 12.4c. The major difference between the two- and three-dimensional hinge supports is that, in the latter case, the hinge must be able to resist forces in the z direction. The reaction force has unknown magnitude, direction, and sense. It may be described in terms of the *three* unknown reaction force components R_x, R_y, and R_z. If the hinge is assumed to be frictionless, the z component of the reaction moment is given by

$$M_z = 0$$

The remaining two components of the reaction moment exerted by the ground on the member have *unknown* magnitudes and senses and are designated M_x and M_y.

Support by a Ball Joint

Figure 12.4d shows a ball joint support, and this method of connection may be thought of as a three-

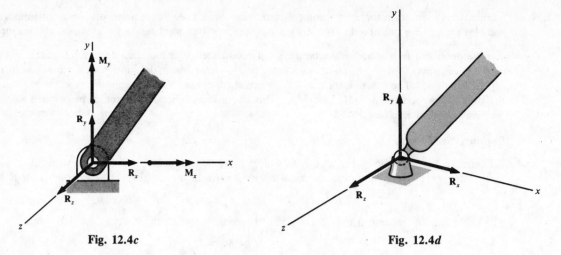

Fig. 12.4c   **Fig. 12.4d**

dimensional "hinge." The reaction force of the ground on the body has three components of unknown magnitude and sense, designated by R_x, R_y, R_z. The reaction moment is zero, so that

$$M_x = M_y = M_z = 0$$

Support by a Rough Surface

The case of support by a rough surface is *statically equivalent* to the case of support by a ball joint, *provided that sliding motion does not occur*. When sliding of the body relative to the surface is impending, the maximum value F of the friction force may be expressed by

$$F = \mu N$$

N is the *compressive* normal force exerted by the foundation on the body, and the friction force acts in the plane of the rough surface. The reaction moment is zero, so that

$$M_x = M_y = M_z = 0$$

Support by a Clamped Connection

The clamped connection is shown in Fig. 12.4e. This is the most general type of three-dimensional support. The components of the reaction force and moment have unknown magnitudes and senses. They form a system of the six unknowns R_x, R_y, R_z and M_x, M_y, M_z.

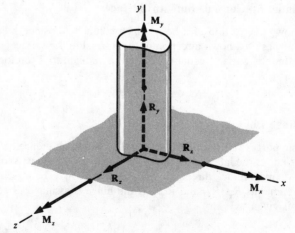

Fig. 12.4e

12.5   (*a*)   State the equilibrium requirements for the general three-dimensional force system.

(*b*)   What new equilibrium requirements are introduced when going from a two-dimensional problem to a three-dimensional problem?

▌ (*a*)   Figure 12.5 shows a symbolic representation of a body which is acted upon by a three-dimensional force

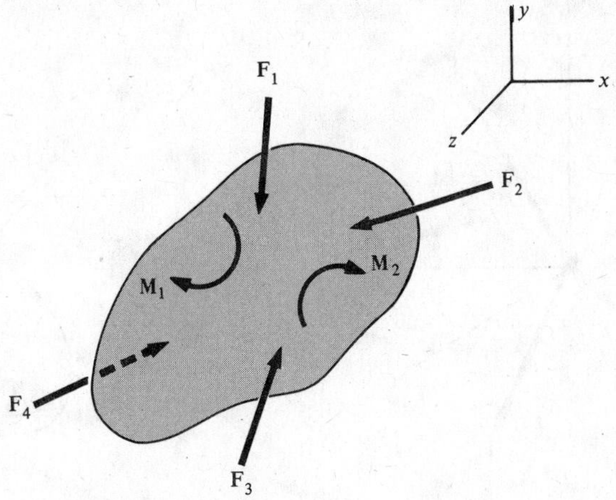

Fig. 12.5

system.   The forces are $\mathbf{F}_1, \mathbf{F}_2, \ldots$ and the moments are $\mathbf{M}_1, \mathbf{M}_2, \ldots$.   These forces and moments include *both the applied loads and the reaction force effects of the foundation on the body*.   The body must separately satisfy the requirements of force and moment equilibrium.   The requirement of force equilibrium is given by

$$\mathbf{F} = 0$$

where $\mathbf{F}$ is the resultant of the applied forces and reaction forces which act on the body.   This equation may be written in terms of its three scalar component equations as

$$\sum F_x = 0 \qquad \sum F_y = 0 \qquad \sum F_z = 0$$

The requirement of moment equilibrium is

$$\mathbf{M} = \mathbf{r} \times \mathbf{F} = 0$$

where $\mathbf{M}$ is the resultant moment of all the applied and reaction moments which act on the body.   This equation may be written in terms of its scalar components as

$$\sum M_x = 0 \qquad \sum M_y = 0 \qquad \sum M_z = 0$$

The most general three-dimensional force system is now seen to require the satisfaction of six independent equations of equilibrium.   In many problems, several of these equations are automatically satisfied.

(*b*)   For the case of a general two-dimensional force system in the *xy* plane, the requirements of equilibrium are

$$\sum F_x = 0 \qquad \sum F_y = 0 \qquad \sum M_z = 0$$

It may be seen that the three-dimensional problem adds a third force equation, $\Sigma F_z = 0$, and the two moment equations $\Sigma M_x = 0$ and $\Sigma M_y = 0$.

12.6   (*a*)   Give the general characteristics, and state the equilibrium requirements, for the concurrent, parallel, and general three-dimensional force systems.

(*b*)   Compare the solutions found by using the scalar equations of equilibrium with those found by using the vector equations of equilibrium.

▌ (*a*)   Two-dimensional force systems were seen to allow a convenient subdivision into four types of systems, with the designations collinear, concurrent, parallel, and general.   A similar classification may be made in the case of three-dimensional force systems, and these cases are shown below.

Concurrent Force System

In the concurrent force system, the lines of action of *all* the applied and reaction forces pass through a common point.   A typical example of such a system is shown in Fig. 12.6*a*.   The *collinear* force system, where all the forces have a common line of action, is a special case of this type of system.   In the concurrent force system, the requirement of moment equilibrium is automatically satisfied.   The remaining requirements for

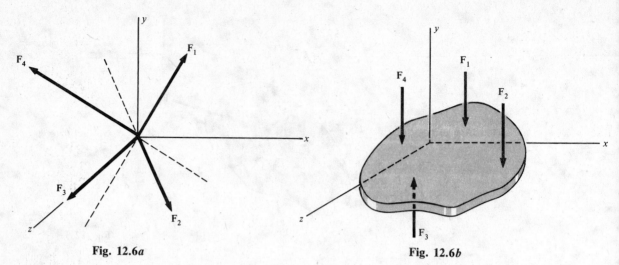

Fig. 12.6a   Fig. 12.6b

equilibrium of this type of force system are

$$\sum F_x = 0 \qquad \sum F_y = 0 \qquad \sum F_z = 0$$

In formal vector notation, the equilibrium requirements are

$$\mathbf{F} = F_x \mathbf{i} + F_y \mathbf{j} + F_z \mathbf{k} = 0$$

where $\mathbf{F}$ is the resultant of all of the applied and reaction forces acting on the body.

Parallel Force System

In the parallel force system, the lines of action of all the applied and reaction forces are parallel to one another. Figure 12.6b shows an example of a parallel force system. The equilibrium requirements for this system, with reference to the coordinate axes shown in the figure, are

$$\sum F_y = 0 \qquad \sum M_x = 0 \qquad \sum M_z = 0$$

It is left as an exercise for the reader to show that the remaining three equations of equilibrium,

$$\sum F_x = 0 \qquad \sum F_z = 0 \qquad \sum M_y = 0$$

are automatically satisfied.

The equilibrium requirements, in formal vector notation, are

$$\mathbf{F} = 0 \qquad \mathbf{M} = 0$$

where $\mathbf{F}$ and $\mathbf{M}$ are the resultants of all of the applied and reaction forces and moments acting on the body. These requirements, for the system shown in Fig. 12.6b, have the forms

$$\mathbf{F} = F_y \mathbf{j} = 0 \qquad \mathbf{M} = M_x \mathbf{i} + M_z \mathbf{k} = 0$$

General Three-Dimensional Force System

If the force system is not one of the cases described above, it is referred to as a general three-dimensional force system. For this case, the requirements for equilibrium are

$$\sum F_x = 0 \qquad \sum F_y = 0 \qquad \sum F_z = 0$$
$$\sum M_x = 0 \qquad \sum M_y = 0 \qquad \sum M_z = 0$$

The above equilibrium requirements, in formal vector notation, are

$$\mathbf{F} = F_x \mathbf{i} + F_y \mathbf{j} + F_z \mathbf{k} = 0 \qquad \mathbf{M} = M_x \mathbf{i} + M_y \mathbf{j} + M_z \mathbf{k} = 0$$

$\mathbf{F}$ and $\mathbf{M}$, as before, are the resultants of all of the applied and reaction forces and moments acting on the body.

(b) Three-dimensional force equilibrium problems may be solved by using either the scalar or vector equations of equilibrium. With the scalar method, particular care must be used when choosing the moment arms for the force component and assigning the correct senses to these moment components. The advantage of this method is that a clear picture of the physical effects is present throughout the solution.

The preliminary steps when using the vector method is to express all forces in terms of unit vectors. A reference point for moments is then chosen, and the position vectors from this point to the forces are then stated in terms of the unit vectors. The vector cross products of the forces and position vectors found above, and the forces, are summed to zero. When using this method, there is no physical interpretation of the problem until the last steps of the solution. It is emphasized, however, that the final equations which result from solution of the vector equations of equilibrium *are exactly the set of scalar equations of equilibrium.*

To allow a comparison of the two techniques, both methods are used in many of the following problems in this chapter. The vector method of solution is probably most useful in problems with complex geometry and in problems involving several forces with different directions. The reader is urged to gain familiarity with both methods of solution.

12.2 CONCURRENT AND PARALLEL FORCE SYSTEMS

12.7 Figure 12.7 shows a ring that is attached to a hook and acted on by two cable forces.

(*a*) Find the components, and the magnitude, of the reaction force exerted by the hook on the ring.

(*b*) Express the result in part (*a*) in formal vector notation.

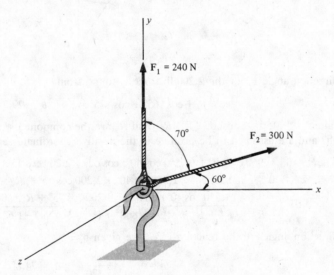

Fig. 12.7

I (*a*) This is a concurrent force system. The third direction angle θ_z is found from

$$\cos^2 \theta_x + \cos^2 \theta_y + \cos^2 \theta_z = 1 \qquad \cos^2 60° + \cos^2 70° + \cos^2 \theta_z = 1 \qquad \theta_z = 37°, \ 143°$$

From consideration of Fig. 12.7, the value $\theta_z = 143°$ is chosen.

The reaction force components R_x, R_y, and R_z exerted by the hook on the ring are chosen to be positive in the positive coordinate senses. For equilibrium of the ring,

$$\sum F_x = 0 \qquad R_x + F_2 \cos \theta_x = 0 \qquad R_x + 300 \cos 60° = 0 \qquad R_x = -150 \text{ N}$$

$$\sum F_y = 0 \qquad R_y + F_1 + F_2 \cos \theta_y = 0 \qquad R_y + 240 + 300 \cos 70° = 0 \qquad R_y = -343 \text{ N}$$

$$\sum F_z = 0 \qquad R_z + F_2 \cos \theta_z = 0 \qquad R_z + 300 \cos 143° = 0 \qquad R_z = 240 \text{ N}$$

The magnitude R of the reaction force is

$$R = \sqrt{R_x^2 + R_y^2 + R_z^2} = \sqrt{(-150)^2 + (-343)^2 + 240^2} = 445 \text{ N}$$

(*b*) The reaction force may be expressed in formal vector notation as

$$\mathbf{R} = -150\mathbf{i} - 343\mathbf{j} + 240\mathbf{k} \qquad \text{N}$$

It is left as an exercise for the reader to show that the above value for the magnitude of R must be equal to the magnitude of the resultant of the two applied forces F_1 and F_2.

12.8 (a) Find the magnitude, direction, and sense of the force exerted by the ground on the eyebolt shown in Fig. 12.8.

(b) Express the result in part (a) in formal vector notation.

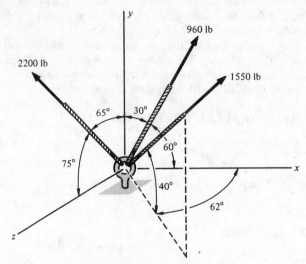

Fig. 12.8

❙ (a) The direction angle θ_x for the 2,200-lb force is found from

$$\cos^2 \theta_x + \cos^2 65° + \cos^2 75° = 1 \qquad \theta_x = 29.7°, \ 150°$$

The value $\theta_x = 150°$ is used. R_x, R_y, and R_z are the components of the force of the ground on the eyebolt, and are assumed to be positive in the positive coordinate senses. For equilibrium,

$$\Sigma F_x = 0 \qquad 960 \cos 60° + (1{,}550 \cos 40°) \cos 62° + 2{,}200 \cos 150° + R_x = 0 \qquad R_x = 868 \ \text{lb}$$
$$\Sigma F_y = 0 \qquad 960 \cos 30° + 1{,}550 \sin 40° + 2{,}200 \cos 65° + R_y = 0 \qquad R_y = -2{,}760 \ \text{lb}$$
$$\Sigma F_z = 0 \qquad (1{,}550 \cos 40°) \sin 62° + 2{,}200 \cos 75° + R_z = 0 \qquad R_z = -1{,}620 \ \text{lb}$$
$$R = \sqrt{R_x^2 + R_y^2 + R_z^2} = \sqrt{868^2 + (-2{,}760)^2 + (-1{,}620)^2} = 3{,}320 \ \text{lb}$$

The direction angles of the reaction force are given by

$$\cos \theta_x = \frac{R_x}{R} = \frac{868}{3{,}320} \qquad \theta_x = 74.8°$$

$$\cos \theta_y = \frac{R_y}{R} = \frac{-2{,}760}{3{,}320} \qquad \theta_y = 146°$$

$$\cos \theta_z = \frac{R_z}{R} = \frac{-1{,}620}{3{,}320} \qquad \theta_z = 119°$$

(b) The force of the ground on the eyebolt may be expressed as

$$\mathbf{R} = 868\mathbf{i} - 2{,}760\mathbf{j} - 1{,}620\mathbf{k} \qquad \text{lb}$$

12.9 Three cables support a weight, as shown in Fig. 12.9a. Find the tensile force in each cable.

❙ Figure 12.9b shows the direction of the cable tensile force T_a. Using $d = \sqrt{2.4^2 + 3^2 + 10^2} = 10.7 \ \text{ft}$, the components of this force are

$$T_{ax} = -\frac{2.4}{10.7} \ T_a = -0.224 T_a \qquad T_{ay} = -\frac{3}{10.7} \ T_a = -0.280 T_a \qquad T_{az} = -\frac{10}{10.7} \ T_a = -0.935 T_a$$

The direction of cable tensile force T_b is seen in Fig. 12.9c. Using $d = \sqrt{3.6^2 + 3^2 + 10^2} = 11.0 \ \text{ft}$, the components of this force have the forms

$$T_{bx} = \frac{3.6}{11.0} \ T_b = 0.327 T_b \qquad T_{by} = -\frac{3}{11.0} \ T_b = -0.273 T_b \qquad T_{bz} = -\frac{10}{11.0} \ T_b = -0.909 T_b$$

Figure 12.9d shows cable force T_c. Using Fig. 12.9d, and $d = \sqrt{1.6^2 + 2^2 + 10^2} = 10.3 \ \text{ft}$, the components

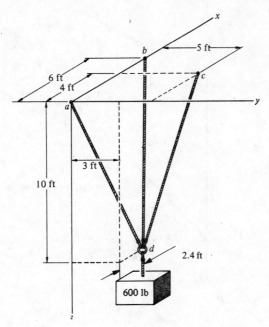

Fig. 12.9a

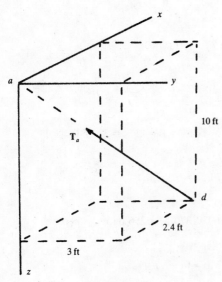

Fig. 12.9b

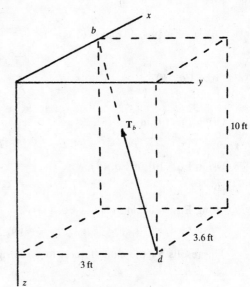

Fig. 12.9c

of cable force T_c are

$$T_{cx} = \frac{1.6}{10.3} T_c = 0.155 T_c \qquad T_{cy} = \frac{2}{10.3} T_c = 0.194 T_c \qquad T_{cz} = -\frac{10}{10.3} T_c = -0.971 T_c$$

For force equilibrium of point d,

$$\Sigma F_x = 0 \qquad T_{ax} + T_{bx} + T_{cx} = 0 \qquad -0.224 T_a + 0.327 T_b + 0.155 T_c = 0 \tag{1}$$

$$\Sigma F_y = 0 \qquad T_{ay} + T_{by} + T_{cy} = 0 \qquad -0.280 T_a - 0.273 T_b + 0.194 T_c = 0 \tag{2}$$

$$\Sigma F_z = 0 \qquad T_{az} + T_{bz} + T_{cz} + 600 = 0 \qquad -0.935 T_a - 0.909 T_b - 0.971 T_c + 600 = 0 \tag{3}$$

Equations (1) through (3) are each divided by the coefficient of the first term, and the signs of Eq. (1) are changed. To avoid loss of accuracy in the subsequent calculations, four significant figures are used, and the results are

$$T_a - 1.460 T_b - 0.6920 T_c = 0 \tag{4}$$
$$-T_a - 0.9750 T_b + 0.6929 T_c = 0 \tag{5}$$
$$-T_a - 0.9722 T_b - 1.039 T_c + 641.7 = 0 \tag{6}$$

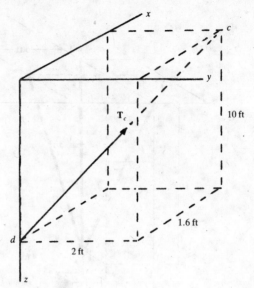

Fig. 12.9d

The sum of Eqs. (4) and (5) is

$$-2.435T_b + 0.0009T_c = 0 \tag{7}$$

The sum of Eqs. (4) and (6) is

$$-2.432T_b - 1.731T_c + 641.7 = 0 \tag{8}$$

From Eq. (7)

$$T_b = 3.7 \times 10^{-4}T_c \tag{9}$$

Using Eq. (9) in Eq. (6),

$$-2.432(3.7 \times 10^{-4}T_c) - 1.731T_c + 641.7 = 0 \qquad T_c = 370.5 \, \text{lb} \tag{10}$$

Using Eq. (10) in Eq. (9),

$$T_b = 3.7 \times 10^{-4}(370.5) = 0.137 \, \text{lb} \tag{11}$$

Using Eqs. (10) and (11) in Eq. (5),

$$T_a = -0.9750(0.137) + 0.6929(370.5) = 256.6 \, \text{lb}$$

The results for the cable forces, to three significant figures, are

$$T_a = 257 \, \text{lb} \qquad T_b = 0.137 \, \text{lb} \approx 0 \qquad T_c = 371 \, \text{lb}$$

The above results will now be checked. Using Eq. (4),

$$257 - 1.460(0.137) - 0.6920(371) \overset{?}{=} 0 \qquad 257 \equiv 257$$

Using Eq. (5),

$$-257 - 0.9750(0.137) + 0.6929(371) \overset{?}{=} 0 \qquad 257 \equiv 257$$

Using Eq. (6),

$$-257 - 0.9722(0.137) - 1.039(371) + 641.7 \overset{?}{=} 0 \qquad 642 \approx 643$$

An interpretation of the result $T_b = 0$ may now be made. Figure 12.9e shows the xy plane with points a and c, and the projection of point d. The x coordinate of point d' is found from the similar triangles $aa'd'$ and $ac'c$ as

$$\frac{x}{3} = \frac{4}{5} \qquad x = 2.4 \, \text{ft}$$

It may be seen that d and d' are coincident points, and thus point d lies in the plane formed by cables ad and cd. For this problem, *cable bd is a zero force member*. A final check on the solution to this problem may be made from consideration of the equilibrium of point d. Figure 12.9f shows the free-body diagram of this

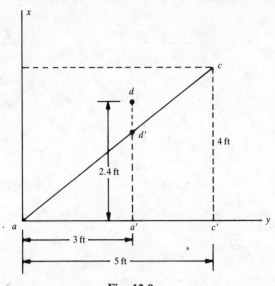

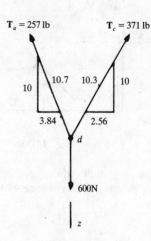

Fig. 12.9e	Fig. 12.9f

point. For equilibrium,

$$\Sigma F_z = 0 \qquad \frac{10}{10.7} T_a + \frac{10}{10.3} T_c \overset{?}{=} 600 \qquad \frac{10}{10.7} (257) + \frac{10}{10.3} (371) \overset{?}{=} 600 \qquad 600 = 600$$

12.10 Find the tensile forces in the three cables that support the weight shown in Fig. 12.10a.

▮ Figure 12.10b shows a view of the xy plane. e is the projection of point d on the xy plane. T_a, T_b, and T_c are the tensile forces in cables ad, bd, and cd, respectively.

Figure 12.10c shows the plane containing forces T_a or T_b and a vertical axis.   From this figure, and using Fig. 12.10b, the components of cable force T_a are found as

$$T_{ax} = -\left(\frac{14.1}{31.3} T_a\right) \cos 45° \qquad T_{ay} = -\left(\frac{14.1}{31.3} T_a\right) \sin 45° \qquad T_{az} = -\frac{28}{31.3} T_a$$

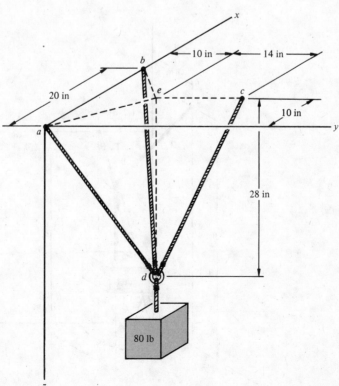

Fig. 12.10a

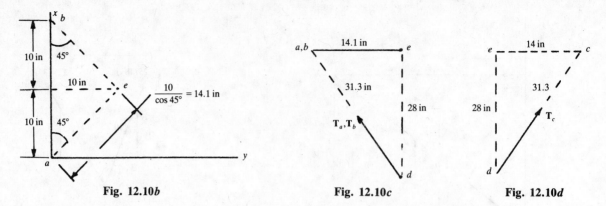

Fig. 12.10b   **Fig. 12.10c**   **Fig. 12.10d**

The magnitudes of the components of T_b have the same forms as those of T_a. The senses of these components are positive in accordance with the positive coordinate senses.

$$T_{bx} = \left(\frac{14.1}{31.3} T_b\right)\cos 45° \qquad T_{by} = -\left(\frac{14.1}{31.3} T_b\right)\sin 45° \qquad T_{bz} = -\frac{28}{31.3} T_b$$

Figure 12.10d shows the plane containing force T_c and a vertical axis. From this figure, and using Fig. 12.10b,

$$T_{cx} = 0 \qquad T_{cy} = \frac{14.1}{31.3} T_c \qquad T_{cz} = -\frac{28}{31.3} T_c$$

For force equilibrium of the system,

$$\Sigma F_x = 0 \qquad -\left(\frac{14.1}{31.3} T_a\right)\cos 45° + \left(\frac{14.1}{31.3} T_b\right)\cos 45° = 0 \qquad T_a = T_b$$

$$\Sigma F_y = 0 \qquad -\left(\frac{14.1}{31.3} T_a\right)\sin 45° - \left(\frac{14.1}{31.3} T_b\right)\sin 45° + \frac{14.1}{31.3} T_c = 0 \qquad T_a = T_b = 0.702 T_c$$

$$\Sigma F_z = 0 \qquad -\frac{28}{31.3} T_a - \frac{28}{31.3} T_b - \frac{28}{31.3} T_c + 80 = 0 \qquad T_c = 37.2 \text{ lb} \qquad T_a = T_b = 0.702(37.2) = 26.1 \text{ lb}$$

12.11   The boom in Fig. 12.11a is connected to the foundation by a ball joint.

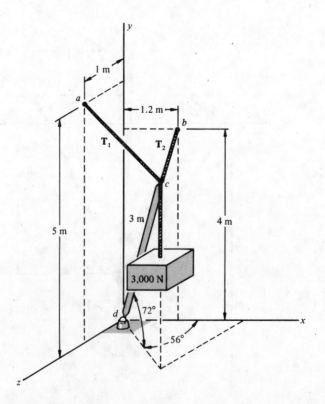

Fig. 12.11a

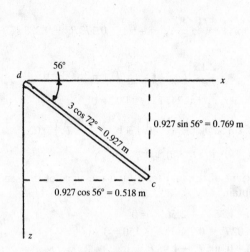

Fig. 12.11b

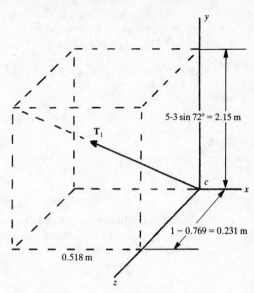

Fig. 12.11c

(a) Find the two cable tensile forces T_1 and T_2.

(b) Find the axial compressive force in the boom.

▌ (a) Figure 12.11b shows a projection of member cd on the xz plane, and Fig. 12.11c shows the direction of cable force T_1. Using Fig. 12.11c,

$$d = \sqrt{0.518^2 + 2.15^2 + 0.231^2} = 2.22 \text{ m}$$

The components of force T_1 acting in the positive coordinate senses are then found as

$$T_{1x} = -\frac{0.518}{2.22}\,T_1 \qquad T_{1y} = \frac{2.15}{2.22}\,T_1 \qquad T_{1z} = \frac{0.231}{2.22}\,T_1$$

Figure 12.11d shows the direction of T_2. Using this figure,

$$d = \sqrt{0.682^2 + 1.15^2 + 0.769^2} = 1.54 \text{ m}$$

The components of force T_2 acting in the positive coordinate senses have the forms

$$T_{2x} = \frac{0.682}{1.54}\,T_2 \qquad T_{2y} = \frac{1.15}{1.54}\,T_2 \qquad T_{2z} = -\frac{0.769}{1.54}\,T_2$$

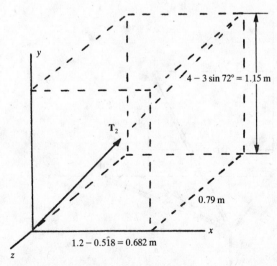

Fig. 12.11d

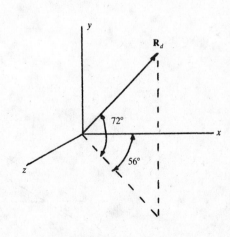

Fig. 12.11e

The forces act only at the ends of the boom, so that the boom is a two-force member. Let R_d be the compressive force exerted by the foundation on the boom, with the direction of the boom shown in Fig. 12.11e. For equilibrium of the boom,

$$\Sigma F_x = 0 \qquad R_d \cos 72° \cos 56° + \left(-\frac{0.518}{2.22} T_1\right) + \left(\frac{0.682}{1.54} T_2\right) = 0 \qquad 0.173 R_d - 0.233 T_1 + 0.443 T_2 = 0 \tag{1}$$

$$\Sigma F_y = 0 \qquad R_d \sin 72° + \left(\frac{2.15}{2.22} T_1\right) + \left(\frac{1.15}{1.54} T_2\right) - 3{,}000 = 0$$
$$0.951 R_d + 0.968 T_1 + 0.747 T_2 - 3{,}000 = 0 \tag{2}$$

$$\Sigma F_z = 0 \qquad R_d \cos 72° \sin 56° + \left(\frac{0.231}{2.22} T_1\right) + \left(-\frac{0.769}{1.54} T_2\right) = 0$$
$$0.256 R_d + 0.104 T_1 - 0.499 T_2 = 0 \tag{3}$$

Each equation is divided by the coefficient of the first term, and the signs of the second and third equations which result are changed, to obtain

$$R_d - 1.35 T_1 + 2.56 T_2 = 0 \tag{4}$$
$$-R_d - 1.02 T_1 - 0.785 T_2 + 3{,}150 = 0 \tag{5}$$
$$-R_d - 0.406 T_1 + 1.95 T_2 = 0 \tag{6}$$

Equations (4) and (6) are added, with the result

$$-1.76 T_1 + 4.51 T_2 = 0 \qquad T_1 = 2.56 T_2 \tag{7}$$

Equations (4) and (5) are added, to obtain

$$-2.37 T_1 + 1.78 T_2 + 3{,}150 = 0 \tag{8}$$

T_1 from Eq. (7) is used in Eq. (8), and T_2 is then found from

$$-2.37(2.56 T_2) + 1.78 T_2 + 3{,}150 = 0 \qquad T_2 = 735 \text{ N}$$

Using Eq. (7),

$$T_1 = 2.56(735) = 1{,}880 \text{ N}$$

I (b) From Eq. (2),

$$0.951 R_d + 0.968(1{,}880) + 0.747(735) - 3{,}000 = 0 \qquad R_d = 664 \text{ N}$$

12.12 Find the forces in the three members of the space truss shown in Fig. 12.12a.

I Each of the three members is a two-force member. Forces F_a and F_b are assumed to be compressive forces, and force F_c is assumed to be a tensile force.

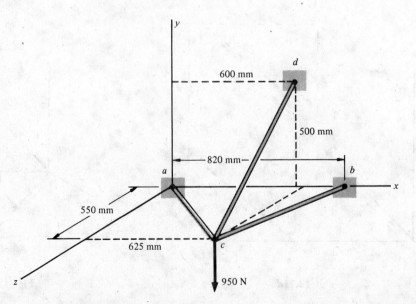

Fig. 12.12a

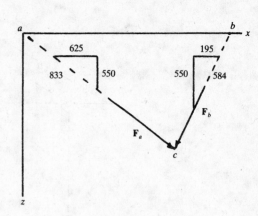

Fig. 12.12b   **Fig. 12.12c**

Figure 12.12b shows the direction of force F_d. Length cd in this figure is given by

$$cd = \sqrt{25^2 + 500^2 + 550^2} = 744 \text{ mm}$$

Figure 12.12c shows the directions of forces F_{ac} and F_{bc}. For equilibrium of point c,

$$\Sigma F_x = 0 \qquad \frac{625}{833} F_a - \frac{195}{584} F_b - \frac{25}{744} F_d = 0 \tag{1}$$

$$\Sigma F_y = 0 \qquad \frac{500}{744} F_d - 950 = 0 \qquad F_d = 1{,}410 \text{ N} \tag{2}$$

$$\Sigma F_z = 0 \qquad \frac{550}{833} F_a + \frac{550}{584} F_b - \frac{550}{744} F_d = 0 \tag{3}$$

The result in Eq. (2) is used in Eqs. (1) and (3), to obtain

$$F_a - 0.445 F_b - 63.1 = 0 \tag{4}$$
$$F_a + 1.43 F_b - 1{,}580 = 0 \tag{5}$$

Equation (5) is subtracted from Eq. (4), with the result

$$-1.88 F_b + 1{,}520 = 0 \qquad F_b = 809 \text{ N}$$

Using Eq. (5),

$$F_a = 1{,}580 - 1.43(809) = 423 \text{ N}$$

12.13 A force P is applied to joint c of the space truss shown in Fig. 12.13a. The direction angles of the line of action of this force are $\theta_x = 44°$, $\theta_y = 70°$, and $\theta_z = 53°$.

(a) If the magnitude of the force in any member may not exceed 1,000 lb, find the maximum allowable value of P.

(b) Do the same as in part (a) if the direction of the line of action of force P is parallel to the y axis.

▌ (a) For uniformity, all truss member forces are assumed to be compressive.
  Figure 12.13b shows the directions of forces F_a and F_b. The x and z components of these forces are

$$F_{ax} = \frac{1.5}{2.83} F_a \qquad F_{az} = \frac{2.4}{2.83} F_a \qquad F_{bx} = -\frac{1.5}{2.83} F_b \qquad F_{bz} = \frac{2.4}{2.83} F_b$$

Figure 12.13c shows the direction of force F_d. The length cd is given by

$$cd = \sqrt{0.3^2 + 3.6^2 + 2.4^2} = 4.34 \text{ ft}$$

The components of F_d have the forms

$$F_{dx} = -\frac{0.3}{4.34} F_d \qquad F_{dy} = \frac{3.6}{4.34} F_d \qquad F_{dz} = \frac{2.4}{4.34} F_d$$

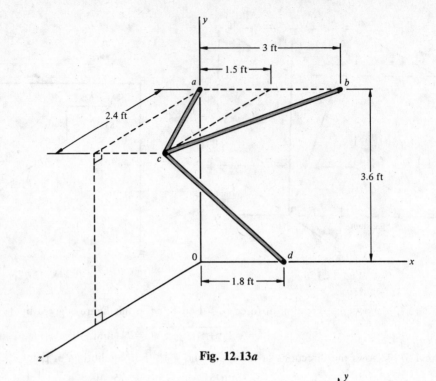

Fig. 12.13a

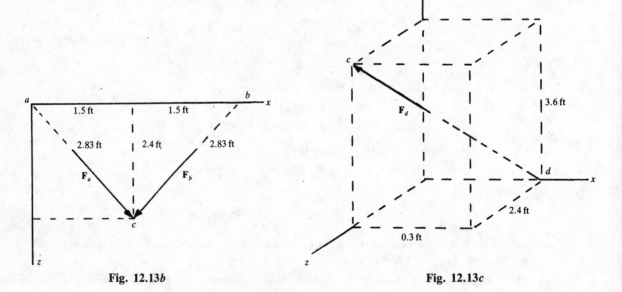

Fig. 12.13b   **Fig. 12.13c**

P is the applied force, with the components

$$P_x = P \cos 44° \qquad P_y = P \cos 70° \qquad P_z = P \cos 53°$$

For equilibrium of joint c,

$$\Sigma F_x = 0 \qquad \frac{1.5}{2.83} F_a - \frac{1.5}{2.83} F_b - \frac{0.3}{4.34} F_d + P \cos 44° = 0 \qquad (1)$$

$$\Sigma F_y = 0 \qquad \frac{3.6}{4.34} F_d + P \cos 70° = 0 \qquad F_d = -0.4123P \qquad (2)$$

$$\Sigma F_z = 0 \qquad \frac{2.4}{2.83} F_a + \frac{2.4}{2.83} F_b + \frac{2.4}{4.34} F_d + P \cos 53° = 0 \qquad (3)$$

To avoid loss of significant figures in the final answers, four significant figures are used to solve the simultaneous equations (1), (2), and (3). Equation (2) is used in Eqs. (1) and (3) to obtain

$$\frac{1.5}{2.83} F_a - \frac{1.5}{2.83} F_b - \frac{0.3}{4.34} (-0.4123P) + P \cos 44° = 0 \qquad (4)$$

$$\frac{2.4}{2.83} F_a + \frac{2.4}{2.83} F_b + \frac{2.4}{4.34} (-0.4123P) + P \cos 53° = 0 \qquad (5)$$

Equation (4) is multiplied by the factor $-2.83/1.5$, and Eq. (5) is multiplied by the factor $2.83/2.4$. The results are

$$-F_a + F_b - 1.411P = 0 \qquad (6)$$
$$F_a + F_b + 0.4408P = 0 \qquad (7)$$

Equations (6) and (7) are added, to obtain

$$2F_b - 0.9702P = 0 \qquad F_b = 0.4851P \qquad (8)$$

The result from Eq. (8) is used in Eq. (4), to obtain

$$F_a = 0.4851P - 1.411P \qquad F_a = -0.9259P \qquad (9)$$

The negative values obtained for F_a and F_d indicate that these forces are tensile forces.

The maximum allowable member force is 1,000 lb. F_a is the member with the maximum force. The minus sign in Eq. (9) is discarded, and

$$F_{aa} = 0.9259P = 1,000$$

$$P_{max} = 1,080 \text{ lb}$$

(b) Force P is assumed to act upward in Fig. 12.13a, in the positive y sense. Forces F_a, F_b, and F_d are assumed to be compressive, with the same functional forms as in part (a).

For equilibrium of joint c,

$$\Sigma F_x = 0 \qquad \frac{1.5}{2.83} F_a - \frac{1.5}{2.83} F_b - \frac{0.3}{4.34} F_d = 0 \qquad (10)$$

$$\Sigma F_y = 0 \qquad \frac{3.6}{4.34} F_d + P = 0 \qquad F_d = -1.206P \qquad (11)$$

$$\Sigma F_z = 0 \qquad \frac{2.4}{2.83} F_a + \frac{2.4}{2.83} F_b + \frac{2.4}{4.34} F_d = 0 \qquad (12)$$

Equation (11) is used in Eqs. (10) and (12), with the results

$$\frac{1.5}{2.83} F_a - \frac{1.5}{2.83} F_b - \frac{0.3}{4.34} (-1.206P) = 0 \qquad (13)$$

$$\frac{2.4}{2.83} F_a + \frac{2.4}{2.83} F_b + \frac{2.4}{4.34} (-1.206P) = 0 \qquad (14)$$

Equation (13) is multiplied by the factor $-2.83/1.5$, and Eq. (14) is multiplied by the factor $2.83/2.4$, with the results

$$-F_a + F_b - 0.1573P = 0 \qquad (15)$$
$$F_a + F_b - 0.7864P = 0 \qquad (16)$$

Equations (15) and (16) are added, to obtain

$$2F_b - 0.9437P = 0 \qquad F_b = 0.4719P \qquad (17)$$

Using Eq. (17) in Eq. (16) gives

$$F_a = -0.4719P + 0.7864P = 0.3145P$$

F_d is a tensile force. The maximum allowable member force is 1,000 lb, and F_d is the maximum member force. Using Eq. (11), with the minus sign discarded,

$$1.206P = 1,000$$

$$P_{max} = 829 \text{ lb}$$

12.14 Figure 12.14a shows a hatch cover which has a weight of 800 N. A cable tensile force T holds the hatch cover in a horizontal plane.

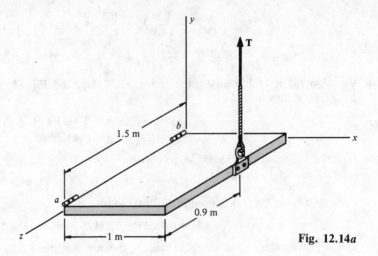

Fig. 12.14a

(a) Find the cable tensile force and the resultant hinge pin forces at a and b, using the scalar equations of equilibrium.

(b) Solve part (a) by using the vector equations of equilibrium.

▌ (a) This is a parallel force system. The free-body diagram of the hatch cover is shown in Fig. 12.14b. For convenience, the hinge pin forces at a and b are shown as acting at the outside ends of the hinges.

The equilibrium requirements are

$$\sum M_z = 0 \qquad -800(0.5) + T(1) = 0 \qquad T = 400 \text{ N}$$

$$\sum M_x = 0 \qquad -R_{ay}(1.5) + 800(0.75) - T(0.6) = 0 \qquad R_{ay} = 240 \text{ N}$$

$$\sum F_y = 0 \qquad R_{ay} + R_{by} - 800 + T = 0 \qquad 240 + R_{by} - 800 + 400 = 0 \qquad R_{by} = 160 \text{ N}$$

As a check on the above computations, a summation of moments, about an axis through hinge a and parallel to the x axis, will be made. The result is

$$R_{by}(1.5) - 800(0.75) + T(0.9) \overset{?}{=} 0 \qquad 160(1.5) - 800(0.75) + 400(0.9) \overset{?}{=} \qquad 0 \equiv 0$$

(b) The four forces acting on the cover are

$$\mathbf{R}_a = R_{ay}\mathbf{j} \qquad \mathbf{R}_b = R_{by}\mathbf{j} \qquad \mathbf{W} = -800\mathbf{j} \text{ N} \qquad \mathbf{T} = T\mathbf{j}$$

Point b is used as the reference point for moments, and the position vectors from point b to the forces are

$$\mathbf{r}_{ba} = 1.5\mathbf{k} \qquad \mathbf{r}_{bb} = 0 \qquad \mathbf{r}_{bW} = 0.5\mathbf{i} + 0.75\mathbf{k} \qquad \mathbf{r}_{bT} = (1)\mathbf{i} + 0.6\mathbf{k}$$

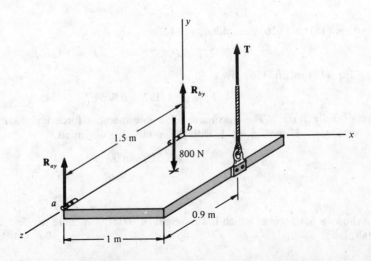

Fig. 12.14b

For moment equilibrium,

$$\mathbf{M}_b = 0 \qquad \mathbf{r}_{ba} \times \mathbf{R}_a + \overbrace{\mathbf{r}_{bb} \times \mathbf{R}_b}^{0} + \mathbf{r}_{bW} \times \mathbf{W} + \mathbf{r}_{bT} \times \mathbf{T} = 0$$

$$\mathbf{M}_b = \begin{vmatrix} \mathbf{i} & \mathbf{j} & \mathbf{k} \\ 0 & 0 & 1.5 \\ 0 & R_{ay} & 0 \end{vmatrix} + 0 + \begin{vmatrix} \mathbf{i} & \mathbf{j} & \mathbf{k} \\ 0.5 & 0 & 0.75 \\ 0 & -800 & 0 \end{vmatrix} + \begin{vmatrix} \mathbf{i} & \mathbf{j} & \mathbf{k} \\ 1 & 0 & 0.6 \\ 0 & T & 0 \end{vmatrix} = 0$$

$$[-R_{ay}(1.5) - (-800)0.75 - 0.6T]\mathbf{i} + [0]\mathbf{j} + [-800(0.5) + T(1)]\mathbf{k} = 0$$

The coefficients of the unit vectors are set equal to zero, with the results

$$-1.5R_{ay} + 800(0.75) - 0.6T = 0 \tag{1}$$
$$-800(0.5) + T = 0 \qquad T = 400 \text{ N}$$

Using the above value in Eq. (1),

$$R_{ay} = 240 \text{ N}$$

For force equilibrium,

$$\mathbf{R}_a + \mathbf{R}_b + \mathbf{W} + \mathbf{T} = 0 \qquad R_{ay}\mathbf{j} + R_{by}\mathbf{j} - 800\mathbf{j} + T\mathbf{j} = 0$$
$$(R_{ay} + R_{by} - 800 + T)\mathbf{j} = 0 \qquad R_{ay} + R_{by} - 800 + T = 0 \qquad R_{by} = 160 \text{ N}$$

12.15 The flat plate shown in Fig. 12.15a has a weight of 0.3 lb/in² of surface area. It rests on three ball supports at locations b, c, and e. Find the magnitudes of the forces exerted by these ball supports on the plate.

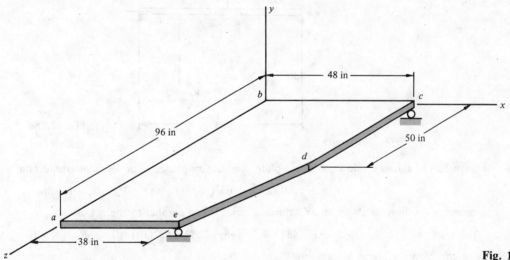

48 in

96 in

50 in

38 in

Fig. 12.15a

❚ Figure 12.15b shows the view of the plate in the xz plane. The centroidal coordinates are found, with area 1 the 48-in by 96-in rectangle, and area 2 the 10-in by 46-in triangle, as

$$x_c = \frac{x_1 A_1 - x_2 A_2}{A_1 - A_2} = \frac{24(48)96 - [38 + \frac{2}{3}(10)]\frac{1}{2}(10)46}{48(96) - \frac{1}{2}(10)46} = 22.9 \text{ in}$$

$$A = A_1 - A_2 = 4,380 \text{ in}^2$$

$$z_c = \frac{z_1 A_1 - z_2 A_2}{A_1 - A_2} = \frac{48(48)96 - [50 + \frac{2}{3}(46)]\frac{1}{2}(10)46}{4,380} = 46.3 \text{ in}$$

The weight of the plate is given by

$$W = 0.3 \frac{\text{lb}}{\text{in}^2} (4,380 \text{ in}^2) = 1,310 \text{ lb}$$

R_b, R_c, and R_e are the vertical reaction forces at the balls, acting in the positive y sense.

Figure 12.15c shows a side view of the plate, and the yz plane. For moment equilibrium about the x axis,

$$\Sigma M_x = 0 \qquad -R_e(96) + 1,310(46.3) = 0 \qquad R_e = 632 \text{ lb}$$

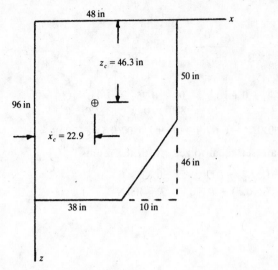

Fig. 12.15b

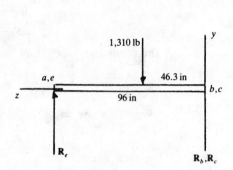

Fig. 12.15c

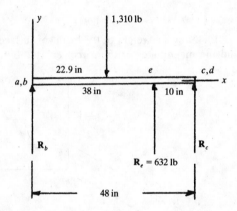

Fig. 12.15d

Figure 12.15d shows a side view of the plate, and the xy plane. For moment equilibrium about the z axis,

$$\Sigma M_z = 0 \qquad -1,310(22.9) + 632(38) + R_c(48) = 0 \qquad R_c = 125 \text{ lb}$$

For moment equilibrium about an axis through point c and parallel to the z axis,

$$\Sigma M_c = 0 \qquad -R_b(48) + 1,310(48 - 22.9) - 632(10) = 0 \qquad R_b = 553 \text{ lb}$$

As a check on the above calculations,

$$\Sigma F_y = 0 \qquad R_b + R_c + R_e \overset{?}{=} 1,310 \text{ lb} \qquad 553 + 125 + 632 \overset{?}{=} 1,310 \qquad 1,310 = 1,310$$

12.16 Figure 12.16a shows a system of four disks mounted on a shaft. An input couple of 2,000 in · lb is applied to disk C, and couples of opposite sense are applied to the remaining three disks.

(a) Find the required magnitude of M_A if the shaft is in equilibrium.

(b) Find the magnitude of the couple transmitted through lengths AB, BC, and CD of the shaft.

(c) Check the results found in parts (a) and (b).

▎ (a) This is a force system consisting of couples in parallel planes. For equilibrium of the entire shaft,

$$\sum M_z = 0 \qquad M_A + 500 - 2,000 + 800 = 0 \qquad M_A = 700 \text{ in} \cdot \text{lb}$$

(b) From physical considerations, it may be concluded that the couple transmitted through the shaft between any two adjacent disks has a *constant* value. Figure 12.16b shows a free-body diagram of disk A and part of shaft length AB. M_{AB} is the magnitude of the couple in length AB, and the sense of this couple is chosen to be that shown in the figure.

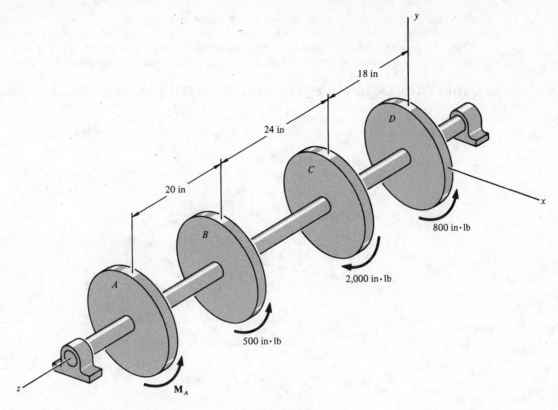

Fig. 12.16a

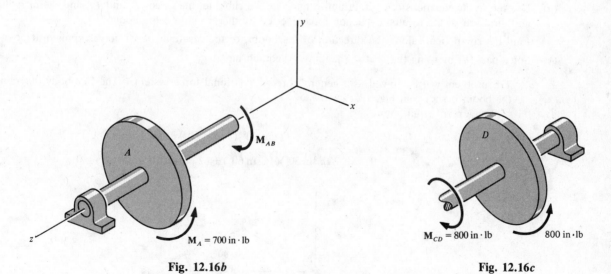

Fig. 12.16b   **Fig. 12.16c**

The equilibrium requirement for the configuration in this figure is

$$\sum M_z = 0 \qquad 700 - M_{AB} = 0 \qquad M_{AB} = 700 \, \text{in} \cdot \text{lb}$$

The remaining values of the couples transmitted through the shaft lengths, using an extension of the free-body diagram of Fig. 12.16b, are

$$\sum M_z = 0 \qquad 700 + 500 + M_{BC} = 0 \qquad M_{BC} = -1{,}200 \, \text{in} \cdot \text{lb}$$

$$\sum M_z = 0 \qquad 700 + 500 - 2{,}000 + M_{CD} = 0 \qquad M_{CD} = 800 \, \text{in} \cdot \text{lb}$$

The actual sense of M_{BC} is clockwise when viewed from the left side in Fig. 12.16a.

(c)   A check on the calculations in parts (a) and (b) is now made.   A free-body diagram of disk D, and a part of shaft CD, is shown in Fig. 12.16c. From Newton's third law, the couple M_{CD} must have a sense *opposite* to the result $M_{CD} = 800$ in · lb, found in part (b) above, when shown in Fig. 12.16c. It may be seen from this figure that the requirement of equilibrium is satisfied.

It may finally be observed that the lengths of the shaft elements do not enter into the problem.

12.3 GENERAL THREE-DIMENSIONAL FORCE SYSTEMS WITH CABLE, HINGE, AND BALL SUPPORTS

12.17 The boom in Fig. 12.17a lies in the xy plane, and it is connected to the ground with a ball joint.

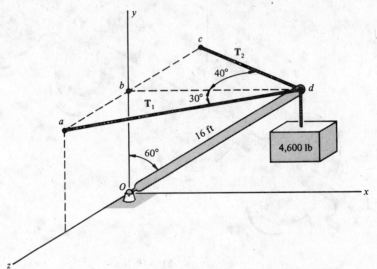

Fig. 12.17a

(a) Use the scalar equations of equilibrium to find the two cable tensile forces T_1 and T_2, and the magnitude and direction of the resultant reaction force exerted by the ground on the boom.

(b) Find the component, along the direction of the boom, of the resultant of the forces applied at point d.

(c) Solve part (a) by using the vector equations of equilibrium.

▌ (a) The problem will be solved as a general three-dimensional force system. The free-body diagram of the boom is shown in Fig. 12.17b.

For moment equilibrium,

$$\sum M_z = 0 \qquad (T_1 \cos 30° + T_2 \cos 40°)(16 \cos 60°) - 4,600(16 \sin 60°) = 0$$

$$\sum M_y = 0 \qquad (-T_1 \sin 30°)(16 \sin 60°) + (T_2 \sin 40°)(16 \sin 60°) = 0$$

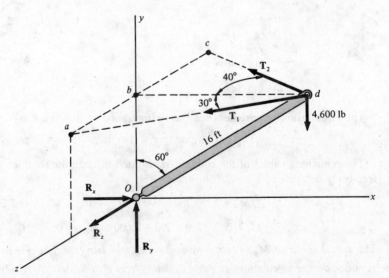

Fig. 12.17b

The solution to the above set of simultaneous equations is

$$T_1 = 5,460 \text{ lb} \qquad T_2 = 4,230 \text{ lb}$$

For force equilibrium,

$$\Sigma F_x = 0 \qquad -T_1 \cos 30° - T_2 \cos 40° + R_x = 0$$
$$-5,460 \cos 30° - 4,230 \cos 40° + R_x = 0 \qquad R_x = 7,970 \text{ lb}$$
$$\Sigma F_y = 0 \qquad R_y - 4,600 = 0 \qquad R_y = 4,600 \text{ lb}$$
$$\Sigma F_z = 0 \qquad T_1 \sin 30° - T_2 \sin 40° + R_z = 0$$
$$5,460 \sin 30° - 4,230 \sin 40° + R_z = 0 \qquad R_z = -11 \text{ lb}$$

From consideration of the magnitudes of R_x and R_y, it may be concluded that

$$R_z \approx 0$$

The resultant R of the reaction forces acting on the boom is

$$R = \sqrt{R_x^2 + R_y^2 + R_z^2} = \sqrt{7,970^2 + 4,600^2} = 9,200 \text{ lb}$$

The direction angles of this force are

$$\cos \theta_x = \frac{R_x}{R} = \frac{7.970}{9,200} \qquad \theta_x = 30° \qquad \cos \theta_y = \frac{R_y}{R} = \frac{4,600}{9,200} \qquad \theta_y = 60°$$

$$\cos \theta_z = 0 \qquad \theta_z = 90°$$

It may be concluded from the above results that *the direction of the resultant of the reaction forces is along the axis of the boom.*

(b) The component F_a along the direction of the boom, of the resultant of the applied forces acting at point d, is

$$F_a = (T_1 \cos 30° + T_2 \cos 40°)(\cos 30°) + 4,600 \cos 60°$$
$$= (5,460 \cos 30° + 4,230 \cos 40°)(\cos 30°) + 4,600 \cos 60°$$
$$= 9,200 \text{ lb}$$

It is left as an exercise for the reader to show that the above force is the *resultant* force acting on the boom at point d. (This may be done by showing that a summation of forces at d, normal to the boom, is equal to zero.)

It may finally be concluded that the system of forces which acts on the boom is effectively a collinear force system. The boom in this problem is actually a two-force member, since the forces are applied at the ends only. A member such as this may be thought of as a *three-dimensional truss member.*

(c) The applied forces which act on the boom are expressed in terms of the unit vectors as

$$T_1 = -(T_1 \cos 30°)i + (T_1 \sin 30°)k = -(0.866 T_1)i + (0.5 T_1)k \qquad \text{lb}$$
$$T_2 = -(T_2 \cos 40°)i - (T_2 \sin 40°)k = -(0.766 T_2)i - (0.643 T_2)k \qquad \text{lb}$$
$$W = -4,600j \qquad \text{lb}$$

The reaction force exerted by the ground on the boom is

$$R = R_x i + R_y j + R_z k \qquad \text{lb}$$

Point 0 in Fig. 12.17b is used as the reference for moments, and the position vector from this point to the force at point d has the form

$$r = (16 \sin 60°)i + (16 \cos 60°)j = 13.9i + 8j \qquad \text{ft}$$

For moment equilibrium,

$$M_0 = r \times F = r \times (T_1 + T_2 + W) = 0$$
$$T_1 + T_2 + W = (-0.866 T_1 - 0.766 T_2)i - 4,600j + (0.5 T_1 - 0.643 T_2)k$$

$$M_0 = r \times F = \begin{vmatrix} i & j & k \\ 13.9 & 8 & 0 \\ -0.866 T_1 - 0.766 T_2 & -4,600 & 0.5 T_1 - 0.643 T_2 \end{vmatrix} = 0$$

$$[8(0.5 T_1 - 0.643 T_2)]i + [-(0.5 T_1 - 0.643 T_2)13.9]j + [-4,600(13.9) - (-0.866 T_1 - 0.766 T_2)8]k = 0$$

The coefficient of **i** is used to obtain

$$0.5T_1 - 0.643T_2 = 0 \qquad T_1 = 1.29T_2 \qquad (1)$$

Using the coefficient of **k**, with Eq. (1), results in

$$-4,600(13.9) + 0.866T_1(8) + 0.766T_2(8) = 0$$
$$T_2 = 4,240 \text{ lb} \approx 4,230 \text{ lb}$$

Using Eq. (1),

$$T_1 = 1.29(4,240) = 5,470 \approx 5,460 \text{ lb}$$

For force equilibrium,

$$\mathbf{T}_1 + \mathbf{T}_2 + \mathbf{W} + \mathbf{R} = 0 \qquad (-0.866T_1 - 0.766T_2 + R_x)\mathbf{i} + (-4,600 + R_y)\mathbf{j} + (0.5T_1 - 0.643T_2 + R_z)\mathbf{k} = 0$$

The coefficients of the unit vectors are set equal to zero, with the results

$$-0.866T_1 - 0.766T_2 + R_x = 0$$
$$R_x = 7,980 \text{ lb} \approx 7,970 \text{ lb} \qquad -4,600 + R_y = 0 \qquad R_y = 4,600 \text{ lb}$$
$$0.5T_1 - 0.643T_2 + R_z = 0$$
$$R_z + 0.5T_1 = 0.643T_2 \qquad R_z + 0.5(5470) = 0.643(4,240) \qquad R_z + 2,740 = 2,730 \qquad R_z \approx 0$$

12.18 The boom in Fig. 12.18a is hinged to move in the yz plane. Find the force in cable bdc and the hinge pin force at a.

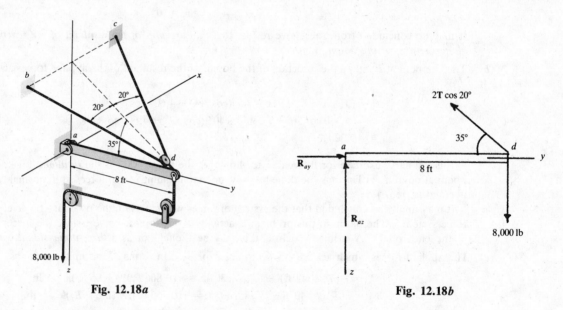

Fig. 12.18a   Fig. 12.18b

▌ T is the tensile force in cable bdc, and a view of the boom in the yz plane is shown in Fig. 12.18b. For equilibrium in the yz plane,

$$\Sigma M_x = 0 \qquad -(2T \cos 20°) \sin 35°(8) + 8,000(8) = 0 \qquad T = 7,420 \text{ lb}$$
$$\Sigma F_y = 0 \qquad R_{ay} - (2T \cos 20°) \cos 35° = 0 \qquad R_{ay} = 11,400 \text{ lb}$$
$$\Sigma F_z = 0 \qquad R_{az} + (2T \cos 20°) \sin 35° - 8,000 = 0 \qquad R_{az} + 8,000 = 8,000 \qquad R_{az} = 0$$

The remaining equilibrium equations $\Sigma F_x = 0$, $\Sigma M_y = 0$, and $\Sigma M_z = 0$ are identically satisfied. Since the forces are applied only at the ends of member ad, this element acts as a two-force member.

12.19 A rigid, weightless boom supports a load, as shown in Fig. 12.19a. The boom is hinged to move in the xy plane only.

(a) Find the tensile force in the cable between the boom and the wall, and the force and moment components exerted by the hinge on the boom.

(b) Find the force acting on the hinge pin normal to the z axis.

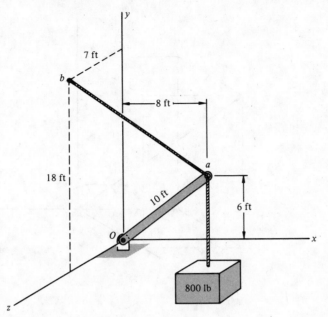

Fig. 12.19a

(c) Find the components of the force and moment exerted by the hinge on the boom, in the directions of the axis of the boom and at right angles to this direction.

(d) Find the results in parts (a) through (c) by using the vector equations of equilibrium.

▌ (a) The free-body diagram of the boom is shown in Fig. 12.19b, and the construction used to find the direction angles of the cable force T is shown in Fig. 12.19c. The magnitude of the dimension d is

$$d = \sqrt{(-8)^2 + 12^2 + 7^2} = 16 \text{ ft}$$

The direction angles are then found as

$$\cos \theta_x = -\tfrac{8}{16} \qquad \theta_x = 120°$$
$$\cos \theta_y = \tfrac{12}{16} \qquad \theta_y = 41.4°$$
$$\cos \theta_z = \tfrac{7}{16} \qquad \theta_z = 64.1°$$

The components of the cable tensile force T are

$$T_x = T \cos \theta_x = -0.5T \tag{1}$$
$$T_y = T \cos \theta_y = 0.75T \tag{2}$$
$$T_z = T \cos \theta_z = 0.438T \tag{3}$$

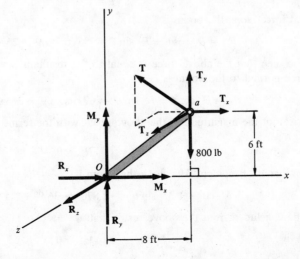

Fig. 12.19b

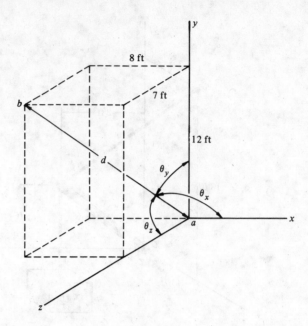

Fig. 12.19c

Because of the use of the direction cosines, the three components T_x, T_y, and T_z above are considered to be quantities *with the senses of the positive coordinate axes*. At the outset of the problem, the component of moment M_z is known to be zero, because of the hinge support.

The equilibrium requirements are

$$\sum M_z = -800(8) - T_x(6) + T_y(8) = 0 \qquad -800(8) - (-0.5T)(6) + 0.75T(8) = 0 \qquad T = 711 \text{ lb}$$

$$\sum F_x = 0 \qquad R_x + T_x = 0 \qquad R_x + (-0.5T) = 0 \qquad R_x = 356 \text{ lb}$$

$$\sum F_y = 0 \qquad R_y + T_y - 800 = 0 \qquad R_y + 0.75T - 800 = 0 \qquad R_y = 267 \text{ lb}$$

$$\sum F_z = 0 \qquad R_z + T_z = 0 \qquad R_z + 0.438T = 0 \qquad R_z = -311 \text{ lb}$$

$$\sum M_x = 0 \qquad M_x + T_z(6) = 0 \qquad M_x + (0.438T)(6) = 0 \qquad M_x = -1,870 \text{ ft} \cdot \text{lb}$$

$$\sum M_y = 0 \qquad M_y - T_z(8) = 0 \qquad M_y - (0.438T)(8) = 0 \qquad M_y = 2,490 \text{ ft} \cdot \text{lb}$$

(*b*) The total force R which acts on the hinge pin, in a direction normal to the z axis, is

$$R = \sqrt{R_x^2 + R_y^2} = \sqrt{356^2 + 267^2} = 445 \text{ lb}$$

(*c*) Figure 12.19*d* shows the components of force and moment exerted by the hinge on the boom. The direction θ of the boom is found from

$$\tan \theta = \tfrac{6}{8} \qquad \theta = 36.9°$$

The total force *along* the boom is

$$R_{x'} = R_x \cos \theta + R_y \sin \theta = 356 \cos 36.9° + 267 \sin 36.9° = 445 \text{ lb}$$

It may be seen that the above force is equal to the resultant hinge pin force found in part (*b*). The total force normal to the boom is

$$R_{y'} = R_y \cos \theta - R_x \sin \theta = 267 \cos 36.9° - 356 \sin 36.9° = -0.2 \text{ lb}$$

Comparison of the magnitude of the above result with the terms used in the computation leads to the conclusion

$$R_{y'} \approx 0$$

The component of moment $M_{x'}$ along the boom is

$$M_{x'} = M_x \cos \theta + M_y \sin \theta = -1,870 \cos 36.9° + 2,490 \sin 36.9° = -0.4 \text{ ft} \cdot \text{lb}$$

It may be concluded from the above result that

$$M_{x'} \approx 0$$

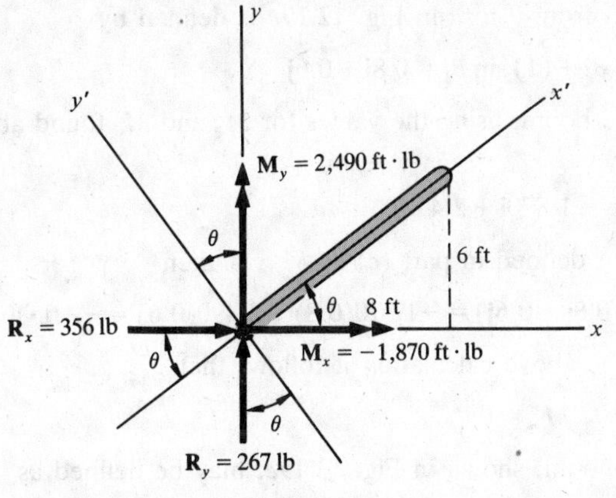

Fig. 12.19d

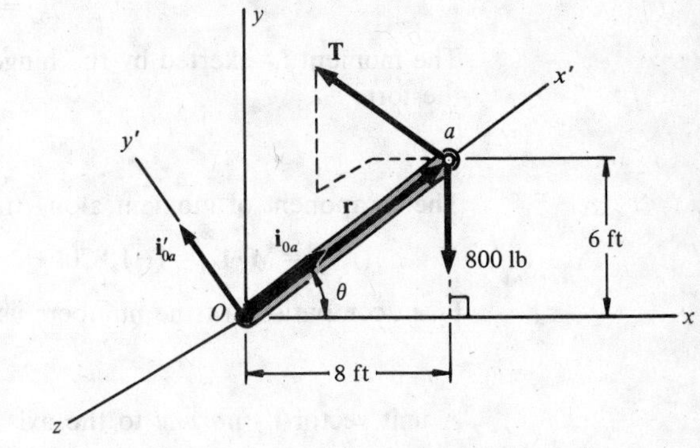

Fig. 12.19e

The above result is expected, since the lines of action of all the forces acting on the boom intersect the x' axis and thus produce no moment about this axis.

The component of moment $M_{y'}$ normal to the boom is

$$M_{y'} = -M_x \sin\theta + M_y \cos\theta = 1{,}870 \sin 36.9° + 2{,}490 \cos 36.9° = 3{,}110 \text{ ft} \cdot \text{lb}$$

The above value would be used to analyze the stresses in the boom, and this topic is treated in the subject strength, or mechanics, of materials.

(*d*) A position vector $\mathbf{r}$ along the boom from 0 to a is shown in Fig. 12.19*e*, and

$$\mathbf{r} = 8\mathbf{i} + 6\mathbf{j}$$

The resultant of all the applied forces which act on the boom at point a is

$$\mathbf{F} = T_x\mathbf{i} + (T_y - 800)\mathbf{j} + T_z\mathbf{k}$$

Using Eqs. (1) through (3) in part (*a*), we can write the above equation as

$$\mathbf{F} = (-0.5T)\mathbf{i} + (0.75T - 800)\mathbf{j} + (0.438T)\mathbf{k}$$

The reaction force $\mathbf{R}$ and moment $\mathbf{M}$ have the forms

$$\mathbf{R} = R_x\mathbf{i} + R_y\mathbf{j} + R_z\mathbf{k} \qquad \mathbf{M} = M_x\mathbf{i} + M_y\mathbf{j} + (0)\mathbf{k}$$

For moment equilibrium about the hinge support,

$$\mathbf{M} + \mathbf{r} \times \mathbf{F} = 0 \qquad M_x\mathbf{i} + M_y\mathbf{j} + (8\mathbf{i} + 6\mathbf{j}) \times [(-0.5T)\mathbf{i} + (0.75T - 800)\mathbf{j} + (0.438T)\mathbf{k}] = 0$$

$$M_x\mathbf{i} + M_y\mathbf{j} + \begin{vmatrix} \mathbf{i} & \mathbf{j} & \mathbf{k} \\ 8 & 6 & 0 \\ -0.5T & 0.75T - 800 & 0.438T \end{vmatrix} = 0$$

$$[M_x + 6(0.438T)]\mathbf{i} + [M_y - 8(0.438T)]\mathbf{j} + [8(0.75T - 800) - 6(-0.5T)]\mathbf{k} = 0$$

The three coefficients of the unit vectors are set equal to zero:

$$8(0.75T - 800) - 6(-0.5T) = 0 \qquad T = 711 \text{ lb}$$
$$M_x + 6(0.438T) = 0 \qquad M_x = -1{,}870 \text{ ft} \cdot \text{lb}$$
$$M_y - 8(0.438T) = 0 \qquad M_y = 2{,}490 \text{ ft} \cdot \text{lb}$$

For force equilibrium of the boom,

$$\mathbf{R} + \mathbf{F} = 0$$

$$R_x\mathbf{i} + R_y\mathbf{j} + R_z\mathbf{k} + (-0.5T)\mathbf{i} + (0.75T - 800)\mathbf{j} + (0.438T)\mathbf{k} = 0$$
$$(R_x - 0.5T)\mathbf{i} + (R_y + 0.75T - 800)\mathbf{j} + (R_z + 0.438T)\mathbf{k} = 0$$

The coefficients of the unit vectors are equated to zero, with the results

$$R_x - 0.5T = 0 \qquad R_x = 356 \text{ lb}$$
$$R_y = 0.75T - 800 = 0 \qquad R_y = 267 \text{ lb}$$
$$R_z + 0.438T = 0 \qquad R_z = -311 \text{ lb}$$

The unit vector $\mathbf{i}_{0a}$ directed along the boom from 0 to a in Fig. 12.19e is defined by

$$\mathbf{i}_{0a} = (1)\cos\theta\mathbf{i} + (1)\sin\theta\mathbf{j} = 0.8\mathbf{i} + 0.6\mathbf{j}$$

The moment $\mathbf{M}$ exerted by the hinge on the boom, using the values for M_x and M_y found above, has the form

$$\mathbf{M} = -1{,}870\mathbf{i} + 2{,}490\mathbf{j}$$

The component of moment along the boom, defined in part (c) as $M_{x'}$, is then

$$M_{x'} = \mathbf{M}\cdot\mathbf{i}_{0a} = (-1{,}870\mathbf{i} + 2{,}490\mathbf{j})\cdot(0.8\mathbf{i} + 0.6\mathbf{j}) = -1{,}870(0.8) + 2{,}490(0.6) = -2\text{ ft}\cdot\text{lb}$$

From comparison of the numbers used in the above calculation it follows that

$$M_{x'} \approx 0$$

A unit vector $\mathbf{i}'_{0a}$ *normal* to the axis of the boom, shown in Fig. 12.19e, may be defined as

$$\mathbf{i}'_{0a} = (-1)\sin\theta\mathbf{i} + (1)\cos\theta\mathbf{j} = -0.6\mathbf{i} + 0.8\mathbf{j}$$

The component of moment normal to the boom is then

$$M_{y'} = \mathbf{M}\cdot\mathbf{i}'_{0a} = (-1{,}870\mathbf{i} + 2{,}490\mathbf{j})\cdot(-0.6\mathbf{i} + 0.8\mathbf{j}) = (-1{,}870)(-0.6) + 2{,}490(0.8) = 3{,}110\text{ ft}\cdot\text{lb}$$

This result agrees with that obtained in part (c). The forces acting along the boom, and normal to this element, may be found from the dot products of the forces $\mathbf{R}$ or $\mathbf{F}$ with the two unit vectors $\mathbf{i}_{0a}$ and $\mathbf{i}'_{0a}$.

12.20 (a) Find the force in cable bc, and the x, y, and z components of the force and moment exerted by the foundation on the link, shown in Fig. 12.20a, by using the scalar equations of equilibrium.

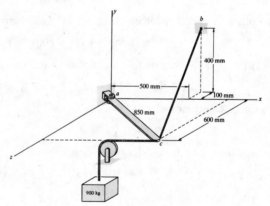

Fig. 12.20a

(b) Find the results in part (a) by using the vector equations of equilibrium.

(c) Find the axial compressive force in the link.

(d) Find the component of moment that acts on the link, in the direction of the axis of the link.

(e) Find the component of moment that acts on the link, in the direction normal to the direction in part (d).

▌ (a) Figure 12.20b shows the direction of the cable tensile force T. From this figure

$$d = \sqrt{102^2 + 400^2 + 700^2} = 813\text{ mm}$$

The components of T are

$$T_x = -\frac{102}{813}T \qquad T_y = \frac{400}{813}T \qquad T_z = -\frac{700}{813}T$$

Figure 12.20c shows the free-body diagram of the link. R_{ax}, R_{ay}, R_{az}, M_x, and M_z are assumed to act in the positive coordinate senses. From the definition of a hinge, $M_y = 0$.
For equilibrium of the link,

$$\sum M_x = 0 \qquad M_x - T_y\left(\frac{600}{1{,}000}\right) = 0 \tag{1}$$

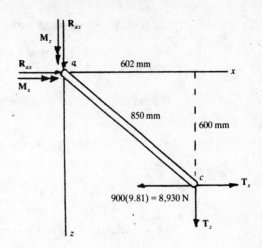

Fig. 12.20b   **Fig. 12.20c**

$$\Sigma M_y = 0 \qquad T_x\left(\frac{600}{1,000}\right) - 8,830\left(\frac{600}{1,000}\right) - T_z\left(\frac{602}{1,000}\right) = 0 \qquad (2)$$

$$\Sigma M_z = 0 \qquad M_z + T_y\left(\frac{602}{1,000}\right) = 0 \qquad (3)$$

$$\Sigma F_x = 0 \qquad R_{ax} + T_x - 8,830 = 0 \qquad (4)$$
$$\Sigma F_y = 0 \qquad R_{ay} + T_y = 0 \qquad (5)$$
$$\Sigma F_z = 0 \qquad R_{az} + T_z = 0 \qquad (6)$$

Using Eq. (2),

$$\left(-\frac{102}{813}\,T\right)\frac{600}{1,000} - 8,830\left(\frac{600}{1,000}\right) - \left(-\frac{700}{813}\,T\right)\frac{602}{1,000} = 0 \qquad T = 12,000\,\text{N}$$

Using Eq. (1),

$$M_x - \left(\frac{400}{813}\,T\right)\frac{600}{1,000} = 0 \qquad M_x = 3,540\,\text{N}\cdot\text{m}$$

Using Eq. (3),

$$M_z + \left(\frac{400}{813}\,T\right)\frac{602}{1,000} = 0 \qquad M_z = -3,550\,\text{N}\cdot\text{m}$$

Using Eq. (4),

$$R_{ax} + \left(-\frac{102}{813}\,T\right) - 8,830 = 0 \qquad R_{ax} = 10,300\,\text{N}$$

Using Eq. (5),

$$R_{ay} + \left(\frac{400}{813}\,T\right) = 0 \qquad R_{ay} = -5,900\,\text{N}$$

Using Eq. (6),

$$R_{az} + \left(-\frac{700}{813}\,T\right) = 0 \qquad R_{az} = 10,300\,\text{N}$$

(b) The forces and moments acting on the link are written in terms of the unit vectors as

$$\mathbf{W} = -8,830\mathbf{i} \qquad \mathbf{T} = -\left(\frac{102}{813}\,T\right)\mathbf{i} + \left(\frac{400}{813}\,T\right)\mathbf{j} - \left(\frac{700}{813}\,T\right)\mathbf{k} \qquad \mathbf{R} = R_{ax}\mathbf{i} + R_{ay}\mathbf{j} + R_{az}\mathbf{k}$$
$$\mathbf{M} = M_x\mathbf{i} + M_z\mathbf{k}$$

Point a in Fig. 12.20b is used as the reference for moments, and the position vector to point c has the form

$$\mathbf{r} = 602\mathbf{i} + 600\mathbf{k} \qquad \text{mm}$$

For moment equilibrium of the link,

$$\mathbf{M} + \mathbf{r} \times (\mathbf{W} + \mathbf{T}) = 0$$

The term $\mathbf{W} + \mathbf{T}$ may be written as

$$\mathbf{W} + \mathbf{T} = \left(-8,830 - \frac{102}{813}\,T\right)\mathbf{i} + \left(\frac{400}{813}\,T\right)\mathbf{j} - \left(\frac{700}{813}\,T\right)\mathbf{k}$$

$$= (-8,830 - 0.125\,T)\mathbf{i} + (0.492\,T)\mathbf{j} - (0.861\,T)\mathbf{k}$$

The moment equation now has the form

$$M_x\mathbf{i} + M_z\mathbf{k} + \begin{vmatrix} \mathbf{i} & \mathbf{j} & \mathbf{k} \\ \dfrac{602}{1,000} & 0 & \dfrac{600}{1,000} \\ -8,830 - 0.125\,T & 0.492\,T & -0.861\,T \end{vmatrix} = 0$$

$$\left[M_x - (0.492\,T)\,\frac{600}{1,000}\right]\mathbf{i} + \frac{1}{1,000}\,[600(-8,830 - 0.125\,T) - (-0.861\,T)602]\mathbf{j} + \left[M_z + (0.492\,T)\,\frac{602}{1,000}\right]\mathbf{k} = 0$$

The coefficients of the unit vectors are equated to zero, with the results

$$600(-8,830 - 0.125\,T) + (0.861\,T)602 = 0 \qquad T = 12,000\ \text{N}$$

$$M_x - (0.492\,T)\,\frac{600}{1,000} = 0 \qquad M_x = 3,540\ \text{N} \cdot \text{m}$$

$$M_z + (0.492\,T)\,\frac{602}{1,000} = 0 \qquad M_z = -3,550\ \text{N} \cdot \text{m}$$

For force equilibrium of the link,

$$\mathbf{W} + \mathbf{T} + \mathbf{R} = 0 \qquad \left(-8,830 - \frac{102}{813}\,T + R_{ax}\right)\mathbf{i} + \left(\frac{400}{813}\,T + R_{ay}\right)\mathbf{j} + \left(-\frac{700}{813}\,T + R_{az}\right)\mathbf{k} = 0$$

The three scalar equations from the above vector equation are

$$R_{ax} - 8,830 - \frac{102}{813}\,T = 0 \qquad R_{ax} = 10,300\ \text{N}$$

$$R_{ay} + \frac{400}{813}\,T = 0 \qquad R_{ay} = -5,900\ \text{N}$$

$$R_{az} - \frac{700}{813}\,T = 0 \qquad R_{az} = 10,300\ \text{N}$$

(c) Figure 12.20d shows the reaction force components R_{ax} and R_{az} acting in the plane of the link. The axial compressive force F_{axial} along the direction of the link is given by

$$F_{\text{axial}} = 10,300\left(\frac{602}{850}\right) + 10,300\left(\frac{600}{850}\right) = 14,600\ \text{N}$$

(d) Figure 12.20e shows the reaction moment components which act in the plane of the link. The reaction moment M_t about the axis of the link is found as

$$M_t = 3,540\left(\frac{602}{850}\right) - 3,550\left(\frac{600}{850}\right) = 0$$

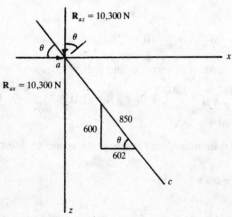

Fig. 12.20d

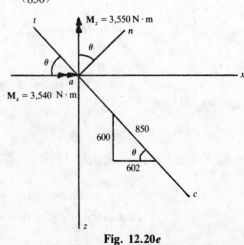

Fig. 12.20e

(e) The reaction moment M_n normal to the axis of the link is found, using Fig. 12.20e, as

$$M_n = 3{,}550\left(\frac{602}{850}\right) + 3{,}540\left(\frac{600}{850}\right) = 5{,}010 \, \text{N} \cdot \text{m}$$

12.21 (a) Use the scalar equations of equilibrium to find the force in the cable, and the components of the reaction force at hinge a, for the boom shown in Fig. 12.21a.

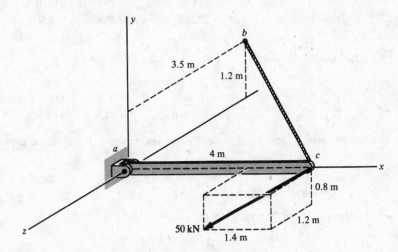

50 kN

Fig. 12.21a

(b) Show the forms of the forces, position vector, and equilibrium equations to solve part (a) by using the vector equations of equilibrium.

(c) Find the magnitude, direction, and sense of the resultant force exerted by the boom on the foundation.

(d) Find the component, along the boom, of the reaction moment exerted by the foundation on the boom.

▌ (a) Figure 12.21b shows the direction of the 50-kN force. The components of this force, using $d = \sqrt{1.4^2 + 0.8^2 + 1.2^2} = 2.01 \, \text{m}$, are

$$F_x = -\frac{1.4}{2.01}(50) = -34.8 \, \text{kN} \qquad F_y = -\frac{0.8}{2.01}(50) = -19.9 \, \text{kN} \qquad F_z = \frac{1.2}{2.01}(50) = 29.9 \, \text{kN}$$

Figure 12.21c shows the direction of cable force T. The components of this force, using $d = \sqrt{4^2 + 1.2^2 + 3.5^2} = 5.45 \, \text{m}$, are

$$T_x = -\frac{4}{5.45}T = -0.734T \qquad T_y = \frac{1.2}{5.45}T = 0.220T \qquad T_z = -\frac{3.5}{5.45}T = -0.642T$$

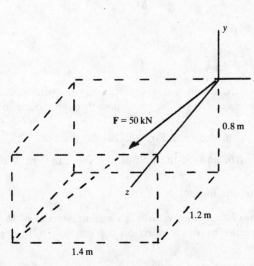

Fig. 12.21b

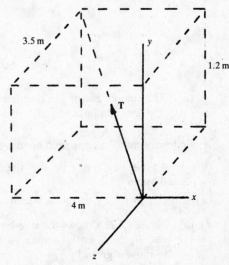

Fig. 12.21c

The reaction force components R_{ax}, R_{ay}, and R_{az} and the reaction moment components M_x and M_y are assumed to act in the positive coordinate senses.

For equilibrium of the link,

$$\Sigma M_z = 0 \qquad F_y(4) + T_y(4) = 0 \qquad F_y = -T_y \qquad\qquad (1)$$
$$-19.9 = -0.220T \qquad T = 90.5 \text{ kN}$$
$$\Sigma M_x = 0 \qquad M_x = 0$$
$$\Sigma M_y = 0 \qquad M_y - F_z(4) - T_z(4) = 0$$
$$M_y = 29.9(4) + (-0.642T)4 = -113 \text{ kN} \cdot \text{m}$$
$$\Sigma F_y = 0 \qquad R_{ay} + F_y + T_y = 0$$

Using Eq. (1),

$$R_{ay} + (-T_y) + T_y = 0 \qquad R_{ay} = 0$$
$$\Sigma F_z = 0 \qquad R_{az} + F_z + T_z = 0 \qquad R_{az} + 29.9 - 0.642T = 0 \qquad R_{az} = 28.2 \text{ kN}$$
$$\Sigma F_x = 0 \qquad R_{ax} + F_x + T_x = 0 \qquad R_{ax} - 34.8 - 0.734T = 0 \qquad R_{ax} = 101 \text{ kN}$$

(b) The 50-kN force $\mathbf{F}$, the cable force $\mathbf{T}$, and the reaction force $\mathbf{R}$ are expressed in terms of the unit vectors as

$$\mathbf{F} = -34.8\mathbf{i} - 19.9\mathbf{j} + 29.9\mathbf{k} \qquad \text{kN}$$
$$\mathbf{T} = -(0.734T)\mathbf{i} + (0.220T)\mathbf{j} - (0.642T)\mathbf{k} \qquad \mathbf{R} = R_{ax}\mathbf{i} + R_{ay}\mathbf{j} + R_{az}\mathbf{k}$$

The reaction moment $\mathbf{M}$ has the form

$$\mathbf{M} = M_x\mathbf{i} + M_y\mathbf{j}$$

Point a is used as the reference for moments, and the position vector from a to c is written as

$$\mathbf{r}_{ac} = 4\mathbf{i} \qquad \text{m}$$

For equilibrium of the link,

$$\mathbf{M}_a = \mathbf{M} + \mathbf{r}_{ac} \times \mathbf{F} + \mathbf{r}_{ac} \times \mathbf{T} = 0 \qquad \mathbf{R} + \mathbf{F} + \mathbf{T} = 0$$

It is left as an exercise for the reader to use the above equations to confirm the results found in part (a).

(c) The components of force exerted by the boom on the foundation are

$$R_{ax} = -101 \text{ kN} \qquad R_{ay} = 0 \qquad R_{az} = -28.2 \text{ kN}$$

The resultant of the above force is

$$R_a = \sqrt{R_{ax}^2 + R_{az}^2} = \sqrt{(-101)^2 + (-28.2)^2} = 105 \text{ kN}$$

The direction of the force is given by

$$\cos \theta_x = -\frac{101}{105} \qquad \theta_x = 164°$$
$$\cos \theta_y = 0 \qquad \theta_y = 90°$$
$$\cos \theta_z = -\frac{28.2}{105} \qquad \theta_z = 106°$$

(d) The component of the reaction moment of the foundation on the boom, along the boom, is $M_x = 0$, since the lines of action of all of the forces acting on the boom pass through the axis of the boom.

12.22 A window of mass 15 kg is held open by a rigid link, as shown in Fig. 12.22a.

(a) Find the forces acting on the hinge pins a and b, and on the link cd, by using the scalar equations of equilibrium.

(b) Solve part (a) by using the vector equations of equilibrium.

▌ (a) Figure 12.22b shows the free-body diagram of the window, with the assumed senses of R_{ay}, R_{az}, R_{by}, and R_{bz}. F_c is the compressive force exerted by member cd on the window. The weight of the window is given by

$$W = 15(9.81) = 147 \text{ N}$$

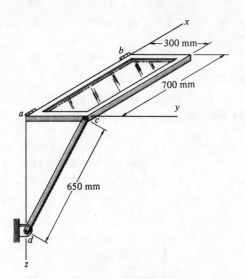

Fig. 12.22a

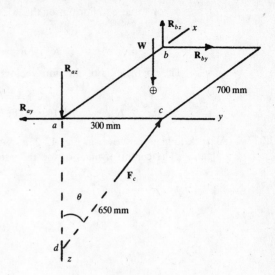

Fig. 12.22b

Angle θ is found from

$$\sin \theta = \frac{300}{650} \qquad \theta = 27.5°$$

For equilibrium of the window,

$$\Sigma M_x = 0 \qquad -(F_c \cos 27.5°)300 + 147(150) = 0 \qquad F_c = 82.9 \text{ N}$$
$$\Sigma M_y = 0 \qquad -147(350) + R_{bz}(700) = 0 \qquad R_{bz} = 73.5 \text{ N}$$
$$\Sigma M_z = 0 \qquad R_{by}(700) = 0 \qquad R_{by} = 0$$
$$\Sigma F_x = 0 \qquad 0 = 0$$
$$\Sigma F_y = 0 \qquad -R_{ay} + F_c \sin 27.5° + R_{by} = 0 \qquad R_{ay} = 38.3 \text{ N}$$
$$\Sigma F_z = 0 \qquad R_{az} - F_c \cos 27.5° + 147 - R_{bz} = 0 \qquad R_{az} + 147 = 73.5 + 73.5 = 147 \qquad R_{az} = 0$$
$$R_a = R_{ay} = 38.3 \text{ N} \qquad R_b = R_{bz} = 73.5 \text{ N}$$

(b) The forces acting on the window are expressed in terms of the unit vectors as

$$\mathbf{R}_a = -R_{ay}\mathbf{j} + R_{az}\mathbf{k} \qquad \mathbf{R}_b = R_{by}\mathbf{j} - R_{bz}\mathbf{k}$$
$$\mathbf{W} = 147\mathbf{k} \qquad \text{N}$$
$$\mathbf{F} = \mathbf{F}_c = (F_c \sin 27.5°)\mathbf{j} - (F_c \cos 27.5°)\mathbf{k} = (0.462F_c)\mathbf{j} - (0.887F_c)\mathbf{k}$$

Point a in Fig. 12.22b is used as the reference point for moments. The position vectors from point a to the forces have the forms

$$\mathbf{r}_{ab} = 700\mathbf{i} \qquad \text{mm} \qquad \mathbf{r}_{aW} = 350\mathbf{i} + 150\mathbf{j} \qquad \text{mm} \qquad \mathbf{r}_{aF} = 300\mathbf{j} \qquad \text{mm}$$

For moment equilibrium of the window,

$$\mathbf{M}_a = \mathbf{r}_{ab} \times \mathbf{R}_b + \mathbf{r}_{aW} \times \mathbf{W} + \mathbf{r}_{aF} \times \mathbf{F} = 0$$

$$\mathbf{M}_a = \begin{vmatrix} \mathbf{i} & \mathbf{j} & \mathbf{k} \\ 700 & 0 & 0 \\ 0 & R_{by} & -R_{bz} \end{vmatrix} + \begin{vmatrix} \mathbf{i} & \mathbf{j} & \mathbf{k} \\ 350 & 150 & 0 \\ 0 & 0 & 147 \end{vmatrix} + \begin{vmatrix} \mathbf{i} & \mathbf{j} & \mathbf{k} \\ 0 & 300 & 0 \\ 0 & 0.462F_c & -0.887F_c \end{vmatrix} = 0$$

$$[150(147) + 300(-0.887F_c)]\mathbf{i} + [-(-R_{bz})700 - 147(350)]\mathbf{j} + [R_{by}(700)]\mathbf{k} = 0$$

Equating the coefficients of the unit vectors to zero results in

$$150(147) + 300(-0.887F_c) = 0 \qquad F_c = 82.9 \text{ N}$$
$$R_{bz}(700) - 147(350) = 0 \qquad R_{bz} = 73.5 \text{ N}$$
$$R_{by}(700) = 0 \qquad R_{by} = 0$$

For force equilibrium of the window,

$$\mathbf{R}_a + \mathbf{R}_b + \mathbf{W} + \mathbf{F} = 0$$

$$(-R_{ay} + R_{by} + 0.462F_c)\mathbf{j} + (R_{az} - R_{bz} + 147 - 0.887F_c)\mathbf{k} = 0$$

The coefficients of the unit vectors are equated to zero to obtain

$$-R_{ay} + 0.462F_c = 0 \qquad R_{ay} = 38.3\,\text{N}$$

$$R_{az} - 73.5 + 147 - 0.887(82.9) = 0 \qquad R_{az} + 147 = 73.5 + 73.5 = 147 \qquad R_{az} = 0$$

12.23 A truck tailgate is shown in Fig. 12.23a. A homogeneous block of specific weight 120 lb/ft³ rests on an 85-lb tailgate. The center of mass of the tailgate is located 12 in from the centerline of the hinges.

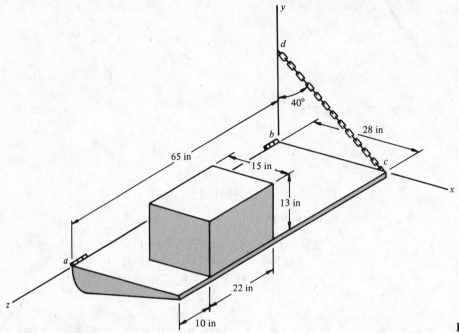

Fig. 12.23a

(a) Using the scalar equations of equilibrium, find the tensile force in the tailgate chain and the resultant forces acting on the hinge pins at a and b.

(b) Express the force and position vectors in terms of the unit vectors and give the vector equations of equilibrium.

▌ (a) Figure 12.23b shows the free-body diagram of the tailgate. The weights W_1 of the tailgate and W_2 of the block are

$$W_1 = 85\,\text{lb} \qquad W_2 = 22(13)15\ \text{in}^3\left(120\ \frac{\text{lb}}{\text{ft}^3}\right)\left(\frac{1\ \text{ft}}{12\ \text{in}}\right)^3 = 298\,\text{lb}$$

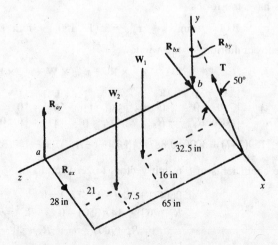

Fig. 12.23b

For equilibrium of the tailgate,

$$\Sigma M_z = 0 \qquad -W_2(20.5) - W_1(12) + (T \sin 50°)28 = 0 \qquad T = 332 \text{ lb}$$
$$\Sigma M_x = 0 \qquad W_1(32.5) + W_2(44) - R_{ay}(65) = 0 \qquad R_{ay} = 244 \text{ lb}$$
$$\Sigma M_y = 0 \qquad R_{ax}(65) = 0 \qquad R_{ax} = 0$$
$$\Sigma F_x = 0 \qquad R_{bx} - T \cos 50° = 0 \qquad R_{bx} = 213 \text{ lb}$$
$$\Sigma F_y = 0 \qquad R_{ay} - W_2 - W_1 - R_{by} + T \sin 50° = 0 \qquad R_{by} = 115 \text{ lb}$$

The resultant forces acting on the hinge pins are

$$R_a = R_{ay} = 244 \text{ lb} \qquad R_b = \sqrt{R_{bx}^2 + R_{by}^2} = \sqrt{213^2 + 115^2} = 242 \text{ lb}$$

(b) The forces are expressed in terms of the unit vectors as

$$W_1 = -85\mathbf{j} \quad \text{lb} \qquad W_2 = -298\mathbf{j} \quad \text{lb}$$
$$T = -(T \cos 50°)\mathbf{i} + (T \sin 50°)\mathbf{j} = -(0.643T)\mathbf{i} + (0.766T)\mathbf{j}$$
$$R_a = R_{ax}\mathbf{i} + R_{ay}\mathbf{j} \qquad R_b = R_{bx}\mathbf{i} - R_{by}\mathbf{j}$$

Using point b as the reference point for moments, the position vectors of the forces have the forms

$$\mathbf{r}_{ba} = 65\mathbf{k} \quad \text{in} \qquad \mathbf{r}_{bW_1} = 12\mathbf{i} + 32.5\mathbf{k} \quad \text{in}$$
$$\mathbf{r}_{bW_2} = 20.5\mathbf{i} + 44\mathbf{k} \quad \text{in} \qquad \mathbf{r}_{bT} = 28\mathbf{i} \quad \text{in}$$

The equilibrium requirements of the system are

$$\mathbf{M}_b = \mathbf{r}_{ba} \times \mathbf{R}_a + \mathbf{r}_{bW_1} \times \mathbf{W}_1 + \mathbf{r}_{bW_2} \times \mathbf{W}_2 + \mathbf{r}_{bT} \times \mathbf{T} = 0$$
$$\mathbf{R}_a + \mathbf{R}_b + \mathbf{W}_1 + \mathbf{W}_2 + \mathbf{T} = 0$$

It is left as an exercise for the reader to solve the above equilibrium equations and verify the results in part (a).

12.24 The plate in Fig. 12.24a has a mass of 350 kg and is supported in the xy plane by the cable cd. Only hinge a can resist forces in the z direction.

(a) Using the vector equations of equilibrium, find the force in the cable, and the reaction forces at hinges a and b.

(b) Do the same as in part (a), if only hinge b can resist forces in the z direction.

▌ (a) Figure 12.24b shows the direction of the cable force T. Using $d = \sqrt{900^2 + 800^2 + 1,200^2} = 1,700$ mm, the components of this force are

$$T_x = -\frac{900}{1,700} T = -0.529T \qquad T_y = \frac{800}{1,700} T = 0.471T \qquad T_z = -\frac{1,200}{1,700} T = -0.706T$$

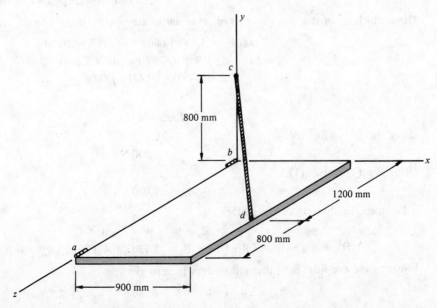

Fig. 12.24a

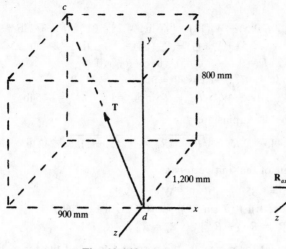

Fig. 12.24b   **Fig. 12.24c**

The above three components are positive when acting in the positive coordinate senses.

The free-body diagram of the plate is shown in Fig. 12.24c, and the weight W of the plate is given by

$$W = 350(9.81) = 3,430 \text{ N}$$

The forces acting on the plate are expressed in terms of the unit vectors, with the results

$$\mathbf{R}_a = R_{ax}\mathbf{i} + R_{ay}\mathbf{j} + R_{az}\mathbf{k} \qquad \mathbf{R}_b = R_{bx}\mathbf{i} + R_{by}\mathbf{j}$$
$$\mathbf{W} = -3,430\mathbf{j} \quad \text{N} \qquad \mathbf{T} = -(0.529\,T)\mathbf{i} + (0.471\,T)\mathbf{j} - (0.706\,T)\mathbf{k}$$

The reference point for moments is chosen to be point a. The position vectors from this point to the forces have the forms

$$\mathbf{r}_{ab} = -2,000\mathbf{k} \quad \text{mm} \qquad \mathbf{r}_{aW} = 450\mathbf{i} - 1,000\mathbf{k} \quad \text{mm} \qquad \mathbf{r}_{aT} = 900\mathbf{i} - 800\mathbf{k} \quad \text{mm}$$

For moment equilibrium of the plate

$$\mathbf{M}_a = \mathbf{r}_{ab} \times \mathbf{R}_b + \mathbf{r}_{aW} \times \mathbf{W} + \mathbf{r}_{aT} \times \mathbf{T} = 0$$

$$\mathbf{M}_a = \begin{vmatrix} \mathbf{i} & \mathbf{j} & \mathbf{k} \\ 0 & 0 & -2,000 \\ R_{bx} & R_{by} & 0 \end{vmatrix} + \begin{vmatrix} \mathbf{i} & \mathbf{j} & \mathbf{k} \\ 450 & 0 & -1,000 \\ 0 & -3,430 & 0 \end{vmatrix} + \begin{vmatrix} \mathbf{i} & \mathbf{j} & \mathbf{k} \\ 900 & 0 & -800 \\ -0.529\,T & 0.471\,T & -0.706\,T \end{vmatrix} = 0$$

$$[-R_{by}(-2,000) - (-3,430)(-1,000) - (0.471\,T)(-800)]\mathbf{i} + [-2,000R_{bx} - 800(-0.529\,T)$$
$$- (-0.706\,T)900]\mathbf{j} + [-3,430(450) + (0.471\,T)900]\mathbf{k} = 0$$

The coefficients of the unit vectors are set equal to zero, with the results

$$2,000R_{by} - 3,430(1,000) + (0.471\,T)800 = 0 \qquad (1)$$
$$-2,000R_{bx} + 800(0.529\,T) + (0.706\,T)900 = 0 \qquad (2)$$
$$-3,430(450) + (0.471\,T)900 = 0 \qquad (3)$$

The solution to Eq. (3) is

$$T = 3,640 \text{ N} \qquad (4)$$

Using Eq. (4) in Eq. (2),

$$R_{bx} = 1,930 \text{ N}$$

Using Eq. (4) in Eq. (1),

$$R_{by} = 1,030 \text{ N}$$

For force equilibrium of the plate,

$$\mathbf{R}_a + \mathbf{R}_b + \mathbf{W} + \mathbf{T} = 0$$
$$(R_{ax} + R_{bx} - 0.529\,T)\mathbf{i} + (R_{ay} + R_{by} - 3,430 + 0.471\,T)\mathbf{j} + (R_{az} - 0.706\,T)\mathbf{k} = 0$$

Equating the coefficients of the unit vectors to zero gives

$$R_{ax} = -4.44 \approx 0 \qquad R_{ay} = 686 \text{ N} \qquad R_{az} = 2,570 \text{ N}$$

(b) It can be shown that the same results are obtained as in part (a), except that

$$R_{az} = 0 \quad \text{and} \quad R_{bz} = 2{,}570 \text{ N}$$

where R_{bz} acts in the positive z sense.

12.25 Plate *acd* shown in Fig. 12.25a has a weight of 1,500 N. It is attached to the foundation by a ball joint at *c*, a hinge at *d*, and a cable. Use the scalar equations of equilibrium to find the force in the cable and the reaction forces acting on the plate at *c* and *d*.

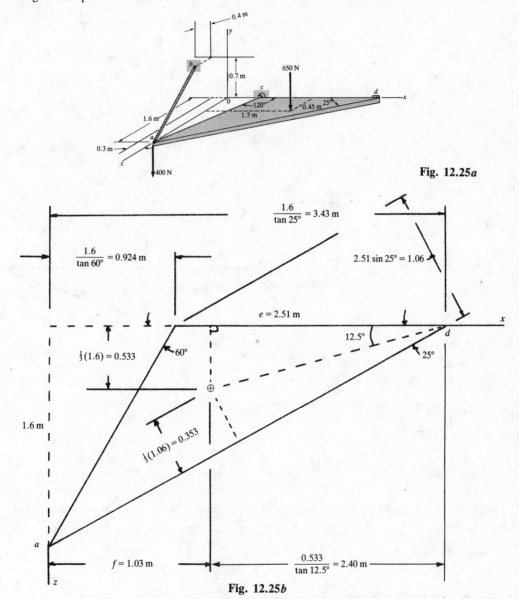

Fig. 12.25a

Fig. 12.25b

▌ A view of the plate in the xz plane is shown in Fig. 12.25b. The dimensions e and f shown in this figure are found as

$$e = 3.43 - 0.924 = 2.51 \text{ m} \qquad f = 3.43 - 2.40 = 1.03 \text{ m}$$

The centroid of the plate area is located at a point that is one-third of the height above the two bases *ad* and *cd*, as shown in Fig. 12.25b.

Figure 12.25c shows the direction of the cable force T. Using $d = \sqrt{0.3^2 + 0.7^2 + 1.2^2} = 1.42 \text{ m}$, the components of the cable force have the forms

$$T_x = -\frac{0.3}{1.42} T = -0.211T \qquad T_y = \frac{0.7}{1.42} T = 0.493T \qquad T_z = -\frac{1.2}{1.42} T = -0.845T$$

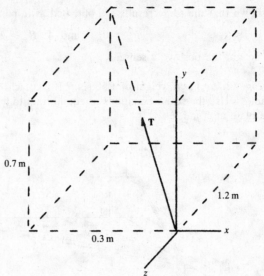

Fig. 12.25c

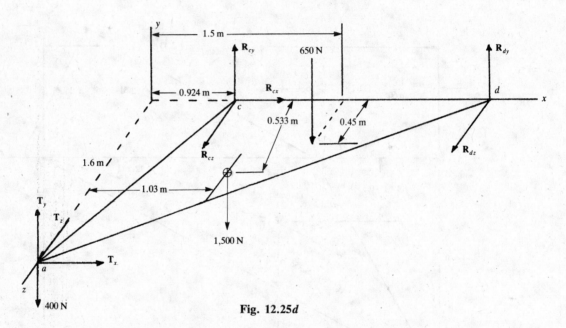

Fig. 12.25d

Figure 12.25d shows the free-body diagram of the plate. For moment equilibrium about the x axis,

$$\Sigma M_x = 0 \qquad 650(0.45) + 1,500(0.533) - T_y(1.6) + 400(1.6) = 0$$

Using $T_y = 0.493T$,

$$T = 2,200 \text{ N}$$

For moment equilibrium about an axis through R_{cy},

$$\Sigma M_{R_{cy}} = 0 \qquad T_z(0.924) + T_x(1.6) - R_{dz}(2.51) = 0$$
$$(-0.845T)0.924 + (-0.211T)1.6 - R_{dz}(2.51) = 0 \qquad R_{dz} = -980 \text{ N}$$

For force equilibrium in the z direction,

$$\Sigma F_z = 0 \qquad T_z + R_{cz} + R_{dz} = 0 \qquad -0.845T + R_{cz} - 980 = 0 \qquad R_{cz} = 2,840 \text{ N}$$

For moment equilibrium about an axis through R_{cz},

$$\Sigma M_{R_{cz}} = 0 \qquad -T_y(0.924) + 400(0.924) - 650(1.5 - 0.924) - 1,500(1.03 - 0.924) + R_{dy}(2.51) = 0$$

Using $T_y = 0.493T$,

$$R_{dy} = 465 \text{ N}$$

For force equilibrium in the x and y directions,

$$\Sigma F_x = 0 \qquad T_x + R_{cx} = 0 \qquad R_{cx} = -T_x = 0.211T = 464 \text{ N}$$

$$\Sigma F_y = 0 \qquad T_y - 400 + R_{cy} - 650 - 1{,}500 + R_{dy} = 0$$

Using $T_y = 0.493T$ and $R_{dy} = 465 \text{ N}$,

$$R_{cy} = 1{,}000 \text{ N}$$

As a check on the above calculations, a moment summation is made about an axis through T_y. The result is

$$-R_{cx}(1.6) - R_{cz}(0.924) \qquad -R_{dz}(3.43) \overset{?}{=} 0$$

$$-R_{dz}(3.43) \overset{?}{=} R_{cx}(1.6) + R_{cz}(0.924) \qquad -(-980)3.43 \overset{?}{=} 464(1.6) + 2{,}840(0.924) \qquad 3{,}360 \approx 3{,}370$$

12.26 The plane vertical plate shown in Fig. 12.26a is attached to the ground by hinges at a and b, and by the link cd. The mass of the plate is 325 kg. Find the hinge pin forces, and the force in the link, if the wind loading on the plate causes a uniform pressure of 690 Pa. Use the scalar equations of equilibrium.

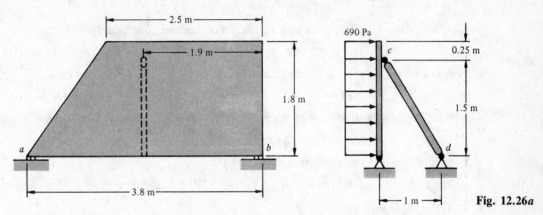

Fig. 12.26a

I The plate area is divided into the two simple shapes shown in Fig. 12.26b. The coordinates of the centroid of this area are given by

$$x_c = \frac{x_1 A_1 + x_2 A_2}{A_1 + A_2} = \frac{\frac{1}{3}(1.8)\frac{1}{2}(1.3)1.8 + (\frac{1.8}{2})(2.5)1.8}{\frac{1}{2}(1.3)1.8 + 2.5(1.8)} = 0.838$$

$$A = A_1 + A_2 = 5.67 \text{ m}^2$$

$$y_c = \frac{y_1 A_1 + y_2 A_2}{A} = \frac{\frac{2}{3}(1.3)\frac{1}{2}(1.3)1.8 + [1.3 + (\frac{2.5}{2})]2.5(1.8)}{5.67} = 2.20 \text{ m}$$

F is the resultant force of the wind on the plate, given by

$$F = pA = 690 \ \frac{\text{N}}{\text{m}^2} (5.67 \text{ m}^2) = 3{,}910 \text{ N}$$

Figure 12.26c shows a side view of the plate. Member cd is a two-force member, with the compressive force F_{cd} acting on it. For equilibrium about the y axis,

$$\Sigma M_y = 0 \qquad -3{,}910(0.838) + \left(\frac{1}{1.80} F_{cd}\right)1.5 = 0 \qquad F_{cd} = 3{,}930 \text{ N}$$

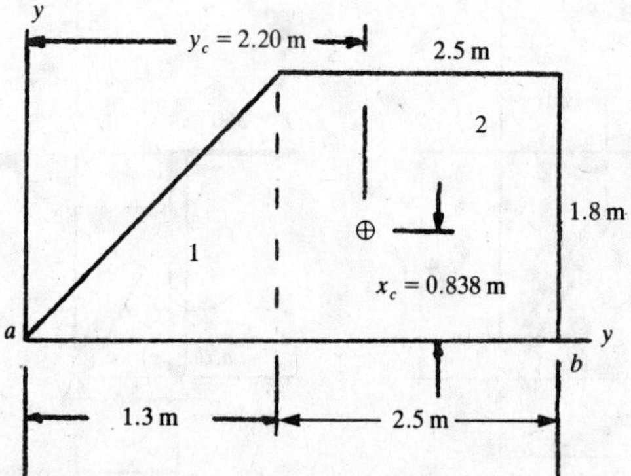

Fig. 12.26b

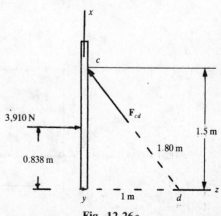

Fig. 12.26c

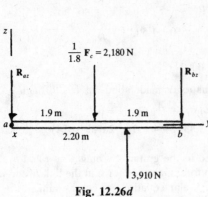

Fig. 12.26d

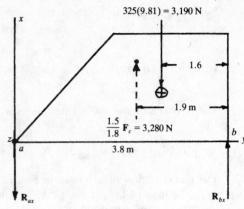

Fig. 12.26e

A top view of the plate is seen in Fig. 12.26d. For moment equilibrium about the x axis,

$$\Sigma M_x = 0 \qquad -2{,}180(1.9) - R_{bx}(3.8) + 3{,}910(2.20) = 0 \qquad R_{bz} = 1{,}170 \text{ N}$$

For force equilibrium in the z direction, using Fig. 12.26d,

$$\Sigma F_z = 0 \qquad -R_{az} - 2{,}180 + 3{,}910 - R_{bz} = 0 \qquad R_{az} = 560 \text{ N}$$

Figure 12.26e shows a front view of the plate. For moment equilibrium about point b,

$$\Sigma M_b = 0 \qquad R_{ax}(3.8) - 3{,}280(1.9) + 3{,}190(1.6) = 0 \qquad R_{ax} = 297 \text{ N}$$

For force equilibrium in the x direction, using Fig. 12.26e,

$$\Sigma F_x = 0 \qquad -R_{ax} + 3{,}280 - 3{,}190 + R_{bx} = 0 \qquad R_{bx} = 207 \text{ N}$$

The resultant forces on the hinge pins are

$$R_a = \sqrt{R_{ax}^2 + R_{az}^2} = \sqrt{297^2 + 560^2} = 634 \text{ N} \qquad R_b = \sqrt{R_{bx}^2 + R_{bz}^2} = \sqrt{207^2 + 1{,}170^2} = 1{,}190 \text{ N}$$

12.27 A 360-lb force acts on corner b of the hinged vertical plate shown in Fig. 12.27a. The link has ball joints at both ends, and only the hinge at a can resist forces in the z direction. The weight of the plate may be neglected.

(a) Use the scalar equations of equilibrium to find the force in the link, and the reaction forces which act normal to the hinge pin axis.

(b) Find the results in part (a) by using the vector equations of equilibrium.

❙ (a) The link has forces applied at its ends only, and is thus a two-force member. The force in this member has the known direction of the member.

The construction used to find the direction of line ce is shown in Fig. 12.27b. The length d of this line is

$$d = \sqrt{1.8^2 + 3^2 + 2^2} = 4.03 \text{ ft}$$

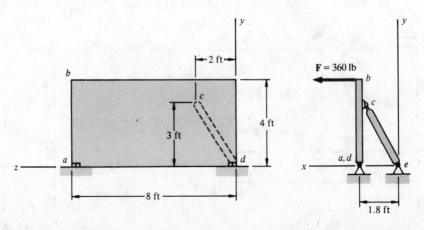

Fig. 12.27a

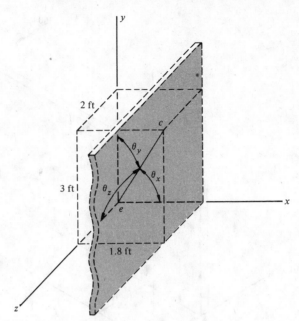

Fig. 12.27b

The direction angles are found from

$$\cos \theta_x = \frac{1.8}{4.03} = 0.447 \qquad \cos \theta_y = \frac{3}{4.03} = 0.744 \qquad \cos \theta_z = \frac{2}{4.03} = 0.496$$

The force in member ce is assumed to be a tensile force. Thus, the force exerted *by* member ce *on* the vertical plate has a sense from c to e. The magnitude of this force is designated R, with the components

$$R_x = -R \cos \theta_x = -0.447R \qquad R_y = -R \cos \theta_y = -0.744R \qquad R_z = -R \cos \theta_z = -0.496R$$

The terms on the right-hand sides of the above three equations *are considered to be quantities which are acting in the positive coordinate senses*.

The free-body diagram of the vertical plate is shown in Fig. 12.27c. The set of forces acting on this member is seen to be a general three-dimensional force system. The reaction force components at the two hinges are assumed to have the senses of the positive coordinate axes.

For moment equilibrium about an axis through the two hinges,

$$\Sigma M_{ad} = 0 \qquad -R_x(3) - 360(4) = 0 \qquad -(-0.447R)(3) - 360(4) = 0 \qquad R = 1{,}070 \text{ lb}$$

Using the above result, the three components of R have the values

$$R_x = -0.447R = -478 \text{ lb} \qquad R_y = -0.744R = -796 \text{ lb} \qquad R_z = -0.496R = -531 \text{ lb}$$

For equilibrium about the x axis in Fig. 12.27c,

$$\Sigma M_x = 0 \qquad R_z(3) - R_y(2) - R_{ay}(8) = 0$$
$$-531(3) - (-796)2 - R_{ay}(8) = 0 \qquad 1{,}590 = 1{,}590 + R_{ay}(8) \qquad R_{ay} = 0$$

The equilibrium requirement about an axis through R_{dy} in Fig. 12.27c is

$$\Sigma M_{R_{dy}} = 0 \qquad R_x(2) + 360(8) + R_{ax}(8) = 0 \qquad -478(2) + 360(8) + R_{ax}(8) = 0 \qquad R_{ax} = -241 \text{ lb}$$

For force equilibrium of the plate,

$$\Sigma F_x = 0 \qquad R_{dx} + R_x + 360 + R_{ax} = 0 \qquad R_{dx} - 478 + 360 - 241 = 0 \qquad R_{dx} = 359 \text{ lb}$$
$$\Sigma F_y = 0 \qquad R_{dy} + R_y + R_{ay} = 0 \qquad R_{dy} - 796 = 0 \qquad R_{dy} = 796 \text{ lb}$$
$$\Sigma F_z = 0 \qquad R_z + R_{az} = 0 \qquad -531 + R_{az} = 0 \qquad R_{az} = 531 \text{ lb}$$

The reaction forces acting on the hinge pins, in the direction normal to these elements, are

$$R_a = \sqrt{R_{ax}^2 + R_{ay}^2} \approx |R_{ax}| = 241 \text{ lb} \qquad R_d = \sqrt{R_{dx}^2 + R_{dy}^2} = \sqrt{359^2 + 796^2} = 873 \text{ lb}$$

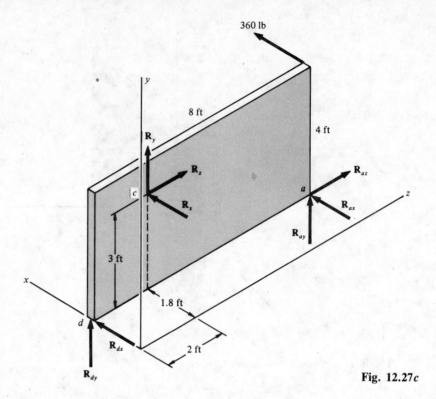

Fig. 12.27c

(b) The forces acting on the plate are expressed in terms of the unit vectors as

$$\mathbf{R}_a = R_{ax}\mathbf{i} + R_{ay}\mathbf{j} + R_{az}\mathbf{k} \qquad \mathbf{R}_d = R_{dx}\mathbf{i} + R_{dy}\mathbf{j}$$
$$\mathbf{R} = R_x\mathbf{i} + R_y\mathbf{j} + R_z\mathbf{k} = -(0.447R)\mathbf{i} - (0.744R)\mathbf{j} - (0.496R)\mathbf{k} \qquad \mathbf{F} = 360\mathbf{i} \quad \text{lb}$$

Point a in Fig. 12.27c is used as the reference point for moments. The position vectors from a to the lines of action of the forces are

$$\mathbf{r}_{ad} = -8\mathbf{k} \quad \text{ft} \qquad \mathbf{r}_{aR} = 3\mathbf{j} - 6\mathbf{k} \quad \text{ft} \qquad \mathbf{r}_{aF} = 4\mathbf{j} \quad \text{ft}$$

For moment equilibrium of the plate,

$$\mathbf{M}_a = \mathbf{r}_{ad} \times \mathbf{R}_d + \mathbf{r}_{aR} \times \mathbf{R} + \mathbf{r}_{aF} \times \mathbf{F} = 0$$

$$\begin{vmatrix} \mathbf{i} & \mathbf{j} & \mathbf{k} \\ 0 & 0 & -8 \\ R_{dx} & R_{dy} & 0 \end{vmatrix} + \begin{vmatrix} \mathbf{i} & \mathbf{j} & \mathbf{k} \\ 0 & 3 & -6 \\ -0.447R & -0.744R & -0.496R \end{vmatrix} + \begin{vmatrix} \mathbf{i} & \mathbf{j} & \mathbf{k} \\ 0 & 4 & 0 \\ 360 & 0 & 0 \end{vmatrix} = 0$$

$$[-R_{dy}(-8) + 3(-0.496R) - (-0.744R)(-6)]\mathbf{i} + [-8R_{dx} - 6(-0.447R)]\mathbf{j} + [-(-0.447R)3 - 360(4)]\mathbf{k} = 0$$

The coefficients of the unit vectors are set equal to zero, with the results

$$8R_{dy} - 3(0.496R) - 6(0.744R) = 0 \qquad -8R_{dx} + 6(0.447R) = 0 \qquad 3(0.447R) - 360(4) = 0$$
$$R = 1{,}070 \text{ lb} \qquad R_{dx} = 359 \text{ lb} \qquad R_{dy} = 796 \text{ lb}$$

For force equilibrium of the plate,

$$\mathbf{R}_a + \mathbf{R}_d + \mathbf{R} + \mathbf{F} = 0 \qquad (R_{ax} + R_{dx} - 0.447R + 360)\mathbf{i} + (R_{ay} + R_{dy} - 0.744R)\mathbf{j} + (R_{az} - 0.496R)\mathbf{k} = 0$$

Equating the coefficients of the unit vectors to zero yields

$$R_{ax} + R_{dx} - 0.447R + 360 = 0 \qquad R_{ay} + R_{dy} - 0.744R = 0 \qquad R_{az} - 0.496R = 0$$
$$R_{ax} = -241 \text{ lb} \qquad R_{ay} \approx 0 \qquad R_{az} = 531 \text{ lb}$$

12.28 Figure 12.28a shows a hinged vertical gate used to dam the end of a channel. The depth of the water is 3.6 ft, and the specific weight of water is 62.4 lb/ft³. The strut bd has ball joints at both ends, and only hinge a can resist forces in the y direction.

(a) Using the scalar equations of equilibrium, find the force exerted by the strut on the gate, and the components of the reaction forces at the two hinges.

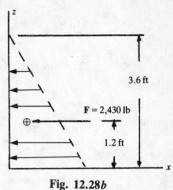

Fig. 12.28a   Fig. 12.28b

(b) Express the force and position vectors in terms of the unit vectors and state the vector equations of equilibrium.

(a) Figure 12.28b shows the distribution of the pressure along the height of the gate. The resultant force of the water on the gate is given by

$$F = h_c \gamma A = \frac{3.6}{2}(62.4)(6)3.6 = 2,430 \text{ lb}$$

Strut bd is a two-force member, with tensile force F_{bd} acting on it. The direction of F_{bd} is shown in Fig. 12.28c. Using $d = \sqrt{4.5^2 + 2^2 + 4^2} = 6.34 \text{ ft}$, the components of this force are

$$F_{bd,x} = \frac{4.5}{6.34}F_{bd} = 0.710F_{bd} \qquad F_{bd,y} = \frac{2}{6.34}F_{bd} = 0.315F_{bd} \qquad F_{bd,z} = -\frac{4}{6.34}F_{bd} = -0.631F_{bd}$$

The above three forces are considered to be positive when acting in the positive coordinate senses. For equilibrium of the gate,

$$\Sigma M_y = 0 \qquad -2,430(1.2) + F_{bd,x}(4) = 0 \qquad -2,430(1.2) + (0.710F_{bd})4 = 0 \qquad F_{bd} = 1,030 \text{ lb}$$
$$\Sigma M_x = 0 \qquad F_{bd,z}(1) - F_{bd,y}(4) + R_{cz}(6) = 0$$
$$-(0.631F_{bd})(1) - (0.315F_{bd})(4) + R_{cz}(6) = 0 \qquad R_{cz} = 325 \text{ lb}$$
$$\Sigma M_z = 0 \qquad -F_{bd,x}(1) + 2,430(3) - R_{cx}(6) = 0$$
$$-(0.710F_{bd})(1) + 2,430(3) - R_{cx}(6) = 0 \qquad R_{cx} = 1,090 \text{ lb}$$

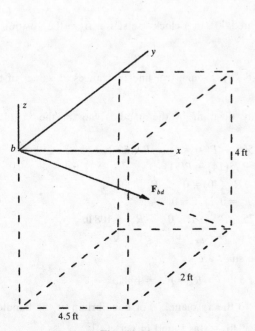

Fig. 12.28c

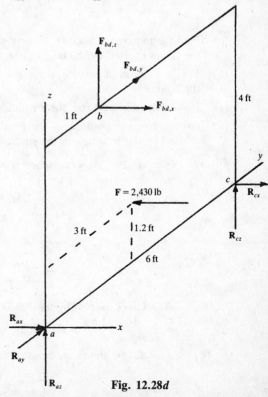

Fig. 12.28d

$$\Sigma F_x = 0 \qquad R_{ax} + F_{bd,x} - 2{,}430 + R_{cx} = 0 \qquad R_{ax} + 0.710F_{bd} - 2{,}430 + R_{cx} = 0 \qquad R_{ax} = 609 \text{ lb}$$
$$\Sigma F_y = 0 \qquad R_{ay} + F_{bd,y} = 0 \qquad R_{ay} + 0.315F_{bd} = 0 \qquad R_{ay} = -324 \text{ lb}$$
$$\Sigma F_z = 0 \qquad R_{az} + F_{bd,z} + R_{cz} = 0 \qquad R_{az} - 0.631F_{bd} + R_{cz} = 0 \qquad R_{az} = 325 \text{ lb}$$

(b) The forces acting on the gate are written in terms of the unit vectors as

$$\mathbf{F}_{bd} = (0.710F_{bd})\mathbf{i} + (0.315F_{bd})\mathbf{j} - (0.631F_{bd})\mathbf{k} \qquad \mathbf{F} = -2{,}430\mathbf{i} \qquad \text{lb}$$
$$\mathbf{R}_c = R_{cx}\mathbf{i} + R_{cz}\mathbf{k} \qquad \mathbf{R}_a = R_{ax}\mathbf{i} + R_{ay}\mathbf{j} + R_{az}\mathbf{k}$$

Point a in Fig. 12.28d is used as the reference point for moments, and the position vectors have the forms

$$\mathbf{r}_{aF} = 3\mathbf{j} + 1.2\mathbf{k} \qquad \text{ft} \qquad \mathbf{r}_{ab} = 1\mathbf{j} + 4\mathbf{k} \qquad \text{ft} \qquad \mathbf{r}_{ac} = 6\mathbf{j} \qquad \text{ft}$$

The equations of equilibrium are

$$\mathbf{M}_a = \mathbf{r}_{aF} \times \mathbf{F} + \mathbf{r}_{ab} \times \mathbf{F}_{bd} + \mathbf{r}_{ac} \times \mathbf{R}_c = 0 \qquad \mathbf{R}_a + \mathbf{R}_c + \mathbf{F} + \mathbf{F}_{bd} = 0$$

It is left as an exercise for the reader to solve the above equations and verify the results found in part (a).

12.29 Figure 12.29a shows a windlass arrangement. The diameter of the drum is 12 in, and force P is normal to the plane of the handle.

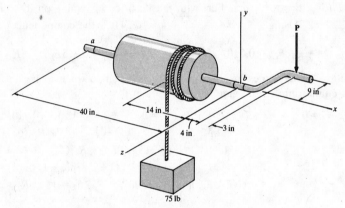

Fig. 12.29a

(a) Find the required value of P, if the system is in equilibrium in the position shown, and the magnitudes of the forces exerted by the bearings on the shaft.

(b) Do the same as in part (a), if the handle is rotated 90°, in a clockwise sense, from the position shown in the figure.

(c) Do the same as part (b), for a rotation of 180°.

(d) For the range of positions of the handle in parts (a) through (c), find the ranges of values of the forces exerted by the bearings on the shaft.

▌ (a) Figure 12.29b shows a schematic free-body diagram of the handle lying in the xz plane. For equilibrium of the handle,

$$\Sigma M_x = 0 \qquad 75(6) - P(9) = 0 \qquad P = 50 \text{ lb}$$
$$\Sigma M_z = 0 \qquad -R_{ay}(40) + 75(14) - P(7) = 0 \qquad R_{ay} = 17.5 \text{ lb}$$
$$\Sigma M_y = 0 \qquad R_{az}(40) = 0 \qquad R_{az} = 0$$
$$\Sigma F_x = 0 \qquad 0 = 0$$
$$\Sigma F_y = 0 \qquad R_{ay} - 75 + R_{by} - P = 0 \qquad R_{by} = 108 \text{ lb}$$
$$\Sigma F_z = 0 \qquad -R_{az} + R_{bz} = 0 \qquad R_{bz} = 0$$

The forces exerted by the bearings on the shaft are

$$R_a = R_{ay} = 17.5 \text{ lb} \qquad R_b = R_{by} = 108 \text{ lb}$$

(b) Figure 12.29c shows the handle rotated 90° to the xy plane. For equilibrium of the handle,

$$\Sigma M_x = 0 \qquad P = 50 \text{ lb} \qquad \text{[as found in part (a)]}$$

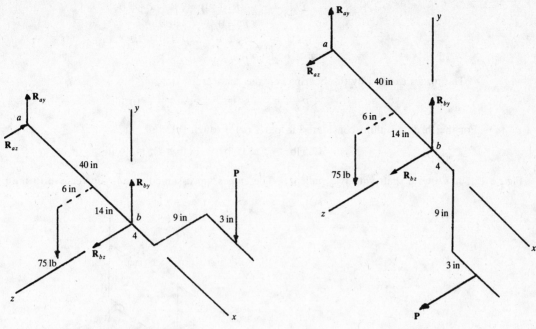

Fig. 12.29b

Fig. 12.29c

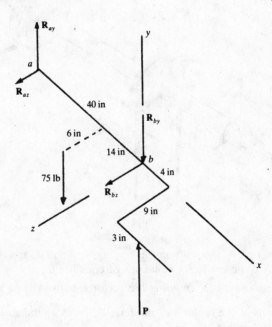

Fig. 12.29d

$$\Sigma M_z = 0 \qquad -R_{ay}(40) + 75(14) = 0 \qquad R_{ay} = 26.3 \text{ lb}$$
$$\Sigma M_y = 0 \qquad R_{az}(40) - P(7) = 0 \qquad R_{az} = 8.75 \text{ lb}$$
$$\Sigma F_x = 0 \qquad 0 = 0$$
$$\Sigma F_y = 0 \qquad R_{ay} - 75 + R_{by} = 0 \qquad R_{by} = 48.7 \text{ lb}$$
$$\Sigma F_z = 0 \qquad R_{az} + R_{bz} + P = 0 \qquad R_{bz} = 58.8 \text{ lb}$$

The forces exerted by the bearings on the shaft are

$$R_a = \sqrt{R_{ay}^2 + R_{az}^2} = \sqrt{26.3^2 + 8.75^2} = 27.7 \text{ lb} \qquad R_b = \sqrt{R_{by}^2 + R_{bz}^2} = \sqrt{48.7^2 + (-58.8)^2} = 76.3 \text{ lb}$$

(c) Figure 12.29d shows the handle rotated 180°. For equilibrium of the handle,

$$\Sigma M_x = 0 \qquad P = 50 \text{ lb} \qquad \text{[as found in part (a)]}$$
$$\Sigma M_z = 0 \qquad -R_{ay}(40) + 75(14) + P(7) = 0 \qquad R_{ay} = 35 \text{ lb}$$

$$\Sigma M_y = 0 \qquad R_{az}(40) = 0 \qquad R_{az} = 0$$
$$\Sigma F_x = 0 \qquad 0 = 0$$
$$\Sigma F_y = 0 \qquad R_{ay} - 75 - R_{by} + P = 0 \qquad R_{by} = 10\,\text{lb}$$
$$\Sigma F_z = 0 \qquad R_{az} + R_{bz} = 0 \qquad R_{bz} = 0$$

The forces exerted by the bearings on the shaft are

$$R_a = R_{ay} = 35\,\text{lb} \qquad R_b = R_{by} = 10\,\text{lb}$$

(d) For the three positions considered in parts (a) through (c),

$$17.5\,\text{lb} \le R_a \le 35\,\text{lb} \qquad 10\,\text{lb} \le R_b \le 108\,\text{lb}$$

12.30 Figure 12.30a shows a shaft with two pulleys. The cables are assumed to not slip relative to the pulleys.

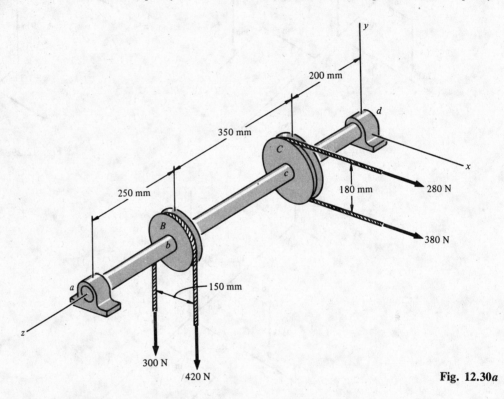

Fig. 12.30a

(a) Find the forces exerted by the shaft on the bearings at a and d, and the torque transmitted through length bc of the shaft, by using the scalar equations of equilibrium.

(b) Solve for the forces in part (a) by using the vector equations of equilibrium.

▌ (a) The force effects of the cables are transferred directly to the shaft, and the free-body diagram of this element is shown in Fig. 12.30b.
The equilibrium requirements are

$$\Sigma M_x = 0 \qquad -R_{ay}(800) + 720(550) = 0 \qquad R_{ay} = 495\,\text{N}$$
$$\Sigma M_y = 0 \qquad -R_{ax}(800) + 660(200) = 0 \qquad R_{ax} = 165\,\text{N}$$
$$\Sigma F_x = 0 \qquad -R_{ax} + 660 - R_{dx} = 0 \qquad R_{dx} = 495\,\text{N}$$
$$\Sigma F_y = 0 \qquad R_{ay} - 720 + R_{dy} = 0 \qquad R_{dy} = 225\,\text{N}$$

The above results can be checked by summing moments about the set of x'y'z' axes shown in Fig. 12.30b. The results are

$$\Sigma M_{x'} = 0 \qquad -720(250) + R_{dy}(800) \overset{?}{=} 0 \qquad 720(250) \overset{?}{=} 225(800) \qquad 1.8 \times 10^5 \equiv 1.8 \times 10^5$$
$$\Sigma M_{y'} = 0 \qquad -660(600) + R_{dx}(800) \overset{?}{=} 0 \qquad 660(600) \overset{?}{=} 495(800) \qquad 3.96 \times 10^5 \equiv 3.96 \times 10^5$$

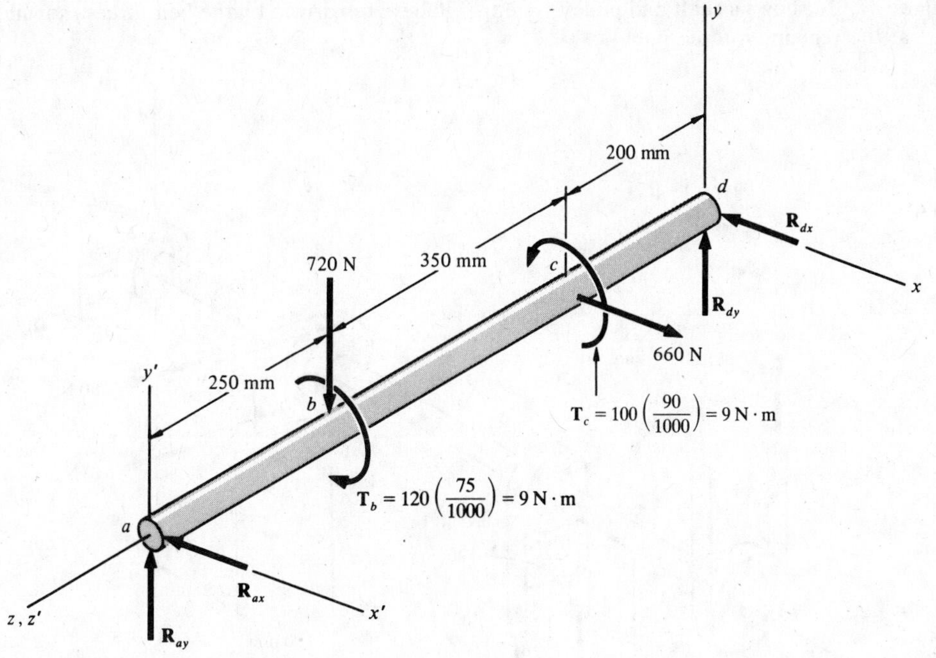

Fig. 12.30b

The forces exerted by the shaft on the bearings at a and d, normal to the direction of the shaft, are

$$R_a = \sqrt{R_{ax}^2 + R_{ay}^2} = \sqrt{165^2 + 495^2} = 522 \text{ N} \qquad R_d = \sqrt{R_{dx}^2 + R_{dy}^2} = \sqrt{495^2 + 225^2} = 544 \text{ N}$$

From consideration of Fig. 12.30b, using $\Sigma M_z = 0$, the torque transmitted through length bc of the shaft is $9 \text{ N} \cdot \text{m}$.

(**b**) The forces acting on the shaft are expressed in terms of the unit vectors as

$$\mathbf{R}_a = -R_{ax}\mathbf{i} + R_{ay}\mathbf{j} \qquad \mathbf{R}_d = -R_{dx}\mathbf{i} + R_{dy}\mathbf{j}$$
$$\mathbf{F}_b = -720\mathbf{j} \quad \text{N} \qquad \mathbf{F}_c = 660\mathbf{i} \quad \text{N}$$

Using d as the reference point for moments, the position vectors are written as

$$\mathbf{r}_{dc} = 200\mathbf{k} \quad \text{mm} \qquad \mathbf{r}_{db} = 550\mathbf{k} \quad \text{mm} \qquad \mathbf{r}_{da} = 800\mathbf{k} \quad \text{mm}$$

For moment equilibrium of the shaft,

$$\mathbf{M}_d = \mathbf{r}_{dc} \times \mathbf{F}_c + \mathbf{r}_{db} \times \mathbf{F}_b + \mathbf{r}_{da} \times \mathbf{R}_a = 0$$

The two couples T_b and T_c, which act on the shaft about the z axis, are not included in the above equations since these two effects are self-canceling when considering the free-body diagram of the entire shaft.

$$\mathbf{M}_d = \begin{vmatrix} \mathbf{i} & \mathbf{j} & \mathbf{k} \\ 0 & 0 & 200 \\ 600 & 0 & 0 \end{vmatrix} + \begin{vmatrix} \mathbf{i} & \mathbf{j} & \mathbf{k} \\ 0 & 0 & 550 \\ 0 & -720 & 0 \end{vmatrix} + \begin{vmatrix} \mathbf{i} & \mathbf{j} & \mathbf{k} \\ 0 & 0 & 800 \\ -R_{ax} & R_{ay} & 0 \end{vmatrix} = 0$$

$$[-(-720)550 - R_{ay}(800)]\mathbf{i} + [200(660) + 800(-R_{ax})]\mathbf{j} + [0]\mathbf{k} = 0$$

Setting the coefficients of the unit vectors equal to zero yields

$$720(550) - R_{ay}(800) = 0 \qquad R_{ay} = 495 \text{ N}$$
$$200(660) - 800R_{ax} = 0 \qquad R_{ax} = 165 \text{ N}$$

For force equilibrium of the shaft,

$$\mathbf{R}_a + \mathbf{R}_d + \mathbf{F}_b + \mathbf{F}_c = 0 \qquad (-R_{ax} - R_{dx} + 660)\mathbf{i} + (R_{ay} + R_{dy} - 720)\mathbf{j} = 0$$

The coefficients of the unit vectors are set equal to zero, with the results

$$-R_{ax} - R_{dx} + 660 = 0 \qquad R_{dx} = 495 \text{ N}$$
$$R_{ay} + R_{dy} - 720 = 0 \qquad R_{dy} = 225 \text{ N}$$

12.31 Figure 12.31a shows a shaft and pulley system. Pulley A is driven by the belt forces shown, and power is taken out at the remaining three pulleys.

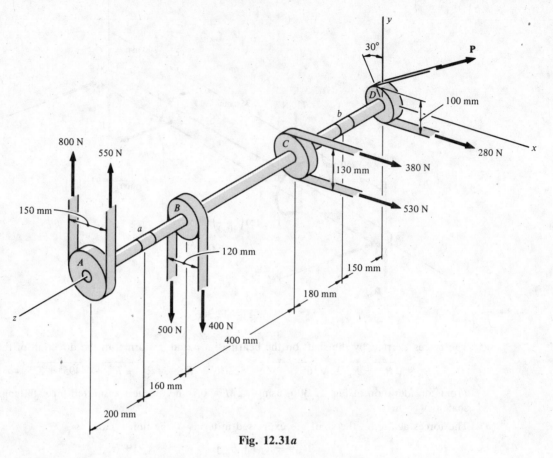

Fig. 12.31a

(a) If the system is in equilibrium, find the value of the belt tensile force P.

(b) Find the forces acting on bearings a and b.

(c) Find the torque transmitted through lengths AB, BC, and CD of the shaft.

▎(a) Using Fig. 12.31a,

$$\Sigma M_z = 0$$

$$(550 - 800)\frac{150}{2} + (500 - 400)\frac{120}{2} + (530 - 380)\frac{130}{2} + (280 - P)\frac{100}{2} = 0 \qquad P = 220 \text{ N}$$

It may be seen that the angle of 30° in Fig. 12.31a does not enter into the above computation.

(b) The belt forces are transferred directly to the shaft, and Fig. 12.31b shows a schematic free-body diagram of the shaft. Using Fig. 12.31a,

$$R_{Dx} = 280 + 220\cos 30° = 471 \text{ N} \qquad R_{Dy} = 220\sin 30° = 110 \text{ N}$$

For moment equilibrium about the line of action of R_{ax},

$$\Sigma M_{R_{ax}} = 0 \qquad -1{,}350(200) - 900(160) + R_{by}(740) + 110(890) = 0 \qquad R_{by} = 427 \text{ N}$$

For force equilibrium in the y direction.

$$\Sigma F_y = 0 \qquad 1{,}350 - R_{ay} - 900 + R_{by} + 110 = 0 \qquad R_{ay} = 987 \text{ N}$$

For moment equilibrium about the line of action of R_{ay},

$$\Sigma M_{R_{ay}} = 0 \qquad -910(560) + R_{bx}(740) - 471(890) = 0 \qquad R_{bx} = 1{,}260 \text{ N}$$

For moment equilibrium about the y axis,

$$\Sigma M_y = 0 \qquad -R_{bx}(150) + 910(330) - R_{ax}(890) = 0 \qquad R_{ax} = 125 \text{ N}$$

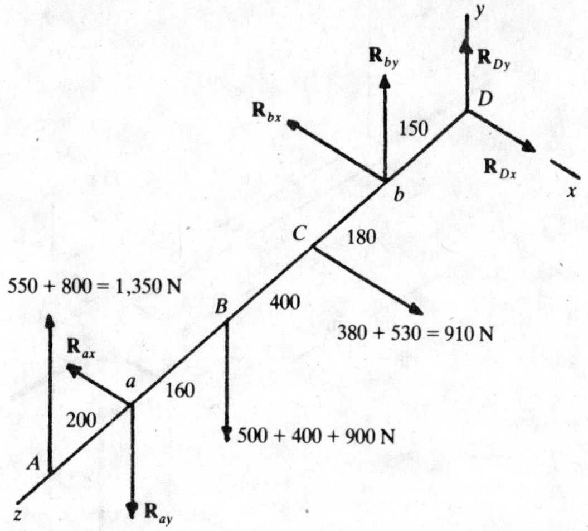

Fig. 12.31b

The resultant bearing forces are

$$R_a = \sqrt{R_{ax}^2 + R_{ay}^2} = \sqrt{125^2 + 987^2} = 995 \text{ N} \qquad R_b = \sqrt{R_{bx}^2 + R_{by}^2} = \sqrt{1{,}260^2 + 427^2} = 1{,}330 \text{ N}$$

As a check on the above calculations,

$$\Sigma F_x = 0 \qquad R_{ax} + R_{bx} \overset{?}{=} 910 + 471 \qquad 125 + 1{,}260 \overset{?}{=} 910 + 471 \qquad 1{,}390 \approx 1{,}380$$

(c) The positive M_z sign convention is used. The torque transmitted through length AB is

$$M_{AB} = \frac{(550 - 800)(150/2)}{1{,}000} = -18.8 \text{ N} \cdot \text{m}$$

The torque transmitted through length BC is

$$M_{BC} = -18.8 + \frac{(500 - 400)(120/2)}{1{,}000} = -12.8 \text{ N} \cdot \text{m}$$

The torque transmitted through length CD is

$$M_{CD} = -12.8 + \frac{(530 - 380)(130/2)}{1{,}000} = -3.05 \text{ N} \cdot \text{m}$$

As a check on the above calculations, the equilibrium of pulley D is tested.

$$\Sigma M_z = 0 \qquad -3.05 + \frac{(280 - 220)(100/2)}{1{,}000} \overset{?}{=} 0 \qquad 3 \approx 3.05$$

12.32 Figure 12.32a shows a bevel gear on a shaft. Only bearing b can resist forces in the x direction. The system is in equililbrium when acted on by the gear tooth forces and the applied torque M_s.

(a) Find the value of the torque M_s and the magnitudes of the forces exerted by the shaft on the bearings, in the direction normal to the shaft, by using the scalar equations of equilibrium.

(b) Find the magnitude, direction, and sense of the force exerted by bearing b on the shaft.

(c) Solve part (a) by using the vector equations of equilibrium.

▌ (a) Figure 12.32b shows a schematic free-body diagram of the shaft and gear assembly.
    For equilibrium of the assembly,

$$\Sigma M_x = 0 \qquad M_s - 200\left(\frac{37.5}{1{,}000}\right) = 0 \qquad M_s = 7.5 \text{ N} \cdot \text{m}$$

$$\Sigma M_y = 0 \qquad -R_{bz}(250) + 25(300) - 70(37.5) = 0 \qquad R_{bz} = 19.5 \text{ N}$$
$$\Sigma M_z = 0 \qquad -R_{by}(250) + 200(300) = 0 \qquad R_{by} = 240 \text{ N}$$
$$\Sigma F_x = 0 \qquad R_{bx} - 70 = 0 \qquad R_{bx} = 70 \text{ N}$$
$$\Sigma F_y = 0 \qquad R_{ay} - R_{by} + 200 = 0 \qquad R_{ay} = 40 \text{ N}$$
$$\Sigma F_z = 0 \qquad R_{az} + R_{bz} - 25 = 0 \qquad R_{az} = 5.5 \text{ N}$$

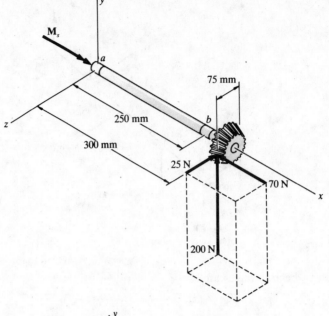

Fig. 12.32*a*

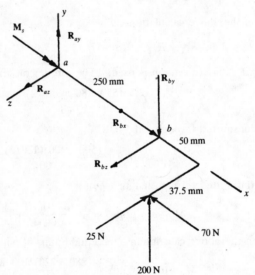

Fig. 12.32*b*

The magnitudes of the forces exerted by the shaft on the bearings, in a direction normal to the shaft axis, are

$$R_{a,n} = \sqrt{R_{ay}^2 + R_{az}^2} = \sqrt{40^2 + 5.5^2} = 40.4 \text{ N} \qquad R_{b,n} = \sqrt{R_{by}^2 + R_{bz}^2} = \sqrt{240^2 + 19.5^2} = 242 \text{ N}$$

(*b*) The magnitude of R_b is given by

$$R_b = \sqrt{R_{bx}^2 + R_{by}^2 + R_{bz}^2} = \sqrt{70^2 + 240^2 + 19.5^2} = 251 \text{ N}$$

The direction of this force is given by

$$\cos \theta = \frac{70}{251} \qquad \theta_x = 73.8°$$

$$\cos \theta_y = \frac{240}{251} \qquad \theta_y = 17.0°$$

$$\cos \theta_z = \frac{19.5}{251} \qquad \theta_z = 85.5°$$

(*c*) The forces, and applied torque M_s, are written in terms of the unit vectors as

$$\mathbf{R}_a = R_{ay}\mathbf{j} + R_{az}\mathbf{k} \qquad \mathbf{R}_b = R_{bx}\mathbf{i} - R_{by}\mathbf{j} + R_{bz}\mathbf{k}$$
$$\mathbf{F} = -70\mathbf{i} + 200\mathbf{j} - 25\mathbf{k} \quad \text{N} \qquad \mathbf{M}_s = M_s\mathbf{i}$$

Using point a as a reference point for moments, the position vectors have the forms

$$\mathbf{r}_{ab} = 250\mathbf{i} \quad \text{mm} \qquad \mathbf{r}_{aF} = 300\mathbf{i} + 37.5\mathbf{k} \quad \text{mm}$$

For moment equilibrium of the gear and shaft assembly,

$$\mathbf{M}_a = \mathbf{r}_{ab} \times \mathbf{R}_b + \mathbf{r}_{aF} \times \mathbf{F} + \mathbf{M}_s = 0$$

$$\mathbf{M}_a = \frac{1}{1,000}\begin{vmatrix} \mathbf{i} & \mathbf{j} & \mathbf{k} \\ 250 & 0 & 0 \\ R_{bx} & -R_{by} & R_{bz} \end{vmatrix} + \frac{1}{1,000}\begin{vmatrix} \mathbf{i} & \mathbf{j} & \mathbf{k} \\ 300 & 0 & 37.5 \\ -70 & 200 & -25 \end{vmatrix} + M_s\mathbf{i} = 0$$

$$\left[-\frac{200(37.5)}{1,000} + M_s\right]\mathbf{i} + \frac{1}{1,000}[-R_{bz}(250) + 37.5(-70) - (-25)300]\mathbf{j}$$

$$+ \frac{1}{1,000}[-R_{by}(250) + 200(300)]\mathbf{k} = 0$$

Setting the coefficients of the unit vectors equal to zero results in

$$-\frac{200(37.5)}{1,000} + M_s = 0 \qquad M_s = 7.5\,\text{N}\cdot\text{m}$$

$$-250R_{bz} - 37.5(70) + 25(300) = 0 \qquad R_{bz} = 19.5\,\text{N}$$

$$-250R_{by} + 200(300) = 0 \qquad R_{by} = 240\,\text{N}$$

For force equilibrium of the assembly,

$$\mathbf{R}_a + \mathbf{R}_b + \mathbf{F} = 0 \qquad (R_{bx} - 70)\mathbf{i} + (R_{ay} - R_{by} + 200)\mathbf{j} + (R_{az} + R_{bz} - 25)\mathbf{k} = 0$$

Equating the coefficients of the unit vectors to zero gives

$$R_{bx} - 70 = 0 \qquad R_{bx} = 70\,\text{N}$$
$$R_{ay} - R_{by} + 200 = 0 \qquad R_{ay} = 40\,\text{N}$$
$$R_{az} + R_{bz} - 25 = 0 \qquad R_{az} = 5.5\,\text{N}$$

12.33 Figure 12.33 shows a pair of helical gears mounted on parallel shafts. The forces shown in the figure are the forces exerted by gear A on gear B. Only bearing a can resist forces in the z direction.

(*a*) Use the vector equations of equilibrium to find the required value of the torque M_s if the system is in equilibrium.

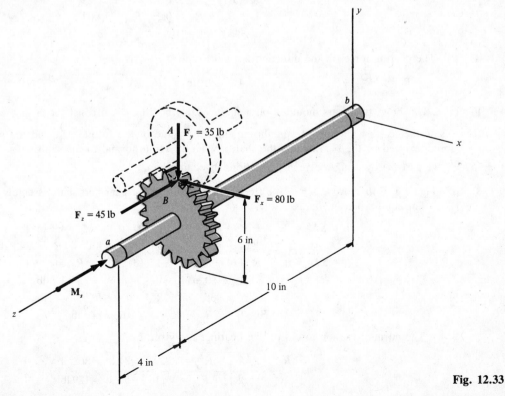

Fig. 12.33

(b) Find the magnitudes of the forces exerted by the shaft on the bearings, in the direction normal to the shaft axis.

▮ (a) The forces and moments acting on the shaft and gear assembly may be written in terms of the unit vectors as

$$\mathbf{F} = -80\mathbf{i} - 35\mathbf{j} - 45\mathbf{k} \qquad \mathbf{R}_a = R_{ax}\mathbf{i} + R_{ay}\mathbf{j} + R_{az}\mathbf{k} \qquad \mathbf{R}_b = R_{bx}\mathbf{i} + R_{by}\mathbf{j} \qquad \mathbf{M} = -M_s\mathbf{k}$$

It follows from the above equations that the unknown components of the bearing forces at a and b are considered to be positive when they act in the positive coordinate senses. The moments will be summed about the origin of the coordinate axes. The position vector $\mathbf{r}_F$ from this point to the point of application of the force $\mathbf{F}$ is

$$\mathbf{r}_F = 3\mathbf{j} + 10\mathbf{k} \qquad \text{in}$$

The position vector $\mathbf{r}_a$ of bearing a is

$$\mathbf{r}_a = 14\mathbf{k} \qquad \text{in}$$

For moment equilibrium about the origin,

$$\mathbf{M}_b = \mathbf{r}_F \times \mathbf{F} + \mathbf{r}_a \times \mathbf{R}_a + \mathbf{M} = 0$$

$$\begin{vmatrix} \mathbf{i} & \mathbf{j} & \mathbf{k} \\ 0 & 3 & 10 \\ -80 & -35 & -45 \end{vmatrix} + \begin{vmatrix} \mathbf{i} & \mathbf{j} & \mathbf{k} \\ 0 & 0 & 14 \\ R_{ax} & R_{ay} & R_{az} \end{vmatrix} - M_s\mathbf{k} = 0$$

$$[3(-45) - (-35)(10) - 14R_{ay}]\mathbf{i} + [10(-80) + 14R_{ax}]\mathbf{j} + [-(-80)3 - M_s]\mathbf{k} = 0$$

The coefficients of the unit vectors are equated to zero, with the results

$$3(-45) - (-35)(10) - 14R_{ay} = 0 \qquad R_{ay} = 15.4 \text{ lb}$$
$$10(-80) + 14R_{ax} = 0 \qquad R_{ax} = 57.1 \text{ lb}$$
$$-(-80)(3) - M_s = 0 \qquad M_s = 240 \text{ in} \cdot \text{lb}$$

For force equilibrium,

$$\mathbf{F} + \mathbf{R}_a + \mathbf{R}_b = 0 \qquad (-80 + R_{ax} + R_{bx})\mathbf{i} + (-35 + R_{ay} + R_{by})\mathbf{j} + (-45 + R_{az})\mathbf{k} = 0$$

The coefficients of the above equation are set equal to zero, with the results

$$-80 + R_{ax} + R_{bx} = 0 \qquad R_{bx} = 22.9 \text{ lb}$$
$$-35 + R_{ay} + R_{by} = 0 \qquad R_{by} = 19.6 \text{ lb}$$
$$-45 + R_{az} = 0 \qquad R_{az} = 45 \text{ lb}$$

(b) The bearing forces in the direction normal to the shaft axis are

$$R_a = \sqrt{R_{ax}^2 + R_{ay}^2} = \sqrt{57.1^2 + 15.4^2} = 59.1 \text{ lb} \qquad R_b = \sqrt{R_{bx}^2 + R_{by}^2} = \sqrt{22.9^2 + 19.6^2} = 30.1 \text{ lb}$$

12.34 Figure 12.34a shows two gears mounted on a shaft. Only bearing a can resist forces in the z direction.

(a) Using the scalar equations of equilibrium, find the value of the torque M_s, and the magnitudes of the forces exerted by the bearings on the shaft, in the direction normal to the shaft.

(b) Solve part (a) by using the vector equations of equilibrium.

▮ (a) Figure 12.34b shows a schematic free-body diagram of the gear and shaft assembly.
For equilibrium of the assembly,

$$\Sigma M_z = 0 \qquad -M_s + 80(3) + 120(1.8) = 0 \qquad M_s = 456 \text{ in} \cdot \text{lb}$$
$$\Sigma M_x = 0 \qquad -R_{ay}(14) - 45(3) + 35(10) + 44(3) = 0 \qquad R_{ay} = 24.8 \text{ lb}$$
$$\Sigma M_y = 0 \qquad R_{ax}(14) - 80(10) - 120(3) = 0 \qquad R_{ax} = 82.9 \text{ lb}$$
$$\Sigma F_x = 0 \qquad R_{ax} - 80 - 120 + R_{bx} = 0 \qquad R_{bx} = 117 \text{ lb}$$
$$\Sigma F_y = 0 \qquad R_{ay} - 35 - 44 + R_{by} = 0 \qquad R_{by} = 54.2 \text{ lb}$$
$$\Sigma F_z = 0 \qquad R_{az} - 45 = 0 \qquad R_{az} = 45 \text{ lb}$$

The normal reaction forces on the bearings are given by

$$R_{a,n} = \sqrt{R_{ax}^2 + R_{ay}^2} = \sqrt{82.9^2 + 24.8^2} = 86.5 \text{ lb}$$
$$R_{b,n} = \sqrt{R_{bx}^2 + R_{by}^2} = \sqrt{117^2 + 54.2^2} = 129 \text{ lb}$$

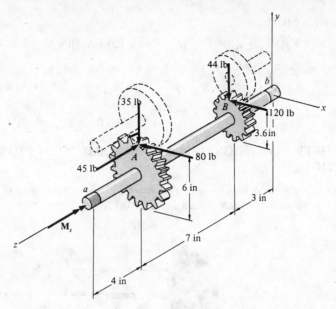

Fig. 12.34a

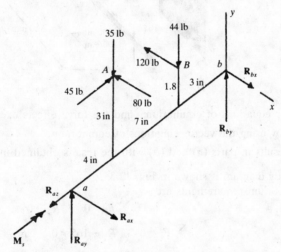

Fig. 12.34b

(**b**) The forces, and the applied torque, are written in terms of the unit vectors as

$$\mathbf{F}_A = -80\mathbf{i} - 35\mathbf{j} - 45\mathbf{k} \quad \text{lb} \qquad \mathbf{F}_B = -120\mathbf{i} - 44\mathbf{j} \quad \text{lb}$$
$$\mathbf{R}_a = R_{ax}\mathbf{i} + R_{ay}\mathbf{j} + R_{az}\mathbf{k} \qquad \mathbf{R}_b = R_{bx}\mathbf{i} + R_{by}\mathbf{j} \qquad \mathbf{M} = -M_s\mathbf{k}$$

Point a is used as the reference for moments, and the position vectors have the forms

$$\mathbf{r}_{aA} = 3\mathbf{j} - 4\mathbf{k} \quad \text{in} \qquad \mathbf{r}_{aB} = 1.8\mathbf{j} - 11\mathbf{k} \quad \text{in} \qquad \mathbf{r}_{ab} = -14\mathbf{k} \quad \text{in}$$

For moment equilibrium of the shaft and gear assembly,

$$\mathbf{M}_a = -M_s\mathbf{k} + \mathbf{r}_{aA} \times \mathbf{F}_A + \mathbf{r}_{aB} \times \mathbf{F}_B + \mathbf{r}_{ab} \times \mathbf{R}_b = 0$$

$$-M_s\mathbf{k} + \begin{vmatrix} \mathbf{i} & \mathbf{j} & \mathbf{k} \\ 0 & 3 & -4 \\ -80 & -35 & -45 \end{vmatrix} + \begin{vmatrix} \mathbf{i} & \mathbf{j} & \mathbf{k} \\ 0 & 1.8 & -11 \\ -120 & -44 & 0 \end{vmatrix} + \begin{vmatrix} \mathbf{i} & \mathbf{j} & \mathbf{k} \\ 0 & 0 & -14 \\ R_{bx} & R_{by} & 0 \end{vmatrix} = 0$$

$$[3(-45) - (-35)(-4) - (-44)(-11) - R_{by}(-14)]\mathbf{i} + [-4(-80) - 11(-120) - 14R_{bx}]\mathbf{j}$$
$$+ [-M_s - (-80)3 - (-120)1.8]\mathbf{k} = 0$$

The coefficients of the unit vectors are set equal to zero, with the results

$$-3(45) - 35(4) - 44(11) + R_{by}(14) = 0 \qquad R_{by} = 54.2 \text{ lb}$$
$$4(80) + 11(120) - 14R_{bx} = 0 \qquad R_{bx} = 117 \text{ lb}$$
$$-M_s + 80(3) + 120(1.8) = 0 \qquad M_s = 456 \text{ in} \cdot \text{lb}$$

For force equilibrium,

$$\mathbf{R}_a + \mathbf{F}_A + \mathbf{F}_B + \mathbf{R}_b = 0 \qquad [R_{ax} - 80 - 120 + R_{bx}]\mathbf{i} + [R_{ay} - 35 - 44 + R_{by}]\mathbf{j} + [R_{az} - 45]\mathbf{k} = 0$$

Equating the coefficients of the unit vectors to zero gives

$$R_{ax} - 80 - 120 + R_{bx} = 0 \qquad R_{ax} = 83 \approx 82.9 \text{ lb}$$
$$R_{ay} - 35 - 44 + R_{by} = 0 \qquad R_{ay} = 24.8 \text{ lb}$$
$$R_{az} - 45 = 0 \qquad R_{az} = 45 \text{ lb}$$

12.4 GENERAL THREE-DIMENSIONAL FORCE SYSTEMS WITH CLAMPED SUPPORTS AND WITH FRICTION

12.35 Figure 12.35a shows a crank arm lying in a horizontal plane and attached to a vertical wall. Someone pushes vertically downward on the handle with a force of 165 N.

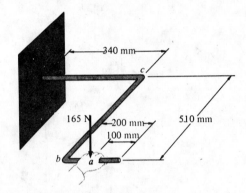

Fig. 12.35a

(a) Using the scalar equations of equilibrium, find the force effects exerted by the wall on the crank arm.

(b) Solve part (a) by using the vector equations of equilibrium.

(c) Compare the results in parts (a) and (b) with the results obtained in Prob. 3.45.

▎ (a) The free-body diagram is shown in Fig. 12.35b.

The equilibrium requirements are

$$\sum F_x = 0 \qquad R_x = 0$$

$$\sum F_y = 0 \qquad R_y - 165 = 0 \qquad R_y = 165 \text{ N}$$

$$\sum F_z = 0 \qquad R_z = 0$$

$$\sum M_x = 0 \qquad M_x + 165\left(\frac{510}{1,000}\right) = 0 \qquad M_x = -84.2 \text{ N} \cdot \text{m}$$

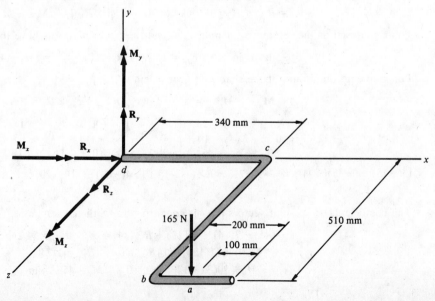

Fig. 12.35b

$$\sum M_y = 0 \qquad M_y = 0$$

$$\sum M_z = 0 \qquad M_z - 165\left(\frac{440}{1,000}\right) = 0 \qquad M_z = 72.6\,\text{N}\cdot\text{m}$$

(b) The applied force, reaction force, and reaction moment are given in terms of the unit vectors by

$$\mathbf{R} = R_x\mathbf{i} + R_y\mathbf{j} + R_z\mathbf{k} \qquad \mathbf{F} = -165\mathbf{j} \quad \text{N} \qquad \mathbf{M} = M_x\mathbf{i} + M_y\mathbf{j} + M_z\mathbf{k}$$

Point d in Fig. 12.35b is used as the reference point for moments. The position vector from this point to force $\mathbf{F}$ has the form

$$\mathbf{r}_{dF} = 440\mathbf{i} + 510\mathbf{k} \qquad \text{mm}$$

For moment equilibrium of the crank arm,

$$\mathbf{M}_d = \mathbf{M} + \mathbf{r}_{dF} \times \mathbf{F} = 0$$

$$\mathbf{M}_d = M_x\mathbf{i} + M_y\mathbf{j} + M_z\mathbf{k} + \frac{1}{1,000}\begin{vmatrix} \mathbf{i} & \mathbf{j} & \mathbf{k} \\ 440 & 0 & 510 \\ 0 & -165 & 0 \end{vmatrix} = 0$$

$$\left[M_x - \frac{(-165)510}{1,000}\right]\mathbf{i} + [M_y]\mathbf{j} + \left[M_z - \frac{165(440)}{1,000}\right]\mathbf{k} = 0$$

Setting the coefficients of the unit vectors equal to zero results in

$$M_x + \frac{165(510)}{1,000} = 0 \qquad M_x = -84.2\,\text{N}\cdot\text{m} \qquad M_y = 0$$

$$M_z - \frac{165(440)}{1,000} = 0 \qquad M_z = 72.6\,\text{N}\cdot\text{m}$$

For force equilibrium,

$$\mathbf{R} + \mathbf{F} = 0 \qquad R_x\mathbf{i} + R_y\mathbf{j} + R_z\mathbf{k} - 165\mathbf{j} = 0$$
$$R_x = 0 \qquad R_y = 165\,\text{N} \qquad R_z = 0$$

(c) The above solutions may be compared with the solution to Problem 3.45. In that problem the 165-N applied force was replaced by a force and two couples at location d, where the crank arm is connected to the wall. The results obtained in that example for the forces and moments exerted by the crank arm on the wall *were the equal and opposite values of the results obtained above*. The resultant of all applied and reaction forces and moments acting on the crank arm is then zero, which is the expected condition of static equilibrium.

12.36 A bracket is clamped at end b to a foundation, as shown in Fig. 12.36. Find the components of the force and moment exerted by the foundation on the bracket.

❚ This problem will be solved by using the vector equations of equilibrium. $\mathbf{R}$ is the reaction force at the support, with components R_x, R_y, and R_z assumed to act in the positive coordinate senses. $\mathbf{M}$ is the reaction moment at the support, with components M_x, M_y, and M_z which are assumed to act in the positive coordinate senses.

The applied forces which act on the bracket, in terms of the unit vectors, are

$$\mathbf{F}_1 = -4,500\mathbf{j} \quad \text{N}$$
$$\mathbf{F}_2 = [(6,000\cos 56°)(\cos 42°)]\mathbf{i} - (6,000\sin 56°)\mathbf{j} + [(6,000\cos 56°)(\sin 42°)]\mathbf{k}$$
$$= 2,490\mathbf{i} - 4,970\mathbf{j} + 2,250\mathbf{k} \quad \text{N}$$
$$\mathbf{F}_3 = 3,200\mathbf{k} \quad \text{N}$$

The reaction moment and force are expressed in terms of the unit vectors as

$$\mathbf{M} = M_x\mathbf{i} + M_y\mathbf{j} + M_z\mathbf{k} \qquad \mathbf{R} = R_x\mathbf{i} + R_y\mathbf{j} + R_z\mathbf{k}$$

Point b is used as the reference for moments, and the position vectors have the forms

$$\mathbf{r}_1 = (30 + 290 - 20)\mathbf{i} - (260 - 36)\mathbf{j} = 300\mathbf{i} - 224\mathbf{j} \qquad \text{mm}$$
$$\mathbf{r}_2 = (30 + 290 - 20)\mathbf{i} + (260 - 18)\mathbf{j} + (290 - 20)\mathbf{k} = 300\mathbf{i} - 242\mathbf{j} + 270\mathbf{k} \qquad \text{mm}$$
$$\mathbf{r}_3 = -30\mathbf{i} - 260\mathbf{j} + 20\mathbf{k} \qquad \text{mm}$$

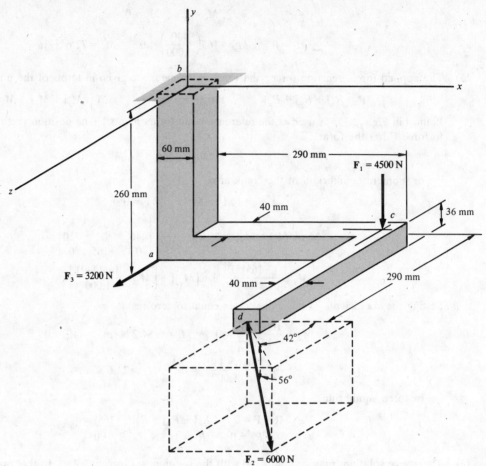

Fig. 12.36

For moment equilibrium of the bracket,

$$\mathbf{M} + \mathbf{r}_1 \times \mathbf{F}_1 + \mathbf{r}_2 \times \mathbf{F}_2 + \mathbf{r}_3 \times \mathbf{F}_3 = 0$$

$$M_x\mathbf{i} + M_y\mathbf{j} + M_z\mathbf{k} + \frac{1}{1,000}\begin{vmatrix} \mathbf{i} & \mathbf{j} & \mathbf{k} \\ 300 & -224 & 0 \\ 0 & -4,500 & 0 \end{vmatrix} + \frac{1}{1,000}\begin{vmatrix} \mathbf{i} & \mathbf{j} & \mathbf{k} \\ 300 & -242 & 270 \\ 2,490 & -4,970 & 2,250 \end{vmatrix} + \frac{1}{1,000}\begin{vmatrix} \mathbf{i} & \mathbf{j} & \mathbf{k} \\ -30 & -260 & 20 \\ 0 & 0 & 3,200 \end{vmatrix} = 0$$

$$\left[M_x - \frac{242(2,250)}{1,000} - \frac{(-4,970)270}{1,000} - \frac{260(3,200)}{1,000} \right]\mathbf{i} + \left[M_y + \frac{270(2,490)}{1,000} - \frac{2,250(300)}{1,000} - \frac{3,200(-30)}{1,000} \right]\mathbf{j}$$

$$+ \left[M_z - \frac{4,500(300)}{1,000} - \frac{4,970(300)}{1,000} - \frac{2,490(-242)}{1,000} \right]\mathbf{k} = 0$$

Setting the coefficients of **i**, **j**, and **k** equal to zero gives

$$M_x = 34.6\,\text{N} \cdot \text{m} \qquad M_y = -93.3\,\text{N} \cdot \text{m} \qquad M_z = 2,240\,\text{N} \cdot \text{m}$$

For force equilibrium of the bracket,

$$\mathbf{F}_1 + \mathbf{F}_2 + \mathbf{F}_3 + \mathbf{R} = 0$$
$$[2,490 + R_x]\mathbf{i} + [-4,500 - 4,970 + R_y]\mathbf{j} + [2,250 + 3,200 + R_z]\mathbf{k} = 0$$

The coefficients of the unit vectors are equated to zero, to obtain

$$R_x = -2,490\,\text{N} \qquad R_y = 9,470\,\text{N} \qquad R_z = -5,450\,\text{N}$$

12.37 Figure 12.37 shows a machine bracket which is clamped at an end. An inclined hole is to be drilled in the part. The drill exerts a force of 40 N, and a couple of 3 N · m, on the bracket. The sense of the couple is clockwise when viewed from the lower right side of the figure. Find the force effects exerted on the clamped end during the drilling operation.

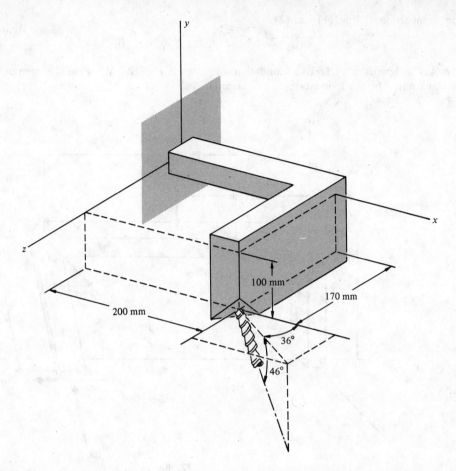

Fig. 12.37

▌ The form of this problem suggests the use of the vector equations of equilibrium. Since the part is clamped, the reaction force and moment have the forms

$$\mathbf{R} = R_x\mathbf{i} + R_y\mathbf{j} + R_z\mathbf{k}$$
$$\mathbf{M}_R = M_x\mathbf{i} + M_y\mathbf{j} + M_z\mathbf{k} \tag{1}$$

All the scalar components are assumed to have the senses of the positive coordinate axes.
The applied force and moment have the forms

$$\mathbf{F} = 40[(-\cos 46° \cos 36°)\mathbf{i} + (\sin 46°)\mathbf{j} - (\cos 46° \sin 36°)\mathbf{k}] = -22.5\mathbf{i} + 28.8\mathbf{j} - 16.3\mathbf{k} \quad \text{N} \tag{2}$$
$$\mathbf{M} = 3[(-\cos 46° \cos 36°)\mathbf{i} + (\sin 46°)\mathbf{j} - (\cos 46° \sin 36°)\mathbf{k}] = -1.69\mathbf{i} + 2.16\mathbf{j} - 1.22\mathbf{k} \quad \text{N·m}$$

The position vector **r** from the origin to the point of application of the drill tip is

$$\mathbf{r} = 200\mathbf{i} - 100\mathbf{j} + 170\mathbf{k} \quad \text{mm}$$

For equilibrium about the origin,

$$\mathbf{M}_R + \mathbf{M} + \mathbf{r} \times \mathbf{F} = 0$$

$$M_x\mathbf{i} + M_y\mathbf{j} + M_z\mathbf{k} - 1.69\mathbf{i} + 2.16\mathbf{j} - 1.22\mathbf{k} + \begin{vmatrix} \mathbf{i} & \mathbf{j} & \mathbf{k} \\ \dfrac{200}{1,000} & \dfrac{-100}{1,000} & \dfrac{170}{1,000} \\ -22.5 & 28.8 & -16.3 \end{vmatrix} = 0$$

$$(M_x - 4.96)\mathbf{i} + (M_y + 1.60)\mathbf{j} + (M_z + 2.29)\mathbf{k} = 0$$

The coefficients of the unit vectors are set equal to zero, and the final solutions are

$$M_x = 4.96 \text{ N·m} \qquad M_y = -1.60 \text{ N·m} \qquad M_z = -2.29 \text{ N·m}$$

For force equilibrium,

$$\mathbf{R} + \mathbf{F} = 0$$

From comparison of Eqs. (1) and (2),

$$R_x = 22.5\,\text{N} \qquad R_y = -28.8\,\text{N} \qquad R_z = 16.3\,\text{N}$$

12.38 A bracket is clamped at end a to a foundation, as shown in Fig. 12.38a. Find the components of the force and moment exerted by the foundation on the bracket.

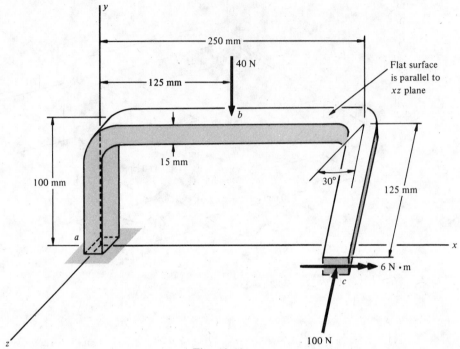

Fig. 12.38a

▮ F_b is the 40-N force, F_c is the 100-N force, and M_c is the 6-N · m moment. R_{ax}, R_{ay}, and R_{az} are the components of the reaction force at a, and M_{ax}, M_{ay}, and M_{az} are the components of the reaction moment at a. All unknown force and moment components are assumed to act in the positive coordinate senses. This problem will be solved by using the vector equations of equilibrium.

Figure 12.38b shows the direction of $\mathbf{F}_c$ and $\mathbf{M}_c$. The forces and moments are expressed in terms of the unit vectors as

$$\mathbf{F}_b = -40\mathbf{j} \quad \text{N} \qquad \mathbf{F}_c = -(100\sin 30°)\mathbf{i} - (100\cos 30°)\mathbf{k} = -50\mathbf{i} - 86.6\mathbf{k} \qquad \text{N}$$
$$\mathbf{R}_a = R_{ax}\mathbf{i} + R_{ay}\mathbf{j} + R_{az}\mathbf{k} \qquad \mathbf{M}_a = M_{ax}\mathbf{i} + M_{ay}\mathbf{j} + M_{az}\mathbf{k}$$
$$\mathbf{M}_c = (6\cos 30°)\mathbf{i} - (6\sin 30°)\mathbf{k} = 5.20\mathbf{i} - 3\mathbf{k} \qquad \text{N} \cdot \text{m}$$

Point a in Fig. 12.38a is used as the reference point for moments, and the position vectors are written as

$$\mathbf{r}_{ab} = 125\mathbf{i} + 100\mathbf{j} \qquad \text{mm}$$
$$\mathbf{r}_{ac} = (250 + 125\sin 30°)\mathbf{i} + (100 - 7.5)\mathbf{j} + (125\cos 30°)\mathbf{k} = 313\mathbf{i} + 92.5\mathbf{j} + 108\mathbf{k} \qquad \text{mm}$$

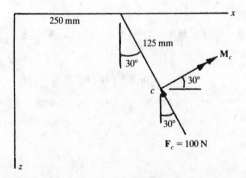

Fig. 12.38b

For moment equilibrium of the bracket,

$$\mathbf{M}_a + \mathbf{r}_{ab} \times \mathbf{F}_b + \mathbf{r}_{ac} \times \mathbf{F}_c + \mathbf{M}_c = 0$$

$$M_{ax}\mathbf{i} + M_{ay}\mathbf{j} + M_{az}\mathbf{k} + \frac{1}{1,000}\begin{vmatrix} \mathbf{i} & \mathbf{j} & \mathbf{k} \\ 125 & 100 & 0 \\ 0 & -40 & 0 \end{vmatrix} + \frac{1}{1,000}\begin{vmatrix} \mathbf{i} & \mathbf{j} & \mathbf{k} \\ 313 & 92.5 & 108 \\ -50 & 0 & -86.6 \end{vmatrix} + (5.20\mathbf{i} - 3\mathbf{k}) = 0$$

$$\left[M_{ax} + \frac{92.5(-86.6)}{1,000} + 5.20\right]\mathbf{i} + \left[M_{ay} + \frac{108(-50)}{1,000} - \frac{(-86.6)313}{1,000}\right]\mathbf{j} + \left[M_{az} - \frac{40(125)}{1,000} - \frac{(-50)92.5}{1,000} - 3\right]\mathbf{k} = 0$$

Setting the coefficients of the unit vectors equal to zero gives

$$M_{ax} = 2.81 \, \text{N} \cdot \text{m} \qquad M_{ay} = -21.7 \, \text{N} \cdot \text{m} \qquad M_{az} = 3.38 \, \text{N} \cdot \text{m}$$

For force equilibrium of the bracket,

$$\mathbf{F}_b + \mathbf{F}_c + \mathbf{R}_a = 0 \qquad (-50 + R_{ax})\mathbf{i} + (-40 + R_{ay})\mathbf{j} + (-86.6 + R_{az})\mathbf{k} = 0$$
$$R_{ax} = 50 \, \text{N} \qquad R_{ay} = 40 \, \text{N} \qquad R_{az} = 86.6 \, \text{N}$$

12.39 The tip of the rigid crank arm shown in Fig. 12.39a rests on a rough surface in the xz plane. The hinge connection at a is frictionless.

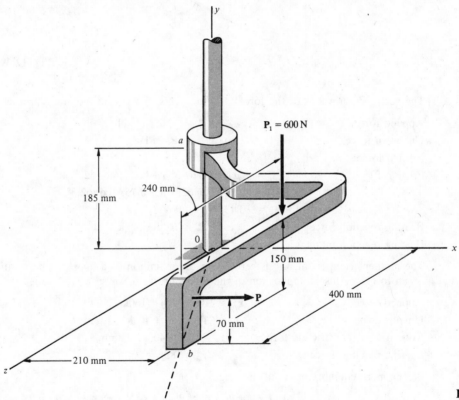

Fig. 12.39a

(a) Find the value of the force P which will cause impending motion of the crank, if the coefficient of friction between the crank and the plane is 0.15, and find the corresponding values of the reaction forces and moments acting on the bearing at a.

(b) Find the value of the bearing reaction force at a, in the direction normal to the y axis.

▌ (a) This problem will be solved using the vector equations of equilibrium. Figure 12.39b shows the xz plane in true view. The contact point between the crank and the plane is at point b, and N is the magnitude of the normal force between these two elements. Since motion is impending, the friction force has its maximum value μN, and the direction of the friction force is along line cc. The sense of the friction force is such as to oppose the impending motion, as shown in Fig. 12.39b.

Angle θ is defined by

$$\tan \theta = \frac{400}{210} \qquad \theta = 62.3°$$

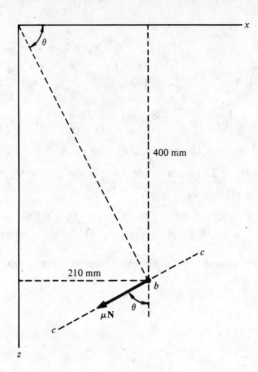

Fig. 12.39b

The vector descriptions of the forces and moments that act on the crank arm are as follows:

Applied force: $\mathbf{P}_1 = -600\mathbf{j}$ N

Unknown force: $\mathbf{P} = P\mathbf{i}$

Normal force: $\mathbf{N} = N\mathbf{j}$

Friction force: $\mathbf{F} = (-\mu N)(\sin 62.3°)\mathbf{i} + (\mu N)(\cos 62.3°)\mathbf{k}$
$$= -0.15N(\sin 62.3°)\mathbf{i} + 0.15N(\cos 62.3°)\mathbf{k}$$
$$= -0.133N\mathbf{i} + 0.0697N\mathbf{k}$$

Bearing reaction force: $\mathbf{R}_a = R_{ax}\mathbf{i} + R_{az}\mathbf{k}$

Bearing reaction moment: $\mathbf{M}_a = M_{ax}\mathbf{i} + M_{az}\mathbf{k}$

The moment component M_{ay} is zero, from the definition of a hinge.   The position vectors from the origin to the points of application of the forces are as follows:

Applied force: $\mathbf{r}_{P_1} = 210\mathbf{i} + 150\mathbf{j} + 160\mathbf{k}$ mm

Unknown force: $\mathbf{r}_P = 210\mathbf{i} + 70\mathbf{j} + 400\mathbf{k}$ mm

Normal force and friction force: $\mathbf{r}_F = 210\mathbf{i} + 400\mathbf{k}$ mm

Bearing reaction force: $\mathbf{r}_a = 185\mathbf{j}$ mm

For moment equilibrium about the origin,

$$\mathbf{M}_a + \mathbf{r}_{P_1} \times \mathbf{P}_1 + \mathbf{r}_P \times \mathbf{P} + \mathbf{r}_F \times (\mathbf{N} + \mathbf{F}) + \mathbf{r}_a \times \mathbf{R}_a = 0$$

$$M_{ax}\mathbf{i} + M_{az}\mathbf{k} + \frac{1}{1000}\begin{vmatrix} \mathbf{i} & \mathbf{j} & \mathbf{k} \\ 210 & 150 & 160 \\ 0 & -600 & 0 \end{vmatrix} + \frac{1}{1000}\begin{vmatrix} \mathbf{i} & \mathbf{j} & \mathbf{k} \\ 210 & 70 & 400 \\ P & 0 & 0 \end{vmatrix}$$

$$+ \frac{1}{1000}\begin{vmatrix} \mathbf{i} & \mathbf{j} & \mathbf{k} \\ 210 & 0 & 400 \\ -0.133N & N & 0.0697N \end{vmatrix} + \frac{1}{1000}\begin{vmatrix} \mathbf{i} & \mathbf{j} & \mathbf{k} \\ 0 & 185 & 0 \\ R_{ax} & 0 & R_{az} \end{vmatrix} = 0$$

$$[-(-600)160 - 400N + 185R_{az} + 1000M_{ax}]\mathbf{i} + [400P + 400(-0.133N) - (0.0697N)210]\mathbf{j}$$
$$+ [-600(210) - 70P + 210N - 185R_{ax} + 1000M_{az}]\mathbf{k} = 0$$

The coefficients are set equal to zero, with the following results:

$$-(-600)160 - 400N + 185R_{az} + 1000M_{ax} = 0 \qquad (1)$$
$$400P - 400(0.133N) - (0.0697N)210 = 0 \qquad (2)$$

$$-600(210) - 70P + 210N - 185R_{ax} + 1000M_{az} = 0 \qquad (3)$$

For force equilibrium,

$$\mathbf{P}_1 + \mathbf{P} + \mathbf{N} + \mathbf{F} + \mathbf{R}_a = 0$$
$$-600\mathbf{j} + P\mathbf{i} + N\mathbf{j} + (-0.133N)\mathbf{i} + (0.0697N)\mathbf{k} + R_{ax}\mathbf{i} + R_{az}\mathbf{k} = 0$$
$$[P - 0.133N + R_{ax}]\mathbf{i} + [-600 + N]\mathbf{j} + [0.0697N + R_{az}]\mathbf{k} = 0$$

The coefficients of these equations are set equal to zero, with the results

$$P - 0.133N + R_{ax} = 0 \qquad (4)$$
$$-600 + N = 0 \qquad (5)$$
$$0.0697N + R_{ax} = 0 \qquad (6)$$

Equations (1) through (6) are a set of six simultaneous equations in the six unknowns N, P, R_{ax}, R_{az}, M_{ax}, and M_{az}. The solutions are

$$N = 600 \text{ N} \qquad R_{ax} = 41.8 \text{ N} \qquad P = 102 \text{ N}$$
$$M_{ax} = 136 \text{ N} \cdot \text{m} \qquad R_{ax} = -22.2 \text{ N} \qquad M_{az} = 3.03 \text{ N} \cdot \text{m}$$

(b) The bearing reaction force at a, in the direction normal to the y axis, is

$$R_a = \sqrt{R_{ax}^2 + R_{az}^2} = \sqrt{(-22.2)^2 + 41.8^2} = 47.3 \text{ N}$$

12.40 The steel vane in Fig. 12.40a weighs 15 lb and is hinged to a fixed vertical rod. The vane fits loosely on the rod. Assume that no moment can be transmitted between the rod and the vane at locations a and b. In addition, the rod may be assumed to be frictionless. The coefficient of friction between tip c on the vane and the xz plane is 0.08.

Use the scalar equations of equilibrium to find the value of P for which motion of the vane is impending and find the forces exerted by the vane on the rod at a and b.

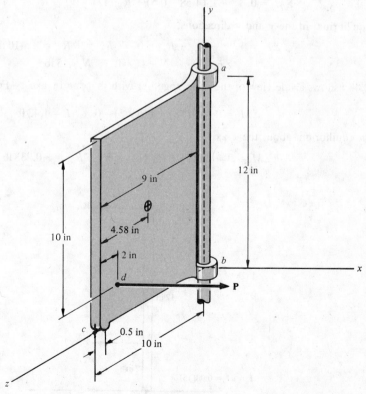

Fig. 12.40a

▮ Figure 12.40b shows a schematic free-body diagram of the vane, and Fig. 12.40c shows a front view of the vane in the xy plane.

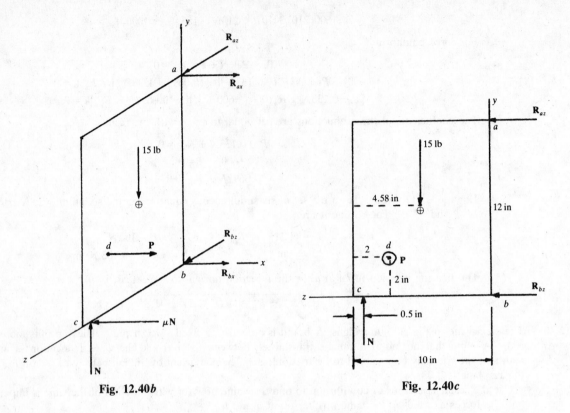

Fig. 12.40b   **Fig. 12.40c**

For moment equilibrium of the vane about point c in this figure,

$$\Sigma M_c = 0 \qquad -15(4.58 - 0.5) + R_{az}(12) = 0 \qquad R_{az} = 5.10 \text{ lb} \tag{1}$$

For force equilibrium in the y and z directions,

$$\Sigma F_z = 0 \qquad -R_{az} - R_{bz} = 0 \qquad R_{bz} = -R_{az} = -5.10 \text{ lb} \tag{2}$$

$$\Sigma F_y = 0 \qquad N - 15 = 0 \qquad N = 15 \text{ lb} \tag{3}$$

Figure 12.40d shows a side view of the vane, together with the x and y axes. For moment equilibrium about the y axis,

$$\Sigma M_y = 0 \qquad P(8) - 1.2(9.5) = 0 \qquad P = 1.43 \text{ lb} \tag{4}$$

For moment equilibrium about the z axis,

$$\Sigma M_b = 0 \qquad P(2) + R_{ax}(12) = 0 \qquad R_{ax} = -0.238 \text{ lb} \tag{5}$$

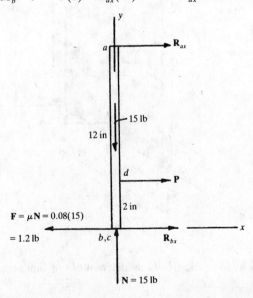

Fig. 12.40d

For force equilibrium in the x direction,

$$\Sigma F_x = 0 \qquad R_{ax} + P + R_{bx} - 1.2 = 0 \tag{6}$$

$$R_{bx} + 1.43 = 0.238 + 1.2 \qquad R_{bx} + 1.43 = 1.44 \qquad R_{bx} \approx 0$$

The forces exerted by the vane on the rod are

$$R_a = \sqrt{R_{ax}^2 + R_{az}^2} = \sqrt{(-0.238)^2 + 5.10^2} = 5.11 \text{ lb} \qquad R_b = R_{bz} = 5.10 \text{ lb}$$

12.41 (a) Solve Prob. 12.40, if the force P has the direction shown in Fig. 12.41a.

(b) The force P that acts on the vane in Prob. 12.40 now has the direction θ shown in Fig. 12.41b. For what limiting value of θ will the vane not move for any value of the applied force P that acts on it?

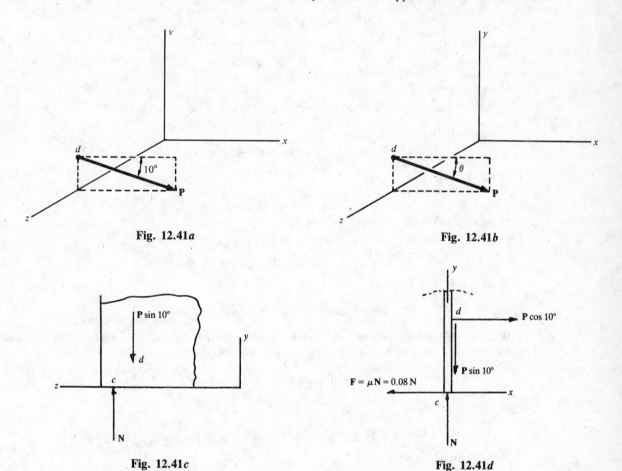

Fig. 12.41a   **Fig. 12.41b**

Fig. 12.41c   **Fig. 12.41d**

▌ (a) Figures 12.41c and d show the details of the components of the applied force P. The equations presented in Prob. 12.40 are now used to solve the problem. The equation numbering below continues from that used in Prob. 12.40.

Using Eq. (1),

$$\Sigma M_c = 0 \qquad -15(4.58 - 0.5) - (P \sin 10°)(2 - 0.5) + R_{az}(12) = 0 \qquad -61.2 - 0.260P + 12R_{az} = 0 \tag{7}$$

Using Eq. (2),

$$-R_{az} - R_{bz} = 0 \tag{8}$$

Using Eq. (3),

$$\Sigma F_y = 0 \qquad N - 15 - P \sin 10° = 0 \tag{9}$$

Using Eq. (4),

$$\Sigma M_y = 0 \qquad (P \cos 10°)8 - (0.08N)9.5 = 0 \tag{10}$$

Using Eq. (5),

$$\Sigma M_b = 0 \qquad -(P\cos 10°)2 - R_{ax}(12) = 0 \tag{11}$$

Using Eq. (6),

$$\Sigma F_x = 0 \qquad R_{ax} + P\cos 10° + R_{bx} - 0.08N = 0 \tag{12}$$

From Eq. (9),

$$N = 15 + P\sin 10° \tag{13}$$

Using Eq. (13) in Eq. (10),

$$(P\cos 10°)8 - 0.08(15 + P\sin 10°)9.5 = 0 \qquad P = 1.47\,\text{lb} \tag{14}$$

Using Eq. (14) in Eq. (7),

$$-61.2 - 0.260(1.47) + 12R_{az} = 0 \qquad R_{az} = 5.13\,\text{lb}$$

Using Eq. (8),

$$R_{bz} = -R_{az} = -5.13\,\text{lb}$$

Using Eq. (14) in Eq. (11),

$$-(1.47\cos 10°)2 - R_{ax}(12) = 0 \qquad R_{ax} = -0.241\,\text{lb}$$

Using Eq. (13),

$$N = 15 + 1.47\sin 10° = 15.3\,\text{lb}$$

Using Eq. (12),

$$-0.241 + 1.47\cos 10° + R_{bx} - 0.08(15.3) = 0 \qquad R_{bx} + 1.45 = 1.47 \qquad R_{bx} \approx 0$$

(b) Using Eqs. (9) and (10), with $10° \to \theta$,

$$N = 15 + P\sin\theta \tag{15}$$
$$(P\cos\theta)8 - (0.08N)9.5 = 0 \tag{16}$$

Using Eq. (15) in Eq. (16),

$$8P\cos\theta - 0.08(9.5)(15 + P\sin\theta) = 0 \qquad P = \frac{11.4}{8\cos\theta - 0.76\sin\theta}$$

When the denominator of the above equation goes to zero, P approaches infinity and the vane will not move for any value of P. The limiting value of θ is found from

$$8\cos\theta - 0.76\sin\theta = 0 \qquad \tan\theta = \frac{8}{0.76} \qquad \theta = 84.6°$$

As a comparison, let

$$\theta = 84° < 84.6° = \theta_{max} \qquad \text{and} \qquad P = \frac{11.4}{8\cos 84° - 0.76\sin 84°} = 142\,\text{lb}$$

From Prob. 12.40, with $\theta = 0$,

$$P = 1.43\,\text{lb}$$

The force P, at $\theta = 84°$, required to cause impending motion of the vane is approximately 100 times greater than this force at $\theta = 0$.

12.42 End c of the vane in Prob. 12.40 rests on a plane surface at a direction θ with the xz plane, as shown in Fig. 12.42a. The coefficient of friction between end c and the plane is 0.08.

(a) For what value of θ is sliding motion of end c impending?

(b) Find the corresponding values of the forces exerted by the vane on the rod at a and b.

❚ (a) Figure 12.42b shows the detail of end c of the vane. Motion is impending when the friction force F satisfies the equation

$$F = 0.08N$$

For moment equilibrium about the y axis, using Figs. 12.40c and 12.42b,

$$\Sigma M_y = 0 \qquad -(0.08N\cos\theta)9.5 + (N\sin\theta)9.5 = 0 \qquad \tan\theta = 0.08 \qquad \theta = 4.57°$$

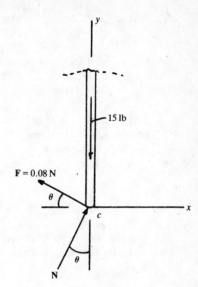

Fig. 12.42*a*   Fig. 12.42*b*

❙ (*b*) Using Eq. (1) in Prob. 12.40,

$$\Sigma M_c = 0 \qquad -15(4.58 - 0.5) + R_{az}(12) = 0 \qquad R_{az} = 5.10 \text{ lb}$$

Using Eq. (2),

$$\Sigma F_z = 0 \qquad -R_{az} - R_{bz} = 0 \qquad R_{bz} - R_{az} = -5.10 \text{ lb}$$

Using Eq. (3),

$$\Sigma F_y = 0 \qquad N \cos \theta + (0.08N) \sin \theta - 15 = 0 \qquad N = 14.95 \approx 15 \text{ lb}$$

Using Eq. (5), with $P = 0$,

$$\Sigma M_b = 0 \qquad R_{ax}(12) = 0 \qquad R_{ax} = 0$$

Using Eq. (6), with $P = 0$,

$$\Sigma F_x = 0 \qquad R_{ax} + R_{bx} + N \sin \theta - (0.08N) \cos \theta = 0 \qquad R_{bx} = 0.001 \approx 0$$
$$R_a = R_{az} = 5.10 \text{ lb} \qquad R_b = R_{bz} = -5.10 \text{ lb}$$

12.43 Give a summary of the basic concepts of analysis of three-dimensional force systems.

❙ In a three-dimensional force system either the physical structure, or the loads, or both, do not lie in a common plane. The direction of a force or moment which is referenced to an *xyz* coordinate system may be expressed in terms of the direction angles θ_x, θ_y, and θ_z. These are the angles between the line of action of the force and the three coordinate axes. The relationship between a force F, or a moment M, and its components along the three coordinate directions is then

$$F_x = F \cos \theta_x \qquad M_x = M \cos \theta_x$$
$$F_y = F \cos \theta_y \qquad M_y = M \cos \theta_y$$
$$F_z = F \cos \theta_z \qquad M_z = M \cos \theta_z$$
$$F = \sqrt{F_x^2 + F_y^2 + F_z^2} \qquad M = \sqrt{M_x^2 + M_y^2 + M_z^2}$$

The direction of a force or moment also may be described by stating the angles between the line of action of the force or moment and fixed reference planes.

The positive senses of the three components of moment about the coordinate axes are clockwise when viewed from the origin of the coordinates.

The reaction force effects for the case of suport by a cable are

$$R_x = R \cos \theta_x \qquad R_y = R \cos \theta_y \qquad R_z = R \cos \theta_z$$

where θ_x, θ_y, and θ_z are the direction angles of the cable. The cable cannot support a moment, and thus

$$M_x = M_y = M_z = 0$$

For support by a plane, or curved, smooth surface, the known direction of the compressive reaction force is normal to the surface. The components of the reaction moment are

$$M_x = M_y = M_z = 0$$

For support by a hinge pin, the component of moment about the axis of the pin is zero. The remaining two components of moment have unknown magnitudes and senses. The unknown reaction force has the components R_x, R_y, and R_z.

A ball joint support cannot transmit a moment. Thus, the components of the reaction moment are

$$M_x = M_y = M_z = 0$$

The components of the unknown reaction force are R_x, R_y, and R_z.

Support by a rough surface is statically equivalent to support by a ball joint, if sliding motion does not occur and the body remains in contact with the surface. When sliding motion is impending, the friction force F of the plane on the body is given by

$$F = \mu N$$

N is the compressive normal force, and the friction force lies in the plane. The moment transmitted is zero, so that

$$M_x = M_y = M_z = 0$$

In the case of support by a clamped connection, both the reaction force and the reaction moment have unknown magnitude, direction, and sense. These two quantities may be expressed symbolically in terms of the six components R_x, R_y, R_z and M_x, M_y, M_z.

For equilibrium of a concurrent force system,

$$\Sigma F_x = 0 \qquad \Sigma F_y = 0 \qquad \Sigma F_z = 0$$

or

$$\mathbf{F} = F_x\mathbf{i} + F_y\mathbf{j} + F_z\mathbf{k} = 0$$

For equilibrium of a parallel force system,

$$\Sigma F_y = 0 \qquad \Sigma M_x = 0 \qquad \Sigma M_z = 0$$

or

$$\mathbf{F} = F_y\mathbf{j} = 0 \qquad \mathbf{M} = M_x\mathbf{i} + M_z\mathbf{k} = 0$$

where y is the direction of all of the forces.

For equilibrium of the general three-dimensional force system,

$$\Sigma F_x = 0 \qquad \Sigma F_y = 0 \qquad \Sigma F_z = 0 \qquad \Sigma M_x = 0 \qquad \Sigma M_y = 0 \qquad \Sigma M_z = 0$$

or

$$\mathbf{F} = F_x\mathbf{i} + F_y\mathbf{j} + F_z\mathbf{k} = 0 \qquad \mathbf{M} = M_x\mathbf{i} + M_y\mathbf{j} + M_z\mathbf{k} = 0$$

Self-Study Review of the Fundamental Definitions, Concepts, and Techniques of Engineering Mechanics: Statics

Note: The selected questions below are repeated from the previous chapters in this text. They contain material which is fundamental to the mastery of the subject statics. The reader is urged to review these questions carefully to make sure that he or she fully understands the definitions, concepts, and techniques. Reference may be made directly, as required, to the solutions which accompany the original problems. The original problem number is shown in parentheses.

INTRODUCTION

13.1   Define the fundamental quantities force and mass. (1.1)

13.2   Describe the four basic units of mechanics. (1.2)

13.3   What is the difference between a gravitational system of units and an absolute system of units? (1.3)

13.4   Describe the U.S. Customary units. (1.4)

13.5   Describe the International System units. (1.5)

13.6   Give the conversion factors for length, force, and mass between USCS and SI units. (1.6)

13.7   Define the basic units of angular measurement. (1.7)

13.8   Express the fundamental relationship between the weight and the mass of a body. (1.8)

13.9   Derive the units of a slug of mass. (1.9)

13.10 Derive the units of a newton of force. (1.10)

13.11 Derive the terms pressure and stress. (1.13)

13.12 Define the material property specific weight and explain how it relates to the weight of a body. (1.14)

13.13 Define the material property density and explain how it relates to the mass of a body. (1.15)

13.14 What is the relationship between specific weight and density? (1.16)

13.15 What is the specific gravity of a material? (1.17)

13.16 State the basic technique used in the conversion of units. (1.25)

13.17 Define the term magnitude. (1.32)

13.18 Define the term direction. (1.33)

13.19 How is the direction of a line defined in three-dimensional space? (1.34)

13.20 Define the term sense. (1.35)

13.21 What is the difference between the term error and the term mistake? (1.41)

13.22 Show two methods of comparing the magnitudes of two different numbers. (1.42)

13.23 Describe the technique of rounding off a number to three significant figures. (1.43)

13.24 How is the solution to a pair of simultaneous equations verified? (1.47)

13.25 Discuss the relationships among the sides and the included angles of a right triangle. (1.48)

13.26 Express the law of cosines for a triangle. (1.49)

13.27 Express the law of sines for a triangle. (1.50)

13.28 What is a scalar quantity? (1.57)

13.29 What is a vector quantity? (1.58)

13.30 Show the graphical construction of a vector. (1.59)

13.31 Define the scalar multiplication operation. (1.62)

OPERATIONS WITH FORCES

13.32 Give a basic definition of the term force. (2.1)

13.33 The magnitudes, directions, and senses of two forces are known. What is the term used to describe the single force which produces the same force effect as that produced by the original two forces? (2.2)

13.34 Show the triangle law of force addition method of summing two forces and compare this law with the parallelogram law of force addition. (2.3)

13.35 If vectors $\mathbf{F}_1$ and $\mathbf{F}_2$ are known, show how the magnitude, direction, and sense of the resultant force $\mathbf{F}$ of the two vectors may be found. (2.4)

13.36 Show how the parallelogram law of force addition method may be extended to cases where more than two forces are to be summed to find a resultant force. (2.6)

13.37 Explain how the triangle law of force addition may be used to add more than two forces in the same plane. (2.7)

13.38 Describe the operation of vector subtraction. (2.9)

13.39 What is meant by the components of a force? (2.11)

13.40 A force which lies in the xy plane has known magnitude F_1 and known direction θ with the x axis.   Find the two components of this force, along the x and y axes, in terms of F and θ. (2.18)

13.41 The description of the components of a force along two perpendicular directions involves the four terms F, θ, F_x, and F_y. If any of the pairs F_x and θ, F_y and θ, or F_x and F_y are given, the force F is uniquely defined.   Show that if the magnitude of the force and one component are given, the force is not uniquely defined. (2.24)

13.42 Describe the method of summing the components to find the resultant of a system of forces. (2.27)

13.43 Show how the method of summing the components to find the resultant of a system of forces can be arranged in a tabular form (2.28)

13.44 Show how the analytic methods developed for finding the resultant of a system of forces in a single plane may be extended to the case where the forces have arbitrary directions in space. (2.39)

13.45 Find the general forms for the x, y, and z components of a force whose line of action is located with respect to fixed reference planes. (2.42)

13.46 Define the unit vectors of an xyz coordinate system. (2.53)

13.47 Show how a vector may be expressed in terms of unit vectors. (2.54)

13.48 Vectors **A** and **B** are known.   Show the operation of vector addition, where **C** is the sum of **A** and **B**, using the unit vectors. (2.56)

13.49 Compare the method of finding a resultant force by summing x, y, and z components with the method using formal vector addition with unit vectors. (2.63)

13.50 Define the vector dot product operation. (2.64)

13.51 Show the general form of the solution for the vector dot product. (2.65)

13.52 Show how the dot product operation may be used to find the component of a vector along a line of known direction. (2.66)

13.53 What is the significance of a dot product that is equal to zero? (2.75)

13.54 Show how the vector dot product may be used to find the angle between two intersecting vectors. (2.77)

OPERATIONS WITH MOMENTS AND COUPLES

13.55 Define the term moment of a force. (3.1)

13.56 For what condition will the moment of a force be zero? (3.2)

13.57 What are the fundamental units of a moment? (3.3)

13.58 Show that a moment is a vector quantity. (3.4)

13.59 Show the graphical representation of a moment using a curved arrow. (3.5)

13.60 Show the graphical representation of a moment using a straight arrow. (3.6)

13.61 Show the relationship between the moment of a force and the sum of the moments of the components of the force. (3.8)

13.62 Describe the general technique for finding the resultant moment of a system of several forces. (3.14)

13.63 Define the term couple. (3.25)

13.64 What is the fundamental difference between a moment and a couple? (3.26)

13.65 Give examples of a couple. (3.27)

13.66 Define the term torque. (3.32)

13.67 Show how to replace a force by a force and a couple. (3.34)

13.68 Show how to replace a force by a force and two couples. (3.43)

13.69 Show the technique for representing a moment in a three-dimensional coordinate system. (3.51)

13.70 Define the vector cross product operation. (3.59)

13.71 Show the general form of the solution for the vector cross product. (3.60)

13.72 Show how the vector cross product can be compactly represented by a 3×3 determinant. (3.61)

13.73 Show the vector cross product definition of the moment of a force. (3.62)

13.74 Compare the method of solving for the moment in a three-dimensional force system by direct application of the fundamental definition $M = Fd$ with the method using the vector cross product definition of moment. (3.66)

FUNDAMENTALS OF FORCE ANALYSIS

13.75 What is the physical interpretation of the term force? (4.1)

13.76 What is the difference between a body force and a surface force? (4.2)

13.77 What is the difference between a tensile force and a compressive force? (4.3)

13.78 Describe weight force. (4.4)

13.79 What is the basic physical effect in engineering mechanics? (4.5)

13.80   What is the difference between a fixed vector and a free vector? (4.16)
13.81   Describe the principle of transmissibility? (4.17)
13.82   Show a collinear force system. (4.18)
13.83   Show a concurrent force system. (4.19)
13.84   Show a parallel force system. (4.20)
13.85   Show a general force system. (4.21)
13.86   What is the difference between coplanar and noncoplanar force systems? (4.22)
13.87   List the types of force systems. (4.23)
13.88   What is a free-body, or equilibrium, diagram? (4.24)
13.89   Explain the difference between the applied forces and moments and the reaction forces and moments which are shown in a free-body diagram. (4.25)
13.90   Give the formal statement of Newton's three laws of motion. (4.26)
13.91   Interpret Newton's laws of motion. (4.27)
13.92   Define the term static equilibrium. (4.28)
13.93   State three uses for the resultant force of a system of forces. (4.29)
13.94   Show the free-body diagram condition for support of a body by a cable. (4.30)
13.95   Show the free-body diagram condition for support of a body by a smooth plane surface (4.13)
13.96   Show the free-body diagram condition for support of a body by a smooth curved surface. (4.33)
13.97   Show the free-body diagram condition for a body which is supported by a ball or roller. (4.34)
13.98   Show the free-body diagram condition for support of a body by a hinge pin. (4.35)
13.99   Show the free-body diagram condition for support of a body by a ball joint. (4.36)
13.100 Show the free-body diagram condition for support of a body by a rough surface. (4.37)
13.101 Show the free-body condition for support of a body by a clamped connection. (4.38)
13.102 How are the senses of the unknown reaction forces and moments in a free-body diagram determined? (4.40)
13.103 Devise a notation for reaction forces and moments in a free-body diagram. (4.41)
13.104 State two general uses for the solutions to problems in static force analysis. (4.53)
13.105 What is the central problem in static force analysis? (4.54)

ANALYSIS OF TWO-DIMENSIONAL FORCE SYSTEMS

13.106 Describe the collinear force system and give the general form for the resultant force. (5.1)
13.107 State the requirement for equilibrium of a body acted on by a collinear force system. (5.4)
13.108 Describe the concurrent force system and give the general forms for the resultant force. (5.8)
13.109 State the requirement for equilibrium of a body acted on by a concurrent force system. (5.10)
13.110 Describe the parallel force system and give the general form for the resultant force. (5.26)
13.111 State the requirement for equilibrium of a body acted on by a parallel force system. (5.30)
13.112 Describe the general two-dimensional force system and give the general form for the resultant force. (5.36)
13.113 State the requirements for equilibrium of a body acted on by a general two-dimensional force system. (5.38)

FORCE ANALYSIS OF PLANE TRUSSES

13.114 Show an example of a plane truss. (6.1a)
13.115 What is the difference between a plane truss and a space truss? (6.1b)
13.116 State the general use of trusses. (6.1c)
13.117 Describe the details of truss construction and the method of application of the loads to the truss. (6.2)
13.118 Discuss the stability of a plane truss. (6.3a)
13.119 State the necessary condition for stability in terms of the number of joints and the number of members. (6.3b)
13.120 What are the basic assumptions used in the force analysis of trusses? (6.4)
13.121 Show that all truss members are two-force members. (6.5)
13.122 Discuss the force transmission through a joint in a truss. (6.6)
13.123 State the two basic methods for solving for the forces in truss members. (6.7)
13.124 Describe how the forces in truss members may be found by using the method of joints, together with the equilibrium requirements for a concurrent force system. (6.9)
13.125 Show how to identify zero force members in a truss. (6.12)
13.126 Show the force transmission through a pulley connection to the joint of a truss. (6.20)
13.127 Describe how the forces in the members of a truss may be found by using the method of joints, together with a closed force triangle or polygon. (6.29)
13.128 Describe how the method of sections may be used to find the forces in selected members of a truss. (6.41)
13.129 Describe the technique for solving for the reaction forces in the free-body diagrams of connected trusses. (6.51)

FORCE ANALYSIS OF PLANE FRAMES AND MACHINES

13.130 Describe the construction, and method of loading, of a plane frame. (7.1)

13.131 Compare the methods of application of loads to a plane frame and to a plane truss. (7.2)

13.132 Show the general form of loading on plane frame members and state the requirements for equilibrium of these members. (7.3)

13.133 Explain how the directions of the forces in two-force members in a plane frame are determined. (7.4)

13.134 Describe the general method for the force analysis of plane frames. (7.5)

13.135 Compare the solutions obtained in the force analyses of plane trusses with those obtained in the force analyses of plane frames. (7.6a)

13.136 What is the central problem in the analysis of plane frame forces? (7.6b)

13.137 Describe the method of analysis for a frame in which
   1. A pin joins three or more members
   2. A pin joins two or more members and a foundation support
   3. A load is applied directly to a pin which joins two or more members

13.138 Give a definition of the term machine. (7.46a)

13.139 Compare the methods of force analysis of frames with the methods of force analysis of machines. (7.46c)

ANALYSIS OF FRICTION FORCES

13.140 Describe the term friction force. (8.1a)

13.141 State the general characteristics of friction forces. (8.1b)

13.142 Give several examples of desirable and undesirable effects of friction forces. (8.1c)

13.143 What is the relationship between the maximum available friction force which may exist between two contacting surfaces and the normal force which is transmitted across these surfaces? (8.4)

13.144 What is the difference between the coefficients of friction for the case where a body is stationary and the case where a body is moving? (8.5)

13.145 What factors affect the value of the coefficient of friction? (8.6)

13.146 Define the term angle of friction. (8.7a)

13.147 How is the maximum value of the angle of friction related to the coefficient of friction? (8.7b)

13.148 What is the mistake most frequently made in the solution of problems involving friction? (8.8)

13.149 Define the term angle of repose. (8.10)

13.150 Show the criteria which may be used to predict whether sliding or tipping of a body occurs. (8.27)

13.151 Show how five distinct regimes of sliding or tipping motion of a body relative to a surface may be identified. (8.28)

13.152 Describe how the friction forces are analyzed in problems where there are multiple sliding surfaces. (8.54)

13.153 Define the term wedge. (8.65a)

13.154 Find the relationship between the force applied to a wedge and the force produced in the direction perpendicular to the applied force. (8.65b)

13.155 Find the general form for the force required to extract, or pull out, a wedge. (8.68)

13.156 What is the relationship between the two belt tensile forces when slipping motion of a belt on a drum is impending? (8.72a)

13.157 Discuss the effects of the values of the coefficient of friction, and the angle of wrap, when slipping motion of a belt on a drum is impending. (8.72b)

13.158 State how friction forces may be used in a braking system. (8.77)

CENTROIDS OF PLANE AREAS AND CURVES

13.159 Show how the model of a thin plate of homogeneous material and constant thickness W may be used to develop a physical interpretation of the concept of a centroid of a plane area. (9.1)

13.160 State the formal definition of the centroidal coordinates of a plane area. (9.2a)

13.161 How is the location of the centroid affected by the placement of the xy axes with respect to the area? (9.2b)

13.162 If a plane area has an axis of symmetry, show that the centroid of the area must lie on this axis. (9.4a)

13.163 Where is the centroid of a plane area located if the area has two axes of symmetry? (9.4b)

13.164 Show how the coordinates of the centroid of a composite area may be found. (9.5)

13.165 Define the first moment of a plane area. (9.14)

13.166 How is the computation of the centroidal coordinates x_c and y_c of a plane area modified if the composite area contains holes or cutouts? (9.17)

13.167 How are the centroidal coordinates of a pattern of rivet holes for the connection of structural members determined? (9.23a)

13.168 What is the major difference between problems that involve finding the centroidal coordinates of a pattern of hole areas and problems that involve finding these coordinates for a single area? (9.23b)

13.169 Show how to organize, in tabular form, the calculations for the case in which a composite area is a complicated shape requiring the description of several elementary areas. (9.26)

13.170 State the formal definition of the centroidal coordinates of a plane curve. (9.33a)

13.171 Show how the coordinates of the centroid of a composite plane curve may be determined. (9.33b)

13.172 How is the computation for the centroidal coordinates simplified if the plane curve has one, or two, axes of symmetry? (9.33c)

13.173 Show how the theorem of Pappus may be used to find the volume generated when a plane area is rotated about an axis. (9.45)

13.174 Show how the theorem of Pappus may be used to find the surface area generated when a plane curve is rotated about an axis. (9.46)

MOMENTS AND PRODUCTS OF INERTIA OF PLANE AREAS AND CURVES

13.175 Show how the model of a plane area A which is positioned with respect to a set of xy coordinate axes, and is divided into the elementary areas A_i with coordinates x_i and y_i, may be used to develop a concept known as the moment of inertia. (10.1a)

13.176 What information does the moment of inertia of a plane area provide about the area? (10.1b)

13.177 What is an alternative term for the moment of inertia of a plane area? (10.1c)

13.178 State the formal definitions of the moments of inertia of a plane area about the x and y axes. (10.2a)

13.179 What are some of the uses of the moment of inertia of a plane area? (10.2b)

13.180 State the fundamental difference between the concepts of the centroid of a plane area and the moment of inertia of a plane area. (10.3)

13.181 What is the definition of the polar moment of inertia of a plane area? (10.6)

13.182 What is meant by the radius of gyration of a plane area? (10.8a)

13.183 State an important useful application of the concept of a radius of gyration. (10.8b)

13.184 Show how the moments of inertia of an area about a set of axes parallel to the centroidal axes may be found if the moments of inertia about the centroidal axes are known. (10.10a)

13.185 How is the moment of inertia determined for a plane area that is not of simple geometric shape? (10.12)

13.186 What is the effect on the computation of the moment of inertia if the plane area has holes or cutouts? (10.21)

13.187 Show how to organize, in tabular format, the calculations for the moments of inertia when the composite area is a complicated shape requiring the description of several elementary areas. (10.26)

13.188 Give the forms for the moments of inertia of a plane curve with respect to the x and y axes. (10.45)

13.189 State the parallel-axis, or transfer, theorems for a plane curve. (10.47)

13.190 Show how the model of a plane area A which is positioned with respect to a set of xy coordinate axes, and divided into the elementary areas A_i with coordinates x_i and y_i, may be used to develop a concept known as the product of inertia. (10.52a)

13.191 State the formal definition of the product of inertia of a plane area. (10.52b)

13.192 What information does the product of inertia of a plane area provide about the area? (10.52c)

13.193 Show how the product of inertia of an area about a set of axes parallel to the centroidal axes may be found if the product of inertia of the area about its centroidal axes is known. (10.54)

13.194 The quantities I_x, I_y, and I_{xy} of an area located with respect to the xy axes are known. A second set of axes, $x'y'$, at direction θ with respect to xy, are placed on the area. Show how the terms $I_{x'}$, $I_{y'}$, and $I_{x'y'}$ may be expressed in terms of I_x, I_y, and I_{xy}. (10.61)

13.195 What is meant by the term principal axes? (10.62a)

13.196 What is the definition of the product of inertia of a plane curve? (10.65a)

13.197 State the parallel-axis, or transfer, theorem for the product of inertia of a plane curve. (10.65b)

DISTRIBUTION OF FORCES ALONG LENGTHS AND OVER AREAS

13.198 What idealization of the physical problem is made when using a concentrated force? (11.1a)

13.199 What is meant by a distributed loading? (11.1b)

13.200 What is the significant difference between loadings that are represented by concentrated forces and loadings that are represented as distributed forces? (11.1c)

13.201 Give the general form for the magnitude of the resultant force of a force that is distributed along a straight length. (11.2a)

13.202 What is the interpretation of the form for the resultant of a force distribution along a straight length? (11.2b)

13.203 Where does the resultant of a force distribution along a straight length act? (11.2c)

13.204 Find the magnitude and position of the resultant force of a force that is distributed uniformly along a length. (11.3a)

13.205 Show a diagram of the replacement of a uniformly distributed force by the resultant force of this distribution. (11.3*b*)

13.206 Show the form of a force which varies uniformly along a length. (11.10*a*)

13.207 Explain how the magnitude and location of the resultant of a force that varies uniformly along a straight length are found. (11.10*b*)

13.208 What common physical problem does a force which varies uniformly along a straight length represent? (11.10*c*)

13.209 Find the magnitude, direction, and location of the resultant force of a uniform pressure acting on a plane area. (11.18)

13.210 Find the pressure at a point in a stationary incompressible liquid. (11.29*a*)

13.211 Find the magnitude and location of the maximum pressure in a stationary incompressible liquid. (11.29*b*)

13.212 What type of force distribution in a stationary incompressible liquid is represented by the variation of pressure with depth? (11.29*c*).

13.213 State several typical examples of the loading produced by a stationary incompressible liquid on a body. (11.29*d*)

13.214 Find the magnitude of the resultant force of a liquid on a submerged plane area. (11.35*a*)

13.215 Find the location of the resultant force acting on a submerged plane area. (11.35*b*)

13.216 What term is used to describe the point where the resultant force on a submerged plane area acts? (11.35*c*)

ANALYSIS OF THREE-DIMENSIONAL FORCE SYSTEMS

13.217 Discuss the fundamental difference between the analysis of two-dimensional force systems and the analysis of three-dimensional force systems. (12.1)

13.218 Show two methods which may be used to define the direction of the line of action of a force or moment in a three-dimensional coordinate system. (12.2)

13.219 Show the sign convention for the sense of a moment or couple in a three-dimensional coordinate system. (12.3)

13.220 Summarize the reaction force and moment requirements in three-dimensional problems for support of a body by a cable, by a plane or curved surface, by a hinge pin or ball joint, and by a clamped connection. (12.4)

13.221 State the equilibrium requirements for the general three-dimensional force system. (12.5*a*)

13.222 What new equilibrium requirements are introduced when going from a two-dimensional problem to a three-dimensional problem? (12.5*b*)

13.223 Give the general characteristics, and state the equilibrium requirements, for the concurrent, parallel, and general three-dimensional force systems. (12.6*a*)

13.224 Discuss the solutions found by using the scalar equations of equilibrium with those found by using the vector equations of equilibrium. (12.6*b*)

APPENDIX A

Appendix B

Appendix C

Index

Note: Each entry in this index is referenced by problem number. The typical listing of a topic is followed by two descriptions. The first—the term *definition*—is broadly understood to include a basic definition, concept, or technique of solution. The second description—*problems*—identifies the problems which directly use this information. In many cases an initial condition of the problem, such as the method of support of the body or the location of the applied loads, is changed to show the effect on the solutions. These types of problems are identified by the statement *problems, with original conditions changed*. All problems in the text are listed in this index, and an asterisk is used to identify problems which have unusually lengthy solutions, or are of a more advanced nature.

The reader is encouraged to review this index and develop facility with its use. It is hoped that this listing has been developed to a sufficient degree of detail to permit rapid identification of specific problems in any desired area of the subject engineering mechanics: statics.

SCHAUM'S SOLVED PROBLEMS SERIES

- ■ Learn the best strategies for solving tough problems in step-by-step detail
- ■ Prepare effectively for exams and save time in doing homework problems
- ■ Use the indexes to quickly locate the types of problems you need the most help solving
- ■ Save these books for reference in other courses and even for your professional library

To order, please check the appropriate box(es) and complete the following coupon.

- ❏ **3000 SOLVED PROBLEMS IN BIOLOGY**
  ORDER CODE 005022-8/**$16.95 406 pp.**

- ❏ **3000 SOLVED PROBLEMS IN CALCULUS**
  ORDER CODE 041523-4/**$19.95 442 pp.**

- ❏ **3000 SOLVED PROBLEMS IN CHEMISTRY**
  ORDER CODE 023684-4/**$20.95 624 pp.**

- ❏ **2500 SOLVED PROBLEMS IN COLLEGE ALGEBRA & TRIGONOMETRY**
  ORDER CODE 055373-4/**$14.95 608 pp.**

- ❏ **2500 SOLVED PROBLEMS IN DIFFERENTIAL EQUATIONS**
  ORDER CODE 007979-x/**$19.95 448 pp.**

- ❏ **2000 SOLVED PROBLEMS IN DISCRETE MATHEMATICS**
  ORDER CODE 038031-7/**$16.95 412 pp.**

- ❏ **3000 SOLVED PROBLEMS IN ELECTRIC CIRCUITS**
  ORDER CODE 045936-3/**$21.95 746 pp.**

- ❏ **2000 SOLVED PROBLEMS IN ELECTROMAGNETICS**
  ORDER CODE 045902-9/**$18.95 480 pp.**

- ❏ **2000 SOLVED PROBLEMS IN ELECTRONICS**
  ORDER CODE 010284-8/**$19.95 640 pp.**

- ❏ **2500 SOLVED PROBLEMS IN FLUID MECHANICS & HYDRAULICS**
  ORDER CODE 019784-9/**$21.95 800 pp.**

- ❏ **1000 SOLVED PROBLEMS IN HEAT TRANSFER**
  ORDER CODE 050204-8/**$19.95 750 pp.**

- ❏ **3000 SOLVED PROBLEMS IN LINEAR ALGEBRA**
  ORDER CODE 038023-6/**$19.95 750 pp.**

- ❏ **2000 SOLVED PROBLEMS IN Mechanical Engineering THERMODYNAMICS**
  ORDER CODE 037863-0/**$19.95 406 pp.**

- ❏ **2000 SOLVED PROBLEMS IN NUMERICAL ANALYSIS**
  ORDER CODE 055233-9/**$20.95 704 pp.**

- ❏ **3000 SOLVED PROBLEMS IN ORGANIC CHEMISTRY**
  ORDER CODE 056424-8/**$22.95 688 pp.**

- ❏ **2000 SOLVED PROBLEMS IN PHYSICAL CHEMISTRY**
  ORDER CODE 041716-4/**$21.95 448 pp.**

- ❏ **3000 SOLVED PROBLEMS IN PHYSICS**
  ORDER CODE 025734-5/**$20.95 752 pp.**

- ❏ **3000 SOLVED PROBLEMS IN PRECALCULUS**
  ORDER CODE 055365-3/**$16.95 385 pp.**

- ❏ **800 SOLVED PROBLEMS IN VECTOR MECHANICS FOR ENGINEERS
  Vol I: STATICS**
  ORDER CODE 056582-1/**$20.95 800 pp.**

- ❏ **700 SOLVED PROBLEMS IN VECTOR MECHANICS FOR ENGINEERS
  Vol II: DYNAMICS**
  ORDER CODE 056687-9/**$20.95 672 pp.**

ASK FOR THE *SCHAUM'S* **SOLVED PROBLEMS SERIES** AT YOUR LOCAL BOOKSTORE
OR CHECK THE APPROPRIATE BOX(ES) ON THE PRECEDING PAGE
AND MAIL WITH THIS COUPON TO:

McGRAW-HILL, INC.
ORDER PROCESSING S-1
PRINCETON ROAD
HIGHTSTOWN, NJ 08520

OR CALL
1-800-338-3987

NAME (PLEASE PRINT LEGIBLY OR TYPE)

ADDRESS (NO P.O. BOXES)

CITY   STATE   ZIP

ENCLOSED IS   ☐ A CHECK   ☐ MASTERCARD   ☐ VISA   ☐ AMEX   (✓ ONE)

ACCOUNT # _____   EXP. DATE _____

SIGNATURE _____

MAKE CHECKS PAYABLE TO MCGRAW-HILL, INC. <u>PLEASE INCLUDE LOCAL SALES TAX AND $1.25 SHIPPING/HANDLING</u>
PRICES SUBJECT TO CHANGE WITHOUT NOTICE AND MAY VARY OUTSIDE THE U.S. FOR THIS
INFORMATION, WRITE TO THE ADDRESS ABOVE OR CALL THE 800 NUMBER.